Linear Algebra in Action
Third Edition

GRADUATE STUDIES
IN MATHEMATICS **232**

Linear Algebra in Action
Third Edition

Harry Dym

AMERICAN MATHEMATICAL SOCIETY
Providence, Rhode Island

EDITORIAL COMMITTEE
Matthew Baker
Marco Gualtieri
Gigliola Staffilani (Chair)
Jeff A. Viaclovsky
Rachel Ward

2020 Mathematics Subject Classification. Primary 15-01, 30-01, 34-01, 39-01, 46E22, 47B32, 52-01, 93-01.

For additional information and updates on this book, visit
www.ams.org/bookpages/gsm-232

Library of Congress Cataloging-in-Publication Data

Names: Dym, H. (Harry), 1938- author.
Title: Linear algebra in action / Harry Dym.
Description: Third edition. | Providence, Rhode Island : American Mathematical Society, [2023] | Series: Graduate studies in mathematics, 1065-7339 ; Volume 232 | Includes bibliographical references and index.
Identifiers: LCCN 2023008107 | ISBN 9781470472061 (hardcover) | ISBN 9781470474195 (paperback) | ISBN 9781470474188 (ebook)
Subjects: LCSH: Algebras, Linear. | AMS: Linear and multilinear algebra; matrix theory – Instructional exposition (textbooks, tutorial papers, etc.). | Functions of a complex variable – Instructional exposition (textbooks, tutorial papers, etc.). | Ordinary differential equations – Instructional exposition (textbooks, tutorial papers, etc.). | Difference and functional equations – Instructional exposition (textbooks, tutorial papers, etc.). | Functional analysis – Linear function spaces and their duals – Hilbert spaces with reproducing kernels (= [proper] functional Hilbert spaces, including de Branges-Rovnyak and other structured spaces | Operator theory – Special classes of linear operators – Operators in reproducing-kernel Hilbert spaces (including de Branges, de Branges-Rovnyak, and other structured spaces). | Convex and discrete geometry – Instructional exposition (textbooks, tutorial papers, etc.). | Systems theory; control – Instructional exposition (textbooks, tutorial papers, etc.).
Classification: LCC QA184.2 .D96 2023 | DDC 512/.5–dc23/eng/20230519
LC record available at https://lccn.loc.gov/2023008107

Copying and reprinting. Individual readers of this publication, and nonprofit libraries acting for them, are permitted to make fair use of the material, such as to copy select pages for use in teaching or research. Permission is granted to quote brief passages from this publication in reviews, provided the customary acknowledgment of the source is given.

Republication, systematic copying, or multiple reproduction of any material in this publication is permitted only under license from the American Mathematical Society. Requests for permission to reuse portions of AMS publication content are handled by the Copyright Clearance Center. For more information, please visit **www.ams.org/publications/pubpermissions**.

Send requests for translation rights and licensed reprints to reprint-permission@ams.org.

© 2023 by the American Mathematical Society. All rights reserved.
The American Mathematical Society retains all rights
except those granted to the United States Government.
Printed in the United States of America.

∞ The paper used in this book is acid-free and falls within the guidelines
established to ensure permanence and durability.
Visit the AMS home page at https://www.ams.org/

10 9 8 7 6 5 4 3 2 1   28 27 26 25 24 23

Dedicated to the memory of:

Irene Lillian Dym, special friend for nigh onto 67 years,

and to the memory of our two oldest sons and our first granddaughter, who were recalled prematurely for no apparent reason:

Jonathan Carol Dym, and he but 44,

David Loren Dym, and he but 57, while playing basketball in the gym named for his older brother Jonathan and his daughter Avital,

Avital Chana Dym, and she but 12.

Yhi zichram baruch

Contents

Preface to the third edition	xvii
Preface to the second edition	xix
Preface to the first edition	xxi
Chapter 1. Prerequisites	1
§1.1. Main definitions	1
§1.2. Mappings	6
§1.3. Triangular matrices	8
§1.4. The binomial formula	9
§1.5. Supplementary notes	10
Chapter 2. Dimension and rank	11
§2.1. The conservation of dimension	11
§2.2. Conservation of dimension for matrices	13
§2.3. What you need to know about rank	13
§2.4. $\{0, 1, \infty\}$	17
§2.5. Block triangular matrices	18
§2.6. Supplementary notes	19
Chapter 3. Gaussian elimination	21
§3.1. Examples	23
§3.2. A remarkable formula	26
§3.3. Extracting a basis	28
§3.4. Augmenting a given set of vectors to form a basis	28

§3.5.	Computing the coefficients in a basis	29
§3.6.	The Gauss-Seidel method	30
§3.7.	Block Gaussian elimination	32
§3.8.	Supplementary notes	34

Chapter 4.   Eigenvalues and eigenvectors — 35

§4.1.	The first step	35
§4.2.	Diagonalizable matrices	37
§4.3.	Invariant subspaces	39
§4.4.	Jordan cells	40
§4.5.	Linear transformations	41
§4.6.	Supplementary notes	42

Chapter 5.   Towards the Jordan decomposition — 43

§5.1.	Direct sums	43
§5.2.	The null spaces of powers of B	48
§5.3.	Verification of Theorem 5.4	50
§5.4.	Supplementary notes	52

Chapter 6.   The Jordan decomposition — 53

§6.1.	The Jordan decomposition	53
§6.2.	Overview	56
§6.3.	Dimension of nullspaces of powers of B	57
§6.4.	Computing J	58
§6.5.	Computing U	58
§6.6.	Two simple examples	60
§6.7.	Real Jordan forms	62
§6.8.	Supplementary notes	62

Chapter 7.   Determinants — 63

§7.1.	Determinants	63
§7.2.	Useful rules for calculating determinants	65
§7.3.	Exploiting block structure	67
§7.4.	Minors	69
§7.5.	Eigenvalues	72
§7.6.	Supplementary notes	74

Chapter 8.	Companion matrices and circulants	75
§8.1.	Companion matrices	75
§8.2.	Circulants	79
§8.3.	Interpolating polynomials	81
§8.4.	An eigenvalue assignment problem	82
§8.5.	Supplementary notes	84
Chapter 9.	Inequalities	85
§9.1.	A touch of convex function theory	85
§9.2.	Four inequalities	91
§9.3.	The Krein-Milman theorem	96
§9.4.	Supplementary notes	97
Chapter 10.	Normed linear spaces	99
§10.1.	Normed linear spaces	99
§10.2.	The vector space of matrices A	101
§10.3.	Evaluating some operator norms	106
§10.4.	Small perturbations	108
§10.5.	Supplementary notes	110
Chapter 11.	Inner product spaces	113
§11.1.	Inner product spaces	113
§11.2.	Gram matrices	116
§11.3.	Adjoints	118
§11.4.	Spectral radius	120
§11.5.	What you need to know about $\|A\|$	122
§11.6.	Supplementary notes	122
Chapter 12.	Orthogonality	125
§12.1.	Orthogonality	125
§12.2.	Projections and direct sums	127
§12.3.	Orthogonal projections	129
§12.4.	The Gram-Schmidt method	133
§12.5.	QR factorization	134
§12.6.	Supplementary notes	134
Chapter 13.	Normal matrices	137
§13.1.	Normal matrices	138
§13.2.	Schur's theorem	140
§13.3.	Commuting normal matrices	142

| §13.4. | Real Hermitian matrices | 144 |
| §13.5. | Supplementary notes | 145 |

Chapter 14. Projections, volumes, and traces — 147
§14.1.	Projection by iteration	147
§14.2.	Computing nonorthogonal projections	149
§14.3.	The general setting	151
§14.4.	Detour on the angle between subspaces	154
§14.5.	Areas, volumes, and determinants	156
§14.6.	Trace formulas	158
§14.7.	Supplementary notes	160

Chapter 15. Singular value decomposition — 161
§15.1.	Singular value decompositions	162
§15.2.	A characterization of singular values	165
§15.3.	Sums and products of singular values	167
§15.4.	Properties of singular values	168
§15.5.	Approximate solutions of linear equations	169
§15.6.	Supplementary notes	170

Chapter 16. Positive definite and semidefinite matrices — 171
§16.1.	A detour on triangular factorization	173
§16.2.	Characterizations of positive definite matrices	176
§16.3.	Square roots	178
§16.4.	Polar forms and partial isometries	180
§16.5.	Some useful formulas	182
§16.6.	Supplementary notes	183

Chapter 17. Determinants redux — 185
§17.1.	Differentiating determinants	185
§17.2.	The characteristic polynomial	188
§17.3.	The Binet-Cauchy formula	189
§17.4.	Inequalities for determinants	190
§17.5.	Some determinant identities	193
§17.6.	Jacobi's determinant identity	194
§17.7.	Sylvester's determinant identity	195
§17.8.	Supplementary notes	198

Chapter 18.	Applications	199
§18.1.	A minimization problem	199
§18.2.	Strictly convex functions and spaces	201
§18.3.	Fitting a line in $\mathbb{R}^2$	204
§18.4.	Fitting a line in $\mathbb{R}^p$	205
§18.5.	Schur complements for semidefinite matrices	207
§18.6.	von Neumann's inequality for contractive matrices	208
§18.7.	Supplementary notes	209
Chapter 19.	Discrete dynamical systems	211
§19.1.	Homogeneous systems	211
§19.2.	Nonhomogeneous systems	214
§19.3.	Second-order difference equations	214
§19.4.	Higher-order difference equations	217
§19.5.	Nonhomogeneous equations	219
§19.6.	Supplementary notes	219
Chapter 20.	Continuous dynamical systems	221
§20.1.	Preliminaries on matrix-valued functions	221
§20.2.	The exponential of a matrix	222
§20.3.	Systems of differential equations	223
§20.4.	Uniqueness	225
§20.5.	Isometric and isospectral flows	226
§20.6.	Nonhomogeneous differential systems	227
§20.7.	Second-order differential equations	228
§20.8.	Higher-order differential equations	229
§20.9.	Wronskians	232
§20.10.	Supplementary notes	233
Chapter 21.	Vector-valued functions	235
§21.1.	Mean value theorems	235
§21.2.	Taylor's formula with remainder	236
§21.3.	Mean value theorem for functions of several variables	237
§21.4.	Mean value theorems for vector-valued functions of several variables	239
§21.5.	Convex minimization problems	241
§21.6.	Supplementary notes	242

Chapter 22.	Fixed point theorems	243
§22.1.	A contractive fixed point theorem	243
§22.2.	A refined contractive fixed point theorem	245
§22.3.	Other fixed point theorems	246
§22.4.	Applications of fixed point theorems	246
§22.5.	Supplementary notes	249
Chapter 23.	The implicit function theorem	251
§23.1.	The inverse function theorem	251
§23.2.	The implicit function theorem	254
§23.3.	Continuous dependence of solutions	257
§23.4.	Roots of polynomials	258
§23.5.	Supplementary notes	261
Chapter 24.	Extremal problems	263
§24.1.	Classical extremal problems	263
§24.2.	Extremal problems with constraints	266
§24.3.	Examples	269
§24.4.	Supplementary notes	273
Chapter 25.	Newton's method	275
§25.1.	Newton's method for scalar functions	275
§25.2.	Newton's method for vector-valued functions	277
§25.3.	Supplementary notes	280
Chapter 26.	Matrices with nonnegative entries	281
§26.1.	A warm-up theorem	282
§26.2.	The Perron-Frobenius theorem	284
§26.3.	Supplementary notes	290
Chapter 27.	Applications of matrices with nonnegative entries	291
§27.1.	Stochastic matrices	291
§27.2.	Behind Google	292
§27.3.	Leslie matrices	294
§27.4.	Minimum matrices	295
§27.5.	Doubly stochastic matrices	297
§27.6.	Inequalities of Ky Fan and von Neuman	298
§27.7.	Supplementary notes	303

| Contents | xiii |

Chapter 28. Eigenvalues of Hermitian matrices	305
§28.1. The Courant-Fischer theorem	305
§28.2. Applications of the Courant-Fischer theorem	307
§28.3. Ky Fan's maximum principle	309
§28.4. The sum of two Hermitian matrices	309
§28.5. On the right-differentiability of eigenvalues	311
§28.6. Sylvester's law of inertia	313
§28.7. Supplementary notes	314

Chapter 29. Singular values redux I	315
§29.1. Sums of singular values	315
§29.2. Majorization	316
§29.3. Norms based on sums of singular values	319
§29.4. Unitarily invariant norms	319
§29.5. Products of singular values	320
§29.6. Eigenvalues versus singular values	322
§29.7. Supplementary notes	323

Chapter 30. Singular values redux II	325
§30.1. Sums of powers of singular values	325
§30.2. Inequalities for singular values in terms of A	328
§30.3. Perturbation of singular values	329
§30.4. Supplementary notes	335

Chapter 31. Approximation by unitary matrices	337
§31.1. Approximation in the Frobenius norm	338
§31.2. Approximation in other norms	339
§31.3. Supplementary notes	344

Chapter 32. Linear functionals	345
§32.1. Linear functionals	345
§32.2. Extensions of linear functionals	348
§32.3. The Minkowski functional	349
§32.4. Separation theorems	351
§32.5. Another path	353
§32.6. Supplementary notes	357

Chapter 33.	A minimal norm problem	359
§33.1.	Dual extremal problems	360
§33.2.	Preliminary calculations	361
§33.3.	Evaluation of (33.1)	364
§33.4.	A numerical example	365
§33.5.	A review	367
§33.6.	Supplementary notes	367
Chapter 34.	Conjugate gradients	369
§34.1.	The recursion	372
§34.2.	Convergence estimates	375
§34.3.	Krylov subspaces	377
§34.4.	The general conjugate gradient method	378
§34.5.	Supplementary notes	379
Chapter 35.	Continuity of eigenvalues	381
§35.1.	Contour integrals of matrix-valued functions	383
§35.2.	Continuous dependence of the eigenvalues	387
§35.3.	Matrices with distinct eigenvalues	389
§35.4.	Supplementary notes	391
Chapter 36.	Eigenvalue location problems	393
§36.1.	Geršgorin disks	393
§36.2.	Spectral radius redux	395
§36.3.	Shifting eigenvalues	399
§36.4.	The Hilbert matrix	400
§36.5.	Fractional powers	401
§36.6.	Supplementary notes	401
Chapter 37.	Matrix equations	403
§37.1.	The equation $X - AXB = C$	403
§37.2.	The Sylvester equation $AX - XB = C$	405
§37.3.	$AX = XB$	409
§37.4.	Special classes of solutions	409
§37.5.	Supplementary notes	411

Chapter 38.	A matrix completion problem	413
§38.1.	Constraints on Ω	414
§38.2.	The central diagonals are specified	417
§38.3.	A moment problem	422
§38.4.	Supplementary notes	422
Chapter 39.	Minimal norm completions	423
§39.1.	A minimal norm completion problem	424
§39.2.	A description of all solutions to the minimal norm completion problem	426
§39.3.	Supplementary notes	428
Chapter 40.	The numerical range	429
§40.1.	The numerical range is convex	429
§40.2.	Eigenvalues versus numerical range	431
§40.3.	The Gauss-Lucas theorem	433
§40.4.	The Heinz inequality	434
§40.5.	Supplementary notes	436
Chapter 41.	Riccati equations	437
§41.1.	Riccati equations	437
§41.2.	Two lemmas	443
§41.3.	The LQR problem	445
§41.4.	Supplementary notes	447
Chapter 42.	Supplementary topics	449
§42.1.	Gaussian quadrature	449
§42.2.	Bezoutians	452
§42.3.	Resultants	456
§42.4.	General QR factorization	457
§42.5.	The QR algorithm	458
§42.6.	Supplementary notes	460
Chapter 43.	Toeplitz, Hankel, and de Branges	461
§43.1.	Reproducing kernel Hilbert spaces	461
§43.2.	de Branges spaces	463
§43.3.	The space of polynomials of degree $\leq n-1$	464
§43.4.	Two subspaces	466

§43.5.	G is a Toeplitz matrix	467
§43.6.	G is a Hankel matrix	470
§43.7.	Supplementary notes	473

Bibliography 475

Notation index 479

Subject index 481

Preface to the third edition

This new edition of *Linear Algebra in Action* is significantly different from the previous edition in both content and style: It includes a number of topics that did not appear in the earlier edition and excludes some that did. In the earlier edition I entered into the proofs of every fact that was used. In this edition I have relegated the proofs of a number of theorems to outside references and have focused instead on their applications, which to my mind has more impact than a proof, especially on a first pass. Moreover, most of the material that is adapted from the previous edition has been rewritten and reorganized.

I have organized this book into short chapters, most of which are a dozen pages or less, because I think this is more amenable to classroom use. I have tried to write this book in the style that most of the mathematicians that I know work in, rather than in the way that they write. In particular, I believe that the discussion of a well-chosen example is often much more helpful than a formal proof, which in many cases is an example hidden by elaborate bookkeeping. A good student will be able to pass to the general setting from the example, and a weaker student will at least have something concrete to focus on.

The book is intended primarily for students who have had at least a little exposure to linear algebra. Nevertheless, the first twelve chapters or so are basically a quick review of the material that is typically offered in a first course, plus a little. The content of this introductory material is dealt with in greater detail in the second edition of this book, which is a useful supplement to this third edition, but the two can be used independently.

A reader who is familiar with the main contents of the first sixteen chapters and Chapter 21 should be able to read any of the other chapters without difficulty, as they are for the most part independent of each other.

The entries **keep in mind**, **warning**, and **notation** appear in the index. The first is to call attention to compilations that I feel are helpful, the second is to call attention to conventions that have been introduced, and the third is to point out the introduction of new notation, most of which is fairly standard, except that a distinction is made between the matrices A^H (the conjugate transpose of A) and A^* (the adjoint of A, which depends upon the inner product).

I extend my thanks to the readers who contributed corrections to the first two editions over the past several years and to Shmuel Aviya for reading and commenting on a number of chapters as they were being prepared for this third edition. I owe a special note of thanks to Dr. Andrei Iacob who carefully copyedited a close to final version of the third edition. It is also a pleasure to thank the staff of the AMS for their friendly help, with extra special thanks to my copyeditor, Arlene O'Sean, for accommodating the author and for her care and devotion to getting things right.

I plan to use the AMS website www.ams.org/bookpages/gsm-232 to supply some supplementary material as well as for sins of omission and commission (and just plain afterthoughts).

I did not add *words of wisdom* at the beginning of each chapter as in the earlier editions. However, I cannot resist repeating two of them:

The first is based on some four score and five years of observing the human scene:

Those who think they know all the answers don't know all the questions.

A Chinese proverb puts it well: *Trust only those who doubt.*

The second is one that I am especially fond of (both the saying and its originator, who was a very special person), though I have never been able to live up to it:

Let's throw everything away; then there will be room for what's left.

–Irene Dym

TAM ACH TEREM NISHLAM
February 14, 2023
Rehovot, Israel

Preface to the second edition

I have an opinion. But I do not agree with it.

Joshua Sobol [**83**]

Most of the chapters in the first edition have been revised, some extensively. The revisions include changes in a number of proofs, to either simplify the argument and/or make the logic clearer, and, on occasion, to sharpen the result.

New short introductory sections on linear programming, extreme points for polyhedra and a Nevanlinna-Pick interpolation problem have been added, as have some very short introductory sections on the mathematics behind Google, Drazin inverses, band inverses, and applications of svd together with a number of new exercises.

I would like to thank the many readers who e-mailed me helpful lists of typographical errors. I owe a special word of thanks to David Kimsey and Motke Porat, whose lists hit double figures. I believe I have fixed all the reported errors and then some.

A couple of oversights in the first edition that came to light (principally the fact that the word Hankel should be removed from the statement and proof of Corollary 21.2; an incomplete definition of a support hyperplane; and a certain fuzziness in the discussion of operator norms and multiplicative norms) have also been fixed.

It is a pleasure to thank the staff of the AMS for being so friendly and helpful; a special note of thanks to my copy/production editor Mike Saitas for his sharp eye and cheerful willingness to accommodate the author and to Mary Medeiros for preparing the indices and her expertise in LaTeX.

The AMS website www.ams.org/bookpages/gsm-78 will be used for sins of omission and commission (and just plain afterthoughts) for the second edition as well as the first.

TAM, ACH TEREM NISHLAM, ...

July 19, 2013

Rehovot, Israel

Preface to the first edition

A foolish consistency is the hobgoblin of little minds,...
                                   Ralph Waldo Emerson, *Self Reliance*

This book is based largely on courses that I have taught at the Feinberg Graduate School of the Weizmann Institute of Science over the past 35 years to graduate students with widely varying levels of mathematical sophistication and interests. The objective of a number of these courses was to present a user-friendly introduction to linear algebra and its many applications. Over the years I wrote and rewrote (and then, more often than not, rewrote some more) assorted sets of notes and learned many interesting things en route. This book is the current end product of that process. The emphasis is on developing a comfortable familiarity with the material. Many lemmas and theorems are made plausible by discussing an example that is chosen to make the underlying ideas transparent in lieu of a formal proof; i.e., I have tried to present the material in the way that most of the mathematicians that I know work rather than in the way they write. The coverage is not intended to be exhaustive (or exhausting), but rather to indicate the rich terrain that is part of the domain of linear algebra and to present a decent sample of some of the tools of the trade of a working analyst that I have absorbed and have found useful and interesting in more than 40 years in the business. To put it another way, I wish someone had taught me this material when I was a graduate student. In those days, in the arrogance of youth, I thought that linear algebra was for boys and girls and that real men and women worked in functional analysis. However, this is but one of many opinions that did not stand the test of time.

In my opinion, the material in this book can (and has been) used on many levels. A core course in classical linear algebra topics can be based on the first six chapters, plus selected topics from Chapters 7–9 and 13. The latter treats difference equations, differential equations, and systems thereof. Chapters 14–16 cover applications to vector calculus, including a proof of the implicit function based on the contractive fixed point theorem, and extremal problems with constraints. Subsequent chapters deal with matrix-valued holomorphic functions, matrix equations, realization theory, eigenvalue location problems, zero location problems, convexity, and matrices with nonnegative entries. I have taken the liberty of straying into areas that I consider significant, even though they are not usually viewed as part of the package associated with linear algebra. Thus, for example, I have added short sections on complex function theory, Fourier analysis, Lyapunov functions for dynamical systems, boundary value problems and more. A number of the applications are taken from control theory.

I have adapted material from many sources. But the one which was most significant for at least the starting point of a number of topics covered in this work is the wonderful book [56] by Lancaster and Tismenetsky.

A number of students read and commented on substantial sections of assorted drafts: Boris Ettinger, Ariel Ginis, Royi Lachmi, Mark Kozdoba, Evgeny Muzikantov, Simcha Rimler, Jonathan Ronen, Idith Segev, and Amit Weinberg. I thank them all, and extend my appreciation to two senior readers: Aad Dijksma and Andrei Iacob for their helpful insightful remarks. A special note of thanks goes to Deborah Smith, my copy editor at the AMS, for her sharp eye and expertise in the world of commas and semicolons.

On the production side, I thank Jason Friedman for typing an early version, and our secretaries Diana Mandelik, Ruby Musrie, Linda Alman, Terry Debesh, all of whom typed selections and to Diana again for preparing all the figures and clarifying numerous mysterious intricacies of LaTeX. I also thank Barbara Beeton of the AMS for helpful advice on AMS LaTeX.

One of the difficulties in preparing a manuscript for a book is knowing when to let go. It is always possible to write it better.[1] Fortunately the

---

[1] Israel Gohberg tells of a conversation with Lev Sakhnovich that took place in Odessa many years ago: Lev: Israel, how is your book with Mark Gregorovic (Krein) progressing? Israel: It's about 85% done. Lev: That's great! Why so sad? Israel: If you would have asked me yesterday, I would have said 95%.

AMS maintains a web page: http://www.ams.org/bookpages/gsm-78, for sins of omission and commission (or just plain afterthoughts).

TAM, ACH TEREM NISHLAM,...

October 18, 2006

Rehovot, Israel

Chapter 1

Prerequisites

We shall assume that the reader is familiar with real and complex vector spaces and the elements of matrix theory and linear transformations. Nevertheless, for the sake of completeness, we begin with a quick review of the main definitions and notations that will be in use in the sequel.

1.1. Main definitions

- **Notation**: The symbols $\mathbb{R}$ and $\mathbb{C}$ will be used to denote the real and complex numbers, respectively. For us, the most important vector spaces are $\mathbb{C}^{p\times q}$, the set of $p \times q$ matrices with complex entries, and $\mathbb{R}^{p\times q}$, the set of $p \times q$ matrices with real entries; $\mathbb{C}^p$ is short for $\mathbb{C}^{p\times 1}$ and $\mathbb{R}^p$ is short for $\mathbb{R}^{p\times 1}$. An element $\mathbf{v}$ in a vector space $\mathcal{V}$ is called a **vector** and is usually printed in boldface. We shall say that $\mathcal{V}$ is a **complex vector space** if $\alpha \mathbf{v} \in \mathcal{V}$ for every $\alpha \in \mathbb{C}$ and every $\mathbf{v} \in \mathcal{V}$. Analogously, we shall say that $\mathcal{V}$ is a **real vector space** if $\alpha \mathbf{v} \in \mathcal{V}$ for every $\alpha \in \mathbb{R}$ and every $\mathbf{v} \in \mathcal{V}$. In both cases, the numbers α will be referred to as **scalars**.

- **Subspaces**: A subspace $\mathcal{M}$ of a vector space $\mathcal{V}$ is a nonempty subset of $\mathcal{V}$ that is **closed under vector addition and scalar multiplication**. In other words, if $\mathbf{x}$ and $\mathbf{y}$ belong to $\mathcal{M}$, then $\mathbf{x} + \mathbf{y} \in \mathcal{M}$ and $\alpha \mathbf{x} \in \mathcal{M}$ for every scalar α. A subspace of a vector space is automatically a vector space in its own right.

- **Span**: If $\mathbf{v_1}, \ldots, \mathbf{v_k}$ is a given set of vectors in a vector space $\mathcal{V}$, then

$$\text{span}\{\mathbf{v_1}, \ldots, \mathbf{v_k}\} = \left\{ \sum_{j=1}^{k} \alpha_j \mathbf{v}_j : \alpha_1, \ldots, \alpha_k \text{ are scalars} \right\}.$$

In words, the span is the set of all **linear combinations** $\alpha_1 \mathbf{v}_1 + \cdots + \alpha_k \mathbf{v}_k$ of the indicated set of vectors, with scalar coefficients $\alpha_1, \ldots, \alpha_k$. Or, to put it another way, span $\{\mathbf{v_1}, \ldots, \mathbf{v_k}\}$ is the smallest subspace of $\mathcal{V}$ that contains the vectors $\mathbf{v_1}, \ldots, \mathbf{v_k}$. It is important to keep in mind that the number of vectors k that were used to define the span is not a good indicator of the size of this space. Thus, for example, if

$$\mathbf{v}_1 = \begin{bmatrix} 1 \\ 2 \\ 1 \end{bmatrix}, \ \mathbf{v}_2 = \begin{bmatrix} 2 \\ 4 \\ 2 \end{bmatrix}, \text{ and } \mathbf{v}_3 = \begin{bmatrix} 3 \\ 6 \\ 3 \end{bmatrix},$$

then

$$\text{span}\{\mathbf{v}_1, \mathbf{v}_2, \mathbf{v}_3\} = \text{span}\{\mathbf{v}_1\}.$$

To clarify the notion of the **size of the span** we need the concept of linear dependence.

- **Linear dependence**: A set of vectors $\{\mathbf{v}_1, \ldots, \mathbf{v}_k\}$ in a vector space $\mathcal{V}$ is said to be **linearly dependent** if there exists a set of scalars $\alpha_1, \ldots, \alpha_k$, not all of which are zero, such that

$$\alpha_1 \mathbf{v}_1 + \cdots + \alpha_k \mathbf{v}_k = \mathbf{0}.$$

This permits us to express one or more of the given vectors in terms of the others. Thus, if $\alpha_1 \neq 0$, then

$$\mathbf{v}_1 = -\frac{\alpha_2}{\alpha_1} \mathbf{v}_2 - \cdots - \frac{\alpha_k}{\alpha_1} \mathbf{v}_k$$

and hence

$$\text{span}\{\mathbf{v}_1, \ldots, \mathbf{v}_k\} = \text{span}\{\mathbf{v}_2, \ldots, \mathbf{v}_k\}.$$

Further reductions are possible if the vectors $\mathbf{v}_2, \ldots, \mathbf{v}_k$ are still linearly dependent.

- **Linear independence**: A set of vectors $\{\mathbf{v}_1, \ldots, \mathbf{v}_k\}$ in a vector space $\mathcal{V}$ is **linearly independent** if the only scalars $\alpha_1, \ldots, \alpha_k$ for which

$$\alpha_1 \mathbf{v}_1 + \cdots + \alpha_k \mathbf{v}_k = \mathbf{0}$$

are $\alpha_1 = \cdots = \alpha_k = 0$. This is just another way of saying that we cannot express one of these vectors as a linear combination of the others. Moreover, if $\{\mathbf{v}_1, \ldots, \mathbf{v}_k\}$ is a set of linearly independent vectors in a vector space $\mathcal{V}$ and if

(1.1) $$\mathbf{v} = \alpha_1 \mathbf{v}_1 + \cdots + \alpha_k \mathbf{v}_k \quad \text{and} \quad \mathbf{v} = \beta_1 \mathbf{v}_1 + \cdots + \beta_k \mathbf{v}_k$$

for some choice of scalars $\alpha_1, \ldots, \alpha_k, \beta_1, \ldots, \beta_k$, then $\alpha_j = \beta_j$ for $j = 1, \ldots, k$.

Exercise 1.1. Show that if (1.1) holds for a linearly independent set of vectors, $\{\mathbf{v}_1, \ldots, \mathbf{v}_k\}$, then $\alpha_j = \beta_j$ for $j = 1, \ldots, k$. Show by example that this conclusion is false if the given set of k vectors is not linearly independent.

- **Basis**: A set of vectors $\{\mathbf{v}_1, \ldots, \mathbf{v}_k\}$ is a **basis** for a vector space $\mathcal{V}$ if:
  (1) $\mathrm{span}\{\mathbf{v}_1, \ldots, \mathbf{v}_k\} = \mathcal{V}$.
  (2) The vectors $\mathbf{v}_1, \ldots, \mathbf{v}_k$ are linearly independent.
  Both of these conditions are essential. The first guarantees that the given set of k vectors is large enough to express every vector $\mathbf{v} \in \mathcal{V}$ as a linear combination of $\mathbf{v}_1, \ldots, \mathbf{v}_k$; the second ensures that we cannot achieve this with less than k vectors.

Exercise 1.2. Let $\mathbf{u}_1, \mathbf{u}_2, \mathbf{u}_3$ be linearly independent vectors in a vector space $\mathcal{U}$ and let $\mathbf{u}_4 = \mathbf{u}_1 + 2\mathbf{u}_2 + \mathbf{u}_3$.

(a) Show that the vectors $\mathbf{u}_1, \mathbf{u}_2, \mathbf{u}_4$ are linearly independent and that span $\{\mathbf{u}_1, \mathbf{u}_2, \mathbf{u}_3\}$ = span $\{\mathbf{u}_1, \mathbf{u}_2, \mathbf{u}_4\}$.

(b) Express the vector $7\mathbf{u}_1 + 13\mathbf{u}_2 + 5\mathbf{u}_3$ as a linear combination of the vectors $\mathbf{u}_1, \mathbf{u}_2, \mathbf{u}_4$. [Note that the coefficients of $\mathbf{u}_1$ and $\mathbf{u}_2$ change.]

- **Dimension**: A nontrivial vector space $\mathcal{V}$ has many bases. However, the number of elements in each basis for $\mathcal{V}$ is exactly the same and is referred to as the **dimension** of $\mathcal{V}$ and will be denoted $\dim \mathcal{V}$. If $\mathcal{U}$ is a subspace of a vector space $\mathcal{V}$, then $\dim \mathcal{U} \leq \dim \mathcal{V}$, with equality if and only if $\mathcal{U} = \mathcal{V}$.

Example 1.1. The $p \times q$ matrices $E_{ij}, i = 1, \ldots, p, j = 1, \ldots, q$, that are defined by setting every entry in E_{ij} equal to zero except for the ij entry, which is set equal to one, form a basis for the vector space $\mathbb{C}^{p \times q}$. Thus, $\dim \mathbb{C}^{p \times q} = pq$. ◇

- **Identity matrix**: The symbol I_n denotes the $n \times n$ matrix $A = [a_{ij}]$, $i, j = 1, \ldots, n$, with $a_{ii} = 1$ for $i = 1, \ldots, n$ and $a_{ij} = 0$ for $i \neq j$. The name stems from the fact that $I_n \mathbf{x} = \mathbf{x}$ for every vector $\mathbf{x} \in \mathbb{C}^n$.

- **Zero matrix**: The symbol $O_{p \times q}$ denotes the matrix in $\mathbb{C}^{p \times q}$ all of whose entries are equal to zero. The subscript $p \times q$ will be dropped if the size is clear from the context.

- **Transposes**: The transpose of a $p \times q$ matrix A is the $q \times p$ matrix A^T whose k'th row is equal to the k'th column of A laid sideways, $k = 1, \ldots, q$. In other words, the ij entry of A is equal to the ji entry of A^T.

- **Hermitian transposes**: The Hermitian transpose A^H of a $p \times q$ matrix A is the same as the transpose A^T of A, except that all the entries in the transposed matrix are replaced by their complex conjugates. Thus, for example,

$$A = \begin{bmatrix} 1 & 3i & 5+i \\ 4 & -i & 6i \end{bmatrix} \implies A^T = \begin{bmatrix} 1 & 4 \\ 3i & -i \\ 5+i & 6i \end{bmatrix} \text{ and } A^H = \begin{bmatrix} 1 & 4 \\ -3i & i \\ 5-i & -6i \end{bmatrix}.$$

It is readily checked that

$$(A^T)^T = A, \quad (AB)^T = B^T A^T, \quad (A^H)^H = A, \quad \text{and} \quad (AB)^H = B^H A^H.$$

- **Permutation matrices**: Every $n \times n$ permutation matrix P is obtained by taking the identity matrix I_n and interchanging some of the rows. Consequently, P can be expressed in terms of the columns $\mathbf{e}_j$, $j = 1, \ldots, n$, of I_n and a one-to-one mapping σ of the set of integers $\{1, \ldots, n\}$ onto itself by the formula

(1.2) $$P = P_\sigma = \sum_{j=1}^n \mathbf{e}_j \mathbf{e}_{\sigma(j)}^T, \quad \text{and hence} \quad I_n = \sum_{j=1}^n \mathbf{e}_j \mathbf{e}_j^T.$$

Thus, for example, if $n = 4$ and $\sigma(1) = 3$, $\sigma(2) = 2$, $\sigma(3) = 4$, and $\sigma(4) = 1$, then

$$P_\sigma = \mathbf{e}_1 \mathbf{e}_3^T + \mathbf{e}_2 \mathbf{e}_2^T + \mathbf{e}_3 \mathbf{e}_4^T + \mathbf{e}_4 \mathbf{e}_1^T = \begin{bmatrix} 0 & 0 & 1 & 0 \\ 0 & 1 & 0 & 0 \\ 0 & 0 & 0 & 1 \\ 1 & 0 & 0 & 0 \end{bmatrix}.$$

- **Orthogonal matrices**: A matrix $V \in \mathbb{R}^{n \times n}$ is said to be an orthogonal matrix if $V^T V = I_n$.

- **Matrix multiplication**: Let $A = [a_{ij}]$ be a $p \times q$ matrix and let $B = [b_{st}]$ be a $q \times r$ matrix. Then the product AB is the $p \times r$ matrix $C = [c_{k\ell}]$ with entries

$$c_{k\ell} = \sum_{j=1}^q a_{kj} b_{j\ell}, \quad k = 1, \ldots, p, \, \ell = 1, \ldots, r.$$

1.1. Main definitions

Notice that $c_{k\ell}$ is the matrix product of the k'th **row** $\vec{\mathbf{a}}_k$ of A with the ℓ'th **column** $\mathbf{b}_\ell$ of B:

$$c_{k\ell} = \vec{\mathbf{a}}_k \mathbf{b}_\ell = \begin{bmatrix} a_{k1} & \cdots & a_{kq} \end{bmatrix} \begin{bmatrix} b_{1\ell} \\ \vdots \\ b_{q\ell} \end{bmatrix}.$$

- Matrix multiplication is **not commutative**, e.g.,

$$A = \begin{bmatrix} 1 & 0 \\ 1 & 0 \end{bmatrix} \text{ and } B = \begin{bmatrix} 0 & 0 \\ 1 & 1 \end{bmatrix} \implies AB = \begin{bmatrix} 0 & 0 \\ 0 & 0 \end{bmatrix} \neq \begin{bmatrix} 0 & 0 \\ 2 & 0 \end{bmatrix} = BA.$$

- Matrix multiplication is **associative**: If $A \in \mathbb{C}^{p \times q}$, $B \in \mathbb{C}^{q \times r}$, and $C \in \mathbb{C}^{r \times s}$, then

$$(AB)C = A(BC).$$

- Matrix multiplication is **distributive**: If $A, A_1, A_2 \in \mathbb{C}^{p \times q}$ and $B, B_1, B_2 \in \mathbb{C}^{q \times r}$, then

$$(A_1 + A_2)B = A_1 B + A_2 B \quad \text{and} \quad A(B_1 + B_2) = AB_1 + AB_2.$$

- If $A \in \mathbb{C}^{p \times q}$ is expressed both as an array of p **row vectors** of width q and as an array of q **column vectors** of height p,

$$A = \begin{bmatrix} \vec{\mathbf{a}}_1 \\ \vdots \\ \vec{\mathbf{a}}_p \end{bmatrix} = \begin{bmatrix} \mathbf{a}_1 & \cdots & \mathbf{a}_q \end{bmatrix},$$

and if $B \in \mathbb{C}^{q \times r}$ is expressed both as an array of q row vectors of width r and as an array of r column vectors of height q,

$$B = \begin{bmatrix} \vec{\mathbf{b}}_1 \\ \vdots \\ \vec{\mathbf{b}}_q \end{bmatrix} = \begin{bmatrix} \mathbf{b}_1 & \cdots & \mathbf{b}_r \end{bmatrix},$$

then the product AB can be expressed in the following three ways:

(1.3) $$AB = \begin{bmatrix} \vec{\mathbf{a}}_1 B \\ \vdots \\ \vec{\mathbf{a}}_p B \end{bmatrix} = \begin{bmatrix} A\mathbf{b}_1 & \cdots & A\mathbf{b}_r \end{bmatrix} = \sum_{s=1}^{q} \mathbf{a}_s \vec{\mathbf{b}}_s.$$

- If $\mathbf{x} \in \mathbb{C}^q$ with entries $x_1, \ldots, x_q$, then

(1.4) $$A\mathbf{x} = \begin{bmatrix} \mathbf{a}_1 & \cdots & \mathbf{a}_q \end{bmatrix} \mathbf{x} = x_1 \mathbf{a}_1 + \cdots + x_q \mathbf{a}_q.$$

- **Inverses**: Let $A \in \mathbb{C}^{p \times q}$. Then A is:
  (1) **Left invertible** if there is a matrix $C \in \mathbb{C}^{q \times p}$ such that $CA = I_q$; in this case C is called a **left inverse** of A and the equation $A\mathbf{x} = \mathbf{b}$ has at most one solution $\mathbf{x}$.

(2) **Right invertible** if there is a matrix $B \in \mathbb{C}^{q \times p}$ such that $AB = I_p$; in this case B is called a **right inverse** of A and the equation $A\mathbf{x} = \mathbf{b}$ has at least one solution $\mathbf{x}$.

If a matrix $A \in \mathbb{C}^{p \times q}$ has both a left inverse C and a right inverse B, then

(1.5) $\quad C = CI_p = C(AB) = (CA)B = I_q B = B \quad \text{and} \quad q = p.$

Thus, if A has both a left and a right inverse, then it has exactly one left inverse and exactly one right inverse. Moreover, they are equal and (as will be shown in Section 2.3) $q = p$. In this instance, we shall say that A is **invertible** and refer to $B = C$ as the **inverse** of A and denote it by A^{-1}.

Exercise 1.3. Show that if $A \in \mathbb{C}^{n \times n}$ and $B \in \mathbb{C}^{n \times n}$ are invertible, then AB is invertible and $(AB)^{-1} = B^{-1}A^{-1}$.

- **Block multiplication**: It is often convenient to express a large matrix as an array of submatrices (i.e., blocks of numbers) rather than as an array of numbers. Then the rules of matrix multiplication still apply (block by block) provided that the block decompositions are compatible. Thus, for example, if

$$A = \begin{bmatrix} A_{11} & A_{12} \\ A_{21} & A_{22} \\ A_{31} & A_{32} \end{bmatrix} \quad \text{and} \quad B = \begin{bmatrix} B_{11} & B_{12} & B_{13} & B_{14} \\ B_{21} & B_{22} & B_{23} & B_{24} \end{bmatrix}$$

with entries $A_{ij} \in \mathbb{C}^{p_i \times q_j}$ and $B_{jk} \in \mathbb{C}^{q_j \times r_k}$, then

$$C = AB = [C_{ij}], \quad i = 1, \ldots, 3, \ j = 1, \ldots, 4,$$

where

$$C_{ij} = A_{i1}B_{1j} + A_{i2}B_{2j}$$

is a $p_i \times r_j$ matrix.

Exercise 1.4. Verify the three ways of writing a matrix product in formula (1.3). [HINT: Show that the ij entries coincide for the first two; then observe that if $I_r = \begin{bmatrix} \mathbf{e}_1 & \cdots & \mathbf{e}_r \end{bmatrix}$, then $AB = AI_r B$ and $I_r = \sum_{s=1}^{r} \mathbf{e}_s \mathbf{e}_s^T$ for the third.]

Exercise 1.5. Show that every permutation matrix $P \in \mathbb{R}^{n \times n}$ is an orthogonal matrix, i.e., $P^T P = PP^T = I_n$. [HINT: Use formula (1.2).]

1.2. Mappings

- **Mappings**: A mapping (or transformation) T from a vector space $\mathcal{U}$ into a vector space $\mathcal{V}$ is a rule that assigns exactly one vector $\mathbf{v} \in \mathcal{V}$ to each $\mathbf{u} \in \mathcal{U}$. In this framework either $\mathcal{U}$ and $\mathcal{V}$ will

1.2. Mappings

both be complex vector spaces or they will both be real vector spaces.

We shall refer to the set

$$\mathcal{N}_T = \{\mathbf{u} \in \mathcal{U} : T\mathbf{u} = \mathbf{0}_\mathcal{V}\}$$

as the **nullspace** (or **kernel**) of T and to the set

$$\mathcal{R}_T = \{T\mathbf{u} : \mathbf{u} \in \mathcal{U}\}$$

as the **range** (or **image**) of T. The subscript $\mathcal{V}$ is added to the symbol $\mathbf{0}$ in the first definition to emphasize that it is the zero vector in $\mathcal{V}$, not in $\mathcal{U}$.

- **Linear mappings**: A mapping T from a vector space $\mathcal{U}$ into a vector space $\mathcal{V}$ is **linear** if for every choice of $\mathbf{u}, \mathbf{v} \in \mathcal{U}$ and every scalar α the following two conditions are met:
  (1) $T(\mathbf{u} + \mathbf{v}) = T\mathbf{u} + T\mathbf{v}$.
  (2) $T(\alpha \mathbf{u}) = \alpha T\mathbf{u}$.
  If T is a linear mapping from a vector space $\mathcal{U}$ into a vector space $\mathcal{V}$, then $\mathcal{N}_T$ is a subspace of $\mathcal{U}$ and $\mathcal{R}_T$ is a subspace of $\mathcal{V}$.

- **The identity**: The special linear transformation from a vector space $\mathcal{U}$ into $\mathcal{U}$ that maps each vector $\mathbf{u} \in \mathcal{U}$ into itself is called the **identity** mapping. It is denoted by the symbol I_n if $\mathcal{U} = \mathbb{C}^n$ or $\mathcal{U} = \mathbb{R}^n$ and by $I_\mathcal{U}$ otherwise, though, more often than not, when the underlying space $\mathcal{U}$ is clear from the context, the subscript $\mathcal{U}$ will be dropped.

If T is a linear mapping of a vector space $\mathcal{U}$ with basis $\{\mathbf{u}_1, \ldots, \mathbf{u}_q\}$ into a vector space $\mathcal{V}$ with basis $\{\mathbf{v}_1, \ldots, \mathbf{v}_p\}$, then there exists a unique set of scalars a_{ij}, $i = 1, \ldots, p$ and $j = 1, \ldots, q$, such that

$$(1.6) \qquad T\mathbf{u}_j = \sum_{i=1}^{p} a_{ij}\mathbf{v}_i \quad \text{for} \quad j = 1, \ldots, q$$

and hence that

$$(1.7) \qquad T\left(\sum_{j=1}^{q} x_j \mathbf{u}_j\right) = \sum_{i=1}^{p} y_i \mathbf{v}_i \iff A \begin{bmatrix} x_1 \\ \vdots \\ x_q \end{bmatrix} = \begin{bmatrix} y_1 \\ \vdots \\ y_p \end{bmatrix}.$$

- **Warning**: If $A \in \mathbb{C}^{p \times q}$, then matrix multiplication defines a linear map that sends $\mathbf{x} \in \mathbb{C}^q$ to $A\mathbf{x} \in \mathbb{C}^p$. Correspondingly, the nullspace of this map,

$$\mathcal{N}_A = \{\mathbf{x} \in \mathbb{C}^q : A\mathbf{x} = \mathbf{0}\}, \quad \text{is a subspace of} \quad \mathbb{C}^q,$$

and the range of this map,

$$\mathcal{R}_A = \{A\mathbf{x} : \mathbf{x} \in \mathbb{C}^q\}, \quad \text{is a subspace of} \quad \mathbb{C}^p.$$

However, if $A \in \mathbb{R}^{p \times q}$, then matrix multiplication also defines a linear map that sends $\mathbf{x} \in \mathbb{R}^q$ to $A\mathbf{x} \in \mathbb{R}^p$; in this setting

$$\mathcal{N}_A = \{\mathbf{x} \in \mathbb{R}^q : A\mathbf{x} = \mathbf{0}\} \quad \text{is a subspace of} \quad \mathbb{R}^q,$$

and the range of this map,

$$\mathcal{R}_A = \{A\mathbf{x} : \mathbf{x} \in \mathbb{R}^q\}, \quad \text{is a subspace of} \quad \mathbb{R}^p.$$

In short, it is important to clarify the space on which A is acting, i.e., the **domain** of A. This will usually be clear from the context.

1.3. Triangular matrices

A matrix $A \in \mathbb{C}^{n \times n}$ with entries a_{ij}, $i, j = 1, \ldots, n$, is said to be:

- **upper triangular** if all its nonzero entries sit either on or above the diagonal, i.e., if $a_{ij} = 0$ when $i > j$;
- **lower triangular** if all its nonzero entries sit either on or below the diagonal, i.e., if A^T is upper triangular;
- **triangular** if it is either upper triangular or lower triangular;
- **diagonal** if $a_{ij} = 0$ when $i \neq j$.

If $A \in \mathbb{C}^{n \times n}$ is a triangular matrix, then:

- A is invertible if and only if all its diagonal entries are nonzero.

If A is an invertible triangular matrix, then:

- A is upper triangular $\iff A^{-1}$ is upper triangular.
- A is lower triangular $\iff A^{-1}$ is lower triangular.

Triangular matrices will be discussed in more detail in Section 2.5. Systems of equations based on a triangular matrix are particularly convenient to work with, even if the matrix is not invertible.

Example 1.2. Let $A \in \mathbb{C}^{4 \times 4}$ be a 4×4 upper triangular matrix with nonzero diagonal entries and let $\mathbf{b}$ be any vector in $\mathbb{C}^4$. Then the vector $\mathbf{x}$ is a solution of the equation

(1.8) $$A\mathbf{x} = \mathbf{b}$$

if and only if

$$\begin{aligned} a_{11}x_1 + a_{12}x_2 + a_{13}x_3 + a_{14}x_4 &= b_1, \\ a_{22}x_2 + a_{23}x_3 + a_{24}x_4 &= b_2, \\ a_{33}x_3 + a_{34}x_4 &= b_3, \\ a_{44}x_4 &= b_4. \end{aligned}$$

Therefore, since the diagonal entries of A are nonzero, it is readily seen that these equations admit exactly one solution, by working from the bottom up:

$$\begin{aligned} x_4 &= a_{44}^{-1} b_4, \\ x_3 &= a_{33}^{-1}(b_3 - a_{34} x_4), \\ x_2 &= a_{22}^{-1}(b_2 - a_{23} x_3 - a_{24} x_4), \\ x_1 &= a_{11}^{-1}(b_1 - a_{12} x_2 - a_{13} x_3 - a_{14} x_4). \end{aligned}$$

Thus, equation (1.8) admits exactly one solution $\mathbf{x}$ for each $\mathbf{b}$.

Let $\mathbf{x}_j$ denote the solution of the equation $A\mathbf{x}_j = \mathbf{e}_j$ for $j = 1, \ldots, 4$ when $\mathbf{e}_j$, $j = 1, \ldots, 4$, denotes the j'th column of the identity matrix I_4. Then the 4×4 matrix $X = [\mathbf{x}_1 \; \mathbf{x}_2 \; \mathbf{x}_3 \; \mathbf{x}_4]$ is a right inverse of A:

$$AX = A[\mathbf{x}_1 \cdots \mathbf{x}_4] = [A\mathbf{x}_1 \cdots A\mathbf{x}_4] = [\mathbf{e}_1 \cdots \mathbf{e}_4] = I_4 \, .$$

Analogous examples can be built for lower triangular $p \times p$ matrices. The only difference is that now it is advantageous to work from the top down. The existence of a left inverse can also be obtained by writing down the requisite equations that must be solved and imitating the preceding arguments. It is easier, however, to take advantage of the fact that $YA = I_p \iff A^T Y^T = I_p$. (In fact, as we shall see shortly, if $A, B \in \mathbb{C}^{p \times p}$, then $AB = I_p \iff BA = I_p$.) ◇

1.4. The binomial formula

The familiar binomial identity

$$(a+b)^m = \sum_{k=0}^{m} \binom{m}{k} a^k b^{m-k}$$

for complex numbers a and b remains valid for square matrices A and B of the same size if they commute:

(1.9) $$(A+B)^m = \sum_{k=0}^{m} \binom{m}{k} A^k B^{m-k} \quad \text{if } AB = BA.$$

It is easy to see that the condition $AB = BA$ is necessary for the formula in (1.9) to hold by comparing both sides when $m = 2$; the sufficiency may be verified by induction.

If $A = \lambda I_n$ and $B \in \mathbb{C}^{n \times n}$, then $AB = BA$ and hence

(1.10) $$(\lambda I_n + B)^m = \sum_{k=0}^{m} \binom{m}{k} \lambda^k B^{m-k}.$$

1.5. Supplementary notes

Most of this chapter is adapted from Chapter 1 of [**30**], which contains more details. A principal difference is that in the present treatment we have simply accepted the fact that if a vector space has a basis with a finite number of vectors, then every other basis of that space will have exactly the same number of vectors. In the somewhat distant future we shall also deal with vector spaces $\mathcal{F}$ of functions that are defined on some reasonable subset Ω of $\mathbb{C}^n$ or $\mathbb{R}^n$. Then vector addition and multiplication by scalars is defined in a natural way: If $f, g \in \mathcal{F}$, $\omega \in \Omega$, and α is a scalar, then

$$(1.11) \qquad (f+g)(\omega) = f(\omega) + g(\omega) \quad \text{and} \quad (\alpha f)(\omega) = \alpha f(\omega).$$

Example 1.3. The set $\mathcal{F}$ of continuous complex- (resp., real-) valued functions $f(x)$ on the interval $0 \leq x \leq 1$ is a complex (resp., real) vector space with respect to the natural rules of vector addition and scalar multiplication that were introduced in (1.11).

Exercise 1.6. Show that the set $\mathcal{F}_0 = \{f \in \mathcal{F} : f(0) = 0 \text{ and } f(1) = 0\}$ is a subspace of the vector space $\mathcal{F}$ considered in the preceding example, but the set $\mathcal{F}_1 = \{f \in \mathcal{F} : f(0) = 0 \text{ and } f(1) = 1\}$ is not a subspace of $\mathcal{F}$.

Chapter 2

Dimension and rank

In this chapter we shall first establish a useful formula (see (2.1)) that we call the conservation of dimension and then explore some of its implications. The last section is devoted to block triangular matrices.

2.1. The conservation of dimension

Theorem 2.1. *Let T be a linear mapping from a finite-dimensional vector space $\mathcal{U}$ into a vector space $\mathcal{V}$ (finite dimensional or not). Then*

(2.1) $$\dim \mathcal{N}_T + \dim \mathcal{R}_T = \dim \mathcal{U}.$$

Proof. Since $\dim \mathcal{R}_T \leq \dim \mathcal{U}$, $\mathcal{R}_T$ is automatically a finite-dimensional space regardless of the dimension of $\mathcal{V}$. Suppose first that $\mathcal{N}_T \neq \{\mathbf{0}\}$ and $\mathcal{R}_T \neq \{\mathbf{0}\}$, and let $\mathbf{u}_1, \ldots, \mathbf{u}_k$ be a basis for $\mathcal{N}_T$, let $\mathbf{v}_1, \ldots, \mathbf{v}_\ell$ be a basis for $\mathcal{R}_T$, and choose vectors $\mathbf{y}_j \in \mathcal{U}$ such that

$$T\mathbf{y}_j = \mathbf{v}_j, \quad j = 1, \ldots, \ell.$$

To verify (2.1), we shall show that the set of vectors $\{\mathbf{u}_1, \ldots, \mathbf{u}_k, \mathbf{y}_1, \ldots, \mathbf{y}_\ell\}$ is a basis for $\mathcal{U}$. The first item of business is to show that the $k + \ell$ vectors in this set are linearly independent. If there exist scalars $\alpha_1, \ldots, \alpha_k$ and $\beta_1, \ldots, \beta_\ell$ such that

(2.2) $$\sum_{i=1}^k \alpha_i \mathbf{u}_i + \sum_{j=1}^\ell \beta_j \mathbf{y}_j = \mathbf{0},$$

then
$$\mathbf{0} = T\mathbf{0} = T\left(\sum_{i=1}^{k}\alpha_i\mathbf{u}_i + \sum_{j=1}^{\ell}\beta_j\mathbf{y}_j\right)$$
$$= \sum_{i=1}^{k}\alpha_i T\mathbf{u}_i + \sum_{j=1}^{\ell}\beta_j T\mathbf{y}_j = \mathbf{0} + \sum_{j=1}^{\ell}\beta_j\mathbf{v}_j.$$

Therefore, $\beta_1 = \cdots = \beta_\ell = 0$ and so too, by (2.2), $\alpha_1 = \cdots = \alpha_k = 0$. This completes the proof of the asserted linear independence.

The next step is to check that

(2.3) $\qquad\qquad \mathrm{span}\{\mathbf{u}_1,\ldots,\mathbf{u}_k,\mathbf{y}_1,\ldots,\mathbf{y}_\ell\} = \mathcal{U}.$

Towards this end, let $\mathbf{w} \in \mathcal{U}$. Then, since
$$T\mathbf{w} = \sum_{j=1}^{\ell}\beta_j\mathbf{v}_j = \sum_{j=1}^{\ell}\beta_j T\mathbf{y}_j$$
for some choice of scalars $\beta_1,\ldots,\beta_\ell$, it follows that
$$T\left(\mathbf{w} - \sum_{j=1}^{\ell}\beta_j\mathbf{y}_j\right) = \mathbf{0}$$
and hence that
$$\mathbf{w} - \sum_{j=1}^{\ell}\beta_j\mathbf{y}_j \in \mathcal{N}_T.$$
Consequently, this vector can be expressed as a linear combination of the vectors $\mathbf{u}_1,\ldots,\mathbf{u}_k$. In other words,
$$\mathbf{w} = \sum_{i=1}^{k}\alpha_i\mathbf{u}_i + \sum_{j=1}^{\ell}\beta_j\mathbf{y}_j$$
for some choice of scalars $\alpha_1,\ldots,\alpha_k$ and $\beta_1,\ldots,\beta_\ell$. But this means that (2.3) holds and hence, in view of the already exhibited linear independence, that
$$\dim \mathcal{U} = k + l = \dim \mathcal{N}_T + \dim \mathcal{R}_T,$$
as claimed.

Suppose next that $\mathcal{N}_T = \{\mathbf{0}\}$ and $\mathcal{R}_T \neq \{\mathbf{0}\}$. Then much the same sort of argument serves to prove that if $\mathbf{v}_1,\ldots,\mathbf{v}_\ell$ is a basis for $\mathcal{R}_T$ and if $\mathbf{y}_j \in \mathcal{U}$ is such that $T\mathbf{y}_j = \mathbf{v}_j$ for $j = 1,\ldots,\ell$, then the vectors $\mathbf{y}_1,\ldots,\mathbf{y}_l$ are linearly independent and span $\mathcal{U}$. Thus, $\dim \mathcal{U} = \dim \mathcal{R}_T = \ell$, and hence formula (2.1) is still in force, since $\dim \mathcal{N}_T = 0$.

It remains only to consider the case $\mathcal{R}_T = \{\mathbf{0}\}$. But then $\mathcal{N}_T = \mathcal{U}$, and formula (2.1) is still valid. $\square$

We shall refer to formula (2.1) as the **principle of conservation of dimension**.

2.2. Conservation of dimension for matrices

One of the main applications of the principle of conservation of dimension is to the particular linear transformation T from $\mathbb{C}^q$ into $\mathbb{C}^p$ that is defined by multiplying each vector $\mathbf{x} \in \mathbb{C}^q$ by a given matrix $A \in \mathbb{C}^{p \times q}$. In this setting:

- $\mathcal{N}_A = \{\mathbf{x} \in \mathbb{C}^q : A\mathbf{x} = \mathbf{0}\}$ is a subspace of $\mathbb{C}^q$.
- $\mathcal{R}_A = \{A\mathbf{x} : \mathbf{x} \in \mathbb{C}^q\}$ is a subspace of $\mathbb{C}^p$.
- The **conservation of dimension** principle translates to

(2.4) $$q = \dim \mathcal{N}_A + \dim \mathcal{R}_A.$$

- The dimension of $\mathcal{R}_A$ is termed the **rank** of A:

$$\operatorname{rank} A = \dim \mathcal{R}_A.$$

Moreover, it is **important to keep in mind** that if $A = \begin{bmatrix} \mathbf{a}_1 & \cdots & \mathbf{a}_q \end{bmatrix}$, then:

- $\mathcal{R}_A = \operatorname{span}\{\mathbf{a}_1, \ldots, \mathbf{a}_q\}$, and the vectors $\mathbf{a}_1, \ldots, \mathbf{a}_q$ are linearly independent if and only if $\mathcal{N}_A = \{\mathbf{0}\}$.

Exercise 2.1. Show that if $A \in \mathbb{C}^{p \times q}$ and $C \in \mathbb{C}^{k \times q}$, then

(2.5) $$\operatorname{rank} \begin{bmatrix} A \\ C \end{bmatrix} = q \iff \mathcal{N}_A \cap \mathcal{N}_C = \{\mathbf{0}\}.$$

Exercise 2.2. Show that if A is a triangular matrix (either upper or lower), then rank A is bigger than or equal to the number of nonzero diagonal entries in A. Give an example of an upper triangular matrix A for which the inequality is strict.

Exercise 2.3. Find the null space $\mathcal{N}_A$ and the range $\mathcal{R}_A$ of the matrix

$$A = \begin{bmatrix} 3 & 1 & 0 & 2 \\ 4 & 1 & 0 & 2 \\ 5 & 2 & 0 & 4 \end{bmatrix} \quad \text{acting on} \quad \mathbb{C}^4$$

and check that the principle of conservation of dimension holds.

2.3. What you need to know about rank

The next theorem is a little on the long side but has the advantage that it plus the **conservation of dimension** for $A \in \mathbb{C}^{p \times q}$,

(2.6) $$q = \dim \mathcal{N}_A + \dim \mathcal{R}_A,$$

contains almost everything you need to know about rank in one location.

Theorem 2.2. *If $A \in \mathbb{C}^{p \times q}$, $B \in \mathbb{C}^{q \times r}$, and $C \in \mathbb{C}^{n \times p}$, then:*

(1) $\operatorname{rank} AB \leq \min\{\operatorname{rank} A, \operatorname{rank} B\}$.
(2) $\mathcal{N}_{A^H A} = \mathcal{N}_A$, $\operatorname{rank} A^H A = \operatorname{rank} A$ *and* $\mathcal{R}_{A^H A} = \mathcal{R}_{A^H}$.
(3) $\operatorname{rank} A = \operatorname{rank} A^H = \operatorname{rank} A^T$.
(4) $\operatorname{rank} A \leq \min\{p, q\}$.
(5) A *is right invertible* $\iff \mathcal{R}_A = \mathbb{C}^p \iff \operatorname{rank} A = p$.
(6) A *is left invertible* $\iff \mathcal{N}_A = \{\mathbf{0}\} \iff \operatorname{rank} A = q$.
(7) *If B is right invertible, then* $\mathcal{R}_{AB} = \mathcal{R}_A$ *and* $\operatorname{rank} AB = \operatorname{rank} A$.
(8) *If C is left invertible, then* $\mathcal{N}_{CA} = \mathcal{N}_A$ *and* $\operatorname{rank} CA = \operatorname{rank} A$.
(9) *If $B \in \mathbb{C}^{q \times p}$, then* $q + \operatorname{rank}(I_p - AB) = p + \operatorname{rank}(I_q - BA)$.

Proof. Since $\mathcal{R}_{AB} \subseteq \mathcal{R}_A$, it is clear that $\operatorname{rank} AB \leq \operatorname{rank} A$. The auxiliary inequality $\operatorname{rank} AB \leq \operatorname{rank} B$ then follows from the fact that if $\{\mathbf{v}_1, \ldots, \mathbf{v}_k\}$ is a basis for $\mathcal{R}_B$, then $\mathcal{R}_{AB} \subseteq \operatorname{span}\{A\mathbf{v}_1, \ldots, A\mathbf{v}_k\}$. Thus, (1) holds.

Suppose next that $\mathbf{x} \in \mathcal{N}_{A^H A}$ and let $\mathbf{y} = A\mathbf{x}$. Then

$$\sum_{j=1}^{p} |y_j|^2 = \mathbf{y}^H \mathbf{y} = (A\mathbf{x})^H A\mathbf{x} = \mathbf{x}^H (A^H A\mathbf{x}) = \mathbf{x}^H \mathbf{0} = 0.$$

Therefore, $A\mathbf{x} = \mathbf{y} = \mathbf{0}$, i.e., $\mathcal{N}_{A^H A} \subseteq \mathcal{N}_A$. Thus, as the inclusion $\mathcal{N}_A \subseteq \mathcal{N}_{A^H A}$ is obvious, the first equality in (2) holds. The second equality in (2) then follows from the principle of conservation of dimension applied first to $A \in \mathbb{C}^{p \times q}$ and then to $A^H A \in \mathbb{C}^{q \times q}$:

$$q = \dim \mathcal{N}_A + \dim \mathcal{R}_A = \dim \mathcal{N}_{A^H A} + \dim \mathcal{R}_{A^H A}.$$

Consequently, $\dim \mathcal{R}_A = \dim \mathcal{R}_{A^H A}$ and hence, in view of (1),

$$\operatorname{rank} A = \operatorname{rank} A^H A \leq \operatorname{rank} A^H.$$

Since the last inequality may be applied to A^H as well as to A and $(A^H)^H = A$, it follows that $\operatorname{rank} A^H \leq \operatorname{rank} A$. Thus, equality must hold. Moreover, $\dim \mathcal{R}_{A^H A} = \dim \mathcal{R}_{A^H}$ and therefore the self-evident inclusion $\mathcal{R}_{A^H A} \subseteq \mathcal{R}_{A^H}$ must be an equality. This completes the proof of (2) and the first equality in (3); the second is left to the reader.

Assertion (4) is immediate from (1) and the observation that $A = I_p A = AI_q$.

Suppose next that $A \in \mathbb{C}^{p \times q}$ is right invertible. Then there exists a matrix $B \in \mathbb{C}^{q \times p}$ such that $AB = I_p$ and hence, by (1) and (4),

$$p = \operatorname{rank} AB \leq \operatorname{rank} A \leq p \implies \operatorname{rank} A = p \implies \mathcal{R}_A = \mathbb{C}^p.$$

Conversely, if $\mathcal{R}_A = \mathbb{C}^p$, then the equations

$$A\mathbf{x}_j = \mathbf{b}_j, \quad j = 1, \ldots, p,$$

2.3. What you need to know about rank

are solvable for every choice of the vectors $\mathbf{b}_j$. If, in particular, $\mathbf{b}_j$ is set equal to the j'th column of the identity matrix I_p, then

$$A \begin{bmatrix} \mathbf{x}_1 & \cdots & \mathbf{x}_p \end{bmatrix} = \begin{bmatrix} \mathbf{b}_1 & \cdots & \mathbf{b}_p \end{bmatrix} = I_p.$$

Thus, $\mathcal{R}_A = \mathbb{C}^p \implies A$ is right invertible, since the $q \times p$ matrix

$$X = \begin{bmatrix} \mathbf{x}_1 & \cdots & \mathbf{x}_p \end{bmatrix}$$

with columns $\mathbf{x}_1, \ldots, \mathbf{x}_p$ is a right inverse of A. Consequently, (5) holds.

Next, (6) follows from (5) and the observation that

$$\begin{aligned} \mathcal{N}_A = \{\mathbf{0}\} &\iff \operatorname{rank} A = q \quad \text{(by (2.4))} \\ &\iff \operatorname{rank} A^H = q \quad \text{(by (3))} \\ &\iff A^H \text{ is right invertible} \quad \text{(by (5))} \\ &\iff A \text{ is left invertible}. \end{aligned}$$

The main step in the justification of (7) is to show that if B is right invertible, then $\mathcal{R}_A \subseteq \mathcal{R}_{AB}$, because the opposite inclusion is valid for every $B \in \mathbb{C}^{q \times p}$, right invertible or not. But, if $BD = I_q$ for some $D \in \mathbb{C}^{r \times q}$, then

$$\mathcal{R}_A = \mathcal{R}_{A(BD)} = \mathcal{R}_{(AB)D} \subseteq \mathcal{R}_{AB} \subseteq \mathcal{R}_A.$$

Therefore, (7) holds, since all these spaces must be equal.

Suppose next that C is left invertible. Then $EC = I_p$ for some $E \in \mathbb{C}^{p \times n}$ and

$$\mathcal{N}_A = \mathcal{N}_{(EC)A} = \mathcal{N}_{E(CA)} \supseteq \mathcal{N}_{CA} \supseteq \mathcal{N}_A.$$

Therefore (8) holds, since all these spaces must be equal.

Item (9) is listed here for ease of access; the proof is postponed to Exercise 3.17. □

It is **important to keep in mind** the following implications of Theorem 2.2:

(1) If $A \in \mathbb{C}^{p \times q}$ is right invertible, then $p \leq q$, i.e., A is either a square matrix or a **fat matrix**.

(2) If $A \in \mathbb{C}^{p \times q}$ is left invertible, then $q \leq p$, i.e., A is either a square matrix or a **thin matrix**.

(3) If $A \in \mathbb{C}^{p \times p}$, then $\mathcal{N}_A = \{\mathbf{0}\} \iff \mathcal{R}_A = \mathbb{C}^p$.

(4) If $A, B \in \mathbb{C}^{p \times p}$, then

$$AB = I_p \iff BA = I_p \iff A \text{ and } B \text{ are both invertible}.$$

The assumption that A and B are square matrices is crucial for the validity of (4):
$$\begin{bmatrix} 1 & 0 \end{bmatrix} \begin{bmatrix} 1 \\ 0 \end{bmatrix} = 1, \quad \text{whereas} \quad \begin{bmatrix} 1 \\ 0 \end{bmatrix} \begin{bmatrix} 1 & 0 \end{bmatrix} = \begin{bmatrix} 1 & 0 \\ 0 & 0 \end{bmatrix}.$$

In a similar vein,
$$\begin{bmatrix} 1 & 0 \end{bmatrix} \begin{bmatrix} 0 \\ 1 \end{bmatrix} = 0, \quad \text{whereas} \quad \begin{bmatrix} 0 \\ 1 \end{bmatrix} \begin{bmatrix} 1 & 0 \end{bmatrix} = \begin{bmatrix} 0 & 0 \\ 1 & 0 \end{bmatrix}.$$

Exercise 2.4. Show that the matrix $\begin{bmatrix} 1 & 0 \end{bmatrix}$ has infinitely many right inverses and no left inverses, whereas the matrix $\begin{bmatrix} 1 \\ 0 \end{bmatrix}$ has infinitely many left inverses and no right inverses.

Exercise 2.5. Show that if $A \in \mathbb{C}^{p \times q}$, $B \in \mathbb{C}^{q \times p}$, and AB is invertible, then $q \geq p$, A is right invertible, and B is left invertible; then show that the converse is false when $q > p$.

Exercise 2.6. Let $A \in \mathbb{C}^{n \times n}$ and suppose that $A^{k-1} \neq O$, but $A^k = O$. Show that
$$\operatorname{rank} \begin{bmatrix} A^{k-1} & A^{k-2} & \cdots & I_n \\ O & A^{k-1} & \cdots & A \\ \vdots & & \ddots & \vdots \\ O & O & \cdots & A^{k-1} \end{bmatrix} = n.$$

[HINT: The given matrix can be expressed as the product of its last block column with its first block row.]

Exercise 2.7. Find a matrix $A \in \mathbb{C}^{p \times q}$ and an invertible matrix $C \in \mathbb{C}^{p \times p}$ such that $\mathcal{R}_{CA} \neq \mathcal{R}_A$. Show that

(2.7) $\quad \mathcal{R}_{CA} = C\mathcal{R}_A \quad$ for any $C \in \mathbb{C}^{m \times p}$, invertible or not.

Exercise 2.8. Let $A \in \mathbb{C}^{4 \times 5}$, let $\mathbf{v}_1, \mathbf{v}_2, \mathbf{v}_3$ be a basis for $\mathcal{R}_A$, and let $V = [\mathbf{v}_1 \ \mathbf{v}_2 \ \mathbf{v}_3]$. Show that $V^H V$ is invertible, that $C = V(V^H V)^{-1} V^H$ is not left invertible, and yet $\mathcal{R}_C = \mathcal{R}_{CA}$.

Exercise 2.9. Let $A \in \mathbb{C}^{p \times q}$, $B \in \mathbb{C}^{q \times r}$. Show that: (1) if A and B are both left invertible, then AB is left invertible; (2) if A and B are both right invertible, then AB is right invertible.

Exercise 2.10. Find a matrix $A \in \mathbb{C}^{p \times q}$ and an invertible matrix $B \in \mathbb{C}^{q \times q}$ such that $\mathcal{N}_{AB} \neq \mathcal{N}_A$. Show that

(2.8) $\quad B\mathcal{N}_{AB} \subseteq \mathcal{N}_A \quad$ for every $B \in \mathbb{C}^{q \times r}$ with equality if B is invertible.

Exercise 2.11. Let $A \in \mathbb{C}^{p \times q}$, $C \in \mathbb{C}^{m \times p}$, and let $\{\mathbf{u}_1, \ldots, \mathbf{u}_k\}$ be a basis for $\mathcal{R}_A$. Show that if C is left invertible, then $\{C\mathbf{u}_1, \ldots, C\mathbf{u}_k\}$ is a basis for $\mathcal{R}_{CA}$.

2.4. $\{0, 1, \infty\}$

Exercise 2.12. Find a pair of matrices $A \in \mathbb{C}^{p \times q}$ and $B \in \mathbb{C}^{p \times p}$ such that B is not left invertible and yet $\{B\mathbf{u}_1, \ldots, B\mathbf{u}_k\}$ is a basis for $\mathcal{R}_{BA}$ for every basis $\{\mathbf{u}_1, \ldots, \mathbf{u}_k\}$ of $\mathcal{R}_A$.

Exercise 2.13. Show that if $A \in \mathbb{C}^{p \times q}$, then $\operatorname{rank} A = \operatorname{rank} A^T$, however, $\operatorname{rank} A^T A$ is not always equal to $\operatorname{rank} A$.

Exercise 2.14. Show that if $A \in \mathbb{C}^{p \times q}$ and $B \in \mathbb{C}^{p \times r}$, then

$$(2.9) \qquad \operatorname{rank} \begin{bmatrix} A & B \end{bmatrix} = p \iff \mathcal{N}_{A^H} \cap \mathcal{N}_{B^H} = \{\mathbf{0}\}.$$

Exercise 2.15. Let $A \in \mathbb{C}^{p \times q}$, $B \in \mathbb{C}^{q \times r}$, and let $\{\mathbf{u}_1, \ldots, \mathbf{u}_k\}$ be a basis for $\mathcal{R}_B$. Show that $\{A\mathbf{u}_1, \ldots, A\mathbf{u}_k\}$ is a basis for $\mathcal{R}_{AB}$ if and only if $\mathcal{R}_B \cap \mathcal{N}_A = \{\mathbf{0}\}$.

Exercise 2.16. Show that if $A \in \mathbb{C}^{p \times q}$ and $B \in \mathbb{C}^{q \times p}$, then $\mathcal{N}_{AB} = \{\mathbf{0}\}$ if and only if $\mathcal{N}_A \cap \mathcal{R}_B = \{\mathbf{0}\}$ and $\mathcal{N}_B = \{\mathbf{0}\}$.

Exercise 2.17. Find a matrix $A \in \mathbb{C}^{p \times q}$ and a vector $\mathbf{b} \in \mathbb{C}^p$ such that $\mathcal{N}_A = \{\mathbf{0}\}$ and yet the equation $A\mathbf{x} = \mathbf{b}$ has no solutions.

Exercise 2.18. Find a basis for $\mathcal{R}_A$ and $\mathcal{N}_A$ if $A = \begin{bmatrix} 1 & 3 & 1 & 8 & 2 \\ 0 & 1 & 2 & 1 & 3 \\ 1 & -2 & 3 & 3 & 1 \\ 1 & 6 & 11 & 5 & 9 \end{bmatrix}$.

[REMARK: A systematic way of tackling such problems will be presented in the next chapter.]

2.4. $\{0, 1, \infty\}$

Theorem 2.3. *The equation $A\mathbf{x} = \mathbf{b}$ has either 0, 1, or infinitely many solutions.*

Proof. There are three possibilities to consider:

(1) $\mathbf{b} \notin \mathcal{R}_A$.
(2) $\mathbf{b} \in \mathcal{R}_A$ and $\mathcal{N}_A = \{\mathbf{0}\}$.
(3) $\mathbf{b} \in \mathcal{R}_A$ and $\mathcal{N}_A \neq \{\mathbf{0}\}$.

In case (1) the equation $A\mathbf{x} = \mathbf{b}$ has no solutions. Suppose next that $\mathbf{b} \in \mathcal{R}_A$ and that $\mathbf{x}_1$ and $\mathbf{x}_2$ are both solutions to the given equation. Then the identities

$$\mathbf{0} = \mathbf{b} - \mathbf{b} = A\mathbf{x}_1 - A\mathbf{x}_2 = A(\mathbf{x}_1 - \mathbf{x}_2)$$

imply that $(\mathbf{x}_1 - \mathbf{x}_2) \in \mathcal{N}_A$. Thus, in case (2) $\mathbf{x}_1 - \mathbf{x}_2 = \mathbf{0}$; i.e., the equation has exactly one solution, whereas in case (3) it has infinitely many solutions: If $A\mathbf{x} = \mathbf{b}$ and $\mathbf{u} \in \mathcal{N}_A$, then $A(\mathbf{x} + \alpha\mathbf{u}) = \mathbf{b}$ for every scalar α. □

Exercise 2.19. Find a system of 5 equations and 3 unknowns that has exactly one solution and a system of 3 equations and 5 unknowns that has no solutions.

Exercise 2.20. Let $n_L = n_L(A)$ and $n_R = n_R(A)$ denote the number of left and right inverses, respectively, of a matrix $A \in \mathbb{C}^{p \times q}$. Show that the combinations $(n_L = 0, n_R = 0)$, $(n_L = 0, n_R = \infty)$, $(n_L = 1, n_R = 1)$, and $(n_L = \infty, n_R = 0)$ are possible.

Exercise 2.21. In the notation of the previous exercise, show that the combinations $(n_L = 0, n_R = 1)$, $(n_L = 1, n_R = 0)$, $(n_L = \infty, n_R = 1)$, $(n_L = 1, n_R = \infty)$, and $(n_L = \infty, n_R = \infty)$ are impossible.

Exercise 2.22. Find the set of all right inverses B to the matrix $A = \begin{bmatrix} A_{11} & A_{12} \end{bmatrix}$ and the set of all left inverses C to A^T when A_{11} is invertible.

2.5. Block triangular matrices

Theorem 2.4. If $A \in \mathbb{C}^{n \times n}$ is a block triangular matrix that is either of the form

$$(2.10) \qquad A = \begin{bmatrix} A_{11} & A_{12} \\ O & A_{22} \end{bmatrix} \quad \text{or the form} \quad A = \begin{bmatrix} A_{11} & O \\ A_{21} & A_{22} \end{bmatrix}$$

with $A_{11} \in \mathbb{C}^{p \times p}$ and $A_{22} \in \mathbb{C}^{q \times q}$, then A is invertible if and only if A_{11} and A_{22} are both invertible. Moreover, if A_{11} and A_{22} are invertible, then

$$(2.11) \qquad \begin{bmatrix} A_{11} & A_{12} \\ O & A_{22} \end{bmatrix}^{-1} = \begin{bmatrix} A_{11}^{-1} & -A_{11}^{-1} A_{12} A_{22}^{-1} \\ O & A_{22}^{-1} \end{bmatrix} \quad \text{and}$$

$$\begin{bmatrix} A_{11} & O \\ A_{21} & A_{22} \end{bmatrix}^{-1} = \begin{bmatrix} A_{11}^{-1} & O \\ -A_{22}^{-1} A_{21} A_{11}^{-1} & A_{22}^{-1} \end{bmatrix}.$$

Proof. Suppose first that A is an invertible block upper triangular matrix. Then there exists a matrix $B \in \mathbb{C}^{n \times n}$ such that $AB = I_n$ and $BA = I_n$. Thus, upon writing B in the block form that is compatible with the block form of A and invoking the first of these formulas, we obtain

$$AB = \begin{bmatrix} A_{11} & A_{12} \\ O & A_{22} \end{bmatrix} \begin{bmatrix} B_{11} & B_{12} \\ B_{21} & B_{22} \end{bmatrix} = \begin{bmatrix} I_p & O_{p \times q} \\ O_{q \times p} & I_q \end{bmatrix}.$$

But this is equivalent to the following four identities:

(1) $A_{11} B_{11} + A_{12} B_{21} = I_p$.

(2) $A_{22} B_{21} = O_{q \times p}$.

(3) $A_{11} B_{12} + A_{12} B_{22} = O_{p \times q}$.

(4) $A_{22} B_{22} = I_q$.

Then, since $A_{22}, B_{22} \in \mathbb{C}^{q \times q}$ and $A_{11}, B_{11} \in \mathbb{C}^{p \times p}$, (4), (2), and (1), considered in that order imply that

A_{22} is invertible, $\quad B_{21} = O_{q \times p}, \quad A_{11} B_{11} = I_p, \quad$ and $\quad A_{11}$ is invertible.

Therefore, the assertion that A invertible implies that A_{11} and A_{22} are invertible is verified. The converse implication is justified by noting that if A_{11} and A_{22} are invertible, then the first block matrix in (2.11) is well-defined, and then checking by direct calculation that the block upper triangular matrix B with these block entries really is the inverse of A. (This is logically correct but leaves open the question of where the entries came from. The answer is that once you know that $B_{22} = A_{22}^{-1}$ and $B_{11} = A_{11}^{-1}$, then the formula $B_{12} = -A_{11}^{-1} A_{12} A_{22}^{-1}$ is obtained from (3).)

The analysis for block lower triangular matrices is similar and is left to the reader. (Another option is to exploit the fact that A is block lower triangular if and only if A^T is block upper triangular.) □

Exercise 2.23. Show that if $A \in \mathbb{C}^{n \times n}$ is an invertible triangular matrix with entries $a_{ij} \in \mathbb{C}$ for $i, j = 1, \ldots, n$, then $a_{ii} \neq 0$ for $i = 1, \ldots, n$. [HINT: Use Theorem 2.4 to show that if the claim is true for $n = k$, then it is also true for $n = k + 1$.]

2.6. Supplementary notes

This chapter is partially adapted from Chapters 1 and 2 in [**30**]. Theorem 2.2 and the **keep in mind** notes are new to this edition. Formula (3.2) in the next chapter exhibits the secret behind the mysterious equality rank A = rank A^T.

The equivalence $\mathcal{N}_A = \{\mathbf{0}\} \iff \mathcal{R}_A = \mathbb{C}^p$ for matrices $A \in \mathbb{C}^{p \times p}$ is a special case of the **Fredholm alternative**, which, in its most provocative form, states that *if the solution to the equation* $A\mathbf{x} = \mathbf{b}$ *is unique, then it exists.*

Chapter 3

Gaussian elimination

Gaussian elimination is a systematic way of passing from a given system of equations
$$A\mathbf{x} = \mathbf{b} \quad \text{to a new system of equations} \quad U\mathbf{x} = \mathbf{c}$$
that is easier to analyze. The passage from the given system to the new system is effected by multiplying both sides of the given system successively on the left by an appropriately chosen sequence of invertible matrices. The restriction to invertible multipliers is essential. Otherwise, the new system may not have the same set of solutions as the given one. In particular, the left multipliers will be either permutation matrices P (which serve to interchange rows) or lower triangular matrices with ones on the diagonal E (which serve to add multiples of one row to another). Thus, for example, if $A \in \mathbb{C}^{3 \times 4}$, then

$$PA = \begin{bmatrix} 0 & 1 & 0 \\ 1 & 0 & 0 \\ 0 & 0 & 1 \end{bmatrix} \begin{bmatrix} a_{11} & a_{12} & a_{13} & a_{14} \\ a_{21} & a_{22} & a_{23} & a_{24} \\ a_{31} & a_{32} & a_{33} & a_{34} \end{bmatrix} = \begin{bmatrix} a_{21} & a_{22} & a_{23} & a_{24} \\ a_{11} & a_{12} & a_{13} & a_{14} \\ a_{31} & a_{32} & a_{33} & a_{34} \end{bmatrix},$$

whereas

$$EA = \begin{bmatrix} 1 & 0 & 0 \\ \alpha & 1 & 0 \\ \beta & 0 & 1 \end{bmatrix} \begin{bmatrix} a_{11} & \cdots & a_{14} \\ a_{21} & \cdots & a_{24} \\ a_{31} & \cdots & a_{34} \end{bmatrix} = \begin{bmatrix} a_{11} & \cdots & a_{14} \\ \alpha a_{11} + a_{21} & \cdots & \alpha a_{14} + a_{24} \\ \beta a_{11} + a_{31} & \cdots & \beta a_{14} + a_{34} \end{bmatrix}.$$

If A is a square matrix, then U will be upper triangular. If A is not square, then U will be an **upper echelon** matrix (which will be defined below).

In practice, Gaussian multiplication is carried out by first forming the augmented matrix $\widetilde{A} = \begin{bmatrix} A & \mathbf{b} \end{bmatrix}$ and then by carrying out the indicated row operations without referring to the corresponding matrix multipliers on the left. This works because if $P_1, \ldots, P_k$ is a sequence of permutation matrices and $E_1, \ldots, E_k$ is a sequence of lower triangular matrices with ones on the diagonal such that

$$(3.1) \qquad E_k P_k \cdots E_1 P_1 \begin{bmatrix} A & \mathbf{b} \end{bmatrix} = \begin{bmatrix} U & \mathbf{c} \end{bmatrix},$$

then the matrix $G = E_k P_k \cdots E_1 P_1$ is invertible and (3.1) ensures that

$$A\mathbf{x} = \mathbf{b} \iff GA\mathbf{x} = G\mathbf{b} \iff U\mathbf{x} = \mathbf{c}.$$

The following notions will prove useful:

- **Upper echelon:** If $U \in \mathbb{C}^{p \times q}$ and $\operatorname{rank} U = k$, then U is an **upper echelon matrix** if: (1) each of the first k rows contains at least one nonzero entry; (2) the first nonzero entry in row i lies to the left of the first nonzero entry in row $i+1$ for $i = 1, \ldots, k-1$; (3) the entries in the remaining $p - k$ rows (if any) are all zero (**because of this special structure, $\operatorname{rank} U = \operatorname{rank} U^T$ for every upper echelon matrix U**). Thus, for example, if

$$V = \begin{bmatrix} 3 & 6 & 2 & 4 & 1 & 0 \\ 0 & 0 & 1 & 0 & 5 & 0 \\ 0 & 0 & 0 & 0 & 2 & 0 \\ 0 & 0 & 0 & 0 & 0 & 0 \end{bmatrix}, \; W = \begin{bmatrix} 0 & 2 & 3 & 1 \\ 0 & 0 & 6 & 0 \\ 0 & 0 & 0 & 0 \\ 0 & 0 & 0 & 0 \end{bmatrix}, \; \text{and} \; X = \begin{bmatrix} 0 & 0 & 0 \\ 4 & 2 & 3 \\ 0 & 0 & 6 \\ 0 & 5 & 0 \end{bmatrix},$$

then V and W are upper echelon matrices, while X is not.

- **Pivots:** The first nonzero entry in each row of an upper echelon matrix is termed a **pivot**. In the preceding display, V has the 3 pivots v_{11}, v_{23}, and v_{35}, and $\operatorname{rank} V = \operatorname{rank} V^T = 3$; W has the 2 pivots w_{12} and w_{23}, and $\operatorname{rank} W = \operatorname{rank} W^T = 2$. Since a byproduct of Gaussian elimination is the fact that if $A \in \mathbb{C}^{p \times q}$ and $\operatorname{rank} A = k \geq 1$, then there exists an invertible matrix $G \in \mathbb{C}^{p \times p}$ such that $GA = U$ is in upper echelon form with k pivots, we now can provide a more transparent proof of the fact that $\operatorname{rank} A = \operatorname{rank} A^T$:

$$(3.2) \qquad \operatorname{rank} A = \operatorname{rank} U = \text{number of pivots} = \operatorname{rank} U^T = \operatorname{rank} A^T.$$

- **Pivot columns:** A column in an upper echelon matrix U will be called a **pivot column** if it contains a pivot. Thus, the first, third, and fifth columns of V are the pivot columns of V, whereas the second and fourth columns of W are the pivot columns of W. Moreover, if $GA = U$, then the columns $\mathbf{a}_{i_1}, \ldots, \mathbf{a}_{i_k}$ of A that correspond in position to the pivot columns $\mathbf{u}_{i_1}, \ldots, \mathbf{u}_{i_k}$ of U will also be called pivot columns (even though the pivots are in U not in A); they form a basis for $\mathcal{R}_A$. The entries $x_{i_1}, \ldots, x_{i_k}$ in $\mathbf{x} \in \mathbb{C}^q$ will be referred to as **pivot variables**.

3.1. Examples

Example 3.1. Consider the equation $A\mathbf{x} = \mathbf{b}$, where

(3.3) $$A = \begin{bmatrix} 0 & 2 & 3 & 1 \\ 1 & 5 & 3 & 4 \\ 2 & 6 & 3 & 2 \end{bmatrix} \quad \text{and} \quad \mathbf{b} = \begin{bmatrix} 1 \\ 2 \\ 1 \end{bmatrix}.$$

1. Construct the augmented matrix

(3.4) $$\widetilde{A} = \begin{bmatrix} A & \mathbf{b} \end{bmatrix} = \begin{bmatrix} 0 & 2 & 3 & 1 & 1 \\ 1 & 5 & 3 & 4 & 2 \\ 2 & 6 & 3 & 2 & 1 \end{bmatrix};$$

it is introduced to ensure that the row operations that are applied to the matrix A are also applied to the vector $\mathbf{b}$.

2. Interchange the first two rows of $\widetilde{A}$ to get a nonzero entry in the upper left-hand corner of the new matrix

$$\begin{bmatrix} 1 & 5 & 3 & 4 & 2 \\ 0 & 2 & 3 & 1 & 1 \\ 2 & 6 & 3 & 2 & 1 \end{bmatrix} = P_1 \widetilde{A}, \quad \text{with } P_1 = \begin{bmatrix} 0 & 1 & 0 \\ 1 & 0 & 0 \\ 0 & 0 & 1 \end{bmatrix}.$$

3. Subtract two times the top row of the matrix $P_1 \widetilde{A}$ from its bottom row to get

$$\begin{bmatrix} 1 & 5 & 3 & 4 & 2 \\ 0 & 2 & 3 & 1 & 1 \\ 0 & -4 & -3 & -6 & -3 \end{bmatrix} = E_1 P_1 \widetilde{A}, \quad \text{where } E_1 = \begin{bmatrix} 1 & 0 & 0 \\ 0 & 1 & 0 \\ -2 & 0 & 1 \end{bmatrix}$$

is chosen to obtain all zeros below the pivot in the first column.

4. Add two times the second row of $E_1 P_1 \widetilde{A}$ to its third row to get

$$\begin{bmatrix} 1 & 5 & 3 & 4 & 2 \\ 0 & 2 & 3 & 1 & 1 \\ 0 & 0 & 3 & -4 & -1 \end{bmatrix} = E_2 E_1 P_1 \widetilde{A} = \begin{bmatrix} U & \mathbf{c} \end{bmatrix},$$

where

$$E_2 = \begin{bmatrix} 1 & 0 & 0 \\ 0 & 1 & 0 \\ 0 & 2 & 1 \end{bmatrix}$$

is chosen to obtain all zeros below the pivot in the second column, $U = E_2 E_1 P_1 A$ is in upper echelon form, and $\mathbf{c} = E_2 E_1 P_1 \mathbf{b}$.

5. Try to solve the new system of equations

(3.5) $$U\mathbf{x} = \begin{bmatrix} 1 & 5 & 3 & 4 \\ 0 & 2 & 3 & 1 \\ 0 & 0 & 3 & -4 \end{bmatrix} \begin{bmatrix} x_1 \\ x_2 \\ x_3 \\ x_4 \end{bmatrix} = \begin{bmatrix} 2 \\ 1 \\ -1 \end{bmatrix}$$

by solving for the pivot variables x_1, x_2, x_3 in terms of x_4, working from the bottom row up:

row 3: $\quad 3x_3 - 4x_4 = -1 \implies 3x_3 = 4x_4 - 1$,
row 2: $\quad 2x_2 + 3x_3 + x_4 = 1 \implies 2x_2 = -3x_3 - x_4 + 1$,
row 1: $\quad x_1 + 5x_2 + 3x_3 + 4x_4 = 2 \implies x_1 = -5x_2 - 3x_3 - 4x_4 + 2$.

Thus, each of the pivot variables x_1, x_2, x_3 can be expressed in terms of the variable x_4: $x_3 = (4x_4 - 1)/3$, $x_2 = (-5x_4 + 2)/2$, and $x_1 = (9x_4 - 4)/2$, i.e.,

$$\mathbf{x} = \begin{bmatrix} x_1 \\ x_2 \\ x_3 \\ x_4 \end{bmatrix} = \begin{bmatrix} -2 \\ 1 \\ -1/3 \\ 0 \end{bmatrix} + x_4 \mathbf{u} \quad \text{with} \quad \mathbf{u} = \begin{bmatrix} 9/2 \\ -5/2 \\ 4/3 \\ 1 \end{bmatrix}$$

is a solution of the system of equations (3.5), or equivalently,

(3.6) $$E_2 E_1 P_1 A \mathbf{x} = E_2 E_1 P_1 \mathbf{b}$$

(with A and $\mathbf{b}$ as in (3.3)) for every choice of x_4. Therefore, $\mathbf{u} \in \mathcal{N}_A$ and, as the matrices E_2, E_1, and P_1 are invertible, $\mathbf{x}$ is a solution of (3.6) if and only if $A\mathbf{x} = \mathbf{b}$, i.e., if and only if $\mathbf{x}$ is a solution of the original equation.

6. Check that the computed solution solves the original system of equations. Strictly speaking, this step is superfluous, because the construction guarantees that every solution of the new system is a solution of the old system, and vice versa. Nevertheless, this is an **extremely important step**, because it gives you a way to check your calculations. ◇

Exercise 3.1. Show that for the matrix A considered in the preceding example $\mathcal{N}_A = \text{span}\{[9/2 \ -5/2 \ 4/3 \ 1]^T\}$.

Example 3.2. Let

$$A = \begin{bmatrix} 0 & 0 & 3 & 4 & 7 \\ 0 & 1 & 0 & 0 & 0 \\ 0 & 2 & 3 & 6 & 8 \\ 0 & 0 & 6 & 8 & 14 \end{bmatrix} \quad \text{and} \quad \mathbf{b} = \begin{bmatrix} b_1 \\ b_2 \\ b_3 \\ b_4 \end{bmatrix}.$$

3.1. Examples

Then a vector $\mathbf{x} \in \mathbb{C}^5$ is a solution of the equation $A\mathbf{x} = \mathbf{b}$ if and only if

$$\begin{bmatrix} 0 & 1 & 0 & 0 & 0 \\ 0 & 0 & 3 & 4 & 7 \\ 0 & 0 & 0 & 2 & 1 \\ 0 & 0 & 0 & 0 & 0 \end{bmatrix} \begin{bmatrix} x_1 \\ x_2 \\ x_3 \\ x_4 \\ x_5 \end{bmatrix} = \begin{bmatrix} b_2 \\ b_1 \\ b_3 - 2b_2 - b_1 \\ b_4 - 2b_1 \end{bmatrix}.$$

The pivots of the upper echelon matrix on the left are in columns 2, 3, and 4. Therefore, upon solving for the pivot variables x_2, x_3, and x_4 in terms of x_1, x_5, and $b_1, \ldots, b_4$ from the bottom row up, we obtain the formulas

$$\begin{aligned} 0 &= b_4 - 2b_1, \\ 2x_4 &= b_3 - 2b_2 - b_1 - x_5, \\ 3x_3 &= b_1 - 4x_4 - 7x_5, \\ &= 3b_1 + 4b_2 - 2b_3 - 5x_5, \\ x_2 &= b_2. \end{aligned}$$

But this is the same as

$$\begin{bmatrix} x_1 \\ x_2 \\ x_3 \\ x_4 \\ x_5 \end{bmatrix} = \begin{bmatrix} x_1 \\ b_2 \\ (-5x_5 + 3b_1 + 4b_2 - 2b_3)/3 \\ (-x_5 + b_3 - 2b_2 - b_1)/2 \\ x_5 \end{bmatrix} = x_1 \begin{bmatrix} 1 \\ 0 \\ 0 \\ 0 \\ 0 \end{bmatrix} + x_5 \begin{bmatrix} 0 \\ 0 \\ -5/3 \\ -1/2 \\ 1 \end{bmatrix}$$

$$+ b_1 \begin{bmatrix} 0 \\ 0 \\ 1 \\ -1/2 \\ 0 \end{bmatrix} + b_2 \begin{bmatrix} 0 \\ 1 \\ 4/3 \\ -1 \\ 0 \end{bmatrix} + b_3 \begin{bmatrix} 0 \\ 0 \\ -2/3 \\ 1/2 \\ 0 \end{bmatrix},$$

i.e., in self-evident notation,

$$\mathbf{x} = x_1 \mathbf{u}_1 + x_5 \mathbf{u}_2 + b_1 \mathbf{u}_3 + b_2 \mathbf{u}_4 + b_3 \mathbf{u}_5.$$

Thus, a vector $\mathbf{b} \in \mathbb{C}^4$ belongs to $\mathcal{R}_A$ if and only if $b_4 = 2b_1$, and, in this case, the displayed vector $\mathbf{x}$ is a solution of the equation $A\mathbf{x} = \mathbf{b}$ for every choice of x_1 and x_5. Therefore, $x_1 \mathbf{u}_1 + x_5 \mathbf{u}_2$ is a solution of the equation $A\mathbf{x} = \mathbf{0}$ for every choice of $x_1, x_5 \in \mathbb{C}$. Thus, $\mathbf{u}_1, \mathbf{u}_2 \in \mathcal{N}_A$ and $\mathcal{R}_A = \text{span}\{\mathbf{a}_2, \mathbf{a}_3, \mathbf{a}_4\}$, the pivot columns of A. Since $\mathbf{u}_1$ and $\mathbf{u}_2$ are linearly independent and $\dim \mathcal{N}_A = 2$ by conservation of dimension, $\{\mathbf{u}_1, \mathbf{u}_2\}$ is a basis for $\mathcal{N}_A$. ◊

3.2. A remarkable formula

One of the surprising dividends of Gaussian elimination is:

Theorem 3.1. *If $A \in \mathbb{C}^{p \times q}$ and $\operatorname{rank} A = r \geq 1$, then there exists a lower triangular $p \times p$ matrix L with ones on the diagonal, an upper echelon matrix $U \in \mathbb{C}^{p \times q}$ with $\operatorname{rank} U = r$, and a $p \times p$ permutation matrix P such that*

$$(3.7) \qquad PA = LU.$$

Discussion. To understand where this formula comes from, suppose that A is a nonzero 4×5 matrix and let $\mathbf{e}_1, \ldots, \mathbf{e}_4$ denote the columns of I_4. Then there exists a choice of permutation matrices P_1, P_2, P_3 and lower triangular matrices

$$E_1 = \begin{bmatrix} 1 & 0 & 0 & 0 \\ a & 1 & 0 & 0 \\ b & 0 & 1 & 0 \\ c & 0 & 0 & 1 \end{bmatrix} = I_4 + \mathbf{u}_1 \mathbf{e}_1^T \quad \text{with} \quad \mathbf{u}_1 = \begin{bmatrix} 0 \\ a \\ b \\ c \end{bmatrix},$$

$$E_2 = \begin{bmatrix} 1 & 0 & 0 & 0 \\ 0 & 1 & 0 & 0 \\ 0 & d & 1 & 0 \\ 0 & e & 0 & 1 \end{bmatrix} = I_4 + \mathbf{u}_2 \mathbf{e}_2^T \quad \text{with} \quad \mathbf{u}_2 = \begin{bmatrix} 0 \\ 0 \\ d \\ e \end{bmatrix},$$

$$E_3 = \begin{bmatrix} 1 & 0 & 0 & 0 \\ 0 & 1 & 0 & 0 \\ 0 & 0 & 1 & 0 \\ 0 & 0 & f & 1 \end{bmatrix} = I_4 + \mathbf{u}_3 \mathbf{e}_3^T \quad \text{with} \quad \mathbf{u}_3 = \begin{bmatrix} 0 \\ 0 \\ 0 \\ f \end{bmatrix},$$

such that

$$(3.8) \qquad E_3 P_3 E_2 P_2 E_1 P_1 A = U$$

is in upper echelon form. The crucial fact is that P_2 is chosen so that it interchanges the second row of $E_1 P_1 A$ with its third or fourth row, if necessary, whereas P_3 interchanges the third and fourth rows, if necessary. Consequently, $\mathbf{e}_i^T P_j = \mathbf{e}_i$ if $i < j$ and

$$P_2 E_1 = P_2(I_4 + \mathbf{u}_1 \mathbf{e}_1^T) = P_2 + (P_2 \mathbf{u}_1)\mathbf{e}_1^T = E_1' P_2,$$

where $E_1' = I_4 + (P_2 \mathbf{u}_1)\mathbf{e}_1^T$ has the same form as E_1. Similarly,

$$P_i E_j = E_j' P_i \quad \text{if} \quad i > j,$$

where E_j' denotes a matrix of the same form as E_j. Thus, in self-evident notation,

$$E_3 P_3 E_2 P_2 E_1 P_1 = E_3 E_2' P_3 E_1' P_2 P_1 = E_3 E_2' E_1'' P_3 P_2 P_1 = EP,$$

which yields (3.7) with $L = E^{-1}$. $\square$

3.2. A remarkable formula

Exercise 3.2. Show that if $A \in \mathbb{C}^{n \times n}$ is an invertible matrix, then there exists a permutation matrix $P \in \mathbb{R}^{n \times n}$ such that

(3.9) $$PA = LDU,$$

where $L \in \mathbb{C}^{n \times n}$ is lower triangular with ones on the diagonal, $U \in \mathbb{C}^{n \times n}$ is upper triangular with ones on the diagonal, and D is a diagonal matrix.

Exercise 3.3. Show that if $L_1 D_1 U_1 = L_2 D_2 U_2$, where L_j, D_j, and U_j are $n \times n$ matrices of the form exhibited in Exercise 3.2, then $L_1 = L_2$, $D_1 = D_2$, and $U_1 = U_2$. [HINT: $L_2^{-1} L_1 D_1 = D_2 U_2 U_1^{-1}$ is both lower and upper triangular.]

Exercise 3.4. Show that there exist a 3×3 permutation matrix P and a lower triangular matrix

$$B = \begin{bmatrix} 1 & 0 & 0 \\ b_{21} & 1 & 0 \\ b_{31} & b_{32} & 1 \end{bmatrix} \quad \text{such that} \quad \begin{bmatrix} 0 & 1 & 0 \\ 1 & 0 & 0 \\ 0 & 0 & 1 \end{bmatrix} \begin{bmatrix} 1 & 0 & 0 \\ \alpha & 1 & 0 \\ \beta & 0 & 1 \end{bmatrix} = BP$$

if and only if $\alpha = 0$.

Exercise 3.5. Find a permutation matrix P such that $PA = LU$, where L is a lower triangular invertible 3×3 matrix and U is an upper triangular invertible 3×3 matrix for the matrix $A = \begin{bmatrix} 0 & 1 & 1 \\ 1 & 1 & 1 \\ 1 & 1 & 0 \end{bmatrix}$.

Exercise 3.6. Find a lower triangular matrix $E_k \in \mathbb{C}^{(k+1) \times (k+1)}$ with ones on the diagonal such that

$$E_k \begin{bmatrix} 1 & \overline{\alpha} & \cdots & \overline{\alpha}^k \\ \alpha & 1 & \cdots & \overline{\alpha}^{k-1} \\ \vdots & & & \vdots \\ \alpha^k & \alpha^{k-1} & \cdots & 1 \end{bmatrix} = \begin{bmatrix} 1 & * & * & \cdots & * & * \\ 0 & \rho & * & \cdots & * & * \\ \vdots & & & & & \vdots \\ 0 & 0 & 0 & \cdots & \rho & * \\ 0 & 0 & 0 & \cdots & 0 & \rho \end{bmatrix}$$

with $\rho = 1 - |\alpha|^2$. [HINT: Compare α times row $j-1$ with row j in the matrix that is being processed.]

Exercise 3.7. Show that if $|\alpha| \neq 1$ and

(3.10) $$\begin{bmatrix} 1 & \overline{\alpha} & \cdots & \overline{\alpha}^k \\ \alpha & 1 & \cdots & \overline{\alpha}^{k-1} \\ \vdots & & & \vdots \\ \alpha^k & \alpha^{k-1} & \cdots & 1 \end{bmatrix} \begin{bmatrix} x_0 \\ x_1 \\ \vdots \\ x_k \end{bmatrix} = \begin{bmatrix} 0 \\ \vdots \\ 0 \\ 1 \end{bmatrix},$$

then $x_k = (1 - |\alpha|^2)^{-1}$. [HINT: Exploit the displayed formula in Exercise 3.6; it is not necessary to compute E_k.]

3.3. Extracting a basis

Let $\{\mathbf{v}_1, \ldots, \mathbf{v}_k\}$ be a set of vectors in $\mathbb{C}^m$. To find a basis for the subspace
$$\mathcal{V} = \operatorname{span}\{\mathbf{v}_1, \ldots, \mathbf{v}_k\} :$$

(1) Let $A = \begin{bmatrix} \mathbf{v}_1 & \cdots & \mathbf{v}_k \end{bmatrix}$.

(2) Use Gaussian elimination to reduce A to an upper echelon matrix U.

(3) The pivot columns of A form a basis for $\mathcal{V}$.

Example 3.3. Let
$$\mathbf{v}_1 = \begin{bmatrix} 1 \\ 3 \\ 1 \end{bmatrix}, \; \mathbf{v}_2 = \begin{bmatrix} 2 \\ 6 \\ 2 \end{bmatrix}, \; \mathbf{v}_3 = \begin{bmatrix} 2 \\ 10 \\ 4 \end{bmatrix}, \; \text{and } \mathbf{v}_4 = \begin{bmatrix} 0 \\ 2 \\ 1 \end{bmatrix}.$$

Then, following the indicated strategy, we apply Gaussian elimination to the matrix
$$A = \begin{bmatrix} 1 & 2 & 2 & 0 \\ 3 & 6 & 10 & 2 \\ 1 & 2 & 4 & 1 \end{bmatrix} \quad \text{to get} \quad U = \begin{bmatrix} 1 & 2 & 2 & 0 \\ 0 & 0 & 4 & 2 \\ 0 & 0 & 0 & 0 \end{bmatrix}.$$

The pivot columns of U are the first and the third. Therefore, by the recipe furnished above, $\dim \mathcal{R}_A = \dim \mathcal{R}_U = 2$ and
$$\mathcal{R}_A = \operatorname{span}\{\mathbf{v}_1, \mathbf{v}_2, \mathbf{v}_3, \mathbf{v}_4\} = \operatorname{span}\{\mathbf{v}_1, \mathbf{v}_3\}.$$

Exercise 3.8. Find a basis for the span of the vectors
$$\begin{bmatrix} 2 \\ 3 \\ 1 \\ 4 \end{bmatrix}, \; \begin{bmatrix} 1 \\ 0 \\ 2 \\ 1 \end{bmatrix}, \; \begin{bmatrix} 0 \\ 3 \\ -3 \\ 2 \end{bmatrix}, \; \begin{bmatrix} 3 \\ -3 \\ 9 \\ 1 \end{bmatrix}.$$

You should **keep in mind** that if you use Gaussian elimination to obtain a basis for $\mathcal{N}_A$, then the pivot columns of A furnish a basis for $\mathcal{R}_A$.

Exercise 3.9. Use Gaussian elimination to find $\mathcal{N}_A$ and $\mathcal{R}_A$ for each of the following choices of the matrix A:
$$\begin{bmatrix} 3 & 1 & 2 & 4 \\ 2 & 1 & 8 & 7 \\ 3 & 2 & 6 & 1 \end{bmatrix}, \; \begin{bmatrix} 1 & 2 & 0 & 2 & 1 \\ -1 & -2 & 1 & 1 & 0 \\ 1 & 2 & -3 & -7 & -2 \end{bmatrix}, \; \begin{bmatrix} 0 & 0 & 8 & 1 \\ 1 & 2 & 4 & 1 \\ 2 & 3 & 0 & 0 \end{bmatrix}.$$

3.4. Augmenting a given set of vectors to form a basis

In future applications we shall need a convenient way to augment a given set of linearly independent vectors with vectors chosen from a second set of linearly independent vectors that span a larger vector space.

Lemma 3.2. *Let $\{\mathbf{w}_1, \ldots, \mathbf{w}_k\}$ be a basis for a subspace $\mathcal{W}$ of a vector space $\mathcal{V} \subseteq \mathbb{C}^n$ with basis $\{\mathbf{v}_1, \ldots, \mathbf{v}_\ell\}$. Then the ℓ pivot columns of the $n \times (k+\ell)$ matrix $A = [\mathbf{w}_1 \cdots \mathbf{w}_k \, \mathbf{v}_1 \cdots \mathbf{v}_\ell]$ will be a basis for $\mathcal{V}$ that includes the vectors $\mathbf{w}_1, \ldots, \mathbf{w}_k$.*

Proof. Since $\operatorname{rank} A = \ell$, the corresponding upper echelon form that is computed by Gaussian elimination will have ℓ pivots. Moreover, it will have k pivots in the first k columns, since the vectors $\mathbf{w}_1, \ldots, \mathbf{w}_k$ are linearly independent. The remaining $\ell - k$ pivots indicate the positions of the vectors from the submatrix $[\mathbf{v}_1 \cdots \mathbf{v}_\ell]$ that together with $\mathbf{w}_1, \ldots, \mathbf{w}_k$ form a basis for $\mathcal{V}$. $\square$

Exercise 3.10. Find a basis for $\mathbb{C}^4$ using the vectors $\mathbf{w}_1, \mathbf{w}_2$ and two of the vectors $\mathbf{v}_1, \ldots, \mathbf{v}_4$, when $\mathbf{w}_1^T = \begin{bmatrix} 1 & 0 & 1 & 0 \end{bmatrix}$, $\mathbf{w}_2^T = \begin{bmatrix} 0 & 1 & 2 & 0 \end{bmatrix}$, $\mathbf{v}_1^T = \begin{bmatrix} 1 & 1 & 1 & 1 \end{bmatrix}$, $\mathbf{v}_2^T = \begin{bmatrix} 0 & 1 & 0 & 1 \end{bmatrix}$, $\mathbf{v}_3^T = \begin{bmatrix} 0 & 0 & 1 & 1 \end{bmatrix}$, and $\mathbf{v}_4^T = \begin{bmatrix} 0 & 0 & 0 & 1 \end{bmatrix}$.

3.5. Computing the coefficients in a basis

Let $\{\mathbf{u}_1, \ldots, \mathbf{u}_k\}$ be a basis for a k-dimensional subspace $\mathcal{U}$ of $\mathbb{C}^n$. Then every vector $\mathbf{b} \in \mathcal{U}$ can be expressed as a unique linear combination of the vectors $\{\mathbf{u}_1, \ldots, \mathbf{u}_k\}$; i.e., there exists a unique set of coefficients $c_1, \ldots, c_k$ such that

$$\mathbf{b} = \sum_{j=1}^{k} c_j \mathbf{u}_j.$$

The problem of computing these coefficients is equivalent to the problem of solving the equation

$$A\mathbf{c} = \mathbf{b},$$

where $A = [\mathbf{u}_1 \cdots \mathbf{u}_k]$ is the $n \times k$ matrix with columns $\mathbf{u}_1, \ldots, \mathbf{u}_k$ and $\mathbf{c} = [c_1 \cdots c_k]^T$. This problem, too, can be solved efficiently by Gaussian elimination.

Exercise 3.11. Let

$$\begin{bmatrix} \mathbf{u}_1 & \mathbf{u}_2 & \mathbf{u}_3 & \mathbf{u}_4 & \mathbf{v} \end{bmatrix} = \begin{bmatrix} 1 & 5 & 7 & 4 & 3 \\ 0 & 2 & 1 & 6 & 1 \\ 0 & 0 & 4 & 2 & 1 \\ 0 & 0 & 0 & 0 & 0 \end{bmatrix}.$$

Show that $\operatorname{span}\{\mathbf{u}_1, \mathbf{u}_2, \mathbf{u}_3\} = \operatorname{span}\{\mathbf{u}_1, \mathbf{u}_2, \mathbf{u}_4\}$ and that $\mathbf{v} = a\mathbf{u}_1 + b\mathbf{u}_2 + c\mathbf{u}_3 = d\mathbf{u}_1 + e\mathbf{u}_2 + f\mathbf{u}_4$, and calculate the coefficients a, b, c, d, e, f in the two representations.

3.6. The Gauss-Seidel method

Gaussian elimination can be used to find the inverse of a $p \times p$ invertible matrix A by solving each of the equations

$$A\mathbf{x}_j = \mathbf{e}_j, \; j = 1, \ldots, p,$$

where the right-hand side $\mathbf{e}_j$ is the j'th column of the identity matrix I_p. Then the formula

$$A \begin{bmatrix} \mathbf{x}_1 & \cdots & \mathbf{x}_p \end{bmatrix} = \begin{bmatrix} \mathbf{e}_1 & \cdots & \mathbf{e}_p \end{bmatrix} = I_p$$

identifies $X = \begin{bmatrix} \mathbf{x}_1 & \cdots & \mathbf{x}_p \end{bmatrix}$ as the inverse A^{-1} of A.

The Gauss-Seidel method is a systematic way of organizing all p of these separate calculations into one more efficient calculation by proceeding as follows, given $A \in \mathbb{C}^{p \times p}$, invertible or not:

1. Construct the $p \times 2p$ augmented matrix

$$\widetilde{A} = \begin{bmatrix} A & I_p \end{bmatrix}.$$

2. Carry out elementary row operations on $\widetilde{A}$ that are designed to bring A into upper echelon form U. This is equivalent to choosing an invertible matrix G so that

$$G \begin{bmatrix} A & I_p \end{bmatrix} = \begin{bmatrix} GA & G \end{bmatrix} = \begin{bmatrix} U & G \end{bmatrix}.$$

3. Observe that U is a $p \times p$ upper triangular matrix with k pivots. If $k < p$, then A is not invertible and the procedure grinds to a halt. If $k = p$, then $u_{ii} \neq 0$ for $i = 1, \ldots, p$. Therefore, by Theorem 2.4, there exists an upper triangular matrix F such that $FU = I_p$ and hence

$$FG \begin{bmatrix} A & I_p \end{bmatrix} = \begin{bmatrix} FGA & FG \end{bmatrix} = \begin{bmatrix} FU & FG \end{bmatrix} = \begin{bmatrix} I_p & FG \end{bmatrix}.$$

Therefore, $A^{-1} = FG$, which is equal to the second block in the $p \times 2p$ matrix on the right in the last display.

To obtain A^{-1} numerically, go on to the next steps.

4. Multiply $\begin{bmatrix} U & G \end{bmatrix}$ on the left by a diagonal matrix $D = \begin{bmatrix} d_{ij} \end{bmatrix}$ with $d_{ii} = (u_{ii})^{-1}$ for $i = 1, \ldots, p$ to obtain

$$D \begin{bmatrix} U & G \end{bmatrix} = \begin{bmatrix} \widetilde{U} & DG \end{bmatrix},$$

where now $\widetilde{U} = DU$ is an upper triangular matrix with ones on the diagonal.

5. Carry out elementary row manipulations on $\begin{bmatrix} \widetilde{U} & DG \end{bmatrix}$ working from the bottom row up that are designed to bring $\widetilde{U}$ to the identity. This is

3.6. The Gauss-Seidel method

equivalent to choosing an upper triangular matrix $\widetilde{F}$ such that $\widetilde{F}\widetilde{U} = I_p$. Then

$$\widetilde{F}\begin{bmatrix}\widetilde{U} & DG\end{bmatrix} = \begin{bmatrix}\widetilde{F}\widetilde{U} & \widetilde{F}DG\end{bmatrix} = \begin{bmatrix}I_p & \widetilde{F}DG\end{bmatrix},$$

and hence, as $\widetilde{F}\widetilde{U} = \widetilde{F}DGA = I_p$, the second block on the right is

$$\widetilde{F}DG = FG = A^{-1}.$$

6. Check! Multiply your candidate for A^{-1} by A to see if you really get the identity matrix I_p as an answer.

Thus, for example, if

$$A = \begin{bmatrix} 1 & 3 & 1 \\ 2 & 8 & 4 \\ 0 & 4 & 7 \end{bmatrix}, \quad \text{then} \quad \widetilde{A} = \begin{bmatrix} 1 & 3 & 1 \\ 2 & 8 & 4 \\ 0 & 4 & 7 \end{bmatrix} \begin{array}{ccc} 1 & 0 & 0 \\ 0 & 1 & 0 \\ 0 & 0 & 1 \end{array},$$

and two steps of Gaussian elimination lead in turn to the forms

$$\widetilde{A}_1 = \begin{bmatrix} 1 & 3 & 1 \\ 0 & 2 & 2 \\ 0 & 4 & 7 \end{bmatrix} \begin{array}{ccc} 1 & 0 & 0 \\ -2 & 1 & 0 \\ 0 & 0 & 1 \end{array}$$

and

$$\widetilde{A}_2 = \begin{bmatrix} 1 & 3 & 1 \\ 0 & 2 & 2 \\ 0 & 0 & 3 \end{bmatrix} \begin{array}{ccc} 1 & 0 & 0 \\ -2 & 1 & 0 \\ 4 & -2 & 1 \end{array}.$$

Next, let

$$\widetilde{A}_3 = \begin{bmatrix} 1 & 0 & 0 \\ 0 & \frac{1}{2} & 0 \\ 0 & 0 & \frac{1}{3} \end{bmatrix} \widetilde{A}_2 = \begin{bmatrix} 1 & 3 & 1 \\ 0 & 1 & 1 \\ 0 & 0 & 1 \end{bmatrix} \begin{array}{ccc} 1 & 0 & 0 \\ -1 & \frac{1}{2} & 0 \\ \frac{4}{3} & -\frac{2}{3} & \frac{1}{3} \end{array},$$

and then subtract the bottom row of $\widetilde{A}_3$ from the second and first rows to obtain

$$\widetilde{A}_4 = \begin{bmatrix} 1 & 3 & 0 \\ 0 & 1 & 0 \\ 0 & 0 & 1 \end{bmatrix} \begin{array}{ccc} -\frac{1}{3} & \frac{2}{3} & -\frac{1}{3} \\ -\frac{7}{3} & \frac{7}{6} & -\frac{1}{3} \\ \frac{4}{3} & -\frac{2}{3} & \frac{1}{3} \end{array}.$$

The next to last step is to subtract three times the second row of $\widetilde{A}_4$ from the first to obtain

$$\widetilde{A}_5 = \begin{bmatrix} 1 & 0 & 0 \\ 0 & 1 & 0 \\ 0 & 0 & 1 \end{bmatrix} \begin{array}{ccc} \frac{20}{3} & -\frac{17}{6} & \frac{2}{3} \\ -\frac{7}{3} & \frac{7}{6} & -\frac{1}{3} \\ \frac{4}{3} & -\frac{2}{3} & \frac{1}{3} \end{array}.$$

The matrix built from the last 3 columns of $\widetilde{A}_5$ is A^{-1}. The final step is to check that this matrix, which is conveniently written as

$$B = \frac{1}{6} \begin{bmatrix} 40 & -17 & 4 \\ -14 & 7 & -2 \\ 8 & -4 & 2 \end{bmatrix},$$

is indeed the inverse of A, i.e., $AB = I_3$.

Exercise 3.12. Find the inverse of the matrix $\begin{bmatrix} 1 & 3 & 2 \\ 2 & 4 & 1 \\ 0 & 4 & 2 \end{bmatrix}$ by the Gauss-Seidel method.

Exercise 3.13. Use the Gauss-Seidel method to show that

(3.11) $$\begin{bmatrix} 1 & 0 & 0 & 0 \\ 1 & 1 & 0 & 0 \\ 1 & 1 & 1 & 0 \\ 1 & 1 & 1 & 1 \end{bmatrix}^{-1} = \begin{bmatrix} 1 & 0 & 0 & 0 \\ -1 & 1 & 0 & 0 \\ 0 & -1 & 1 & 0 \\ 0 & 0 & -1 & 1 \end{bmatrix}.$$

To compute the **right inverses** of a matrix $A \in \mathbb{C}^{p \times n}$ with rank $A = p$ and $q = n - p \geq 1$, use Gaussian elimination to find the pivot columns of A and then choose a permutation matrix $P \in \mathbb{R}^{n \times n}$ such that $AP = \begin{bmatrix} B_{11} & B_{12} \end{bmatrix}$ and B_{11} is invertible. Then express $X \in \mathbb{C}^{n \times p}$ in block form with blocks $X_{11} \in \mathbb{C}^{p \times p}$ and $X_{21} \in \mathbb{C}^{q \times p}$ and observe that

$$APX = I_p \iff \begin{bmatrix} B_{11} & B_{12} \end{bmatrix} \begin{bmatrix} X_{11} \\ X_{21} \end{bmatrix} = I_p \iff X_{11} = B_{11}^{-1}(I_p - B_{12}X_{21}).$$

Consequently the matrix PX with $X_{11} = B_{11}^{-1}(I_p - B_{12}X_{21})$ is a right inverse of A for every choice of the pq entries in X_{21}.

The **left inverses** of a matrix $A \in \mathbb{C}^{n \times q}$ with rank $A = q$ and $n > q$ are obtained easily from the right inverses of A^T.

3.7. Block Gaussian elimination

Gaussian elimination can also be carried out in block matrices provided that appropriate range conditions are fulfilled. Thus, for example, if

$$A = \begin{bmatrix} A_{11} & A_{12} & A_{13} \\ A_{21} & A_{22} & A_{23} \\ A_{31} & A_{32} & A_{33} \end{bmatrix}$$

is a block matrix and if there exists a pair of matrices K_1 and K_2 such that

(3.12) $$A_{21} = K_1 A_{11} \quad \text{and} \quad A_{31} = K_2 A_{11},$$

3.7. Block Gaussian elimination

then
$$\begin{bmatrix} I & O & O \\ -K_1 & I & O \\ -K_2 & O & I \end{bmatrix} A = \begin{bmatrix} A_{11} & A_{12} & A_{13} \\ O & -K_1 A_{12} + A_{22} & -K_1 A_{13} + A_{23} \\ O & -K_2 A_{12} + A_{32} & -K_2 A_{13} + A_{33} \end{bmatrix}.$$

This operation is the block matrix analogue of clearing the first column in conventional Gaussian elimination. The implementation of such a step depends critically on the existence of matrices K_1 and K_2 that fulfill the conditions in (3.12). If A_{11} is invertible, then clearly $K_1 = A_{21} A_{11}^{-1}$ and $K_2 = A_{31} A_{11}^{-1}$ meet the requisite conditions. However, matrices K_1 and K_2 that satisfy the conditions in (3.12) may exist even if A_{11} is not invertible.

Lemma 3.3. *Let $A \in \mathbb{C}^{p \times q}$, $B \in \mathbb{C}^{p \times r}$, and $C \in \mathbb{C}^{r \times q}$. Then:*

(1) *There exists a matrix $K \in \mathbb{C}^{r \times q}$ such that $A = BK$ if and only if $\mathcal{R}_A \subseteq \mathcal{R}_B$.*

(2) *There exists a matrix $L \in \mathbb{C}^{p \times r}$ such that $A = LC$ if and only if $\mathcal{R}_{A^H} \subseteq \mathcal{R}_{C^H}$.*

Proof. Suppose first that $\mathcal{R}_A \subseteq \mathcal{R}_B$ and let $\mathbf{e}_j$ denote the j'th column of I_q. Then the presumed range inclusion implies that $A\mathbf{e}_j = B\mathbf{u}_j$ for some vector $\mathbf{u}_j \in \mathbb{C}^r$ for $j = 1, \ldots, q$ and hence that $A = A \begin{bmatrix} \mathbf{e}_1 & \cdots & \mathbf{e}_q \end{bmatrix} = B \begin{bmatrix} \mathbf{u}_1 & \cdots & \mathbf{u}_q \end{bmatrix} = BK$, with $K = \begin{bmatrix} \mathbf{u}_1 & \cdots & \mathbf{u}_q \end{bmatrix}$. This proves half of (1). The other half is easy and is left to the reader together with (2). □

Under appropriate assumptions, a double application of block Gaussian elimination (once on the left and once on the right) is applicable and leads to useful factorization formulas for square matrices.

Theorem 3.4. *If $A_{11} \in \mathbb{C}^{p \times p}$, $A_{12} \in \mathbb{C}^{p \times q}$, $A_{21} \in \mathbb{C}^{q \times p}$, $A_{22} \in \mathbb{C}^{q \times q}$ and the range conditions*

$$(3.13) \qquad \mathcal{R}_{A_{12}} \subseteq \mathcal{R}_{A_{11}} \quad \text{and} \quad \mathcal{R}_{A_{21}^H} \subseteq \mathcal{R}_{A_{11}^H}$$

are in force, then there exists a pair of matrices $K \in \mathbb{C}^{p \times q}$ and $L \in \mathbb{C}^{q \times p}$ such that

$$(3.14) \qquad A_{12} = A_{11} K \quad \text{and} \quad A_{21} = L A_{11}$$

and hence

$$(3.15) \qquad \begin{bmatrix} A_{11} & A_{12} \\ A_{21} & A_{22} \end{bmatrix} = \begin{bmatrix} I_p & O \\ L & I_q \end{bmatrix} \begin{bmatrix} A_{11} & O \\ O & A_{22} - L A_{11} K \end{bmatrix} \begin{bmatrix} I_p & K \\ O & I_q \end{bmatrix}.$$

Proof. Lemma 3.3 guarantees the existence of a pair of matrices $L \in \mathbb{C}^{q \times p}$ and $K \in \mathbb{C}^{p \times q}$ that meet the conditions in (3.14). Thus,

$$\begin{bmatrix} I_p & O \\ -L & I_q \end{bmatrix} \begin{bmatrix} A_{11} & A_{12} \\ A_{21} & A_{22} \end{bmatrix} = \begin{bmatrix} A_{11} & A_{12} \\ O & A_{22} - L A_{12} \end{bmatrix}$$

and
$$\begin{bmatrix} A_{11} & A_{12} \\ O & A_{22} - LA_{12} \end{bmatrix} \begin{bmatrix} I_p & -K \\ O & I_q \end{bmatrix} = \begin{bmatrix} A_{11} & O \\ O & A_{22} - LA_{12} \end{bmatrix},$$
which in turn leads easily to formula (3.15). $\square$

The right-hand side of (3.15) is the product of block lower, block diagonal, and block upper triangular matrices. The next exercise reverses the order.

Exercise 3.14. Let $A \in \mathbb{C}^{n \times n}$ be a four-block matrix with entries $A_{11} \in \mathbb{C}^{p \times p}$, $A_{12} \in \mathbb{C}^{p \times q}$, $A_{21} \in \mathbb{C}^{q \times p}$, $A_{22} \in \mathbb{C}^{q \times q}$, where $n = p + q$. Show that if the range conditions

(3.16) $$\mathcal{R}_{A_{12}^H} \subseteq \mathcal{R}_{A_{22}^H} \quad \text{and} \quad \mathcal{R}_{A_{21}} \subseteq \mathcal{R}_{A_{22}}$$

are in force, then A admits a factorization of the form

(3.17) $$\begin{bmatrix} A_{11} & A_{12} \\ A_{21} & A_{22} \end{bmatrix} = \begin{bmatrix} I_p & M \\ O & I_q \end{bmatrix} \begin{bmatrix} A_{11} - MA_{22}N & O \\ O & A_{22} \end{bmatrix} \begin{bmatrix} I_p & O \\ N & I_q \end{bmatrix}.$$

Exercise 3.15. Show that if A_{11} is invertible, then formula (3.15) can be reexpressed as
(3.18)
$$A = \begin{bmatrix} I_p & O \\ A_{21}A_{11}^{-1} & I_q \end{bmatrix} \begin{bmatrix} A_{11} & O \\ O & A_{22} - A_{21}A_{11}^{-1}A_{12} \end{bmatrix} \begin{bmatrix} I_p & A_{11}^{-1}A_{12} \\ O & I_q \end{bmatrix}.$$

Exercise 3.16. Show that if A_{22} is invertible, then formula (3.17) can be reexpressed as
(3.19)
$$A = \begin{bmatrix} I_p & A_{12}A_{22}^{-1} \\ O & I_q \end{bmatrix} \begin{bmatrix} A_{11} - A_{12}A_{22}^{-1}A_{21} & O \\ O & A_{22} \end{bmatrix} \begin{bmatrix} I_p & O \\ A_{22}^{-1}A_{21} & I_q \end{bmatrix}.$$

Exercise 3.17. Verify item (9) of Theorem 2.2. [HINT: Exploit formulas (3.18) and (3.19) for the block matrix with $A_{11} = I_p$, $A_{22} = I_q$ and A and B in the corners.]

3.8. Supplementary notes

This chapter is partially adapted from Chapters 2 and 3 of [**30**]. The matrices $A_{22} - A_{21}A_{11}^{-1}A_{12}$ in formula (3.18) and $A_{11} - A_{12}A_{22}^{-1}A_{21}$ in formula (3.19) are referred to as the **Schur complements** of A_{11} and A_{22}, respectively. They make many computations transparent. If A is invertible and $B = A^{-1}$ is expressed in compatible four-block form, then

(3.20) $$\begin{aligned} A \text{ and } A_{11} \text{ invertible} &\implies B_{22} = (A_{22} - A_{21}A_{11}^{-1}A_{12})^{-1}, \\ A \text{ and } A_{22} \text{ invertible} &\implies B_{11} = (A_{11} - A_{12}A_{22}^{-1}A_{21})^{-1}. \end{aligned}$$

Chapter 4

Eigenvalues and eigenvectors

- A point $\lambda \in \mathbb{C}$ is said to be an **eigenvalue** of a matrix $A \in \mathbb{C}^{n \times n}$ if there exists a nonzero vector $\mathbf{u} \in \mathbb{C}^n$ such that $A\mathbf{u} = \lambda \mathbf{u}$, i.e., if
$$\mathcal{N}_{(A-\lambda I_n)} \neq \{\mathbf{0}\}.$$
Every nonzero vector $\mathbf{u} \in \mathcal{N}_{(A-\lambda I_n)}$ is said to be an **eigenvector** of A corresponding to the eigenvalue λ.

4.1. The first step

Theorem 4.1. *If $A \in \mathbb{C}^{n \times n}$, then:*

(1) *A has at least one eigenvalue $\lambda \in \mathbb{C}$.*

(2) *A has at most n distinct eigenvalues.*

(3) *Eigenvectors corresponding to distinct eigenvalues are linearly independent.*

Proof. Let $\mathbf{u}$ be a nonzero vector in $\mathbb{C}^n$. Then, since the set of $n+1$ vectors
$$\mathbf{u}, A\mathbf{u}, \ldots, A^n \mathbf{u}$$
is linearly dependent, there exists a set of complex numbers $c_0, \ldots, c_n$, not all of which are zero, such that
$$c_0 \mathbf{u} + c_1 A\mathbf{u} + \cdots + c_n A^n \mathbf{u} = \mathbf{0}.$$
Let $k = \max\{j : c_j \neq 0\}$. Then, by the fundamental theorem of algebra, the polynomial
$$p(x) = c_0 + c_1 x + \cdots + c_n x^n = c_0 + c_1 x + \cdots + c_k x^k$$

can be factored as a product of k polynomials of degree one with roots $\mu_1, \ldots, \mu_k \in \mathbb{C}$:
$$p(x) = c_k(x - \mu_k) \cdots (x - \mu_1).$$
Moreover, the same holds true for polynomials in A (e.g., $x^2 + 5x + 6 = (x+3)(x+2) \implies A^2 + 5A + 6I_n = (A+3I_n)(A+2I_n)$). Correspondingly,
$$\begin{aligned} c_0 \mathbf{u} + \cdots + c_n A^n \mathbf{u} &= c_0 \mathbf{u} + \cdots + c_k A^k \mathbf{u} \\ &= c_k(A - \mu_k I_n) \cdots (A - \mu_2 I_n)(A - \mu_1 I_n) \mathbf{u} = \mathbf{0}. \end{aligned}$$

This in turn implies that there are k possibilities:

(1) $(A - \mu_1 I_n)\mathbf{u} = \mathbf{0}$.
(2) $(A - \mu_1 I_n)\mathbf{u} \neq \mathbf{0}$ and $(A - \mu_2 I_n)(A - \mu_1 I_n)\mathbf{u} = \mathbf{0}$.
$\vdots$
(k) $(A - \mu_{k-1} I_n) \cdots (A - \mu_1 I_n)\mathbf{u} \neq \mathbf{0}$ and $(A - \mu_k I_n) \cdots (A - \mu_1 I_n)\mathbf{u} = \mathbf{0}$.

In the first case, μ_1 is an eigenvalue and $\mathbf{u}$ is an eigenvector.
In the second case, $\mathbf{w}_1 = (A - \mu_1 I_n)\mathbf{u}$ is a nonzero vector in $\mathbb{C}^n$ and $A\mathbf{w}_1 = \mu_2 \mathbf{w}_1$. Therefore, $(A - \mu_1 I_n)\mathbf{u}$ is an eigenvector of A corresponding to the eigenvalue μ_2.
$\vdots$
In the k'th case, $\mathbf{w}_{k-1} = (A - \mu_{k-1} I_n) \cdots (A - \mu_1 I_n)\mathbf{u}$ is a nonzero vector in $\mathbb{C}^n$ and $A\mathbf{w}_{k-1} = \mu_k \mathbf{w}_{k-1}$. Therefore, $\mathbf{w}_{k-1}$ is an eigenvector of A corresponding to the eigenvalue μ_k. This completes the verification of (1).

Items (2) and (3) of Theorem 4.1 are stated here for perspective; they will be justified in the proof of Theorem 4.2. (But clearly, (2) follows from (3).) □

Notice that the proof **does not guarantee the existence of real eigenvalues** for A even if $A \in \mathbb{R}^{n \times n}$, because the polynomial $p(x) = c_0 + c_1 x + \cdots + c_k x^k$ may have only complex roots $\mu_1, \ldots, \mu_k$ even if the coefficients $c_1, \ldots, c_k$ are real.

Exercise 4.1. Show that if $A = \begin{bmatrix} 0 & -1 \\ 1 & 0 \end{bmatrix}$, then the equation $A\mathbf{x} = \mu\mathbf{x}$ has a nonzero solution $\mathbf{x} \in \mathbb{C}^2$ if and and only if $\mu = \pm i$.

If $\lambda_1, \ldots, \lambda_k$ are distinct eigenvalues of a matrix $A \in \mathbb{C}^{n \times n}$, then:

- The number
$$\gamma_j = \dim \mathcal{N}_{(A - \lambda_j I_n)}$$
is termed the **geometric multiplicity** of the eigenvalue λ_j, $j = 1, \ldots, k$.

- The number
$$\alpha_j = \dim \mathcal{N}_{(A-\lambda_j I_n)^n}$$
is termed the **algebraic multiplicity** of the eigenvalue λ_j, $j = 1, \ldots, k$.
- The inclusions

(4.1) $$\mathcal{N}_{(A-\lambda_j I_n)} \subseteq \mathcal{N}_{(A-\lambda_j I_n)^2} \subseteq \cdots \subseteq \mathcal{N}_{(A-\lambda_j I_n)^n}$$

guarantee that

(4.2) $$\gamma_j \leq \alpha_j \quad \text{for} \quad j = 1, \ldots, k,$$

and hence (as $\alpha_1 + \cdots + \alpha_k = n$ by Theorem 5.4) that

(4.3) $$k \leq \gamma_1 + \cdots + \gamma_k \leq \alpha_1 + \cdots + \alpha_k = n.$$

- The set

(4.4) $$\sigma(A) = \{\lambda \in \mathbb{C} : \mathcal{N}_{(A-\lambda I_n)} \neq \{\mathbf{0}\}\}$$

is called the **spectrum** of A. Clearly, $\sigma(A) = \{\lambda_1, \ldots, \lambda_k\}$, the set of all the distinct eigenvalues of the matrix A in $\mathbb{C}$.
- A nonzero vector $\mathbf{u} \in \mathbb{C}^n$ is said to be a **generalized eigenvector** of order k for the matrix $A \in \mathbb{C}^{n \times n}$ corresponding to the eigenvalue $\lambda \in \mathbb{C}$ if $\mathbf{u} \in \mathcal{N}_{(A-\lambda I_n)^k}$ but $\mathbf{u} \notin \mathcal{N}_{(A-\lambda I_n)^{k-1}}$.

4.2. Diagonalizable matrices

A matrix $A \in \mathbb{C}^{n \times n}$ is said to be **similar** to a matrix $B \in \mathbb{C}^{n \times n}$ if there exists an invertible matrix $U \in \mathbb{C}^{n \times n}$ such that $A = UBU^{-1}$; A is said to be **diagonalizable** if it is similar to a diagonal matrix D, i.e., if

(4.5) $$A = UDU^{-1}.$$

Theorem 4.2. *Let $A \in \mathbb{C}^{n \times n}$ and suppose that A has exactly k distinct eigenvalues $\lambda_1, \ldots, \lambda_k \in \mathbb{C}$ with geometric multiplicities $\gamma_1, \ldots, \gamma_k$, respectively. Then:*

(1) *There exists an $n \times (\gamma_1 + \cdots + \gamma_k)$ matrix $U = \begin{bmatrix} U_1 & \cdots & U_k \end{bmatrix}$ with blocks $U_j \in \mathbb{C}^{n \times \gamma_j}$ such that $\operatorname{rank} U = \gamma_1 + \cdots + \gamma_k$ and*
$$AU = UD, \quad \text{with} \quad D = \operatorname{diag}\{\lambda_1 I_{\gamma_1}, \ldots, \lambda_k I_{\gamma_k}\}.$$

(2) $k \leq \gamma_1 + \cdots + \gamma_k \leq n$.

(3) *A is diagonalizable if and only if $\gamma_1 + \cdots + \gamma_k = n$.*

(4) *A is diagonalizable if $k = n$.*

Discussion. To ease the exposition, suppose that $k = 3$ and let

$$r = \gamma_1 \text{ and } \{\mathbf{u}_1, \ldots, \mathbf{u}_r\} \text{ be a basis for } \mathcal{N}_{(A-\lambda_1 I_n)},$$
$$s = \gamma_2 \text{ and } \{\mathbf{v}_1, \ldots, \mathbf{v}_s\} \text{ be a basis for } \mathcal{N}_{(A-\lambda_2 I_n)},$$
$$t = \gamma_3 \text{ and } \{\mathbf{w}_1, \ldots, \mathbf{w}_t\} \text{ be a basis for } \mathcal{N}_{(A-\lambda_3 I_n)},$$

and set

$$U_1 = \begin{bmatrix} \mathbf{u}_1 & \cdots & \mathbf{u}_r \end{bmatrix}, \, U_2 = \begin{bmatrix} \mathbf{v}_1 & \cdots & \mathbf{v}_s \end{bmatrix}, \, U_3 = \begin{bmatrix} \mathbf{w}_1 & \cdots & \mathbf{w}_t \end{bmatrix}.$$

Then

$$A \begin{bmatrix} U_1 & U_2 & U_3 \end{bmatrix} = \begin{bmatrix} AU_1 & AU_2 & AU_3 \end{bmatrix} = \begin{bmatrix} U_1(\lambda_1 I_r) & U_2(\lambda_2 I_s) & U_3(\lambda_3 I_t) \end{bmatrix}$$
$$= \begin{bmatrix} U_1 & U_2 & U_3 \end{bmatrix} \begin{bmatrix} \lambda_1 I_r & O & O \\ O & \lambda_2 I_s & O \\ O & O & \lambda_3 I_t \end{bmatrix}.$$

To verify (1), it remains to show that $\text{rank}\, U = \gamma_1 + \gamma_2 + \gamma_3$. Towards this end, suppose that

$$\sum_{i=1}^{r} a_i \mathbf{u}_i + \sum_{j=1}^{s} b_j \mathbf{v}_j + \sum_{k=1}^{t} c_k \mathbf{w}_k = \mathbf{0}$$

for some set of coefficients $a_1, \ldots, a_r, b_1, \ldots, b_s, c_1, \ldots, c_t \in \mathbb{C}$ and, to simplify the bookkeeping, let

$$\mathbf{u} = \sum_{i=1}^{r} a_i \mathbf{u}_i, \quad \mathbf{v} = \sum_{j=1}^{s} b_j \mathbf{v}_j, \quad \text{and} \quad \mathbf{w} = \sum_{k=1}^{t} c_k \mathbf{w}_k.$$

Then

$$A\mathbf{u} = \lambda_1 \mathbf{u}, \; A\mathbf{v} = \lambda_2 \mathbf{v}, \; A\mathbf{w} = \lambda_3 \mathbf{w} \quad \text{and} \quad \mathbf{u} + \mathbf{v} + \mathbf{w} = \mathbf{0}.$$

Thus,

$$\mathbf{0} = (A - \lambda_1 I_n)(\mathbf{u} + \mathbf{v} + \mathbf{w}) = (\lambda_2 - \lambda_1)\mathbf{v} + (\lambda_3 - \lambda_1)\mathbf{w}$$

and

$$\mathbf{0} = (A - \lambda_2 I_n)((\lambda_2 - \lambda_1)\mathbf{v} + (\lambda_3 - \lambda_1)\mathbf{w}) = (\lambda_3 - \lambda_2)(\lambda_3 - \lambda_1)\mathbf{w}.$$

Since the eigenvalues are distinct, the last three displays, starting from the third, clearly imply that $\mathbf{w} = \sum_{k=1}^{t} c_k \mathbf{w}_k = \mathbf{0}$, $\mathbf{v} = \sum_{j=1}^{s} b_j \mathbf{v}_j = \mathbf{0}$, and $\mathbf{u} = \sum_{i=1}^{r} a_i \mathbf{u}_i = \mathbf{0}$. Therefore, since each of these three sums is a linear combination of linearly independent vectors, all the coefficients are equal to zero. Consequently, (1) holds for $k = 3$. The argument is easily adapted for arbitrary positive integers k.

Items (2) and (3) are immediate from (1); (4) is immediate from (2). $\square$

Diagonalizable matrices are pleasant to work with. In particular, formula (4.5) implies that

$$A^2 = (UDU^{-1})(UDU^{-1}) = UD^2U^{-1}, \ A^3 = UD^3U^{-1}, \ldots, \ A^k = UD^kU^{-1}.$$

The advantage is that the powers $D^2, D^3, \ldots, D^k$ are easy to compute:

$$(4.6) \qquad D = \begin{bmatrix} \mu_1 & & \\ & \ddots & \\ & & \mu_n \end{bmatrix} \Longrightarrow D^k = \begin{bmatrix} \mu_1^k & & \\ & \ddots & \\ & & \mu_n^k \end{bmatrix}.$$

Exercise 4.2. Show that if a matrix $A \in \mathbb{C}^{n \times n}$ is diagonalizable, i.e., if $A = UDU^{-1}$ with $D = \text{diag}\{\lambda_1, \ldots, \lambda_n\}$, and if

$$U = \begin{bmatrix} \mathbf{u}_1 & \cdots & \mathbf{u}_n \end{bmatrix} \quad \text{and} \quad (U^{-1})^T = \begin{bmatrix} \mathbf{v}_1 & \cdots & \mathbf{v}_n \end{bmatrix}, \text{ then:}$$

(1) $A^k = UD^kU^{-1} = \sum_{j=1}^n \lambda_j^k \mathbf{u}_j \mathbf{v}_j^T$.

(2) $(A - \lambda I_n)^{-1} = U(D - \lambda I_n)^{-1} U^{-1} = \sum_{j=1}^n (\lambda_j - \lambda)^{-1} \mathbf{u}_j \mathbf{v}_j^T$, if $\lambda \notin \sigma(A)$.

Exercise 4.3. Show that the matrices

$$A = \begin{bmatrix} 1 & -1 \\ 1 & 1 \end{bmatrix} \quad \text{and} \quad A = \begin{bmatrix} 2 & -1 \\ 3 & -1 \end{bmatrix}$$

have no real eigenvalues, i.e., $\sigma(A) \cap \mathbb{R} = \emptyset$ in both cases.

Exercise 4.4. Show that although the following upper triangular matrices

$$\begin{bmatrix} 2 & 0 & 0 \\ 0 & 2 & 0 \\ 0 & 0 & 2 \end{bmatrix}, \ \begin{bmatrix} 2 & 1 & 0 \\ 0 & 2 & 0 \\ 0 & 0 & 2 \end{bmatrix}, \ \begin{bmatrix} 2 & 1 & 0 \\ 0 & 2 & 1 \\ 0 & 0 & 2 \end{bmatrix}$$

have the same diagonal, $\dim \mathcal{N}_{(A-2I_3)}$ is equal to three for the first, two for the second, and one for the third. Calculate $\mathcal{N}_{(A-2I_3)^j}$ for $j = 1, 2, 3, 4$ for each of the three choices of A.

Exercise 4.5. Show that if $A \in \mathbb{C}^{n \times n}$ is a triangular matrix with entries a_{ij}, then $\sigma(A) = \bigcup_{i=1}^n \{a_{ii}\}$.

4.3. Invariant subspaces

A subspace $\mathcal{U}$ of $\mathbb{C}^n$ is **invariant** under A if $A\mathbf{u} \in \mathcal{U}$ for every vector $\mathbf{u} \in \mathcal{U}$. The simplest invariant subspaces are the one-dimensional ones.

Exercise 4.6. Show that if $A \in \mathbb{C}^{n \times n}$, then a nonzero vector $\mathbf{u} \in \mathbb{C}^n$ is an eigenvector of A if and only if the one-dimensional subspace $\{\alpha \mathbf{u} : \alpha \in \mathbb{C}\}$ is invariant under A.

Example 4.1. If $A \in \mathbb{C}^{n \times n}$, then
$$\mathcal{N}_{(A-\lambda I_n)} = \{\mathbf{u} \in \mathbb{C}^{n \times n} : A\mathbf{u} = \lambda \mathbf{u}\}$$
and
$$\mathcal{R}_{(A-\lambda I_n)} = \{(A - \lambda I_n)\mathbf{u} : \mathbf{u} \in \mathbb{C}^n\}$$
are both invariant under A:
$$(A - \lambda I_n)\mathbf{u} = \mathbf{0} \implies (A - \lambda I_n)A\mathbf{u} = A(A - \lambda I_n)\mathbf{u} = \mathbf{0}$$
and
$$\mathbf{u} = (A - \lambda I_n)\mathbf{v} \implies A\mathbf{u} = A(A - \lambda I_n)\mathbf{v} = (A - \lambda I_n)A\mathbf{v}\,.$$

Theorem 4.3. *Let $A \in \mathbb{C}^{n \times n}$ and let $\mathcal{U}$ be a subspace of $\mathbb{C}^n$ that is invariant under A, i.e., $\mathbf{u} \in \mathcal{U} \implies A\mathbf{u} \in \mathcal{U}$. Then:*

(1) *Either $\mathcal{U} = \{\mathbf{0}\}$ or there exist a nonzero vector $\mathbf{u} \in \mathcal{U}$ and a number $\lambda \in \mathbb{C}$ such that $A\mathbf{u} = \lambda \mathbf{u}$.*

(2) *If $\mathbf{u}_1, \ldots, \mathbf{u}_k \in \mathcal{U}$ are eigenvectors of A corresponding to distinct eigenvalues $\lambda_1, \ldots, \lambda_k$, then $k \leq \dim \mathcal{U}$.*

Proof. If $\mathbf{u}$ is a nonzero vector in $\mathcal{U}$ and $\dim \mathcal{U} = m$, then, since the $m+1$ vectors $\mathbf{u}, A\mathbf{u}, \ldots, A^m\mathbf{u}$ all belong to $\mathcal{U}$, they must be linearly dependent. Therefore, the same argument that was used to justify item (1) of Theorem 4.1 serves to justify item (1) of this theorem too. Item (2) then follows from the fact that $k = \dim\{\mathrm{span}\{\mathbf{u}_1, \ldots, \mathbf{u}_k\}\} \leq \dim \mathcal{U}$. $\square$

4.4. Jordan cells

Not all matrices are diagonalizable: The criterion $\gamma_1 + \cdots + \gamma_k = n$ established in Theorem 4.2 may not be satisfied. Thus, for example, if

$$A = \begin{bmatrix} 2 & 1 & 0 \\ 0 & 2 & 1 \\ 0 & 0 & 2 \end{bmatrix}, \quad \text{then the nullspace of} \quad A - \lambda I_3 = \begin{bmatrix} 2-\lambda & 1 & 0 \\ 0 & 2-\lambda & 1 \\ 0 & 0 & 2-\lambda \end{bmatrix}$$

will be nonzero if and only if $\lambda = 2$. In this case $\dim \mathcal{N}_{(A-2I_3)} = 1$ and $\dim \mathcal{N}_{(A-\lambda I_3)} = 0$ if $\lambda \neq 2$. More elaborate examples may be constructed by taking larger matrices of the same form or by putting such blocks together.

The matrix A in the last display is of the form

$$(4.7) \qquad C_\mu^{(p)} = \mu I_p + \sum_{j=1}^{p-1} \mathbf{e}_j \mathbf{e}_{j+1}^T \qquad \text{where } I_p = \begin{bmatrix} \mathbf{e}_1 & \cdots & \mathbf{e}_p \end{bmatrix},$$

with $\mu = 2$ and $p = 3$. Matrices of the form (4.7) are called **Jordan cells**.

Exercise 4.7. Let $A = C_\mu^{(p)}$ be a Jordan cell of size $p \times p$ with μ on the diagonal. Show that $\mathcal{N}_{(A-\lambda I_p)} \neq \{\mathbf{0}\}$ if and only if $\lambda = \mu$ and that $\dim \mathcal{N}_{(A-\mu I_p)^k} = k$ for $k = 1, \ldots, p$.

Nevertheless, the news is not all bad. There is a more general factorization formula than (4.5) in which the matrix D is replaced by a block diagonal matrix $J = \text{diag}\{B_{\lambda_1}, \ldots, B_{\lambda_k}\}$, where B_{λ_j} is an $\alpha_j \times \alpha_j$ matrix that is block diagonal with γ_j Jordan cells as blocks. Thus, it is an $\alpha_j \times \alpha_j$ upper triangular matrix with λ_j on the diagonal and the columns of U are generalized eigenvectors of A. This representation will be developed in the next two chapters.

4.5. Linear transformations

In this section we shall reformulate some of the main conclusions that were obtained for matrices in the language of linear transformations.

Let T be a linear transformation (mapping) from a vector space $\mathcal{V}$ into itself. Then a subspace $\mathcal{M}$ of $\mathcal{V}$ is said to be **invariant** under T if $T\mathbf{v} \in \mathcal{M}$ whenever $\mathbf{v} \in \mathcal{M}$. A number $\lambda \in \mathbb{C}$ is an **eigenvalue** of T if there exists a nonzero vector $\mathbf{v} \in \mathcal{V}$ such that $T\mathbf{v} = \lambda\mathbf{v}$. Such a vector $\mathbf{v}$ is said to be an **eigenvector** of T. Thus, every nonzero vector in $\mathcal{N}_{(T-\lambda I_\mathcal{V})}$ is an eigenvector of T.

Let
$$\gamma_j = \dim \mathcal{N}_{(T-\lambda_j I_\mathcal{V})}.$$

The next two theorems paraphrase the main conclusions that have already been obtained for matrices. The proofs, which are much the same as for their matrix counterparts, are left to the reader.

Theorem 4.4. *If T is a linear transformation from an n-dimensional complex vector space $\mathcal{V}$ into itself, then:*

(1) *T has at least one eigenvalue $\lambda \in \mathbb{C}$.*

(2) *Eigenvectors $\mathbf{v}_1, \ldots, \mathbf{v}_k$ of T that correspond to distinct eigenvalues $\lambda_1, \ldots, \lambda_k$ are linearly independent.*

Moreover, if T has exactly k distinct eigenvalues $\lambda_1, \ldots, \lambda_k$ with geometric multiplicities $\gamma_j = \dim \mathcal{N}_{(T-\lambda_j I_\mathcal{V})}$, then:

(3) *$k \leq \gamma_1 + \cdots + \gamma_k \leq n$.*

(4) *The space $\mathcal{V}$ admits a basis of eigenvectors of T if and only if $\gamma_1 + \cdots + \gamma_k = n$.*

Theorem 4.5. *Let T be a linear transformation from a complex vector space $\mathcal{V}$ into itself and let $\mathcal{M}$ be a finite-dimensional subspace of $\mathcal{V}$ that is invariant under T. Then either $\mathcal{M} = \{\mathbf{0}\}$ or there exist a nonzero vector $\mathbf{w} \in \mathcal{M}$ and a number $\lambda \in \mathbb{C}$ such that*
$$T\mathbf{w} = \lambda\mathbf{w}.$$

Exercise 4.8. Show that if T is a linear transformation from a complex vector space $\mathcal{V}$ into itself, then the vector spaces $\mathcal{N}_{(T-\lambda I)}$ and $\mathcal{R}_{(T-\lambda I)}$ are both invariant under T for each choice of $\lambda \in \mathbb{C}$.

Exercise 4.9. The set $\mathcal{V}$ of polynomials $p(t)$ with complex coefficients is a complex vector space with respect to the natural rules of vector addition and scalar multiplication. Let $Tp = p''(t) + tp'(t)$ and $Sp = p''(t) + t^2 p'(t)$. Show that the subspace $\mathcal{U}_k$ of $\mathcal{V}$ of polynomials $p(t) = c_0 + c_1 t + \cdots + c_k t^k$ of degree less than or equal to k is invariant under T, but not under S. Find a nonzero polynomial $p \in \mathcal{U}_3$ and a number $\lambda \in \mathbb{C}$ such that $Tp = \lambda p$.

Exercise 4.10. Let T be a linear transformation from a real vector space $\mathcal{V}$ into itself and let $\mathcal{U}$ be a two-dimensional subspace of $\mathcal{V}$ with basis $\{\mathbf{u}_1, \mathbf{u}_2\}$. Show that if $T\mathbf{u}_1 = \mathbf{u}_2$ and $T\mathbf{u}_2 = -\mathbf{u}_1$, then $T^2\mathbf{u} + \mathbf{u} = \mathbf{0}$ for every vector $\mathbf{u} \in \mathcal{U}$, but that there are no one-dimensional subspaces of $\mathcal{U}$ that are invariant under T. Why? [HINT: One-dimensional subspace of $\mathcal{U}$ are of the form $\{\alpha(c_1\mathbf{u}_1 + c_2\mathbf{u}_2) : \alpha \in \mathbb{R}\}$ for some choice of $c_1, c_2 \in \mathbb{R}$ with $|c_1| + |c_2| > 0$.]

4.6. Supplementary notes

This chapter is adapted from Chapter 4 of [**30**]. As noted there, the proof of Theorem 4.1 (which avoids the use of determinants) was influenced by a conversation with Sheldon Axler at the Holomorphic Functions Session at MSRI, Berkeley, in 1995 and his paper [**7**].

Chapter 5

Towards the Jordan decomposition

The main result in this chapter is Theorem 5.4. To understand what it says, we need to understand direct sums of subspaces.

5.1. Direct sums

Let $\mathcal{U}$ and $\mathcal{V}$ be subspaces of a vector space $\mathcal{Y}$ and let
$$\mathcal{U} + \mathcal{V} = \{\mathbf{u} + \mathbf{v} : \mathbf{u} \in \mathcal{U} \text{ and } \mathbf{v} \in \mathcal{V}\}.$$
Clearly, the **sum** $\mathcal{U} + \mathcal{V}$ is a subspace of $\mathcal{Y}$ with respect to the rules of vector addition and scalar multiplication that are inherited from the vector space $\mathcal{Y}$, since $\mathcal{U} + \mathcal{V}$ is closed under vector addition and scalar multiplication.

If $\mathcal{U}$ and $\mathcal{V}$ are finite dimensional, then the sum $\mathcal{U} + \mathcal{V}$ is said to be a **direct sum** if
$$\dim \mathcal{U} + \dim \mathcal{V} = \dim(\mathcal{U} + \mathcal{V}).$$
Direct sums are denoted by the symbol $\dotplus$, i.e., $\mathcal{U} \dotplus \mathcal{V}$ rather than $\mathcal{U} + \mathcal{V}$.

Analogously, if the $\mathcal{U}_j$, $j = 1, \ldots, k$, are subspaces of a vector space $\mathcal{Y}$, then the sum

(5.1) $\quad \mathcal{U}_1 + \cdots + \mathcal{U}_k = \{\mathbf{u}_1 + \cdots + \mathbf{u}_k : \mathbf{u}_i \in \mathcal{U}_i \text{ for } i = 1, \ldots, k\}$

is a subspace of $\mathcal{Y}$ with respect to the rules of vector addition and scalar multiplication that are inherited from the vector space $\mathcal{Y}$. If these k subspaces are finite dimensional, the sum $\mathcal{U} = \mathcal{U}_1 + \cdots + \mathcal{U}_k$ is said to be a **direct sum** if

(5.2) $\quad \dim \mathcal{U}_1 + \cdots + \dim \mathcal{U}_k = \dim\{\mathcal{U}_1 + \cdots + \mathcal{U}_k\}.$

If the sum $\mathcal{U}$ is direct, then we write
$$\mathcal{U} = \mathcal{U}_1 \dotplus \cdots \dotplus \mathcal{U}_k.$$

Example 5.1. If $\{\mathbf{u}_1, \ldots, \mathbf{u}_n\}$ is a basis for $\mathcal{U}$ and $1 < k < \ell < n$, then
$$\mathcal{U} = \operatorname{span}\{\mathbf{u}_1, \ldots, \mathbf{u}_{k-1}\} \dotplus \operatorname{span}\{\mathbf{u}_k, \ldots, \mathbf{u}_{\ell-1}\} \dotplus \operatorname{span}\{\mathbf{u}_\ell, \ldots, \mathbf{u}_n\}.$$

Theorem 5.1. *If $\mathcal{U}$ and $\mathcal{V}$ are finite-dimensional subspaces of a vector space $\mathcal{Y}$, then*

(5.3) $$\dim(\mathcal{U} + \mathcal{V}) = \dim \mathcal{U} + \dim \mathcal{V} - \dim(\mathcal{U} \cap \mathcal{V})$$

and hence

(5.4) $$\text{the sum } \mathcal{U} + \mathcal{V} \text{ is direct} \iff \mathcal{U} \cap \mathcal{V} = \{\mathbf{0}\}.$$

Proof. If $\mathcal{U} \cap \mathcal{V} = \mathcal{U}$ or $\mathcal{U} \cap \mathcal{V} = \mathcal{V}$, then it is easily seen that formula (5.3) holds, since $\mathcal{U} + \mathcal{V} = \mathcal{V}$ in the first case and $\mathcal{U} + \mathcal{V} = \mathcal{U}$ in the second.

Suppose next that $\mathcal{U} \cap \mathcal{V}$ is a nonzero proper subspace of $\mathcal{U}$ and of $\mathcal{V}$ and that $\{\mathbf{w}_1, \ldots, \mathbf{w}_k\}$ is a basis for $\mathcal{U} \cap \mathcal{V}$. Then there exists a family of vectors $\{\mathbf{u}_1, \ldots, \mathbf{u}_r\}$ and a family of vectors $\{\mathbf{v}_1, \ldots, \mathbf{v}_s\}$ such that $\{\mathbf{w}_1, \ldots, \mathbf{w}_k, \mathbf{u}_1, \ldots, \mathbf{u}_r\}$ is a basis for $\mathcal{U}$ and $\{\mathbf{w}_1, \ldots, \mathbf{w}_k, \mathbf{v}_1, \ldots, \mathbf{v}_s\}$ is a basis for $\mathcal{V}$. It is clear that
$$\operatorname{span}\{\mathbf{w}_1, \ldots, \mathbf{w}_k, \mathbf{u}_1, \ldots, \mathbf{u}_r, \mathbf{v}_1, \ldots, \mathbf{v}_s\} = \mathcal{U} + \mathcal{V}.$$
Moreover, if $a_1, \ldots, a_k, b_1, \ldots, b_r, c_1, \ldots, c_s$ are scalars, then
$$\sum_{i=1}^k a_i \mathbf{w}_i + \sum_{j=1}^r b_j \mathbf{u}_j + \sum_{\ell=1}^s c_\ell \mathbf{v}_\ell = \mathbf{0} \implies \sum_{i=1}^k a_i \mathbf{w}_i + \sum_{j=1}^r b_j \mathbf{u}_j = -\sum_{\ell=1}^s c_\ell \mathbf{v}_\ell$$
and hence $c_1 = \cdots = c_s = 0$, since the last equality implies that $\sum_{\ell=1}^s c_s \mathbf{v}_\ell$ belongs to $\mathcal{U} \cap \mathcal{V} = \operatorname{span}\{\mathbf{w}_1, \ldots, \mathbf{w}_k\}$. Thus, also $a_1 = \cdots = a_k = b_1 = \cdots = b_s = 0$, since $\{\mathbf{w}_1, \ldots, \mathbf{w}_k, \mathbf{u}_1, \ldots, \mathbf{u}_r\}$ is a basis for $\mathcal{U}$. Consequently, the full family of $k + r + s$ vectors is linearly independent and so is a basis for $\mathcal{U} + \mathcal{V}$. Therefore,
$$\dim(\mathcal{U} + \mathcal{V}) = k + r + s = (k+r) + (k+s) - k,$$
which serves to verify (5.3) when $\mathcal{U} \cap \mathcal{V}$ is a nonzero proper subspace of both $\mathcal{U}$ and $\mathcal{V}$.

The verification of (5.3) when $\mathcal{U} \cap \mathcal{V} = \{\mathbf{0}\}$ is similar, but easier: If $\{\mathbf{u}_1, \ldots, \mathbf{u}_s\}$ is a basis for $\mathcal{U}$ and $\{\mathbf{v}_1, \ldots, \mathbf{v}_t\}$ is a basis for $\mathcal{V}$, then clearly $\operatorname{span}\{\mathbf{u}_1, \ldots, \mathbf{u}_s, \mathbf{v}_1, \ldots, \mathbf{v}_t\} = \mathcal{U} + \mathcal{V}$ and this set of $s + t$ vectors is linearly independent since
$$\sum_{i=1}^s b_i \mathbf{u}_i + \sum_{j=1}^t c_j \mathbf{v}_j = \mathbf{0} \implies \sum_{i=1}^s b_i \mathbf{u}_i = -\sum_{j=1}^t c_j \mathbf{v}_j$$

and hence both sides of the last equality belong to $\mathcal{U} \cap \mathcal{V} = \{\mathbf{0}\}$. Therefore, (5.3) holds in this case also.

Finally, the characterization (5.4) is immediate from the definition of a direct sum and formula (5.3). □

We remark that the characterization (5.4) has the advantage of being applicable to infinite-dimensional spaces. However, it does not have simple analogues for sums of three or more subspaces; see, e.g., Example 5.2.

Example 5.2. If

$$\mathcal{U} = \mathrm{span}\left\{\begin{bmatrix}1\\1\\0\end{bmatrix}\right\}, \quad \mathcal{V} = \mathrm{span}\left\{\begin{bmatrix}0\\-1\\1\end{bmatrix}\right\}, \quad \text{and} \quad \mathcal{W} = \mathrm{span}\left\{\begin{bmatrix}1\\0\\1\end{bmatrix}\right\},$$

then $\mathcal{U} \cap \mathcal{V} = \{\mathbf{0}\}$, $\mathcal{U} \cap \mathcal{W} = \{\mathbf{0}\}$, and $\mathcal{V} \cap \mathcal{W} = \{\mathbf{0}\}$ but the sum $\mathcal{U} + \mathcal{V} + \mathcal{W}$ is not direct. ◇

Exercise 5.1. Let $\mathcal{Y}$ be a finite-dimensional vector space. Show that if $\mathcal{Y} = \mathcal{U} \dotplus \mathcal{V}$ and $\mathcal{V} = \mathcal{X} \dotplus \mathcal{W}$, then $\mathcal{Y} = \mathcal{U} \dotplus \mathcal{X} \dotplus \mathcal{W}$.

If the vector space $\mathcal{Y}$ admits the **direct sum decomposition**

$$\mathcal{Y} = \mathcal{U} \dotplus \mathcal{V},$$

then the subspace $\mathcal{V}$ is said to be a **complementary space** to the subspace $\mathcal{U}$ and the subspace $\mathcal{U}$ is said to be a complementary space to the subspace $\mathcal{V}$.

Exercise 5.2. Let T be a linear transformation from a real vector space $\mathcal{V}$ into itself and let $\mathcal{U}$ be a two-dimensional subspace of $\mathcal{V}$ with basis $\{\mathbf{u}_1, \mathbf{u}_2\}$. Show that if $T\mathbf{u}_1 = \mathbf{u}_1 + 2\mathbf{u}_2$ and $T\mathbf{u}_2 = 2\mathbf{u}_1 + \mathbf{u}_2$, then $\mathcal{U}$ is the direct sum of two one-dimensional spaces that are each invariant under T.

Lemma 5.2. *Let $\mathcal{U}$ and $\mathcal{V}$ be subspaces of a vector space $\mathcal{Y}$ such that $\mathcal{U} \dotplus \mathcal{V} = \mathcal{Y}$. Then every vector $\mathbf{y} \in \mathcal{Y}$ can be expressed as a sum of the form $\mathbf{y} = \mathbf{u} + \mathbf{v}$ for exactly one pair of vectors $\mathbf{u} \in \mathcal{U}$ and $\mathbf{v} \in \mathcal{V}$.*

Proof. Suppose that $\mathbf{y} = \mathbf{u}_1 + \mathbf{v}_1 = \mathbf{u}_2 + \mathbf{v}_2$ with $\mathbf{u}_1, \mathbf{u}_2 \in \mathcal{U}$ and $\mathbf{v}_1, \mathbf{v}_2 \in \mathcal{V}$. Then the vector $\mathbf{u}_1 - \mathbf{u}_2 = \mathbf{v}_2 - \mathbf{v}_1$ belongs to $\mathcal{U} \cap \mathcal{V}$ and is therefore equal to $\mathbf{0}$. Thus, $\mathbf{u}_1 = \mathbf{u}_2$ and $\mathbf{v}_1 = \mathbf{v}_2$. □

Lemma 5.3. *Let $\mathcal{U}, \mathcal{V},$ and $\mathcal{W}$ be subspaces of a vector space $\mathcal{Y}$ such that*

$$\mathcal{U} \dotplus \mathcal{V} = \mathcal{Y} \quad \text{and} \quad \mathcal{U} \subseteq \mathcal{W}.$$

Then

(5.5) $$\mathcal{W} = \mathcal{U} \dotplus (\mathcal{W} \cap \mathcal{V}).$$

Proof. Clearly,
$$\mathcal{U} + (\mathcal{W} \cap \mathcal{V}) = (\mathcal{W} \cap \mathcal{U}) + (\mathcal{W} \cap \mathcal{V}) \subseteq \mathcal{W} + \mathcal{W} = \mathcal{W}.$$

To establish the opposite inclusion, let $\mathbf{w} \in \mathcal{W}$. Then, since $\mathcal{W} \subseteq \mathcal{Y}$ and $\mathcal{Y} = \mathcal{U} \dotplus \mathcal{V}$, $\mathbf{w} = \mathbf{u} + \mathbf{v}$ for exactly one pair of vectors $\mathbf{u} \in \mathcal{U}$ and $\mathbf{v} \in \mathcal{V}$. Moreover, under the added assumption that $\mathcal{U} \subseteq \mathcal{W}$, it follows that both $\mathbf{u}$ and $\mathbf{v} = \mathbf{w} - \mathbf{u}$ belong to $\mathcal{W}$. Therefore, $\mathbf{u} \in \mathcal{W} \cap \mathcal{U}$ and $\mathbf{v} \in \mathcal{W} \cap \mathcal{V}$, and hence
$$\mathcal{W} \subseteq (\mathcal{W} \cap \mathcal{U}) + (\mathcal{W} \cap \mathcal{V}).$$

Therefore, (5.5) holds. □

Example 5.3. If $I_2 = \begin{bmatrix} \mathbf{e}_1 & \mathbf{e}_2 \end{bmatrix}$, $\mathcal{U} = \mathrm{span}\{\mathbf{e}_1\}$, $\mathcal{V} = \mathrm{span}\{\mathbf{e}_2\}$, and $\mathcal{W} = \mathrm{span}\{\mathbf{e}_1 + \mathbf{e}_2\}$, then $\mathbb{C}^2 = \mathcal{U} \dotplus \mathcal{V}$, but $(\mathcal{W} \cap \mathcal{U}) + (\mathcal{W} \cap \mathcal{V}) = \{\mathbf{0}\} \neq \mathcal{W}$. What went wrong? ◇

The next theorem is the first step towards the verification of the general formula $A = UJU^{-1}$ alluded to earlier.

Theorem 5.4. *Let $A \in \mathbb{C}^{n \times n}$ and suppose that A has exactly k distinct eigenvalues, $\lambda_1, \ldots, \lambda_k \in \mathbb{C}$. Then*
$$(5.6) \qquad \mathbb{C}^n = \mathcal{N}_{(A - \lambda_1 I_n)^n} \dotplus \cdots \dotplus \mathcal{N}_{(A - \lambda_k I_n)^n}.$$

The proof of this theorem will be furnished in Section 5.3. At this point we shall focus on its implications.

Corollary 5.5. *If $A \in \mathbb{C}^{n \times n}$ has exactly k distinct eigenvalues, $\lambda_1, \ldots, \lambda_k \in \mathbb{C}$, and the matrices $U_j \in \mathbb{C}^{n \times \alpha_j}$, $j = 1, \ldots, k$, are constructed so that the columns of U_j are a basis for $\mathcal{N}_{(A - \lambda_j I_n)^n}$, then:*

(1) *The $n \times n$ matrix $U = \begin{bmatrix} U_1 & \cdots & U_k \end{bmatrix}$ is invertible.*
(2) *There exists a set of matrices $G_j \in \mathbb{C}^{\alpha_j \times \alpha_j}$ such that $AU_j = U_j G_j$ for $j = 1, \ldots, k$.*
(3) $A = UGU^{-1}$ *with* $G = \mathrm{diag}\{G_1, \ldots, G_k\}$.

Proof. It suffices to verify (2), since (1) follows easily from Theorem 5.4 and (3) follows easily from (2). Towards this end, let $\mu = \lambda_j$ and $m = \alpha_j$ for any choice of j in the set $\{1, \ldots, k\}$ and suppose that the columns of the matrix $V = [\mathbf{v}_1, \ldots, \mathbf{v}_m]$ are a basis for $\mathcal{N}_{(A - \mu I_n)^n}$. Then $(A - \mu I_n)^n \mathbf{v}_i = \mathbf{0}$ for $i = 1, \ldots, m$. Therefore,
$$(A - \mu I_n)^n A \mathbf{v}_i = A[(A - \mu I_n)^n \mathbf{v}_i] = A\mathbf{0} = \mathbf{0},$$

i.e., $\mathbf{v}_i \in \mathcal{N}_{(A - \mu I_n)^n} \implies A\mathbf{v}_i \in \mathcal{N}_{(A - \mu I_n)^n}$. But this means that $A\mathbf{v}_i$ is a linear combination of the columns of V, i.e., $A\mathbf{v}_i = V\mathbf{g}_i$ for some vector

$\mathbf{g}_i \in \mathbb{C}^m$. Thus,

$$AV = A\begin{bmatrix} \mathbf{v}_1 & \cdots & \mathbf{v}_m \end{bmatrix} = \begin{bmatrix} A\mathbf{v}_1 & \cdots & A\mathbf{v}_m \end{bmatrix} = \begin{bmatrix} V\mathbf{g}_1 & \cdots & V\mathbf{g}_m \end{bmatrix}$$
$$= VG \text{ with } G = \begin{bmatrix} \mathbf{g}_1 & \cdots & \mathbf{g}_m \end{bmatrix}.$$

This serves to justify the formula $AU_j = U_j G_j$ for $j = 1, \ldots, k$. Therefore (2) holds. □

Example 5.4. Let $A \in \mathbb{C}^{9 \times 9}$ and suppose that A has exactly three distinct eigenvalues λ_1, λ_2, and λ_3 with algebraic multiplicities $\alpha_1 = 4$, $\alpha_2 = 2$, and $\alpha_3 = 3$, respectively. Let $\{\mathbf{v}_1, \mathbf{v}_2, \mathbf{v}_3, \mathbf{v}_4\}$ be any basis for $\mathcal{N}_{(A-\lambda_1 I_9)^9}$, let $\{\mathbf{w}_1, \mathbf{w}_2\}$ be any basis for $\mathcal{N}_{(A-\lambda_2 I_9)^9}$, and let $\{\mathbf{x}_1, \mathbf{x}_2, \mathbf{x}_3\}$ be any basis for $\mathcal{N}_{(A-\lambda_3 I_9)^9}$. Then, since each of the spaces $\mathcal{N}_{(A-\lambda_j I_9)^9}$, $j = 1, 2, 3$, is invariant under multiplication by the matrix A,

$$\begin{aligned} A[\mathbf{v}_1 \ \mathbf{v}_2 \ \mathbf{v}_3 \ \mathbf{v}_4] &= [\mathbf{v}_1 \ \mathbf{v}_2 \ \mathbf{v}_3 \ \mathbf{v}_4]G_1, \\ A[\mathbf{w}_1 \ \mathbf{w}_2] &= [\mathbf{w}_1 \ \mathbf{w}_2]G_2, \\ A[\mathbf{x}_1 \ \mathbf{x}_2 \ \mathbf{x}_3] &= [\mathbf{x}_1 \ \mathbf{x}_2 \ \mathbf{x}_3]G_3 \end{aligned}$$

for some choice of $G_1 \in \mathbb{C}^{4 \times 4}$, $G_2 \in \mathbb{C}^{2 \times 2}$, and $G_3 \in \mathbb{C}^{3 \times 3}$. In other notation, upon setting

$$V = [\mathbf{v}_1 \ \mathbf{v}_2 \ \mathbf{v}_3 \ \mathbf{v}_4], \ W = [\mathbf{w}_1 \ \mathbf{w}_2], \text{ and } X = [\mathbf{x}_1 \ \mathbf{x}_2 \ \mathbf{x}_3],$$

we can write the preceding three sets of equations together as

$$A[V \ W \ X] = [V \ W \ X] \begin{bmatrix} G_1 & O & O \\ O & G_2 & O \\ O & O & G_3 \end{bmatrix}$$

or, equivalently, upon setting $U = [V \ W \ X]$, as

$$(5.7) \qquad A = U \begin{bmatrix} G_1 & O & O \\ O & G_2 & O \\ O & O & G_3 \end{bmatrix} U^{-1},$$

since the matrix $U = [V \ W \ X]$ is invertible, thanks to Theorem 5.4. ◊

Formula (5.7) is the best that can be achieved in Example 5.4 if we only know $\alpha_1, \alpha_2, \alpha_3$. Our ultimate objective is to obtain a factorization in which each of the blocks G_1, G_2, G_3 is itself a block diagonal matrix of Jordan cells. To achieve this, the corresponding bases $\{\mathbf{v}_1, \mathbf{v}_2, \mathbf{v}_3, \mathbf{v}_4\}$, $\{\mathbf{w}_1 \ \mathbf{w}_2\}$, and $\{\mathbf{x}_1 \ \mathbf{x}_2 \ \mathbf{x}_3\}$ must be chosen appropriately. This will be taken up in the next chapter.

5.2. The null spaces of powers of B

The next few lemmas focus on the null spaces $\mathcal{N}_{B^j}$ for $j = 0, 1, \ldots$ when $B \in \mathbb{C}^{n \times n}$. They will play a role in the proof of Theorem 5.11. However, to verify the first assertion of that theorem, only (1) and (2) of Lemma 5.6 and (1)–(3) of Lemma 5.8 are needed.

Lemma 5.6. *If $B \in \mathbb{C}^{n \times n}$, then:*

  (1) $\mathcal{N}_B \subseteq \mathcal{N}_{B^2} \subseteq \mathcal{N}_{B^3} \subseteq \cdots$.
  (2) *If $\mathcal{N}_{B^j} = \mathcal{N}_{B^{j+1}}$ for some integer $j \geq 1$, then $\mathcal{N}_{B^{j+1}} = \mathcal{N}_{B^{j+2}}$.*
  (3) *If $j \geq 1$ is an integer, then $\mathcal{N}_{B^j} \neq \{\mathbf{0}\} \iff \mathcal{N}_{B^{j+1}} \neq \{\mathbf{0}\}$.*

Proof. The proof is left to the reader. □

Exercise 5.3. Verify the assertions in Lemma 5.6.

Lemma 5.7. *If $B \in \mathbb{C}^{n \times n}$ and $\mathbf{u} \in \mathbb{C}^n$ belongs to $\mathcal{N}_{B^k}$ for some positive integer k, then the vectors*

$$\mathbf{u}, B\mathbf{u}, \ldots, B^{k-1}\mathbf{u} \text{ are linearly independent} \iff B^{k-1}\mathbf{u} \neq \mathbf{0}.$$

Proof. Suppose first that $B^{k-1}\mathbf{u} \neq \mathbf{0}$ and that

$$\alpha_0 \mathbf{u} + \alpha_1 B\mathbf{u} + \cdots + \alpha_{k-1} B^{k-1}\mathbf{u} = \mathbf{0}$$

for some choice of coefficients $\alpha_0, \ldots, \alpha_{k-1} \in \mathbb{C}$. Then, since $\mathbf{u} \in \mathcal{N}_{B^k}$,

$$B^{k-1}(\alpha_0 \mathbf{u} + \alpha_1 B\mathbf{u} + \cdots + \alpha_{k-1} B^{k-1}\mathbf{u}) = \alpha_0 B^{k-1}\mathbf{u} = \mathbf{0},$$

which clearly implies that $\alpha_0 = 0$. Similarly, the identity

$$B^{k-2}(\alpha_1 B\mathbf{u} + \cdots + \alpha_{k-1} B^{k-1}\mathbf{u}) = \alpha_1 B^{k-1}\mathbf{u} = \mathbf{0}$$

implies that $\alpha_1 = 0$. Continuing in this vein it is readily seen that $B^{k-1}\mathbf{u} \neq \mathbf{0} \Longrightarrow$ the vectors $\mathbf{u}, B\mathbf{u}, \ldots, B^{k-1}\mathbf{u}$ are linearly independent. Thus, as the converse is self-evident, the proof is complete. □

Lemma 5.8. *If $B \in \mathbb{C}^{n \times n}$, then:*

  (1) $\mathcal{N}_{B^k} \subseteq \mathcal{N}_{B^n}$ *and* $\mathcal{R}_{B^k} \supseteq \mathcal{R}_{B^n}$ *for every positive integer k.*
  (2) $\mathbb{C}^n = \mathcal{N}_{B^n} \dotplus \mathcal{R}_{B^n}$.
  (3) $\mathcal{N}_{(B+\lambda I_n)^n} \subseteq \mathcal{R}_{B^n}$ *for every point $\lambda \in \mathbb{C} \setminus \{0\}$.*
  (4) $\mathcal{R}_B \cap \mathcal{N}_B = \{\mathbf{0}\} \Longrightarrow \mathcal{R}_B \cap \mathcal{N}_{B^k} = \{\mathbf{0}\}$ *for $k = 1, 2, \ldots$.*
  (5) $\mathbb{C}^n = \mathcal{N}_B \dotplus \mathcal{R}_B \iff \mathcal{N}_B = \mathcal{N}_{B^n}$.

Proof. If $\mathcal{N}_{B^k} = \{\mathbf{0}\}$, then the first assertion in (1) is clear. If $B^k \mathbf{u} = \mathbf{0}$ for some nonzero vector $\mathbf{u} \in \mathbb{C}^n$ and some positive integer k, let j be the smallest positive integer such that $B^{j-1}\mathbf{u} \neq \mathbf{0}$ and $B^j \mathbf{u} = \mathbf{0}$. Then, in view of Lemma 5.7, the vectors $\mathbf{u}, B\mathbf{u}, \ldots, B^{j-1}\mathbf{u}$ are linearly independent.

5.2. The null spaces of powers of B

Therefore, $j \leq n$, and hence $B^n\mathbf{u} = B^{n-j}(B^j\mathbf{u}) = B^{n-j}\mathbf{0} = \mathbf{0}$, i.e., $B^k\mathbf{u} = \mathbf{0} \implies B^n\mathbf{u} = \mathbf{0}$. This justifies the first assertion in (1); the second is left to the reader.

To verify (2), suppose first that $\mathbf{u} \in \mathcal{N}_{B^n} \cap \mathcal{R}_{B^n}$. Then $B^n\mathbf{u} = \mathbf{0}$ and $\mathbf{u} = B^n\mathbf{v}$ for some vector $\mathbf{v} \in \mathbb{C}^n$. Therefore,

$$\mathbf{0} = B^n\mathbf{u} = B^{2n}\mathbf{v}.$$

But, then (1) ensures that

$$\mathbf{u} = B^n\mathbf{v} = \mathbf{0}.$$

Thus, the sum is direct. Therefore, as $\{\mathcal{R}_{B^n} \dotplus \mathcal{N}_{B^n}\} \subseteq \mathbb{C}^n$ and

$$\dim\{\mathcal{R}_{B^n} \dotplus \mathcal{N}_{B^n}\} = \dim \mathcal{R}_{B^n} + \dim \mathcal{N}_{B^n} = n$$

by the principle of conservation of dimension, (2) holds.

Suppose next that $(B + \lambda I_n)^n\mathbf{u} = \mathbf{0}$ for some nonzero vector $\mathbf{u} \in \mathbb{C}^n$ and some nonzero number $\lambda \in \mathbb{C}$. Then, since λI_n commutes with B, we can invoke the binomial theorem for matrices to obtain

$$\mathbf{0} = \sum_{j=0}^{n} \binom{n}{j} \lambda^{n-j} B^j \mathbf{u} = \lambda^n \mathbf{u} + \sum_{j=1}^{n} \binom{n}{j} \lambda^{n-j} B^j \mathbf{u}.$$

Therefore,

(5.8) $\qquad \mathbf{u} = Bp(B)\mathbf{u}, \quad \text{where} \quad p(B) = -\sum_{j=1}^{n} \binom{n}{j} \lambda^{-j} B^{j-1}$

is a polynomial in B. Thus, as $Bp(B) = p(B)B$,

$$\mathbf{u} = Bp(B)\mathbf{u} = (Bp(B))^2\mathbf{u} = \cdots = (Bp(B))^n\mathbf{u} = B^n p(B)^n \mathbf{u} \in \mathcal{R}_{B^n},$$

as claimed in (3).

We now turn to (4): If $\mathcal{R}_B \cap \mathcal{N}_B = \{\mathbf{0}\}$ and $\mathbf{x} \in \mathcal{R}_B \cap \mathcal{N}_{B^k}$ for some positive integer $k \geq 2$, then $B^{k-1}\mathbf{x} \in \mathcal{R}_B \cap \mathcal{N}_B = \{\mathbf{0}\}$ and hence $\mathcal{R}_B \cap \mathcal{N}_{B^k} \subseteq \mathcal{R}_B \cap \mathcal{N}_{B^{k-1}} \subseteq \cdots \subseteq \mathcal{R}_B \cap \mathcal{N}_B$.

Finally, if $\mathbb{C}^n = \mathcal{N}_B \dotplus \mathcal{R}_B$, then $\mathcal{N}_B \cap \mathcal{R}_B = \{\mathbf{0}\}$ and hence, in view of Lemma 5.3 and (4),

$$\mathcal{N}_{B^n} = (\mathcal{N}_B \cap \mathcal{N}_{B^n}) \dotplus (\mathcal{R}_B \cap \mathcal{N}_{B^n}) = \mathcal{N}_B \dotplus (\mathcal{R}_B \cap \mathcal{N}_{B^n})$$
$$= \mathcal{N}_B \dotplus (\mathcal{R}_B \cap \mathcal{N}_B) = \mathcal{N}_B.$$

It remains to show that if $\mathcal{N}_{B^n} = \mathcal{N}_B$, then $\mathbb{C}^n = \mathcal{N}_B \dotplus \mathcal{R}_B$. This is left to the reader as an exercise. $\square$

Exercise 5.4. Show that if $B \in \mathbb{C}^{n \times n}$, then $\mathcal{N}_{(B+\lambda I_n)^k} \subseteq \mathcal{R}_{B^k}$ for every point $\lambda \in \mathbb{C} \setminus \{0\}$ and every positive integer k.

Exercise 5.5. Show that if $\mathcal{N}_{B^n} = \mathcal{N}_B$, then $\mathbb{C}^n = \mathcal{N}_B \dotplus \mathcal{R}_B$, to complete the proof of (5) in Lemma 5.8.

Remark 5.9. The last lemma may be exploited to give a quick proof of the fact that *generalized eigenvectors corresponding to distinct eigenvalues are automatically linearly independent*. To verify this, let
$$(A - \lambda_j I_n)^n \mathbf{u}_j = 0, \ j = 1, \ldots, k,$$
for some set of distinct eigenvalues $\lambda_1, \ldots, \lambda_k$ and suppose that
$$c_1 \mathbf{u}_1 + \cdots + c_k \mathbf{u}_k = \mathbf{0}.$$
Then
$$-c_1 \mathbf{u}_1 = c_2 \mathbf{u}_2 + \cdots + c_k \mathbf{u}_k$$
and, since $-c_1 \mathbf{u}_1 \in \mathcal{N}_{(A-\lambda_1 I_n)^n}$ and, by (3) of Lemma 5.8, $c_2 \mathbf{u}_2 + \cdots + c_k \mathbf{u}_k \in \mathcal{R}_{(A-\lambda_1 I_n)^n}$, both sides of the last displayed equality must equal zero, thanks to (2) of Lemma 5.8. Therefore, $c_1 = 0$ and
$$c_2 \mathbf{u}_2 + \cdots + c_k \mathbf{u}_k = \mathbf{0}.$$
To complete the verification, just keep on going.

Exercise 5.6. Show that if $B \in \mathbb{C}^{n \times n}$, $\lambda \in \mathbb{C}$, and $\lambda \neq 0$, then

(5.9) $$\mathcal{N}_{B^j} \subseteq \mathcal{R}_{(B+\lambda I_n)^k}$$

for every choice of $j, k \in \{1, \ldots, n\}$. [HINT: The proof of (3) in Lemma 5.8.]

5.3. Verification of Theorem 5.4

Lemma 5.8 guarantees that

(5.10) $$\mathbb{C}^n = \mathcal{N}_{(A-\lambda I_n)^n} \dotplus \mathcal{R}_{(A-\lambda I_n)^n}$$

for every point $\lambda \in \mathbb{C}$. The next step is to obtain an analogous direct sum decomposition for $\mathcal{R}_{(A-\lambda I_n)^n}$.

Lemma 5.10. *If* $A \in \mathbb{C}^{n \times n}$, $\lambda_1, \lambda_2 \in \mathbb{C}$, *and* $\lambda_1 \neq \lambda_2$, *then*

(5.11) $$\mathcal{R}_{(A-\lambda_1 I_n)^n} = \mathcal{N}_{(A-\lambda_2 I_n)^n} \dotplus (\mathcal{R}_{(A-\lambda_1 I_n)^n} \cap \mathcal{R}_{(A-\lambda_2 I_n)^n}).$$

Proof. Since $\mathbb{C}^n = \mathcal{N}_{(A-\lambda_2 I_n)^n} \dotplus \mathcal{R}_{(A-\lambda_2 I_n)^n}$ and $\mathcal{N}_{(A-\lambda_2 I_n)^n} \subseteq \mathcal{R}_{(A-\lambda_1 I_n)^n}$ when $\lambda_1 \neq \lambda_2$ by Lemma 5.8, the assertion follows from Lemma 5.3 (with $\mathcal{U} = \mathcal{N}_{(A-\lambda_2 I_n)^n}$, $\mathcal{V} = \mathcal{R}_{(A-\lambda_2 I_n)^n}$, and $\mathcal{W} = \mathcal{R}_{(A-\lambda_1 I_n)^n}$). □

The first assertion in the next theorem is used in the proof of Theorem 5.4; the second assertion, which serves to characterize diagonalizable matrices, is included for added perspective.

5.3. Verification of Theorem 5.4

Theorem 5.11. *If $A \in \mathbb{C}^{n \times n}$ has k distinct eigenvalues $\lambda_1, \ldots, \lambda_k \in \mathbb{C}$ with geometric multiplicities $\gamma_1, \ldots, \gamma_k$ and algebraic multiplicities $\alpha_1, \ldots, \alpha_k$, respectively, then*

(5.12) $$\mathcal{R}_{(A-\lambda_1 I_n)^n} \cap \mathcal{R}_{(A-\lambda_2 I_n)^n} \cap \cdots \cap \mathcal{R}_{(A-\lambda_k I_n)^n} = \{\mathbf{0}\} \,.$$

However,

(5.13) $$\mathcal{R}_{(A-\lambda_1 I_n)} \cap \mathcal{R}_{(A-\lambda_2 I_n)} \cap \cdots \cap \mathcal{R}_{(A-\lambda_k I_n)} = \{\mathbf{0}\} \iff \gamma_j = \alpha_j$$
$$\text{for } j = 1, \ldots, k \,.$$

Proof. Let $\mathcal{M}$ denote the intersection of the k subspaces on the left-hand side of the asserted identity (5.12). Then it is readily checked that $\mathcal{M}$ is invariant under A; i.e., if $\mathbf{u} \in \mathcal{M}$, then $A\mathbf{u} \in \mathcal{M}$, because each of the subspaces $\mathcal{R}_{(A-\lambda_j I_n)^n}$ is invariant under A: if $\mathbf{u} \in \mathcal{R}_{(A-\lambda_j I_n)^n}$, then $\mathbf{u} = (A - \lambda_j I_n)^n \mathbf{v}_j$ for some vector $\mathbf{v}_j \in \mathbb{C}^n$ for each choice of $j = 1, \ldots, n$ and hence

$$A\mathbf{u} = A(A - \lambda_j I_n)^n \mathbf{v}_j = (A - \lambda_j I_n)^n A\mathbf{v}_j \in \mathcal{R}_{(A-\lambda_j I_n)^n} \quad \text{for } j = 1, \ldots, k \,.$$

Thus, if $\mathcal{M} \neq \{\mathbf{0}\}$, then, by Theorem 4.3, there exist a number $\lambda \in \mathbb{C}$ and a nonzero vector $\mathbf{v} \in \mathcal{M}$ such that $A\mathbf{v} - \lambda\mathbf{v} = 0$. But this means that λ is equal to one of the eigenvalues of A, say λ_t, i.e., $\mathbf{v} \in \mathcal{N}_{(A-\lambda_t I_n)}$. But this in turn implies that

$$\mathbf{v} \in \mathcal{N}_{(A-\lambda_t I_n)^n} \cap \mathcal{R}_{(A-\lambda_t I_n)^n} = \{\mathbf{0}\} \,.$$

Therefore, $\mathcal{M} = \{\mathbf{0}\}$. This completes the proof of (5.12).

To verify the implication $\implies$ in (5.13), observe first that, in view of Lemma 5.8 and the inclusions $\mathcal{N}_B \subseteq \mathcal{N}_{B^n}$ and $\mathcal{R}_{B^n} \subseteq \mathcal{R}_B$:

(a) $\mathcal{N}_{A-\lambda_1 I_n} \subseteq \mathcal{R}_{(A-\lambda_2 I_n)} \cap \cdots \cap \mathcal{R}_{(A-\lambda_k I_n)}$.

(b) $\mathcal{R}_{(A-\lambda_1 I_n)} \cap \mathcal{N}_{(A-\lambda_1 I_n)} \subseteq \mathcal{R}_{(A-\lambda_1 I_n)} \cap \cdots \cap \mathcal{R}_{(A-\lambda_k I_n)}$.

(c) $\mathcal{R}_{(A-\lambda_1 I_n)} \cap \cdots \cap \mathcal{R}_{(A-\lambda_k I_n)} = \{\mathbf{0}\} \implies \mathcal{R}_{(A-\lambda_1 I_n)} \cap \mathcal{N}_{(A-\lambda_1 I_n)} = \{\mathbf{0}\}$.

Therefore, $\mathcal{R}_{(A-\lambda_1 I_n)} \dotplus \mathcal{N}_{(A-\lambda_1 I_n)} = \mathbb{C}^n$ and hence, by (5) of Lemma 5.8, $\alpha_1 = \gamma_1$, and, by analogous arguments, $\alpha_j = \gamma_j$ for $j = 2, \ldots, k$ also.

Conversely, if $\alpha_j = \gamma_j$ for $j = 1, \ldots, k$, then $\mathcal{N}_{(A-\lambda_1 I_n)} = \mathcal{N}_{(A-\lambda_1 I_n)^n}$. Therefore, $\mathcal{R}_{(A-\lambda_1 I_n)} = \mathcal{R}_{(A-\lambda_1 I_n)^n}$ and hence the implication $\impliedby$ in (5.13) follows from (5.12). $\square$

We are now ready to prove Theorem 5.4.

Proof of Theorem 5.4. Let us suppose that $k \geq 3$. Then, by Lemmas 5.8 and 5.10,
$$\mathbb{C}^n = \mathcal{N}_{(A-\lambda_1 I_n)^n} \dotplus \mathcal{R}_{(A-\lambda_1 I_n)^n}$$
and
$$\mathcal{R}_{(A-\lambda_1 I_n)^n} = \mathcal{N}_{(A-\lambda_2 I_n)^n} \dotplus \mathcal{R}_{(A-\lambda_1 I_n)^n} \cap \mathcal{R}_{(A-\lambda_2 I_n)^n}.$$
Therefore,
$$(5.14) \quad \mathbb{C}^n = \mathcal{N}_{(A-\lambda_1 I_n)^n} \dotplus \mathcal{N}_{(A-\lambda_2 I_n)^n} \dotplus \mathcal{R}_{(A-\lambda_1 I_n)^n} \cap \mathcal{R}_{(A-\lambda_2 I_n)^n}.$$
Moreover, since $\mathbb{C}^n = \mathcal{N}_{(A-\lambda_3 I_n)^n} \dotplus \mathcal{R}_{(A-\lambda_3 I_n)^n}$ and
$$\mathcal{N}_{(A-\lambda_3 I_n)^n} \subseteq \mathcal{R}_{(A-\lambda_1 I_n)^n} \cap \mathcal{R}_{(A-\lambda_2 I_n)^n}$$
by Lemma 5.8, the supplementary formula
$$\mathcal{R}_{(A-\lambda_1 I_n)^n} \cap \mathcal{R}_{(A-\lambda_2 I_n)^n}$$
$$= \mathcal{N}_{(A-\lambda_3 I_n)^n} \dotplus \mathcal{R}_{(A-\lambda_1 I_n)^n} \cap \mathcal{R}_{(A-\lambda_2 I_n)^n} \cap \mathcal{R}_{(A-\lambda_3 I_n)^n}$$
follows from Lemma 5.3 with $\mathcal{U} = \mathcal{N}_{(A-\lambda_3 I_n)^n}$, $\mathcal{V} = \mathcal{R}_{(A-\lambda_3 I_n)^n}$, and $\mathcal{W} = \mathcal{R}_{(A-\lambda_1 I_n)^n} \cap \mathcal{R}_{(A-\lambda_2 I_n)^n}$. When substituted into formula (5.14), this yields
$$(5.15) \quad \mathbb{C}^n = \mathcal{N}_{(A-\lambda_1 I_n)^n} \dotplus \cdots \dotplus \mathcal{N}_{(A-\lambda_j I_n)^n} \dotplus \mathcal{R}_{(A-\lambda_1 I_n)^n} \cap \cdots \cap \mathcal{R}_{(A-\lambda_j I_n)^n}$$
for $j = 3$. To complete the proof, just keep on going until you run out of eigenvalues (i.e., until $j = k$ in (5.15)) and then invoke (5.12). □

5.4. Supplementary notes

This chapter is mostly adapted from Chapter 4 in [**30**]. However, (5.13) does not appear there.

Chapter 6

The Jordan decomposition

The decomposition (5.6) ensures that if $A \in \mathbb{C}^{n \times n}$ has k distinct eigenvalues $\lambda_1, \ldots, \lambda_k$ with algebraic multiplicities $\alpha_1, \ldots, \alpha_k$, then there exist an invertible matrix $U \in \mathbb{C}^{n \times n}$ and a block diagonal matrix $G = \mathrm{diag}\{G_1, \ldots, G_k\}$ with blocks G_j of size $\alpha_j \times \alpha_j$, $j = 1, \ldots, k$, such that $A = UGU^{-1}$. Our next objective is to establish the Jordan decomposition theorem, which is essentially an algorithm for choosing the basis for each of the spaces $\mathcal{N}_{(A-\lambda_j I_n)^n}$ to obtain a nice form for G_j.

6.1. The Jordan decomposition

Theorem 6.1. *If $A \in \mathbb{C}^{n \times n}$ has k distinct eigenvalues $\lambda_1, \ldots, \lambda_k$ with geometric multiplicities $\gamma_1, \ldots, \gamma_k$ and algebraic multiplicities $\alpha_1, \ldots, \alpha_k$, respectively, then there exists an invertible matrix $U \in \mathbb{C}^{n \times n}$ such that*

$$AU = UJ,$$

where:

(1) $J = \mathrm{diag}\{B_{\lambda_1}, \ldots, B_{\lambda_k}\}$.

(2) B_{λ_j} is an $\alpha_j \times \alpha_j$ block diagonal matrix that is built out of γ_j Jordan cells $C_{\lambda_j}^{(\cdot)}$ of the form (4.7).

(3) The number of Jordan cells $C_{\lambda_j}^{(m)}$ of size $m \times m$ in B_{λ_j} is equal to the number of columns of height m in the array of α_j symbols $\times$ that is constructed by placing $\kappa_i = \dim \mathcal{N}_{(A-\lambda_j I_n)^i} - \dim \mathcal{N}_{(A-\lambda_j I_n)^{i-1}}$

symbols in row i for $i = 1, 2, \ldots$ (Lemma 6.4 guarantees that $\kappa_1 \geq \kappa_2 \geq \cdots$):

(6.1)
$$\begin{array}{cccccc} \times & \times & \times & \times & \cdots & \times \\ \times & \times & \cdots & \times & & \\ \times & \cdots & \times & & & \\ \vdots & & & & & \end{array} \qquad \begin{array}{l} \kappa_1 \text{ symbols} \\ \kappa_2 \text{ symbols} \\ \kappa_3 \text{ symbols} \\ \vdots \end{array}$$

(4) The columns of U are generalized eigenvectors of the matrix A.

(5) $(A - \lambda_1 I_n)^{\alpha_1} \cdots (A - \lambda_k I_n)^{\alpha_k} = O$.

(6) If $\nu_j = \min\{i : \dim \mathcal{N}_{(A-\lambda_j I_n)^i} = \dim \mathcal{N}_{(A-\lambda_j I_n)^n}\}$, then $\nu_j \leq \alpha_j$ for $j = 1, \ldots, k$ and $(A - \lambda_1 I_n)^{\nu_1} \cdots (A - \lambda_k I_n)^{\nu_k} = O$.

Discussion. Corollary 5.5 guarantees that there exist an invertible matrix $U \in \mathbb{C}^{n \times n}$ and a block diagonal matrix $G = \mathrm{diag}\{G_1, \ldots, G_k\}$ with $G_j \in \mathbb{C}^{\alpha_j \times \alpha_j}$ for $j = 1, \ldots, k$ such that $AU = UG$. This result was obtained by choosing $U_j \in \mathbb{C}^{n \times \alpha_j}$ so that the columns of U_j are a basis for $\mathcal{N}_{(A-\lambda_j I_n)^n}$ and then setting $U = \begin{bmatrix} U_1 & \cdots & U_k \end{bmatrix}$.

It remains to show that it is possible to choose the U_j, i.e., the basis of $\mathcal{N}_{(A-\lambda_j I_n)^n}$, in such a way that G_j will be a block diagonal matrix made up of γ_j Jordan cells. This will be done in two steps.

The first step rests on Lemma 5.7, which guarantees that if $\mathbf{u} \in \mathcal{N}_{B^m}$ and $B^{m-1}\mathbf{u} \neq \mathbf{0}$, then the vectors $\mathbf{u}, B\mathbf{u}, \ldots, B^{m-1}\mathbf{u}$ are linearly independent and hence that

$$\mathrm{rank}\begin{bmatrix} B^{m-1}\mathbf{u} & \cdots & \mathbf{u} \end{bmatrix} = m.$$

The reason for stacking the vectors in this order is that then

$$B\begin{bmatrix} B^{m-1}\mathbf{u} & \cdots & \mathbf{u} \end{bmatrix} = \begin{bmatrix} B^m\mathbf{u} & \cdots & B\mathbf{u} \end{bmatrix} = \begin{bmatrix} \mathbf{0} & B^{m-1}\mathbf{u} & \cdots & B\mathbf{u} \end{bmatrix}$$
$$= \begin{bmatrix} B^{m-1}\mathbf{u} & \cdots & \mathbf{u} \end{bmatrix} C_0^{(m)},$$

where $C_0^{(m)}$ is the $m \times m$ Jordan cell with 0 on the diagonal. Thus, if $B = A - \mu I_n$, then

$$A\begin{bmatrix} B^{m-1}\mathbf{u} & \cdots & \mathbf{u} \end{bmatrix} = \mu \begin{bmatrix} B^{m-1}\mathbf{u} & \cdots & \mathbf{u} \end{bmatrix} + \begin{bmatrix} B^{m-1}\mathbf{u} & \cdots & \mathbf{u} \end{bmatrix} C_0^{(m)}$$
$$= \begin{bmatrix} B^{m-1}\mathbf{u} & \cdots & \mathbf{u} \end{bmatrix} \mu I_m + \begin{bmatrix} B^{m-1}\mathbf{u} & \cdots & \mathbf{u} \end{bmatrix} C_0^{(m)}$$
$$= \begin{bmatrix} B^{m-1}\mathbf{u} & \cdots & \mathbf{u} \end{bmatrix} C_\mu^{(m)},$$

which is of the form

(6.2)
$$A\begin{bmatrix} \mathbf{x}_1 & \cdots & \mathbf{x}_m \end{bmatrix} = \begin{bmatrix} \mathbf{x}_1 & \cdots & \mathbf{x}_m \end{bmatrix} C_\mu^{(m)}.$$

6.1. The Jordan decomposition

Thus, $\mathbf{x}_1 = B^{m-1}\mathbf{u} = (A - \mu I_n)^{m-1}\mathbf{u}$ is an eigenvector of A and $\mathbf{u}$ is a nonzero vector in $\mathcal{N}_{(A-\mu I_n)^m}$ and $\mathbf{x}_j = B^{m-j}\mathbf{u}$ is a generalized eigenvector of A of order j for $j = 2, \ldots, m$.

The second step is to show that it is possible to choose γ_j such chains of vectors in $\mathcal{N}_{(A-\lambda_j I_n)^{\alpha_j}}$ that are linearly independent and span the space. We shall present an algorithm that achieves this in Section 6.5.

These two steps and the algorithm referred to above serve to justify the first four assertions; (5) and (6) then drop out easily from the formulas $A = UJU^{-1}$ and $A - \mu I_n = U(J - \mu I_n)U^{-1}$ upon taking the structure of J into account; see Exercise 6.1 for the underlying idea. □

A set of vectors $\{\mathbf{x}_1, \ldots, \mathbf{x}_m\}$ with $\mathbf{x}_1 \neq 0$ for which (6.2) holds for some $\mu \in \mathbb{C}$ is called a **Jordan chain**.

Remark 6.2. Item (5) is the **Cayley-Hamilton theorem**:

$$(6.3) \quad (\lambda - \lambda_1)^{\alpha_1} \cdots (\lambda - \lambda_k)^{\alpha_k} = \lambda^n + \sum_{j=0}^{n-1} c_j \lambda^j \implies A^n = -\sum_{j=0}^{n-1} c_j A^j.$$

In view of (6), the polynomial $p(\lambda) = (\lambda - \lambda_1)^{\nu_1} \cdots (\lambda - \lambda_k)^{\nu_k}$ is referred to as the **minimal polynomial** for A. Moreover, the number ν_j is the size of the largest Jordan cell in B_{λ_j}.

Exercise 6.1. Show that if $A = UJU^{-1}$ and $J = \text{diag}\{C_{\lambda_1}^{(4)}, C_{\lambda_2}^{(3)}\}$ with $\lambda_1 \neq \lambda_2$, then $(A - \lambda_1 I_7)^4 (A - \lambda_2 I_7)^3 = O$. [HINT: $(J - \lambda_1 I_7)^4 (J - \lambda_2 I_7)^3 = O$.]

Exercise 6.2. Show that if $\{\mathbf{x}_1, \ldots, \mathbf{x}_m\}$ is a set of vectors in $\mathbb{C}^n$ such that (6.2) holds for some $\mu \in \mathbb{C}$ and $\mathbf{x}_1 \neq 0$, then:

(1) $A\mathbf{x}_1 = \mu \mathbf{x}_1$ and $A\mathbf{x}_j = \mu \mathbf{x}_j + \mathbf{x}_{j-1}$ for $j = 2, \ldots, m$ if $m > 1$.

(2) $\mathbf{x}_j = B^{m-j} \mathbf{x}_m$ for $j = 1, \ldots, m$ with $B = A - \mu I_n$.

Exercise 6.3. Show that if $A = UC_\alpha^{(n)}U^{-1}$, then $\dim \mathcal{N}_{(A-\lambda I_n)} = 1$ if $\lambda = \alpha$ and 0 otherwise.

Exercise 6.4. Calculate $\dim \mathcal{N}_{(A-\lambda I_p)^t}$ for every $\lambda \in \mathbb{C}$ and $t = 1, 2, \ldots$ for the 17×17 matrix $A = UJU^{-1}$ when $J = \text{diag}\{B_{\lambda_1}, B_{\lambda_2}, B_{\lambda_3}\}$, the points $\lambda_1, \lambda_2, \lambda_3$ are distinct, $B_{\lambda_1} = C_{\lambda_1}^{(4)}$, $B_{\lambda_2} = \text{diag}\{C_{\lambda_2}^{(3)}, C_{\lambda_2}^{(3)}\}$, and $B_{\lambda_3} = \text{diag}\{C_{\lambda_3}^{(4)}, C_{\lambda_3}^{(2)}, C_{\lambda_3}^{(1)}\}$. Then check that the number of Jordan cells in B_{λ_j} of size $i \times i$ is equal to the number of columns of height i in the array that is built via the instructions in item (3) of Theorem 6.1.

Exercise 6.5. Show that if $A \in \mathbb{C}^{n \times n}$ has exactly k distinct eigenvalues $\lambda_1, \ldots, \lambda_k$ in $\mathbb{C}$ with algebraic multiplicities $\alpha_1, \ldots, \alpha_k$, then
$$\mathcal{N}_{(A-\lambda_1 I_n)^{\alpha_1}} \dotplus \cdots \dotplus \mathcal{N}_{(A-\lambda_k I_n)^{\alpha_k}} = \mathbb{C}^n.$$
Is it possible to reduce the powers further? Explain your answer.

Exercise 6.6. Let $\mathbf{u}, \mathbf{v} \in \mathbb{C}^n$ and $B \in \mathbb{C}^{n \times n}$ be such that $B^4 \mathbf{u} = 0$, $B^3 \mathbf{v} = 0$, and the pair of vectors $B^3 \mathbf{u}$ and $B^2 \mathbf{v}$ are linearly independent. Show that the seven vectors $\mathbf{u}, B\mathbf{u}, B^2\mathbf{u}, B^3\mathbf{u}, \mathbf{v}, B\mathbf{v}$, and $B^2 \mathbf{v}$ are linearly independent.

Exercise 6.7. Calculate $(C_\mu^{(4)})^{100}$. [HINT: $C_\mu^{(4)} = \mu I_4 + C_0^{(4)}$.]

6.2. Overview

The calculation of the matrices U and J in the representation $A = UJU^{-1}$ presented in Theorem 6.1 can be conveniently divided into three parts:

(1) Obtain the distinct eigenvalues $\lambda_1, \ldots, \lambda_k$ of the matrix A and their algebraic multiplicities $\alpha_1, \ldots, \alpha_k$. These are the points $\lambda \in \mathbb{C}$ at which $\mathcal{N}_{(\lambda I_n - A)} \neq \{\mathbf{0}\}$ and $\alpha_j = \dim \mathcal{N}_{(\lambda_j I_n - A)^n}$. As will be explained in the next chapter, the eigenvalues may also be characterized as the distinct roots $\lambda_1, \ldots, \lambda_k$ of the **characteristic polynomial**
$$p(\lambda) = \det(\lambda I_n - A) = a_0 + a_1 \lambda + \cdots + a_{n-1} \lambda^{n-1} + \lambda^n$$
that are obtained by writing it in factored form as
$$p(\lambda) = (\lambda - \lambda_1)^{\alpha_1} \cdots (\lambda - \lambda_k)^{\alpha_k},$$
which also displays the algebraic multiplicities.

(2) Compute $J = \text{diag}\{B_{\lambda_1} \ldots, B_{\lambda_k}\}$ by calculating
$$\dim \mathcal{N}_{(A - \lambda_j I_n)^i} \quad \text{for} \quad i = 1, \ldots, \alpha_j - 1$$
for each of the distinct eigenvalues $\lambda_1, \ldots, \lambda_k$, in order to obtain the sizes of the Jordan cells in B_{λ_j} from the algorithm in (3) of Theorem 6.1. The dimensions of these null spaces specify J up to order.

(3) Construct a basis of $\mathcal{N}_{(A - \lambda_j I_n)^{\alpha_j}}$ made up of γ_j **Jordan chains**, one Jordan chain for each Jordan cell in B_{λ_j}, in order to obtain blocks U_j such that $AU_j = U_j B_{\lambda_j}$, $j = 1, \ldots, k$. Then $U = \begin{bmatrix} U_1 & \cdots & U_k \end{bmatrix}$.

Remark 6.3. The information in (1) is enough to guarantee a factorization of A of the form $A = UGU^{-1}$, where $G = \text{diag}\{G_1, \ldots, G_k\}$ for some choice of $G_j \in \mathbb{C}^{\alpha_j \times \alpha_j}$, $j = 1, \ldots, k$. The information in (2) serves to determine the number of Jordan cells in J and their sizes, but not the order in which

6.3. Dimension of nullspaces of powers of B

Lemma 6.4. *If $B \in \mathbb{C}^{n \times n}$, then*

(6.4) $\quad \dim \mathcal{N}_{B^{j+1}} - \dim \mathcal{N}_{B^j} \leq \dim \mathcal{N}_{B^j} - \dim \mathcal{N}_{B^{j-1}} \quad \text{for } j = 1, 2, \ldots.$

Proof. Fix an integer $j \geq 1$ and assume that

$$\mathcal{N}_{B^j} = \mathcal{N}_{B^{j-1}} \dotplus \text{span}\{\mathbf{v}_1, \ldots, \mathbf{v}_s\} \quad \text{and} \quad \mathcal{N}_{B^{j+1}} = \mathcal{N}_{B^j} \dotplus \text{span}\{\mathbf{w}_1, \ldots, \mathbf{w}_t\},$$

where the vectors $\{\mathbf{v}_1, \ldots, \mathbf{v}_s\}$ and $\{\mathbf{w}_1, \ldots, \mathbf{w}_t\}$ are linearly independent and $\mathcal{N}_{B^0} = \{\mathbf{0}\}$. The proof amounts to verifying that:

(1) $\text{span}\{B^j \mathbf{w}_1, \ldots, B^j \mathbf{w}_t\}$ is a t-dimensional subspace of $\mathcal{N}_B$.

(2) $\text{span}\{B^{j-1} \mathbf{v}_1, \ldots, B^{j-1} \mathbf{v}_s\}$ is an s-dimensional subspace of $\mathcal{N}_B$.

(3) $\text{span}\{B^j \mathbf{w}_1, \ldots, B^j \mathbf{w}_t\} \subseteq \text{span}\{B^{j-1} \mathbf{v}_1, \ldots, b^{j-1} \mathbf{v}_s\}$.

Clearly $B^j \mathbf{w}_i \in \mathcal{N}_B$ for $i = 1, \ldots, t$, since $B(B^j \mathbf{w}_i) = B^{j+1} \mathbf{w}_i = \mathbf{0}$. Moreover, the vectors $\{B^j \mathbf{w}_1, \ldots, B^j \mathbf{w}_t\}$ are linearly independent, because

$$\delta_1 B^j \mathbf{w}_1 + \cdots + \delta_t B^j \mathbf{w}_t = \mathbf{0} \implies B^j(\delta_1 \mathbf{w}_1 + \cdots + \delta_t \mathbf{w}_t) = \mathbf{0}$$
$$\implies \delta_1 \mathbf{w}_1 + \cdots + \delta_t \mathbf{w}_t \in \mathcal{N}_{B^j} \cap \text{span}\{\mathbf{w}_1, \ldots, \mathbf{w}_t\} = \{\mathbf{0}\}.$$

Therefore, the coefficients $\delta_1 = \cdots = \delta_t = 0$, since $\{\mathbf{w}_1, \ldots, \mathbf{w}_t\}$ is a linearly independent set of vectors. This completes the proof of (1). The proof of (2) is similar and is left to the reader.

Next, since $B\mathbf{w}_i \in \mathcal{N}_{B^j}$, it follows that

$$B\mathbf{w}_i = \mathbf{u} + \beta_1 \mathbf{v}_1 + \cdots + \beta_s \mathbf{v}_s$$

for some choice of $\mathbf{u} \in \mathcal{N}_{B^{j-1}}$ and $\beta_1, \ldots, \beta_s \in \mathbb{C}$. Thus,

$$B^j \mathbf{w}_i = B^{j-1} \mathbf{u} + B^{j-1}(\beta_1 \mathbf{v}_1 + \cdots + \beta_s \mathbf{v}_s) = \mathbf{0} + \beta_1 B^{j-1} \mathbf{v}_1 + \cdots + \beta_s B^{j-1} \mathbf{v}_s.$$

Therefore,

$$\text{span}\{B^j \mathbf{w}_1, \ldots, B^j \mathbf{w}_t\} \subseteq \text{span}\{B^{j-1} \mathbf{v}_1, \ldots, B^{j-1} \mathbf{v}_s\},$$

which justifies (3) and so too that $t \leq s$. Consequently,

$$\dim \mathcal{N}_{B^{j+1}} - \dim \mathcal{N}_{B^j} = t \leq s = \dim \mathcal{N}_{B^j} - \dim \mathcal{N}_{B^{j-1}},$$

as claimed. $\square$

6.4. Computing J

To illustrate the construction of J, suppose that $A \in \mathbb{C}^{n \times n}$ has k distinct eigenvalues $\lambda_1, \ldots, \lambda_k$ with geometric multiplicities $\gamma_1, \ldots, \gamma_k$ and algebraic multiplicities $\alpha_1, \ldots, \alpha_k$, respectively. To construct the Jordan blocks associated with λ_1, let $B = A - \lambda_1 I_n$ for short and suppose for the sake of definiteness that $\gamma_1 = 6$, $\alpha_1 = 15$, and, to be more concrete, suppose that

$$\dim \mathcal{N}_B = 6, \ \dim \mathcal{N}_{B^2} = 10, \ \dim \mathcal{N}_{B^3} = 13, \ \text{and} \ \dim \mathcal{N}_{B^4} = 15.$$

These numbers are chosen to meet the inequalities in (6.4) but are otherwise completely arbitrary.

To see what to expect, construct an array of ×'s with 6 in the first row, $10 - 6 = 4$ in the second row, $13 - 10 = 3$ in the third row, and $15 - 13 = 2$ in the fourth row:

$$\begin{array}{cccccc} \times & \times & \times & \times & \times & \times \\ \times & \times & \times & \times & & \\ \times & \times & \times & & & \\ \times & \times & & & & \end{array}$$

There will be one Jordan cell for each column; the size of the cell is equal to the height of the column: two cells of size 4, one cell of size 3, one cell of size 2, and two cells of size 1.

Exercise 6.8. Find an 11×11 matrix B such that $\dim \mathcal{N}_B = 4$, $\dim \mathcal{N}_{B^2} = 7$, $\dim \mathcal{N}_{B^3} = 9$, $\dim \mathcal{N}_{B^4} = 10$, and $\dim \mathcal{N}_{B^5} = 11$. [HINT: Use the array.]

6.5. Computing U

The Jordan decomposition theorem amounts to showing that there exist γ_j Jordan chains in $\mathcal{N}_{(A-\lambda_j I_n)^n}$ such the vectors in these chains form a basis for $\mathcal{N}_{(A-\lambda_j I_n)^n}$. To illustrate the main ideas underlying the proof of this theorem, suppose, for example, that $B = A - \mu I_n$,

$\{\mathbf{u}_1, \ldots, \mathbf{u}_r\}$ is a basis for $\mathcal{N}_B$,

$\{\mathbf{u}_1, \ldots, \mathbf{u}_r; \mathbf{v}_1, \ldots, \mathbf{v}_s\}$ is a basis for $\mathcal{N}_{B^2}$,

$\{\mathbf{u}_1, \ldots, \mathbf{u}_r; \mathbf{v}_1, \ldots, \mathbf{v}_s; \mathbf{w}_1, \ldots, \mathbf{w}_t\}$ is a basis for $\mathcal{N}_{B^3}$,

and $\mathcal{N}_{B^3} = \mathcal{N}_{B^4}$. Then

$$\mathcal{N}_B = \mathrm{span}\{\mathbf{u}_1, \ldots, \mathbf{u}_r\},$$
$$\mathcal{N}_{B^2} = \mathcal{N}_B \dotplus \mathrm{span}\{\mathbf{v}_1, \ldots, \mathbf{v}_s\}, \ \text{and}$$
$$\mathcal{N}_{B^3} = \mathcal{N}_{B^2} \dotplus \mathrm{span}\{\mathbf{w}_1, \ldots, \mathbf{w}_t\}.$$

Since $\mathbf{w}_i \in \mathcal{N}_{B^3}$ and $B^2 \mathbf{w}_i \neq \mathbf{0}$, Lemma 5.7 guarantees that $\{B^2 \mathbf{w}_i, B \mathbf{w}_i, \mathbf{w}_i\}$ is a set of three linearly independent vectors for $i = 1, \ldots, t$. Analogously,

6.5. Computing U

as $\mathbf{v}_j \in \mathcal{N}_{B^2}$ and $B\mathbf{v}_j \neq \mathbf{0}$, the same lemma guarantees that $\{B\mathbf{v}_j, \mathbf{v}_j\}$ is a set of two linearly independent vectors for $j = 1, \ldots, s$.

Next, consider the following three sets of chains:

(1) $\{B^2\mathbf{w}_1, B\mathbf{w}_1, \mathbf{w}_1\}, \ldots, \{B^2\mathbf{w}_t, B\mathbf{w}_t, \mathbf{w}_t\}$ (t chains of length 3),
(2) $\{B\mathbf{v}_1, \mathbf{v}_1\}, \ldots, \{B\mathbf{v}_s, \mathbf{v}_s\}$ (s chains of length 2),
(3) $\{\mathbf{u}_1\}, \ldots, \{\mathbf{u}_r\}$ (r chains of length 1).

Exercise 6.9. Show that the $3t$ vectors in (1) are linearly independent and the $2s$ vectors in (2) are linearly independent.

Since the r vectors in (3) are a basis for $\mathcal{N}_B$ they must also be linearly independent. However, the full collection of $3t + 2s + r$ vectors cannot be linearly independent, since they all belong to $\mathcal{N}_{B^3}$ and $\dim \mathcal{N}_{B^3} = r + s + t$.

The next step is to show that it is possible to select a subset of the exhibited set of $3t + 2s + r$ chains to form a basis for $\mathcal{N}_{B^3}$. The proof of Lemma 6.4 ensures that

(6.5) $\quad t = \dim \mathrm{span}\{B^2\mathbf{w}_1, \ldots, B^2\mathbf{w}_t\}, \quad s = \dim \mathrm{span}\{B\mathbf{v}_1, \ldots, B\mathbf{v}_s\},$

and

(6.6) $\quad \mathrm{span}\{B^2\mathbf{w}_1, \ldots, B^2\mathbf{w}_t\} \subseteq \mathrm{span}\{B\mathbf{v}_1, \ldots, B\mathbf{v}_s\}$
$\subseteq \mathrm{span}\{\mathbf{u}_1, \ldots, \mathbf{u}_r\}.$

The information in (6.5) and (6.6) is **super important**. It is the key to the whole construction. In view of (6.5) and (6.6), $t \leq s \leq r$ and:

(a) The vectors $B^2\mathbf{w}_1, \ldots, B^2\mathbf{w}_t$, together with an appropriately chosen set of $s - t$ vectors from the set $\{B\mathbf{v}_1, \ldots, B\mathbf{v}_s\}$ and an appropriately chosen set of $r - s$ vectors from the set $\{\mathbf{u}_1, \ldots, \mathbf{u}_r\}$ form a basis for $\mathcal{N}_B$.

(b) The set of $3t + 2(s-t) + 1(r-s) = t + s + r$ vectors in the chains corresponding to the vectors selected in (a) are linearly independent. Therefore, they form a good basis for $\mathcal{N}_{B^3} = \mathcal{N}_{B^n}$.

If $\mu = \lambda_j$ and the columns of U_j are constructed from the chains selected in (b), then $AU_j = U_j B_{\lambda_j}$ in which $B_{\lambda_j} \in \mathbb{C}^{\alpha_j \times \alpha_j}$ is itself a block diagonal matrix with $r = \delta_j$ Jordan cells, one for each of the selected chains: t Jordan cells $C_{\lambda_j}^{(3)}$, $s - t$ Jordan cells $C_{\lambda_j}^{(2)}$, and $r - s$ Jordan cells $C_{\lambda_j}^{(1)}$.

- Thus, in order to describe the sizes of the Jordan cells in the matrix B_{λ_j} corresponding to a good choice of the block U_j, it is only necessary to know the numbers r, s, and t, or, equivalently, the dimensions of the spaces $\mathcal{N}_B, \mathcal{N}_{B^2}, \ldots$.

It is helpful to record this information graphically as a stacked array of r symbols in the first row, s symbols in the second, t symbols in the third, which suffices for the example under consideration:

$$\begin{array}{llllllll} \times & \cdots & \times & \times & \times & \times & \times & \times \quad (r \text{ entries})\\ \times & \cdots & \times & \times & \times & & & \quad (s \text{ entries})\\ \times & \cdots & \times & & & & & \quad (t \text{ entries}).\end{array}$$

The columns are in one-to-one correspondence with the Jordan cells: t columns of height 3; $s-t$ columns of height 2; and $r-s$ columns of height 1.

6.6. Two simple examples

Example 6.1. Let

$$A = \begin{bmatrix} 3 & 0 & 0 & 0 & 1\\ 0 & 3 & 0 & 1 & 0\\ 0 & 0 & 3 & 0 & 0\\ 0 & 0 & 0 & 3 & 0\\ 0 & 1 & 0 & 0 & 3 \end{bmatrix}, \quad B = A - 3I_5,$$

and let $\mathbf{e}_j$ denote the j'th column of I_5. It is then readily checked that $\mathcal{N}_B = \mathrm{span}\{\mathbf{e}_1, \mathbf{e}_3\}$, $\mathcal{N}_{B^2} = \mathcal{N}_B \dotplus \mathrm{span}\{\mathbf{e}_5\}$, $\mathcal{N}_{B^3} = \mathcal{N}_{B^2} \dotplus \mathrm{span}\{\mathbf{e}_2\}$, and $\mathcal{N}_{B^4} = \mathcal{N}_{B^3} \dotplus \mathrm{span}\{\mathbf{e}_4\}$. The algorithm is to consider the array

$$\begin{array}{llll} B^3\mathbf{e}_4 & B^2\mathbf{e}_2 & B\mathbf{e}_5 & \mathbf{e}_1 \quad \mathbf{e}_3\\ B^2\mathbf{e}_4 & B\mathbf{e}_2 & \mathbf{e}_5 & \\ B\mathbf{e}_4 & \mathbf{e}_2 & & \\ \mathbf{e}_4 & & & \end{array}$$

Since

$$\mathrm{span}\{B^3\mathbf{e}_4\} \subseteq \mathrm{span}\{B^2\mathbf{e}_2\} \subseteq \mathrm{span}\{B\mathbf{e}_5\} \subseteq \mathrm{span}\{\mathbf{e}_1, \mathbf{e}_3\}$$

and equality prevails in the first two inclusions, the second and third columns in the array can be deleted. Furthermore, as $B^3\mathbf{e}_4 = \mathbf{e}_1$, the fourth column should also be deleted. Consequently, if $U_1 = \begin{bmatrix} B^3\mathbf{e}_4 & B^2\mathbf{e}_4 & B\mathbf{e}_4 & \mathbf{e}_4 \end{bmatrix}$ and $U_2 = \begin{bmatrix} \mathbf{e}_3 \end{bmatrix}$, then $U = \begin{bmatrix} U_1 & U_2 \end{bmatrix}$ is invertible and

$$A \begin{bmatrix} U_1 & U_2 \end{bmatrix} = \begin{bmatrix} U_1 & U_2 \end{bmatrix} \begin{bmatrix} C_3^{(4)} & O\\ O & C_3^{(1)} \end{bmatrix}.$$

Example 6.2. Let $A \in \mathbb{C}^{n \times n}$, $\lambda_1 \in \sigma(A)$, $B = A - \lambda_1 I_n$ and suppose that

$$\dim \mathcal{N}_B = 2, \quad \dim \mathcal{N}_{B^2} = 4, \text{ and } \dim \mathcal{N}_{B^j} = 5 \text{ for } j = 3, \ldots, n.$$

6.6. Two simple examples

The given information guarantees the existence of five linearly independent vectors $\mathbf{a}_1$, $\mathbf{a}_2$, $\mathbf{b}_1$, $\mathbf{b}_2$, and $\mathbf{c}_1$ such that

$$\mathcal{N}_B = \mathrm{span}\{\mathbf{a}_1, \mathbf{a}_2\},$$
$$\mathcal{N}_{B^2} = \mathcal{N}_B \dotplus \mathrm{span}\{\mathbf{b}_1, \mathbf{b}_2\},$$
$$\mathcal{N}_{B^3} = \mathcal{N}_{B^2} \dotplus \mathrm{span}\{\mathbf{c}_1\},$$

i.e., $\{\mathbf{a}_1, \mathbf{a}_2\}$ is a basis for $\mathcal{N}_B$, $\{\mathbf{a}_1, \mathbf{a}_2, \mathbf{b}_1, \mathbf{b}_2\}$ is a basis for $\mathcal{N}_{B^2}$, and $\{\mathbf{a}_1, \mathbf{a}_2, \mathbf{b}_1, \mathbf{b}_2, \mathbf{c}_1\}$ is a basis for $\mathcal{N}_{B^3}$. By the list that is justified in the proof of Lemma 6.4,

$\mathrm{span}\{B^2\mathbf{c}_1\}$ is a one-dimensional subspace of $\mathrm{span}\{B\mathbf{b}_1, B\mathbf{b}_2\}$ and

$\mathrm{span}\{B\mathbf{b}_1, B\mathbf{b}_2\}$ is a two-dimensional subspace of (and so equal to) $\mathcal{N}_B$.

If $B^2\mathbf{c}_1$ and $B\mathbf{b}_1$ are linearly independent, then the strategy is to build the array

$$\begin{array}{cc} B^2\mathbf{c}_1 & B\mathbf{b}_1 \\ B\mathbf{c}_1 & \mathbf{b}_1 \\ \mathbf{c}_1 & \end{array}$$

and then to check that the five vectors in these two columns are linearly independent. To this end, suppose that there exist scalars β_1, β_2, δ_1, δ_2, and δ_3 such that

$$\beta_1 \mathbf{b}_1 + \beta_2 B\mathbf{b}_1 + \delta_1 \mathbf{c}_1 + \delta_2 B\mathbf{c}_1 + \delta_3 B^2\mathbf{c}_1 = \mathbf{0}$$
$$\implies \beta_1 B\mathbf{b}_1 + \delta_1 B\mathbf{c}_1 + \delta_2 B^2\mathbf{c}_1 = \mathbf{0}$$
$$\implies \delta_1 B^2\mathbf{c}_1 = \mathbf{0},$$

where the implications in the second and third row are obtained by multiplying the preceding row through by B.

The third row implies that $\delta_1 = 0$, since $B^2\mathbf{c}_1 \ne \mathbf{0}$; then the second row implies that $\beta_1 = \delta_2 = 0$, since the vectors $B^2\mathbf{c}_1$ and $B\mathbf{b}_1$ are presumed to be linearly independent in the case at hand. Then the first row implies that $\beta_2 = \delta_3 = 0$ for the same reason. Thus, the set of vectors $\{B^2\mathbf{c}_1, B\mathbf{c}_1, \mathbf{c}_1, B\mathbf{b}_1, \mathbf{b}_1\}$ is a basis for $\mathcal{N}_{B^3}$. Moreover, since

$$B[B^2\mathbf{c}_1 \quad B\mathbf{c}_1 \quad \mathbf{c}_1 \quad B\mathbf{b}_1 \quad \mathbf{b}_1] = [B^3\mathbf{c}_1 \quad B^2\mathbf{c}_1 \quad B\mathbf{c}_1 \quad B^2\mathbf{b}_1 \quad B\mathbf{b}_1]$$
$$= [\mathbf{0} \quad B^2\mathbf{c}_1 \quad B\mathbf{c}_1 \quad \mathbf{0} \quad B\mathbf{b}_1]$$
$$= [B^2\mathbf{c}_1 \quad B\mathbf{c}_1 \quad \mathbf{c}_1 \quad B\mathbf{b}_1 \quad \mathbf{b}_1]N,$$

where $N = \mathrm{diag}\left\{C_0^{(3)}, C_0^{(2)}\right\}$, it is now readily seen that the vectors

$$\mathbf{u}_1 = B^2\mathbf{c}_1, \quad \mathbf{u}_2 = B\mathbf{c}_1, \quad \mathbf{u}_3 = \mathbf{c}_1, \quad \mathbf{u}_4 = B\mathbf{b}_1, \quad \text{and} \quad \mathbf{u}_5 = \mathbf{b}_1$$

are linearly independent and that

(6.7) $$A[\mathbf{u}_1 \cdots \mathbf{u}_5] = [\mathbf{u}_1 \cdots \mathbf{u}_5] \begin{bmatrix} C^{(3)}_{\lambda_1} & O \\ O & C^{(2)}_{\lambda_1} \end{bmatrix}.$$

Similar conclusions prevail with $\mathbf{b}_2$ in place of $\mathbf{b}_1$ if $B^2\mathbf{c}_1$ and $B\mathbf{b}_1$ are linearly dependent. ◇

Exercise 6.10. Show that if $N = \sum_{j=1}^{p-1} \mathbf{e}_j \mathbf{e}_{j+1}^T$ is the $p \times p$ matrix with ones on the first super diagonal and zeros elsewhere, then $\alpha I_p + \beta N$ is similar to $\alpha I_p + N$ if $\beta \neq 0$.

6.7. Real Jordan forms

If $A \in \mathbb{R}^{n \times n}$, then the Jordan decomposition $A = UJU^{-1}$ can be reexpressed in terms of real matrices. Thus, for example, if $A \in \mathbb{R}^{4 \times 4}$ and

$$A \begin{bmatrix} \mathbf{u}_1 & \mathbf{u}_2 \end{bmatrix} = \begin{bmatrix} \mathbf{u}_1 & \mathbf{u}_2 \end{bmatrix} \begin{bmatrix} \omega & 1 \\ 0 & \omega \end{bmatrix}$$

for a pair of nonzero vectors $\mathbf{u}_1, \mathbf{u}_2 \in \mathbb{C}^4$ and a point $\omega \in \mathbb{C} \setminus \mathbb{R}$, then

$$A \begin{bmatrix} \mathbf{u}_1 & \mathbf{u}_2 & \overline{\mathbf{u}_1} & \overline{\mathbf{u}_2} \end{bmatrix} = \begin{bmatrix} \mathbf{u}_1 & \mathbf{u}_2 & \overline{\mathbf{u}_1} & \overline{\mathbf{u}_2} \end{bmatrix} \begin{bmatrix} \omega & 1 & 0 & 0 \\ 0 & \omega & 0 & 0 \\ 0 & 0 & \overline{\omega} & 1 \\ 0 & 0 & 0 & \overline{\omega} \end{bmatrix}.$$

Consequently, upon writing $\mathbf{u}_j = \mathbf{x}_j + i\mathbf{y}_j$ with $\mathbf{x}_j, \mathbf{y}_j \in \mathbb{R}^4$ for $j = 1, 2$, and $\omega = r\cos\theta + ir\sin\theta$ with $r > 0$ and $\theta \in [0, 2\pi)$, it is readily checked that there exists an invertible matrix $V \in \mathbb{C}^{4 \times 4}$ such that

$$\begin{bmatrix} \mathbf{u}_1 & \mathbf{u}_2 & \overline{\mathbf{u}_1} & \overline{\mathbf{u}_2} \end{bmatrix} V = \begin{bmatrix} \mathbf{x}_1 & \mathbf{y}_1 & \mathbf{x}_2 & \mathbf{y}_2 \end{bmatrix}$$

and subsequently that

$$A \begin{bmatrix} \mathbf{x}_1 & \mathbf{y}_1 & \mathbf{x}_2 & \mathbf{y}_2 \end{bmatrix}$$

$$= \begin{bmatrix} \mathbf{x}_1 & \mathbf{y}_1 & \mathbf{x}_2 & \mathbf{y}_2 \end{bmatrix} \begin{bmatrix} r\cos\theta & r\sin\theta & 1 & 0 \\ -r\sin\theta & r\cos\theta & 0 & 1 \\ 0 & 0 & r\cos\theta & r\sin\theta \\ 0 & 0 & -r\sin\theta & r\cos\theta \end{bmatrix}.$$

Exercise 6.11. Justify the transition from the complex Jordan decomposition to the real Jordan decomposition and en route show that the complex Jordan form is similar to the real Jordan form.

6.8. Supplementary notes

This chapter is adapted from Chapters 4 and 6 in [30].

Chapter 7

Determinants

In this chapter we shall develop the theory of determinants axiomatically and shall then briefly survey a number of their properties. Some more advanced topics on determinants will be considered in Chapter 17.

7.1. Determinants

Let Σ_n denote the set of all the $n!$ one-to-one mappings σ of the set of integers $\{1,\ldots,n\}$ onto itself and let $\mathbf{e}_i$ denote the i'th column of the identity matrix I_n. Then the formula

$$P_\sigma = \sum_{i=1}^n \mathbf{e}_i \mathbf{e}_{\sigma(i)}^T = \begin{bmatrix} \mathbf{e}_{\sigma(1)}^T \\ \vdots \\ \mathbf{e}_{\sigma(n)}^T \end{bmatrix}$$

that was introduced earlier defines a one-to-one correspondence between the set of all $n \times n$ permutation matrices P_σ and the set Σ_n. A permutation matrix $P_\sigma \in \mathbb{R}^{n \times n}$ with $n \geq 2$ is said to be **simple** if σ interchanges exactly two of the integers in the set $\{1,\ldots,n\}$, i.e., if and only if it can be expressed as

$$P = \sum_{j \in \Lambda} \mathbf{e}_j \mathbf{e}_j^T + \mathbf{e}_{i_1} \mathbf{e}_{i_2}^T + \mathbf{e}_{i_2} \mathbf{e}_{i_1}^T,$$

where $\Lambda = \{1,\ldots,n\} \setminus \{i_1, i_2\}$, $i_1, i_2 \in \{1,\ldots,n\}$, and $i_1 \neq i_2$.

Exercise 7.1. Show that: (1) if $P \in \mathbb{R}^{n \times n}$ is a permutation matrix, then $PP^T = I_n$; (2) if P is simple, then $P = P^T$.

Theorem 7.1. *There is exactly one way of assigning a number $d(A) \in \mathbb{C}$ to each matrix $A \in \mathbb{C}^{n \times n}$ that meets the following three requirements:*

1°. $d(I_n) = 1$.

2°. $d(PA) = -d(A)$ *for every simple permutation matrix P.*

3°. $d(A)$ *is a multilinear functional of the rows of A; i.e., it is linear in each row separately.*

Discussion. The first two of these requirements are easily understood. The third is perhaps best visualized by example. Thus, if

$$A = \begin{bmatrix} a_{11} & a_{12} & a_{13} \\ a_{21} & a_{22} & a_{23} \\ a_{31} & a_{32} & a_{33} \end{bmatrix} = \begin{bmatrix} \vec{\mathbf{a}}_1 \\ \vec{\mathbf{a}}_2 \\ \vec{\mathbf{a}}_3 \end{bmatrix} \quad \text{and} \quad I_3 = \begin{bmatrix} \mathbf{e}_1 & \mathbf{e}_2 & \mathbf{e}_3 \end{bmatrix},$$

then

$$\vec{\mathbf{a}}_1 = \sum_{i=1}^{3} a_{1i} \mathbf{e}_i^T, \quad \vec{\mathbf{a}}_2 = \sum_{j=1}^{3} a_{2j} \mathbf{e}_j^T, \quad \text{and} \quad \vec{\mathbf{a}}_3 = \sum_{k=1}^{3} a_{3k} \mathbf{e}_k^T.$$

Therefore, by successive applications of rule **3°**,

$$d(A) = \sum_{i=1}^{3} a_{1i} d\left(\begin{bmatrix} \vec{\mathbf{e}}_i \\ \vec{\mathbf{a}}_2 \\ \vec{\mathbf{a}}_3 \end{bmatrix} \right) = \sum_{i=1}^{3} a_{1i} \left\{ \sum_{j=1}^{3} a_{2j} d\left(\begin{bmatrix} \vec{\mathbf{e}}_i \\ \vec{\mathbf{e}}_j \\ \vec{\mathbf{a}}_3 \end{bmatrix} \right) \right\}$$

$$= \sum_{i=1}^{3} a_{1i} \left\{ \sum_{j=1}^{3} a_{2j} \left[\sum_{k=1}^{3} a_{3k} d\left(\begin{bmatrix} \vec{\mathbf{e}}_i \\ \vec{\mathbf{e}}_j \\ \vec{\mathbf{e}}_k \end{bmatrix} \right) \right] \right\},$$

which is an explicit formula for $d(A)$ in terms of the entries a_{st} in the matrix A and the 27 numbers $d\left(\begin{bmatrix} \vec{\mathbf{e}}_i \\ \vec{\mathbf{e}}_j \\ \vec{\mathbf{e}}_k \end{bmatrix} \right)$. But, the second rule in Theorem 7.1 implies that if two rows of $A \in \mathbb{C}^{n \times n}$ are identical, then $d(A) = 0$. Consequently only $6 = 3!$ of these 27 numbers are not equal to 0 and the last expression simplifies to

$$d(A) = \sum_{\sigma \in \Sigma_3} a_{1\sigma(1)} a_{2\sigma(2)} a_{3\sigma(3)} d\left(\begin{bmatrix} \mathbf{e}_{\sigma(1)}^T \\ \mathbf{e}_{\sigma(2)}^T \\ \mathbf{e}_{\sigma(3)}^T \end{bmatrix} \right),$$

where, as noted earlier, Σ_n denotes the set of all the $n!$ one-to-one mappings of the set $\{1, \ldots, n\}$ onto itself.

Analogously, if $A \in \mathbb{C}^{n \times n}$ and $\mathbf{e}_j$ now denotes the j'th column of I_n, then

(7.1) $\qquad d(A) = \displaystyle\sum_{\sigma \in \Sigma_n} a_{1\sigma(1)} \cdots a_{n\sigma(n)} d(P_\sigma) \quad \text{with} \quad P_\sigma = \begin{bmatrix} \mathbf{e}_{\sigma(1)}^T \\ \vdots \\ \mathbf{e}_{\sigma(n)}^T \end{bmatrix}.$

If P_σ is equal to the product of k simple permutations, then
$$d(P_\sigma) = (-1)^k d(I_n) = (-1)^k \quad (= (-1)^{\text{sign}\,\sigma} \text{ in the usual notation}). \qquad \square$$

The unique number $d(A)$ that is determined by the three conditions in Theorem 7.1 is called the **determinant** of A and will be denoted
$$\det(A) \quad \text{or} \quad \det A \quad \text{or} \quad |A|,$$
from now on. It is clear from formula (7.1) that if $A \in \mathbb{C}^{n \times n}$, then $\det A$ is **a continuous function of the** n^2 **entries in** A.

Exercise 7.2. Use the three rules in Theorem 7.1 to show that if $A \in \mathbb{C}^{2 \times 2}$, then $\det A = a_{11}a_{22} - a_{12}a_{21}$, and if $A \in \mathbb{C}^{3 \times 3}$, then $\det A$ is equal to
$$a_{11}a_{22}a_{33} - a_{11}a_{23}a_{32} + a_{12}a_{23}a_{31} - a_{12}a_{21}a_{33} + a_{13}a_{21}a_{32} - a_{13}a_{22}a_{31}.$$

7.2. Useful rules for calculating determinants

Theorem 7.2. *The determinant of a matrix $A \in \mathbb{C}^{n \times n}$ satisfies the following rules:*

- **4°.** *If two rows of A are identical, then $\det A = 0$.*
- **5°.** *If B is the matrix that is obtained by adding a multiple of one row of A to another row of A, then $\det B = \det A$.*
- **6°.** *If A has a row in which all the entries are equal to zero, then $\det A = 0$.*
- **7°.** *If two nonzero rows of A are linearly dependent, then $\det A = 0$.*
- **8°.** *If $A \in \mathbb{C}^{n \times n}$ is either upper triangular or lower triangular, then*
$$\det A = a_{11} \cdots a_{nn}.$$
- **9°.** *If $A \in \mathbb{C}^{n \times n}$, then A is invertible if and only if $\det A \neq 0$.*
- **10°.** *If $A, B \in \mathbb{C}^{n \times n}$, then $\det(AB) = \det A \times \det B = \det(BA)$.*
- **11°.** *If $A \in \mathbb{C}^{n \times n}$ and A is invertible, then $\det A^{-1} = (\det A)^{-1}$.*
- **12°.** *If $A \in \mathbb{C}^{n \times n}$, then $\det A = \det A^T$.*
- **13°.** *If $A \in \mathbb{C}^{n \times n}$, then rules $\mathbf{3°}$ to $\mathbf{7°}$ remain valid if the word rows is replaced by the word columns and the row interchange in rule $\mathbf{2°}$ is replaced by a column interchange.*

Proof. We shall discuss rules $\mathbf{8°}$, $\mathbf{9°}$, $\mathbf{10°}$, and $\mathbf{12°}$ and leave the other rules to the reader.

To illustrate $\mathbf{8°}$, observe that, in view of rules $\mathbf{3°}$, $\mathbf{5°}$, and $\mathbf{1°}$,

$$\begin{vmatrix} a_{11} & a_{12} & a_{13} \\ 0 & a_{22} & a_{23} \\ 0 & 0 & a_{33} \end{vmatrix} = a_{33} \begin{vmatrix} a_{11} & a_{12} & a_{13} \\ 0 & a_{22} & a_{23} \\ 0 & 0 & 1 \end{vmatrix} = a_{33} \begin{vmatrix} a_{11} & a_{12} & 0 \\ 0 & a_{22} & 0 \\ 0 & 0 & 1 \end{vmatrix}$$

$$= a_{33}a_{22} \begin{vmatrix} a_{11} & a_{12} & 0 \\ 0 & 1 & 0 \\ 0 & 0 & 1 \end{vmatrix} = a_{33}a_{22} \begin{vmatrix} a_{11} & 0 & 0 \\ 0 & 1 & 0 \\ 0 & 0 & 1 \end{vmatrix} = a_{33}a_{22}a_{11}.$$

The computation for triangular matrices in $\mathbb{C}^{n \times n}$ is much the same.

The verification of $\mathbf{9°}$ rests on the fact that the two basic steps of Gaussian elimination applied to a matrix $A \in \mathbb{C}^{n \times n}$, i.e., (1) permuting rows and (2) adding a multiple of one row to another, preserve both the rank of A and (in view of $\mathbf{2°}$ and $\mathbf{5°}$) $|\det A|$. More precisely, there exists a permutation matrix $P \in \mathbb{C}^{n \times n}$ (which is a product of simple permutations) and an upper echelon matrix $U \in \mathbb{C}^{n \times n}$ such that

$$\det PA = \det U \quad \text{and} \quad \operatorname{rank} A = \operatorname{rank} U.$$

Thus, as U is automatically upper triangular (since it is square in this application),

$$|\det A| = |\det U| = |u_{11} \cdots u_{nn}|.$$

But this serves to justify $\mathbf{9°}$, since

$$A \text{ is invertible} \iff U \text{ is invertible}$$

and

$$U \text{ is invertible} \iff u_{11} \cdots u_{nn} \neq 0.$$

To verify $\mathbf{10°}$, observe first that if $\det B = 0$, then the asserted identities are immediate from rule $\mathbf{9°}$, since B, AB, and BA are then all noninvertible matrices.

If $\det B \neq 0$, set

$$\varphi(A) = \frac{\det(AB)}{\det B}$$

and check that $\varphi(A)$ meets rules $\mathbf{1°}$–$\mathbf{3°}$. Then

$$\varphi(A) = \det A,$$

since there is only one functional that meets these three conditions, i.e.,

$$\det(AB) = \det A \times \det B,$$

as claimed. Now, having this last formula for every choice of A and B, invertible or not, we can interchange the roles of A and B to obtain
$$\det(BA) = \det B \times \det A = \det A \times \det B = \det(AB).$$
To verify **12°**, we first invoke the formula $EPA = U$ that summarizes Gaussian elimination to obtain the equalities
$$\det PA = \det EPA = \det U = \det U^T = \det A^T P^T E^T = \det A^T P^T,$$
since E is lower triangular with ones on the diagonal and U is triangular. The formula in **12°** emerges by multiplying through by $\det P$, since $(\det P)^2 = 1$ and $\det P^T P = \det I_n = 1$ for permutation matrices P. □

Exercise 7.3. Show that if $\det B \neq 0$, then the functional $\varphi(A) = \dfrac{\det(AB)}{\det B}$ meets conditions **1°**–**3°**. [HINT: To verify **3°**, observe that if $\vec{a}_1, \ldots, \vec{a}_n$ designate the rows of A, then the rows of AB are $\vec{a}_1 B, \ldots, \vec{a}_n B$.]

Exercise 7.4. Calculate the determinants of the following matrices by Gaussian elimination:
$$\begin{bmatrix} 1 & 3 & 2 & 1 \\ 0 & 4 & 1 & 6 \\ 0 & 0 & 2 & 1 \\ 1 & 1 & 0 & 4 \end{bmatrix}, \begin{bmatrix} 1 & 0 & 1 & 0 \\ 0 & 1 & 0 & 1 \\ 1 & 0 & 0 & 1 \\ 0 & 1 & 1 & 0 \end{bmatrix}, \begin{bmatrix} 1 & 3 & 2 & 4 \\ 0 & 2 & 1 & 6 \\ 0 & 0 & 3 & 0 \\ 0 & 0 & 1 & 2 \end{bmatrix}, \begin{bmatrix} 0 & 0 & 0 & 4 \\ 1 & 2 & 3 & 1 \\ 0 & 0 & 1 & 1 \\ 0 & 1 & 2 & 6 \end{bmatrix}.$$
[HINT: If k simple permutations were used to pass from A to U, then $\det A = (-1)^k \det U$.]

Exercise 7.5. Calculate the determinants of the matrices in the previous exercise by rules **1°** to **13°**.

Exercise 7.6. Calculate $\det A$ when $A = \begin{bmatrix} 1 & \overline{\alpha} & \overline{\alpha}^2 & \overline{\alpha}^3 \\ \alpha & 1 & \overline{\alpha} & \overline{\alpha}^2 \\ \alpha^2 & \alpha & 1 & \overline{\alpha} \\ \alpha^3 & \alpha^2 & \alpha & 1 \end{bmatrix}$. [HINT: Use Gaussian elimination to find a lower triangular matrix E with ones on the diagonal such that $U = EA$ is an upper echelon matrix.]

7.3. Exploiting block structure

The calculation of determinants is often simplified by taking advantage of block structure.

Lemma 7.3. *If $A \in \mathbb{C}^{n \times n}$ is a block triangular matrix, i.e., if either*
$$A = \begin{bmatrix} A_{11} & A_{12} \\ O & A_{22} \end{bmatrix} \quad \text{or} \quad A = \begin{bmatrix} A_{11} & O \\ A_{21} & A_{22} \end{bmatrix}$$
with square blocks $A_{11} \in \mathbb{C}^{p \times p}$ and $A_{22} \in \mathbb{C}^{q \times q}$, then
$$\det A = \det A_{11} \det A_{22}.$$

Proof. In view of Theorem 6.1, $A_{11} = V_1 J_1 V_1^{-1}$ and $A_{22} = V_2 J_2 V_2^{-1}$, where J_1 and J_2 are in Jordan form. Thus, if A is block upper triangular, then

$$A = \begin{bmatrix} A_{11} & A_{12} \\ O & A_{22} \end{bmatrix} = \begin{bmatrix} V_1 & O \\ O & V_2 \end{bmatrix} \begin{bmatrix} J_1 & V_1^{-1} A_{12} V_2 \\ O & J_2 \end{bmatrix} \begin{bmatrix} V_1 & O \\ O & V_2 \end{bmatrix}^{-1}.$$

Therefore, since J_1 and J_2 are upper triangular, the middle matrix on the right is also upper triangular and

$$\det A = \det J_1 \det J_2 = \det A_{11} \det A_{22},$$

as claimed. The proof for block lower triangular matrices is similar. □

Exercise 7.7. Compute $\det A$ when $A = \begin{bmatrix} 3 & 1 & 4 & 6 & 7 & 8 \\ 0 & 2 & 5 & 1 & 9 & 4 \\ 0 & 0 & 1 & 1 & 1 & 1 \\ 0 & 0 & 0 & 4 & 0 & 0 \\ 0 & 0 & 0 & 1 & 1 & 0 \\ 0 & 0 & 0 & 2 & 1 & 1 \end{bmatrix}$. [REMARK: This should not take more than about 30 seconds.]

Exercise 7.8. Let $A = \begin{bmatrix} A_{11} & A_{12} \\ A_{21} & A_{22} \end{bmatrix} \in \mathbb{C}^{n \times n}$ be a four-block matrix with entries $A_{11} \in \mathbb{C}^{p \times p}$, $A_{12} \in \mathbb{C}^{p \times q}$, $A_{21} \in \mathbb{C}^{q \times p}$, $A_{22} \in \mathbb{C}^{q \times q}$, where $n = p + q$. Show that:

(7.2) if A_{11} is invertible, then $\det A = \det A_{11} \det (A_{22} - A_{21} A_{11}^{-1} A_{12})$;

(7.3) if A_{22} is invertible, then $\det A = \det A_{22} \det (A_{11} - A_{12} A_{22}^{-1} A_{21})$.

[HINT: Use the Schur complement formulas (3.18) and (3.19).]

Exercise 7.9. Show that if $A \in \mathbb{C}^{p \times q}$ and $B \in \mathbb{C}^{q \times p}$, then

$$\det(I_p - AB) = \det(I_q - BA)$$

and

$$q + \mathrm{rank}(I_p - AB) = p + \mathrm{rank}(I_q - BA).$$

[HINT: Imbed A and B appropriately in a $(p+q) \times (p+q)$ matrix and then invoke formulas (7.2) and (7.3).]

Exercise 7.10. Show that if $A \in \mathbb{C}^{p \times q}$ and $B \in \mathbb{C}^{q \times p}$, then

(7.4) $\lambda^q \det(\lambda I_p - AB) = \lambda^p \det(\lambda I_q - BA)$ for every $\lambda \in \mathbb{C}$.

Exercise 7.11. Show that if $\mathbf{u}, \mathbf{v} \in \mathbb{C}^n$, then $\det (I_n - \mathbf{u}\mathbf{v}^H) \ne 0$ if and only if $1 - \mathbf{v}^H \mathbf{u} \ne 0$ and then compute $(I_n - \mathbf{u}\mathbf{v}^H)^{-1}$ when this condition is in force. [HINT: $(I_n - \mathbf{u}\mathbf{v}^H)^{-1}$ must be of the form $(I_n + \kappa \mathbf{u}\mathbf{v}^H)^{-1}$ for some choice of $\kappa \in \mathbb{C}$.]

7.4. Minors

Exercise 7.12. Let $A \in \mathbb{C}^{n \times n}$ be invertible and let $\mathbf{u}, \mathbf{v} \in \mathbb{C}^n$. Show that the matrix $A + \mathbf{u}\mathbf{v}^H$ is invertible if and only if $1 + \mathbf{v}^H A^{-1} \mathbf{u} \neq 0$ and that if this condition is in force, then

$$(A + \mathbf{u}\mathbf{v}^H)^{-1} = A^{-1} - \frac{A^{-1}\mathbf{u}\mathbf{v}^H A^{-1}}{1 + \mathbf{v}^H A^{-1} \mathbf{u}}.$$

[HINT: Exploit Exercise 7.11.]

Exercise 7.13. Verify the formula

$$\det(I_n - \mathbf{e}_k \mathbf{e}_j^T + \mathbf{u}\mathbf{v}^H) = (1 + \mathbf{v}^H \mathbf{u})(1 - \mathbf{e}_j^T \mathbf{e}_k) + u_j \overline{v_k},$$

where $\mathbf{u}, \mathbf{v} \in \mathbb{C}^n$, $u_j = \mathbf{e}_j^T \mathbf{u}$, $v_k = \mathbf{e}_k^T \mathbf{v}$, and $\mathbf{e}_i$ denotes the i'th column of I_n, $i = 1, \ldots, n$. [HINT: First verify the formula under the condition $1 + \mathbf{v}^H \mathbf{u} \neq 0$, and then use the fact that the determinant of a matrix $A \in \mathbb{C}^{n \times n}$ is a continuous function of the entries a_{ij} in A.]

Exercise 7.14. Calculate the determinant of the matrix $\begin{bmatrix} O_{p \times q} & I_p \\ I_q & O_{q \times p} \end{bmatrix}$.

7.4. Minors

The *ij* **minor** $A_{\{i;j\}}$ of a matrix $A \in \mathbb{C}^{n \times n}$ is defined as the determinant of the $(n-1) \times (n-1)$ matrix that is obtained by deleting the i'th row and the j'th column of A. Thus, for example, if

$$A = \begin{bmatrix} 1 & 3 & 1 \\ 2 & 0 & 4 \\ 1 & 1 & 2 \end{bmatrix}, \text{ then } A_{\{1;2\}} = \det \begin{bmatrix} 2 & 4 \\ 1 & 2 \end{bmatrix}.$$

Exercise 7.15. Show that if $A \in \mathbb{C}^{n \times n}$, then $A_{\{i;j\}} = (-1)^{i+j} \det \widetilde{A}$, where $\widetilde{A}$ denotes the matrix A with its i'th row replaced by $\mathbf{e}_j^T$. [HINT: $\det \widetilde{A}$ does not change if a_{kj} is replaced by 0 when $k \neq i$. Then by $i - 1 + j - 1$ row and column interchanges the one in the ij position moves to the 11 position. Now invoke Lemma 7.3.]

Theorem 7.4. *If A is an $n \times n$ matrix, then $\det A$ can be expressed as an* **expansion along the i'th row**:

(7.5) $$\det A = \sum_{j=1}^n a_{ij}(-1)^{i+j} A_{\{i;j\}}$$

for each choice of i, $i = 1, \ldots, n$, and as an **expansion along the j'th column**:

(7.6) $$\det A = \sum_{i=1}^n a_{ij}(-1)^{i+j} A_{\{i;j\}}$$

for each choice of j, $j = 1, \ldots, n$.

Discussion. Let $\mathbf{e}_j$ denote the j'th column of I_n. Then (7.5) follows from the fact that $\mathbf{e}_i^T A = \sum_{j=1}^n a_{ij} \mathbf{e}_j^T$ and Exercise 7.15.

The verification of formula (7.6) rests on analogous decompositions for the columns of A. □

• **Moral:** Formulas (7.5) and (7.6) yield $2n$ different ways of calculating the determinant of an $n \times n$ matrix, one for each row and one for each column, respectively. It is usually advantageous to expand along the row or column with the most zeros.

Exercise 7.16. Evaluate the determinant of the 4×4 matrix

$$A = \begin{bmatrix} 5 & 2 & 3 & 1 \\ 3 & 0 & 0 & 2 \\ 1 & 1 & 0 & 1 \\ 0 & 2 & 0 & 1 \end{bmatrix}$$

twice; first begin by expanding in minors along the third column and then again by expanding in minors along the fourth column.

Theorem 7.5. *If $A \in \mathbb{C}^{n \times n}$ and if $C \in \mathbb{C}^{n \times n}$ denotes the matrix with entries*

$$c_{ij} = (-1)^{i+j} A_{\{j;i\}}, \ i,j = 1,\ldots,n,$$

then:

(1) $AC = CA = \det A \cdot I_n$.

(2) *If $\det A \neq 0$, then A is invertible and $A^{-1} = \dfrac{1}{\det A} C$.*

Discussion. If A is a 3×3 matrix, then this theorem states that

(7.7) $\begin{bmatrix} a_{11} & a_{12} & a_{13} \\ a_{21} & a_{22} & a_{23} \\ a_{31} & a_{32} & a_{33} \end{bmatrix} \begin{bmatrix} A_{\{1;1\}} & -A_{\{2;1\}} & A_{\{3;1\}} \\ -A_{\{1;2\}} & A_{\{2;2\}} & -A_{\{3;2\}} \\ A_{\{1;3\}} & -A_{\{2;3\}} & A_{\{3;3\}} \end{bmatrix} = \det A \cdot I_3$.

This formula may be verified by three simple sets of calculations. The first set is based on the formula

(7.8) $\det \begin{bmatrix} x & y & z \\ a_{21} & a_{22} & a_{23} \\ a_{31} & a_{32} & a_{33} \end{bmatrix} = x A_{\{1;1\}} - y A_{\{1;2\}} + z A_{\{1;3\}}$

and the observation that:

• if $(x,y,z) = (a_{11}, a_{12}, a_{13})$, then the left-hand side of (7.8) is equal to $\det A$;

• if $(x,y,z) = (a_{21}, a_{22}, a_{23})$, then, by rule **4°**, the left-hand side of (7.8) is equal to 0;

7.4. Minors

- if $(x, y, z) = (a_{31}, a_{32}, a_{33})$, then, by rule **4°**, the left-hand side of (7.8) is equal to 0.

These three evaluations can be recorded in the following more revealing way:

$$\begin{bmatrix} a_{11} & a_{12} & a_{13} \\ a_{21} & a_{22} & a_{23} \\ a_{31} & a_{32} & a_{33} \end{bmatrix} \begin{bmatrix} A_{\{1;1\}} \\ -A_{\{1;2\}} \\ A_{\{1;3\}} \end{bmatrix} = \det A \begin{bmatrix} 1 \\ 0 \\ 0 \end{bmatrix}.$$

The next set of calculations uses the formula

$$\det \begin{bmatrix} a_{11} & a_{12} & a_{13} \\ x & y & z \\ a_{31} & a_{32} & a_{33} \end{bmatrix} = -x A_{\{2;1\}} + y A_{\{2;2\}} - z A_{\{2;3\}}$$

to verify that

$$\begin{bmatrix} a_{11} & a_{12} & a_{13} \\ a_{21} & a_{22} & a_{23} \\ a_{31} & a_{32} & a_{33} \end{bmatrix} \begin{bmatrix} -A_{\{2;1\}} \\ A_{\{2;2\}} \\ -A_{\{2;3\}} \end{bmatrix} = \det A \begin{bmatrix} 0 \\ 1 \\ 0 \end{bmatrix}.$$

The final step in the verification of (7.7) is to substitute x, y, and z for a_{31}, a_{32}, and a_{33}, respectively, in order to obtain analogues of the formulas obtained in the first two steps and then to combine these results appropriately. □

Exercise 7.17. Formulate and verify the analogue of formula (7.7) for 4×4 matrices.

Exercise 7.18. Show that if $A \in \mathbb{C}^{3 \times 3}$ is invertible and $A\mathbf{x} = \mathbf{b}$, then

$$x_1 = \frac{\det \begin{bmatrix} b_1 & a_{12} & a_{13} \\ b_2 & a_{22} & a_{23} \\ b_3 & a_{32} & a_{33} \end{bmatrix}}{\det A}, \quad x_2 = \frac{\det \begin{bmatrix} a_{11} & b_1 & a_{13} \\ a_{21} & b_2 & a_{23} \\ a_{31} & b_3 & a_{33} \end{bmatrix}}{\det A}$$

and state and verify the analogous formula for x_3. [REMARK: This is an example of **Cramer's rule**.]

Exercise 7.19. Show that if $A, B \in \mathbb{C}^{n \times n}$ and $AB = I_n$, then for every $\lambda \in \mathbb{C}$

(7.9) $$b_{11} + b_{21}\lambda + \cdots + b_{n1}\lambda^{n-1} = \frac{\det \begin{bmatrix} 1 & \lambda & \cdots & \lambda^{n-1} \\ a_{21} & a_{22} & \cdots & a_{2n} \\ \vdots & & & \vdots \\ a_{n1} & a_{n2} & \cdots & a_{nn} \end{bmatrix}}{\det A}.$$

Exercise 7.20. Compute the inverse of the matrix $A = \begin{bmatrix} 1 & 2 & 2 \\ 2 & 1 & 2 \\ 1 & x & 0 \end{bmatrix}$ for those values of x for which A is invertible. [HINT: Exploit formula (7.7).]

7.5. Eigenvalues

Determinants play a useful role in calculating the eigenvalues of a matrix $A \in \mathbb{C}^{n \times n}$. In particular, if $A = UJU^{-1}$, where J is in Jordan form, then

$$\det(\lambda I_n - A) = \det(\lambda I_n - UJU^{-1}) = \det\left(U(\lambda I_n - J)U^{-1}\right).$$

Therefore, by rules **10°**, **11°**, and **8°**, applied in that order,

$$\det(\lambda I_n - A) = \det(\lambda I_n - J) = (\lambda - j_{11})(\lambda - j_{22}) \cdots (\lambda - j_{nn}),$$

where j_{ii}, $i = 1, \ldots, n$, are the diagonal entries of J. The polynomial

(7.10) $$p(\lambda) = \det(\lambda I_n - A)$$

is termed the **characteristic polynomial** of A. In particular, **a number λ is an eigenvalue of the matrix A if and only if** $p(\lambda) = 0$. Thus, for example, to find the eigenvalues of the matrix

$$A = \begin{bmatrix} 1 & 2 \\ 2 & 1 \end{bmatrix},$$

look for the roots of the polynomial

$$\det(\lambda I_2 - A) = (\lambda - 1)^2 - 2^2 = \lambda^2 - 2\lambda - 3.$$

This leads readily to the conclusion that the eigenvalues of the given matrix A are $\lambda_1 = 3$ and $\lambda_2 = -1$. Moreover, if $J = \text{diag}\{3, -1\}$, then

$$\begin{aligned} A^2 - 2A - 3I_2 &= U(J^2 - 2J - 3I)U^{-1} = U(J - 3I_2)(J + I_2)U^{-1} \\ &= U \begin{bmatrix} 0 & 0 \\ 0 & -4 \end{bmatrix} \begin{bmatrix} 4 & 0 \\ 0 & 0 \end{bmatrix} U^{-1} = U \begin{bmatrix} 0 & 0 \\ 0 & 0 \end{bmatrix} U^{-1}, \end{aligned}$$

which yields the far from obvious conclusion

$$A^2 - 2A - 3I_2 = O.$$

The argument propogates: If $\lambda_1, \ldots, \lambda_k$ denote the distinct eigenvalues of A and if α_i denotes the algebraic multiplicity of the eigenvalue λ_i, $i = 1, \ldots, k$, then the characteristic polynomial can be written in the more revealing form

(7.11) $$p(\lambda) = (\lambda - \lambda_1)^{\alpha_1}(\lambda - \lambda_2)^{\alpha_2} \cdots (\lambda - \lambda_k)^{\alpha_k}.$$

Thus,

$$\begin{aligned} p(A) &= (A - \lambda_1 I_n)^{\alpha_1}(A - \lambda_2 I_n)^{\alpha_2} \cdots (A - \lambda_k I_n)^{\alpha_k} \\ &= U(J - \lambda_1 I_n)^{\alpha_1}(J - \lambda_2 I_n)^{\alpha_2} \cdots (J - \lambda_k I_n)^{\alpha_k} U^{-1} \\ &= O. \end{aligned}$$

7.5. Eigenvalues

This serves to justify the **Cayley-Hamilton** theorem that was referred to in Remark 6.2. In more striking terms, it states that

(7.12)
$$\det(\lambda I_n - A) = a_0 + \cdots + a_{n-1}\lambda^{n-1} + \lambda^n$$
$$\implies a_0 I_n + \cdots + a_{n-1}A^{n-1} + A^n = O.$$

Exercise 7.21. Show that if $J = \mathrm{diag}\{C_{\lambda_1}^{(5)}, C_{\lambda_2}^{(3)}, C_{\lambda_3}^{(2)}\}$, then
$$(J - \lambda_1 I_{10})^5 (J - \lambda_2 I_{10})^3 (J - \lambda_3 I_{10})^2 = O.$$

Exercise 7.22. Show that if $\lambda_1 = \lambda_2$ in Exercise 7.21, then
$$(J - \lambda_1 I_{10})^5 (J - \lambda_3 I_{10})^2 = O.$$

Exercise 7.22 illustrates the fact that if ν_j, $j = 1, \ldots, k$, denotes the size of the largest Jordan cell in the matrix J with λ_j on its diagonal, then $p(A) = 0$ holds for the possibly lower-degree polynomial
$$p_{\min}(\lambda) = (\lambda - \lambda_1)^{\nu_1} (\lambda - \lambda_2)^{\nu_2} \cdots (\lambda - \lambda_k)^{\nu_k},$$
which is the **minimal polynomial** referred to in Remark 6.2:
$$\begin{aligned} p_{\min}(A) &= (A - \lambda_1 I_n)^{\nu_1} (A - \lambda_2 I_n)^{\nu_2} \cdots (A - \lambda_k I_n)^{\nu_k} \\ &= U(J - \lambda_1 I_n)^{\nu_1} (J - \lambda_2 I_n)^{\nu_2} \cdots (J - \lambda_k I_n)^{\nu_k} U^{-1} \\ &= O. \end{aligned}$$

The Jordan decomposition $A = UJU^{-1}$ leads easily to a number of useful formulas for determinants and traces, where the **trace** of an $n \times n$ matrix A is defined as the sum of its diagonal elements:

(7.13)
$$\mathrm{trace}\, A = a_{11} + a_{22} + \cdots + a_{nn}.$$

Theorem 7.6. *If $A \in \mathbb{C}^{n \times n}$ has k distinct eigenvalues $\lambda_1, \ldots, \lambda_k$ with algebraic multiplicities $\alpha_1, \ldots, \alpha_k$, then*

(7.14)
$$\det A = \lambda_1^{\alpha_1} \lambda_2^{\alpha_2} \cdots \lambda_k^{\alpha_k}$$

and

(7.15)
$$\mathrm{trace}\, A = \alpha_1 \lambda_1 + \alpha_2 \lambda_2 + \cdots + \alpha_k \lambda_k.$$

Moreover, if $f(\lambda)$ is a polynomial, then

(7.16)
$$\det(\lambda I_n - f(A)) = (\lambda - f(\lambda_1))^{\alpha_1} \cdots (\lambda - f(\lambda_k))^{\alpha_k},$$

(7.17)
$$\det f(A) = f(\lambda_1)^{\alpha_1} f(\lambda_2)^{\alpha_2} \cdots f(\lambda_k)^{\alpha_k},$$

and

(7.18)
$$\mathrm{trace}\, f(A) = \alpha_1 f(\lambda_1) + \alpha_2 f(\lambda_2) + \cdots + \alpha_k f(\lambda_k).$$

Proof. The verification of formulas (7.15) and (7.18) depends upon the fact that

$$\text{trace}\,(AB) = \sum_{i=1}^{n}\sum_{j=1}^{n} a_{ij} b_{ji} = \text{trace}\,(BA)\,. \tag{7.19}$$

Thus, in particular,

$$\text{trace}\,A = \text{trace}\,(UJU^{-1}) = \text{trace}\,(JU^{-1}U) = \text{trace}\,J\,, \tag{7.20}$$

which leads easily to (7.15); the verification of (7.18) is similar but is based on the formula $f(A) = Uf(J)U^{-1}$. The rest is left to the reader. $\square$

Corollary 7.7 (Spectral mapping principle). *If $A \in \mathbb{C}^{n \times n}$ and $f(\lambda)$ is a polynomial, then*

$$\lambda \in \sigma(A) \iff f(\lambda) \in \sigma(f(A))\,. \tag{7.21}$$

Exercise 7.23. Verify Corollary 7.7, but show by example that the multiplicities may change. [HINT: The key is (7.16).]

7.6. Supplementary notes

This chapter is partially adapted in abbreviated form from Chapter 5 of [**30**]. As noted earlier, additional properties of determinants will be considered in Chapter 17. Formula (7.13) is a special case of a general formula for the trace that will be presented in Section 14.6. The formula in Exercise 7.12 is known as the **Sherman-Morrison formula**.

A byproduct of formula (7.4) is that:

$$\text{if } A, B \in \mathbb{C}^{n \times n}, \text{ then } \sigma(AB) \setminus \{\mathbf{0}\} = \sigma(BA) \setminus \{\mathbf{0}\}. \tag{7.22}$$

Moreover, since $\det A$ is a homogeneous polynomial of degree n in its variables, $\partial \det A / \partial a_{ij} = (-1)^{i+j} A_{\{i;j\}}$ and hence formulas (7.5) and (7.6) can be expressed as

$$\det A = \sum_{j=1}^{n} a_{ij} \frac{\partial \det A}{\partial a_{ij}} \quad \text{and} \quad \det A = \sum_{i=1}^{n} a_{ij} \frac{\partial \det A}{\partial a_{ij}}\,, \tag{7.23}$$

respectively.

Chapter 8

Companion matrices and circulants

A matrix $A \in \mathbb{C}^{n \times n}$ of the form

$$(8.1) \qquad A = \begin{bmatrix} 0 & 1 & 0 & \cdots & 0 \\ 0 & 0 & 1 & \cdots & 0 \\ \vdots & & & \ddots & \vdots \\ 0 & 0 & 0 & \cdots & 1 \\ -a_0 & -a_1 & -a_2 & \cdots & -a_{n-1} \end{bmatrix}$$

is called a **companion matrix**. In this chapter some of the special properties of companion matrices are developed and then applied to study a class of matrices called circulants. Sections 8.3 and 8.4 treat advanced applications that can be postponed to a later reading without loss of continuity.

8.1. Companion matrices

Theorem 8.1. *If the companion matrix $A \in \mathbb{C}^{n \times n}$ in (8.1) has k distinct eigenvalues $\lambda_1, \ldots, \lambda_k$ with geometric multiplicities $\gamma_1, \ldots, \gamma_k$ and algebraic multiplicities $\alpha_1, \ldots, \alpha_k$, respectively, then:*

(1) $\det(\lambda I_n - A) = a_0 + a_1 \lambda + \cdots + a_{n-1} \lambda^{n-1} + \lambda^n$.

(2) $\gamma_j = 1$ for $j = 1, \ldots, k$.

(3) A is similar to the Jordan matrix $J = \mathrm{diag}\{C_{\lambda_1}^{(\alpha_1)}, \ldots, C_{\lambda_k}^{(\alpha_k)}\}$.

(4) A is invertible if and only if $a_0 \neq 0$.

Proof. The formula in (1) is obtained by expanding in minors along the last column of $\lambda I_n - A$ and taking advantage of the structure; see Exercises

8.1 and 8.2. Next, since $\dim \mathcal{R}_{(A-\lambda I_n)} \geq n - 1$ for every point $\lambda \in \mathbb{C}$, it follows that $\dim \mathcal{N}_{(A-\lambda_j I_n)} = 1$ for $j = 1, \ldots, k$. Therefore (2) holds, and hence there is exactly one Jordan cell $C_{\lambda_j}^{(\alpha_j)}$ for each distinct eigenvalue λ_j, i.e., (3) holds. Assertion (4) holds because rank $A = n \iff a_0 \neq 0$. □

Exercise 8.1. Verify the formula in (1) in Theorem 8.1 for $n = 2$ and $n = 3$.

Exercise 8.2. Justify the formula for $p_n(\lambda) = \det(\lambda I_n - A)$ in (1) in Theorem 8.1 for an arbitrary positive integer n by induction after first verifying the identity

(8.2) $\quad p_n(\lambda) = (a_{n-1} + \lambda)\lambda^{n-1} + p_{n-1}(\lambda) - \lambda^{n-1} \quad$ for $n \geq 2$.

[HINT: Expand in minors along the last column. The key observation when $n = 4$, for example, is that the 34 minor

$$\det \begin{bmatrix} \lambda & -1 & 0 \\ 0 & \lambda & -1 \\ a_0 & a_1 & a_2 \end{bmatrix} = \det \begin{bmatrix} \lambda & -1 & 0 \\ 0 & \lambda & -1 \\ a_0 & a_1 & a_2 + \lambda \end{bmatrix} - \det \begin{bmatrix} \lambda & -1 & 0 \\ 0 & \lambda & 0 \\ 0 & 0 & \lambda \end{bmatrix}.]$$

Our next objective is to analyze the Jordan decomposition of companion matrices. To warm up, we begin with an example.

Example 8.1. If λ is an eigenvalue of a 3×3 companion matrix A with characteristic polynomial $f(\lambda) = a_0 + a_1\lambda + a_2\lambda^2 + \lambda^3$, then

$$\lambda \begin{bmatrix} x_1 \\ x_2 \\ x_3 \end{bmatrix} = \begin{bmatrix} 0 & 1 & 0 \\ 0 & 0 & 1 \\ -a_0 & -a_1 & -a_2 \end{bmatrix} \begin{bmatrix} x_1 \\ x_2 \\ x_3 \end{bmatrix} = \begin{bmatrix} x_2 \\ x_3 \\ -(a_0 x_1 + a_1 x_2 + a_2 x_3) \end{bmatrix}.$$

Consequently, $x_2 = \lambda x_1$, $x_3 = \lambda x_2 = \lambda^2 x_1$, and

$$-(a_0 x_1 + a_1 x_2 + a_2 x_3) = -(a_0 + a_1\lambda + a_2\lambda^2)x_1 = -f(\lambda)x_1 + \lambda^3 x_1 = \lambda^3 x_1,$$

since $f(\lambda) = 0$ when $\lambda \in \sigma(A)$. Thus, $\begin{bmatrix} x_1 & x_2 & x_3 \end{bmatrix}^T = x_1 \begin{bmatrix} 1 & \lambda & \lambda^2 \end{bmatrix}^T$, and, if A has three distinct eigenvalues $\lambda_1, \lambda_2, \lambda_3$, then

$$A \begin{bmatrix} 1 & 1 & 1 \\ \lambda_1 & \lambda_2 & \lambda_3 \\ \lambda_1^2 & \lambda_2^2 & \lambda_3^2 \end{bmatrix} = \begin{bmatrix} 1 & 1 & 1 \\ \lambda_1 & \lambda_2 & \lambda_3 \\ \lambda_1^2 & \lambda_2^2 & \lambda_3^2 \end{bmatrix} \begin{bmatrix} \lambda_1 & 0 & 0 \\ 0 & \lambda_2 & 0 \\ 0 & 0 & \lambda_3 \end{bmatrix}.$$

Analogous formulas hold for $n \times n$ companion matrices with n distinct eigenvalues. This and more follows from the evaluations in the next lemma.

Lemma 8.2. If $A \in \mathbb{C}^{n \times n}$ is a companion matrix with characteristic polynomial $f(\lambda) = a_0 + \cdots + a_{n-1}\lambda^{n-1} + \lambda^n$ and $\mathbf{v}(\lambda)^T = \begin{bmatrix} 1 & \lambda & \cdots & \lambda^{n-1} \end{bmatrix}$, then

(8.3) $\quad A\mathbf{v}(\lambda) = \lambda\mathbf{v}(\lambda) - f(\lambda)\mathbf{e}_n,$

8.1. Companion matrices

where $\mathbf{e}_n$ denotes the n'th column of I_n, and

(8.4) $\quad A\dfrac{\mathbf{v}^{(j)}(\lambda)}{j!} = \lambda\dfrac{\mathbf{v}^{(j)}(\lambda)}{j!} + \dfrac{\mathbf{v}^{(j-1)}(\lambda)}{(j-1)!} - \dfrac{f^{(j)}(\lambda)}{j!}\mathbf{e}_n \quad for \ j = 1, 2, \ldots .$

Proof. By direct computation

$$A\mathbf{v}(\lambda) = \begin{bmatrix} \lambda \\ \vdots \\ \lambda^{n-1} \\ -(a_0 + a_1\lambda + \cdots + a_{n-1}\lambda^{n-1}) \end{bmatrix} = \begin{bmatrix} \lambda \\ \vdots \\ \lambda^{n-1} \\ \lambda^n \end{bmatrix} - \begin{bmatrix} 0 \\ \vdots \\ 0 \\ f(\lambda) \end{bmatrix},$$

which coincides with (8.3). The formulas in (8.4) are obtained by differentiating both sides of (8.3) k times with respect to λ to first verify the formula

(8.5) $\quad A\mathbf{v}^{(k)}(\lambda) = \lambda\mathbf{v}^{(k)}(\lambda) + k\mathbf{v}^{(k-1)}(\lambda) - f^{(k)}(\lambda)\mathbf{e}_n$

for $k = 1, 2, \ldots$ by **Leibniz's rule**

$$(gh)^{(k)} = \sum_{j=0}^{k} \binom{k}{j} g^{(j)} h^{(k-j)}$$

for the derivative of a product (or by induction) and then dividing both sides by $k!$. $\square$

Exercise 8.3. Use formulas (8.3) and (8.4) to give another proof of formula (1) in Theorem 8.1. [HINT: $\det \begin{bmatrix} \dfrac{\mathbf{v}(\lambda)}{0!} & \dfrac{\mathbf{v}^{(1)}(\lambda)}{1!} & \cdots & \dfrac{\mathbf{v}^{(n-1)}(\lambda)}{(n-1)!} \end{bmatrix} = 1$.]

Exercise 8.4. Show that if $f(\lambda) = a_0 + a_1\lambda + a_2\lambda^2 + \lambda^3 = (\lambda - \mu)^3$, then

$$\begin{bmatrix} 0 & 1 & 0 \\ 0 & 0 & 1 \\ -a_0 & -a_1 & -a_2 \end{bmatrix} V = V \begin{bmatrix} \mu & 1 & 0 \\ 0 & \mu & 1 \\ 0 & 0 & \mu \end{bmatrix}, \quad \text{where} \quad V = \begin{bmatrix} 1 & 0 & 0 \\ \mu & 1 & 0 \\ \mu^2 & 2\mu & 1 \end{bmatrix}.$$

[HINT: Invoke (8.3), (8.4), and the evaluations $f(\mu) = f'(\mu) = f''(\mu) = 0$.]

Theorem 8.3. *If $J = \mathrm{diag}\{C_{\lambda_1}^{(\alpha_1)}, \ldots, C_{\lambda_k}^{(\alpha_k)}\}$ is an $n \times n$ matrix in Jordan form based on k distinct eigenvalues $\lambda_1, \ldots, \lambda_k$, then J is similar to the companion matrix A in (8.1) with entries a_j equal to the coefficients of the polynomial $a_0 + a_1\lambda + \cdots + a_{n-1}\lambda^{n-1} + \lambda^n = (\lambda - \lambda_1)^{\alpha_1} \cdots (\lambda - \lambda_k)^{\alpha_k}$: If*

(8.6) $\quad V_j = \begin{bmatrix} \dfrac{\mathbf{v}(\lambda_j)}{0!} & \dfrac{\mathbf{v}^{(1)}(\lambda_j)}{1!} & \cdots & \dfrac{\mathbf{v}^{(\alpha_j-1)}(\lambda_j)}{(\alpha_j-1)!} \end{bmatrix}$

for $j = 1, \ldots, k$ with $\mathbf{v}(\lambda)^T = \begin{bmatrix} 1 & \lambda & \cdots & \lambda^{n-1} \end{bmatrix}$, then the $n \times n$ matrix

(8.7) $\quad V = \begin{bmatrix} V_1 & \cdots & V_k \end{bmatrix}$

is invertible and

(8.8) $$A = VJV^{-1}.$$

If A has n distinct eigenvalues $\lambda_1, \ldots, \lambda_n$, then $J = \text{diag}\{\lambda_1, \ldots, \lambda_n\}$ and

(8.9) $$V = \begin{bmatrix} 1 & \cdots & 1 \\ \lambda_1 & \cdots & \lambda_n \\ \vdots & & \vdots \\ \lambda_1^{n-1} & \cdots & \lambda_n^{n-1} \end{bmatrix}.$$

Proof. In view of formulas (8.3) and (8.4),

(8.10) $$AV_j = V_j(\lambda_j I_{\alpha_j} + C_0^{(\alpha_j)}) = V_j C_{\lambda_j}^{(\alpha_j)} \quad \text{for} \quad j = 1, \ldots, k.$$

Consequently, $AV = VJ$ with V as in (8.7) and, as $\text{rank}\, V_j = \alpha_j$ for $j = 1, \ldots, k$ and $\text{rank}\, V = \text{rank}\, V_1 + \cdots + \text{rank}\, V_k = n$, V is invertible. Thus, (8.8) holds.

Formula (8.9) is obtained from (8.6) and (8.7) when $k = n$. □

A matrix $V \in \mathbb{C}^{n \times n}$ of the form (8.9) is called a **Vandermonde** matrix, whereas a matrix of the form (8.7) with V_j as in (8.6) is called a **generalized Vandermonde** matrix.

Corollary 8.4. *The Vandermonde matrix V defined by formula (8.9) is invertible if and only if the points $\lambda_1, \ldots, \lambda_n$ are distinct.*

Example 8.2. If
$$f(\lambda) = (\lambda - \alpha)^3 (\lambda - \beta)^2 = a_0 + a_1 \lambda + \cdots + a_4 \lambda^4 + \lambda^5$$
and $\alpha \neq \beta$, then

$$\begin{bmatrix} 0 & 1 & 0 & 0 & 0 \\ 0 & 0 & 1 & 0 & 0 \\ 0 & 0 & 0 & 1 & 0 \\ 0 & 0 & 0 & 0 & 1 \\ -a_0 & -a_1 & -a_2 & -a_3 & -a_4 \end{bmatrix} \begin{bmatrix} 1 & 0 & 0 & 1 & 0 \\ \alpha & 1 & 0 & \beta & 1 \\ \alpha^2 & 2\alpha & 1 & \beta^2 & 2\beta \\ \alpha^3 & 3\alpha^2 & 3\alpha & \beta^3 & 3\beta^2 \\ \alpha^4 & 4\alpha^3 & 6\alpha^2 & \beta^4 & 4\beta^3 \end{bmatrix}$$

$$= \begin{bmatrix} 1 & 0 & 0 & 1 & 0 \\ \alpha & 1 & 0 & \beta & 1 \\ \alpha^2 & 2\alpha & 1 & \beta^2 & 2\beta \\ \alpha^3 & 3\alpha^2 & 3\alpha & \beta^3 & 3\beta^2 \\ \alpha^4 & 4\alpha^3 & 6\alpha^2 & \beta^4 & 4\beta^3 \end{bmatrix} \begin{bmatrix} \alpha & 1 & 0 & 0 & 0 \\ 0 & \alpha & 1 & 0 & 0 \\ 0 & 0 & \alpha & 0 & 0 \\ 0 & 0 & 0 & \beta & 1 \\ 0 & 0 & 0 & 0 & \beta \end{bmatrix}.$$

Exercise 8.5. Compute the determinant of the Vandermonde matrix V given by formula (8.9) when $n = 4$. [HINT: Let $f(x)$ denote the value of the determinant when λ_4 is replaced by x and observe that $f(x)$ is a polynomial of degree ≤ 3 and $f(\lambda_1) = f(\lambda_2) = f(\lambda_3) = 0$.]

8.2. Circulants

Exercise 8.6. Find an invertible matrix U and a matrix J in Jordan form such that $A = UJU^{-1}$ if $A \in \mathbb{C}^{6 \times 6}$ is a companion matrix, $\det(\lambda I_6 - A) = (\lambda - \lambda_1)^4 (\lambda - \lambda_2)^2$, and $\lambda_1 \neq \lambda_2$.

If $A \in \mathbb{C}^{n \times n}$ has k distinct eigenvalues $\lambda_1, \ldots, \lambda_k$ with geometric multiplicities $\gamma_1 = \cdots = \gamma_k = 1$, then A is similar to a companion matrix. The next three exercises serve to indicate the possibilities when this condition fails.

Exercise 8.7. Show that if $A \in \mathbb{C}^{n \times n}$ is similar to the Jordan matrix

$$J = \mathrm{diag}\{C_{\lambda_1}^{(4)}, C_{\lambda_1}^{(2)}, C_{\lambda_2}^{(3)}, C_{\lambda_2}^{(1)}, C_{\lambda_3}^{(3)}\} \quad \text{with 3 distinct eigenvalues},$$

then A is also similar to the block diagonal matrix $\mathrm{diag}\{A_1, A_2\}$ based on a pair of companion matrices with characteristic polynomials $f_1(\lambda)$ and $f_2(\lambda)$, and find the polynomials.

Exercise 8.8. Find a Jordan form J for the matrix

$$A = \begin{bmatrix} 0 & 1 & 0 & 0 & 0 \\ 0 & 0 & 1 & 0 & 0 \\ 8 & -12 & 6 & 0 & 0 \\ -1 & 1 & 0 & 0 & 1 \\ -4 & 1 & 0 & -4 & 4 \end{bmatrix}.$$

[HINT: You may find the formula $x^3 - 6x^2 + 12x - 8 = (x-2)^3$ useful.]

Exercise 8.9. Find an invertible matrix U such that $AU = UJ$ for the matrices A and J considered in Exercise 8.8.

8.2. Circulants

A matrix $A \in \mathbb{C}^{n \times n}$ of the form

$$(8.11) \qquad A = g(P) = a_0 I_n + a_1 P + \cdots + a_{n-1} P^{n-1}$$

based on the polynomial $g(\lambda) = a_0 + a_1 \lambda + \cdots + a_{n-1} \lambda^{n-1}$ and the $n \times n$ permutation matrix

$$(8.12) \qquad P = \sum_{j=1}^{n-1} \mathbf{e}_j \mathbf{e}_{j+1}^T + \mathbf{e}_n \mathbf{e}_1^T \quad \text{(which is also a companion matrix)}$$

is called a **circulant**. To illustrate more graphically, if $n = 5$, then

$$P = \begin{bmatrix} 0 & 1 & 0 & 0 & 0 \\ 0 & 0 & 1 & 0 & 0 \\ 0 & 0 & 0 & 1 & 0 \\ 0 & 0 & 0 & 0 & 1 \\ 1 & 0 & 0 & 0 & 0 \end{bmatrix} \quad \text{and} \quad A = \begin{bmatrix} a_0 & a_1 & a_2 & a_3 & a_4 \\ a_4 & a_0 & a_1 & a_2 & a_3 \\ a_3 & a_4 & a_0 & a_1 & a_2 \\ a_2 & a_3 & a_4 & a_0 & a_1 \\ a_1 & a_2 & a_3 & a_4 & a_0 \end{bmatrix}.$$

Circulants have very nice properties:

(8.13)
if $A \in \mathbb{C}^{n \times n}$ and $B \in \mathbb{C}^{n \times n}$ are circulants, then $AB = BA$,
A^H is a circulant, and hence $A^H A = A A^H$.

Moreover, circulants are diagonalizable and, as will be spelled out below in Theorem 8.5, it is easy to compute their Jordan decomposition.

Exercise 8.10. Show that if a permutation matrix $P \in \mathbb{R}^{n \times n}$ is also a companion matrix, then $PP^H = I_n$ and $P^n = I_n$.

Exercise 8.11. Verify the assertions in (8.13). [HINT: Exploit Exercise 8.10.]

Exercise 8.12. Show that there exist permutation matrices $P \in \mathbb{R}^{n \times n}$ such that $P^n \neq I_n$ if $n \geq 3$.

Theorem 8.5. *If $A \in \mathbb{C}^{n \times n}$ is the circulant that is defined in terms of the polynomial $g(\lambda) = a_0 + a_1 \lambda + \cdots + a_{n-1} \lambda^{n-1}$ and the permutation matrix P by formula (8.11), then*

(8.14)
$$A = \frac{1}{n} V D V^H, \quad \text{where} \quad D = \text{diag}\{g(\lambda_1), \ldots, g(\lambda_n)\},$$
$$\lambda_j = \exp(2\pi i (j/n)) \quad \text{for} \quad j = 1, \ldots, n,$$

V is the Vandermonde matrix defined by formula (8.9), and $V^H V = n I_n$.

Proof. The permutation matrix P defined by (8.12) is a companion matrix. Thus, in view of Theorem 8.1, $\det(\lambda I_n - P) = \lambda^n - 1$ and hence P has n distinct eigenvalues

$$\lambda_j = \zeta^j \text{ for } j = 1, \ldots, n, \text{ with } \zeta = e^{i 2\pi/n}.$$

Therefore, by Theorem 8.3, $P = V \Delta V^{-1}$ with $\Delta = \text{diag}\{\lambda_1, \ldots, \lambda_n\}$ and V as in (8.9). Consequently, $A = g(P) = V g(\Delta) V^{-1} = V D V^{-1}$. Moreover, since

(8.15)
$$1 + \overline{\lambda_j} \lambda_k + \cdots + (\overline{\lambda_j} \lambda_k)^{n-1} = \begin{cases} 0 & \text{if } j \neq k, \\ n & \text{if } j = k, \end{cases}$$

$V^H V = n I_n$. □

Exercise 8.13. Verify (8.15).

Exercise 8.14. Show that if P denotes the permutation matrix defined by formula (8.12), then $\mathcal{N}_{(P - \lambda_j I_n)} = \text{span}\{\mathbf{v}_j\}$, where $\mathbf{v}_j$ denotes the j'th column of the Vandermonde matrix V with $\lambda_j = \exp(2\pi i j / n)$ and justify the formula

(8.16)
$$P = V \Delta V^{-1} \quad \text{where} \quad \Delta = \text{diag}\{\lambda_1, \ldots, \lambda_n\}.$$

Remark 8.6. If A and V are defined as in Theorem 8.5, but the eigenvalues $\lambda_j = \exp(2\pi i(j/n))$ of A are indexed $j = 0, \ldots, n-1$, then

$$(8.17) \qquad V = \begin{bmatrix} 1 & 1 & \cdots & 1 \\ 1 & \omega & \cdots & \omega^{n-1} \\ \vdots & & & \vdots \\ 1 & \omega^{n-1} & \cdots & \omega^{(n-1)^2} \end{bmatrix}.$$

The matrix $F_n = n^{-1/2} V$ is called the **Fourier matrix**.

Exercise 8.15. Show that if $n \geq 4$, then $\sigma(F_n) = \{1, i, -1, -i\}$. [HINT: First show that $F_n^4 = I_n$.]

8.3. Interpolating polynomials

The next theorem is a useful byproduct of the properties of Vandermonde matrices that were developed in Section 8.1.

Theorem 8.7. If $\{\alpha_0, \ldots, \alpha_n\}$ and $\{\beta_0, \ldots, \beta_n\}$ are two sets of points in $\mathbb{C}$ and $\alpha_i \neq \alpha_j$ when $i \neq j$, then there exists exactly one polynomial $p(\lambda) = c_0 + c_1 \lambda + \cdots + c_n \lambda^n$ of degree n such that $p(\alpha_i) = \beta_i$ for $i = 0, \ldots, n$. The coefficients c_j of this polynomial are specified by (8.18) below.

Proof. Let $p(\lambda) = c_0 + c_1 \lambda + \cdots + c_n \lambda^n$ be a polynomial of degree n. Then $p(\alpha_j) = \beta_j$ for $j = 0, \ldots, n$ if and only if

$$(8.18) \qquad \begin{bmatrix} 1 & \alpha_0 & \cdots & \alpha_0^n \\ 1 & \alpha_1 & \cdots & \alpha_1^n \\ \vdots & \cdots & & \vdots \\ 1 & \alpha_n & \cdots & \alpha_n^n \end{bmatrix} \begin{bmatrix} c_0 \\ c_1 \\ \vdots \\ c_n \end{bmatrix} = \begin{bmatrix} \beta_0 \\ \beta_1 \\ \vdots \\ \beta_n \end{bmatrix}.$$

Since the points $\alpha_0, \ldots, \alpha_n$ are distinct, the matrix in (8.18) is the transpose of an invertible Vandermonde matrix. Therefore, there is only one set of coefficients $c_0, \ldots, c_n$, for which (8.18) holds. □

The same circle of ideas applied to generalized Vandermonde matrices allows us to specify derivatives.

Example 8.3. If $\lambda_1 \neq \lambda_2$, then for each choice of $\beta_0, \ldots, \beta_4 \in \mathbb{C}$, there exists exactly one polynomial $p(\lambda) = c_0 + c_1 \lambda + \cdots + c_4 \lambda^4$ of degree four such that $p(\lambda_1) = \beta_0$, $p'(\lambda_1) = \beta_1$, $p''(\lambda_1)/2! = \beta_2$, $p(\lambda_2) = \beta_3$, and $p'(\lambda_2) = \beta_4$. The coefficients of this polynomial are the solutions of the equation

$$\begin{bmatrix} 1 & \lambda_1 & \lambda_1^2 & \lambda_1^3 & \lambda_1^4 \\ 0 & 1 & 2\lambda_1 & 3\lambda_1^2 & 4\lambda_1^3 \\ 0 & 0 & 1 & 3\lambda_1 & 6\lambda_1^2 \\ 1 & \lambda_2 & \lambda_2^2 & \lambda_2^3 & \lambda_2^4 \\ 0 & 1 & 2\lambda_1 & 3\lambda_2^2 & 4\lambda_2^3 \end{bmatrix} \begin{bmatrix} c_0 \\ c_1 \\ \vdots \\ c_4 \end{bmatrix} = \begin{bmatrix} \beta_0 \\ \beta_1 \\ \vdots \\ \beta_4 \end{bmatrix}.$$

This equation has exactly one solution because the matrix on the left is the transpose of an invertible generalized Vandermonde matrix. ◇

8.4. An eigenvalue assignment problem

Let

(8.19) $$K_f = \begin{bmatrix} 0 & 1 & 0 & \cdots & 0 \\ 0 & 0 & 1 & \cdots & 0 \\ \vdots & & & \ddots & \vdots \\ 0 & 0 & 0 & \cdots & 1 \\ -a_0 & -a_1 & -a_2 & \cdots & -a_{n-1} \end{bmatrix}$$

denote the companion matrix based on the polynomial

(8.20) $$f(\lambda) = a_0 + a_1 \lambda + \cdots + a_{n-1}\lambda^{n-1} + \lambda^n$$

and let H_f denote the invertible **Hankel** matrix

(8.21) $$H_f = \begin{bmatrix} a_1 & a_2 & \cdots & a_{n-1} & a_n \\ a_2 & a_3 & \cdots & a_n & 0 \\ \vdots & & & & \vdots \\ a_n & 0 & \cdots & 0 & 0 \end{bmatrix} \quad \text{with } a_n = 1$$

based on the coefficients of the same polynomial.

Lemma 8.8. *The product $H_f K_f$ of the matrices K_f and H_f defined by formulas (8.19) and (8.21) is symmetric:*

(8.22) $$K_f^T H_f = H_f K_f$$

Proof. It turns out to be convenient to express K_f in terms of the $n \times n$ Jordan cell $C_0^{(n)}$ with 0 on the diagonal, the vector $\mathbf{u}^T = \begin{bmatrix} a_0 & \cdots & a_{n-1} \end{bmatrix}$, and the standard basis vectors $\mathbf{e}_1, \ldots, \mathbf{e}_n$ of $\mathbb{C}^n$ as $K_f = C_0^{(n)} - \mathbf{e}_n \mathbf{u}^T$. Then

$$H_f K_f = H_f(C_0^{(n)} - \mathbf{e}_n \mathbf{u}^T) = \begin{bmatrix} 0 & a_1 & \cdots & a_{n-1} \\ \mathbf{0} & & B & \end{bmatrix} - \mathbf{e}_1 \mathbf{u}^T = \begin{bmatrix} -a_0 & \mathbf{0}^T \\ \mathbf{0} & B \end{bmatrix},$$

8.4. An eigenvalue assignment problem

where
$$B = \begin{bmatrix} a_2 & a_3 & \cdots & a_{n-1} & a_n \\ a_3 & a_4 & \cdots & a_n & 0 \\ \vdots & & & & \vdots \\ a_{n-1} & a_n & & 0 & 0 \\ a_n & 0 & \cdots & 0 & 0 \end{bmatrix} = B^T \quad \text{with } a_n = 1.$$

Thus, $H_f K_f = (H_f K_f)^T = K_f^T H_f$, since H_f is symmetric. $\square$

Theorem 8.9. *If $A \in \mathbb{C}^{n \times n}$ has k distinct eigenvalues $\lambda_1, \ldots, \lambda_k$, with geometric multiplicities $\gamma_1, \ldots, \gamma_k$ and algebraic multiplicities $\alpha_1, \ldots, \alpha_k$, respectively, then the following statements are equivalent:*

(1) $\gamma_1 = \cdots = \gamma_k = 1$.

(2) *A is similar to the companion matrix K_f based on the polynomial $f(\lambda) = \det(\lambda I_n - A)$.*

(3) *A is similar to K_f^T.*

(4) *There exists a vector $\mathbf{b} \in \mathbb{C}^n$ such that the matrix*
$$\mathfrak{C} = [\mathbf{b} \ A\mathbf{b} \ \cdots \ A^{n-1}\mathbf{b}]$$
*is invertible. (In other terminology, $\mathbf{b}$ is a **cyclic vector** for A.)*

Proof. If (1) is in force, then the matrix J in the Jordan decomposition of A is of the form $J = \mathrm{diag}\{C_{\lambda_1}^{(\alpha_1)}, \ldots, C_{\lambda_k}^{(\alpha_k)}\}$. Thus, in view of Theorem 8.3, J is similar to the companion matrix K_f based on the characteristic polynomial of A. Therefore, (1) $\Longrightarrow$ (2). The implications (2) $\Longrightarrow$ (1) and (2) $\Longleftrightarrow$ (3) are justified by Theorem 8.1 and Lemma 8.8, respectively.

Suppose next that (3) holds. Then there exists an invertible matrix $U \in \mathbb{C}^{n \times n}$ with columns $\mathbf{u}_1, \ldots, \mathbf{u}_n$ such that
$$[\mathbf{u}_1 \ \cdots \ \mathbf{u}_n] K_f^T = A [\mathbf{u}_1 \ \cdots \ \mathbf{u}_n],$$
i.e.,
$$\mathbf{u}_2 = A\mathbf{u}_1, \quad \mathbf{u}_3 = A\mathbf{u}_2 = A^2\mathbf{u}_1, \cdots, \mathbf{u}_n = A\mathbf{u}_{n-1} = A^{n-1}\mathbf{u}_1.$$

Thus, as U is invertible, (4) holds with $\mathbf{b} = \mathbf{u}_1$.

Conversely, if (4) holds, then
$$\mathfrak{C} K_f^T = [A\mathbf{b} \ \cdots \ A^{n-1}\mathbf{b} \ \mathbf{c}] \quad \text{with } \mathbf{c} = -(a_0\mathbf{b} + \cdots + a_{n-1}A^{n-1}\mathbf{b}) = A^n\mathbf{b},$$
by the Cayley-Hamilton theorem. Therefore, $\mathfrak{C} K_f^T = A\mathfrak{C}$ and, as $\mathfrak{C}$ is invertible, (4) $\Longrightarrow$ (3). $\square$

A basic problem in control theory amounts to shifting the eigenvalues of a given matrix A to preassigned values, or a preassigned region, by an appropriately chosen additive perturbation of the matrix, which in practice is implemented by feedback. Since the eigenvalues of A are the roots of its characteristic polynomial, this corresponds to shifting the polynomial $f(\lambda) = \det(\lambda I_n - A)$ to a polynomial $g(\lambda) = c_0 + \cdots + c_{n-1}\lambda^{n-1} + \lambda^n$ with suitably chosen roots.

Theorem 8.10. *If $A \in \mathbb{C}^{n \times n}$ has k distinct eigenvalues $\lambda_1, \ldots, \lambda_k$, with geometric multiplicities $\gamma_1 = \cdots = \gamma_k = 1$, then for every polynomial $g(\lambda) = c_0 + c_1\lambda + \cdots + c_{n-1}\lambda^{n-1} + \lambda^n$, there exists a pair of vectors $\mathbf{b}, \mathbf{u} \in \mathbb{C}^n$ such that*
$$\det(\lambda I_n - A - \mathbf{b}\mathbf{u}^T) = g(\lambda).$$

Proof. Let K_f denote the companion matrix based on the polynomial $f(\lambda) = \det(\lambda I_n - A)$ and let H_f denote the invertible Hankel matrix based on the coefficients of $f(\lambda)$ that is defined in (8.21). Then, in view of Theorem 8.9, there exists an invertible matrix $\mathfrak{C} = \begin{bmatrix} \mathbf{b} & A\mathbf{b} & \cdots & A^{n-1}\mathbf{b} \end{bmatrix}$ such that $A\mathfrak{C} = \mathfrak{C}K_f^T$. Thus, as $K_f^T H_f = H_f K_f$,
$$(A + \mathbf{b}\mathbf{u}^T)\mathfrak{C}H_f = \mathfrak{C}K_f^T H_f + \mathbf{b}\mathbf{u}^T\mathfrak{C}H_f = \mathfrak{C}H_f K_f + \mathbf{b}\mathbf{u}^T\mathfrak{C}H_f.$$
Therefore, since $K_g - K_f = \mathbf{e}_n\mathbf{w}^T$ with
$$\mathbf{w}^T = \begin{bmatrix} (a_0 - c_0) & (a_1 - c_1) & \cdots & (a_{n-1} - c_{n-1}) \end{bmatrix},$$
$$(A + \mathbf{b}\mathbf{u}^T)\mathfrak{C}H_f = \mathfrak{C}H_f K_g \iff \mathbf{b}\mathbf{u}^T\mathfrak{C}H_f = \mathfrak{C}H_f(K_g - K_f)$$
$$\iff \mathbf{b}\mathbf{u}^T\mathfrak{C}H_f = \mathfrak{C}H_f\mathbf{e}_n\mathbf{w}^T$$
$$\iff \mathbf{b}\mathbf{u}^T\mathfrak{C}H_f = \mathbf{b}\mathbf{w}^T \iff \mathbf{u}^T = \mathbf{w}^T(\mathfrak{C}H_f)^{-1}.$$
(The passage to the last line uses the identities $H_f\mathbf{e}_n = \mathbf{e}_1$ and $\mathfrak{C}\mathbf{e}_1 = \mathbf{b}$.) $\square$

Exercise 8.16. Show that if $A \in \mathbb{C}^{n \times n}$ is similar to a companion matrix K_f: $A = UK_fU^{-1}$, $U = \begin{bmatrix} \mathbf{u}_1 & \cdots & \mathbf{u}_n \end{bmatrix}$, $\mathfrak{C} = \begin{bmatrix} \mathbf{u}_1 & A\mathbf{u}_1 & \cdots & A^{n-1}\mathbf{u}_1 \end{bmatrix}$, and H_f is the Hankel matrix defined in (8.21), then $A = (\mathfrak{C}H_f)K_f(\mathfrak{C}H_f)^{-1}$.

Exercise 8.17. Let $A = \begin{bmatrix} 1 & 1 & 0 \\ 0 & 1 & 1 \\ 0 & 0 & 1 \end{bmatrix}$ and let $\mathbf{b} = \begin{bmatrix} 0 \\ 0 \\ 1 \end{bmatrix}$. Find a vector $\mathbf{u} \in \mathbb{C}^3$ such that $\sigma(A + \mathbf{b}\mathbf{u}^H) = \{2, 3, 4\}$.

8.5. Supplementary notes

Companion matrices play a significant role in Chapters 19, 20, and 42. The restriction in Theorem 8.10 that the matrix A is similar to a companion matrix will be relaxed in Section 36.3.

Chapter 9

Inequalities

In this chapter we shall establish a number of inequalities for future use. We begin, however, with a brief introduction to the theory of convex functions, because they play a useful role in verifying these inequalities (and many, many others). In the last section we discuss ever so briefly a finite-dimensional version of another basic tool in convex analysis: the Krein-Milman theorem.

9.1. A touch of convex function theory

A subset Q of a vector space is said to be a **convex set** if $t\mathbf{a} + (1-t)\mathbf{b} \in Q$ for every pair of vectors $\mathbf{a}, \mathbf{b} \in Q$ and every t in the interval $0 \le t \le 1$. This is the same as saying that if $\mathbf{a}$ and $\mathbf{b}$ belong to Q, then every point on the line segment between $\mathbf{a}$ and $\mathbf{b}$ also belongs to Q.

A real-valued function $f(\mathbf{x})$ defined on a convex set Q is said to be **convex** if

$$(9.1) \quad f(t\mathbf{a} + (1-t)\mathbf{b}) \le tf(\mathbf{a}) + (1-t)f(\mathbf{b}) \quad \text{for every choice of } \mathbf{a}, \mathbf{b} \in Q$$

and every choice of t in the interval $0 \le t \le 1$; $f(x)$ is said to be a **strictly convex function** if the inequality in (9.1) is strict whenever $\mathbf{a} \ne \mathbf{b}$ and $0 < t < 1$.

Lemma 9.1 (Jensen's inequality). *If f is a real-valued convex function that is defined on a convex set Q and if $\mathbf{x}_1, \ldots, \mathbf{x}_n \in Q$ and $t_1, \ldots, t_n$ are positive numbers such that $t_1 + \cdots + t_n = 1$, then*

$$(9.2) \quad \sum_{i=1}^{n} t_i \mathbf{x}_i \in Q \quad \text{and} \quad f\left(\sum_{i=1}^{n} t_i \mathbf{x}_i\right) \le \sum_{i=1}^{n} t_i f(\mathbf{x}_i).$$

If f is strictly convex, then equality holds if and only if $x_1 = \cdots = x_n$.

Proof. To justify (9.2), we shall show that if it is valid for $n = k$ for some positive integer $k \geq 2$, then it is also valid for $n = k+1$. Since (9.2) is valid for $n = 2$ by definition, this will complete the justification. Towards this end, let $\mathbf{x}_1, \ldots, \mathbf{x}_{k+1} \in Q$ and let $t_1, \ldots, t_{k+1}$ be positive numbers such that $t_1 + \cdots + t_{k+1} = 1$. Then

$$t_1 \mathbf{x}_1 + \cdots + t_k \mathbf{x}_k + t_{k+1} \mathbf{x}_{k+1} = (1 - t_{k+1})\mathbf{u} + t_{k+1} \mathbf{x}_{k+1},$$

where $\mathbf{u} = \tau_1 \mathbf{x}_1 + \cdots + \tau_k \mathbf{x}_k$ and $\tau_j = (1 - t_{k+1})^{-1} t_j$ for $j = 1, \ldots, k$. Thus, as $\tau_1 + \cdots + \tau_k = 1$, the presumed validity of (9.2) for $n = k$ ensures that $\mathbf{u} \in Q$. Therefore, by the identity in the last display, $t_1 \mathbf{x}_1 + \cdots + t_k \mathbf{x}_k + t_{k+1} \mathbf{x}_{k+1} \in Q$ and

$$\begin{aligned}
f(t_1 \mathbf{x}_1 + \cdots + t_{k+1} \mathbf{x}_{k+1}) &= f((1 - t_{k+1})\mathbf{u} + t_{k+1} \mathbf{x}_{k+1}) \\
&\leq (1 - t_{k+1}) f(\mathbf{u}) + t_{k+1} f(\mathbf{x}_{k+1}) \\
&\leq (1 - t_{k+1})(\tau_1 f(\mathbf{x}_1) + \cdots + \tau_k f(\mathbf{x}_k)) \\
&\quad + t_{k+1} f(\mathbf{x}_{k+1}) \\
&= t_1 f(\mathbf{x}_1) + \cdots + t_{k+1} f(\mathbf{x}_{k+1}),
\end{aligned}$$

as needed.

Suppose next that f is strictly convex and equality holds in (9.2) for $n = k + 1$. Then the inequalities in the preceding display are all equalities. Therefore,

$$\mathbf{x}_{k+1} = \mathbf{u} \quad \text{and} \quad \tau_1 f(\mathbf{x}_1) + \cdots + \tau_k f(\mathbf{x}_k) = f(\tau_1 \mathbf{x}_1 + \cdots + \tau_k \mathbf{x}_k).$$

Repeating this procedure $k - 2$ times, we obtain three positive numbers s_1, s_2, s_3 such that $s_1 + s_2 + s_3 = 1$ and

$$\mathbf{x}_3 = s'_1 \mathbf{x}_1 + s'_2 \mathbf{x}_2 \quad \text{and} \quad f(s'_1 \mathbf{x}_1 + s'_2 \mathbf{x}_2) = s'_1 f(\mathbf{x}_1) + s'_2 f(\mathbf{x}_2),$$

where $s'_j = s_j/(1 - s_3)$ for $j = 1, 2$. Thus, as f is strictly convex, $\mathbf{x}_1 = \mathbf{x}_2$ and hence $\mathbf{x}_3 = \mathbf{x}_1$ and, as $\mathbf{x}_{j+1}$ is a convex combination of $\mathbf{x}_1, \ldots, \mathbf{x}_j$ for $j = 2, \ldots, k$, it follows that $\mathbf{x}_1 = \cdots = \mathbf{x}_{k+1}$. The converse is self-evident. □

For the rest of this section we shall restrict our attention to convex functions $f(x)$ of one real variable that are defined on convex subsets Q of $\mathbb{R}$.

Exercise 9.1. Show that if Q is a convex subset of $\mathbb{R}$, then Q must be an interval, i.e., the set of points between a pair of points $\alpha \in \mathbb{R}$ and $\beta \in \mathbb{R}$, where one or both of these points may or may not belong to Q. [HINT: Let $\alpha = \inf \{x \in \mathbb{R} : x \in Q\}$ and $\beta = \sup \{x \in \mathbb{R} : x \in Q\}$.]

9.1. A touch of convex function theory

Lemma 9.2. *If $f(x)$ is defined on an open subinterval Q of $\mathbb{R}$, then:*

(1) *f is convex if and only if*

(9.3) $$\frac{f(c) - f(a)}{c - a} \leq \frac{f(b) - f(a)}{b - a} \leq \frac{f(b) - f(c)}{b - c}$$

for every set of three points a, b, c in Q with $a < c < b$.

(2) *f is strictly convex if and only if*

(9.4) $$\frac{f(c) - f(a)}{c - a} < \frac{f(b) - f(a)}{b - a} < \frac{f(b) - f(c)}{b - c}$$

for every set of three points a, b, c in Q with $a < c < b$.

Proof. Suppose first that f is strictly convex. Then the inequality

$$f(c) < tf(a) + (1-t)f(b) \quad \text{for } a < b, \ c = ta + (1-t)b, \text{ and } 0 < t < 1$$

implies that

$$f(c) - f(a) < (1-t)(f(b) - f(a)) \quad \text{and} \quad t(f(b) - f(a)) < f(b) - f(c).$$

The inequalities in (9.4) are obtained by noting that

$$c = ta + (1-t)b \implies t = \frac{b - c}{b - a} \quad \text{and} \quad 1 - t = \frac{c - a}{b - a}.$$

Conversely, if (9.4) holds for $a < c < b$, then $c = ta + (1-t)b$ for some $t \in (0, 1)$ and

$$f(c) < \left(1 - \frac{c - a}{b - a}\right) f(a) + \frac{c - a}{b - a} f(b) = tf(a) + (1-t)f(b),$$

and hence f is strictly convex.

The preceding argument is easily adapted to show that f is convex if and only if (9.3) holds for every choice of points $a, b, c \in Q$ with $a < c < b$; the details are left to the reader. $\square$

Exercise 9.2. Show that the inequalities in (9.3) hold for every choice of points $a, b, c \in Q$ with $a < c < b$ if and only if f is convex.

Theorem 9.3. *If f is a convex function that is defined on an open interval $(\alpha, \beta) \subseteq \mathbb{R}$, then f is automatically continuous and the one-sided derivatives*

$$f^+(x) = \lim_{\varepsilon \downarrow 0} \frac{f(x + \varepsilon) - f(x)}{\varepsilon} \quad \text{and} \quad f^-(x) = \lim_{\varepsilon \downarrow 0} \frac{f(x) - f(x - \varepsilon)}{\varepsilon}$$

exist at every point $x \in (\alpha, \beta)$.

Proof. We first show that each point $a \in (\alpha, \beta)$ has a right derivative $f^+(a)$. Choose points a', a and a sequence $\varepsilon_1 > \varepsilon_2 > \cdots > 0$ such that

$$\alpha < a' < a < \cdots < a + \varepsilon_{j+1} < a + \varepsilon_j < \cdots < a + \varepsilon_1 < \beta$$

and let
$$\mu_j = \frac{f(a+\varepsilon_j) - f(a)}{\varepsilon_j}.$$
Then, by repeated applications of the first inequality in (9.3), we obtain
$$\mu_1 \geq \mu_2 \geq \cdots \geq \mu_j \geq \mu_{j+1} \geq \cdots \geq \frac{f(a) - f(a')}{a - a'}.$$
A monotonely decreasing sequence of finite numbers that is bounded below must tend to a limit, and that limit (which must be the same regardless of how the points ε_j are chosen) is $f^+(a)$. A similar argument based on the second inequality in (9.3) shows that $(f(b) - f(b - \varepsilon_j))/\varepsilon_j$ is a monotonely increasing sequence of numbers that is bounded above and hence that f has a left derivative $f^-(b)$ at each point $b \in (\alpha, \beta)$.

Thus, as f has both a left and a right derivative at every point in (α, β), it must be continuous. The details are left to the reader. □

Exercise 9.3. Show that if f is a convex function on $(\alpha, \beta) \subseteq \mathbb{R}$, then f has a left derivative at every point in (α, β).

Exercise 9.4. Show that if f is a convex function on $(\alpha, \beta) \subseteq \mathbb{R}$, then f is continuous. [HINT: If $\varepsilon > 0$, then
$$|f(c+\varepsilon) - f(c)| \leq \varepsilon |\varepsilon^{-1}(f(c+\varepsilon) - f(c)) - f^+(c)| + \varepsilon |f^+(c)|.]$$

Our next objective is to establish a practical way of checking whether or not a given function $f(x)$ of one variable is convex, at least for functions in the class $\mathcal{C}^2(Q)$, where
$$\mathcal{C}^k(Q) = \{f \text{ with } k \text{ continuous derivatives in the open interval } Q \subseteq \mathbb{R}\}.$$
To attain this objective we need a preliminary result.

Theorem 9.4. *Let $Q = (\alpha, \beta)$ be an open subinterval of $\mathbb{R}$ and let $f \in \mathcal{C}^1(Q)$. Then:*

(1) *$f(x)$ is convex on Q if and only if $f'(x) \leq f'(y)$ for every pair of points $x, y \in Q$ with $x < y$.*

(2) *$f(x)$ is strictly convex on Q if and only if $f'(x) < f'(y)$ for every pair of points $x, y \in Q$ with $x < y$.*

Moreover, if f is convex and

(9.5) $\qquad f'(x) = 0$ *for some $x \in Q$, then $f(x) \leq f(y)$ for every $y \in Q$;*

if f is strictly convex and

(9.6) $\qquad f'(x_1) = f'(x_2) = 0$ *for $x_1, x_2 \in Q$, then $x_1 = x_2$.*

9.1. A touch of convex function theory

Proof. Suppose first that $f(x)$ is strictly convex on Q and let $a < c < b$ be three points in Q. Then, upon letting $c \downarrow a$ in the first inequality in (9.4) and $c \uparrow b$ in the second inequality in (9.4), it is readily seen (with the help of Exercise 9.5, to justify strict inequalities) that

$$f'(a) < \frac{f(b) - f(a)}{b - a} \quad \text{and} \quad \frac{f(b) - f(a)}{b - a} < f'(b).$$

Conversely, if $a < c < b$ are three points in Q and $f'(x)$ is strictly increasing in Q, then, by the mean value theorem (see Theorem 21.1 for the statement),

$$\frac{f(c) - f(a)}{c - a} = f'(\xi) \quad \text{for some point} \quad \xi \in (a, c)$$

and

$$\frac{f(b) - f(c)}{b - c} = f'(\eta) \quad \text{for some point} \quad \eta \in (c, b).$$

Therefore, since $f'(\eta) - f'(\xi) > 0$, it follows that

$$(f(c) - f(a))(b - c) < (f(b) - f(c))(c - a).$$

But this in turn implies that

$$f(c)(b - a) < (b - c)f(a) + (c - a)f(b),$$

which, upon setting $c = ta + (1 - t)b$ for any choice of $t \in (0, 1)$, is easily seen to be equivalent to the requisite condition for strict convexity. This completes the proof of (2).

Suppose next that f is convex on Q and that $a, b \in Q$ and $f'(a) = 0$. Then, as the inequality $f(a + t(b - a)) \leq f(a) + t(f(b) - f(a))$ is valid for every $t \in (0, 1)$,

$$f'(a) = \lim_{t \downarrow 0} \frac{f(a + t(b - a)) - f(a)}{t} \leq f(b) - f(a)$$

and hence (9.5) holds. Thus, if f is strictly convex and $f(a_1) \leq f(b)$ and $f(a_2) \leq f(b)$ for every point $b \in Q$, then $f(a_1) = f(a_2)$ and hence if $a_1 \neq a_2$ and $0 < t < 1$, then $f(ta_1 + (1-t)a_2) < tf(a_1) + (1-t)f(a_2) = f(a_1)$, which is not possible. Therefore, (9.6) holds. The verification of (1) is left to the reader. $\square$

Exercise 9.5. Show that if f is strictly convex on $(\alpha, \beta) \subseteq \mathbb{R}$ and if $\alpha < a < a + \varepsilon_1 < b < \beta$ and $\varepsilon_1 > \varepsilon_2 > \cdots$ tends to zero, then

$$\frac{f(a + \varepsilon_{j+1}) - f(a)}{\varepsilon_{j+1}} < \frac{f(a + \varepsilon_j) - f(a)}{\varepsilon_j} < \frac{f(b) - f(a)}{b - a}$$
$$< \frac{f(b) - f(b - \varepsilon_j)}{\varepsilon_j} < \frac{f(b) - f(b - \varepsilon_{j+1})}{\varepsilon_{j+1}}.$$

Exercise 9.6. Show that if $Q = (\alpha, \beta)$ is an open subinterval of $\mathbb{R}$ and $f \in C^2(Q)$, then

(9.7) $\quad f$ is convex on Q if and only if $\quad f''(x) \geq 0 \quad$ for every $x \in Q$.

Exercise 9.7. Show that if $Q = (\alpha, \beta)$ is an open subinterval of $\mathbb{R}$ and if $f \in C^2(Q)$, then

(9.8) $\quad f$ is strictly convex on Q if $\quad f''(x) > 0 \quad$ for every $x \in Q$.

Exercise 9.8. Show that x^4 is strictly convex on $\mathbb{R}$ even though $f''(0) = 0$.

Exercise 9.9. Show that x^r is convex on $(0, \infty)$ if and only if $r \geq 1$ or $r \leq 0$ and that $-f(x)$ is convex on $(0, \infty)$ if and only if $0 \leq r \leq 1$.

Exercise 9.10. Show that e^x is strictly convex on $\mathbb{R}$ and that $-\ln x$ and x^r with $1 < r < \infty$ are strictly convex on $(0, \infty)$.

Exercise 9.11. Show that if $f(x)$ is convex on an interval Q and $g(y)$ is a nondecreasing convex function on an interval Q' that contains $\{f(x) : x \in Q\}$, then $g(f(x))$ is convex on Q.

Exercise 9.12. Show that if $f \in C^{(1)}(0, \infty) \bigcap C([0, \infty))$ is convex and $f(0) = 0$, then $\int_0^x f(s)ds \leq x^2 f'(x)/2$ for every $x > 0$. [HINT: Exploit (9.7).]

Exercise 9.13. Show that if $a_1, \ldots, a_n$ and $t_1, \ldots, t_n$ are two sequences of positive numbers and $t_1 + \cdots + t_n = 1$, then

(9.9) $$\left(\sum_{j=1}^n t_j a_j\right)^r \leq \sum_{j=1}^n t_j a_j^r \quad \text{when } r \geq 1.$$

Example 9.1. If $a_1, \ldots, a_n$ and $t_1, \ldots, t_n$ are two sequences of positive numbers such that $t_1 + \cdots + t_n = 1$, then

(9.10) $\quad a_1^{t_1} \cdots a_n^{t_n} \leq t_1 a_1 + \cdots + t_n a_n, \quad$ with equality $\iff a_1 = \cdots = a_n$.

In particular, the **geometric mean** of a given set of positive numbers $a_1, \ldots, a_n$ is less than or equal to its **arithmetic mean**, i.e.,

(9.11) $$(a_1 a_2 \cdots a_n)^{1/n} \leq \frac{a_1 + a_2 + \cdots + a_n}{n}.$$

Since $-\ln x$ is a strictly convex function of x on the interval $(0, \infty)$, Lemma 9.1 ensures that

$$-\ln(t_1 a_1 + \cdots + t_n a_n) \leq -(t_1 \ln a_1 + \cdots + t_n \ln a_n) = -\ln(a_1^{t_1} \cdots a_n^{t_n}),$$

with equality if and only if $a_1 = \cdots = a_n$. But this is easily seen to be equivalent to (9.10). ◇

Exercise 9.14. Show that if $A \in \mathbb{C}^{n \times n}$ has nonnegative eigenvalues, then

(9.12) $$(\det A)^{1/n} \leq \frac{\text{trace } A}{n}.$$

9.2. Four inequalities

If $s > 1$ and $t > 1$, then it is readily checked that

(9.13)
$$\frac{1}{s} + \frac{1}{t} = 1 \iff (s-1)(t-1) = 1$$
$$\iff (s-1)t = s$$
$$\iff (t-1)s = t \, .$$

Lemma 9.5. *If $c > 0$, $d > 0$, $s > 1$, $t > 1$, and $(s-1)(t-1) = 1$, then*

(9.14) $$cd \leq \frac{c^s}{s} + \frac{d^t}{t}, \quad \text{with equality if and only if} \quad c^s = d^t \, .$$

Proof. This is just (9.10) with $n = 2$, $t_1 = 1/s$, $t_2 = 1/t$, $c = a_1^{t_1}$, and $d = a_2^{t_2}$; it is treated in detail in the discussion of Example 9.1. $\square$

Theorem 9.6 (Hölder). *If $s > 1$, $t > 1$, and $(s-1)(t-1) = 1$ and if $\mathbf{a}, \mathbf{b} \in \mathbb{C}^n$ with components $a_1, \ldots, a_n$ and $b_1, \ldots, b_n$, respectively, then*

(9.15) $$\sum_{k=1}^n |a_k b_k| \leq \left\{ \sum_{k=1}^n |a_k|^s \right\}^{1/s} \left\{ \sum_{k=1}^n |b_k|^t \right\}^{1/t} .$$

Moreover, equality will prevail in (9.15) if and only

(9.16) $$\begin{array}{l} \text{either } \mathbf{a} = \mathbf{0} \\ \text{or } \mathbf{a} \neq \mathbf{0} \text{ and } |b_k|^t = \mu |a_k|^s \quad \text{for } k = 1, \ldots, n \text{ and some } \mu \geq 0 \, . \end{array}$$

Proof. Suppose first that $\mathbf{a} \neq \mathbf{0}$ and $\mathbf{b} \neq \mathbf{0}$ and let

$$\alpha_k = \frac{a_k}{\left\{\sum_{j=1}^n |a_j|^s\right\}^{1/s}} \quad \text{and} \quad \beta_k = \frac{b_k}{\left\{\sum_{j=1}^n |b_j|^t\right\}^{1/t}} .$$

Then

$$\sum_{k=1}^n |\alpha_k|^s = 1 \quad \text{and} \quad \sum_{k=1}^n |\beta_k|^t = 1 ,$$

and hence, in view of Lemma 9.5,

(9.17) $$\sum_{k=1}^n |\alpha_k \beta_k| \leq \sum_{k=1}^n \frac{|\alpha_k|^s}{s} + \sum_{k=1}^n \frac{|\beta_k|^t}{t} = \frac{1}{s} + \frac{1}{t} = 1 \, .$$

This yields the desired inequality because

$$\sum_{k=1}^n |\alpha_k \beta_k| = \frac{\sum_{k=1}^n |a_k b_k|}{\left(\sum_{j=1}^n |a_j|^s\right)^{1/s} \left(\sum_{j=1}^n |b_j|^t\right)^{1/t}} \, .$$

Moreover, equality will prevail in (9.15) if and only if it prevails in (9.17), i.e., if and only if

$$(9.18) \qquad |\alpha_i \beta_i| = \frac{|\alpha_i|^s}{s} + \frac{|\beta_i|^t}{t} \quad \text{for} \quad i = 1, \ldots, n.$$

If $\alpha_k \beta_k \neq 0$, then Lemma 9.5 implies that (9.18) holds for $i = k$ if and only if

$$|\alpha_k|^s = |\beta_k|^t \iff |b_k|^t = \mu |a_k|^s \text{ with } \mu = \frac{\sum_{j=1}^n |b_j|^t}{\sum_{j=1}^n |a_j|^s}.$$

On the other hand, if $\alpha_k \beta_k = 0$, then equality holds in (9.18) for $i = k$ if and only if $\alpha_k = 0$ and $\beta_k = 0$, i.e., if and only if $a_k = 0$ and $|b_k|^t = \mu |a_k|^s$. This completes the proof when $\mathbf{a} \neq \mathbf{0}$ and $\mathbf{b} \neq \mathbf{0}$.

This leaves two cases to consider: (1) $\mathbf{a} \neq \mathbf{0}$ and $\mathbf{b} = \mathbf{0}$; (2) $\mathbf{a} = \mathbf{0}$. Equality holds in (9.15) for both, and both are covered by (9.16). □

Exercise 9.15. Show that if $a, b \in \mathbb{C}$, then

$$(9.19) \quad |a+b| = |a| + |b| \iff \begin{array}{l} \text{either } a = 0 \\ \text{or } a \neq 0 \text{ and } b = \mu a \text{ for some } \mu \geq 0 \end{array}.$$

[HINT: If $a \neq 0$, then $b/a = re^{i\theta}$ and (9.19) holds $\iff |1 + re^{i\theta}| = 1 + r$.]

Exercise 9.16. Show that if $\mathbf{c} \in \mathbb{C}^n$ with entries $c_1, \ldots, c_n$ and $\mathbf{c} \neq \mathbf{0}$, then

$$(9.20) \qquad \left| \sum_{j=1}^n c_j \right| = \sum_{j=1}^n |c_j| \iff c_j = e^{i\theta} |c_j|$$

for $j = 1, \ldots, n$ and some $\theta \in [0, 2\pi)$.

[HINT: Suppose $c_k \neq 0$ and deduce from (9.20) that $|c_k + c_j| = |c_k| + |c_j|$ for $j = 1, \ldots, n$ and hence, in view of Exercise 9.15, that $c_j = t_j c_k = t_j |c_k| e^{i\theta} = |c_j| e^{i\theta}$ with $t_j \geq 0$ for some $\theta \in [0, 2\pi)$ and $j = 1, \ldots, n$.]

The next exercise supplements Theorem 9.6.

Exercise 9.17. Show that if, in the setting of Theorem 9.6, $\sum_{k=1}^n |a_k| > 0$, then

$$(9.21) \qquad \left| \sum_{k=1}^n a_k b_k \right| = \left\{ \sum_{k=1}^n |a_k|^s \right\}^{1/s} \left\{ \sum_{k=1}^n |b_k|^t \right\}^{1/t}$$

if and only if $|b_k|^t = \mu |a_k|^s$ for some choice of $\mu \geq 0$ and $a_k b_k = e^{i\theta} |a_k b_k|$ for $k = 1, \ldots, n$ and some choice of $\theta \in [0, 2\pi)$. [HINT: Exploit the implications of Theorem 9.6 and Exercises 9.15 and 9.16.]

9.2. Four inequalities

If $s = t = 2$, then (9.15) reduces to the well-known Cauchy-Schwarz inequality:

Theorem 9.7 (Cauchy-Schwarz). *If $\mathbf{a}, \mathbf{b} \in \mathbb{C}^n$ with entries $a_1, \ldots, a_n$ and $b_1, \ldots, b_n$, respectively, then*

$$(9.22) \qquad \left| \sum_{k=1}^n a_k b_k \right| \leq \left(\sum_{k=1}^n |a_k|^2 \right)^{1/2} \left(\sum_{k=1}^n |b_k|^2 \right)^{1/2},$$

with equality if and only if

$$(9.23) \qquad \text{either } \mathbf{a} = \mathbf{0}, \text{ or } \mathbf{a} \neq \mathbf{0} \text{ and } \mathbf{b} = \beta \overline{\mathbf{a}} \text{ for some } \beta \in \mathbb{C}.$$

Proof. The inequality is immediate from (9.15) with $s = t = 2$.

Let $f(\mathbf{a}, \mathbf{b})$ denote the right-hand side of (9.22). If equality holds in (9.22), then

$$f(\mathbf{a}, \mathbf{b}) = \left| \sum_{k=1}^n a_k b_k \right| \leq \sum_{k=1}^n |a_k b_k| \leq f(\mathbf{a}, \mathbf{b}).$$

Therefore, equality prevails throughout, and hence, if $\mathbf{a} \neq \mathbf{0}$, then, in view of (9.20) and Theorem 9.6, there exists a $\theta \in [0, 2\pi)$ and a $\mu \geq 0$ such that

$$a_k b_k = e^{i\theta} |a_k b_k| \quad \text{and} \quad |b_k| = \mu |a_k| \quad \text{for } k = 1, \ldots, n.$$

Thus, if $a_k \neq 0$, then

$$b_k = e^{i\theta} \frac{|a_k b_k|}{a_k} = e^{i\theta} \mu \frac{|a_k|^2}{a_k} = e^{i\theta} \mu \overline{a_k}.$$

Thus equality in (9.22) implies that (9.23) is in force. Since the converse implication is self-evident, the proof is complete. □

Exercise 9.18. Show that if $\alpha, \beta \in \mathbb{R}$ and $\theta \in [0, 2\pi)$, then $\alpha \cos\theta + \beta \sin\theta \leq \sqrt{\alpha^2 + \beta^2}$ and that the upper bound is achieved for some choice of θ.

Theorem 9.8 (Minkowski). *If $\mathbf{a} \in \mathbb{C}^n$ with entries $a_1, \ldots, a_n$, $\mathbf{b} \in \mathbb{C}^n$ with entries $b_1, \ldots, b_n$, and $1 \leq s < \infty$, then*

$$(9.24) \qquad \left\{ \sum_{k=1}^n |a_k + b_k|^s \right\}^{1/s} \leq \left\{ \sum_{k=1}^n |a_k|^s \right\}^{1/s} + \left\{ \sum_{k=1}^n |b_k|^s \right\}^{1/s}.$$

Moreover, if $1 < s < \infty$, then equality holds in (9.24) if and only if either $\mathbf{a} = \mathbf{0}$, or $\mathbf{a} \neq \mathbf{0}$ and $\mathbf{b} = \mu \mathbf{a}$ for some $\mu \geq 0$. If $s = 1$, then equality holds in (9.24) if and only if for each index k, either $a_k = 0$, or $a_k \neq 0$ and $b_k = \mu_k a_k$ for some $\mu_k \geq 0$.

Proof. The inequality for $s = 1$ is an immediate consequence of the fact that for every pair of complex numbers a and b, $|a + b| \leq |a| + |b|$; the conditions for equality follow from Exercise 9.15. The proof for $1 < s < \infty$ is divided into parts.

1. Verification of the inequality (9.24) for $1 < s < \infty$.

If $s > 1$, then

(9.25)
$$\sum_{k=1}^{n} |a_k + b_k|^s = \sum_{k=1}^{n} |a_k + b_k|^{s-1} |a_k + b_k|$$
$$\leq \sum_{k=1}^{n} |a_k + b_k|^{s-1} (|a_k| + |b_k|).$$

By Hölder's inequality,

(9.26)
$$\sum_{k=1}^{n} |a_k + b_k|^{s-1} |a_k| \leq \left\{ \sum_{k=1}^{n} |a_k + b_k|^{(s-1)t} \right\}^{1/t} \left\{ \sum_{k=1}^{n} |a_k|^s \right\}^{1/s}$$

and

(9.27)
$$\sum_{k=1}^{n} |a_k + b_k|^{s-1} |b_k| \leq \left\{ \sum_{k=1}^{n} |a_k + b_k|^{(s-1)t} \right\}^{1/t} \left\{ \sum_{k=1}^{n} |b_k|^s \right\}^{1/s}$$

for $t = s/(s-1)$. Since $(s-1)t = s$, the last three inequalities imply that

$$\sum_{k=1}^{n} |a_k + b_k|^s \leq \left\{ \sum_{k=1}^{n} |a_k + b_k|^s \right\}^{1/t} \left\{ \left(\sum_{k=1}^{n} |a_k|^s \right)^{1/s} + \left(\sum_{k=1}^{n} |b_k|^s \right)^{1/s} \right\}.$$

Now, if

$$\sum_{k=1}^{n} |a_k + b_k|^s > 0,$$

then we can divide both sides of the last inequality by $\left\{ \sum_{k=1}^{n} |a_k + b_k|^s \right\}^{1/t}$ to obtain the desired inequality (9.24). On the other hand, if $\sum_{k=1}^{n} |a_k + b_k|^s = 0$, then the inequality (9.24) is self-evident.

2. Verification of the conditions for equality in (9.24) for $1 < s < \infty$.

Clearly equality holds in (9.24) if $\mathbf{a} = \mathbf{0}$. Suppose next that $\mathbf{a} \neq \mathbf{0}$ and let $\mathbf{c} \in \mathbb{C}^n$ with components $c_1 = a_1 + b_1, \ldots, c_n = a_n + b_n$. A necessary condition for (9.24) to hold when $\mathbf{a} \neq \mathbf{0}$ is that $\mathbf{c} \neq \mathbf{0}$. Moreover, if equality

9.2. Four inequalities

holds in (9.24), then, in view of (9.25)–(9.27),

$$\sum_{k=1}^n |c_k|^s \le \sum_{k=1}^n |c_k|^{s-1}|a_k| + \sum_{k=1}^n |c_k|^{s-1}|b_k|$$

$$\le \left\{\sum_{k=1}^n |c_k|^s\right\}^{1/t} \left\{\left(\sum_{k=1}^n |a_k|^s\right)^{1/s} + \left(\sum_{k=1}^n |b_k|^s\right)^{1/s}\right\}$$

$$= \left\{\sum_{k=1}^n |c_k|^s\right\}^{1/t} \left\{\left(\sum_{k=1}^n |c_k|^s\right)^{1/s}\right\} = \sum_{k=1}^n |c_k|^s.$$

Thus, if equality holds in (9.24), then it also holds in (9.25)–(9.27).

Then Theorem 9.6 ensures that equality holds in (9.26) if and only if $|a_k|^s = \alpha |c_k|^{(s-1)t}$ for $k = 1,\ldots,n$ and some $\alpha \ge 0$. Similarly, equality holds in (9.27) if and only if $|b_k|^s = \beta |c_k|^{(s-1)t}$ for $k = 1,\ldots,n$ and some $\beta \ge 0$. Moreover, since $\mathbf{a} \ne \mathbf{0}$ by assumption, $\alpha > 0$. Thus, as $(s-1)t = s$,

$$|b_k|^s = \beta |c_k|^s = \frac{\beta}{\alpha}\alpha|c_k|^s = \frac{\beta}{\alpha}|a_k|^s \quad \text{for } k = 1,\ldots,n.$$

Therefore,

(9.28) $\qquad |b_k| = \mu|a_k| \quad \text{for } k = 1,\ldots,n \text{ and some } \mu \ge 0.$

Equality in (9.25) implies that $|c_k|^{s-1}|a_k + b_k| = |c_k|^{s-1}(|a_k| + |b_k|)$ for $k = 1,\ldots,n$. If $a_k = 0$, then this is automatically so. If $a_k \ne 0$, then the already established equality $|a_k|^s = \alpha|c_k|^s$ ensures that $c_k \ne 0$ and hence that $|a_k + b_k| = |a_k| + |b_k|$. But then (9.19) ensures that $b_k = t_k a_k$ for some $t_k \ge 0$. Thus, in view of (9.28), $a_k = 0 \implies b_k = 0$ and $a_k \ne 0 \implies t_k = \mu$. Consequently, $b_k = \mu a_k$ in both cases; i.e., if $\mathbf{a} \ne \mathbf{0}$, then $\mathbf{b} = \mu\mathbf{a}$ for some $\mu \ge 0$. $\square$

Remark 9.9. The inequality (9.14) is a special case of a more general statement that is usually referred to as **Young's inequality**: *If $b_1,\ldots,b_n$ and $p_1,\ldots,p_n$ are positive numbers and $\frac{1}{p_1} + \cdots + \frac{1}{p_n} = 1$, then*

(9.29) $\qquad b_1 \cdots b_n \le \frac{b_1^{p_1}}{p_1} + \cdots + \frac{b_n^{p_n}}{p_n},$

which is equivalent to (9.10).

9.3. The Krein-Milman theorem

A point $\mathbf{c}$ in a convex set Q is said to be an **extreme point** of Q if

$$0 < t < 1, \ \mathbf{a}, \mathbf{b} \in Q, \text{ and } \mathbf{c} = t\mathbf{a} + (1-t)\mathbf{b} \implies \mathbf{a} = \mathbf{b} = \mathbf{c}.$$

Theorem 9.10 (Krein-Milman). *If Q is a closed convex subset of $\mathbb{R}^n$ such that $Q \subset \{\mathbf{x} \in \mathbb{R}^n : \mathbf{x}^H \mathbf{x} \leq R\}$ for some $R < \infty$, then every vector in Q is a convex combination of the extreme points of Q.*

Proof. See, e.g., Section 22.8 in [30]. □

In fact, since $Q \subset \mathbb{R}^n$, every vector in Q is a convex combination of a finite number of extreme points of Q. An argument of **Carathéodory** serves to bound the number of extreme points needed:

Theorem 9.11. *If Q is a closed convex subset of $\mathbb{R}^n$ such that $Q \subset \{\mathbf{x} \in \mathbb{R}^n : \mathbf{x}^H \mathbf{x} \leq R\}$ for some $R < \infty$, then every vector in Q is a convex combination of at most $n+1$ extreme points of Q.*

Proof. If $\mathbf{u} \in Q$, then, in view of Theorem 9.10, $\mathbf{u} = \sum_{j=1}^{k} t_j \mathbf{v}_j$ is a convex combination of k extreme points of Q, $\mathbf{v}_1, \ldots, \mathbf{v}_k$. If $k > n+1$, then the vectors $\begin{bmatrix} 1 & \mathbf{v}_1 \end{bmatrix}^T, \ldots, \begin{bmatrix} 1 & \mathbf{v}_k \end{bmatrix}^T$ are linearly dependent. Therefore, there exist a set of numbers $s_1, \ldots, s_k$, not all of which are zero, such that

$$\sum_{j=1}^{k} s_j \begin{bmatrix} 1 \\ \mathbf{v}_j \end{bmatrix} = \mathbf{0}_{n+1}, \quad \text{i.e., } \sum_{j=1}^{k} s_j = 0 \text{ and } \sum_{j=1}^{k} s_j \mathbf{v}_j = \mathbf{0}_n.$$

Since we may assume that $t_j > 0$ for $j = 1, \ldots, k$, there exists a number $\mu > 0$ such that $t_j - \mu s_j \geq 0$ for $j = 1, \ldots, k$, with equality for at least one index j. Thus, as $\sum_{j=1}^{k}(t_j - \mu s_j) = 1$,

$$\mathbf{u} = \sum_{j=1}^{k}(t_j - \mu s_j)\mathbf{v}_j$$

is a convex combination of at most $k-1$ extreme points. The argument can be repeated until a representation with at most $n+1$ extreme points is obtained. □

Exercise 9.19. Show that if $\mathbf{v}_1, \ldots, \mathbf{v}_k$ are linearly independent vectors in $\mathbb{R}^n$ and $Q = \mathbf{v}_1 + \{\sum_{j=2}^{k} t_j \mathbf{v}_j : t_j \geq 0 \text{ and } \sum_{j=2}^{k} t_j = 1\}$, then Q is a convex combination of $k-1$ extreme points.

Exercise 9.20. Show that if $f(\mathbf{x})$ is a real-valued linear function on $\mathbb{R}^n$ (i.e., $f(c\mathbf{u} + \mathbf{v}) = cf(\mathbf{u}) + f(\mathbf{v})$ for $\mathbf{u}, \mathbf{v} \in \mathbb{R}^n$ and $c \in \mathbb{R}$) and Q is as in Exercise 9.19 with extreme points $\mathbf{u}_1, \ldots, \mathbf{u}_{k-1}$, then $\max\{f(\mathbf{u}) : \mathbf{u} \in Q\} = \max_j f(\mathbf{u}_j)$.

9.4. Supplementary notes

For general versions of the Krein-Milman theorem see, e.g., Bollobás [12]. Exercise 9.19 illustrates a refinement of Theorem 9.11 for convex subsets of $\mathbb{R}^n$ having (appropriately defined) dimension less than n; see, e.g., Theorem 8.11 in Simon [70]. Exercise 9.12 was posed by Andrica [4].

Chapter 10

Normed linear spaces

This chapter introduces normed linear spaces and surveys a number of their basic properties. In the final section it is shown that left invertibility and right invertibility of a matrix are preserved under small perturbations, though rank is not.

10.1. Normed linear spaces

A vector space $\mathcal{U}$ is said to be a **normed linear space** if there exists a number $\varphi(\mathbf{x})$ assigned to each vector $\mathbf{x} \in \mathcal{U}$ such that for every choice of $\mathbf{x}, \mathbf{y} \in \mathcal{U}$ and every scalar α the following four conditions are met:

(1) $\varphi(\mathbf{x}) \geq 0$.
(2) $\varphi(\mathbf{x}) = 0$ if and only if $\mathbf{x} = 0$.
(3) $\varphi(\alpha \mathbf{x}) = |\alpha| \varphi(\mathbf{x})$.
(4) $\varphi(\mathbf{x} + \mathbf{y}) \leq \varphi(\mathbf{x}) + \varphi(\mathbf{y})$.

Every function $\varphi(\mathbf{x})$ that meets these four conditions is called a **norm** and is usually denoted by the symbol $\|\mathbf{x}\|$, or by the symbol $\|\mathbf{x}\|_{\mathcal{U}}$ if it is desired to clarify the space under consideration. The inequality in (4) is called the **triangle inequality**.

Lemma 10.1. *Let $\mathcal{U}$ be a normed linear space with norm $\varphi(\mathbf{x})$. Then*

$$|\varphi(\mathbf{x}) - \varphi(\mathbf{y})| \leq \varphi(\mathbf{x} - \mathbf{y})$$

for every choice of $\mathbf{x}$ and $\mathbf{y}$ in $\mathcal{U}$ (and hence φ is continuous).

Proof. The triangle inequality implies that
$$\varphi(\mathbf{x}) = \varphi(\mathbf{x} - \mathbf{y} + \mathbf{y}) \leq \varphi(\mathbf{x} - \mathbf{y}) + \varphi(\mathbf{y}).$$
Therefore,
$$\varphi(\mathbf{x}) - \varphi(\mathbf{y}) \leq \varphi(\mathbf{x} - \mathbf{y})$$
and, since $\mathbf{x}$ and $\mathbf{y}$ may be interchanged, the supplementary inequality
$$\varphi(\mathbf{y}) - \varphi(\mathbf{x}) \leq \varphi(\mathbf{y} - \mathbf{x}) = \varphi(\mathbf{x} - \mathbf{y})$$
also holds. Thus,
$$-\varphi(\mathbf{x} - \mathbf{y}) \leq \varphi(\mathbf{x}) - \varphi(\mathbf{y}) \leq \varphi(\mathbf{x} - \mathbf{y}),$$
which is equivalent to the stated inequality. $\square$

In the special case that $\mathcal{U} = \mathbb{C}^n$ or $\mathcal{U} = \mathbb{R}^n$ the classical norms are

(10.1) $$\|\mathbf{x}\|_s = \begin{cases} \max\{|x_j| : 1 \leq j \leq n\} & \text{if } s = \infty, \\ \left\{\sum_{j=1}^n |x_j|^s\right\}^{1/s} & \text{if } 1 \leq s < \infty. \end{cases}$$

Exercise 10.1. Verify that $\|\mathbf{x}\|_s$ defines a norm on $\mathbb{C}^n$ for each choice of s, $1 \leq s \leq \infty$. [HINT: Use Minkowski's inequality (9.24) to justify the triangle inequality when $1 < s < \infty$.]

Exercise 10.2. Let $\mathcal{U}$ be a vector space with basis $\mathbf{u}_1, \ldots, \mathbf{u}_n$. Show that for each choice of s in the interval $1 \leq s < \infty$ the formula
$$\varphi\left(\sum_{j=1}^n x_j \mathbf{u}_j\right) = \left\{\sum_{j=1}^n |x_j|^s\right\}^{1/s}$$
defines a norm on $\mathcal{U}$. [HINT: See the hint in Exercise 10.1.]

The next exercise illustrates a special case of the general principle that **in a finite-dimensional normed linear space all norms are equivalent,** i.e., if $\varphi(\mathbf{u})$ and $\psi(\mathbf{u})$ are norms in a finite-dimensional normed linear space $\mathcal{U}$, then there exists a pair of positive constants α and β such that

(10.2) $$\alpha\psi(\mathbf{u}) \leq \varphi(\mathbf{u}) \leq \beta\psi(\mathbf{u}).$$

Exercise 10.3. Show that if $\mathbf{x} \in \mathbb{C}^n$, then
$$\frac{1}{n}\|\mathbf{x}\|_1 \leq \|\mathbf{x}\|_\infty \leq \|\mathbf{x}\|_2 \leq \|\mathbf{x}\|_1 \leq n\|\mathbf{x}\|_\infty.$$

Exercise 10.4. Show that if $\mathbf{x} \in \mathbb{C}^n$, $s \geq 1$, and $t \geq 0$, then

(10.3) $$\|\mathbf{x}\|_1 \geq \|\mathbf{x}\|_s \geq \|\mathbf{x}\|_{s+t} \geq \|\mathbf{x}\|_\infty \quad \text{for each vector } \mathbf{x} \in \mathbb{C}^n.$$

[HINT: If $y_j = (\|\mathbf{x}\|_s)^{-1}|x_j|$, then $0 \leq y_j \leq 1$ and $\sum_{j=0}^n y_j^{s+t} \leq \sum_{j=0}^n y_j^s = 1$.]

10.2. The vector space of matrices A

Exercise 10.5. Show that

(10.4) $$\sum_{j=1}^{n} |a_j|^t \geq \left(\sum_{j=1}^{n} |a_j|\right)^t \quad \text{if } 0 < t < 1.$$

[HINT: $0 < t < 1$ if and only if $1 < 1/t < \infty$.]

Exercise 10.6. Show that $\lim_{s \uparrow \infty} \|\mathbf{x}\|_s = \|\mathbf{x}\|_\infty$ for each vector $\mathbf{x} \in \mathbb{C}^n$. [HINT: $\|\mathbf{x}\|_s \leq n^{1/s} \|\mathbf{x}\|_\infty$.]

Exercise 10.7. Show that $\|\mathbf{x}\|_3$ is equivalent to $\|\mathbf{x}\|_1$. [HINT: Exploit the preceding exercises.]

The **most important norms** in $\mathbb{C}^n$ and $\mathbb{R}^n$ are $\|\mathbf{x}\|_1$, $\|\mathbf{x}\|_2$, and $\|\mathbf{x}\|_\infty$; the choice $s = 2$ yields the familiar Euclidean norm:

(10.5) $$\|\mathbf{x}\|_2 = \left\{\sum_{j=1}^{n} |x_j|^2\right\}^{1/2}.$$

Remark 10.2. Even though all norms on a finite-dimensional normed linear space are equivalent in the sense noted above, particular choices may be most appropriate for certain applications. Thus, for example, if the entries u_i in a vector $\mathbf{u} \in \mathbb{R}^n$ denote deviations from a navigational path (such as a channel through shallow waters) at successive increments of time, it's important to keep $\|\mathbf{u}\|_\infty$ small. If $\mathbf{a}, \mathbf{b} \in \mathbb{R}^2$, then although $\|\mathbf{a} - \mathbf{b}\|_2$ is equal to the usual Euclidean distance between the points $\mathbf{a}$ and $\mathbf{b}$, the norm $\|\mathbf{a} - \mathbf{b}\|_1$ might give a better indication of the driving distance.

10.2. The vector space of matrices A

There are many ways to define a norm on the vector space $\mathbb{C}^{p \times q}$. One could for example view a matrix $A \in \mathbb{C}^{p \times q}$ as a funny way to record the pq entries a_{ij} of a vector in the space $\mathbb{C}^{pq}$. Then

(10.6) $$\|A\|_s = \begin{cases} \{\sum_{i=1}^{p} \sum_{j=1}^{q} |a_{ij}|^s\}^{1/s} & \text{if } 1 \leq s < \infty, \\ \max\{|a_{ij}| : i = 1, \ldots, p, \ j = 1, \ldots, q\} & \text{if } s = \infty. \end{cases}$$

We shall, however, be primarily interested in other norms which reflect the action of A as a linear transformation.

We begin with a preliminary estimate:

Lemma 10.3. *If $A \in \mathbb{C}^{p \times q}$, $\mathbf{u} \in \mathbb{C}^q$, and $1 \leq s \leq \infty$, then*

(10.7) $$\|A\mathbf{u}\|_t \leq \left\{\sum_{i=1}^{p}\left(\sum_{j=1}^{q}|a_{ij}|\right)^t\right\}^{1/t} \|\mathbf{u}\|_s \quad \text{for } 1 \leq t < \infty$$

and

(10.8) $$\|A\mathbf{u}\|_\infty \leq \max_i \left\{ \sum_{j=1}^q |a_{ij}| \right\} \|\mathbf{u}\|_s \leq q\|A\|_\infty \|\mathbf{u}\|_s.$$

Proof. If $1 \leq t < \infty$, then

$$\left| \sum_{j=1}^q a_{ij} u_j \right|^t \leq \left(\sum_{j=1}^q |a_{ij} u_j| \right)^t \leq \left(\sum_{j=1}^q |a_{ij}| \right)^t \|\mathbf{u}\|_s^t,$$

since $|u_j| \leq \|\mathbf{u}\|_s$ when $1 \leq s \leq \infty$. But this leads easily to (10.7). The verification of (10.8) is left to the reader. $\square$

Exercise 10.8. Show that if $A \in \mathbb{C}^{p \times q}$, $\mathbf{u} \in \mathbb{C}^q$, and $1 \leq t \leq \infty$, then

(10.9) $$\|A\mathbf{u}\|_t \leq \|A\|_t \|\mathbf{u}\|_{t'},$$

where $t' = t/(t-1)$ if $1 < t < \infty$, $t' = \infty$ if $t = 1$, and $t' = 1$ if $t = \infty$. [HINT: Hölder's inequality is the key.]

Lemma 10.3 ensures that there exists a finite positive number $\gamma_{t,s}$ such that

(10.10) $\quad \|A\mathbf{u}\|_t \leq \gamma_{t,s} \|\mathbf{u}\|_s \quad$ for every vector $\mathbf{u} \in \mathbb{C}^q$ and $1 \leq s \leq \infty$.

Thus, the function $f(\mathbf{u}) = \|A\mathbf{u}\|_t$ is continuous in $\mathbb{C}^q$ with respect to the norm $\|\cdot\|_s$. Therefore it attains its maximum value on the set $\{\mathbf{u} \in \mathbb{C}^q : \|\mathbf{u}\|_s = 1\}$, since this is a closed bounded set in a finite-dimensional space. Consequently, for every pair of numbers $1 \leq s, t \leq \infty$, there exists at least one vector $\mathbf{u}_{\max} \in \mathbb{C}^q$ with $\|\mathbf{u}_{\max}\|_s = 1$ such that

(10.11) $\quad \|A\mathbf{u}\|_t \leq \|A\mathbf{u}_{\max}\|_t \quad$ for every $\mathbf{u} \in \mathbb{C}^q$ with $\|\mathbf{u}\|_s = 1$.

Let $\|A\|_{s,t} = \|A\mathbf{u}_{\max}\|_t$. Then

(10.12) $\quad \|A\|_{s,t} = \max\{\|A\mathbf{u}\|_t : \mathbf{u} \in \mathbb{C}^q \text{ and } \|\mathbf{u}\|_s = 1\}$

for every pair of numbers $1 \leq s, t \leq \infty$. (Some evaluations of $\|A\|_{s,t}$ are furnished in Section 10.3.)

Theorem 10.4. *If $A \in \mathbb{C}^{p \times q}$ and $1 \leq s, t \leq \infty$, then the number $\|A\|_{s,t}$ that is defined by formula (10.12) defines a norm on $\mathbb{C}^{p \times q}$. Moreover,*

(10.13) $\quad \|A\mathbf{u}\|_t \leq \|A\|_{s,t} \|\mathbf{u}\|_s \quad$ *for every vector* $\mathbf{u} \in \mathbb{C}^q$,

and $\|A\|_{s,t}$ may also be evaluated by each of the following two supplementary recipes:

(10.14) $\quad \|A\|_{s,t} \;=\; \max\{\|A\mathbf{u}\|_t : \mathbf{u} \in \mathbb{C}^q \text{ and } \|\mathbf{u}\|_s \leq 1\}$

(10.15) $\quad\qquad\;\; = \max\left\{ \dfrac{\|A\mathbf{u}\|_t}{\|\mathbf{u}\|_s} : \mathbf{u} \in \mathbb{C}^q \text{ and } \mathbf{u} \neq \mathbf{0} \right\}.$

Proof. Let α, β, and γ denote the right-hand sides of (10.12), (10.14), and (10.15), respectively, and note that if $\mathbf{u}$ is a nonzero vector in $\mathbb{C}^q$ and $\mathbf{v} = \mathbf{u}/\|\mathbf{u}\|_s$, then, as $\|\mathbf{v}\|_s = 1$,

$$\|A\mathbf{u}\|_t = \frac{\|A\mathbf{u}\|_t}{\|\mathbf{u}\|_s}\|\mathbf{u}\|_s = \|A\mathbf{v}\|_t \|\mathbf{u}\|_s \le \alpha \|\mathbf{u}\|_s \,.$$

This clearly implies that $\gamma \le \alpha$ and also serves to justify (10.13), since the inequality $\|A\mathbf{u}\|_t \le \alpha \|\mathbf{u}\|_s$ is also valid for $\mathbf{u} = \mathbf{0}$.

On the other hand, if $\|\mathbf{u}\|_s \le 1$ and $\mathbf{u} \ne \mathbf{0}$, then

$$\|A\mathbf{u}\|_t \le \frac{\|A\mathbf{u}\|_t}{\|\mathbf{u}\|_s} \le \gamma,$$

which implies that $\beta \le \gamma$, and hence, in view of the already established bound $\gamma \le \alpha$ and the self-evident inequality $\alpha \le \beta$, that $\alpha \le \beta \le \gamma \le \alpha$. Thus, $\alpha = \beta = \gamma$.

We still need to verify that $\|A\|_{s,t}$ really defines a norm on $\mathbb{C}^{p \times q}$. The conditions $\|A\|_{s,t} \ge 0$ and $\|\alpha A\|_{s,t} = |\alpha|\,\|A\|_{s,t}$ for $\alpha \in \mathbb{C}$ are self-evident. Moreover, if $\|A\|_{s,t} = 0$, then, in view of the bound (10.13), $\|A\mathbf{u}\|_t = 0$ for every vector $\mathbf{u} \in \mathbb{C}^q$. Therefore $A\mathbf{u} = \mathbf{0}$ for every vector $\mathbf{u} \in \mathbb{C}^q$ and hence $\|A\|_{s,t} = 0 \implies A = O$.

It remains only to check the triangle inequality. But if also $B \in \mathbb{C}^{p \times q}$, then

$$\|(A+B)\mathbf{u}\|_t \le \|A\mathbf{u}\|_t + \|B\mathbf{u}\|_t \le \|A\|_{s,t}\|\mathbf{u}\|_s + \|B\|_{s,t}\|\mathbf{u}\|_s\,,$$

which implies that $\|(A+B)\|_{s,t} \le \|A\|_{s,t} + \|B\|_{s,t}$, as needed to complete the proof. $\square$

To put the preceding discussion into context, we define the **operator norm** of a linear transformation T from a finite-dimensional normed linear space $\mathcal{U}$ into a normed linear space $\mathcal{V}$ by the formula

(10.16) $\qquad \|T\|_{\mathcal{U},\mathcal{V}} = \max\{\|T\mathbf{u}\|_{\mathcal{V}} : \mathbf{u} \in \mathcal{U} \text{ and } \|\mathbf{u}\|_{\mathcal{U}} = 1\}\,.$

Thus, $\|A\|_{s,t}$ is the **operator norm** for the linear transformation that sends vectors $\mathbf{u}$ in the normed linear space $\mathbb{C}^q$ with norm $\|\mathbf{u}\|_s$ to vectors $\mathbf{v} = A\mathbf{u}$ in the normed linear space $\mathbb{C}^p$ with norm $\|\mathbf{v}\|_t$.

The assumption that $\mathcal{U}$ is finite dimensional is essential to ensure the existence of a vector $\mathbf{u}_{\max} \in \mathcal{U}$ with $\|\mathbf{u}_{\max}\|_{\mathcal{U}} = 1$ such that $\|T\|_{\mathcal{U},\mathcal{V}} = \|T\mathbf{u}_{\max}\|_{\mathcal{V}}$ (or, in the setting discussed above, that (10.11) holds); see Exercise 10.9.

Exercise 10.9. The vector space $\mathcal{U}$ of **infinite** column vectors $\mathbf{u}$ with entries $u_1, u_2, \ldots$ that meet the constraint $\sum_{j=1}^{\infty} |u_j|^2 < \infty$ is a normed linear space with norm $\|\mathbf{u}\| = \left\{\sum_{i=1}^{\infty} |u_i|^2\right\}^{1/2}$. Let T denote the operator that maps $\mathbf{u}$ into the vector $T\mathbf{u}$ with components $a_1 u_1, a_2 u_2, \ldots$, where $0 < a_j < 1$ for all j and $\lim_{j \uparrow \infty} a_j = 1$. Show that $\sup\{\|T\mathbf{u}\| : \|\mathbf{u}\| = 1\} = 1$, but that there does not exist a nonzero vector $\mathbf{u} \in \mathcal{U}$ such that $\|T\mathbf{u}\| = \|\mathbf{u}\|$.

Exercise 10.10. Show that if T is a linear transformation from a finite-dimensional normed linear space $\mathcal{U}$ into a normed linear space $\mathcal{V}$, then

(10.17)
$$\begin{aligned}\|T\|_{\mathcal{U},\mathcal{V}} &= \max\{\|T\mathbf{u}\|_{\mathcal{V}} : \mathbf{u} \in \mathcal{U} \text{ and } \|\mathbf{u}\|_{\mathcal{U}} \leq 1\} \\ &= \max\left\{\frac{\|T\mathbf{u}\|_{\mathcal{V}}}{\|\mathbf{u}\|_{\mathcal{U}}} : \mathbf{u} \in \mathcal{U} \text{ and } \mathbf{u} \neq \mathbf{0}\right\}.\end{aligned}$$

Exercise 10.11. Show that if $A \in \mathbb{C}^{p \times 1}$, then $\|A\|_{s,t} = \|A\|_t$.

The next theorem gives useful bounds.

Theorem 10.5. *If $A \in \mathbb{C}^{p \times q}$, $u \in \{1, \ldots, p\}$, and $v \in \{1, \ldots, q\}$, then*

(10.18)
$$|a_{uv}| \leq \|A\|_{s,t} \leq \sum_{i=1}^{p}\sum_{j=1}^{q} |a_{ij}| \quad \text{for } 1 \leq s, t \leq \infty,$$

i.e.,

(10.19)
$$\|A\|_\infty \leq \|A\|_{s,t} \leq \|A\|_1 \quad \text{for } 1 \leq s, t \leq \infty.$$

Proof. Let E_{ij} denote the $p \times q$ matrix with a 1 in the ij place and 0's elsewhere. Then

$$\|A\|_{s,t} = \left\|\sum_{i=1}^{p}\sum_{j=1}^{q} a_{ij} E_{ij}\right\|_{s,t} \leq \sum_{i=1}^{p}\sum_{j=1}^{q} |a_{ij}| \|E_{ij}\|_{s,t}.$$

To evaluate $\|E_{ij}\|_{s,t}$, let $\mathbf{e}_1, \ldots, \mathbf{e}_p$ denote the columns of I_p and let $\mathbf{f}_1, \ldots, \mathbf{f}_q$ denote the columns of I_q. Then, since $E_{ij} = \mathbf{e}_i \mathbf{f}_j^T$,

$$\|E_{ij}\mathbf{u}\|_t = \|\mathbf{e}_i \mathbf{f}_j^T \mathbf{u}\|_t = |u_j| \|\mathbf{e}_i\|_t = |u_j| \leq \|\mathbf{u}\|_s$$

for every choice of s, $1 \leq s \leq \infty$. Moreover, equality is achieved by choosing $\mathbf{u} = \mathbf{f}_j$. Thus, $\|E_{ij}\|_{s,t} = 1$. This establishes the upper bound in (10.18).

On the other hand if $\mathbf{a}_v$ denotes the v'th column of A, then

$$\|A\|_{s,t} \geq \|A\mathbf{f}_v\|_t = \|\mathbf{a}_v\|_t \geq |a_{uv}| \quad \text{for } 1 \leq s, t \leq \infty$$

and every choice of $u \in \{1, \ldots, p\}$. □

10.2. The vector space of matrices A

Corollary 10.6. *If $A, B \in \mathbb{C}^{p \times q}$, $u \in \{1, \ldots, p\}$, and $v \in \{1, \ldots, q\}$, then*

(10.20) $\quad |a_{uv} - b_{uv}| \leq \|A - B\|_{s,t} \leq \sum_{i=1}^{p} \sum_{j=1}^{q} |a_{ij} - b_{ij}| \quad \text{for } 1 \leq s, t \leq \infty.$

The bounds in (10.18) are not the best possible; however, they have the advantage of simplicity and, as is spelled out in (10.20), they clearly display the fact that $\|A - B\|_{s,t}$ is small if and only if the entries in A are close to the entries in B.

We have special interest in norms $\|A\|$ on matrices A that have two extra properties:

(10.21) $\qquad\qquad \|AB\| \leq \|A\| \, \|B\| \quad \text{and} \quad \|I_n\| = 1 \, .$

Theorem 10.7. *The norm $\|A\|_{s,s}$, $1 \leq s \leq \infty$, meets both of the conditions in (10.21). The norm $\|A\|_s$ never meets both of these conditions: If $A, B \in \mathbb{C}^{n \times n}$ and $n > 1$, then*

(10.22)
$$\|AB\|_s \leq \|A\|_s \|B\|_s \iff 1 \leq s \leq 2 \quad \text{and} \quad \|I_p\|_s = 1 \iff s = \infty \, .$$

Proof. If $A \in \mathbb{C}^{p \times q}$, $B \in \mathbb{C}^{q \times k}$, and $\mathbf{u} \in \mathbb{C}^k$, then, by successive applications of the bound (10.13),

$$\|(AB)\mathbf{u}\|_t = \|A(B\mathbf{u})\|_t \leq \|A\|_{s,t} \|B\mathbf{u}\|_s \leq \|A\|_{s,t} \|B\|_{r,s} \|\mathbf{u}\|_r \, ,$$

which implies that

(10.23) $\qquad\qquad \|AB\|_{r,t} \leq \|A\|_{s,t} \|B\|_{r,s} \quad \text{for } 1 \leq r, s, t \leq \infty$

and hence that $\|AB\|_{s,s} \leq \|A\|_{s,s} \|B\|_{s,s}$. Thus, as $\|I_n\|_{s,s} = 1$, both of the conditions in (10.21) are met. The verification of (10.22) is left to the reader; see Exercises 10.12, 10.13, and 10.14. $\square$

Exercise 10.12. Show that if $1 \leq s \leq \infty$ and $n > 1$, then $\|I_n\|_s = 1$ if and only if $s = \infty$.

Exercise 10.13. Show that if $A \in \mathbb{C}^{p \times q}$, $B \in \mathbb{C}^{q \times r}$, and $1 \leq s \leq 2$, then $\|AB\|_s \leq \|A\|_s \|B\|_s$. [HINT: If $1 < s \leq 2$ and $s^{-1} + t^{-1} = 1$, then $t \geq s$ and hence $\|\mathbf{u}\|_t \leq \|\mathbf{u}\|_s$.]

Exercise 10.14. Show that if $A = \begin{bmatrix} a & a \\ a & a \end{bmatrix}$ with $a > 0$, then $\|A^2\|_s > \|A\|_s^2$ when $2 < s \leq \infty$ and $\|A^2\|_s < \|A\|_s^2 < 2\|A^2\|_s$ when $1 \leq s < 2$.

Exercise 10.15. Show that if $A \in \mathbb{C}^{p \times q}$, then

(10.24) $\qquad \max_i \left\{ \sum_{j=1}^{q} |a_{ij}|^2 \right\}^{1/2} \leq \|A\|_{2,2} \leq \left\{ \sum_{i=1}^{p} \sum_{j=1}^{q} |a_{ij}|^2 \right\}^{1/2} .$

Exercise 10.16. Show that if $A = \begin{bmatrix} a & b \\ 0 & c \end{bmatrix} \in \mathbb{R}^{2\times 2}$ and $d^2 = a^2 + b^2 + c^2$, then

$$\max\{\|A\mathbf{x}\|_2^2 : \mathbf{x} \in \mathbb{R}^2 \text{ and } \|\mathbf{x}\|_2 = 1\} = \frac{d^2 + \sqrt{d^4 - 4a^2c^2}}{2}.$$

[HINT: If $\mathbf{u} \in \mathbb{R}^2$ and $\|\mathbf{u}\|_2 = 1$, then $\mathbf{u}^T = [\cos\theta \quad \sin\theta]$. To finish, refer to Exercise 9.18.]

10.3. Evaluating some operator norms

The next lemma lists a number of cases for which it is possible to evaluate $\|A\|_{s,t}$ precisely.

Lemma 10.8. *If $A \in \mathbb{C}^{p\times q}$, then:*

(1) $\|A\|_{1,1} = \max_j \{\sum_{i=1}^p |a_{ij}|\}$.

(2) $\|A\|_{\infty,\infty} = \max_i \{\sum_{j=1}^q |a_{ij}|\}$.

(3) $\|A\|_{2,2} = s_1$, where s_1^2 is the largest eigenvalue of the matrix $A^H A$.

(4) $\|A\|_{1,\infty} = \max_{i,j} |a_{ij}|$.

(5) $\|A\|_{2,\infty} = \max_i \{(\sum_{j=1}^q |a_{ij}|^2)^{1/2}\}$.

(6) $\|A\|_{1,2} = \max_j \{(\sum_{i=1}^p |a_{ij}|^2)^{1/2}\}$.

Discussion. To obtain the first formula, observe that

$$\|A\mathbf{x}\|_1 = \sum_{i=1}^p \left|\sum_{j=1}^q a_{ij} x_j\right| \le \sum_{j=1}^q \left(\sum_{i=1}^p |a_{ij}|\right) |x_j|$$

and hence that

(10.25) $$\|A\mathbf{x}\|_1 \le \max_j \left\{\sum_{i=1}^p |a_{ij}|\right\} \|\mathbf{x}\|_1.$$

This establishes the inequality

(10.26) $$\|A\|_{1,1} \le \max_j \left\{\sum_{i=1}^p |a_{ij}|\right\}.$$

To obtain equality, it suffices to exhibit a vector $\mathbf{x} \in \mathbb{C}^q$ such that $\mathbf{x} \ne 0$ and equality prevails in formula (10.25). Suppose that the maximum in (10.26)

10.3. Evaluating some operator norms

is achieved when $j = k$. Then for the vector $\mathbf{u}$ with $u_k = 1$ and all other coordinates equal to zero, we obtain $\|\mathbf{u}\|_1 = 1$ and

$$\|A\|_{1,1} \geq \|A\mathbf{u}\|_1 = \sum_{i=1}^{p}\left|\sum_{j=1}^{q} a_{ij}u_j\right| = \sum_{i=1}^{p}|a_{ik}| = \max_{j}\left\{\sum_{i=1}^{p}|a_{ij}|\right\}.$$

This completes the proof of the first formula.

Next, to obtain the second formula, observe that

$$\left|\sum_{j=1}^{q} a_{ij}x_j\right| \leq \sum_{j=1}^{q}|a_{ij}||x_j| \leq \sum_{j=1}^{q}|a_{ij}|\|\mathbf{x}\|_\infty$$

and hence that

(10.27) $$\|A\mathbf{x}\|_\infty = \max_{i}\left\{\left|\sum_{j=1}^{q} a_{ij}x_j\right|\right\} \leq \max_{i}\left\{\sum_{j=1}^{q}|a_{ij}|\right\}\|\mathbf{x}\|_\infty,$$

i.e.,

(10.28) $$\|A\|_{\infty,\infty} \leq \max_{i}\left\{\sum_{j=1}^{q}|a_{ij}|\right\}.$$

To obtain equality in (10.28), it suffices to exhibit a vector $\mathbf{x} \in \mathbb{C}^q$ such that $\mathbf{x} \neq 0$ and equality prevails in (10.27). Suppose that the maximum in (10.28) is attained at $i = k$ and that it is not equal to zero, and let $\mathbf{u}$ be the vector in $\mathbb{C}^q$ with entries

$$u_j = \begin{cases} \overline{a_{kj}}/|a_{kj}| & \text{if } a_{kj} \neq 0, \\ 0 & \text{if } a_{kj} = 0. \end{cases}$$

Then $\|\mathbf{u}\|_\infty = 1$ and

$$\|A\|_{\infty,\infty} \geq \|A\mathbf{u}\|_\infty \geq \left|\sum_{j=1}^{q} a_{kj}u_j\right| = \sum_{j=1}^{q}|a_{kj}| = \max_{i}\left\{\sum_{j=1}^{q}|a_{ij}|\right\}.$$

This completes the proof of the second assertion if $A \neq O_{p \times q}$. However, if $A = O_{p \times q}$, then the asserted formula is self-evident.

We shall postpone the proof of the third assertion to Lemma 15.3 and leave the remaining assertions to the reader. $\square$

Exercise 10.17. Compute the maximum eigenvalue of the matrix $A^H A$ when $A = \begin{bmatrix} a & b \\ 0 & c \end{bmatrix} \in \mathbb{R}^{2 \times 2}$ and show that it is equal to the maximum that was calculated in Exercise 10.16.

10.4. Small perturbations

In subsequent chapters we shall work mostly (though not exclusively) with the norm $\|A\|_{2,2}$. But in this section (and this section only), just to give some idea of the possibilities, we shall let $\|A\|_\blacklozenge$ denote the operator norm for $A \in \mathbb{C}^{p \times q}$ that is defined by the formula

$$\|A\|_\blacklozenge = \max\{\|A\mathbf{x}\|_\diamond : \mathbf{x} \in \mathbb{C}^q \text{ and } \|\mathbf{x}\|_\diamond = 1\},$$

where $\|\mathbf{x}\|_\diamond$ is any norm on $\mathbb{C}^q$. You can, if it makes you more comfortable, replace $\|A\|_\blacklozenge$ by $\|A\|_{2,2}$ and $\|\mathbf{x}\|_\diamond$ by $\|\mathbf{x}\|_2$.

Lemma 10.9. *If $X \in \mathbb{C}^{p \times p}$ and $\|X\|_\blacklozenge < 1$, then $I_p - X$ is invertible.*

Proof. Let $\mathbf{u} \in \mathbb{C}^p$ and $(I_p - X)\mathbf{u} = \mathbf{0}$; then $\|\mathbf{u}\|_\diamond = \|X\mathbf{u}\|_\diamond \leq \|X\|_\blacklozenge \|\mathbf{u}\|_\diamond$. Therefore, $(1 - \|X\|_\blacklozenge)\|\mathbf{u}\|_\diamond \leq 0$, which implies that $\mathbf{u} = \mathbf{0}$ and hence that the nullspace of $I_p - X$ is equal to $\{\mathbf{0}\}$. Therefore, $I_p - X$ is invertible. □

Theorem 10.10. *If $A, B \in \mathbb{C}^{p \times p}$, A is invertible, and*

$$\|A - B\|_\blacklozenge < \{\|A^{-1}\|_\blacklozenge\}^{-1},$$

then B is invertible.

Proof. Since A is invertible,

$$B = A - (A - B) = A(I_p - A^{-1}(A - B)),$$

which will be invertible if $\|A^{-1}(A - B)\|_\blacklozenge < 1$ by Lemma 10.9. But, if $\|A - B\|_\blacklozenge < \{\|A^{-1}\|_\blacklozenge\}^{-1}$, then $\|A^{-1}(A - B)\|_\blacklozenge \leq \|A^{-1}\|_\blacklozenge \|A - B\|_\blacklozenge < 1$. □

Theorem 10.10 ensures that **invertibilty is preserved under small perturbations**. But more is true:

(1) If B is close to A, then the eigenvalues of B will be close to the eigenvalues of A; see Theorem 35.7.

(2) **Left and right invertibility are also preserved under small perturbations**. We shall justify this indirectly by first exploring the behavior of the rank, which is not necessarily preserved under small perturbations.

Lemma 10.11. *If $A \in \mathbb{C}^{p \times q}$ and B is a submatrix of A, then $\|B\|_\blacklozenge \leq \|A\|_\blacklozenge$.*

Discussion. Suppose that $A \in \mathbb{C}^{6 \times 5}$ and

$$B = \begin{bmatrix} a_{21} & a_{23} & a_{24} \\ a_{41} & a_{43} & a_{44} \end{bmatrix}$$

and let $\mathbf{e}_i$ denote the i'th column of I_6 for $i = 1, \ldots, 6$ and $\mathbf{f}_j$ denote the j'th column of I_5 for $j = 1, \ldots, 5$. Then

$$B = E^T A F, \quad \text{where} \quad E = \begin{bmatrix} \mathbf{e}_2 & \mathbf{e}_4 \end{bmatrix} \quad \text{and} \quad F = \begin{bmatrix} \mathbf{f}_1 & \mathbf{f}_3 & \mathbf{f}_4 \end{bmatrix}.$$

10.4. Small perturbations

Thus, $\|B\|_\blacklozenge = \|E^T AF\|_\blacklozenge \leq \|E^T\|_\blacklozenge \|A\|_\blacklozenge \|F\|_\blacklozenge$. But,

$$\begin{aligned}
\|F\|_\blacklozenge &= \max\{\|F\mathbf{x}\|_\diamond : \mathbf{x} \in \mathbb{C}^3 \text{ and } \|\mathbf{x}\|_\diamond = 1\} \\
&= \max\{\|I_5\mathbf{x}\|_\diamond : \mathbf{x} \in \mathbb{C}^5, \quad x_2 = x_5 = 0, \text{ and } \|\mathbf{x}\|_\diamond = 1\} \\
&\leq \max\{\|I_5\mathbf{x}\|_\diamond : \mathbf{x} \in \mathbb{C}^5 \text{ and } \|\mathbf{x}\|_\diamond = 1\} = \|I_5\|_\blacklozenge = 1.
\end{aligned}$$

By similar considerations, $\|E^T\|_\blacklozenge \leq 1$. Therefore, $\|B\|_\blacklozenge \leq \|A\|_\blacklozenge$ in this example. The verification of this inequality in the general setting is essentially the same; only the bookkeeping is a little more elaborate. □

Exercise 10.18. Show that in the setting of the preceding discussion,
$$\|F\|_\blacklozenge = \|E^T\|_\blacklozenge = 1.$$

Exercise 10.19. Show by direct calculation that if $E \in \mathbb{R}^{n \times k}$ is a submatrix of I_n that is obtained by discarding $n - k$ columns of I_n for some choice of k, $1 \leq k \leq n$, then $\|E\|_\blacklozenge = 1$ and $\|E^T\|_\blacklozenge = 1$.

Theorem 10.12. *If $A \in \mathbb{C}^{p \times q}$, then:*

(1) $\operatorname{rank} A = r \implies$ *there exists an invertible $r \times r$ submatrix of A.*

(2) *If there exists a $k \times k$ invertible submatrix of A, then $\operatorname{rank} A \geq k$.*

Proof. If $A = \begin{bmatrix} \mathbf{a}_1 & \cdots & \mathbf{a}_q \end{bmatrix}$ and $\operatorname{rank} A = r$, then there exists a submatrix $B \in \mathbb{C}^{p \times r}$ of A with $\operatorname{rank} B = r$. Therefore, $\operatorname{rank} B^T = r$ and there exists an $r \times r$ submatrix C of B^T with $\operatorname{rank} C = r$. Thus C^T is an invertible $r \times r$ submatrix of A. This completes the proof of (1).

Suppose next that there exists a $k \times k$ invertible submatrix of A. Then the k columns of A that overlap the columns of this submatrix are linearly independent. Thus, $\operatorname{rank} A \geq k$. □

Theorem 10.13. *If $A \in \mathbb{C}^{p \times q}$ and $\operatorname{rank} A = r$, then there exists an $\varepsilon > 0$ such that if $B \in \mathbb{C}^{p \times q}$ and $\|A - B\|_\blacklozenge < \varepsilon$, then $\operatorname{rank} B \geq r$.*

Proof. If $A \in \mathbb{C}^{p \times q}$ and $\operatorname{rank} A = r$, then there exist a $p \times r$ submatrix E of I_p and a $q \times r$ submatrix F of I_q such that the $r \times r$ submatrix $E^T AF$ of A is invertible. Moreover, since

$$\|E^T AF - E^T BF\|_\blacklozenge = \|E^T(A - B)F\|_\blacklozenge \leq \|A - B\|_\blacklozenge,$$

Theorem 10.10 ensures that $E^T BF$ is invertible if

$$\|A - B\|_\blacklozenge < \frac{1}{\|(E^T AF)^{-1}\|_\blacklozenge}.$$

Thus, if this condition is met, then, in view of Theorem 10.12, $\operatorname{rank} B \geq r$, as claimed. □

Example 10.1. If
$$A = \begin{bmatrix} 0 & 1 \\ 0 & 1 \end{bmatrix} \quad \text{and} \quad B = \begin{bmatrix} 0 & 1 \\ \alpha & 1 \end{bmatrix} \quad \text{with } \alpha \neq 0,$$
then $\|A - B\|_\blacklozenge = |\alpha|$ and hence there exists a matrix B with rank $B = 2$ in the set $\{X \in \mathbb{C}^{p \times q} : \|A - X\|_\blacklozenge < \varepsilon\}$ for every $\varepsilon > 0$, no matter how small, whereas rank $A = 1$. ◇

Corollary 10.14. *If $A, B \in \mathbb{C}^{p \times q}$, then there exists an $\varepsilon > 0$ such that:*

(1) *A left invertible and $\|A - B\|_\blacklozenge < \varepsilon \Longrightarrow B$ is left invertible.*

(2) *A right invertible and $\|A - B\|_\blacklozenge < \varepsilon \Longrightarrow B$ is right invertible.*

Proof. This depends upon the fact that A is right invertible (resp., left invertible) if and only if rank $A = p$ (resp., rank $A = q$). Consequently, if A is right invertible and $B - A$ is small enough, then
$$p \geq \min\{p, q\} \geq \operatorname{rank} B \geq \operatorname{rank} A = p.$$
Thus, rank $B = p$ and (1) holds; the justification of (2) is similar. □

Theorem 10.15. *If $A \in \mathbb{C}^{p \times p}$ and $\varepsilon > 0$, then there exists a diagonalizable matrix $B \in \mathbb{C}^{p \times p}$ such that $\|B - A\|_\blacklozenge < \varepsilon$.*

Proof. If $A = UJU^{-1}$, choose $B = U(J + D)U^{-1}$, where D is a diagonal matrix that is chosen so that the diagonal entries of $J + D$ are all distinct and $\|UDU^{-1}\|_\blacklozenge < \varepsilon$. □

Theorem 10.15 implies that the set of complex diagonalizable matrices is **dense** in $\mathbb{C}^{p \times p}$; however, it is not an **open** set: If $A \in \mathbb{C}^{p \times p}$ is diagonalizable and $\varepsilon > 0$, then $\{B \in \mathbb{C}^{p \times p} : \|B - A\|_\circ < \varepsilon\}$ will contain nondiagonalizable matrices. The simplest example is
$$A = \begin{bmatrix} 1 & 0 \\ 0 & 1 \end{bmatrix} \quad \text{and} \quad \begin{bmatrix} 1 & \alpha \\ 0 & 1 \end{bmatrix} \quad \text{with} \quad 0 < |\alpha| < \varepsilon.$$

Exercise 10.20. Show that if $A \in \mathbb{C}^{n \times n}$ and $\|A\|_\infty < 1/(n+1)$, then the matrix $B = I_n - A$ is invertible and $\|B\|_\infty < 3/2$.

10.5. Supplementary notes

This chapter is partially adapted from Chapter 7 of [**30**]. For a proof of the equivalence of norms in a finite-dimensional space, see, e.g., Section 7.3 of [**30**].

A sequence of vectors $\mathbf{u}_1, \mathbf{u}_2, \ldots$ in a normed linear space $\mathcal{U}$ is a **Cauchy sequence** if for every $\varepsilon > 0$ there exists a positive integer N such that $\|\mathbf{u}_{n+k} - \mathbf{u}_n\|_\mathcal{U} < \varepsilon$ for $n \geq N$ and all positive integers k. A normed linear space is a **Banach space** if every Cauchy sequence tends to a limit in the

10.5. Supplementary notes

space. In a finite-dimensional normed linear space every Cauchy sequence tends to a limit in the space. Thus, every finite-dimensional normed linear space is a Banach space. Moreover, in a finite-dimensional normed linear space, every infinite sequence of vectors $\mathbf{u}_1, \mathbf{u}_2, \ldots$ with $\|\mathbf{u}_j\|_{\mathcal{U}} \leq K < \infty$ has a convergent subsequence; and if S, T are linear transformations from $\mathcal{U}$ into $\mathcal{U}$, then $ST = I \iff TS = I$. Both of these properties fail in infinite-dimensional spaces such as the space $\mathcal{U}$ considered in Exercise 10.9.

Chapter 11

Inner product spaces

The first three sections of this chapter are devoted to inner product spaces, Gram matrices, and the adjoint of a linear transformation, respectively. We then consider the spectral radius of a matrix (which could just as well have been presented in Chapter 10) and finally present a list of what you need to know about the operator norm $\|A\|_{2,2}$ of a matrix $A \in \mathbb{C}^{p \times q}$.

11.1. Inner product spaces

A vector space $\mathcal{U}$ is said to be an **inner product space** if there is a number $\langle \mathbf{u}, \mathbf{v} \rangle_\mathcal{U} \in \mathbb{C}$ associated with every pair of vectors $\mathbf{u}, \mathbf{v} \in \mathcal{U}$ such that:

(1) $\langle \mathbf{u} + \mathbf{w}, \mathbf{v} \rangle_\mathcal{U} = \langle \mathbf{u}, \mathbf{v} \rangle_\mathcal{U} + \langle \mathbf{w}, \mathbf{v} \rangle_\mathcal{U}$ for every $\mathbf{w} \in \mathcal{U}$.

(2) $\langle \alpha \mathbf{u}, \mathbf{v} \rangle_\mathcal{U} = \alpha \langle \mathbf{u}, \mathbf{v} \rangle_\mathcal{U}$ for every scalar α.

(3) $\langle \mathbf{u}, \mathbf{v} \rangle_\mathcal{U} = \overline{\langle \mathbf{v}, \mathbf{u} \rangle_\mathcal{U}}$.

(4) $\langle \mathbf{u}, \mathbf{u} \rangle_\mathcal{U} \geq 0$ with equality if and only if $\mathbf{u} = \mathbf{0}$.

The number $\langle \mathbf{u}, \mathbf{v} \rangle_\mathcal{U}$ is termed the **inner product**. Items (1) and (2) imply that the inner product is linear in the first entry and hence, in particular, that

$$2\langle \mathbf{0}, \mathbf{v} \rangle_\mathcal{U} = \langle 2\mathbf{0}, \mathbf{v} \rangle_\mathcal{U} = \langle \mathbf{0}, \mathbf{v} \rangle_\mathcal{U},$$

which implies that $\langle \mathbf{0}, \mathbf{v} \rangle_\mathcal{U} = 0$. Item (3) serves to guarantee that the inner product is additive in the second entry, i.e.,

$$\langle \mathbf{u}, \mathbf{v} + \mathbf{w} \rangle_\mathcal{U} = \langle \mathbf{u}, \mathbf{v} \rangle_\mathcal{U} + \langle \mathbf{u}, \mathbf{w} \rangle_\mathcal{U}; \quad \text{however,} \quad \langle \mathbf{u}, \beta \mathbf{v} \rangle_\mathcal{U} = \overline{\beta} \langle \mathbf{u}, \mathbf{v} \rangle_\mathcal{U}.$$

When the underlying inner product space $\mathcal{U}$ is clear from the context, the inner product is often denoted simply as $\langle \mathbf{u}, \mathbf{v} \rangle$ instead of $\langle \mathbf{u}, \mathbf{v} \rangle_\mathcal{U}$.

A **Hilbert space** is an inner product space $\mathcal{U}$ in which every Cauchy sequence tends to a limit in $\mathcal{U}$. A finite-dimensional inner product space is automatically a Hilbert space.

Exercise 11.1. Let $\mathcal{U}$ be an inner product space and let $\mathbf{u} \in \mathcal{U}$. Show that
$$\langle \mathbf{u}, \mathbf{v} \rangle = 0 \quad \text{for every} \quad \mathbf{v} \in \mathcal{U} \iff \mathbf{u} = \mathbf{0}$$
and (consequently)
$$\langle \mathbf{u}_1, \mathbf{v} \rangle = \langle \mathbf{u}_2, \mathbf{v} \rangle \quad \text{for every} \quad \mathbf{v} \in \mathcal{U} \iff \mathbf{u}_1 = \mathbf{u}_2 \,.$$

The **notation** $\langle \mathbf{x}, \mathbf{y} \rangle_{\mathrm{st}}$, which is defined for $\mathbf{x}, \mathbf{y} \in \mathbb{C}^n$ by the formula

(11.1) $$\langle \mathbf{x}, \mathbf{y} \rangle_{\mathrm{st}} = \mathbf{y}^H \mathbf{x} = \sum_{i=1}^{n} \overline{y_i} x_i \,,$$

will be used on occasion to denote the **standard inner product** on $\mathbb{C}^n$. The conjugation in this formula can be dropped if $\mathbf{x}, \mathbf{y} \in \mathbb{R}^n$. It is important to bear in mind that there are many other inner products that can be imposed on $\mathbb{C}^n$:

Exercise 11.2. Show that if $B \in \mathbb{C}^{p \times q}$ and $\operatorname{rank} B = q$, then the formula

(11.2) $$\langle \mathbf{x}, \mathbf{y} \rangle = (B\mathbf{y})^H B \mathbf{x}$$

defines an inner product on $\mathbb{C}^q$.

Exercise 11.3. Show that the formula $\langle A, B \rangle = \operatorname{trace}(B^H A)$ defines an inner product on the space $\mathcal{U} = \mathbb{C}^{p \times q}$ and then find a basis for $\mathbb{C}^{p \times q}$ that is orthonormal with respect to this inner product.

The norm $\|A\|_2 = (\langle A, A \rangle)^{1/2}$ based on the inner product introduced in Exercise 11.3 is called the **Frobenius norm**. Although it looks totally new, the identities

(11.3) $$\sum_{i=1}^{p} \sum_{j=1}^{q} a_{ij} \overline{b_{ij}} = \operatorname{trace} B^H A = \operatorname{trace} A B^H \quad \text{for } A, B \in \mathbb{C}^{p \times q}$$

show that it really is just the standard inner product for vectors in $\mathbb{C}^{pq}$ that are displayed as matrices.

Exercise 11.4. Let $\mathcal{U}$ denote the set of continuous complex-valued functions $f(t)$ on the finite closed interval $[a, b]$.

(a) Show that $\mathcal{U}$ is a complex vector space with respect to the natural rules of addition and multiplication by constants. Identify the zero element.

(b) Show that $\mathcal{U}$ is a normed linear space with respect to the norm $\|f\| = \left\{ \int_a^b |f(t)|^2 dt \right\}^{1/2}$.

(c) Show that $\mathcal{U}$ is an inner product space with respect to the inner product $\langle f, g \rangle = \int_a^b f(t)\overline{g(t)}dt$.

Lemma 11.1 (The Cauchy-Schwarz inequality for inner products).
Let $\mathcal{U}$ be an inner product space with inner product $\langle \mathbf{u}, \mathbf{v} \rangle$ for every pair of vectors $\mathbf{u}, \mathbf{v} \in \mathcal{U}$. Then

(11.4) $$|\langle \mathbf{u}, \mathbf{v} \rangle| \leq (\langle \mathbf{u}, \mathbf{u} \rangle)^{1/2} (\langle \mathbf{v}, \mathbf{v} \rangle)^{1/2},$$

with equality if and only if either (1) $\mathbf{v} = \mathbf{0}$ *and* $\mathbf{u}$ *is arbitrary or* (2) $\mathbf{v} \neq \mathbf{0}$ *and* $\mathbf{u} = \lambda \mathbf{v}$ *for some* $\lambda \in \mathbb{C}$.

Proof. The proof rests essentially on the fact that the inequality

$$0 \leq \langle \mathbf{u} - \lambda \mathbf{v}, \mathbf{u} - \lambda \mathbf{v} \rangle = \langle \mathbf{u}, \mathbf{u} \rangle - \overline{\lambda} \langle \mathbf{u}, \mathbf{v} \rangle - \lambda \langle \mathbf{v}, \mathbf{u} \rangle + |\lambda|^2 \langle \mathbf{v}, \mathbf{v} \rangle$$

is valid for every choice of $\lambda \in \mathbb{C}$. If $\mathbf{v} \neq \mathbf{0}$, then $\langle \mathbf{v}, \mathbf{v} \rangle > 0$ and we may set

$$\lambda = \frac{\langle \mathbf{u}, \mathbf{v} \rangle}{\langle \mathbf{v}, \mathbf{v} \rangle} \quad \text{to obtain} \quad \langle \mathbf{u} - \lambda \mathbf{v}, \mathbf{u} - \lambda \mathbf{v} \rangle = \frac{\langle \mathbf{u}, \mathbf{u} \rangle \langle \mathbf{v}, \mathbf{v} \rangle - |\langle \mathbf{u}, \mathbf{v} \rangle|^2}{\langle \mathbf{v}, \mathbf{v} \rangle},$$

which clearly justifies the inequality (11.4) and in addition shows that if equality prevails when $\mathbf{v} \neq \mathbf{0}$, then $\mathbf{u} + \lambda \mathbf{v} = \mathbf{0}$. On the other hand, if $\mathbf{v} = \mathbf{0}$, then equality holds in (11.4) for every vector $\mathbf{u} \in \mathcal{U}$.

It remains only to show that if $\mathbf{v} = \mathbf{0}$ or $\mathbf{v} \neq \mathbf{0}$ and $\mathbf{u} - \lambda \mathbf{v} = \mathbf{0}$, then equality holds in (11.4). But this is easy and is left to the reader. □

The condition for equality in the Cauchy-Schwarz inequality should not be overlooked; it is useful:

Example 11.1. If $\mathbf{x} \in \mathbb{C}^n$ with components x_i, $i = 1, \ldots, n$, then

(11.5) $\quad \|\mathbf{x}\|_1 \leq \sqrt{n}\|\mathbf{x}\|_2 \quad$ with equality if and only if $\quad |x_1| = \cdots = |x_n|$.

Discussion. Let $\mathbf{v}, \mathbf{a} \in \mathbb{C}^n$ with components $v_i = |x_i|$ and $a_i = 1$ for $i = 1, \ldots, n$, respectively. Then

$$\|\mathbf{x}\|_1 = \sum_{i=1}^n |x_i| = |\langle \mathbf{v}, \mathbf{a} \rangle| \leq \|\mathbf{v}\|_2 \|\mathbf{a}\|_2 = \sqrt{n} \|\mathbf{x}\|_2,$$

with equality if and only if $\mathbf{v} = \mu \mathbf{a}$ for some constant $\mu \geq 0$. ◇

Exercise 11.5. Show that if $f(t)$ and $g(t)$ are continuous complex-valued functions $f(t)$ on the finite closed interval $[a, b]$, then

$$\left| \int_a^b f(t)\overline{g(t)}dt \right|^2 \leq \int_a^b |f(t)|^2 dt \int_a^b |g(t)|^2 dt$$

with equality if and only if either $f(t) \equiv 0$ (i.e., $f(t) = 0$ for every point $t \in [a, b]$) or $f(t) \not\equiv 0$ and $g(t) = \beta f(t)$ for some point $\beta \in \mathbb{C}$.

Since an inner product space $\mathcal{U}$ is automatically a normed linear space with respect to the norm $\|\mathbf{u}\| = \{\langle \mathbf{u}, \mathbf{u}\rangle\}^{1/2}$, it is natural to ask whether or not the converse is true: Is every normed linear space automatically an inner product space? The answer is no, because the norm induced by the inner product has an extra property: It satisfies the **parallelogram law**:

(11.6) $$\|\mathbf{u}+\mathbf{v}\|^2 + \|\mathbf{u}-\mathbf{v}\|^2 = 2\|\mathbf{u}\|^2 + 2\|\mathbf{v}\|^2.$$

It can be shown that every normed linear space for which (11.6) holds is an inner product space and that the inner product is then specified in terms of the norm by the **polarization identity**

(11.7) $$\langle \mathbf{u}, \mathbf{v}\rangle = \frac{1}{4}\sum_{k=1}^{4} i^k \|\mathbf{u}+i^k\mathbf{v}\|^2 \quad \text{for complex spaces}$$

and

(11.8) $$\langle \mathbf{u}, \mathbf{v}\rangle = \frac{1}{4}\{\|\mathbf{u}+\mathbf{v}\|^2 - \|\mathbf{u}-\mathbf{v}\|^2\} \quad \text{for real spaces}.$$

11.2. Gram matrices

Let $\mathbf{v}_1, \ldots, \mathbf{v}_k$ be a set of vectors in an inner product space $\mathcal{U}$. Then the $k \times k$ matrix G with entries

(11.9) $$g_{ij} = \langle \mathbf{v}_j, \mathbf{v}_i\rangle_{\mathcal{U}} \quad \text{for} \quad i, j = 1, \ldots, k$$

is called the **Gram matrix** of the given set of vectors. It is easy to see that $G = G^H$ and, in terms of notation that will be discussed in Chapter 16,

(11.10) $$G \succeq O, \quad \text{i.e.,} \quad \langle G\mathbf{x}, \mathbf{x}\rangle_{\text{st}} \geq 0 \text{ for every } \mathbf{x} \in \mathbb{C}^k.$$

The notation $G \succ O$ signifies that $G \succeq O$ and G is invertible, i.e.,

(11.11) $$G \succ O \iff \langle G\mathbf{x}, \mathbf{x}\rangle_{\text{st}} > 0 \text{ for every nonzero } \mathbf{x} \in \mathbb{C}^k.$$

Lemma 11.2. *Let $\mathcal{U}$ be an inner product space and let G denote the Gram matrix of a set of vectors $\mathbf{v}_1, \ldots, \mathbf{v}_k$ in $\mathcal{U}$. Then $G \succ O$ if and only if the vectors $\mathbf{v}_1, \ldots, \mathbf{v}_k$ are linearly independent.*

Proof. Let $\mathbf{c}, \mathbf{d} \in \mathbb{C}^k$ with components $c_1, \ldots, c_k$ and $d_1, \ldots, d_k$, respectively, and let $\mathbf{v} = \sum_{j=1}^{k} c_j \mathbf{v}_j$ and $\mathbf{w} = \sum_{i=1}^{k} d_i \mathbf{v}_i$. Then

(11.12) $$\langle \mathbf{v}, \mathbf{w}\rangle_{\mathcal{U}} = \mathbf{d}^H G \mathbf{c} = \langle G\mathbf{c}, \mathbf{d}\rangle_{\text{st}}.$$

If G is invertible and $\sum_{j=1}^{k} c_j \mathbf{v}_j = \mathbf{0}$ for some choice of $c_1, \ldots, c_k \in \mathbb{C}$, then, in view of formula (11.12),

$$0 = \left\langle \sum_{j=1}^{k} c_j \mathbf{v}_j, \sum_{i=1}^{k} d_i \mathbf{v}_i \right\rangle_{\mathcal{U}} = \mathbf{d}^H G \mathbf{c} = \langle G\mathbf{c}, \mathbf{d}\rangle_{\text{st}}$$

for every choice of $d_1,\ldots,d_k \in \mathbb{C}$. Therefore, $G\mathbf{c} = \mathbf{0}$, which in turn implies that $\mathbf{c} = \mathbf{0}$, since G is invertible. Thus, the vectors $\mathbf{v}_1,\ldots,\mathbf{v}_k$ are linearly independent.

Suppose next that the vectors $\mathbf{v}_1,\ldots,\mathbf{v}_k$ are linearly independent and that $\mathbf{c} \in \mathcal{N}_G$. Then, by formula (11.12),
$$\left\langle \sum_{j=1}^k c_j \mathbf{v}_j, \sum_{i=1}^k c_i \mathbf{v}_i \right\rangle_\mathcal{U} = \mathbf{c}^H G \mathbf{c} = 0.$$
Therefore, $\sum_{j=1}^k c_j \mathbf{v}_j = \mathbf{0}$ and hence, in view of the presumed linear independence, $c_1 = \cdots = c_k = 0$. Thus, G is invertible. □

Exercise 11.6. Verify the assertions in (11.10).

Exercise 11.7. Verify formula (11.12).

Lemma 11.2 can be strengthened:

Theorem 11.3. *If G is the Gram matrix of a set of vectors $\mathbf{v}_1,\ldots,\mathbf{v}_k$ in an inner product space $\mathcal{U}$ and $\mathcal{V} = \mathrm{span}\{\mathbf{v}_1,\ldots,\mathbf{v}_k\}$, then*

(11.13) $$\mathrm{rank}\, G = \dim \mathrm{span}\{\mathbf{v}_1,\ldots,\mathbf{v}_k\} = \dim \mathcal{V}.$$

Proof. Suppose first that $\dim \mathcal{R}_G = r$, with $1 \leq r < k$. Then, there exists a permutation matrix P such that the first r columns of GP are a basis for $\mathcal{R}_G$. Consequently, the first r columns of $P^H G P$ are a basis for $\mathcal{R}_{P^H GP}$. Thus, if $P^H G P$ is written in block form as $[A_{ij}]$, $i,j = 1,2$, with $A_{11} = A_{11}^H \in \mathbb{C}^{r \times r}$, $A_{21} = A_{12}^H \in \mathbb{C}^{(k-r) \times r}$, and $A_{22} = A_{22}^H \in \mathbb{C}^{(k-r) \times (k-r)}$, then there exists a matrix $B \in \mathbb{C}^{r \times (k-r)}$ such that the following implications hold:
$$\begin{bmatrix} A_{12} \\ A_{22} \end{bmatrix} = \begin{bmatrix} A_{11} \\ A_{21} \end{bmatrix} B \implies \begin{bmatrix} A_{11} \\ A_{21} \end{bmatrix} = \begin{bmatrix} I_r \\ B^H \end{bmatrix} A_{11} \implies \mathrm{rank}\begin{bmatrix} A_{11} \\ A_{21} \end{bmatrix} = \mathrm{rank}\, A_{11}.$$
Thus, as A_{11} can be identified as the Gram matrix of a set $\{\mathbf{v}_{i_1},\ldots,\mathbf{v}_{i_r}\}$ of r linearly independent vectors,
$$\mathrm{rank}\, G = \dim \mathrm{span}\{\mathbf{v}_{i_1},\ldots,\mathbf{v}_{i_r}\} \leq \dim \mathcal{V}.$$

Conversely, if $\dim \mathcal{V} = r$ and $\{\mathbf{v}_{i_1},\ldots,\mathbf{v}_{i_r}\}$ is a basis for $\mathcal{V}$, then the Gram matrix of this set of vectors is invertible. Thus, as this Gram matrix is a submatrix of G, it follows that
$$r = \dim \mathcal{V} \leq \mathrm{rank}\, G.$$
This completes the proof. □

Exercise 11.8. Show that if the second and fourth columns of a matrix $A \in \mathbb{C}^{4 \times 4}$ are linearly independent and if $A = A^H$ and $\mathrm{rank}\, A = 2$, then $\begin{bmatrix} a_{22} & a_{24} \\ a_{42} & a_{44} \end{bmatrix}$ is invertible.

Exercise 11.9. Show that if $\{\mathbf{v}_1,\ldots,\mathbf{v}_r\}$ is a basis for the space $\mathcal{V}$ introduced in the proof of Theorem 11.3 and $1 \leq r < k$, then there exists a matrix $A \in \mathbb{C}^{s \times r}$ with $s = k - r$ such that the columns of $\begin{bmatrix} A & I_s \end{bmatrix}^T$ are linearly independent and belong to $\mathcal{N}_G$.

Theorem 11.4. *If $\mathbb{C}^n$ is equipped with an inner product $\langle \mathbf{x}, \mathbf{y}\rangle_\mathcal{U}$ and G is the Gram matrix with entries $g_{ij} = \langle \mathbf{e}_j, \mathbf{e}_i\rangle_\mathcal{U}$ based on the columns of I_n, then (in terms of the notation (11.11)), $G \succ O$ and*

(11.14) $\qquad \langle \mathbf{x}, \mathbf{y}\rangle_\mathcal{U} = \langle G\mathbf{x}, \mathbf{y}\rangle_{\mathrm{st}} \quad$ *for every choice of* $\mathbf{x}, \mathbf{y} \in \mathbb{C}^n$.

Proof. The proof is easy and is left to the reader. □

11.3. Adjoints

If T is a linear transformation from a finite-dimensional inner product space $\mathcal{U}$ into a finite-dimensional inner product space $\mathcal{V}$, then (as is spelled out in Exercises 11.10 and 11.11 below) there exists exactly one transformation T^* from $\mathcal{V}$ into $\mathcal{U}$ such that

(11.15) $\qquad \langle T\mathbf{u}, \mathbf{v}\rangle_\mathcal{V} = \langle \mathbf{u}, T^*\mathbf{v}\rangle_\mathcal{U} \quad$ for every choice of $\mathbf{u} \in \mathcal{U}$ and $\mathbf{v} \in \mathcal{V}$.

The transformation T^* is called the **adjoint** of T; it is automatically linear and depends upon the inner products. Moreover, if T_1 and T_2 are linear transformations from $\mathcal{U}$ into $\mathcal{V}$ and $\alpha \in \mathbb{C}$, then:

(1) $(\alpha T_1 + T_2)^* = \overline{\alpha} T_1^* + T_2^*, \quad (T^*)^* = T, \quad$ and $\quad \|T\|_{\mathcal{U},\mathcal{V}} = \|T^*\|_{\mathcal{V},\mathcal{U}}$.
(2) $\|T\|_{\mathcal{U},\mathcal{V}} = \max\{|\langle T\mathbf{u}, \mathbf{v}\rangle_\mathcal{V}| : \mathbf{u} \in \mathcal{U}, \mathbf{v} \in \mathcal{V}, \text{ and } \|\mathbf{u}\|_\mathcal{U} = \|\mathbf{v}\|_\mathcal{V} = 1\}$.
(3) If $\dim \mathcal{U} = \dim \mathcal{V}$, then $T^*T = I_\mathcal{U} \iff TT^* = I_\mathcal{V}$.

Discussion. The first two equalities in (1) are easy consequences of the definition of the adjoint; the third follows from (2). To verify (2), observe first that the Cauchy-Schwarz inequality ensures that the right-hand side of the asserted equality in (2) is always $\leq \|T\|_{\mathcal{U},\mathcal{V}}$. To achieve equality, fix a vector $\mathbf{u}_0 \in \mathbb{C}^q$ such that $\|\mathbf{u}_0\|_\mathcal{U} = 1$ and $\|T\mathbf{u}_0\|_\mathcal{V} = \|T\|_{\mathcal{U},\mathcal{V}}$ and choose $\mathbf{v} = T\mathbf{u}_0/\|T\mathbf{u}_0\|_\mathcal{V}$.

To justify (3), it suffices to verify the implication $\Longrightarrow$. If $T^*T = I_\mathcal{U}$, then $\dim \mathcal{U} = \dim \mathcal{R}_T$, since $\mathcal{N}_T = \{\mathbf{0}\}$. Thus, if also $\dim \mathcal{U} = \dim \mathcal{V}$, then T maps $\mathcal{U}$ onto $\mathcal{V}$. Therefore, for each vector $\mathbf{v} \in \mathcal{V}$, there exists exactly one vector $\mathbf{u} \in \mathcal{U}$ such that $T\mathbf{u} = \mathbf{v}$. Consequently, $T^*\mathbf{v} = T^*T\mathbf{u} = \mathbf{u}$, and hence $TT^*\mathbf{v} = T\mathbf{u} = \mathbf{v}$ for every vector $\mathbf{v} \in \mathcal{V}$. □

Example 11.2. The **most important example** for us is when $\mathcal{U} = \mathbb{C}^q$ and $\mathcal{V} = \mathbb{C}^p$. In keeping with the definition for linear transformations, we shall say that the matrix $A^* \in \mathbb{C}^{q \times p}$ is the **adjoint** of a matrix $A \in \mathbb{C}^{p \times q}$ if

(11.16) $\qquad \langle A\mathbf{u}, \mathbf{v}\rangle_\mathcal{V} = \langle \mathbf{u}, A^*\mathbf{v}\rangle_\mathcal{U} \quad$ for every choice of $\mathbf{u} \in \mathbb{C}^q$ and $\mathbf{v} \in \mathbb{C}^p$.

11.3. Adjoints

In view of Theorem 11.4, (11.16) can be expressed in terms of the Gram matrices B and C of the columns of I_q in $\mathcal{U}$ and the columns of I_p in $\mathcal{V}$ as

(11.17) $\quad \langle CA\mathbf{u}, \mathbf{v} \rangle_{\text{st}} = \langle B\mathbf{u}, A^*\mathbf{v} \rangle_{\text{st}} \quad$ for every $\mathbf{u} \in \mathbb{C}^q$ and $\mathbf{v} \in \mathbb{C}^p$.

But, as $\langle X\mathbf{u}, \mathbf{v} \rangle_{\text{st}} = \mathbf{v}^H X \mathbf{u} = (X^H \mathbf{v})^H \mathbf{u} = \langle \mathbf{u}, X^H \mathbf{v} \rangle_{\text{st}}$,

(11.18) $\qquad\qquad$ (11.17) holds if and only if $A^* = B^{-1} A^H C$.

If $\mathbb{C}^q$ and $\mathbb{C}^p$ are both equipped with the standard inner product, then $B = I_q$, $C = I_p$ and hence $A^* = A^H$. $\qquad\qquad\qquad\qquad\qquad\quad\diamond$

Exercise 11.10. Let T be a linear transformation from an inner product space $\mathcal{U}$ with basis $\{\mathbf{u}_1, \ldots, \mathbf{u}_q\}$ and Gram matrix $G_\mathcal{U}$ into an inner product space $\mathcal{V}$ with basis $\{\mathbf{v}_1, \ldots, \mathbf{v}_p\}$ and Gram matrix $G_\mathcal{V}$ that is defined in terms of the entries a_{ij} of a matrix $A \in \mathbb{C}^{p \times q}$ by the formula $T\mathbf{u}_i = \sum_{k=1}^{p} a_{ki} \mathbf{v}_k$ for $i = 1, \ldots, q$, and let S be a linear transformation from $\mathcal{U}$ into $\mathcal{V}$ that is defined in terms of the entries b_{ij} of a matrix $B \in \mathbb{C}^{q \times p}$ by the formula $S\mathbf{v}_j = \sum_{k=1}^{q} b_{kj} \mathbf{u}_k$ for $j = 1, \ldots, p$. Show that $\langle T\mathbf{u}, \mathbf{v} \rangle_\mathcal{V} = \langle \mathbf{u}, S\mathbf{v} \rangle_\mathcal{U}$ for every choice of $\mathbf{u} \in \mathcal{U}$ and $\mathbf{v} \in \mathcal{V}$ if and only if $B = G_\mathcal{U}^{-1} A^H G_\mathcal{V}$.

Exercise 11.11. Use Exercise 11.10 to show that if T is a linear transformation from a finite-dimensional inner product space $\mathcal{U}$ into a finite-dimensional inner product space $\mathcal{V}$, then there exists at least one transformation T^* from $\mathcal{V}$ into $\mathcal{U}$ such that (11.15) holds and then show that if S is any transformation from $\mathcal{V}$ into $\mathcal{U}$ such that $\langle T\mathbf{u}, \mathbf{v} \rangle_\mathcal{V} = \langle \mathbf{u}, S\mathbf{v} \rangle_\mathcal{U}$ for every choice of $\mathbf{u} \in \mathcal{U}$ and $\mathbf{v} \in \mathcal{V}$, then $S = T^*$.

Example 11.3. If $\mathcal{U} = \mathbb{C}^{p \times q}$ is equipped with the inner product $\langle A, B \rangle_\mathcal{U} = \text{trace}\, B^H A$, $\mathcal{V} = \mathbb{C}^p$ is equipped with the standard inner product, and $\mathbf{u} \in \mathbb{C}^q$, then the adjoint T^* of the linear transformation T from $\mathcal{U}$ into $\mathcal{V}$ that is defined by the formula $TA = A\mathbf{u}$ for every $A \in \mathcal{U}$ must satisfy the identity

$$\langle TA, \mathbf{v} \rangle_\mathcal{V} = \langle A, T^*\mathbf{v} \rangle_\mathcal{U}$$

for every choice of $A \in \mathbb{C}^{p \times q}$ and $\mathbf{v} \in \mathbb{C}^p$. Thus,

$$\begin{aligned} \langle TA, \mathbf{v} \rangle_\mathcal{V} &= \langle A\mathbf{u}, \mathbf{v} \rangle_\mathcal{V} = \mathbf{v}^H A \mathbf{u} = \text{trace}\{\mathbf{v}^H A \mathbf{u}\} \\ &= \text{trace}\{\mathbf{u}\mathbf{v}^H A\} = \langle A, \mathbf{v}\mathbf{u}^H \rangle_\mathcal{U}, \end{aligned}$$

i.e., $\langle A, T^*\mathbf{v} \rangle_\mathcal{U} = \langle A, \mathbf{v}\mathbf{u}^H \rangle_\mathcal{U}$ for every $\mathbf{v} \in \mathbb{C}^p$. Therefore, $T^*\mathbf{v} = \mathbf{v}\mathbf{u}^H$ for every $\mathbf{v} \in \mathcal{V}$. $\qquad\qquad\qquad\qquad\qquad\qquad\qquad\qquad\qquad\qquad\quad\diamond$

Exercise 11.12. Let $\mathcal{U} = \mathbb{C}^n$ equipped with the inner product $\langle \mathbf{u}, \mathbf{v} \rangle_\mathcal{U} = \sum_{j=1}^{n} j \overline{v_j} u_j$ for vectors $\mathbf{u}, \mathbf{v} \in \mathbb{C}^n$ with components $u_1, \ldots, u_n$ and $v_1, \ldots, v_n$, respectively. Find the adjoint A^* of a matrix $A \in \mathbb{C}^{n \times n}$ with respect to this inner product.

11.4. Spectral radius

The next theorem provides a remarkable connection between the growth of the numbers $\|A^n\|$ and the **spectral radius**

(11.19) $$r_\sigma(A) = \max\{|\lambda| : \lambda \in \sigma(A)\} \quad \text{for} \quad A \in \mathbb{C}^{p\times p}.$$

Theorem 11.5. *If $A \in \mathbb{C}^{p\times p}$ and $\|A\| = \|A\|_{2,2}$, then*

(11.20) $$\lim_{n\uparrow\infty} \|A^n\|^{1/n} = r_\sigma(A);$$

i.e., the indicated limit exists and is equal to the spectral radius of A.

Proof. To verify (11.20), it suffices to justify the inequalities

(11.21) $$r_\sigma(A) \leq \|A^n\|^{1/n} \leq r_\sigma(A)(1 + \kappa_n),$$

for every positive integer n, where $\kappa_n \geq 0$ tends to zero as $n \uparrow \infty$.

The lower bound is easy: If $A\mathbf{x} = \lambda \mathbf{x}$ for some nonzero vector $\mathbf{x} \in \mathbb{C}^p$, then, since $A^n \mathbf{x} = \lambda^n \mathbf{x}$, it is readily seen that

$$|\lambda^n|\|\mathbf{x}\|_2 = \|A^n \mathbf{x}\|_2 \leq \|A^n\|\|\mathbf{x}\|_2$$

and hence that $|\lambda^n| \leq \|A^n\|$ for every $\lambda \in \sigma(A)$. Therefore, $r_\sigma(A) \leq \|A^n\|^{1/n}$ for every positive integer n.

To verify the upper bound in (11.21), we first invoke the Jordan decomposition theorem, which ensures that there exists an invertible matrix $U \in \mathbb{C}^{p\times p}$ such that $A = UJU^{-1}$. Therefore,

$$\|A^n\| = \|UJ^n U^{-1}\| \leq \|U\|\|J^n\|\|U^{-1}\|$$

and

(11.22) $$\|A^n\|^{1/n} \leq \|U\|^{1/n}\|J^n\|^{1/n}\|U^{-1}\|^{1/n} = \|J^n\|^{1/n}(1 + \varepsilon_n),$$

where

$$(1 + \varepsilon_n) = \{\|U\|\|U^{-1}\|\}^{1/n} \geq \|UU^{-1}\|^{1/n} = 1$$

and $\varepsilon_n \downarrow 0$ as $n \uparrow \infty$.

To obtain an upper bound on $\|J^n\|$, it suffices to obtain an upper bound on $\|(C_\mu^{(k)})^n\|$ for every Jordan cell $C_\mu^{(k)} = \mu I_k + N$ (with $N = C_0^{(k)}$) that appears in J. But, if $\mu \in \sigma(A)$ and $n > p$, then, as $N^k = O$,

$$\begin{aligned}
\|(C_\mu^{(k)})^n\| &= \left\|\sum_{j=0}^{n} \binom{n}{j} \mu^{n-j} N^j\right\| = \left\|\sum_{j=0}^{k-1} \binom{n}{j} \mu^{n-j} N^j\right\| \\
&\leq \sum_{j=0}^{k-1} \binom{n}{j} |\mu|^{n-j} \leq \sum_{j=0}^{k-1} n^j (r_\sigma(A))^{n-j} \leq \sum_{j=0}^{p-1} n^j (r_\sigma(A))^{n-j} \\
&\leq (r_\sigma(A))^n (n/r_\sigma(A))^p \quad \text{if } n \geq 2r_\sigma(A) \text{ and } p \geq 1.
\end{aligned}$$

Therefore,

(11.23) $$\|J^n\|^{1/n} \leq r_\sigma(A)(1+\delta_n),$$

where

$$1+\delta_n = \{n/r_\sigma(A)\}^{p/n} = \exp\left\{\frac{p}{n}\ln[n/r_\sigma(A)]\right\} \to 1 \quad \text{as} \quad n \uparrow \infty.$$

The bounds (11.22) and (11.23) imply that $\|A^n\|^{1/n} \leq r_\sigma(A)(1+\delta_n)(1+\varepsilon_n)$. Therefore, $0 \leq \|A\|^{1/n} - r_\sigma(A) \leq r_\sigma(A)(\varepsilon_n + \delta_n + \varepsilon_n\delta_n)$, which serves to complete the proof, since $\varepsilon_n + \delta_n + \varepsilon_n\delta_n$ tends to 0 as $n \uparrow \infty$. □

Theorem 11.6. *If $A, B \in \mathbb{C}^{p \times p}$ and $AB = BA$, then:*

(1) $\sigma(A+B) \subseteq \sigma(A) + \sigma(B)$.

(2) $r_\sigma(A+B) \leq r_\sigma(A) + r_\sigma(B)$.

(3) $r_\sigma(AB) \leq r_\sigma(A)r_\sigma(B)$.

Proof. Let **u** be an eigenvector of $A + B$ corresponding to the eigenvalue μ. Then

$$(A+B)\mathbf{u} = \mu\mathbf{u}$$

and hence, since $BA = AB$,

$$(A+B)B\mathbf{u} = B(A+B)\mathbf{u} = \mu B\mathbf{u};$$

that is to say, $\mathcal{N}_{(A+B-\mu I_p)}$ is a nonzero subspace of $\mathbb{C}^p$ that is invariant under B. Therefore, by Theorem 4.5, there exists an eigenvector **v** of B in this null space, i.e., there exists a nonzero vector $\mathbf{v} \in \mathbb{C}^p$ such that

$$(A+B)\mathbf{v} = \mu\mathbf{v} \text{ and } B\mathbf{v} = \beta\mathbf{v}$$

for some $\beta \in \mathbb{C}$. But this in turn implies that $\beta \in \sigma(B)$ and

$$A\mathbf{v} = (\mu - \beta)\mathbf{v},$$

i.e., the number $\alpha = \mu - \beta$ is an eigenvalue of A. Thus we have shown that

$$\mu \in \sigma(A+B) \Longrightarrow \mu = \alpha + \beta, \quad \text{where} \quad \alpha \in \sigma(A) \text{ and } \beta \in \sigma(B).$$

Therefore, (1) holds. Moreover, (2) is an immediate consequence of (1) and the definition of spectral radius; (3) is left to the reader as an exercise. □

Exercise 11.13. Verify the third assertion in Theorem 11.6.

Exercise 11.14. Verify the second assertion in Theorem 11.6 by estimating $\|(A+B)^n\|$ with the aid of the binomial theorem. [REMARK: This is not as easy as the proof furnished above, but it has the advantage of being applicable in wider circumstances.]

Exercise 11.15. Show that if $A, B \in \mathbb{C}^{n \times n}$, then $r_\sigma(AB) = r_\sigma(BA)$, even if $AB \neq BA$. [HINT: Recall formula (7.4).]

Exercise 11.16. Show that if $A = \begin{bmatrix} 1 & 1 \\ 0 & 0 \end{bmatrix}$ and $B = \begin{bmatrix} 1 & 0 \\ 1 & 0 \end{bmatrix}$, then
$$r_\sigma(AB) > r_\sigma(A)\, r_\sigma(B) \quad \text{and} \quad r_\sigma(A+B) > r_\sigma(A) + r_\sigma(B)\,.$$

Exercise 11.17. Show that if $A, B \in \mathbb{C}^{n \times n}$, then $r_\sigma(A+B) \leq r_\sigma(A) + \|B\|$, even if the two matrices do not commute.

11.5. What you need to know about $\|A\|$

- **Warning:** From now on, we adopt the convention that, unless specified otherwise, $\|A\| = \|A\|_{2,2}$ for matrices $A \in \mathbb{C}^{p \times q}$ and $\|\mathbf{x}\| = \|\mathbf{x}\|_2$ for vectors $\mathbf{x} \in \mathbb{C}^q$, for every choice of the positive integers p and q. The main properties of $\|A\|$ are:

  (1) $\|AB\| \leq \|A\|\,\|B\|$ for $B \in \mathbb{C}^{q \times r}$.
  (2) $\|I_p\| = 1$.
  (3) The inequality $\|A\mathbf{u}\|_2 \leq \|A\|\,\|\mathbf{u}\|_2$ is in force for every $\mathbf{u} \in \mathbb{C}^q$.
  (4) $\|A\| = \max\{|\langle A\mathbf{x}, \mathbf{y}\rangle| : \mathbf{x} \in \mathbb{C}^q,\ \mathbf{y} \in \mathbb{C}^p,\ \text{and } \|\mathbf{x}\|_2 = \|\mathbf{y}\|_2 = 1\}$.
  (5) $\|A\| = \|A^H\|$ (this is an easy consequence of the formula in (4)).
  (6) $\|(A^H A)^k\| = \|A\|^{2k}$ and $\|A(A^H A)^k\| = \|A\|^{2k+1}$ for $k = 1, 2, \ldots$.
  (7) If $V \in \mathbb{C}^{n \times p}$, $V^H V = I_p$, $U \in \mathbb{C}^{r \times q}$, and $U^H U = I_q$, then $\|V A U^H\| = \|A\|$.

Exercise 11.18. Verify the first formula in (6) when $k = 1$. [HINT: $\|A^H A\| \geq \max\{\langle A^H A \mathbf{x}, \mathbf{x}\rangle : \|\mathbf{x}\| = 1\} = \max\{\|A\mathbf{x}\|^2 : \|\mathbf{x}\| = 1\} = \|A\|^2$.]

Exercise 11.19. Verify the second formula in (6) for $k = 1$. [HINT: $\|A\|^4 = \|A^H A A^H A\| \leq \|A^H\|\,\|A A^H A\| \leq \|A\|^4$.]

Exercise 11.20. Show that if $\|(A^H A)^k\| = \|A\|^{2k}$ for some integer $k \geq 2$, then $\|(A^H A)^{k-1}\| = \|A\|^{2(k-1)}$ and $\|A(A^H A)^{k-1}\| = \|A\|^{2k-1}$. [HINT: $\|A\|^{2k} = \|A^H A (A^H A)^{k-1}\| \leq \|A\|^2\, \|(A^H A)^{k-1}\| \leq \|A\|^{2k}$, for the first.]

Exercise 11.21. Verify the formulas in (6) for $k = 1, 2, \ldots$. [HINT: Exploit the implications in Exercises 11.18–11.20.]

Exercise 11.22. Verify the formula in (7).

11.6. Supplementary notes

This chapter is partially adapted from Chapter 8 of [30]. Formula (11.20) is valid in a much wider context than was considered here; see, e.g., Chapter 18 of Rudin [66]. Lemma 42.1 and the surrounding discussion is a good supplement to the section on adjoints. Example 11.3 is adapted from the monograph by Borwein and Lewis [13].

11.6. Supplementary notes

The next few exercises deal with variations of the polarization identities exhibited in (11.7) and (11.8).

Exercise 11.23. Show that if $G \in \mathbb{C}^{n \times n}$, then

$$(11.24) \quad \mathbf{v}^H G \mathbf{u} = \frac{1}{4} \sum_{k=1}^{4} i^k (\mathbf{u} + i^k \mathbf{v})^H G (\mathbf{u} + i^k \mathbf{v}) \quad \text{for every } \mathbf{u}, \mathbf{v} \in \mathbb{C}^n.$$

Exercise 11.24. Show that if $G \in \mathbb{C}^{n \times n}$, then

$$(11.25) \quad \mathbf{x}^H G \mathbf{x} = 0 \text{ for every } \mathbf{x} \in \mathbb{C}^n \implies G = O.$$

[HINT: Exploit Exercise 11.23.]

Exercise 11.25. Show that if $G \in \mathbb{R}^{n \times n}$, then

$$(11.26) \quad \mathbf{v}^H (G + G^T) \mathbf{u} = \frac{1}{2} \{ (\mathbf{u} + \mathbf{v})^T G (\mathbf{u} + \mathbf{v}) - (\mathbf{u} - \mathbf{v})^T G (\mathbf{u} - \mathbf{v}) \}$$

for every choice of $\mathbf{u}, \mathbf{v} \in \mathbb{R}^n$.

Exercise 11.26. Show that if $G \in \mathbb{R}^{n \times n}$, then

$$(11.27) \quad \mathbf{x}^H G \mathbf{x} = 0 \text{ for every } \mathbf{x} \in \mathbb{R}^n \text{ does not imply that } G = O.$$

However,

$$(11.28) \quad G = G^T \text{ and } \mathbf{x}^H G \mathbf{x} = 0 \text{ for every } \mathbf{x} \in \mathbb{R}^n \implies G = O.$$

Chapter 12

Orthogonality

In this chapter we shall discuss orthogonality in inner product spaces.

Recall that the cosine of the angle θ between the line segment running from $\mathbf{0}$ to $\mathbf{a} = (a_1, a_2, a_3)$ and the line segment running from $\mathbf{0}$ to $\mathbf{b} = (b_1, b_2, b_3)$ for a pair of points $\mathbf{a}$ and $\mathbf{b}$ in the first octant in $\mathbb{R}^3$ is

$$(12.1) \qquad \cos\theta = \frac{\|\mathbf{a}\|^2 + \|\mathbf{b}\|^2 - \|\mathbf{b}-\mathbf{a}\|^2}{2\|\mathbf{a}\|\|\mathbf{b}\|} = \frac{\sum_{i=1}^{3} a_i b_i}{\sqrt{\sum_{i=1}^{3} a_i^2}\sqrt{\sum_{i=1}^{3} b_i^2}}.$$

Thus, in terms of the standard inner product $\langle \mathbf{a}, \mathbf{b} \rangle = \sum_{i=1}^{3} a_i b_i$ in $\mathbb{R}^3$,

$$\langle \mathbf{a}, \mathbf{b} \rangle = 0 \iff \cos\theta = 0 \iff \theta = \pi/2 \iff \mathbf{a} \perp \mathbf{b}.$$

12.1. Orthogonality

Formula (12.1) (the law of cosines) serves to motivate the following definitions in an inner product space $\mathcal{U}$ with inner product $\langle \mathbf{u}, \mathbf{v} \rangle_\mathcal{U}$:

- **Orthogonal vectors**: A pair of vectors $\mathbf{u}$ and $\mathbf{v}$ in $\mathcal{U}$ is said to be **orthogonal** if $\langle \mathbf{u}, \mathbf{v} \rangle_\mathcal{U} = 0$.
- **Orthogonal family**: A set of nonzero vectors $\{\mathbf{u}_1, \ldots, \mathbf{u}_k\}$ in $\mathcal{U}$ is said to be an **orthogonal family** if

$$\langle \mathbf{u}_i, \mathbf{u}_j \rangle_\mathcal{U} = 0 \quad \text{when } i \neq j.$$

The assumption that none of the vectors $\mathbf{u}_1, \ldots, \mathbf{u}_k$ are equal to $\mathbf{0}$ serves to guarantee that they are automatically linearly independent.

- **Orthonormal family**: A set of vectors $\mathbf{u}_1, \ldots, \mathbf{u}_k$ in $\mathcal{U}$ is said to be an **orthonormal family** if:
  (1) $\langle \mathbf{u}_i, \mathbf{u}_j \rangle_{\mathcal{U}} = 0$ for $i, j = 1, \ldots, k$ and $i \neq j$ and
  (2) $\|\mathbf{u}_i\|_{\mathcal{U}}^2 = \langle \mathbf{u}_i, \mathbf{u}_i \rangle_{\mathcal{U}} = 1$ for $i = 1, \ldots, k$.
- **Orthogonal decomposition**: A pair of subspaces $\mathcal{V}$ and $\mathcal{W}$ of $\mathcal{U}$ is said to form an **orthogonal decomposition** of $\mathcal{U}$ if:
  (1) $\mathcal{V} + \mathcal{W} = \mathcal{U}$ and
  (2) $\langle \mathbf{v}, \mathbf{w} \rangle_{\mathcal{U}} = 0$ for every $\mathbf{v} \in \mathcal{V}$ and $\mathbf{w} \in \mathcal{W}$.
  Orthogonal decompositions will be indicated by the symbol
  $$\mathcal{U} = \mathcal{V} \oplus \mathcal{W}.$$
- **Orthogonal complement**: If $\mathcal{V}$ is a subspace of an inner product space $\mathcal{U}$, then the set

(12.2) $$\mathcal{V}^{\perp} = \{\mathbf{u} \in \mathcal{U} : \langle \mathbf{u}, \mathbf{v} \rangle_{\mathcal{U}} = 0 \quad \text{for every} \quad \mathbf{v} \in \mathcal{V}\}$$

  is referred to as the **orthogonal complement** of $\mathcal{V}$ in $\mathcal{U}$. It is a subspace of $\mathcal{U}$.

- **Orthonormal expansions**: An orthonormal expansion of a vector $\mathbf{u} \in \mathcal{U}$ is a linear combination of vectors $\mathbf{u} = \sum_{j=1}^{k} c_j \mathbf{u}_j$ wherein the vectors $\mathbf{u}_1, \ldots, \mathbf{u}_k$ are **orthonormal**. The advantage of orthonormal expansions is that the computation of the coefficients $c_1, \ldots, c_k$ and the evaluation of $\langle \mathbf{u}, \mathbf{u} \rangle_{\mathcal{U}}$ is now easy:

(12.3) $$\langle \mathbf{u}, \mathbf{u}_i \rangle_{\mathcal{U}} = \sum_{j=1}^{k} c_j \langle \mathbf{u}_j, \mathbf{u}_i \rangle_{\mathcal{U}} = c_i \quad \text{for } i = 1, \ldots, k,$$

and

(12.4) $$\langle \mathbf{u}, \mathbf{u} \rangle_{\mathcal{U}} = \left\langle \sum_{j=1}^{k} c_j \mathbf{u}_j, \sum_{i=1}^{k} c_i \mathbf{u}_i \right\rangle_{\mathcal{U}} = \sum_{i=1}^{k} \overline{c_i} \left\langle \sum_{j=1}^{k} c_j \mathbf{u}_j, \mathbf{u}_i \right\rangle_{\mathcal{U}}$$
$$= \sum_{i=1}^{k} |c_i|^2.$$

Moreover, if $\mathbf{w} = \sum_{j=1}^{k} d_j \mathbf{u}_j$, then, by much the same sort of calculation,

(12.5) $$\langle \mathbf{u}, \mathbf{w} \rangle_{\mathcal{U}} = \sum_{i=1}^{n} \overline{d_i} c_i.$$

Exercise 12.1. Show that every orthogonal sum decomposition is a direct sum decomposition and give an example of a direct sum decomposition that is not an orthogonal decomposition.

Exercise 12.2. Show that if $\{\mathbf{u}_1,\ldots,\mathbf{u}_k\}$ is an orthogonal family of nonzero vectors in an inner product space $\mathcal{U}$, then $\mathbf{u}_1,\ldots,\mathbf{u}_k$ are linearly independent.

Exercise 12.3. Show that if $A \in \mathbb{C}^{p \times q}$ and $\mathbb{C}^q$ and $\mathbb{C}^p$ are both equipped with the standard inner product, then

(12.6) $$\mathbb{C}^q = \mathcal{R}_{A^H} \oplus \mathcal{N}_A \quad \text{and} \quad \mathbb{C}^p = \mathcal{R}_A \oplus \mathcal{N}_{A^H}.$$

12.2. Projections and direct sums

Recall that the sum $\mathcal{V} + \mathcal{W} = \{\mathbf{v} + \mathbf{w} : \mathbf{v} \in \mathcal{V} \text{ and } \mathbf{w} \in \mathcal{W}\}$ of a pair of subspaces $\mathcal{V}$ and $\mathcal{W}$ of a vector space $\mathcal{U}$ is direct if $\mathcal{V} \cap \mathcal{W} = \{\mathbf{0}\}$. In this section we shall establish a correspondence between the decomposition $\mathcal{U} = \mathcal{V} \dotplus \mathcal{W}$ of a vector space $\mathcal{U}$ as a direct sum and a special class of linear transformations from $\mathcal{U}$ into $\mathcal{U}$ that are called projections.

- **Projections:** A linear transformation T of a vector space $\mathcal{U}$ into itself is said to be a **projection** if $T^2 = T$.

Lemma 12.1. *If a linear transformation T from a vector space $\mathcal{U}$ into itself is a projection, then*

(12.7) $$\mathcal{U} = \mathcal{R}_T \dotplus \mathcal{N}_T.$$

Proof. Let $\mathbf{x} \in \mathcal{U}$. Then clearly

$$\mathbf{x} = T\mathbf{x} + (I - T)\mathbf{x}$$

and $T\mathbf{x} \in \mathcal{R}_T$. Moreover, $(I - T)\mathbf{x} \in \mathcal{N}_T$, since

$$T(I - T)\mathbf{x} = (T - T^2)\mathbf{x} = (T - T)\mathbf{x} = \mathbf{0}.$$

Thus,

$$\mathcal{U} = \mathcal{R}_T + \mathcal{N}_T.$$

The sum is direct because

$$\mathbf{y} \in \mathcal{R}_T \iff \mathbf{y} = T\mathbf{y} \quad \text{and} \quad \mathbf{y} \in \mathcal{N}_T \iff T\mathbf{y} = \mathbf{0}.$$

(The key to the first equivalence is $\mathbf{y} = T\mathbf{x} \implies T\mathbf{y} = T^2\mathbf{x} = T\mathbf{x} = \mathbf{y}$.) □

Lemma 12.1 exhibits $\mathcal{U}$ as the direct sum of the spaces $\mathcal{V} = \mathcal{R}_T$ and $\mathcal{W} = \mathcal{N}_T$ that are defined in terms of a given projection T. Conversely, the complementary spaces in any given direct sum decomposition $\mathcal{U} = \mathcal{V} \dotplus \mathcal{W}$ may be identified as the range and null space, respectively, of a projection

T, i.e., $\mathcal{V} = \mathcal{R}_T$ and $\mathcal{W} = \mathcal{N}_T$:

Lemma 12.2. *Let $\mathcal{V}$ and $\mathcal{W}$ be subspaces of a vector space $\mathcal{U}$ and suppose that $\mathcal{U} = \mathcal{V} \dotplus \mathcal{W}$. Then:*

(1) *For every vector $\mathbf{u} \in \mathcal{U}$ there exists exactly one vector $\mathbf{v} \in \mathcal{V}$ such that $\mathbf{u} - \mathbf{v} \in \mathcal{W}$.*

The transformation T that maps $\mathbf{u} \in \mathcal{U}$ into the unique vector $\mathbf{v} \in \mathcal{V}$ considered in (1) enjoys the following properties:

(2) $\mathcal{R}_T = \mathcal{V}$ *and* $T\mathbf{v} = \mathbf{v}$ *for every vector* $\mathbf{v} \in \mathcal{V}$; $\mathcal{N}_T = \mathcal{R}_{(I-T)} = \mathcal{W}$ *and* $(I - T)\mathbf{w} = \mathbf{w}$ *for every vector* $\mathbf{w} \in \mathcal{W}$.

(3) T *is linear and* $T^2 = T$.

Proof. The first assertion is immediate from the definition of a direct sum decomposition.

To verify (2), suppose first that $\mathbf{v} \in \mathcal{V}$. Then, since $\mathbf{v} = \mathbf{v} + \mathbf{0}$ and $\mathbf{0} \in \mathcal{W}$, it follows that $\mathbf{v} = T\mathbf{v}$ and hence that $\mathcal{V} \subseteq \mathcal{R}_T$. Thus, as $\mathcal{R}_T \subseteq \mathcal{V}$ by definition, the first equality in (2) must hold. To get the second, let $\mathbf{u} \in \mathcal{N}_T$ and write $\mathbf{u} = \mathbf{v} + \mathbf{w}$ with $\mathbf{v} \in \mathcal{V}$ and $\mathbf{w} \in \mathcal{W}$. Then, since $\mathbf{0} = T\mathbf{u} = \mathbf{v}$, it follows that $\mathcal{N}_T \subseteq \mathcal{W}$. Thus, as the opposite inclusion follows from the equality $\mathbf{w} = \mathbf{0} + \mathbf{w}$, the proof of (2) is complete.

Suppose next that $\mathbf{u}_1 = \alpha \mathbf{v}_1 + \mathbf{w}_1$ and $\mathbf{u}_2 = \mathbf{v}_2 + \mathbf{w}_2$ with $\mathbf{v}_1, \mathbf{v}_2 \in \mathcal{V}$, $\mathbf{w}_1, \mathbf{w}_2 \in \mathcal{W}$, and a scalar α. Then, as $\alpha \mathbf{v}_1 + \mathbf{v}_2 \in \mathcal{V}$ and $\mathbf{w}_1 + \mathbf{w}_2 \in \mathcal{W}$,

$$T(\alpha \mathbf{u}_1 + \mathbf{u}_2) = \alpha \mathbf{v}_1 + \mathbf{v}_2 = \alpha T\mathbf{u}_1 + T\mathbf{u}_2.$$

Thus, T is linear. The equality $T^2 = T$ is immediate from (2). $\square$

• **Notation:** We shall use the symbol $P_\mathcal{V}^\mathcal{W}$ to denote the projection onto the subspace $\mathcal{V}$ with respect to the decomposition $\mathcal{V} \dotplus \mathcal{W}$ in order to emphasize that this projection **depends upon both** $\mathcal{V}$ **and the complementary space** $\mathcal{W}$.

Exercise 12.4. Let $\{\mathbf{v}, \mathbf{w}\}$ be a basis for a vector space $\mathcal{U}$. Find the projection of the vector $\mathbf{u} = 2\mathbf{v} + 3\mathbf{w}$ onto the space $\mathcal{V}$ with respect to each of the following direct sum decompositions: $\mathcal{U} = \mathcal{V} \dotplus \mathcal{W}$ and $\mathcal{U} = \mathcal{V} \dotplus \mathcal{W}_1$, when $\mathcal{V} = \mathrm{span}\{\mathbf{v}\}$, $\mathcal{W} = \mathrm{span}\{\mathbf{w}\}$, and $\mathcal{W}_1 = \mathrm{span}\{\mathbf{w} + \mathbf{v}\}$.

Exercise 12.5. Let

$$\begin{bmatrix} \mathbf{u}_1 & \mathbf{u}_2 & \mathbf{u}_3 & \mathbf{u}_4 & \mathbf{u}_5 & \mathbf{u}_6 \end{bmatrix} = \begin{bmatrix} 1 & 1 & 1 & 2 & 3 & 4 \\ 2 & 0 & 4 & 1 & 5 & 0 \\ 1 & 1 & 1 & 0 & 1 & 0 \\ 0 & 1 & -1 & 0 & -1 & 1 \end{bmatrix},$$

and let $\mathcal{U} = \mathrm{span}\{\mathbf{u}_1, \mathbf{u}_2, \mathbf{u}_3, \mathbf{u}_4\}$, $\mathcal{V} = \mathrm{span}\{\mathbf{u}_1, \mathbf{u}_2, \mathbf{u}_3\}$, $\mathcal{W}_1 = \mathrm{span}\{\mathbf{u}_4\}$, and $\mathcal{W}_2 = \mathrm{span}\{\mathbf{u}_5\}$.

(a) Find a basis for the vector space $\mathcal{V}$.

(b) Show that $\mathcal{U} = \mathcal{V} \dotplus \mathcal{W}_1$ and $\mathcal{U} = \mathcal{V} \dotplus \mathcal{W}_2$.

(c) Find the projection of the vector $\mathbf{u}_6$ onto the space $\mathcal{V}$ with respect to each of the two direct sum decompositions defined in (b).

Exercise 12.6. Show that if $A \in \mathbb{R}^{n \times n}$ and $\operatorname{rank} A = r$, then A is a projection (i.e., $A^2 = A$) if and only if $\det(\lambda I_n - A) = (\lambda - 1)^r \lambda^{n-r}$ and A is diagonalizable.

12.3. Orthogonal projections

- **Orthogonal projections:** A linear transformation T of an inner product space $\mathcal{U}$ into itself is an **orthogonal projection** if

(12.8) $\qquad T^2 = T \quad \text{and} \quad \mathcal{N}_T \text{ is orthogonal to } \mathcal{R}_T$

(hence $\mathcal{U} = \mathcal{R}_T \oplus \mathcal{N}_T$, which is a stronger condition than (12.7)).

Exercise 12.7. Let $\mathbf{u}_1$ and $\mathbf{u}_2$ be a pair of orthonormal vectors in an inner product space $\mathcal{U}$ and let α be a scalar. Show that the transformation T that is defined by the formula $T\mathbf{u} = \langle \mathbf{u}, \mathbf{u}_1 + \alpha \mathbf{u}_2 \rangle_\mathcal{U} \mathbf{u}_1$ is a projection but is not an orthogonal projection unless $\alpha = 0$.

Theorem 12.3. *If a linear transformation T from an inner product space $\mathcal{U}$ into itself is a projection, then T is an orthogonal projection if and only if*

(12.9) $\qquad \langle T\mathbf{x}, \mathbf{y} \rangle_\mathcal{U} = \langle \mathbf{x}, T\mathbf{y} \rangle_\mathcal{U} \quad \text{for every choice of } \mathbf{x}, \mathbf{y} \in \mathcal{U}.$

Proof. Suppose first that T is an orthogonal projection, i.e.,

(12.10) $\qquad \langle \mathbf{v}, \mathbf{w} \rangle_\mathcal{U} = 0 \quad \text{for every choice of } \mathbf{v} \in \mathcal{R}_T \text{ and } \mathbf{w} \in \mathcal{N}_T.$

Then, since $\mathcal{R}_{(I-T)} = \mathcal{N}_T$,

$$\langle T\mathbf{x}, \mathbf{y} \rangle_\mathcal{U} = \langle T\mathbf{x}, T\mathbf{y} + (I-T)\mathbf{y} \rangle_\mathcal{U} = \langle T\mathbf{x}, T\mathbf{y} \rangle_\mathcal{U}$$
$$= \langle T\mathbf{x} + (I-T)\mathbf{x}, T\mathbf{y} \rangle_\mathcal{U} = \langle \mathbf{x}, T\mathbf{y} \rangle_\mathcal{U}$$

for every choice of $\mathbf{x}, \mathbf{y} \in \mathcal{U}$. Thus, (12.10) implies (12.9).

Conversely, if $\mathbf{v} \in \mathcal{R}_T$, $\mathbf{w} \in \mathcal{N}_T$, and (12.9) is in force, then

$$\langle \mathbf{v}, \mathbf{w} \rangle = \langle T\mathbf{v}, \mathbf{w} \rangle = \langle \mathbf{v}, T\mathbf{w} \rangle = \langle \mathbf{v}, \mathbf{0} \rangle = 0.$$

Therefore, (12.9) implies (12.10), as needed to complete the proof. $\square$

Exercise 12.8. Show that if $T^2 = T$ and (12.9) holds, then

(12.11) $\qquad \langle T\mathbf{x}, \mathbf{y} \rangle_\mathcal{U} = \langle T\mathbf{x}, T\mathbf{y} \rangle_\mathcal{U} \quad \text{for every choice of } \mathbf{x}, \mathbf{y} \in \mathcal{U}.$

The next result is an analogue of Lemma 12.2 for orthogonal projections that also includes a recipe for calculating the projection. It is formulated in terms of one subspace $\mathcal{V}$ of the underlying inner product space $\mathcal{U}$ rather than in terms of a pair of complementary subspaces $\mathcal{V}$ and $\mathcal{W}$, because the second space $\mathcal{W}$ is specified as the **orthogonal complement** $\mathcal{V}^\perp$ of $\mathcal{V}$, i.e.,

$$\mathcal{U} = \mathcal{V} \oplus \mathcal{V}^\perp. \tag{12.12}$$

Since $\mathcal{V} \cap \mathcal{V}^\perp = \{\mathbf{0}\}$, Lemma 12.2 guarantees the existence of exactly one linear transformation T that maps $\mathcal{U}$ onto $\mathcal{V}$ such that $\mathbf{u} - T\mathbf{u} \in \mathcal{V}^\perp$ and further guarantees that $T^2 = T$, $\mathcal{R}_T = \mathcal{V}$, and $\mathcal{N}_T = \mathcal{V}^\perp$. We shall refer to this transformation as the **orthogonal projection** of $\mathcal{U}$ onto $\mathcal{V}$ and denote it by the symbol $\Pi_\mathcal{V}$.

Theorem 12.4. *If $\mathcal{U}$ is an inner product space, $\mathcal{V}$ is a subspace of $\mathcal{U}$ with basis $\{\mathbf{v}_1, \ldots, \mathbf{v}_k\}$, $\mathbf{u} \in \mathcal{U}$, and $G \in \mathbb{C}^{(k+1) \times (k+1)}$ is the Gram matrix of the set of vectors $\{\mathbf{v}_1, \ldots, \mathbf{v}_k, \mathbf{u}\}$, then the orthogonal projection $\Pi_\mathcal{V}$ of $\mathcal{U}$ onto $\mathcal{V}$ is given by the formula*

$$\Pi_\mathcal{V} \mathbf{u} = \sum_{j=1}^k (G_{11}^{-1} G_{12})_j \mathbf{v}_j, \quad \text{where} \quad G = \begin{bmatrix} G_{11} & G_{12} \\ G_{21} & G_{22} \end{bmatrix}, \tag{12.13}$$

$$(G_{11})_{ij} = \langle \mathbf{v}_j, \mathbf{v}_i \rangle_\mathcal{U} \text{ for } i, j = 1, \ldots, k,$$
$$(G_{12})_i = \langle \mathbf{u}, \mathbf{v}_i \rangle_\mathcal{U} \text{ for } i = 1, \ldots, k,$$
$$G_{21} = G_{12}^H, \quad \text{and} \quad G_{22} = \langle \mathbf{u}, \mathbf{u} \rangle_\mathcal{U}.$$

Moreover,

$$\|\mathbf{u} - \mathbf{v}\|_\mathcal{U}^2 \geq \|\mathbf{u}\|_\mathcal{U}^2 - \|\Pi_\mathcal{V} \mathbf{u}\|_\mathcal{U}^2 = G_{22} - G_{21} G_{11}^{-1} G_{12} \tag{12.14}$$

for every vector $\mathbf{v} \in \mathcal{V}$, with equality if and only if $\mathbf{v} = \Pi_\mathcal{V} \mathbf{u}$.

If $\mathcal{U} = \mathbb{C}^n$ is endowed with the inner product $\langle \mathbf{x}, \mathbf{y} \rangle_\mathcal{U} = \langle B\mathbf{x}, \mathbf{y} \rangle_{\text{st}} = \mathbf{y}^H B \mathbf{x}$ (based on any $B \in \mathbb{C}^{n \times n}$ for which $\mathbf{x}^H B \mathbf{x} > 0$ for every $\mathbf{x} \neq \mathbf{0}$) and $V = [\mathbf{v}_1 \; \cdots \; \mathbf{v}_k]$, then $G_{11} = V^H B V$, $G_{12} = V^H B \mathbf{u}$, $G_{22} = \mathbf{u}^H B \mathbf{u}$, and

$$\Pi_\mathcal{V} \mathbf{u} = V(V^H B V)^{-1} V^H B \mathbf{u} = V G_{11}^{-1} G_{12} \quad \text{for every } \mathbf{u} \in \mathbb{C}^n. \tag{12.15}$$

Proof. The vector $\mathbf{u} - \sum_{j=1}^k c_j \mathbf{v}_j$ belongs to $\mathcal{V}^\perp$ if and only if

$$\left\langle \left(\mathbf{u} - \sum_{j=1}^k c_j \mathbf{v}_j \right), \mathbf{v}_i \right\rangle_\mathcal{U} = 0 \quad \text{for} \quad i = 1, \ldots, k,$$

or, equivalently, in terms of the entries in G, if and only if

$$\langle \mathbf{u}, \mathbf{v}_i \rangle_\mathcal{U} = \sum_{j=1}^k (G_{11})_{ij} c_j \quad \text{for} \quad i = 1, \ldots, k.$$

12.3. Orthogonal projections

But this in turn is the same as saying that the vector $\mathbf{c} \in \mathbb{C}^k$ with components $c_1, \ldots, c_k$ is a solution of the vector equation $G_{12} = G_{11}\mathbf{c}$. Since G_{11} is invertible by Lemma 11.2, $\Pi_{\mathcal{V}}\mathbf{u}$ is uniquely specified by formula (12.13).

Next, since $(\mathbf{u} - \Pi_{\mathcal{V}}\mathbf{u}) \in \mathcal{V}^\perp$ and $(\Pi_{\mathcal{V}}\mathbf{u} - \mathbf{v}) \in \mathcal{V}$, it is readily seen that
$$
(12.16) \quad \|\mathbf{u} - \mathbf{v}\|_{\mathcal{U}}^2 = \|\mathbf{u} - \Pi_{\mathcal{V}}\mathbf{u} + \Pi_{\mathcal{V}}\mathbf{u} - \mathbf{v}\|_{\mathcal{U}}^2 = \|\mathbf{u} - \Pi_{\mathcal{V}}\mathbf{u}\|_{\mathcal{U}}^2 + \|\Pi_{\mathcal{V}}\mathbf{u} - \mathbf{v}\|_{\mathcal{U}}^2
$$
$$
\geq \|\mathbf{u} - \Pi_{\mathcal{V}}\mathbf{u}\|_{\mathcal{U}}^2 = \|\mathbf{u}\|_{\mathcal{U}}^2 - \|\Pi_{\mathcal{V}}\mathbf{u}\|_{\mathcal{U}}^2
$$
for every $\mathbf{v} \in \mathcal{V}$, which serves to justify (12.14), modulo a straightforward calculation.

Finally, (12.15) follows from (12.13), since $G_{11} = V^H B V$ and $G_{12} = V^H B \mathbf{u}$ in the given setting. $\square$

Exercise 12.9. Show that if $\mathbf{u}_1, \ldots, \mathbf{u}_n$ are linearly independent vectors in an inner product space $\mathcal{U}$ and G_j is the Gram matrix for $\{\mathbf{u}_1, \ldots, \mathbf{u}_j\}$, then
$$
\min\{\|\mathbf{u}_k - \mathbf{u}\|_{\mathcal{U}} : \mathbf{u} \in \text{span}\{\mathbf{u}_1, \ldots, \mathbf{u}_{k-1}\}\} = (\det G_k / \det G_{k-1})^{1/2}
$$
for $k = 2, \ldots, n$.

In the future we shall usually denote the orthogonal projection of an inner product space $\mathcal{U}$ onto a subspace $\mathcal{V}$ by $\Pi_{\mathcal{V}}$. Here, there is no danger of going astray because it is understood that the projection is with respect to the decomposition $\mathcal{U} = \mathcal{V} \oplus \mathcal{V}^\perp$.

If the vectors $\mathbf{v}_1, \ldots, \mathbf{v}_k$ that are specified in Theorem 12.4 are orthonormal in $\mathcal{U}$, then the formulas simplify, because $G_{11} = I_k$, and the orthogonal projection $\Pi_{\mathcal{V}}\mathbf{u}$ of a vector $\mathbf{u} \in \mathcal{U}$ onto $\mathcal{V}$ is given by the formula

$$
(12.17) \quad \Pi_{\mathcal{V}}\mathbf{u} = \langle \mathbf{u}, \mathbf{v}_1 \rangle_{\mathcal{U}} \mathbf{v}_1 + \cdots + \langle \mathbf{u}, \mathbf{v}_k \rangle_{\mathcal{U}} \mathbf{v}_k.
$$

Correspondingly,

$$
(12.18) \quad \|\Pi_{\mathcal{V}}\mathbf{u}\|_{\mathcal{U}}^2 = \sum_{j=1}^{k} |\langle \mathbf{u}, \mathbf{v}_j \rangle_{\mathcal{U}}|^2 \quad \text{for every vector } \mathbf{u} \in \mathcal{U}
$$

and **(Bessel's inequality)**

$$
(12.19) \quad \sum_{j=1}^{k} |\langle \mathbf{u}, \mathbf{v}_j \rangle_{\mathcal{U}}|^2 \leq \|\mathbf{u}\|_{\mathcal{U}}^2, \quad \text{with equality if and only if } \mathbf{u} \in \mathcal{V}.
$$

Moreover, **the coefficients $c_j = \langle \mathbf{u}, \mathbf{v}_j \rangle_{\mathcal{U}}$, $j = 1, \ldots, k$, computed in (12.17) do not change if the space $\mathcal{V}$ is enlarged by adding more orthonormal vectors.**

To this point the analysis in this section is applicable to any inner product space. Thus, for example, we may choose $\mathcal{U}$ equal to the set of continuous

complex-valued functions on the interval $[0,1]$, with inner product
$$\langle f, g \rangle_{\mathcal{U}} = \int_0^1 f(t)\overline{g(t)}dt.$$
Then it is readily checked that the set of functions
$$\varphi_j(t) = e^{j2\pi i t}, \quad j = 1, \ldots, k,$$
is an orthonormal family in $\mathcal{U}$ for any choice of the integer k. Consequently,
$$\sum_{j=1}^k \left| \int_0^1 f(t)\overline{\varphi_j(t)}dt \right|^2 \leq \int_0^1 |f(t)|^2 dt,$$
by (12.19).

Exercise 12.10. Show that no matter how large you choose k, the family $\varphi_j(t) = e^{j2\pi i t}$, $j = 1, \ldots, k$, is not a basis for the space of continuous complex-valued functions $\mathcal{U}$ considered just above.

Exercise 12.11. Let $A = \begin{bmatrix} \mathbf{a}_1 & \mathbf{a}_2 & \mathbf{a}_3 & \mathbf{a}_4 \end{bmatrix} \in \mathbb{C}^{4 \times 4}$, let $G = A^H A$, and let $\Pi_{\mathcal{V}}$ denote the orthogonal projection onto $\mathcal{V} = \text{span}\{\mathbf{a}_1, \mathbf{a}_2, \mathbf{a}_3\}$. Show that:

(a) G is a Gram matrix.

(b) If $\mathbf{a}_1, \mathbf{a}_2, \mathbf{a}_3$ are linearly independent and H denotes the 3×3 Gram matrix for these 3 vectors, then the Schur complement
$$g_{44} - \begin{bmatrix} g_{41} & g_{42} & g_{43} \end{bmatrix} H^{-1} \begin{bmatrix} g_{14} \\ g_{24} \\ g_{34} \end{bmatrix} = \|\mathbf{a}_4 - \Pi_{\mathcal{V}} \mathbf{a}_4\|^2.$$

Exercise 12.12. Show that the choice $\lambda = \langle \mathbf{u}, \mathbf{v} \rangle / \langle \mathbf{v}, \mathbf{v} \rangle$ when $\mathbf{v} \neq \mathbf{0}$ in the proof of Lemma 11.1 (the Cauchy-Schwarz inequality) minimizes $\{\|\mathbf{u} - \alpha \mathbf{v}\| : \alpha \in \mathbb{C}\}$.

Exercise 12.13. Verify directly that if $V \in \mathbb{C}^{n \times k}$ with $1 \leq k < n$ and rank $V = k$, then $V^H V$ is invertible and $\Pi_{\mathcal{V}} = V(V^H V)^{-1} V^H$ is an orthogonal projection in the space $\mathbb{C}^n$ equipped with the standard inner product. Show that $\mathcal{R}_{\Pi_{\mathcal{V}}} = \mathcal{R}_V$ and $\mathcal{N}_{\Pi_{\mathcal{V}}} = \mathcal{N}_{V^H}$. [HINT: The fact that $\mathbf{v}_j = V\mathbf{e}_j$, where $\mathbf{e}_j$ is the j'th column vector of I_k, may be helpful.]

Exercise 12.14. Show that the norm of the projection that is defined in Exercise 12.7 is equal to $(1 + |\alpha|^2)^{1/2}$.

Exercise 12.15. Show that if $P \in \mathbb{C}^{n \times n}$, rank $P \geq 1$, and $P^2 = P$, then:

(a) $\|P\| = 1$ if $\mathcal{R}_P$ is orthogonal to $\mathcal{N}_P$.

(b) $\|P\|$ can be very large if $\mathcal{R}_P$ is not orthogonal to $\mathcal{N}_P$.

Exercise 12.16. Find the orthogonal projection of the vector $\mathbf{u}_6$ onto the space $\mathcal{V}$ in the setting of Exercise 12.5.

Exercise 12.17. Let $\mu_1 < \cdots < \mu_k$ and let $p(x) = c_0 + c_1 x + \cdots + c_n x^n$ be a polynomial of degree $n \geq k$ with coefficients $c_0, \ldots, c_n \in \mathbb{R}$. Show that if $p(\mu_j) = \beta_j$ for $j = 1, \ldots, k$, then

$$\sum_{j=0}^{n} c_j^2 \geq \mathbf{b}^T (V^T V)^{-1} \mathbf{b}, \quad \text{where} \quad V = \begin{bmatrix} 1 & \cdots & 1 \\ \mu_1 & & \mu_k \\ \vdots & & \vdots \\ \mu_1^n & & \mu_k^n \end{bmatrix} \quad \text{and} \quad \mathbf{b} = \begin{bmatrix} \beta_1 \\ \vdots \\ \beta_k \end{bmatrix},$$

and find a polynomial that achieves the exhibited minimum. [HINT: First verify the fact that $\|V(V^T V)^{-1} V^T\| = 1$.]

12.4. The Gram-Schmidt method

Let $\{\mathbf{u}_1, \ldots, \mathbf{u}_k\}$ be a set of linearly independent vectors in an inner product space $\mathcal{U}$. The Gram-Schmidt method is a procedure for finding a set of orthonormal vectors $\{\mathbf{v}_1, \ldots, \mathbf{v}_k\}$ such that

$$\mathcal{V}_j = \mathrm{span}\{\mathbf{v}_1, \ldots, \mathbf{v}_j\} = \mathrm{span}\{\mathbf{u}_1, \ldots, \mathbf{u}_j\} \quad \text{for } j = 1, \ldots, k.$$

The steps of this procedure may be expressed in terms of the orthogonal projection $\Pi_{\mathcal{V}_j}$ of $\mathcal{U}$ onto $\mathcal{V}_j$ as follows:

$$\mathbf{v}_1 = \mathbf{u}_1 / \rho_1 \quad \text{with } \rho_1 = \|\mathbf{u}_1\|_\mathcal{U} > 0,$$

$$\mathbf{v}_{j+1} = \frac{\mathbf{u}_{j+1} - \Pi_{\mathcal{V}_j} \mathbf{u}_{j+1}}{\rho_{j+1}} = \frac{\mathbf{u}_{j+1} - [\langle \mathbf{u}_{j+1}, \mathbf{v}_1 \rangle_\mathcal{U} \mathbf{v}_1 + \cdots + \langle \mathbf{u}_{j+1}, \mathbf{v}_j \rangle_\mathcal{U} \mathbf{v}_j]}{\rho_{j+1}}$$

with

$$\rho_{j+1} = \|\mathbf{u}_{j+1} - [\langle \mathbf{u}_{j+1}, \mathbf{v}_1 \rangle_\mathcal{U} \mathbf{v}_1 + \cdots + \langle \mathbf{u}_{j+1}, \mathbf{v}_j \rangle_\mathcal{U} \mathbf{v}_j]\|_\mathcal{U} > 0$$

for $j = 1, \ldots, k$. It is easily checked that the vectors constructed this way are orthonormal and that $\mathcal{V}_j = \mathrm{span}\{\mathbf{u}_1, \ldots, \mathbf{u}_j\}$ for $j = 1, \ldots, k-1$.

To see the pattern underlying this construction more clearly, note that

(12.20)
$$\mathbf{v}_1 = \mathbf{u}_1 / \rho_1,$$
$$\mathbf{v}_2 = [\mathbf{u}_2 - \langle \mathbf{u}_2, \mathbf{v}_1 \rangle_\mathcal{U} \mathbf{v}_1] / \rho_2,$$
$$\mathbf{v}_3 = [\mathbf{u}_3 - \langle \mathbf{u}_3, \mathbf{v}_1 \rangle_\mathcal{U} \mathbf{v}_1 - \langle \mathbf{u}_3, \mathbf{v}_2 \rangle_\mathcal{U} \mathbf{v}_2] / \rho_3.$$

Exercise 12.18. Find a set of orthonormal vectors $\{\mathbf{y}_1, \mathbf{y}_2, \mathbf{y}_3\}$ in $\mathbb{C}^4$ such that $\mathrm{span}\{\mathbf{y}_1, \mathbf{y}_2\} = \mathrm{span}\{\mathbf{u}_1, \mathbf{u}_2\}$, $\mathrm{span}\{\mathbf{y}_1, \mathbf{y}_2, \mathbf{y}_3\} = \mathrm{span}\{\mathbf{u}_1, \mathbf{u}_2, \mathbf{u}_4\}$, and $\mathrm{span}\{\mathbf{y}_1\} = \mathrm{span}\{\mathbf{u}_1\}$, for the vectors $\mathbf{u}_1$, $\mathbf{u}_2$, and $\mathbf{u}_4$ defined in Exercise 12.5.

Exercise 12.19. Find a set of three polynomials $p_0(t) = a$, $p_1(t) = b + ct$, and $p_3(t) = d + et + ft^2$ with $a, b, c, d, e, f \in \mathbb{R}$ so that they form an orthonormal set with respect to the real inner product $\langle f, g \rangle = \int_0^2 f(t) g(t) dt$.

12.5. QR factorization

Lemma 12.5. *If $A \in \mathbb{C}^{p \times q}$ and $\operatorname{rank} A = q$, then there exist exactly one matrix $Q \in \mathbb{C}^{p \times q}$ with $Q^H Q = I_q$ and exactly one upper triangular matrix $R \in \mathbb{C}^{q \times q}$ with positive entries on the diagonal such that $A = QR$.*

Proof. The existence of at least one factorization of the indicated form is a consequence of the Gram-Schmidt procedure. Thus, for example, if $k = 3$, then (12.20) can be reexpressed as

$$\begin{bmatrix} \mathbf{u}_1 & \mathbf{u}_2 & \mathbf{u}_3 \end{bmatrix} = \begin{bmatrix} \mathbf{v}_1 & \mathbf{v}_2 & \mathbf{v}_3 \end{bmatrix} \begin{bmatrix} \rho_1 & \langle \mathbf{u}_2, \mathbf{v}_1 \rangle & \langle \mathbf{u}_3, \mathbf{v}_1 \rangle \\ 0 & \rho_2 & \langle \mathbf{u}_3, \mathbf{v}_2 \rangle \\ 0 & 0 & \rho_3 \end{bmatrix}.$$

To verify the asserted uniqueness, suppose that there were two such factorizations: $A = Q_1 R_1$ and $A = Q_2 R_2$. Then

$$R_1^H R_1 = R_1^H Q_1^H Q_1 R_1 = A^H A = R_2^H Q_2^H Q_2 R_2 = R_2^H R_2$$

and hence

$$R_1 (R_2)^{-1} = (R_1^H)^{-1} R_2^H.$$

Therefore, since the left-hand side of the last equality is upper triangular and the right-hand side is lower triangular, it follows that $D = R_1 (R_2)^{-1}$ is a diagonal matrix with positive diagonal entries d_{jj} for $j = 1, \ldots, q$ and, as

$$R_2^H R_2 = R_1^H R_1 = R_2^H D^H D R_2,$$

that $D^H D = I_q$. Thus, $|d_{jj}|^2 = 1$ and hence as $d_{jj} > 0$, $D = I_q$, $R_1 = R_2$, and $Q_1 = Q_2$, as claimed. $\square$

Exercise 12.20. Show that if $A \in \mathbb{C}^{n \times n}$ is invertible, then there exists an invertible lower triangular matrix C such that the columns of CA are orthonormal.

12.6. Supplementary notes

The factorization $A = QR$ established in Lemma 12.5 for matrices $A \in \mathbb{C}^{p \times q}$ with $\operatorname{rank} A = q$ is called the **QR factorization** of A. There is a very beautiful formula for the columns of Q when $p = q$ and A is invertible:

Theorem 12.6. *The columns $\mathbf{q}_1, \ldots, \mathbf{q}_n$ of the matrix Q in the QR factorization of an invertible matrix $A \in \mathbb{C}^{n \times n}$ can be expressed in terms of the columns $\mathbf{a}_1, \ldots, \mathbf{a}_n$ of A by the formula*

$$(12.21) \qquad \mathbf{q}_k = \left(\frac{\det G_k}{\det G_{k-1}} \right)^{1/2} A_k G_k^{-1} \mathbf{f}_k \quad \text{for } k = 2, \ldots, n,$$

where $A_k = \begin{bmatrix} \mathbf{a}_1 & \cdots & \mathbf{a}_k \end{bmatrix}$, $G_k = A_k^H A_k$ is the Gram matrix of the columns of A_k, and $\mathbf{f_k} = \begin{bmatrix} 0 & \cdots & 0 & 1 \end{bmatrix}^T \in \mathbb{R}^k$.

Proof. Since R is upper triangular, it is readily checked that $A_k = Q_k R_{[1,k]}$, where $Q_k = \begin{bmatrix} \mathbf{q}_1 & \cdots & \mathbf{q}_k \end{bmatrix}$, $R_{[1,k]}$ denotes the upper left $k \times k$ corner of R, and hence that $Q_k = A_k(R_{[1,k]})^{-1}$. Therefore, $\mathbf{q}_k = A_k(R_{[1,k]})^{-1}\mathbf{f}_k$. However, as $Q_k^H Q_k = I_k$, $G_k = A_k^H A_k = (R_{[1,k]})^H R_{[1,k]}$. Consequently,

$$(R_{[1,k]})^{-1}\mathbf{f}_k = G_k^{-1}(R_{[1,k]})^H \mathbf{f}_k = G_k^{-1}\mathbf{f}_k r_{kk},$$

since $r_{kk} > 0$. The final formula (12.21) is obtained by noting that
$$r_{kk}^{-2} = \mathbf{f}_k^H R_{[1,k]}^{-1}(R_{[1,k]}^{-1})^H \mathbf{f}_k = \mathbf{f}_k^H G_k^{-1} \mathbf{f}_k = G_{\{k;k\}}/\det G_k = \det G_{k-1}/\det G_k.$$

□

Chapter 13

Normal matrices

A matrix $A \in \mathbb{C}^{n \times n}$ is said to be

- **normal** if $A^H A = AA^H$,
- **Hermitian** if $A^H = A$,
- **skew-Hermitian** if $A^H = -A$,
- **unitary** if $A^H A = I_n$ and $AA^H = I_n$ (but keep Exercise 13.1 in mind). A real unitary matrix is also called an **orthogonal matrix**. Permutation matrices are orthogonal matrices.

A matrix $A \in \mathbb{C}^{p \times q}$ is said to be

- **isometric** if $A^H A = I_q$.

Warning: The definitions of normal, unitary, and isometric matrices provided above are linked to the standard inner product. In particular, the term isometric stems from the fact that if $A \in \mathbb{C}^{p \times q}$ and $A^H A = I_q$, then $\langle A\mathbf{x}, A\mathbf{x}\rangle_{\mathrm{st}} = \langle \mathbf{x}, A^H A \mathbf{x}\rangle_{\mathrm{st}} = \langle \mathbf{x}, \mathbf{x}\rangle_{\mathrm{st}}$, i.e., the norm is preserved: $\|A\mathbf{x}\|_{\mathrm{st}} = \|\mathbf{x}\|_{\mathrm{st}}$. If $A \in \mathbb{C}^{p \times q}$, $\mathcal{U} = \mathbb{C}^q$, and $\mathcal{V} = \mathbb{C}^p$ are equipped with arbitrary inner products, then

$$\langle A\mathbf{x}, A\mathbf{x}\rangle_{\mathcal{V}} = \langle \mathbf{x}, \mathbf{x}\rangle_{\mathcal{U}} \iff A^* A = I_q,$$

i.e., A^H should be replaced by A^*. Correspondingly, A is **isometric** if $A^* A = I_q$; if $p = q$, then A is **normal** if $A^* A = AA^*$ and **unitary** if $A^* A = I_p$; see Example 11.2 and Theorem 13.10.

Exercise 13.1. Show that if $A \in \mathbb{C}^{p \times q}$, then

(13.1)
$$\begin{aligned}
A^H A = I_q \text{ and } AA^H = I_p &\iff A^H A = I_q \text{ and } p = q \\
&\iff A^H A = I_q \text{ and } \mathcal{R}_A = \mathbb{C}^p \\
&\iff A^H A = I_q \text{ and } A \text{ is invertible}.
\end{aligned}$$

137

Exercise 13.2. Show that if $A \in \mathbb{C}^{n \times n}$ is a cyclic matrix, then $A^H A = AA^H$.

13.1. Normal matrices

The main result of this section is Theorem 13.1. As a byproduct of this theorem we shall see that the class of $n \times n$ normal matrices is the largest class of matrices $A \in \mathbb{C}^{n \times n}$ that admit a factorization of the form

(13.2) $\quad A = UDU^H$, with $U \in \mathbb{C}^{n \times n}$ unitary and $D \in \mathbb{C}^{n \times n}$ diagonal.

Exercise 13.3. Show that a matrix $A \in \mathbb{C}^{n \times n}$ admits a factorization of the form (13.2) with $D \in \mathbb{R}^{n \times n}$ if and only if $A = A^H$.

Theorem 13.1. *If $A \in \mathbb{C}^{n \times n}$, then there exists an orthonormal family of eigenvectors $\{\mathbf{u}_1, \ldots, \mathbf{u}_n\}$ of A in $\mathbb{C}^n$ (equipped with the standard inner product) if and only if $A^H A = AA^H$. Moreover:*

(1) *If $A = A^H$, then $A^H A = AA^H$ and $\sigma(A) \subset \mathbb{R}$.*
(2) *If $A = -A^H$, then $A^H A = AA^H$ and $\sigma(A) \subset i\mathbb{R}$.*
(3) *If $A^H A = I_n$, then $A^H A = AA^H$ and $\sigma(A) \subset \{\lambda \in \mathbb{C} : |\lambda| = 1\}$.*

Proof. If $\{\mathbf{u}_1, \ldots, \mathbf{u}_n\}$ is a family of eigenvectors of A in $\mathbb{C}^n$ that are orthonormal in the standard inner product and $A\mathbf{u}_j = \lambda_j \mathbf{u}_j$ for $j = 1, \ldots, n$, then the matrix $U = \begin{bmatrix} \mathbf{u}_1 & \cdots & \mathbf{u}_n \end{bmatrix}$ is unitary (i.e., $U^H U = I_n$) and

$$AU = UD \quad \text{with} \quad D = \operatorname{diag}\{\lambda_1, \ldots, \lambda_n\}.$$

Therefore, $A = UDU^H$, $A^H = UD^H U^H$, and, as $D^H D = DD^H$,

$$A^H A = (UD^H U^H)(UDU^H) = UD^H DU^H$$
$$= UDD^H U^H = (UDU^H)(UD^H U^H) = AA^H.$$

Thus, $A^H A = AA^H$; the same argument shows that if $A \in \mathbb{C}^{n \times n}$ admits a factorization of the form (13.2), then $A^H A = AA^H$.

The verification of the converse and assertions (1)–(3) is broken into steps.

1. *If $A \in \mathbb{C}^{n \times n}$ and $A^H A = AA^H$, then $\|(A - \lambda I_n)\mathbf{u}\| = \|(A^H - \overline{\lambda} I_n)\mathbf{u}\|$ for every vector $\mathbf{u} \in \mathbb{C}^n$ and every point $\lambda \in \mathbb{C}$.*

This is an easy consequence of the fact that

(13.3) $\quad (A^H - \overline{\lambda} I_n)(A - \lambda I_n) = (A - \lambda I_n)(A^H - \overline{\lambda} I_n) \quad$ for every point $\lambda \in \mathbb{C}$.

13.1. Normal matrices

2. If $A \in \mathbb{C}^{n \times n}$ and $A^H A = AA^H$, then $\mathcal{N}_{(A-\lambda I_n)^2} = \mathcal{N}_{(A-\lambda I_n)}$ for every point $\lambda \in \mathbb{C}$.

If $\mathbf{u} \in \mathcal{N}_{(A-\lambda I_n)^2}$ and $\mathbf{v} = (A - \lambda I_n)\mathbf{u}$, then, in view of step 1,
$$0 = \|(A - \lambda I_n)\mathbf{v}\| = \|(A^H - \overline{\lambda} I_n)\mathbf{v}\| \implies (A^H - \overline{\lambda} I_n)\mathbf{v} = \mathbf{0}.$$
Therefore,
$$0 = \langle (A^H - \overline{\lambda} I_n)(A - \lambda I_n)\mathbf{u}, \mathbf{u} \rangle = \|(A - \lambda I_n)\mathbf{u}\|^2,$$
which implies that $\mathcal{N}_{(A-\lambda I_n)^2} \subseteq \mathcal{N}_{(A-\lambda I_n)}$. Since the opposite inclusion is self-evident, this completes the proof of this step.

3. If $A \in \mathbb{C}^{n \times n}$ and $A^H A = AA^H$, then there exists an orthonormal family of eigenvectors $\{\mathbf{u}_1, \ldots, \mathbf{u}_n\}$ of A.

Suppose that $\mathcal{U}_k = \text{span}\{\mathbf{u}_1, \ldots, \mathbf{u}_k\}$ is the span of k orthonormal eigenvectors of A for some positive integer k with $k < n$ and let $\mathcal{U}_k^\perp$ denote the orthogonal complement of $\mathcal{U}$ with respect to the standard inner product. Then $\mathcal{U}_k^\perp \neq \{\mathbf{0}\}$. The key observation is that $\mathcal{U}_k^\perp$ is invariant under A^H, i.e., if $\mathbf{v} \in \mathcal{U}_k^\perp$, then $A^H \mathbf{v} \in \mathcal{U}_k^\perp$:
$$\langle \mathbf{u}_j, A^H \mathbf{v} \rangle = \langle A\mathbf{u}_j, \mathbf{v} \rangle = \lambda_j \langle \mathbf{u}_j, \mathbf{v} \rangle = 0 \quad \text{for } j = 1, \ldots, k.$$
Consequently, there exists an eigenvector $\mathbf{w}$ of A^H in $\mathcal{U}_k^\perp$ with $\|\mathbf{w}\| = 1$. But if $A^H \mathbf{w} = \beta \mathbf{w}$, then, in view of step 1, $A\mathbf{w} = \overline{\beta} \mathbf{w}$, i.e., $\mathbf{w}$ is an eigenvector of A that is orthogonal to $\mathcal{U}_k$. Therefore, $\mathcal{U}_{k+1} = \text{span}\{\mathbf{u}_1, \ldots, \mathbf{u}_k, \mathbf{w}\}$ is an orthonormal family of $k+1$ eigenvectors of A. If $k+1 = n$, then we are finished. If $k+1 < n$, then we repeat the procedure $n - (k+1)$ more times.

4. *Verification of* (1)–(3).

It is easily seen that if $A^H = \pm A$, or if $A \in \mathbb{C}^{n \times n}$ and $A^H A = I_n$, then $A^H A = AA^H$ (see Exercise 13.1 for help with the last assertion).

If $A\mathbf{u}_j = \lambda_j \mathbf{u}_j$, $\|\mathbf{u}_j\| = 1$, and $A = A^H$, then
$$\lambda_j \langle \mathbf{u}_j, \mathbf{u}_j \rangle = \langle A\mathbf{u}_j, \mathbf{u}_j \rangle = \langle \mathbf{u}_j, A\mathbf{u}_j \rangle = \langle \mathbf{u}_j, \lambda_j \mathbf{u}_j \rangle = \overline{\lambda_j} \langle \mathbf{u}_j, \mathbf{u}_j \rangle.$$
Therefore, (1) holds.

If $A\mathbf{u}_j = \lambda_j \mathbf{u}_j$, $\|\mathbf{u}_j\| = 1$, and $A^H A = I_n$, then
$$|\lambda_j|^2 \langle \mathbf{u}_j, \mathbf{u}_j \rangle = \langle A\mathbf{u}_j, A\mathbf{u}_j \rangle = \langle \mathbf{u}_j, A^H A \mathbf{u}_j \rangle = \langle \mathbf{u}_j, \mathbf{u}_j \rangle.$$
Therefore, (3) holds; (2) is left to the reader as an exercise. $\square$

Exercise 13.4. Show that if $A \in \mathbb{C}^{n \times n}$ and $A^H = -A$, then $\sigma(A) \subset i\mathbb{R}$.

Exercise 13.5. Show that if $A \in \mathbb{R}^{n \times n}$, $A^T = -A$, and n is odd, then $\det A = 0$.

Exercise 13.6. Show that if $A \in \mathbb{C}^{n \times n}$ and $A^H A = AA^H$, then $r_\sigma(A) = \|A\|$.

Exercise 13.7. Find the Jordan decomposition of the matrix
$$A = \begin{bmatrix} 0 & 0 & 9 \\ 0 & 1 & 0 \\ 0 & 0 & 1 \end{bmatrix}.$$

[HINT: First check that A is a projection.]

Exercise 13.8. Show that if $A \in \mathbb{C}^{n \times n}$ and $A^2 = A$, then A is an orthogonal projection if and only if A is normal.

Exercise 13.9. Show that if $A \in \mathbb{C}^{p \times q}$, then

(13.4) $\quad A^H A = I_q \iff \langle A\mathbf{x}, A\mathbf{x}\rangle_{\mathrm{st}} = \langle \mathbf{x}, \mathbf{x}\rangle_{\mathrm{st}} \quad$ for every $\quad \mathbf{x} \in \mathbb{C}^q$.

[HINT: Use the polarization identities to check that $\langle B\mathbf{x}, \mathbf{x}\rangle = \langle \mathbf{x}, \mathbf{x}\rangle$ for all vectors $\mathbf{x} \in \mathbb{C}^q$ if and only if $\langle B\mathbf{x}, \mathbf{y}\rangle = \langle \mathbf{x}, \mathbf{y}\rangle$ for all vectors $\mathbf{x}, \mathbf{y} \in \mathbb{C}^q$.]

Some elementary facts to keep in mind.
(1) If $U \in \mathbb{C}^{n \times n}$ and $U^H U = I_n$, then the columns of U form an orthonormal basis of $\mathbb{C}^n$ with respect to the standard inner product.
(2) If $U \in \mathbb{R}^{n \times n}$ and $U^H U = I_n$, then the columns of U form an orthonormal basis of $\mathbb{R}^n$ with respect to the standard inner product.
(3) If $A \in \mathbb{R}^{n \times n}$, then A is Hermitian if and only if it is symmetric. But this is not true if $A \in \mathbb{C}^{n \times n}$. Thus, for example,
$$A = \begin{bmatrix} 1 & i \\ i & 1 \end{bmatrix} = A^T \neq A^H, \quad \text{whereas} \quad B = \begin{bmatrix} 1 & i \\ -i & 1 \end{bmatrix} = B^H \neq B^T.$$
(4) (13.4)

Exercise 13.10. Show that if $A \in \mathbb{C}^{p \times q}$, then $\mathcal{R}_A \cap \mathcal{N}_{A^H} = \{\mathbf{0}\}$ and hence that if $A = A^H$, then $p = q$ and $\mathcal{R}_A \cap \mathcal{N}_A = \{\mathbf{0}\}$.

13.2. Schur's theorem

Theorem 13.1 ensures that every normal matrix is unitarily equivalent to a diagonal matrix, i.e., it admits a factorization of the form exhibited in (13.2). A theorem of Issai Schur states that every square matrix is unitarily

13.2. Schur's theorem

equivalent to a triangular matrix:

Theorem 13.2 (Schur). *If $A \in \mathbb{C}^{n \times n}$, then there exist a matrix $V \in \mathbb{C}^{n \times n}$ and an upper triangular matrix T such that $V^H V = I_n$ and*

(13.5) $$A = VTV^H$$

is upper triangular. Moreover, V can be chosen so that the diagonal entries of T coincide with the eigenvalues of A, repeated according to their algebraic multiplicity.

Proof. The Jordan decomposition theorem guarantees that there exists an invertible matrix $U \in \mathbb{C}^{n \times n}$ such that $A = UJU^{-1}$. Since U is invertible, it has a QR decomposition $U = QR$ with a unitary factor $Q \in \mathbb{C}^{n \times n}$ and an upper triangular invertible factor $R \in \mathbb{C}^{n \times n}$ with positive entries on the diagonal. Thus,

$$A = QRJR^{-1}Q^{-1} = Q(RJR^{-1})Q^H.$$

Moreover, upon writing

$$R = D_1 + X_1, \quad J = D_0 + X_0, \text{ and } R^{-1} = D_2 + X_2,$$

where D_j is diagonal and X_j is strictly upper triangular (i.e., upper triangular with zero entries on the diagonal), it is readily checked that

$$RJR^{-1} = (D_1 + X_1)(D_0 + X_0)(D_2 + X_2) = D_1 D_0 D_2 + X_3,$$

where X_3 is strictly upper triangular and hence as

$$D_1 D_0 D_2 = D_0 D_1 D_2 \text{ and } D_1 D_2 = I_n,$$

the diagonal entries of RJR^{-1} coincide with the diagonal entries of J, which run through the eigenvalues of A, repeated according to their algebraic multiplicity. Thus, (13.5) holds with $V = Q$ and $T = RJR^{-1}$. □

Theorem 13.3. *If $A \in \mathbb{C}^{n \times n}$ with eigenvalues $\mu_1, \ldots, \mu_n$, repeated according to their algebraic multiplicity, then*

(13.6) $$\sum_{j=1}^{n} |\mu_j|^2 \leq \sum_{i,j=1}^{n} |a_{ij}|^2,$$

with equality if and only if $A^H A = A A^H$.

Proof. By Schur's theorem, there exists a unitary matrix U such that $U^H A U = T$ is upper triangular and $t_{ii} = \mu_i$ for $i = 1, \ldots, n$. Consequently,

$$\begin{aligned}\operatorname{trace}\{A^H A\} &= \operatorname{trace}\{UT^H U^H UTU^H\} = \operatorname{trace}\{UT^H TU^H\} \\ &= \operatorname{trace}\{T^H TU^H U\} = \operatorname{trace}\{T^H T\},\end{aligned}$$

and hence the identity
$$\sum_{i,j=1}^n |t_{ij}|^2 = \sum_{i,j=1}^n |a_{ij}|^2$$
is in force for the entries t_{ij} of T and a_{ij} of A. Therefore,

(13.7) $$\sum_{i=1}^n |\mu_i|^2 = \sum_{i=1}^n |t_{ii}|^2 \leq \sum_{i,j=1}^n |t_{ij}|^2 = \text{trace}\, A^H A = \sum_{i,j=1}^n |a_{ij}|^2,$$

with equality if and only if $t_{ij} = 0$ when $i \neq j$, i.e., if and only if T is a diagonal matrix. But then $A^H A = UT^H T U^H = UTT^H U^H = AA^H$, since diagonal matrices commute. Consequently, equality in (13.7) implies that A is normal. The opposite implication is left to the reader. □

Exercise 13.11. Show that if $A \in \mathbb{C}^{n \times n}$, then $|\det A| \leq \|A\|_\infty^n n^{n/2}$. [HINT: Use (13.6) and (9.11).]

13.3. Commuting normal matrices

Lemma 13.4. *If $A, B \in \mathbb{C}^{n \times n}$ and $B^H B = BB^H$, then*

(13.8) $$AB = BA \iff AB^H = B^H A.$$

Proof. If $AB = BA$, then $AB^m = B^m A$ for every positive integer m. Therefore, $A\varphi(B) = \varphi(B)A$ for every polynomial φ. Moreover, since $B^H B = BB^H$, B admits a representation of the form $B = WDW^H$ with $D \in \mathbb{C}^{n \times n}$ diagonal and $W \in \mathbb{C}^{n \times n}$ unitary. Consequently,
$$\varphi(B) = \varphi(WDW^H) = W\varphi(D)W^H \quad \text{and} \quad B^H = WD^H W^H.$$
Thus, $\varphi(B) = B^H \iff \varphi(D) = D^H$. If D has k distinct entries $\lambda_1, \ldots, \lambda_k$ with $k \geq 2$ and $\varphi(\lambda) = a_0 + a_1 \lambda + \cdots + a_{k-1}\lambda^{k-1}$, then

$$\varphi(\lambda_j) = \overline{\lambda_j} \;\text{ for } j = 1, \ldots, k \iff \begin{bmatrix} 1 & \lambda_1 & \cdots & \lambda_1^{k-1} \\ 1 & \lambda_2 & \cdots & \lambda_2^{k-1} \\ \vdots & & & \vdots \\ 1 & \lambda_k & \cdots & \lambda_k^{k-1} \end{bmatrix} \begin{bmatrix} a_0 \\ a_1 \\ \vdots \\ a_{k-1} \end{bmatrix} = \begin{bmatrix} \overline{\lambda_1} \\ \overline{\lambda_2} \\ \vdots \\ \overline{\lambda_k} \end{bmatrix}.$$

Thus, as the Vandermonde matrix in the preceding display is invertible, the polynomial $\varphi(\lambda) = \sum_{j=0}^{k-1} a_j \lambda^j$ maps λ_j to $\overline{\lambda_j}$. Consequently, $\varphi(B) = B^H$ and
$$AB = BA \implies A\varphi(B) = \varphi(B)A \implies AB^H = B^H A.$$
This serves to complete the proof, since the implication $\Longleftarrow$ follows from the implication $\Longrightarrow$ and the fact that $(B^H)^H = B$. □

Exercise 13.12. Show that if $A, B, C \in \mathbb{C}^{n \times n}$, $B^H B = BB^H$, and $C^H C = CC^H$, then $AB = CA \iff AB^H = C^H A$.

13.3. Commuting normal matrices

Theorem 13.5. *If $A \in \mathbb{C}^{n\times n}$ and $B \in \mathbb{C}^{n\times n}$ are both normal matrices, then $AB = BA$ if and only if there exists a unitary matrix $U \in \mathbb{C}^{n\times n}$ that diagonalizes both A and B.*

Proof. Suppose first that there exists a unitary matrix $U \in \mathbb{C}^{n\times n}$ such that $D_A = U^H A U$ and $D_B = U^H B U$ are both diagonal matrices. Then, since $D_A D_B = D_B D_A$,

$$\begin{aligned} AB &= U D_A U^H U D_B U^H = U D_A D_B U^H \\ &= U D_B D_A U^H = U D_B U^H U D_A U^H = BA \,. \end{aligned}$$

Suppose next that $AB = BA$ and A has k distinct eigenvalues $\lambda_1, \ldots, \lambda_k$ with geometric multiplicities $\gamma_1, \ldots, \gamma_k$, respectively. Then there exists a set of k isometric matrices $U_1 \in \mathbb{C}^{n\times \gamma_1}, \ldots, U_k \in \mathbb{C}^{n\times \gamma_k}$ such that $U = \begin{bmatrix} U_1 & \cdots & U_k \end{bmatrix}$ is unitary, $AU_j = \lambda_j U_j$, and $A^H U_j = \overline{\lambda_j} U_j$. Therefore,

$$ABU_j = BAU_j = \lambda_j BU_j \quad \text{for} \quad j = 1, \ldots, k,$$

which implies that the columns of the matrix BU_j belong to $\mathcal{N}_{(A - \lambda_j I_n)}$ and hence, since the columns of U_j form a basis for that space, that there exists a set of $\gamma_j \times \gamma_j$ matrices C_j such that

$$BU_j = U_j C_j \quad \text{for} \quad j = 1, \ldots, k.$$

Since $AB = BA \implies AB^H = B^H A$, by Lemma 13.4, analogous reasoning yields a set of $\gamma_j \times \gamma_j$ matrices D_j such that

$$B^H U_j = U_j D_j \quad \text{for} \quad j = 1, \ldots, k.$$

Moreover,

$$C_j^H = (U_j^H B U_j)^H = U_j^H B^H U_j = D_j \quad \text{for } j = 1, \ldots, k$$

and

$$C_j^H C_j = U_j^H B^H U_j C_j = U_j^H B^H B U_j = U_j^H B B^H U_j = D_j^H D_j = C_j C_j^H$$

for $j = 1, \ldots, k$, i.e., the C_j are all normal matrices. Therefore, upon writing

$$C_j = W_j \Delta_j W_j^H \quad \text{for} \quad j = 1, \ldots, k,$$

with W_j unitary and Δ_j diagonal, and setting

$$V_j = U_j W_j \quad \text{for} \quad j = 1, \ldots, k \quad \text{and} \quad W = \text{diag}\{W_1, \ldots, W_k\},$$

it is readily seen that

$$AV_j = AU_j W_j = \lambda_j U_j W_j = \lambda_j V_j \quad \text{for} \quad j = 1, \ldots, k$$

and

$$BV_j = BU_j W_j = U_j C_j W_j = U_j W_j \Delta_j = V_j \Delta_j \quad \text{for} \quad j = 1, \ldots, k.$$

Thus, the matrix $V = \begin{bmatrix} V_1 & \cdots & V_k \end{bmatrix} = UW$ is a unitary matrix that serves to diagonalize both A and B. $\square$

Remark 13.6. The proof of Theorem 13.5 simplifies when $k = n$, because then every eigenvector of A is also an eigenvector of B.

13.4. Real Hermitian matrices

In this section we shall show that if $A = A^H$ and $A \in \mathbb{R}^{n \times n}$, then the unitary matrix U in the formula $A = UDU^H$ may also be chosen to belong to $\mathbb{R}^{n \times n}$.

Theorem 13.7. *If $A = A^H$ and $A \in \mathbb{R}^{n \times n}$, then there exist an orthogonal matrix $Q \in \mathbb{R}^{n \times n}$ and a real diagonal matrix $D \in \mathbb{R}^{n \times n}$ such that*

(13.9) $$A = QDQ^T.$$

Proof. Let $\mu \in \sigma(A)$ and let $\mathbf{u}_1, \ldots, \mathbf{u}_\ell$ be a basis for the nullspace of the matrix $B = A - \mu I_n$. Then, since $B \in \mathbb{R}^{n \times n}$, the real and imaginary parts of the vectors $\mathbf{u}_j$ also belong to $\mathcal{N}_B$: If $\mathbf{u}_j = \mathbf{x}_j + i\mathbf{y}_j$ with $\mathbf{x}_j$ and $\mathbf{y}_j$ in $\mathbb{R}^n$ for $j = 1, \ldots, \ell$, then

$$B(\mathbf{x}_j + i\mathbf{y}_j) = \mathbf{0} \implies B\mathbf{x}_j = \mathbf{0} \quad \text{and} \quad B\mathbf{y}_j = \mathbf{0} \quad \text{for } j = 1, \ldots, \ell.$$

Moreover, if

$$U = \begin{bmatrix} \mathbf{u}_1 & \cdots & \mathbf{u}_\ell \end{bmatrix}, \quad X = \begin{bmatrix} \mathbf{x}_1 & \cdots & \mathbf{x}_\ell \end{bmatrix}, \quad \text{and} \quad Y = \begin{bmatrix} \mathbf{y}_1 & \cdots & \mathbf{y}_\ell \end{bmatrix},$$

then $U = X + iY$ and the formula

$$U = \begin{bmatrix} X & Y \end{bmatrix} \begin{bmatrix} I_\ell \\ iI_\ell \end{bmatrix}$$

implies that

$$\operatorname{rank} \begin{bmatrix} X & Y \end{bmatrix} \geq \operatorname{rank} U = \ell.$$

Therefore, ℓ of the columns in $\begin{bmatrix} X & Y \end{bmatrix}$ form a basis for $\mathcal{N}_B$, and an orthonormal basis of ℓ vectors in $\mathbb{R}^n$ for $\mathcal{N}_B$ may be found by invoking the Gram-Schmidt procedure.

If A has k distinct eigenvalues $\lambda_1, \ldots, \lambda_k$, let Q_i, $i = 1, \ldots, k$, denote the $n \times \gamma_i$ matrix that is obtained by stacking the vectors that are obtained by applying the procedure described above to $B_i = A - \lambda_i I_n$ for $i = 1, \ldots, k$. Then, $AQ_i = \lambda_i Q_i$ and

$$A \begin{bmatrix} Q_1 & \cdots & Q_k \end{bmatrix} = \begin{bmatrix} Q_1 & \cdots & Q_k \end{bmatrix} D, \text{ where } D = \operatorname{diag}\{\lambda_1 I_{\gamma_1}, \ldots, \lambda_k I_{\gamma_k}\}.$$

Moreover, the matrix $Q = \begin{bmatrix} Q_1 & \cdots & Q_k \end{bmatrix}$ is an orthogonal matrix, since the columns in Q_j form an orthonormal basis for $\mathcal{N}_{(A - \lambda_j I_n)}$ and the columns in Q_i are orthogonal to the columns in Q_j if $i \neq j$. □

Lemma 13.8. *If $A \in \mathbb{R}^{p \times q}$, then*

$$\max \{\|A\mathbf{x}\|_{\operatorname{st}} : \mathbf{x} \in \mathbb{C}^q \quad \text{and} \quad \|\mathbf{x}\|_{\operatorname{st}} = 1\}$$
$$= \max \{\|A\mathbf{x}\|_{\operatorname{st}} : \mathbf{x} \in \mathbb{R}^q \quad \text{and} \quad \|\mathbf{x}\|_{\operatorname{st}} = 1\}.$$

Proof. Since $A \in \mathbb{R}^{p \times q}$, $A^H A$ is a real $q \times q$ Hermitian matrix. Therefore, $A^H A = QDQ^T$, where $Q \in \mathbb{R}^{q \times q}$ is orthogonal, $D = \mathrm{diag}\{\lambda_1, \ldots, \lambda_q\} \in \mathbb{R}^{q \times q}$, and $\lambda_j \geq 0$ for $j = 1, \ldots, q$. Thus, if $\mathbf{x} \in \mathbb{C}^q$, $\mathbf{y} = Q^T \mathbf{x}$, and $\delta = \max\{\lambda_j : j = 1, \ldots, q\}$, then $\delta \geq 0$ and

$$\begin{aligned}
\|A\mathbf{x}\|_{\mathrm{st}}^2 &= \langle A^H A \mathbf{x}, \mathbf{x} \rangle_{\mathrm{st}} = \langle QDQ^T \mathbf{x}, \mathbf{x} \rangle_{\mathrm{st}} \\
&= \langle DQ^T \mathbf{x}, Q^T \mathbf{x} \rangle_{\mathrm{st}} = \langle D\mathbf{y}, \mathbf{y} \rangle_{\mathrm{st}} \\
&= \sum_{j=1}^n \lambda_j |y_j|^2 \leq \delta \sum_{j=1}^n |y_j|^2 \\
&= \delta \|\mathbf{y}\|_{\mathrm{st}}^2 = \delta \|Q^T \mathbf{x}\|_{\mathrm{st}}^2 = \delta \|\mathbf{x}\|_{\mathrm{st}}^2 \, .
\end{aligned}$$

Consequently,

$$\max\{\|A\mathbf{x}\|_{\mathrm{st}} : \mathbf{x} \in \mathbb{C}^n \quad \text{and} \quad \|\mathbf{x}\|_{\mathrm{st}} = 1\} \leq \sqrt{\delta} \, .$$

However, if $\delta = \lambda_k$, then it is readily seen that this maximum can be attained by choosing $\mathbf{x} = Q\mathbf{e}_k$, the k'th column of Q. But this proves the claim, since $Q\mathbf{e}_k \in \mathbb{R}^q$. $\square$

13.5. Supplementary notes

This chapter is partially adapted from Chapter 9 in [**30**]. The inequality in Exercise 13.11 is due to Hadamard; the simple proof based on (13.6) is due to Schur.

Lemma 13.4 and Exercise 13.12 are finite-dimensional versions of theorems due to Fuglede and Putnam, respectively.

Theorem 13.1 is a special case of a general result for linear transformations T from a finite-dimensional inner product space $\mathcal{U}$ into itself. In this setting, T is said to be **normal** if $T^*T = TT^*$; **selfadjoint** if $T = T^*$; and **unitary** if $T^*T = I$ (here, too, just as in Exercise 13.1, $T^*T = I \iff TT^* = I$ when $\mathcal{U}$ is a finite-dimensional inner product space).

Theorem 13.9. *If T is a linear transformation from an n-dimensional inner product space $\mathcal{U}$ into itself, then:*

(1) *There exists an orthonormal basis $\{\mathbf{u}_1, \ldots, \mathbf{u}_n\}$ of $\mathcal{U}$ of eigenvectors of T if and only if T is normal.*

(2) *If T is selfadjoint, then the eigenvalues of T are all real.*

(3) *If T is unitary, then the eigenvalues of T all have absolute value equal to one.*

The proof is much the same as the proof of Theorem 13.1. If T is normal, then every vector $\mathbf{u} \in \mathcal{U}$ can be expressed in terms of the orthonormal eigenvectors $\mathbf{u}_1, \ldots, \mathbf{u}_n$ of T as $\mathbf{u} = \sum_{j=1}^n \langle \mathbf{u}, \mathbf{u}_j \rangle_{\mathcal{U}} \mathbf{u}_j$. Moreover, if $T\mathbf{u}_j = \lambda_j \mathbf{u}_j$ for $j = 1, \ldots, n$, then $T^* \mathbf{u}_j = \overline{\lambda_j} \mathbf{u}_j$ for $j = 1, \ldots, n$. Correspondingly,

$$(13.10) \qquad T\mathbf{u} = \sum_{j=1}^n \lambda_j \langle \mathbf{u}, \mathbf{u}_j \rangle_{\mathcal{U}} \mathbf{u}_j \quad \text{and} \quad T^*\mathbf{u} = \sum_{j=1}^n \overline{\lambda_j} \langle \mathbf{u}, \mathbf{u}_j \rangle_{\mathcal{U}} \mathbf{u}_j$$

for every vector $\mathbf{u} \in \mathcal{U}$. It is easy to extract (2) and (3) from (13.10). Nevertheless, it is helpful to keep the following simple alternate argument in mind:

If $T\mathbf{u}_j = \lambda_j \mathbf{u}_j$ for $j = 1, \ldots, n$ and $T = T^*$, then

$$(13.11) \qquad \begin{aligned} \lambda_j \langle \mathbf{u}_j, \mathbf{u}_i \rangle_{\mathcal{U}} &= \langle T\mathbf{u}_j, \mathbf{u}_i \rangle_{\mathcal{U}} = \langle \mathbf{u}_j, T^*\mathbf{u}_i \rangle_{\mathcal{U}} \\ &= \langle \mathbf{u}_j, T\mathbf{u}_i \rangle_{\mathcal{U}} = \langle \mathbf{u}_j, \lambda_i \mathbf{u}_i \rangle_{\mathcal{U}} = \overline{\lambda_i} \langle \mathbf{u}_j, \mathbf{u}_i \rangle_{\mathcal{U}}, \end{aligned}$$

i.e., $(\lambda_j - \overline{\lambda_i}) \langle \mathbf{u}_j, \mathbf{u}_i \rangle_{\mathcal{U}} = 0$. Consequently, $\lambda_i = \overline{\lambda_i}$ for $i = 1, \ldots, n$ and $\langle \mathbf{u}_j, \mathbf{u}_i \rangle_{\mathcal{U}} = 0$ for $i, j = 1, \ldots, n$ if $i \neq j$. For additional details, see, e.g., Section 8.12 of [**30**].

Thus, if $T : \mathbf{x} \in \mathbb{C}^n \mapsto A\mathbf{x} \in \mathbb{C}^n$, then $T^* : \mathbf{x} \in \mathbb{C}^n \mapsto A^*\mathbf{x} \in \mathbb{C}^n$ and:

Theorem 13.10. *If $\mathcal{U} = \mathbb{C}^n$ equipped with the inner product $\langle \mathbf{x}, \mathbf{y} \rangle_{\mathcal{U}} = \langle G\mathbf{x}, \mathbf{y} \rangle_{\mathrm{st}}$ for some $G \in \mathbb{C}^{n \times n}$ with $G \succ O$, then $A^* = G^{-1} A^H G$ and*

$$(13.12) \qquad \begin{aligned} A^*A = AA^* &\iff \text{there exist } W, D \in \mathbb{C}^{n \times n} \text{ with } D \text{ diagonal such} \\ &\text{that } AW = WD \quad \text{and} \quad W^H G W = G \ (i.e., W^*W = I_n). \end{aligned}$$

Moreover, $A^* = A \implies D \in \mathbb{R}^{n \times n}$ and $A^*A = I_n \implies |d_{ii}| = 1$.

Chapter 14

Projections, volumes, and traces

The first three sections of this chapter deal with projections; the fourth is a short detour on the angle between subspaces; the fifth gives a geometric interpretation to determinants; the sixth develops trace formulas.

14.1. Projection by iteration

In this section we consider the problem of computing the orthogonal projection of a vector $\mathbf{u}$ in a finite-dimensional inner product space $\mathcal{U}$ onto the intersection $\mathcal{V} \cap \mathcal{W}$ of a pair of subspaces $\mathcal{V}$ and $\mathcal{W}$ of $\mathcal{U}$ in terms of the orthogonal projections $\Pi_\mathcal{V}$ of $\mathcal{U}$ onto $\mathcal{V}$ and $\Pi_\mathcal{W}$ of $\mathcal{U}$ onto $\mathcal{W}$.

Lemma 14.1. *If $\mathcal{V}$ and $\mathcal{W}$ are subspaces of a finite-dimensional inner product space $\mathcal{U}$ and $\Pi_\mathcal{V}$, $\Pi_\mathcal{W}$, and $\Pi_{\mathcal{V} \cap \mathcal{W}}$ denote the orthogonal projections onto $\mathcal{V}$, $\mathcal{W}$, and $\mathcal{V} \cap \mathcal{W}$, respectively, then*

(14.1) $$\lim_{k \uparrow \infty} \|\Pi_{\mathcal{V} \cap \mathcal{W}} - (\Pi_\mathcal{V} \Pi_\mathcal{W})^k\| = 0.$$

Moreover,

(14.2) $$\|\Pi_\mathcal{V} \Pi_\mathcal{W}\| < 1 \iff \mathcal{V} \cap \mathcal{W} = \{\mathbf{0}\}.$$

Proof. The proof is divided into steps; the first serves to justify (14.2).

1. $\mathcal{V} \cap \mathcal{W} \neq \{\mathbf{0}\} \iff \|\Pi_\mathcal{V} \Pi_\mathcal{W}\| = 1$.

If $\|\Pi_\mathcal{V} \Pi_\mathcal{W}\| = 1$ and $\|\Pi_\mathcal{V} \Pi_\mathcal{W} \mathbf{x}\| = \|\Pi_\mathcal{V} \Pi_\mathcal{W}\| \|\mathbf{x}\|$ for some nonzero vector $\mathbf{x} \in \mathbb{C}^n$, then $\|\Pi_\mathcal{W}\| = \|\Pi_\mathcal{V}\| = 1$ and

(14.3) $\|\mathbf{x}\| = \|\Pi_\mathcal{V} \Pi_\mathcal{W} \mathbf{x}\| \leq \|\Pi_\mathcal{V}\| \|\Pi_\mathcal{W} \mathbf{x}\| = \|\Pi_\mathcal{W} \mathbf{x}\| \leq \|\Pi_\mathcal{W}\| \|\mathbf{x}\| = \|\mathbf{x}\|$.

147

Consequently, equality holds throughout (14.3). In particular,
$$\|(I_n - \Pi_\mathcal{W})\mathbf{x}\|^2 = \|\mathbf{x}\|^2 - \|\Pi_\mathcal{W}\mathbf{x}\|^2 = 0\,,$$
which implies that $\mathbf{x} = \Pi_\mathcal{W}\mathbf{x} \in \mathcal{W}$; and
$$\|(I_n - \Pi_\mathcal{V})\mathbf{x}\|^2 = \|\mathbf{x}\|^2 - \|\Pi_\mathcal{V}\mathbf{x}\|^2 = \|\mathbf{x}\|^2 - \|\Pi_\mathcal{V}\Pi_\mathcal{W}\mathbf{x}\|^2 = 0\,,$$
which implies that $\mathbf{x} = \Pi_\mathcal{V}\mathbf{x} \in \mathcal{V}$. Therefore, $\mathbf{x} \in \mathcal{V} \cap \mathcal{W}$, i.e., $\|\Pi_\mathcal{V}\Pi_\mathcal{W}\| = 1 \implies \mathcal{V} \cap \mathcal{W} \neq \{\mathbf{0}\}$.

Conversely, if $\mathbf{x} \in \mathcal{V} \cap \mathcal{W}$ and $\mathbf{x} \neq \mathbf{0}$, then $\Pi_\mathcal{W}\mathbf{x} = \mathbf{x}$ and $\Pi_\mathcal{V}\Pi_\mathcal{W}\mathbf{x} = \Pi_\mathcal{V}\mathbf{x} = \mathbf{x}$. Therefore,
$$\|\mathbf{x}\| = \|\Pi_\mathcal{V}\Pi_\mathcal{W}\mathbf{x}\| \leq \|\Pi_\mathcal{V}\Pi_\mathcal{W}\|\,\|\mathbf{x}\| \leq \|\mathbf{x}\|\,,$$
i.e., $\mathcal{V} \cap \mathcal{W} \neq \{\mathbf{0}\} \implies \|\Pi_\mathcal{V}\Pi_\mathcal{W}\| = 1$.

2. *Verification of* (14.1).

Let $\mathcal{X} = \mathcal{V} \cap \mathcal{W}$ and consider the orthogonal decompositions
$$\mathcal{V} = \mathcal{X} \oplus \mathcal{V}_1 \quad \text{and} \quad \mathcal{W} = \mathcal{X} \oplus \mathcal{W}_1\,.$$
Then, in self-evident notation, the orthogonal projections
$$\Pi_\mathcal{V} = \Pi_\mathcal{X} + \Pi_{\mathcal{V}_1}, \quad \Pi_\mathcal{W} = \Pi_\mathcal{X} + \Pi_{\mathcal{W}_1}$$
and, as $\Pi_\mathcal{X}\Pi_{\mathcal{V}_1} = \Pi_{\mathcal{V}_1}\Pi_\mathcal{X} = O$ and $\Pi_\mathcal{X}\Pi_{\mathcal{W}_1} = \Pi_{\mathcal{W}_1}\Pi_\mathcal{X} = O$,
$$\Pi_\mathcal{V}\Pi_\mathcal{W} = (\Pi_\mathcal{X} + \Pi_{\mathcal{V}_1})(\Pi_\mathcal{X} + \Pi_{\mathcal{W}_1}) = (\Pi_\mathcal{X})^2 + \Pi_{\mathcal{V}_1}\Pi_{\mathcal{W}_1} = \Pi_\mathcal{X} + \Pi_{\mathcal{V}_1}\Pi_{\mathcal{W}_1}\,.$$
Since $\Pi_\mathcal{X}\Pi_{\mathcal{V}_1}\Pi_{\mathcal{W}_1} = O = \Pi_{\mathcal{V}_1}\Pi_{\mathcal{W}_1}\Pi_\mathcal{X}$, we can invoke the binomial formula to obtain
$$(\Pi_\mathcal{V}\Pi_\mathcal{W})^k = (\Pi_\mathcal{X} + \Pi_{\mathcal{V}_1}\Pi_{\mathcal{W}_1})^k = \sum_{j=0}^{k} \binom{k}{j} (\Pi_\mathcal{X})^{k-j} (\Pi_{\mathcal{V}_1}\Pi_{\mathcal{W}_1})^j$$
$$= (\Pi_\mathcal{X})^k + (\Pi_{\mathcal{V}_1}\Pi_{\mathcal{W}_1})^k = \Pi_\mathcal{X} + (\Pi_{\mathcal{V}_1}\Pi_{\mathcal{W}_1})^k$$
for $k = 1, 2, \ldots$. Therefore,
$$\|\Pi_{\mathcal{V} \cap \mathcal{W}} - (\Pi_\mathcal{V}\Pi_\mathcal{W})^k\| = \|(\Pi_{\mathcal{V}_1}\Pi_{\mathcal{W}_1})^k\| \leq \|\Pi_{\mathcal{V}_1}\Pi_{\mathcal{W}_1}\|^k\,,$$
which tends to zero as $k \uparrow \infty$, since
$$\mathcal{V}_1 \cap \mathcal{W}_1 \subseteq \mathcal{V} \cap \mathcal{W} = \mathcal{X} \implies \mathcal{V}_1 \cap \mathcal{W}_1 \subseteq \mathcal{V}_1 \cap \mathcal{X} = \{\mathbf{0}\}\,,$$
and hence, in view of (14.2), $\|\Pi_{\mathcal{V}_1}\Pi_{\mathcal{W}_1}\| < 1$. □

Exercise 14.1. Show that if $\mathcal{V}$ is a subspace of a finite-dimensional inner product space $\mathcal{U}$, $\Pi_\mathcal{V}$ is an orthogonal projection, and $\mathbf{u} \in \mathcal{U}$, then

(14.4) $$\|\Pi_\mathcal{V}\mathbf{u}\|_\mathcal{U} = \|\mathbf{u}\|_\mathcal{U} \iff \Pi_\mathcal{V}\mathbf{u} = \mathbf{u}.$$

Exercise 14.2. Show that if the columns of the matrices $V = \begin{bmatrix} X & V_1 \end{bmatrix}$, $W = \begin{bmatrix} X & W_1 \end{bmatrix}$, and X form orthonormal bases for the subspaces $\mathcal{V}$, $\mathcal{W}$, and $\mathcal{V} \cap \mathcal{W}$, respectively, then $\|V_1^H W_1\| < 1$ and

(14.5) $$\|\Pi_{\mathcal{V} \cap \mathcal{W}} - (\Pi_\mathcal{V} \Pi_\mathcal{W})^k\| = \|(\Pi_{\mathcal{V}_1} \Pi_{\mathcal{W}_1})^k\| = \|(V_1 V_1^H W_1 W_1^H)^k\|.$$

[HINT: $\Pi_\mathcal{V} \Pi_\mathcal{W} = (XX^H + V_1 V_1^H)(XX^H + W_1 W_1^H) = Y + Z$, where $Y = XX^H$, $Z = V_1 V_1^H W_1 W_1^H$, and $YZ = ZY = O$.]

Exercise 14.3. Show that in the setting of Exercise 14.2,

(14.6) $$\|(\Pi_{\mathcal{V}_1} \Pi_{\mathcal{W}_1})^k\| = \|V_1^H W_1\|^{2k-1} = \|\Pi_{\mathcal{V}_1} \Pi_{\mathcal{W}_1}\|^{2k-1}.$$

[HINT: $(\Pi_{\mathcal{V}_1} \Pi_{\mathcal{W}_1})^k = V_1 A (A^H A)^{k-1} W_1^H$ with $A = V_1^H W_1$; and Section 11.5.]

14.2. Computing nonorthogonal projections

A formula for the orthogonal projection $\Pi_\mathcal{V}$ of an inner product space $\mathcal{U}$ onto a subspace $\mathcal{V}$ was presented in Theorem 12.4. This is the projection that is defined by the orthogonal sum decomposition $\mathcal{U} = \mathcal{V} \oplus \mathcal{V}^\perp$. In this section we shall obtain a formula for the projection $P_\mathcal{V}^\mathcal{W}$ of $\mathcal{U}$ onto $\mathcal{V}$ with respect to the direct sum decomposition $\mathcal{U} = \mathcal{V} \dotplus \mathcal{W}$.

To carry out the computations, it is again convenient to draw upon some facts from the theory of positive definite matrices that will be justified in Chapter 16: If $G \in \mathbb{C}^{k \times k}$ and $\mathbf{x}^H G \mathbf{x} > 0$ for every nonzero $\mathbf{x} \in \mathbb{C}^k$, then G is **positive definite** and we indicate this by writing $G \succ O$. Moreover, if $G \in \mathbb{C}^{k \times k}$ and $G \succ O$, then $G = G^H$ and there exists exactly one matrix $F \in \mathbb{C}^{k \times k}$ such that $F \succ O$ and $F^2 = G$. The symbol $G^{1/2}$ is used to denote this matrix.

To warm up, we first consider the special case in which $\mathcal{U}$ is a subspace of $\mathbb{C}^n$.

Theorem 14.2. *If $V \in \mathbb{C}^{n \times p}$, $W \in \mathbb{C}^{n \times q}$, $A = \begin{bmatrix} V & W \end{bmatrix}$, and $\operatorname{rank} A = p + q$, then the Gram matrix $G = A^H A$ is positive definite and the space $\mathcal{U} = \mathcal{R}_A$ is the direct sum of the spaces $\mathcal{V} = \mathcal{R}_V$ and $\mathcal{W} = \mathcal{R}_W$. Moreover,*

(14.7) $$P_\mathcal{V}^\mathcal{W} = \begin{bmatrix} V & O \end{bmatrix} G^{-1} A^H$$

and

(14.8) $$\|P_\mathcal{V}^\mathcal{W}\|^2 = \frac{1}{1 - \|\Pi_\mathcal{V}\Pi_\mathcal{W}\|^2}\,.$$

Proof. Let $\mathbf{u} \in \mathcal{U}$. Then, as $\operatorname{rank} V = p$, $\operatorname{rank} W = q$, and $\mathcal{V} \cap \mathcal{W} = \{\mathbf{0}\}$, there exists exactly one vector $\mathbf{a} \in \mathbb{C}^p$ and exactly one vector $\mathbf{b} \in \mathbb{C}^q$ such that $\mathbf{u} = V\mathbf{a} + W\mathbf{b}$. Thus, $P_\mathcal{V}^\mathcal{W} \mathbf{u} = V\mathbf{a}$. In terms of the blocks G_{ij}, $i,j = 1,2$, of the Gram matrix G,

$$V^H \mathbf{u} = V^H V \mathbf{a} + V^H W \mathbf{b} = G_{11} \mathbf{a} + G_{12} \mathbf{b}$$

and

$$W^H \mathbf{u} = W^H V \mathbf{a} + W^H W \mathbf{b} = G_{21} \mathbf{a} + G_{22} \mathbf{b}\,.$$

Moreover, since $G \succ O$, the matrices G_{11}, G_{22}, and $G_{11} - G_{12} G_{22}^{-1} G_{21}$ are all positive definite. Thus,

(14.9) $$\mathbf{a} = G_{11}^{-1}(V^H \mathbf{u} - G_{12} \mathbf{b}) \quad \text{and} \quad \mathbf{b} = G_{22}^{-1}(W^H \mathbf{u} - G_{21} \mathbf{a})\,,$$

and hence

$$(G_{11} - G_{12} G_{22}^{-1} G_{21})\mathbf{a} = V^H \mathbf{u} - G_{12} G_{22}^{-1} W^H \mathbf{u} = \begin{bmatrix} I_p & -G_{12} G_{22}^{-1} \end{bmatrix} A^H \mathbf{u}\,.$$

Formula (14.7) drops out easily upon invoking the identity

(14.10) $$\begin{bmatrix} I_p & O \end{bmatrix} G^{-1} = (G_{11} - G_{12} G_{22}^{-1} G_{21})^{-1} \begin{bmatrix} I_p & -G_{12} G_{22}^{-1} \end{bmatrix}\,,$$

which is readily obtained from the Schur complement identity (3.19).

To verify (14.8), observe first that

$$\|P_\mathcal{V}^\mathcal{W}\|^2 = \|P_\mathcal{V}^\mathcal{W}(P_\mathcal{V}^\mathcal{W})^H\| = \left\| \begin{bmatrix} V & O \end{bmatrix} G^{-1} A^H A G^{-1} \begin{bmatrix} V^H \\ O \end{bmatrix} \right\|$$

$$= \|V(G_{11} - G_{12} G_{22}^{-1} G_{21})^{-1} V^H\|$$

$$= \|V G_{11}^{-1/2}(I_p - G_{11}^{-1/2} G_{12} G_{22}^{-1} G_{21} G_{11}^{-1/2})^{-1} G_{11}^{-1/2} V^H\|$$

$$= \|(I_p - G_{11}^{-1/2} G_{12} G_{22}^{-1} G_{21} G_{11}^{-1/2})^{-1}\|\,.$$

Thus, if $\operatorname{rank} G_{12} = r$, then the isometric factor V_1 in the singular value decomposition $G_{11}^{-1/2} G_{12} G_{22}^{-1/2} = V_1 S_1 U_1^H$ is of size $p \times r$. If $r < p$, then there exists a matrix V_2 such that $\begin{bmatrix} V_1 & V_2 \end{bmatrix}$ is unitary and hence

$$\|P_\mathcal{V}^\mathcal{W}\|^2 = \|(I_p - V_1 S_1 U_1^H U_1 S_1 V_1^H)^{-1}\|$$

$$= \left\| \left(\begin{bmatrix} V_1 & V_2 \end{bmatrix} \begin{bmatrix} I_r - S_1^2 & O \\ O & I_{p-r} \end{bmatrix} \begin{bmatrix} V_1^H \\ V_2^H \end{bmatrix} \right)^{-1} \right\|$$

$$= \left\| \begin{bmatrix} I_r - S_1^2 & O \\ O & I_{p-r} \end{bmatrix}^{-1} \right\| = \frac{1}{1 - s_1^2}\,,$$

where
$$s_1 = \|G_{11}^{-1/2}G_{12}G_{22}^{-1/2}\| = \|VG_{11}^{-1}G_{12}G_{22}^{-1}W^H\| = \|\Pi_{\mathcal{V}}\Pi_{\mathcal{W}}\|.$$

Since the same conclusion holds when $r = p$, the proof is complete. □

Exercise 14.4. Verify formula (14.10).

Exercise 14.5. Show that in the setting of Theorem 14.2,

(14.11) $\qquad P_{\mathcal{V}}^{\mathcal{W}} = (I_n - \Pi_{\mathcal{V}}\Pi_{\mathcal{W}})^{-1}\Pi_{\mathcal{V}}(I_n - \Pi_{\mathcal{V}}\Pi_{\mathcal{W}}).$

[HINT: Invoke (14.9) and the identities $VG_{11}^{-1}V^H = \Pi_{\mathcal{V}}$, $VG_{11}^{-1}G_{12}W^H = \Pi_{\mathcal{V}}\Pi_{\mathcal{W}}$, and $VG_{11}^{-1}G_{12}G_{22}^{-1}G_{21} = \Pi_{\mathcal{V}}\Pi_{\mathcal{W}}V$.]

Exercise 14.6. Show that if $V \in \mathbb{C}^{n \times p}$ and $W \in \mathbb{C}^{n \times q}$ are isometric matrices, then

(14.12) $\qquad \min\{\|V\mathbf{a} - W\mathbf{b}\|^2 : \|\mathbf{a}\| = \|\mathbf{b}\| = 1\} = 2(1 - \|\Pi_{\mathcal{V}}\Pi_{\mathcal{W}}\|).$

[HINT: First show that the left-hand side of (14.12) is $\geq 2(1 - \|V^H W\|)$.]

Exercise 14.7. Show that if $A = \begin{bmatrix} V & W \end{bmatrix}$ with $V \in \mathbb{C}^{n \times p}$, $W \in \mathbb{C}^{n \times q}$, and $G = A^H A$, then

$$\operatorname{rank} W = q \implies \|V\mathbf{a} - \Pi_{\mathcal{W}}V\mathbf{a}\|^2 = \langle (G_{11} - G_{12}G_{22}^{-1}G_{21})\mathbf{a}, \mathbf{a}\rangle$$

and

$$\operatorname{rank} V = p \implies \|W\mathbf{a} - \Pi_{\mathcal{V}}W\mathbf{a}\|^2 = \langle (G_{22} - G_{21}G_{11}^{-1}G_{12})\mathbf{a}, \mathbf{a}\rangle.$$

14.3. The general setting

In this section we obtain formulas for $P_{\mathcal{V}}^{\mathcal{W}}$ and $\|P_{\mathcal{V}}^{\mathcal{W}}\|_{\{\mathcal{U},\mathcal{U}\}}$ when $\mathcal{V}$ and $\mathcal{W}$ are subspaces of a finite-dimensional inner product space $\mathcal{U}$. To distinguish between the two inner products that will be in play, the subscript $\mathcal{U}$ is added to all norms and inner products in $\mathcal{U}$; the standard inner product and norm are not marked.

Theorem 14.3. If $\mathcal{U} = \mathcal{V} \dotplus \mathcal{W}$, $\{\mathbf{v}_1, \ldots, \mathbf{v}_p\}$ is a basis for $\mathcal{V}$, $\{\mathbf{w}_1, \ldots, \mathbf{w}_q\}$ is a basis for $\mathcal{W}$, and $n = p + q$, then the Gram matrix $G \in \mathbb{C}^{n \times n}$ of the vectors $\{\mathbf{v}_1 \ldots, \mathbf{v}_p, \mathbf{w}_1, \ldots, \mathbf{w}_q\}$, G_{11}, the Gram matrix for $\{\mathbf{v}_1, \ldots, \mathbf{v}_p\}$, G_{22}, the Gram matrix for $\{\mathbf{w}_1, \ldots, \mathbf{w}_q\}$, and $G_{11} - G_{12}G_{22}^{-1}G_{21}$ are all positive definite. Moreover,

(14.13) $\qquad P_{\mathcal{V}}^{\mathcal{W}}\mathbf{u} = \sum_{s=1}^{p} a_s \mathbf{v}_s,$

where

(14.14) $\quad \begin{bmatrix} a_1 & \cdots & a_p \end{bmatrix}^T = \mathbf{a} = (G_{11} - G_{12}G_{22}^{-1}G_{21})^{-1}(\mathbf{c} - G_{12}G_{22}^{-1}\mathbf{d})$,

$$\mathbf{c} = \begin{bmatrix} \langle \mathbf{u}, \mathbf{v}_1 \rangle_{\mathcal{U}}, \ldots, \langle \mathbf{u}, \mathbf{v}_p \rangle_{\mathcal{U}} \end{bmatrix}^T \quad \text{and} \quad \mathbf{d} = \begin{bmatrix} \langle \mathbf{u}, \mathbf{w}_1 \rangle_{\mathcal{U}}, \ldots, \langle \mathbf{u}, \mathbf{w}_q \rangle_{\mathcal{U}} \end{bmatrix}^T,$$

and

(14.15) $\quad \|P_{\mathcal{V}}^{\mathcal{W}}\|_{\{\mathcal{U},\mathcal{U}\}}^2 = \dfrac{1}{1 - \|G_{11}^{-1/2} G_{12} G_{22}^{-1/2}\|^2} = \dfrac{1}{1 - \|\Pi_{\mathcal{V}}\Pi_{\mathcal{W}}\|_{\{\mathcal{U},\mathcal{U}\}}^2}$.

Proof. Since the sum is direct, G, G_{11}, G_{22}, and $G_{11} - G_{12}G_{22}^{-1}G_{21}$ are all positive definite (see (3.19) for the latter). The rest of the proof is divided into parts.

1. *Verification of* (14.13) *and* (14.14).

If

(14.16) $$\mathbf{u} = \sum_{s=1}^{p} a_s \mathbf{v}_s + \sum_{t=1}^{q} b_t \mathbf{w}_t,$$

then

$$\langle \mathbf{u}, \mathbf{v}_i \rangle_{\mathcal{U}} = \sum_{s=1}^{p} a_s \langle \mathbf{v}_s, \mathbf{v}_i \rangle_{\mathcal{U}} + \sum_{t=1}^{q} b_t \langle \mathbf{w}_t, \mathbf{v}_i \rangle_{\mathcal{U}} = \sum_{s=1}^{p} (G_{11})_{is} a_s + \sum_{t=1}^{q} (G_{12})_{it} b_t$$

for $i = 1, \ldots, p$ and

$$\langle \mathbf{u}, \mathbf{w}_i \rangle_{\mathcal{U}} = \sum_{s=1}^{p} a_s \langle \mathbf{v}_s, \mathbf{w}_i \rangle_{\mathcal{U}} + \sum_{t=1}^{q} b_t \langle \mathbf{w}_t, \mathbf{w}_i \rangle_{\mathcal{U}} = \sum_{s=1}^{p} (G_{21})_{is} a_s + \sum_{t=1}^{q} (G_{22})_{it} b_t$$

for $i = 1, \ldots, q$. Therefore,

$$\mathbf{c} = G_{11}\mathbf{a} + G_{12}\mathbf{b} \quad \text{and} \quad \mathbf{d} = G_{21}\mathbf{a} + G_{22}\mathbf{b},$$

with $\mathbf{b} = \begin{bmatrix} b_1 & \cdots & b_q \end{bmatrix}^T$; (14.13) and (14.14) are then obtained by straightforward computation. The details are left to the reader.

2. *Verification of* (14.15).

Let $Q = G_{11}^{-1/2} G_{12} G_{22}^{-1/2}$. Then, since

$$I_q - Q^H Q = G_{22}^{-1/2}(G_{22} - G_{21}G_{11}^{-1}G_{21})G_{22}^{-1/2} \succ O$$

it follows that $\|Q\| < 1$ and hence that $\|Q\| = \cos\theta$ for some $\theta \in (0, \pi/2]$. Thus, in view of formulas (14.16) and (14.14), it is readily checked that

$$\|P_{\mathcal{V}}^{\mathcal{W}}\mathbf{u}\|_{\mathcal{U}}^2 = \|G_{11}^{1/2}\mathbf{a}\|^2 \quad \text{and} \quad \|\mathbf{u}\|_{\mathcal{U}}^2 = \|G_{11}^{1/2}\mathbf{a}\|^2 + \|G_{22}^{1/2}\mathbf{b}\|^2 + 2\Re\langle G_{12}\mathbf{b}, \mathbf{a}\rangle,$$

14.3. The general setting

which, upon setting $\mathbf{x} = G_{11}^{1/2}\mathbf{a}$ and $\mathbf{y} = G_{22}^{1/2}\mathbf{b}$, implies that

$$\frac{\|P_\mathcal{V}^\mathcal{W}\mathbf{u}\|_\mathcal{U}^2}{\|\mathbf{u}\|_\mathcal{U}^2} = \frac{\|\mathbf{x}\|^2}{\|\mathbf{x}\|^2 + \|\mathbf{y}\|^2 + 2\Re\langle Q\mathbf{y}, \mathbf{x}\rangle} \leq \frac{\|\mathbf{x}\|^2}{\|\mathbf{x}\|^2 + \|\mathbf{y}\|^2 - 2\|Q\|\|\mathbf{y}\|\|\mathbf{x}\|}$$

$$= \frac{\|\mathbf{x}\|^2}{\|\mathbf{x}\|^2 + \|\mathbf{y}\|^2 - 2\cos\theta\|\mathbf{y}\|\|\mathbf{x}\|} = \frac{\|\mathbf{x}\|^2}{(\|\mathbf{y}\| - \|\mathbf{x}\|\cos\theta)^2 + \|\mathbf{x}\|^2\sin^2\theta}$$

$$\leq \frac{1}{\sin^2\theta}.$$

This supplies the bound $\|P_\mathcal{V}^\mathcal{W}\|_{\{\mathcal{U},\mathcal{U}\}} \leq 1/\sin\theta$. To complete the proof it remains to show first that this upper bound can be achieved (by choosing $\mathbf{y} = \cos\theta\,\mathbf{u}_1$ and $\mathbf{x} = -\mathbf{v}_1$, where $\mathbf{u}_1$ and $\mathbf{v}_1$ are the first columns in the isometric factors U_1 and V_1 in the singular value decomposition $Q = V_1 S_1 U_1^H$, respectively, and $\cos\theta = s_1(Q) = \|Q\|$) and then to verify that $\|\Pi_\mathcal{V}\Pi_\mathcal{W}\|_{\{\mathcal{U},\mathcal{U}\}} = \|Q\|$. (See Exercises 14.9 and 14.10.) □

Exercise 14.8. Show that if $\mathcal{W}$ is orthogonal to $\mathcal{V}$, then formula (14.14) reduces to $\mathbf{a} = G_{11}^{-1}\mathbf{c}$.

Exercise 14.9. Show that in the setting of Theorem 14.3,

$$\|\Pi_\mathcal{V}\Pi_\mathcal{W}\|_{\{\mathcal{U},\mathcal{U}\}} = \max\left\{\|\Pi_\mathcal{V}\Pi_\mathcal{W}\mathbf{w}\|_\mathcal{U} : \mathbf{w} \in \mathcal{W} \text{ and } \|\mathbf{w}\|_\mathcal{U} = 1\right\}.$$

Exercise 14.10. Show that in the setting of Theorem 14.3,

$$\|\Pi_\mathcal{V}\Pi_\mathcal{W}\|_{\{\mathcal{U},\mathcal{U}\}} = \|G_{11}^{-1/2} G_{12} G_{22}^{-1/2}\|.$$

Theorem 14.4. *In the setting and notation of Theorem 14.3, the projection $P_\mathcal{V}^\mathcal{W}$ can be expressed in terms of the orthogonal projections $\Pi_\mathcal{V}$ and $\Pi_\mathcal{W}$ by the formula*

$$(14.17) \qquad P_\mathcal{V}^\mathcal{W}\mathbf{u} = (I_\mathcal{U} - \Pi_\mathcal{V}\Pi_\mathcal{W})^{-1} \Pi_\mathcal{V} (I_\mathcal{U} - \Pi_\mathcal{V}\Pi_\mathcal{W})\mathbf{u}.$$

Proof. The proof rests on the evaluations

$$\Pi_\mathcal{V}\mathbf{w}_i = \sum_{s=1}^{p} (G_{11}^{-1} G_{12})_{si} \mathbf{v}_s \quad \text{for } i = 1, \ldots, q,$$

$$\Pi_\mathcal{W}\mathbf{v}_j = \sum_{t=1}^{q} (G_{22}^{-1} G_{21})_{tj} \mathbf{w}_t \quad \text{for } j = 1, \ldots, p,$$

$$\Pi_\mathcal{V}\mathbf{u} = \sum_{s=1}^{p} (\mathbf{a} + G_{11}^{-1} G_{12}\mathbf{b})_s \mathbf{v}_s$$

and the formula $P_\mathcal{V}^\mathcal{W}\mathbf{u} = \sum_{s=1}^{p} a_s \mathbf{v}_s$, where $a_1, \ldots, a_p$ are the entries in the vector $\mathbf{a}$ that is specified in (14.14). The proof amounts to checking the following sequence of equalities:

$$(I_{\mathcal{U}} - \Pi_{\mathcal{V}}\Pi_{\mathcal{W}}) P_{\mathcal{V}}^{\mathcal{W}} \mathbf{u} = (I_{\mathcal{U}} - \Pi_{\mathcal{V}}\Pi_{\mathcal{W}}) \sum_{s=1}^{p} a_s \mathbf{v}_s$$

$$= \sum_{s=1}^{p} ((I_p - G_{11}^{-1} G_{12} G_{22}^{-1} G_{21}) \mathbf{a})_s \mathbf{v}_s$$

$$= \Pi_{\mathcal{V}} \mathbf{u} - \sum_{s=1}^{p} (G_{11}^{-1} G_{12} (G_{22}^{-1} G_{21} \mathbf{a} + \mathbf{b}))_s \mathbf{v}_s$$

$$= \Pi_{\mathcal{V}} \mathbf{u} - \Pi_{\mathcal{V}} \Pi_{\mathcal{W}} \mathbf{u} = \Pi_{\mathcal{V}} (I_{\mathcal{U}} - \Pi_{\mathcal{V}} \Pi_{\mathcal{W}}) \mathbf{u},$$

which yields (14.17), since $I_{\mathcal{U}} - \Pi_{\mathcal{V}}\Pi_{\mathcal{W}}$ is invertible. □

Notice that formulas (14.13) and (14.17) for $P_{\mathcal{V}}^{\mathcal{W}} \mathbf{u}$ depend upon both $\mathcal{V}$ and $\mathcal{W}$. But, if $\mathcal{V}$ is orthogonal to $\mathcal{W}$, then $G_{12} = O$ and $\Pi_{\mathcal{V}}\Pi_{\mathcal{W}} = O$ and hence $P_{\mathcal{V}}^{\mathcal{W}} \mathbf{u} = \Pi_{\mathcal{V}} \mathbf{u}$, which depends only on $\mathcal{V}$.

Exercise 14.11. Let

$$A = \begin{bmatrix} \mathbf{a}_1 & \mathbf{a}_2 & \mathbf{a}_3 & \mathbf{a}_4 \end{bmatrix} = \begin{bmatrix} 1 & 1 & 1 & 1 \\ 0 & 1 & 1 & 1 \\ 0 & 0 & 1 & 1 \\ 0 & 0 & 0 & 1 \end{bmatrix}, \quad \mathbf{u} = \begin{bmatrix} 1 \\ 2 \\ 3 \\ 4 \end{bmatrix},$$

$\mathcal{V} = \mathrm{span}\{\mathbf{a}_1, \mathbf{a}_2\}$, $\mathcal{W}_1 = \mathrm{span}\{\mathbf{a}_3, \mathbf{a}_4\}$, and $\mathcal{W}_2 = \mathrm{span}\{\mathbf{a}_1 + \mathbf{a}_3, \mathbf{a}_2 + \mathbf{a}_4\}$.

(a) Show that $\mathbb{C}^4 = \mathcal{V} \dotplus \mathcal{W}_1$ and $\mathbb{C}^4 = \mathcal{V} \dotplus \mathcal{W}_2$, i.e., both sums are direct.

(b) Find the projection of the vector $\mathbf{u}$ onto $\mathcal{V}$ with respect to the decomposition $\mathbb{C}^4 = \mathcal{V} \dotplus \mathcal{W}_1$.

(c) Find the projection of the vector $\mathbf{u}$ onto $\mathcal{V}$ with respect to the decomposition $\mathbb{C}^4 = \mathcal{V} \dotplus \mathcal{W}_2$.

(d) Compute the orthogonal projection of $\mathbf{u}$ onto $\mathcal{V}$.

[REMARK: Exploit the fact that A is block triangular to compute A^{-1}.]

14.4. Detour on the angle between subspaces

The symbol $\cos\theta$ was introduced as a convenient shorthand for the norm of the operator Q in step 2 of the proof of Theorem 14.3. But there is more to this choice of notation than the fact that $\|Q\| \leq 1$. The story begins with a definition:

The **angle** θ between a pair of subspaces $\mathcal{V}$ and $\mathcal{W}$ of a finite-dimensional inner product space $\mathcal{U}$ is defined in the interval $[0, \pi/2]$ by the formula

(14.18) $\quad \cos\theta = \max\{|\langle \mathbf{v}, \mathbf{w} \rangle| : \mathbf{v} \in \mathcal{V}, \ \mathbf{w} \in \mathcal{W}, \ \text{and} \ \ \|\mathbf{v}\| = \|\mathbf{w}\| = 1\}.$

14.4. Detour on the angle between subspaces

The next few exercises are designed to give just a little additional insight. They can be skipped without loss of continuity.

Exercise 14.12. Show that the angle θ defined in (14.18) can also be expressed in terms of the orthogonal projectors $\Pi_\mathcal{V}$ and $\Pi_\mathcal{W}$ by the formula $\cos\theta = \|\Pi_\mathcal{V}\Pi_\mathcal{W}\|$.

Exercise 14.13. Show that in the setting of Exercise 14.12,

$$(14.19) \quad \min\{\|\Pi_\mathcal{V}\mathbf{w}\|^2 : \mathbf{w} \in \mathcal{W} \quad \text{and} \quad \|\mathbf{w}\| = 1\} = 1 - \|(I - \Pi_\mathcal{V})\Pi_\mathcal{W}\|^2$$

and

$$(14.20) \quad \min\{\|(I - \Pi_\mathcal{V})\mathbf{w}\|^2 : \mathbf{w} \in \mathcal{W} \quad \text{and} \quad \|\mathbf{w}\| = 1\} = 1 - \|\Pi_\mathcal{V}\Pi_\mathcal{W}\|^2.$$

[HINT: $\|\mathbf{w}\|^2 = \|\Pi_\mathcal{V}\mathbf{w}\|^2 + \|(I - \Pi_\mathcal{V})\mathbf{w}\|^2.$]

In view of (14.18) and (14.20),

$$(14.21) \quad \sin^2\theta = \min\{\|(I - \Pi_\mathcal{V})\mathbf{w}\|^2 : \mathbf{w} \in \mathcal{W} \quad \text{and} \quad \|\mathbf{w}\| = 1\}.$$

Exercise 14.14. Show that if $\mathcal{V}$ and $\mathcal{W}$ are subspaces of a finite-dimensional inner product space $\mathcal{U}$, then

$$(14.22) \quad \mathbf{w} \in \mathcal{W} \implies \min\{\|\mathbf{v} - \mathbf{w}\| : \mathbf{v} \in \mathcal{V}\} = \|(I - \Pi_\mathcal{V})\mathbf{w}\|$$

and

$$(14.23) \quad \mathbf{v} \in \mathcal{V} \implies \min\{\|\mathbf{v} - \mathbf{w}\| : \mathbf{w} \in \mathcal{W}\} = \|(I - \Pi_\mathcal{W})\mathbf{v}\|.$$

Exercise 14.15. Show that if, in the notation of Exercise 14.14, we set $d(\mathbf{v}; \mathcal{W}) = \min\{\|\mathbf{v} - \mathbf{w}\| : \mathbf{w} \in \mathcal{W}\}$ (which is a reasonable measure of the distance from a point $\mathbf{v} \in \mathcal{V}$ to $\mathcal{W}$), then

$$(14.24) \quad \max\{d(\mathbf{v}, \mathcal{W}) : \mathbf{v} \in \mathcal{V} \quad \text{and} \quad \|\mathbf{v}\| = 1\} = \|(I - \Pi_\mathcal{W})\Pi_\mathcal{V}\|.$$

Exercise 14.16. Show that

$$(14.25) \quad \|(\Pi_\mathcal{V} - \Pi_\mathcal{W})\mathbf{u}\|^2 = \|(\Pi_\mathcal{V}(I - \Pi_\mathcal{W}))\mathbf{u}\|^2 + \|((I - \Pi_\mathcal{V})\Pi_\mathcal{W})\mathbf{u}\|^2$$

for every vector $\mathbf{u} \in \mathcal{U}$ and

$$(14.26) \quad \|\Pi_\mathcal{V} - \Pi_\mathcal{W}\| = \max\{\|\Pi_\mathcal{V}(I - \Pi_\mathcal{W})\|, \|(I - \Pi_\mathcal{V})\Pi_\mathcal{W}\|\}.$$

[HINT: To obtain (14.26) from (14.25), let $\mathbf{u} = \cos\theta\mathbf{w} + \sin\theta\mathbf{x}$, where $\mathbf{w} \in \mathcal{W}$, $\mathbf{x}$ is orthogonal to $\mathcal{W}$, and $\|\mathbf{w}\| = \|\mathbf{x}\| = 1$.]

14.5. Areas, volumes, and determinants

To warm up, consider the following:

Exercise 14.17. Let $\mathbf{a}, \mathbf{b} \in \mathbb{R}^2$, let $V = [\mathbf{a} \quad \mathbf{b}]$ denote the 2×2 matrix with columns $\mathbf{a}$ and $\mathbf{b}$ in the first quadrant, and let $G = V^H V$ denote the Gram matrix of the given vectors. Then the area of the parallelogram generated by $\mathbf{a}$ and $\mathbf{b}$ is equal to $(\det G)^{1/2}$.

Lemma 14.5. *Let* $\mathbf{a}, \mathbf{b} \in \mathbb{R}^n$, *let* $V = [\mathbf{a} \quad \mathbf{b}]$ *denote the* $n \times 2$ *matrix with columns* $\mathbf{a}$ *and* $\mathbf{b}$, *and let* $G = V^H V$ *denote the Gram matrix of the given vectors. Then the area of the parallelogram generated by* $\mathbf{a}$ *and* $\mathbf{b}$ *is equal to* $(\det G)^{1/2}$.

Proof. Suppose first that $\mathbf{a}$ and $\mathbf{b}$ are linearly independent and let $\mathcal{A} = \mathrm{span}\{\mathbf{a}\}$. Since $\mathrm{span}\{\mathbf{a}\} = \mathrm{span}\{-\mathbf{a}\}$ and $\det G$ does not change if $\mathbf{a}$ is replaced by $-\mathbf{a}$, we can suppose without loss of generality that the angle θ between $\mathbf{a}$ and $\mathbf{b}$ is between 0 and $\pi/2$. Then $\mathbf{b}$ admits the orthogonal decomposition $\mathbf{b} = \Pi_\mathcal{A} \mathbf{b} + (I - \Pi_\mathcal{A})\mathbf{b}$ and the area α of the parallelogram is equal to $\|\mathbf{a}\| \|\mathbf{b}\| \sin \theta = \|\mathbf{a}\| \|(I - \Pi_\mathcal{A})\mathbf{b}\|$. Consequently,

$$\alpha^2 = \|(I - \Pi_\mathcal{A})\mathbf{b}\|^2 \|\mathbf{a}\|^2$$
$$= \langle (I - \Pi_\mathcal{A})\mathbf{b}, (I - \Pi_\mathcal{A})\mathbf{b} \rangle \|\mathbf{a}\|^2$$
$$= \langle (I - \Pi_\mathcal{A})\mathbf{b}, \mathbf{b} \rangle \|\mathbf{a}\|^2$$
$$= \|\mathbf{b}\|^2 \|\mathbf{a}\|^2 - \langle \Pi_\mathcal{A}\mathbf{b}, \mathbf{b} \rangle \|\mathbf{a}\|^2 \,.$$

Thus, as
$$\Pi_\mathcal{A} \mathbf{b} = \mathbf{a}(\mathbf{a}^H \mathbf{a})^{-1} \mathbf{a}^H \mathbf{b} = \langle \mathbf{b}, \mathbf{a} \rangle \|\mathbf{a}\|^{-2} \mathbf{a}$$

by formula (12.15),

(14.27) $$\alpha^2 = \|\mathbf{a}\|^2 \|\mathbf{b}\|^2 - |\langle \mathbf{a}, \mathbf{b} \rangle|^2 \,.$$

To complete the proof, observe that the Gram matrix

$$G = V^H V = \begin{bmatrix} \mathbf{a}^H \\ \mathbf{b}^H \end{bmatrix} [\mathbf{a} \quad \mathbf{b}]$$

$$= \begin{bmatrix} \mathbf{a}^H \mathbf{a} & \mathbf{a}^H \mathbf{b} \\ \mathbf{b}^H \mathbf{a} & \mathbf{b}^H \mathbf{b} \end{bmatrix} = \begin{bmatrix} \|\mathbf{a}\|^2 & \langle \mathbf{b}, \mathbf{a} \rangle \\ \langle \mathbf{a}, \mathbf{b} \rangle & \|\mathbf{b}\|^2 \end{bmatrix}$$

and hence that

(14.28) $$\det G = \|\mathbf{a}\|^2 \|\mathbf{b}\|^2 - |\langle \mathbf{a}, \mathbf{b} \rangle|^2 = \alpha^2 \,.$$

This completes the proof of the asserted formula when $\mathbf{a}$ and $\mathbf{b}$ are linearly independent. Formula (14.28) remains valid, however, even if $\mathbf{a}$ and $\mathbf{b}$ are linearly dependent because then both sides are equal to zero. $\square$

14.5. Areas, volumes, and determinants

As a byproduct of the proof of the last lemma we obtain the formula
$$\text{(14.29)} \qquad |\langle \mathbf{a}, \mathbf{b} \rangle|^2 = \|\mathbf{a}\|^2 \|\mathbf{b}\|^2 - \alpha^2 .$$

This yields another proof of the Cauchy-Schwarz inequality for vectors in $\mathbb{R}^n$:
$$|\langle \mathbf{a}, \mathbf{b} \rangle| \leq \|\mathbf{a}\| \|\mathbf{b}\|$$
with equality if and only if the area is equal to zero, i.e., if and only if $\mathbf{a}$ and $\mathbf{b}$ are linearly dependent.

Formula (14.28) is a special case of the formula for the volume of the **parallelepiped**
$$\mathcal{P}(\mathbf{v}_1, \ldots, \mathbf{v}_k) = \left\{ t_1 \mathbf{v}_1 + \cdots + t_k \mathbf{v}_k : t_j \geq 0 \text{ for } j = 1, \ldots, k \text{ and } \sum_{j=1}^{k} t_j = 1 \right\}$$
generated by the vectors $\mathbf{v}_1, \ldots, \mathbf{v}_k$ in $\mathbb{R}^n$ that is defined inductively by the formula
$$\text{(14.30)} \qquad \operatorname{vol} \mathcal{P}(\mathbf{v}_1, \ldots, \mathbf{v}_{j+1}) = \operatorname{vol} \mathcal{P}(\mathbf{v}_1, \ldots, \mathbf{v}_j) \|(I_n - \Pi_{\mathcal{V}_j}) \mathbf{v}_{j+1}\|,$$
where $\Pi_{\mathcal{V}_j}$ denotes the orthogonal projection of $\mathbb{R}^n$ onto
$$\mathcal{V}_j = \operatorname{span}\{\mathbf{v}_1, \ldots, \mathbf{v}_j\}.$$

Theorem 14.6. *If*
$$G = \begin{bmatrix} G_{11} & G_{12} \\ G_{21} & G_{22} \end{bmatrix} \quad \text{with } G_{11} \in \mathbb{R}^{(k-1) \times (k-1)} \text{ and } G_{22} \in \mathbb{R}$$
is the Gram matrix of a set of vectors $\{\mathbf{v}_1, \ldots, \mathbf{v}_k\}$ in $\mathbb{R}^n$, then:

(1) *The volume of the parallelepiped generated by these vectors is given by the formula*
$$\text{(14.31)} \qquad \operatorname{vol} \mathcal{P}(\mathbf{v}_1, \ldots, \mathbf{v}_k) = (\det G)^{1/2} .$$

(2) *If G_{11} is invertible,*
$$\text{(14.32)} \qquad \|(I_n - \Pi_{\mathcal{V}_{k-1}}) \mathbf{v}_k\|^2 = G_{22} - G_{21} G_{11}^{-1} G_{12} .$$

Proof. If the vectors $\mathbf{v}_1, \ldots, \mathbf{v}_{k-1}$ are linearly dependent, then $\det G_{11} = 0$ and $\det G = 0$. Thus, in view of (14.30) with $j = k-1$, both sides of (14.31) are equal to zero.

If the vectors $\mathbf{v}_1, \ldots, \mathbf{v}_{k-1}$ are linearly independent and
$$V = \begin{bmatrix} \mathbf{v}_1 & \cdots & \mathbf{v}_{k-1} \end{bmatrix},$$
then $V^H V = G_{11}$, $V^H \mathbf{v}_k = G_{12} = G_{21}^H$, $\mathbf{v}_k^H \mathbf{v}_k = G_{22}$, and
$$\begin{aligned}
\|(I_n - \Pi_{\mathcal{V}_{k-1}}) \mathbf{v}_k\|^2 &= \langle (I_n - V(V^H V)^{-1} V^H) \mathbf{v}_k, \mathbf{v}_k \rangle \\
&= \mathbf{v}_k^H \mathbf{v}_k - \mathbf{v}_k^H V(V^H V)^{-1} V^H \mathbf{v}_k = G_{22} - G_{21} G_{11}^{-1} G_{12},
\end{aligned}$$

which justifies (2) and so too the formula

$$\{\operatorname{vol}\mathcal{P}(\mathbf{v}_1,\ldots,\mathbf{v}_k)\}^2 = \{\operatorname{vol}\mathcal{P}(\mathbf{v}_1,\ldots,\mathbf{v}_{k-1})\}^2 \, (G_{22} - G_{21}G_{11}^{-1}G_{12})$$

for every $k \geq 3$. Thus, if (14.31) holds for $k-1$ vectors, then the first factor on the right is equal to $\det G_{11}$ and hence, as $(G_{22} - G_{21}G_{11}^{-1}G_{12}) \in \mathbb{R}$ in this case,

$$\{\operatorname{vol}\mathcal{P}(\mathbf{v}_1,\ldots,\mathbf{v}_k)\}^2 = \det G_{11} \det(G_{22} - G_{21}G_{11}^{-1}G_{12}) = \det G,$$

by (7.3). Consequently, if (14.31) holds for k, then it holds for $k+1$. Thus, as (14.31) holds for $k=2$, it holds for all positive integers $k \geq 2$. □

14.6. Trace formulas

The **trace** of a linear transformation T that maps an inner product space $\mathcal{U}$ into itself is defined by the formula

$$(14.33) \qquad \operatorname{trace} T = \sum_{j=1}^{n} \langle T\mathbf{u}_j, \mathbf{u}_j \rangle_{\mathcal{U}},$$

where $\{\mathbf{u}_1,\ldots,\mathbf{u}_n\}$ is any orthonormal basis for $\mathcal{U}$.

The fact that the sum in (14.33) is independent of the choice of the orthonormal basis will be established in Lemma 14.7. This is perhaps less surprising if you keep in mind that if $\mathcal{U} = \mathbb{C}^n$, $A \in \mathbb{C}^{n\times n}$, $I_n = \begin{bmatrix} \mathbf{e}_1 & \cdots & \mathbf{e}_n \end{bmatrix}$, and $\{\mathbf{u}_1,\ldots,\mathbf{u}_n\}$ is an orthonormal basis for $\mathbb{C}^n$, then $U = \begin{bmatrix} \mathbf{u}_1 & \cdots & \mathbf{u}_n \end{bmatrix}$ is unitary and, in view of (7.13) and (7.19),

$$\sum_{j=1}^{n} \langle A\mathbf{u}_j, \mathbf{u}_j \rangle = \operatorname{trace} U^H A U = \operatorname{trace} A = \sum_{j=1}^{n} a_{jj} = \sum_{j=1}^{n} \langle A\mathbf{e}_j, \mathbf{e}_j \rangle.$$

Thus, formula (7.13) is consistent with formula (14.33).

Lemma 14.7. *Let T be a linear transformation from an inner product space $\mathcal{U}$ into itself and let $\{\mathbf{u}_1,\ldots,\mathbf{u}_n\}$ and $\{\mathbf{w}_1,\ldots,\mathbf{w}_n\}$ be any two orthonormal bases for $\mathcal{U}$. Then*

$$(14.34) \qquad \sum_{j=1}^{n} \langle T\mathbf{u}_j, \mathbf{u}_j \rangle_{\mathcal{U}} = \sum_{j=1}^{n} \langle T\mathbf{w}_j, \mathbf{w}_j \rangle_{\mathcal{U}}.$$

Proof. Since

$$T\mathbf{u}_j = \sum_{i=1}^{n} \langle T\mathbf{u}_j, \mathbf{w}_i \rangle_{\mathcal{U}} \mathbf{w}_i \quad \text{and} \quad \mathbf{u}_j = \sum_{k=1}^{n} \langle \mathbf{u}_j, \mathbf{w}_k \rangle_{\mathcal{U}} \mathbf{w}_k,$$

formula (12.5) ensures that

$$\sum_{j=1}^{n} \langle T\mathbf{u}_j, \mathbf{u}_j \rangle_{\mathcal{U}} = \sum_{j=1}^{n} \left\{ \sum_{i=1}^{n} \langle T\mathbf{u}_j, \mathbf{w}_i \rangle_{\mathcal{U}} \overline{\langle \mathbf{u}_j, \mathbf{w}_i \rangle_{\mathcal{U}}} \right\}$$

$$= \sum_{i=1}^{n} \left\{ \sum_{j=1}^{n} \langle \mathbf{w}_i, \mathbf{u}_j \rangle_{\mathcal{U}} \overline{\langle T^*\mathbf{w}_i, \mathbf{u}_j \rangle_{\mathcal{U}}} \right\}$$

$$= \sum_{i=1}^{n} \langle \mathbf{w}_i, T^*\mathbf{w}_i \rangle_{\mathcal{U}} = \sum_{i=1}^{n} \langle T\mathbf{w}_i, \mathbf{w}_i \rangle_{\mathcal{U}},$$

as claimed. $\square$

Exercise 14.18. Show that in the setting of Lemma 14.7

(14.35) $$\sum_{j=1}^{n} \|T\mathbf{u}_j\|_{\mathcal{U}}^2 = \sum_{j=1}^{n} \|T\mathbf{w}_j\|_{\mathcal{U}}^2 = \sum_{j=1}^{n} \|T^*\mathbf{u}_j\|_{\mathcal{U}}^2.$$

Lemma 14.8. *Let $\mathcal{U}$ and $\mathcal{V}$ be finite-dimensional inner product spaces and let T be a linear transformation from $\mathcal{U}$ into $\mathcal{V}$ and let S be a linear transformation from $\mathcal{V}$ into $\mathcal{U}$. Then*

(14.36) $$\operatorname{trace} ST = \operatorname{trace} TS.$$

Proof. Let $\mathbf{u}_1, \ldots, \mathbf{u}_q$ be an orthonormal basis for $\mathcal{U}$ and let $\mathbf{v}_1, \ldots, \mathbf{v}_p$ be an orthonormal basis for $\mathcal{V}$. Then, in view of the definition of the adjoint of a linear transformation and (12.5),

$$\sum_{j=1}^{q} \langle ST\mathbf{u}_j, \mathbf{u}_j \rangle_{\mathcal{U}} = \sum_{j=1}^{q} \langle T\mathbf{u}_j, S^*\mathbf{u}_j \rangle_{\mathcal{V}} = \sum_{j=1}^{q} \left\{ \sum_{i=1}^{p} \langle T\mathbf{u}_j, \mathbf{v}_i \rangle_{\mathcal{V}} \overline{\langle S^*\mathbf{u}_j, \mathbf{v}_i \rangle_{\mathcal{V}}} \right\}$$

$$= \sum_{i=1}^{p} \left\{ \sum_{j=1}^{q} \langle S\mathbf{v}_i, \mathbf{u}_j \rangle_{\mathcal{U}} \overline{\langle T^*\mathbf{v}_i, \mathbf{u}_j \rangle_{\mathcal{U}}} \right\}$$

$$= \sum_{i=1}^{p} \langle S\mathbf{v}_i, T^*\mathbf{v}_i \rangle_{\mathcal{U}} = \sum_{i=1}^{p} \langle TS\mathbf{v}_i, \mathbf{v}_i \rangle_{\mathcal{V}},$$

which justifies (14.36). $\square$

Corollary 14.9. *If T is a linear transformation from a finite-dimensional inner product space $\mathcal{U}$ into an inner product space $\mathcal{V}$ and if $\{\mathbf{u}_1, \ldots, \mathbf{u}_q\}$ is an orthonormal basis for $\mathcal{U}$ and $\{\mathbf{v}_1, \ldots, \mathbf{v}_p\}$ is an orthonormal basis for $\mathcal{R}_T$, then*

(14.37) $$\sum_{j=1}^{q} \|T\mathbf{u}_j\|_{\mathcal{V}}^2 = \sum_{i=1}^{p} \|T^*\mathbf{v}_i\|_{\mathcal{U}}^2.$$

Proof. This is immediate from Lemma 14.8, upon viewing T as a transformation onto the finite-dimensional subspace $\mathcal{R}_T$ of $\mathcal{V}$ and then setting $S = T^*$. $\square$

14.7. Supplementary notes

This chapter was partially adapted from Chapters 8 and 9 of [30]. Section 14.1 was motivated by a report by Klaus Diepold [24] on the design of sensors for cars. A formula for $\|P_\mathcal{V}^\mathcal{W}\|$ appears in the paper [53] by Krein and Spitkovsky. Exercises 14.14–14.16 are adapted from the discussion in the book [39] by Glazman and Ljubic of (in their terminology) the aperture between subspaces.

Chapter 15

Singular value decomposition

In this chapter we introduce the singular value decomposition for matrices $A \in \mathbb{C}^{p \times q}$. It is convenient to first review some facts about matrices $A \in \mathbb{C}^{p \times q}$ that preserve norm: $\|A\mathbf{u}\| = \|\mathbf{u}\|$ for every vector $\mathbf{u} \in \mathbb{C}^q$.

- **Warning:** Recall that, unless explicitly indicated otherwise, $\|\mathbf{u}\| = \sqrt{\langle \mathbf{u}, \mathbf{u} \rangle}$ and $\langle \mathbf{u}, \mathbf{v} \rangle = \langle \mathbf{u}, \mathbf{v} \rangle_{\text{st}}$ for vectors $\mathbf{u}, \mathbf{v} \in \mathbb{C}^n$ and $\|A\| = \|A\|_{2,2}$ for matrices A.

- A matrix $A \in \mathbb{C}^{p \times q}$ is said to be **isometric** if $A^H A = I_q$. The name stems from the fact that

(15.1) $\quad A^H A = I_q \iff \langle A\mathbf{x}, A\mathbf{x} \rangle = \langle \mathbf{x}, \mathbf{x} \rangle \quad$ for every $\mathbf{x} \in \mathbb{C}^q$.

If $A \in \mathbb{C}^{p \times q}$ is isometric, then $\operatorname{rank} A = q$ and hence $p \geq q$. Thus, there are two possibilities:
(1) If $p = q$, then $A^H A = I_p \implies A A^H = I_p$.
(2) If $p > q$, then $A^H A = I_p \implies A A^H = A(A^H A)^{-1} A^H = \Pi_{\mathcal{R}_A}$, the orthogonal projection of $\mathbb{C}^p$ onto $\mathcal{R}_A$.

In case (1), A and A^H are both **isometric**; in case (2), A is an isometry and A^H is a **partial isometry**, i.e., it is isometric on the orthogonal complement of $\mathcal{N}_{A^H}$ (i.e., on $\mathcal{R}_A$).

- A matrix $A \in \mathbb{C}^{p \times q}$ is said to be **unitary** if it is both isometric and invertible, i.e., if

$$A^H A = I_q \quad \text{and} \quad A A^H = I_p, \quad \text{which is only possible if } q = p.$$

If $U_1 \in \mathbb{C}^{n \times p}$ is isometric and $n - p = q \geq 1$, then there exists a second isometric matrix $U_2 \in \mathbb{C}^{n \times q}$ such that $U = \begin{bmatrix} U_1 & U_2 \end{bmatrix}$ is a unitary matrix.

Thus,

$$I_n = U^H U = \begin{bmatrix} U_1^H \\ U_2^H \end{bmatrix} \begin{bmatrix} U_1 & U_2 \end{bmatrix} = \begin{bmatrix} U_1^H U_1 & U_1^H U_2 \\ U_2^H U_1 & U_2^H U_2 \end{bmatrix} = \begin{bmatrix} I_p & O_{p \times q} \\ O_{q \times p} & I_q \end{bmatrix},$$

$$I_n = UU^H = \begin{bmatrix} U_1 & U_2 \end{bmatrix} \begin{bmatrix} U_1^H \\ U_2^H \end{bmatrix} = U_1 U_1^H + U_2 U_2^H = \Pi_{\mathcal{R}_{U_1}} + \Pi_{\mathcal{R}_{U_2}},$$

and U_1^H and U_2^H are both partial isometries. In particular,

$$\langle \mathbf{x}, \mathbf{x} \rangle = \langle (U_1 U_1^H + U_2 U_2^H)\mathbf{x}, \mathbf{x} \rangle = \langle U_1^H \mathbf{x}, U_1^H \mathbf{x} \rangle + \langle U_2^H \mathbf{x}, U_2^H \mathbf{x} \rangle$$
$$\geq \langle U_1^H \mathbf{x}, U_1^H \mathbf{x} \rangle, \quad \text{with equality if and only if } \mathbf{x} \in \mathcal{R}_{U_1}.$$

Exercise 15.1. Justify the equivalence in (15.1). [HINT: Exploit Exercise 11.24.]

Exercise 15.2. Show that if $A \in \mathbb{C}^{p \times p}$, then $A^H A = I_p \iff AA^H = I_p$ and hence **every isometric matrix $A \in \mathbb{C}^{p \times p}$ is automatically unitary**.

15.1. Singular value decompositions

The statement of the next theorem is rather long. The main fact to focus on at a first pass is (15.2). The remaining assertions can be absorbed later, as needed.

Theorem 15.1. *If $A \in \mathbb{C}^{p \times q}$ and $\operatorname{rank} A = r$, with $r \geq 1$, then there exist a pair of isometric matrices*

$$U_1 = \begin{bmatrix} \mathbf{u}_1 & \cdots & \mathbf{u}_r \end{bmatrix} \in \mathbb{C}^{q \times r} \quad \text{and} \quad V_1 = \begin{bmatrix} \mathbf{v}_1 & \cdots & \mathbf{v}_r \end{bmatrix} \in \mathbb{C}^{p \times r}$$

and a diagonal matrix

$$S_1 = \operatorname{diag}\{s_1, \ldots, s_r\} \in \mathbb{R}^{r \times r} \quad \text{with} \quad s_1 \geq \cdots \geq s_r > 0$$

such that

$$(15.2) \qquad A = \sum_{j=1}^r s_j \mathbf{v}_j \mathbf{u}_j^H = V_1 S_1 U_1^H.$$

Moreover, $\mathcal{R}_A = \mathcal{R}_{V_1}$, $\mathcal{N}_A = \mathcal{R}_{U_2}$, $\mathcal{R}_{A^H} = \mathcal{R}_{U_1}$, $\mathcal{N}_{A^H} = \mathcal{R}_{V_2}$, and:

(1) $V_1 V_1^H = \Pi_{\mathcal{R}_A}$, *the orthogonal projection of $\mathbb{C}^p$ onto $\mathcal{R}_A$, and $V_1 U_1^H$ maps $\mathcal{R}_{A^H}$ isometrically onto $\mathcal{R}_A$.*

(2) $U_1 U_1^H = \Pi_{\mathcal{R}_{A^H}}$, *the orthogonal projection of $\mathbb{C}^q$ onto $\mathcal{R}_{A^H}$, and $U_1 V_1^H$ maps $\mathcal{R}_A$ isometrically onto $\mathcal{R}_{A^H}$.*

(3) *If $r < p$ and $V_2 \in \mathbb{C}^{p \times (p-r)}$ is any isometric matrix such that $V = \begin{bmatrix} V_1 & V_2 \end{bmatrix}$ is unitary, then $V_2 V_2^H = \Pi_{\mathcal{N}_{A^H}}$, the orthogonal projection of $\mathbb{C}^p$ onto $\mathcal{N}_{A^H}$.*

15.1. Singular value decompositions

(4) If $r < q$ and $U_2 \in \mathbb{C}^{q \times (q-r)}$ is any isometric matrix such that $U = \begin{bmatrix} U_1 & U_2 \end{bmatrix}$ is unitary, then $U_2 U_2^H = \Pi_{\mathcal{N}_A}$, the orthogonal projection of $\mathbb{C}^q$ onto $\mathcal{N}_A$.

(5) $A^H = \sum_{j=1}^r s_j \mathbf{u}_j \mathbf{v}_j^H = U_1 S_1 V_1^H$.

(6) If also $X_1 \in \mathbb{C}^{q \times r}$ and $Y_1 \in \mathbb{C}^{p \times r}$ are isometric matrices such that $A = V_1 S_1 U_1^H = Y_1 S_1 X_1^H$, then $V_1 U_1^H = Y_1 X_1^H$.

Proof. Let $A \in \mathbb{C}^{p \times q}$ with $\operatorname{rank} A = r \geq 1$. Then, since $A^H A \in \mathbb{C}^{q \times q}$ and $(A^H A)^H = A^H A$, there exists a unitary matrix $U = \begin{bmatrix} \mathbf{u}_1 & \cdots & \mathbf{u}_q \end{bmatrix}$ such that
$$A^H A U = U D \quad \text{with } D = \operatorname{diag}\{\mu_1, \ldots, \mu_q\} \in \mathbb{R}^{q \times q}.$$
Thus, $\operatorname{rank} A = \operatorname{rank} A^H A = \operatorname{rank} D$ and
$$\mu_j = \mu_j \langle \mathbf{u}_j, \mathbf{u}_j \rangle = \langle A^H A \mathbf{u}_j, \mathbf{u}_j \rangle = \langle A \mathbf{u}_j, A \mathbf{u}_j \rangle \geq 0.$$
Consequently, upon setting $\mu_j = s_j^2$ with $s_j \geq 0$ and rearranging the indices, if need be, so that $s_1 \geq s_2 \geq \cdots \geq s_q$, the preceding formula for $A^H A$ can be expressed as

$$(15.3) \quad A^H A U = U \begin{bmatrix} s_1^2 & 0 & \cdots & 0 \\ 0 & s_2^2 & \cdots & 0 \\ \vdots & & \ddots & \vdots \\ 0 & 0 & \cdots & s_q^2 \end{bmatrix} \quad \text{with } s_1 \geq s_2 \geq \cdots \geq s_q \geq 0.$$

Formula (15.3) implies that $\|A\mathbf{u}_j\| = s_j$ for $j = 1, \ldots, q$ and
$$\operatorname{rank} A = r \iff s_r > 0 \text{ and } s_j = 0 \text{ for } j > r \text{ (if } r < q\text{)}.$$
Let
$$\mathbf{v}_j = s_j^{-1} A \mathbf{u}_j \quad \text{for } j = 1, \ldots, r.$$
Then
$$\langle \mathbf{v}_j, \mathbf{v}_k \rangle = \langle s_j^{-1} A \mathbf{u}_j, s_k^{-1} A \mathbf{u}_k \rangle = \{s_j s_k\}^{-1} \langle A^H A \mathbf{u}_j, \mathbf{u}_k \rangle$$
$$= s_j s_k^{-1} \langle \mathbf{u}_j, \mathbf{u}_k \rangle = \begin{cases} 1 & \text{if } j = k, \\ 0 & \text{if } j \neq k. \end{cases}$$
Thus,
$$AU_1 = A \begin{bmatrix} \mathbf{u}_1 & \cdots & \mathbf{u}_r \end{bmatrix} = \begin{bmatrix} A\mathbf{u}_1 & \cdots & A\mathbf{u}_r \end{bmatrix} = \begin{bmatrix} s_1 \mathbf{v}_1 & \cdots & s_r \mathbf{v}_r \end{bmatrix} = V_1 S_1.$$
If $r = q$, then $U_1 = U$ is unitary and $A = V_1 S_1 U_1^H$. If $r < q$ and $U_2 = \begin{bmatrix} \mathbf{u}_{r+1} & \cdots & \mathbf{u}_q \end{bmatrix}$, then $AU_2 = O$ and
$$V_1 S_1 U_1^H = AU_1 U_1^H = A(U_1 U_1^H + U_2 U_2^H) = AUU^H = A.$$

This completes the justification of (15.2). Assertions (1)–(5) are left to the reader to fill in at his/her leisure; (6) is covered by Corollary 16.8 in the next chapter. □

The numbers $s_1 \geq \cdots \geq s_q \geq 0$ in the decomposition of $A^H A$ in (15.3) are termed the **singular values** of A; formula (15.2) is called the **singular value decomposition** of A, or the **svd** of A for short.

Exercise 15.3. Show that if $A \in \mathbb{R}^{p \times q}$ with rank $A = r \geq 1$, then (15.2) holds with $V_1 \in \mathbb{R}^{p \times r}$ and $U_1 \in \mathbb{R}^{q \times r}$. [HINT: Keep Theorem 13.7 in mind.]

Remark 15.2. In the setting and terminology of Theorem 15.1 formula (15.2) can be expressed in a number of different ways:

(1) If $r < q$ and $U = \begin{bmatrix} U_1 & U_2 \end{bmatrix}$ is unitary, then

$$(15.4) \qquad A = V_1 \begin{bmatrix} S_1 & O_{r \times (q-r)} \end{bmatrix} U^H.$$

(2) If $r < p$ and $V = \begin{bmatrix} V_1 & V_2 \end{bmatrix}$ is unitary, then

$$(15.5) \qquad A = V \begin{bmatrix} S_1 \\ O_{(p-r) \times r} \end{bmatrix} U_1^H.$$

(3) If $r < \min\{p, q\}$ and $U = \begin{bmatrix} U_1 & U_2 \end{bmatrix} \in \mathbb{C}^{q \times q}$ and $V = \begin{bmatrix} V_1 & V_2 \end{bmatrix} \in \mathbb{C}^{p \times p}$ are both unitary, then

$$(15.6) \qquad A = V \begin{bmatrix} S_1 & O_{r \times (q-r)} \\ O_{(p-r) \times r} & O_{(p-r) \times (q-r)} \end{bmatrix} U^H.$$

Formula (15.6) may seem particularly attractive because of the presence of unitary factors. However, formula (15.2) works just as well because $\|V_1 B U_1^H\| = \|B\|$ for every $B \in \mathbb{C}^{r \times r}$. Moreover, there is the added advantage that S_1 is invertible.

Exercise 15.4. Show that if $A \in \mathbb{C}^{p \times q}$, then

$p > q \implies \begin{bmatrix} A & O_{p \times (p-q)} \end{bmatrix}$ has the same nonzero singular values as A,

$p < q \implies \begin{bmatrix} A \\ O_{(q-p) \times q} \end{bmatrix}$ has the same nonzero singular values as A.

If $A \in \mathbb{C}^{p \times q}$ is expressed in the form (15.2), then the matrix

$$(15.7) \qquad A^\dagger = U_1 S_1^{-1} V_1^H$$

is called the **Moore-Penrose inverse** of A. It can be characterized as the only matrix in $\mathbb{C}^{q \times p}$ that meets the four conditions $A^\dagger A A^\dagger = A^\dagger$, $A A^\dagger A = A$, $(A^\dagger A)^H = A^\dagger A$, and $(A A^\dagger)^H = A A^\dagger$.

Exercise 15.5. Show that if $A = V_1 S_1 U_1^H$ and $A^\dagger = U_1 S_1^{-1} V_1^H$, then $A^\dagger A A^\dagger = A^\dagger$, $A A^\dagger A = A$, $(A^\dagger A)^H = A^\dagger A$, and $(A A^\dagger)^H = A A^\dagger$.

Exercise 15.6. Redo Exercises 11.18–11.21 using the singular value decomposition (15.2).

15.2. A characterization of singular values

Lemma 15.3. *If $A \in \mathbb{C}^{p \times q}$ and rank $A \geq 1$, then $\|A\| = s_1$.*

Proof. By formula (15.2), $A = V_1 S_1 U_1^H$, with isometric factors $V_1 \in \mathbb{C}^{p \times r}$ and $U_1 \in \mathbb{C}^{q \times r}$. Therefore, $\|A\| = \|V_1 S_1 U_1^H\| = \|S_1\|$. Thus, as

$$\|S_1 \mathbf{x}\|^2 = \sum_{j=1}^{r} |s_j x_j|^2 \leq s_1^2 \sum_{j=1}^{r} |x_j|^2 = s_1^2 \|\mathbf{x}\|^2$$

for every vector $\mathbf{x} = \begin{bmatrix} x_1 & \cdots & x_r \end{bmatrix}^T$ in $\mathbb{C}^r$, $\|S_1\| \leq s_1$. Since equality is attained by choosing $x_1 = 1$ and $x_j = 0$ for $j = 2, \ldots, r$, $\|A\| = s_1$. $\square$

The next result extends Lemma 15.3 and serves to characterize all the singular values of $A \in \mathbb{C}^{p \times q}$ in terms of a problem of best approximation.

Theorem 15.4. *If $A \in \mathbb{C}^{p \times q}$, then its singular values $s_1, \ldots, s_q$ can be characterized as solutions of the following extremal problem:*

$$(15.8) \qquad s_{k+1} = \min \left\{ \|A - B\| : B \in \mathbb{C}^{p \times q} \quad \text{and} \quad \text{rank } B \leq k \right\},$$

for $k = 0, \ldots, q - 1$.

Proof. If $k = 0$ in (15.8), then $B = O$ and the assertion follows from Lemma 15.3. If rank $A = r$ and $k \geq r$, then the minimum is achieved by choosing $B = A$ and then (15.8) simply confirms the fact that $s_i = 0$ if $i > r$. Suppose therefore that rank $A = r$ with $r \geq 1$ and rank $B \leq k$ with $k < r$; and let $U_1 = \begin{bmatrix} \mathbf{u}_1 & \cdots & \mathbf{u}_r \end{bmatrix} \in \mathbb{C}^{q \times r}$ and $V_1 = \begin{bmatrix} \mathbf{v}_1 & \cdots & \mathbf{v}_r \end{bmatrix} \in \mathbb{C}^{p \times r}$ be the isometric matrices in the singular value decomposition (15.2) of A. Then, since rank $B \begin{bmatrix} \mathbf{u}_1 & \cdots & \mathbf{u}_{k+1} \end{bmatrix} \leq k$, there exists a unit vector $\mathbf{d} \in \mathbb{C}^{k+1}$ such that $B \begin{bmatrix} \mathbf{u}_1 & \cdots & \mathbf{u}_{k+1} \end{bmatrix} \mathbf{d} = \mathbf{0}$. Thus, upon setting

$$\mathbf{x} = \begin{bmatrix} \mathbf{u}_1 & \cdots & \mathbf{u}_{k+1} \end{bmatrix} \mathbf{d} = \begin{bmatrix} \mathbf{u}_1 & \cdots & \mathbf{u}_r \end{bmatrix} \begin{bmatrix} \mathbf{d} \\ \mathbf{0} \end{bmatrix} = U_1 \begin{bmatrix} \mathbf{d} \\ \mathbf{0} \end{bmatrix} \quad \text{with } \mathbf{0} \in \mathbb{C}^{r-k-1},$$

it is readily seen that if $\mathbf{d}^T = \begin{bmatrix} d_1 & \cdots & d_{k+1} \end{bmatrix}$, then

$$A\mathbf{x} = V_1 S_1 U_1^H \mathbf{x} = V_1 S_1 U_1^H U_1 \begin{bmatrix} \mathbf{d} \\ \mathbf{0} \end{bmatrix} = \sum_{j=1}^{k+1} d_j s_j \mathbf{v}_j$$

and hence (since $B\mathbf{x} = \mathbf{0}$ and $\|\mathbf{x}\| = 1$) that

$$\|A - B\|^2 \geq \|A\mathbf{x} - B\mathbf{x}\|^2 = \|A\mathbf{x}\|^2 = \left\| \sum_{j=1}^{k+1} d_j s_j \mathbf{v}_j \right\|^2$$

$$= \sum_{j=1}^{k+1} |d_j|^2 s_j^2 \geq s_{k+1}^2 \sum_{j=1}^{k+1} |d_j|^2 = s_{k+1}^2 \, .$$

Let δ_k denote the right-hand side of (15.8). Then, as the last inequality is valid for every choice of $B \in \mathbb{C}^{p \times q}$ with rank $B \leq k$, it follows that $\delta_k \geq s_{k+1}$.

To complete the proof, it suffices to check that if

$$
(15.9) \qquad B = \begin{bmatrix} \mathbf{v}_1 & \cdots & \mathbf{v}_k \end{bmatrix} \begin{bmatrix} s_1 & 0 & \cdots & 0 \\ 0 & s_2 & \cdots & 0 \\ \vdots & & \ddots & \\ 0 & 0 & & s_k \end{bmatrix} \begin{bmatrix} \mathbf{u}_1 & \cdots & \mathbf{u}_k \end{bmatrix}^H ,
$$

then rank $B \leq k$ and $\|A - B\| = s_{k+1}$. Therefore, $\delta_k \leq s_{k+1}$. $\square$

The usefulness of the singular value decomposition stems in large part from the fact the matrix $B = \sum_{j=1}^{k} s_j \mathbf{v}_j \mathbf{u}_j^H$ in (15.9) is the **best approximation** to A in the matrix norm $\|\ \|_{2,2}$ in the set of $p \times q$ matrices with rank $\leq k$. It is also the best approximation in the norm

$$
(15.10) \qquad \|A\|_2 = \left(\sum_{i,j} |a_{ij}|^2 \right)^{1/2} = \left(\operatorname{trace} A^H A \right)^{1/2},
$$

which is also referred to as the **Frobenius norm**; see (15.11) and (15.13) below.

Theorem 15.5. *If $A \in \mathbb{C}^{p \times q}$ and rank $A = r \geq 1$, with nonzero singular values $s_1, \ldots, s_r$, then*

$$
(15.11) \qquad \sum_{j=k+1}^{r} s_j^2 = \min \left\{ \|A - B\|_2^2 : B \in \mathbb{C}^{p \times q} \text{ and } \operatorname{rank} B \leq k \right\},
$$

for $k = 0, \ldots, r - 1$.

Proof. Let $A = V_1 S_1 U_1^H$ be the singular value decomposition of A with $V_1 \in \mathbb{C}^{p \times r}$ and $U_1 \in \mathbb{C}^{q \times r}$ isometric, $S_1 = \operatorname{diag}\{s_1, \ldots, s_r\}$, $s_1 \geq \cdots \geq s_r > 0$, and let $C = V_1^H B U_1$. Then rank $C \leq$ rank $B \leq k$ and

$$
\|A - B\|_2^2 \geq \|V_1 S_1 U_1^H - V_1 V_1^H B U_1 U_1^H\|_2^2
$$
$$
= \|S_1 - V_1^H B U_1\|_2^2 = \|S_1 - C\|_2^2
$$
$$
= \sum_{i=1}^{r} |s_i - c_{ii}|^2 + \sum_{\substack{i,j=1 \\ i \neq j}}^{r} |c_{ij}|^2 \geq \sum_{i=1}^{r} |s_i - c_{ii}|^2
$$

(15.12)

$$
\geq \min \left\{ \sum_{i=1}^{r} |s_i - c_{ii}|^2 : C \text{ is diagonal and } \operatorname{rank} C \leq k \right\}
$$
$$
= \sum_{j=k+1}^{r} s_j^2 .
$$

Therefore, setting δ_k equal to the right-hand side of (15.11), we see that $\delta_k \geq \sum_{j=k+1}^{r} s_j^2$. On the other hand, the specific choice (in the notation of (15.9))

$$(15.13) \qquad B = \sum_{j=1}^{k} s_j \mathbf{v}_j \mathbf{u}_j^H \implies \|A - B\|_2^2 = \sum_{j=k+1}^{r} s_j^2 \geq \delta_k,$$

which completes the proof. $\square$

The last evaluation in the proof of Theorem 15.5 serves to identify $B = \sum_{j=1}^{k} s_j \mathbf{v}_j \mathbf{u}_j^H$ as the **best approximation** to A (in the matrix norm $\|\ \|_2$) in the set of $p \times q$ matrices with rank $\leq k$.

Exercise 15.7. Justify the first inequality in the first line of (15.12) by showing that $\|V_1 S_1 U_1^H - B\|_2^2 = \|S_1 - V_1^H B U_1\|_2^2 + \|V_1^H B U_2\|_2^2 + \|V_2^H B\|_2^2$, where $\begin{bmatrix} V_1 & V_2 \end{bmatrix}$ and $\begin{bmatrix} U_1 & U_2 \end{bmatrix}$ are both unitary matrices. [HINT: $\|A\|_2^2 = \operatorname{trace} A^H A$.]

Exercise 15.8. Let $A \in \mathbb{C}^{p \times q}$ and suppose that $s_1(A) \leq 1$. Show that the matrix $I_p - AB$ is invertible for every choice of $B \in \mathbb{C}^{q \times p}$ with $s_1(B) \leq 1$ if and only if $s_1(A) < 1$.

15.3. Sums and products of singular values

The formula

$$(15.14) \qquad s_1 = \max\{|\langle A\mathbf{x}, \mathbf{y}\rangle| : \mathbf{x} \in \mathbb{C}^q, \mathbf{y} \in \mathbb{C}^p, \text{ and } \mathbf{x}^H \mathbf{x} = \mathbf{y}^H \mathbf{y} = 1\}$$

for matrices $A \in \mathbb{C}^{p \times q}$ has a natural generalization:

$$(15.15) \qquad \begin{aligned} s_1 + \cdots + s_k = \max \{ &|\operatorname{trace}(Y^H A X)| : \\ & X \in \mathbb{C}^{q \times k},\ Y \in \mathbb{C}^{p \times k}, \text{ and } X^H X = I_k = Y^H Y \}, \end{aligned}$$

since the inner products in these two formulas are equal to $\operatorname{trace} \mathbf{y}^H A \mathbf{x}$ and $\operatorname{trace} Y^H A X$, respectively. To put it another way, formula (15.14) coincides with (15.15) when $k = 1$.

In a similar vein, the formula

$$(15.16) \quad s_1^2 \cdots s_k^2 = \max\left\{\det\{W^H A^H A W\} : W \in \mathbb{C}^{q \times k} \text{ and } W^H W = I_k\right\}$$

for the singular values $s_1, \ldots, s_k$ of $A \in \mathbb{C}^{p \times q}$ is a natural generalization of the formula

$$(15.17) \qquad s_1^2 = \max\left\{\det\{\mathbf{w}^H A^H A \mathbf{w}\} : \mathbf{w} \in \mathbb{C}^q \text{ and } \mathbf{w}^H \mathbf{w} = 1\right\}.$$

Proofs of (15.15) and (15.16) will be furnished in Chapter 29.

15.4. Properties of singular values

The next theorem summarizes a number of important properties of singular values that are good to **keep in mind**. The verification of the theorem will make use of the extremal characterizations (15.15) and (15.16), even though they have not been justified yet.

Theorem 15.6. *If $A \in \mathbb{C}^{p \times q}$, $B \in \mathbb{C}^{m \times p}$, and $C \in \mathbb{C}^{q \times n}$, then:*

(1) $s_j(A) = s_j(A^H)$ *for* $j \leq \operatorname{rank} A$.

(2) $s_j(BA) \leq \|B\| s_j(A)$, *with equality if $B \in \mathbb{C}^{p \times p}$ is unitary (and hence also $\|B\| = 1$).*

(3) $s_j(AC) \leq s_j(A) \|C\|$, *with equality if $C \in \mathbb{C}^{q \times q}$ is unitary (and hence also $\|C\| = 1$).*

(4) $\prod_{j=1}^{k} s_j(AC) \leq \prod_{j=1}^{k} s_j(A) \prod_{j=1}^{k} s_j(C)$.

(5) $\sum_{j=1}^{k} s_j(A+B) \leq \sum_{j=1}^{k} s_j(A) + \sum_{j=1}^{k} s_j(B)$ *when also $B \in \mathbb{C}^{p \times q}$.*

Proof. Item (1) follows from Theorem 15.4, since $\|A - B\| = \|A^H - B^H\|$ and $\operatorname{rank} B = \operatorname{rank} B^H$. To verify (2), observe that if $B \in \mathbb{C}^{m \times p}$ and $j > 1$, then, in view of Theorem 15.4,

(15.18)
$$\begin{aligned} s_j(BA) &= \min\{\|BA - D\| : D \in \mathbb{C}^{m \times q} \text{ and } \operatorname{rank} D \leq j-1\} \\ &\leq \min\{\|BA - BE\| : E \in \mathbb{C}^{p \times q} \text{ and } \operatorname{rank} E \leq j-1\} \\ &\leq \|B\| \min\{\|A - E\| : E \in \mathbb{C}^{p \times q} \text{ and } \operatorname{rank} E \leq j-1\} \\ &= \|B\| \, s_j(A) \,. \end{aligned}$$

Therefore (2) holds. The proof of (3) is similar and is left to the reader.

Next, a double application of the formula

(15.19) $\quad s_1^2 \cdots s_k^2 \det\{W^H W\} = \max\left\{\det\{W^H A^H AW\} : W \in \mathbb{C}^{q \times k}\right\},$

which stems from (15.16), yields the inequalities

$$\begin{aligned} \det\{V^H C^H A^H ACV\} &\leq s_1(A)^2 \cdots s_k(A)^2 \det\{V^H C^H CV\} \\ &\leq s_1(A)^2 \cdots s_k(A)^2 s_1(C)^2 \cdots s_k(C)^2 \det\{V^H V\} \end{aligned}$$

for every matrix $V \in \mathbb{C}^{n \times k}$. The inequality (4) then follows from (15.16) (with V in place of W and AC in place of A).

15.5. Approximate solutions of linear equations

Finally, the justification of (5) rests on (15.15) and the observation that if $V \in \mathbb{C}^{n \times k}$ is isometric, then

$$|\text{trace}\{V^H(A+B)V\}| = |\text{trace}\{V^HAV\} + \text{trace}\{V^HBV\}|$$
$$\leq |\text{trace}\{V^HAV\}| + |\text{trace}\{V^HBV\}|$$
$$\leq \sum_{j=1}^{k} s_j(A) + \sum_{j=1}^{k} s_j(B).$$

Item (5) is then obtained by maximizing the left-hand side of this inequality over all isometric matrices $V \in \mathbb{C}^{n \times k}$ and invoking (15.15) once more. □

Exercise 15.9. Show that $\sum_{j=0}^{k} s_j(A)$ defines a norm on $\mathbb{C}^{p \times q}$ when $1 \leq k \leq q$. [HINT: Use (15.15) to obtain the triangle inequality.]

Exercise 15.10. Let $\lambda_1(A), \ldots, \lambda_n(A)$ denote the eigenvalues of $A \in \mathbb{C}^{n \times n}$, repeated according to their algebraic multiplicity, and let $s_1(A), \ldots, s_n(A)$ denote the singular values of A. Show that $\sum_{j=1}^{n} |\lambda_j(A)|^2 \leq \sum_{j=1}^{n} s_j(A)^2$. [HINT: Theorem 13.3 is the key.]

Exercise 15.11. Show that if $A, B \in \mathbb{C}^{n \times n}$ and $AB = (AB)^H$, then $\text{trace}\{(AB)^HAB\} \leq \text{trace}\{(BA)^HBA\}$. [HINT: Use formula (7.4) and Exercise 15.10.]

Exercise 15.12. Justify $s_j(A) = s_j(A^H)$ for $A \in \mathbb{C}^{p \times q}$ and $j \leq \text{rank } A$ by comparing $\det(\lambda I_q - A^HA)$ with $\det(\lambda I_p - AA^H)$.

15.5. Approximate solutions of linear equations

If $A \in \mathbb{C}^{p \times q}$ and $\mathbf{b} \in \mathbb{C}^p$, then the equation $A\mathbf{x} = \mathbf{b}$ has a solution $\mathbf{x} \in \mathbb{C}^q$ if and only if $\mathbf{b} \in \mathcal{R}_A$. However, if $\mathbf{b} \notin \mathcal{R}_A$, then a reasonable strategy is to look for vectors $\mathbf{x} \in \mathbb{C}^q$ that minimize $\|A\mathbf{x} - \mathbf{b}\|$. (There may be many.) Since this problem is only of interest if $\mathbf{b} \notin \mathcal{R}_A$, it suffices to focus on the case where $\text{rank } A = r$ with $1 \leq r < p$.

Theorem 15.7. *If $A \in \mathbb{C}^{p \times q}$, $\text{rank } A = r$, $1 \leq r < p$, and, in terms of the notation introduced earlier for the singular value decomposition of A, $A = V_1 S_1 U_1^H$, $A^\dagger = U_1 S_1^{-1} V_1^H$, where $\begin{bmatrix} V_1 & V_2 \end{bmatrix}$ and $\begin{bmatrix} U_1 & U_2 \end{bmatrix}$ are unitary, then:*

(1) $\min\{\|A\mathbf{x} - \mathbf{b}\| : \mathbf{x} \in \mathbb{C}^q\} = \|(I_q - \Pi_{\mathcal{R}_A})\mathbf{b}\| = \|V_2 V_2^H \mathbf{b}\| = \|V_2^H \mathbf{b}\|$.

(2) $\|A\mathbf{x} - \mathbf{b}\| = \|V_2^H \mathbf{b}\| \iff \mathbf{x} = A^\dagger \mathbf{b} + U_2 \mathbf{c}$ *for some* $\mathbf{c} \in \mathbb{C}^{q-r}$, *with* $U_2 = O$ *if* $q = r$.

Proof. Let $\mathbf{b} = \mathbf{b}_1 + \mathbf{b}_2$ with $\mathbf{b}_1 \in \mathcal{R}_A$ and $\mathbf{b}_2$ orthogonal to $\mathcal{R}_A$. Then $\mathbf{b}_2 = (I_p - \Pi_{\mathcal{R}_A})\mathbf{b} = V_2 V_2^H \mathbf{b}$ and

$$\|A\mathbf{x} - \mathbf{b}\|^2 = \|A\mathbf{x} - \mathbf{b}_1\|^2 + \|\mathbf{b}_2\|^2.$$

Therefore, (1) holds and
$$Ax = b_1 \iff V_1 S_1 U_1^H x = V_1 V_1^H b$$
$$\iff S_1 U_1^H x = V_1^H b$$
$$\iff U_1^H x = S_1^{-1} V_1^H b.$$

Thus, upon writing $x = U_1 a + U_2 c$, it is easily seen that the last equality is met if and only if
$$a = S_1^{-1} V_1^H b,$$
i.e., if and only if
$$x = U_1 S_1^{-1} V_1^H b + U_2 c = A^\dagger b + U_2 c.$$
Thus, (2) holds. $\square$

The matrix $A^\dagger = U_1 S_1^{-1} V_1^H$ is called the **Moore-Penrose inverse** of the matrix A with singular value decomposition $A = V_1 S_1 U_1^H$.

Exercise 15.13. Show that if $A \in \mathbb{C}^{p \times q}$, then rank $A^\dagger =$ rank A and

(15.20) $AA^\dagger A = A$, $A^\dagger A A^\dagger = A^\dagger$, $A^\dagger A = (A^\dagger A)^H$, and $AA^\dagger = (AA^\dagger)^H$.

Exercise 15.14. Show that $AA^\dagger = \Pi_{\mathcal{R}_A}$, $A^\dagger A = \Pi_{\mathcal{R}_{A^H}}$, and hence that rank $A = p \iff AA^\dagger = I_p$ and rank $A = q \iff A^\dagger A = I_q$.

Exercise 15.15. In the setting of Theorem 15.7, show that if $r < q$, then

(15.21) $$x = A^\dagger b = U_1 S_1^{-1} V_1^H b = \sum_{j=1}^{r} \frac{\langle b, v_j \rangle}{s_j} u_j$$

is the vector of smallest norm in $\mathbb{C}^q$ that minimizes $\|Ax - b\|$.

Exercise 15.16. Show that if $A \in \mathbb{C}^{p \times q}$ and rank $A = q$, then $A^H A$ is invertible and $Ax = \Pi_{\mathcal{R}_A} b$ if and only if $x = (A^H A)^{-1} A^H b$.

15.6. Supplementary notes

The monograph [43] of Gohberg and Krein is an excellent source of supplementary information on singular value decompositions in a Hilbert space setting. Theorem 15.5 was presented in a 1936 paper [35] by Eckart and Young. In 1960 Mirsky observed [62] that the same conclusions hold for any norm $\varphi(A)$ on $\mathbb{C}^{p \times q}$ if $\varphi(A) = \varphi(VAU)$ when V and U are unitary; see also Golub, Hoffman, and Stewart [44] for additional developments and references.

If $A \in \mathbb{C}^{p \times q}$, then there is exactly one matrix $A^\dagger \in \mathbb{C}^{q \times p}$ that meets the four conditions in (15.20); see, e.g., Section 11.2 of [30].

Chapter 16

Positive definite and semidefinite matrices

A matrix $A \in \mathbb{C}^{n\times n}$ is said to be **positive semidefinite** if

(16.1) $\qquad \langle A\mathbf{x}, \mathbf{x}\rangle \geq 0$ for every $\mathbf{x} \in \mathbb{C}^n$;

it is said to be **positive definite** if

(16.2) $\qquad \langle A\mathbf{x}, \mathbf{x}\rangle > 0$ for every nonzero vector $\mathbf{x} \in \mathbb{C}^n$.

Correspondingly, $A \in \mathbb{C}^{n\times n}$ is said to be **negative semidefinite** if $-A$ is positive semidefinite, and it is said to be **negative definite** if $-A$ is positive definite.

If $A \in \mathbb{C}^{n\times n}$ and $B \in \mathbb{C}^{n\times n}$, then the **notation**

$A \succeq B$ (resp., $A \succ B$) means that $A - B$ is positive semidefinite (resp., $A - B$ is positive definite).

Lemma 16.1. *If $A \in \mathbb{C}^{n\times n}$ and $A \succeq O$, then:*

(1) *A is automatically Hermitian.*

(2) *The eigenvalues of A are nonnegative numbers.*

Moreover,

(16.3) $\quad A \succ O \iff A = A^H$ *and the eigenvalues of A are all positive*

(16.4) $\qquad \iff A \succeq O$ *and* $\det A > 0$.

Proof. If $A \succeq O$, then

$$\langle A\mathbf{x}, \mathbf{x}\rangle = \overline{\langle A\mathbf{x}, \mathbf{x}\rangle} = \langle \mathbf{x}, A\mathbf{x}\rangle$$

171

for every $\mathbf{x} \in \mathbb{C}^n$. Therefore, by a straightforward calculation,

$$4\langle A\mathbf{x}, \mathbf{y}\rangle = \sum_{k=1}^{4} i^k \langle A(\mathbf{x} + i^k \mathbf{y}), (\mathbf{x} + i^k \mathbf{y})\rangle$$

$$= \sum_{k=1}^{4} i^k \langle (\mathbf{x} + i^k \mathbf{y}), A(\mathbf{x} + i^k \mathbf{y})\rangle = 4\langle \mathbf{x}, A\mathbf{y}\rangle ;$$

i.e., $\langle A\mathbf{x}, \mathbf{y}\rangle = \langle \mathbf{x}, A\mathbf{y}\rangle$ for every choice of $\mathbf{x}, \mathbf{y} \in \mathbb{C}^n$. Therefore, (1) holds.

Next, let $\mathbf{x}$ be an eigenvector of A corresponding to the eigenvalue λ. Then

$$\lambda \langle \mathbf{x}, \mathbf{x}\rangle = \langle A\mathbf{x}, \mathbf{x}\rangle \geq 0.$$

Therefore $A \succeq O \implies \lambda \geq 0$ and $A \succ O \implies \lambda > 0$, since $\langle \mathbf{x}, \mathbf{x}\rangle > 0$. Thus, (2) and, in view of (1), the implication $\implies$ in (16.3) hold. The implication $\impliedby$ in (16.3) follows from the fact that $A = A^H \implies A = WDW^H$, in which W is unitary and $D = \mathrm{diag}\{\lambda_1, \ldots, \lambda_n\}$; the verification of (16.4) is left to the reader. □

• **Warning:** The conclusions of Lemma 16.1 are not true under the less restrictive constraint

$$\langle A\mathbf{x}, \mathbf{x}\rangle \geq 0 \text{ for every } \mathbf{x} \in \mathbb{R}^n.$$

Thus, for example, if

$$A = \begin{bmatrix} 2 & -2 \\ 0 & 2 \end{bmatrix} \quad \text{and} \quad \mathbf{x} = \begin{bmatrix} x_1 \\ x_2 \end{bmatrix},$$

then

$$\langle A\mathbf{x}, \mathbf{x}\rangle = (x_1 - x_2)^2 + x_1^2 + x_2^2 > 0$$

for every nonzero vector $\mathbf{x} \in \mathbb{R}^n$. However, A is clearly not Hermitian. The next lemma serves to clarify this example.

Lemma 16.2. *If $A \in \mathbb{R}^{n \times n}$, then*

$$\langle A\mathbf{u}, \mathbf{u}\rangle \geq 0 \text{ for every } \mathbf{u} \in \mathbb{C}^n \iff \begin{cases} \langle A\mathbf{x}, \mathbf{x}\rangle \geq 0 \text{ for every } \mathbf{x} \in \mathbb{R}^n \\ \text{and } A = A^T. \end{cases}$$

Proof. If the conditions on the right hold and $\mathbf{u} = \mathbf{x} + i\mathbf{y}$ with $\mathbf{x}, \mathbf{y} \in \mathbb{R}^n$, then

$$\langle A(\mathbf{x} + i\mathbf{y}), (\mathbf{x} + i\mathbf{y})\rangle = \langle A\mathbf{x}, \mathbf{x}\rangle - i\langle A\mathbf{x}, \mathbf{y}\rangle + i\langle A\mathbf{y}, \mathbf{x}\rangle + \langle A\mathbf{y}, \mathbf{y}\rangle$$
$$= \langle A\mathbf{x}, \mathbf{x}\rangle + \langle A\mathbf{y}, \mathbf{y}\rangle \geq 0,$$

since $\langle A\mathbf{y}, \mathbf{x}\rangle = \langle \mathbf{y}, A\mathbf{x}\rangle = \langle A\mathbf{x}, \mathbf{y}\rangle$ when $A = A^T \in \mathbb{R}^{n \times n}$ and $\mathbf{x}, \mathbf{y} \in \mathbb{R}^n$. Thus, the conditions on the left hold. The converse implication is justified by Lemma 16.1. □

Exercise 16.1. Show that if $A \in \mathbb{C}^{n \times n}$ and $A = A^H$ with eigenvalues $\lambda_1 \geq \cdots \geq \lambda_n$, then $\lambda_1 I_n - A \succeq O$ **(even if $\lambda_1 \leq 0$)**.

Exercise 16.2. Show that if $V \in \mathbb{C}^{n \times k}$ and $\operatorname{rank} V = k$, then
$$A \succ O \Longrightarrow V^H A V \succ O,$$
but the converse implication is not true if $k < n$.

Exercise 16.3. Show that if $A \in \mathbb{C}^{n \times n}$ with entries a_{ij}, $i, j = 1, \ldots, n$, and $A \succeq O$, then $|a_{ij}|^2 \leq a_{ii} a_{jj}$.

Exercise 16.4. Show that if $A \in \mathbb{C}^{n \times n}$, $n = p + q$, and
$$A = \begin{bmatrix} A_{11} & A_{12} \\ A_{21} & A_{22} \end{bmatrix},$$
where $A_{11} \in \mathbb{C}^{p \times p}$, $A_{22} \in \mathbb{C}^{q \times q}$, then
$$A \succ O \iff A_{11} \succ O, \quad A_{21} = A_{12}^H, \quad \text{and} \quad A_{22} - A_{21} A_{11}^{-1} A_{12} \succ O.$$

Exercise 16.5. Let $U \in \mathbb{C}^{n \times n}$ be unitary and let $A \in \mathbb{C}^{n \times n}$. Show that if $A \succ O$ and $AU \succ O$, then $U = I_n$. [HINT: Consider $\langle AU\mathbf{x}, \mathbf{x} \rangle$ for eigenvectors $\mathbf{x}$ of U.]

16.1. A detour on triangular factorization

The **notation**

$$(16.5) \quad A_{[j,k]} = \begin{bmatrix} a_{jj} & \cdots & a_{jk} \\ \vdots & \ddots & \vdots \\ a_{kj} & \cdots & a_{kk} \end{bmatrix} \quad \text{for} \quad A \in \mathbb{C}^{n \times n} \quad \text{and} \quad 1 \leq j \leq k \leq n$$

will be convenient.

The **trade secret** behind the factorization formulas that will be considered below is that if $B, L, U \in \mathbb{C}^{n \times n}$, L is lower triangular, and U is upper triangular, then

$$(16.6) \quad \begin{array}{l} (LB)_{[1,k]} = L_{[1,k]} B_{[1,k]} \quad \text{and} \quad (BU)_{[1,k]} = B_{[1,k]} U_{[1,k]}, \\ (BL)_{[k,n]} = B_{[k,n]} L_{[k,n]} \quad \text{and} \quad (UB)_{[k,n]} = U_{[k,n]} B_{[k,n]} \end{array}$$

for $k = 1, \ldots, n$.

Exercise 16.6. Let $P_k = \operatorname{diag}\{I_k, O_{(n-k) \times (n-k)}\}$. Show that

(a) $A \in \mathbb{C}^{n \times n}$ is upper triangular $\iff AP_k = P_k A P_k$ for $k = 1, \ldots, n$.
(b) $A \in \mathbb{C}^{n \times n}$ is lower triangular $\iff P_k A = P_k A P_k$ for $k = 1, \ldots, n$.

We shall say that a matrix $A \in \mathbb{C}^{n \times n}$ admits an LU (resp., UL) factorization if there exist a lower triangular matrix $L \in \mathbb{C}^{n \times n}$ and an upper triangular matrix $U \in \mathbb{C}^{n \times n}$ such that $A = LU$ (resp., $A = UL$).

Theorem 16.3. *If $A \in \mathbb{C}^{n \times n}$, then:*

(1) *A admits an LU factorization with invertible triangular factors L and U $\iff \det A_{[1,k]} \neq 0$ for $k = 1, \ldots, n$.*

(2) *A admits a UL factorization with invertible triangular factors L and U $\iff \det A_{[k,n]} \neq 0$ for $k = 1, \ldots, n$.*

(3) *If $\det A_{[1,k]} \neq 0$ for $k = 1, \ldots, n$, then $A = LDU$ for exactly one lower triangular matrix L with ones on the diagonal, one upper triangular matrix U with ones on the diagonal, and one diagonal matrix D.*

(4) *If $\det A_{[k,n]} \neq 0$ for $k = 1, \ldots, n$, then $A = UDL$ for exactly one lower triangular matrix L with ones on the diagonal, one upper triangular matrix U with ones on the diagonal, and one diagonal matrix D.*

Proof. The proof is divided into steps.

1. *Verification of* (1):

Suppose first that $A = LU$ with invertible factors L and U. Then, by the first formula in (16.6), $A_{[1,k]} = L_{[1,k]} U_{[1,k]}$. Moreover, since L and U are triangular matrices, $L_{[1,k]}$ and $U_{[1,k]}$ are also invertible for $k = 1, \ldots, n$. Thus, $A_{[1,k]}$ is invertible for $k = 1, \ldots, n$.

Suppose next that $A_{[1,k]}$ is invertible for $k = 1, \ldots, n$ and let $X \in \mathbb{C}^{n \times n}$ be the lower triangular matrix with entries x_{ij} for $i \geq j$ that are specified by the formulas

(16.7) $\quad \begin{bmatrix} x_{k1} & \cdots & x_{kk} \end{bmatrix} = \begin{bmatrix} 0 & \cdots & 0 & 1 \end{bmatrix} (A_{[1,k]})^{-1} \quad$ for $k = 1, \ldots, n$,

with the understanding that $x_{11} = 1/a_{11}$. Now, let $Y = XA$. Then, by the first formula in (16.6),

$$Y_{[1,k]} = X_{[1,k]} A_{[1,k]} \quad \text{for } k = 1, \ldots, n,$$

and hence, in view of (16.7),

$$\begin{bmatrix} y_{k1} & \cdots & y_{kk} \end{bmatrix} = \begin{bmatrix} x_{k1} & \cdots & x_{kk} \end{bmatrix} A_{[1,k]} = \begin{bmatrix} 0 & \cdots & 0 & 1 \end{bmatrix}.$$

Thus, Y is upper triangular with $y_{jj} = 1$ for $j = 1, \ldots, n$. Therefore, Y and $X = YA^{-1}$ are invertible and $A = LU$ with $L = X^{-1}$ and $U = Y$.

2. *Verification of* (2):

If $A = UL$ and U and L are both invertible, then, as U and L are triangular, $U_{[k,n]}$ and $L_{[k,n]}$ are both invertible for $k = 1, \ldots, n$. Thus, as $A_{[k,n]} = U_{[k,n]} L_{[k,n]}$ for $k = 1, \ldots, n$ by the fourth formula in (16.6), $A_{[k,n]}$ is also invertible for $k = 1, \ldots, n$.

16.1. A detour on triangular factorization

Suppose next that $A_{[k,n]}$ is invertible for $k = 1, \ldots, n$ and let $X \in \mathbb{C}^{n \times n}$ be the lower triangular matrix with entries x_{ij} for $i \geq j$ that are specified by the formulas

$$(16.8) \qquad \begin{bmatrix} x_{kk} \\ \vdots \\ x_{kn} \end{bmatrix} = (A_{[k,n]})^{-1} \begin{bmatrix} 1 \\ 0 \\ \vdots \\ 0 \end{bmatrix} \quad \text{for } k = 1, \ldots, n,$$

with the understanding that $x_{nn} = 1/a_{nn}$, and let $Y = AX$. Then, by the third formula in (16.6), $Y_{[k,n]} = A_{[k,n]} X_{[k,n]}$ for $k = 1, \ldots, n$ and hence

$$\begin{bmatrix} y_{kk} \\ \vdots \\ y_{kn} \end{bmatrix} = A_{[k,n]} \begin{bmatrix} x_{kk} \\ \vdots \\ x_{kn} \end{bmatrix} = \begin{bmatrix} 1 \\ 0 \\ \vdots \\ 0 \end{bmatrix}.$$

Thus, Y is upper triangular with $y_{jj} = 1$ for $j = 1, \ldots, n$. Therefore, Y and $X = A^{-1}Y$ are invertible and $A = UL$ with $L = X^{-1}$ and $U = Y$.

3. *Verification of* (3) *and* (4):

To verify (3), suppose that an invertible matrix A admits a pair of factorizations $A = L_1 D_1 U_1 = L_2 D_2 U_2$ in which the diagonal entries of the triangular factors are all equal to one. Then the identity $L_2^{-1} L_1 D_1 = D_2 U_2 U_1^{-1}$ implies that $D_1 = D_2$ and that $L_2^{-1} L_1 D_1$ is both upper and lower triangular and hence is a diagonal matrix, which must be equal to D_1. Therefore, $L_1 = L_2$ and $U_1 = U_2$.

The verification of (4) is left to the reader; it is similar to the verification of (3). □

Remark 16.4. Formulas (16.7) and (16.8) serve to make the proof of Theorem 16.3 efficient, but mysterious.

To explain where the first of these two formulas comes from, we first observe that if $A = LU$ is invertible, then L and U are invertible and the diagonal matrix $\Delta = \text{diag}\{u_{11}, \ldots, u_{nn}\}$ based on the diagonal entries of U is invertible. Therefore, $Y = \Delta^{-1} U$ is an upper triangular matrix with diagonal entries $y_{jj} = 1$ for $j = 1, \ldots, n$ and

$$A = LU \iff L^{-1} A = U \iff \Delta^{-1} L^{-1} A = \Delta^{-1} U \iff XA = Y,$$

with $X = \Delta^{-1} L^{-1}$. Thus A admits an LU factorization if and only if there exist a lower triangular matrix $X \in \mathbb{C}^{n \times n}$ and an upper triangular matrix $Y \in \mathbb{C}^{n \times n}$ with $y_{jj} = 1$ for $j = 1, \ldots, n$ such that $XA = Y$ (because then X is invertible and $A = LY$ with $L = X^{-1}$).

It is remarkable that the awesome looking nonlinear matrix equation $XA = Y$, which is a system of n^2 equations with $(n^2 + n)/2$ unknown entries x_{ij} with $1 \leq j \leq i \leq n$ in X and $(n^2 - n)/2$ unknown entries y_{ij} with $1 \leq i < j \leq n$ in Y, is tractable. But, in self-evident notation,

(16.9)
$$Y_{[1,k]} = (XA)_{[1,k]} = \begin{bmatrix} I_k & O \end{bmatrix} \begin{bmatrix} X_{11} & O \\ X_{21} & X_{22} \end{bmatrix} \begin{bmatrix} A_{11} & A_{12} \\ A_{21} & A_{22} \end{bmatrix} \begin{bmatrix} I_k \\ O \end{bmatrix}$$
$$= X_{11}A_{11} = X_{[1,k]}A_{[1,k]}$$

for $k = 1, \ldots, n$. It is now easily seen that when $\det A_{[1,k]} \neq 0$ for $k = 1, \ldots, n$, then (16.7) is just the bottom row of (16.9).

The motivation for (16.8) is similar.

Exercise 16.7. Verify item (4) in Theorem 16.3.

Exercise 16.8. Show that if $A \in \mathbb{C}^{n \times n}$ is invertible and $A^{-1} = B$, then $A_{[1,k]}$ is invertible for $k = 1, \ldots, n$ if and only if $B_{[k,n]}$ is invertible for $k = 1, \ldots, n$.

Exercise 16.9. Show that if $A \in \mathbb{C}^{n \times n}$ and $A_{[1,k]}$ is invertible for $k = 1, \ldots, n$, then formula (16.7) implies that $x_{kk} = \det A_{[1,k-1]}/\det A_{[1,k]}$ for $k = 2, \ldots, n$, whereas, if $A_{[k,n]}$ is invertible for $k = 1, \ldots, n$, then (16.8) implies that $x_{kk} = \det A_{[k+1,n]}/\det A_{[k,n]}$ for $k = 1, \ldots, n-1$.

Exercise 16.10. Show that the matrix $A = \begin{bmatrix} 1 & -1 \\ 1 & 0 \end{bmatrix}$ admits an LU factorization but does not admit a UL factorization and find matrices L, D, U such that $A = LDU$ with L lower (resp., U upper) triangular with ones on the diagonal and D diagonal.

Exercise 16.11. Let $A \in \mathbb{C}^{n \times n}$ be a Vandermonde matrix with columns $\mathbf{v}(\lambda_1), \ldots, \mathbf{v}(\lambda_n)$ based on n distinct points $\lambda_1, \ldots, \lambda_n$. Show that A admits an LU factorization, but may not admit a UL factorization.

Exercise 16.12. Show that if $A \in \mathbb{C}^{n \times n}$ and $A^2 = A$, then A is an orthogonal projection if and only if $A \succeq O$.

16.2. Characterizations of positive definite matrices

There are a number of different characterizations of positive definite matrices:

Theorem 16.5. *If $A \in \mathbb{C}^{n \times n}$, then the following statements are equivalent:*

(1) $A \succ O$.

(2) $A = A^H$ and the eigenvalues, $\lambda_1, \ldots, \lambda_n$, of A are all positive; i.e., $\lambda_j > 0$ for $j = 1, \ldots, n$.

16.2. Characterizations of positive definite matrices

(3) $A = A^H$ and $\det A_{[1,k]} > 0$ for $k = 1, \ldots, n$.

(4) $A = LL^H$, where L is a lower triangular invertible matrix.

(5) $A = A^H$ and $\det A_{[k,n]} > 0$ for $k = 1, \ldots, n$.

(6) $A = UU^H$, where U is an upper triangular invertible matrix.

Proof. Lemma 16.1 ensures that (1) $\implies$ (2). Conversely, if (2) is in force, then $A = VDV^H$ with $V \in \mathbb{C}^{n \times n}$ unitary and $D \succ O$ and diagonal. Therefore,

$$\langle A\mathbf{x}, \mathbf{x} \rangle = \langle DV^H \mathbf{x}, V^H \mathbf{x} \rangle > 0$$

for every nonzero vector $\mathbf{x} \in \mathbb{C}^n$. Thus, (2) $\iff$ (1).

Next, it is clear that (1) $\implies$ (3) and hence, in view of Theorem 16.3, that A admits exactly one factorization of the form $A = L_1 D U_1$, where L_1 is lower triangular with ones on the diagonal, U_1 is upper triangular with ones on the diagonal, and D is a diagonal matrix. Since $A = A^H$, $L_1 D U_1 = U_1^H D^H L_1^H$. Consequently, $U_1 = L_1^H$ and $D = D^H$. Moreover, as

$$A = L_1 D L_1^H \implies A_{[1,k]} = (L_1)_{[1,k]} D_{[1,k]} (L_1^H)_{[1,k]}$$
$$\implies \det A_{[1,k]} = |\det (L_1)_{[1,k]}|^2 \det D_{[1,k]} = \det D_{[1,k]}$$

for $k = 1, \ldots, n$, the diagonal entries in the matrix $D = \mathrm{diag}\{\lambda_1, \ldots, \lambda_n\}$ are positive. Thus, D admits a square root

$$D^{1/2} = \mathrm{diag}\{\sqrt{\lambda_1}, \ldots, \sqrt{\lambda_n}\}$$

and hence (4) holds with $L = L_1 D^{1/2}$. Since the implication (4) $\implies$ (1) is clear, the implications (1) $\implies$ (3) $\implies$ (4) $\implies$ (1) are justified.

To complete the proof, it suffices to check that (1) $\implies$ (5) $\implies$ (6) $\implies$ (1). The details are left to the reader. $\square$

Exercise 16.13. Show that if $A \in \mathbb{C}^{n \times n}$, then $A \succ O$ if and only if $A = V^H V$ for some invertible matrix $V \in \mathbb{C}^{n \times n}$.

The next three exercises are formulated in terms of the matrix
(16.10)

$$Z_n = \sum_{j=1}^n \mathbf{e}_j \mathbf{e}_{n+1-j}^T = \begin{bmatrix} 0 & 0 & \cdots & 0 & 1 \\ 0 & 0 & \cdots & 1 & 0 \\ \vdots & & & & \vdots \\ 1 & 0 & \cdots & 0 & 0 \end{bmatrix}, \quad \text{where } \begin{bmatrix} \mathbf{e}_1 & \cdots & \mathbf{e}_n \end{bmatrix} = I_n.$$

Exercise 16.14. Show that $Z_n^H = Z_n$ and $Z_n^H Z_n = I_n$, i.e., Z_n is both Hermitian and unitary.

Exercise 16.15. Show that $U \in \mathbb{C}^{n \times n}$ is an invertible upper triangular matrix if and only if $Z_n U Z_n$ is an invertible lower triangular matrix and then use this information to verify the equivalence of (4) and (6) in Theorem 16.5.

Exercise 16.16. Show that if $p \geq 1$, $q \geq 1$, and $p + q = n$, then
$$\begin{bmatrix} O & Z_q \\ Z_p & O \end{bmatrix} \begin{bmatrix} A_{11} & A_{12} \\ A_{21} & A_{22} \end{bmatrix} \begin{bmatrix} O & Z_p \\ Z_q & O \end{bmatrix} = \begin{bmatrix} Z_q A_{22} Z_q & Z_q A_{21} Z_p \\ Z_p A_{12} Z_q & Z_p A_{11} Z_p \end{bmatrix}$$
and then use this identity to verify the equivalence of (3) and (5) in Theorem 16.5.

16.3. Square roots

Theorem 16.6. *If $A \in \mathbb{C}^{n \times n}$ and $A \succeq O$, then there is exactly one matrix $B \in \mathbb{C}^{n \times n}$ such that $B \succeq O$ and $B^2 = A$.*

Proof. If $A \in \mathbb{C}^{n \times n}$ and $A \succeq O$, then there exists a unitary matrix U and a diagonal matrix
$$D = \mathrm{diag}\{d_{11}, \ldots, d_{nn}\}$$
with nonnegative entries such that $A = UDU^H$. Therefore, upon setting
$$D^{1/2} = \mathrm{diag}\{d_{11}^{1/2}, \ldots, d_{nn}^{1/2}\},$$
it is readily checked that the matrix $B = UD^{1/2}U^H$ is again positive semidefinite and
$$B^2 = (UD^{1/2}U^H)(UD^{1/2}U^H) = UDU^H = A.$$
This completes the proof of the existence of at least one positive semidefinite square root of A.

Suppose next that there are two positive semidefinite square roots of A, B_1 and B_2. Then there exist a pair of unitary matrices U_1 and U_2 and a pair of diagonal matrices $D_1 = \mathrm{diag}\{\alpha_1, \ldots, \alpha_n\}$ with $\alpha_1 \geq \cdots \geq \alpha_n \geq 0$ and $D_2 = \mathrm{diag}\{\beta_1, \ldots, \beta_n\}$ with $\beta_1 \geq \cdots \geq \beta_n \geq 0$ such that
$$B_1 = U_1 D_1 U_1^H \quad \text{and} \quad B_2 = U_2 D_2 U_2^H.$$
Thus, as
$$U_1 D_1^2 U_1^H = B_1^2 = A = B_2^2 = U_2 D_2^2 U_2^H,$$
it follows that
$$U_2^H U_1 D_1^2 = D_2^2 U_2^H U_1$$
and hence, upon setting $W = U_2^H U_1$, that
$$W D_1^2 - D_2 W D_1 = D_2^2 W - D_2 W D_1.$$
But this in turn implies that the matrix
$$X = W D_1 - D_2 W$$

16.3. Square roots

with entries x_{ij} for $i,j = 1,\ldots,n$ is a solution of the equation
$$XD_1 + D_2 X = O$$
and hence that
$$x_{ij}\alpha_j + \beta_i x_{ij} = 0 \quad \text{for } i,j = 1,\ldots,n.$$
Thus, $x_{ij} = 0$ if $\alpha_j + \beta_i > 0$. On the other hand, if $\alpha_j + \beta_i = 0$, then $\alpha_j = \beta_i = 0$ and so $x_{ij} = w_{ij}\alpha_j - \beta_i w_{ij} = 0$ in this case also. Therefore, $X = O$ is the only solution of the equation $XD_1 + D_2 X = O$. Consequently,
$$U_2^H U_1 D_1 - D_2 U_2^H U_1 = X = O;$$
i.e.,
$$B_1 = U_1 D_1 U_1^H = U_2 D_2 U_2^H = B_2,$$
as claimed. $\square$

If $A \succeq O$, the symbol $A^{1/2}$ will be used to denote the unique $n \times n$ matrix $B \succeq O$ with $B^2 = A$. Correspondingly, B will be referred to as the **square root** of A. The restriction that $B \succeq O$ is essential to ensure uniqueness. Thus, for example,
$$\begin{bmatrix} O & A \\ A^{-1} & O \end{bmatrix} \begin{bmatrix} O & A \\ A^{-1} & O \end{bmatrix} = I_{2n},$$
for every invertible matrix $A \in \mathbb{C}^{n \times n}$.

Exercise 16.17. Show that if $A \in \mathbb{C}^{n \times n}$, then
$$\begin{bmatrix} A & A \\ A & A \end{bmatrix} \succeq O \iff A \succeq O.$$

Exercise 16.18. Show that if $A \in \mathbb{C}^{n \times n}$, then
$$\begin{bmatrix} A^2 & A \\ A & I_n \end{bmatrix} \succeq O \iff A = A^H.$$

Exercise 16.19. Show that if $A, G \in \mathbb{C}^{n \times n}$, $G \succ O$, and $GA = A^H G$, then $\sigma(A) \subset \mathbb{R}$ and $A = A^*$ with respect to an appropriately defined inner product.

Exercise 16.20. Show that if $A, B \in \mathbb{C}^{n \times n}$ and $A \succeq B \succ O$, then $B^{-1} \succeq A^{-1} \succ O$. [HINT: $A - B \succ O \implies A^{-1/2} B A^{-1/2} \prec I_n$.]

Exercise 16.21. Show that if $A, B \in \mathbb{C}^{n \times n}$ and if $A \succeq O$ and $B \succeq O$, then trace $AB \geq 0$ (even if $AB \not\succeq O$).

16.4. Polar forms and partial isometries

A matrix $A \in \mathbb{C}^{p \times q}$ is said to be a **partial isometry** if $A^H A \mathbf{x} = \mathbf{x}$ for every vector $\mathbf{x} \in \mathbb{C}^q$ that is orthogonal to $\mathcal{N}_A$. Since $\mathbb{C}^q = \mathcal{N}_A \oplus \mathcal{R}_{A^H}$, $A \in \mathbb{C}^{p \times q}$ is a partial isometry if and only if $A^H A A^H \mathbf{y} = A^H \mathbf{y}$ for every $\mathbf{y} \in \mathbb{C}^p$. Thus:

(1) $A \in \mathbb{C}^{p \times q}$ is an isometry if $A^H A = I_q$.

(2) $A \in \mathbb{C}^{p \times q}$ is a partial isometry if $A^H A A^H = A^H$.

Exercise 16.22. Show that if $A \in \mathbb{C}^{p \times q}$ is a partial isometry, then it is an isometry if and only if rank $A = q$.

Theorem 16.7. *If $A \in \mathbb{C}^{p \times q}$, then there exists exactly one partial isometry $B \in \mathbb{C}^{p \times q}$ and one positive semidefinite matrix $P \in \mathbb{C}^{q \times q}$ such that $A = BP$ and $\mathcal{N}_B = \mathcal{N}_P$. In this factorization, P is **the** positive semidefinite square root of $A^H A$.*

Proof. If B and P meet the stated conditions, then $\mathbb{C}^q = \mathcal{N}_B \oplus \mathcal{R}_{B^H} = \mathcal{N}_P \oplus \mathcal{R}_{P^H}$ and $P = P^H$. Therefore, $\mathcal{R}_{B^H} = \mathcal{R}_P$ and hence

$$B^H B P = P \quad \text{and} \quad A^H A = P B^H B P = P^2\,.$$

Thus, P is the one and only positive semidefinite square root of $A^H A$. If $C \in \mathbb{C}^{p \times q}$ is a partial isometry such that $A = CP$ and $\mathcal{N}_C = \mathcal{N}_P$, then

$$CP\mathbf{x} = A\mathbf{x} = BP\mathbf{x} \text{ for every } \mathbf{x} \in \mathbb{C}^q \quad \text{and} \quad C\mathbf{y} = B\mathbf{y} \text{ for every } \mathbf{y} \in \mathcal{N}_P\,.$$

Therefore, $C = B$. $\square$

Corollary 16.8. *If $A \in \mathbb{C}^{p \times q}$ and rank $A = r \geq 1$ and A admits a pair of singular value decompositions $A = V_1 S_1 U_1^H = Y_1 S_1 X_1^H$, with isometric factors $V_1, Y_1 \in \mathbb{C}^{p \times r}$ and $U_1, X_1 \in \mathbb{C}^{q \times r}$, then $V_1 U_1^H = Y_1 X_1^H$.*

Proof. If $A \in \mathbb{C}^{p \times q}$ and $A = V_1 S_1 U_1^H$ with isometric factors $V_1 \in \mathbb{C}^{p \times r}$ and $U_1 \in \mathbb{C}^{q \times r}$ and $S_1 = \text{diag}\{s_1, \ldots, s_r\} \succ O$, then

(16.11) $\quad A = BP \quad$ with $B = V_1 U_1^H \in \mathbb{C}^{p \times q}$ and $P = U_1 S_1 U_1^H \in \mathbb{C}^{q \times q}$.

The asserted uniqueness follows from Theorem 16.7, since $B^H B B^H = B^H$, $P \succeq O$, and $\mathcal{N}_B = \mathcal{N}_P$. $\square$

Exercise 16.23. Show that the factors V_1 and U_1 in the factorization $A = V_1 S_1 U_1^H$ are not unique. [HINT: Diagonal matrices commute.]

The factorization BP in (16.11) is called the **right polar form** of A.

Exercise 16.24. Show that if $A \in \mathbb{C}^{p \times q}$ and rank $A = r \geq 1$, then A admits exactly one **left polar form** $A = QC$ in which $Q \succeq O$ is a square root of $A A^H$ and C is a partial isometry with $\mathcal{R}_C = \mathcal{R}_Q$.

16.4. Polar forms and partial isometries

Exercise 16.25. Show that if $P \in \mathbb{C}^{n \times n}$ is a positive semidefinite matrix and $Y_1, Y_2 \in \mathbb{C}^{n \times n}$ are such that $Y_1 P = Y_2 P$ and $\mathcal{N}_{Y_1} = \mathcal{N}_{Y_2} = \mathcal{N}_P$, then $Y_1 = Y_2$.

Theorem 16.9. *If $A \in \mathbb{C}^{p \times q}$ and* rank $A = r \geq 1$*, then*

(16.12) $\qquad A^H A = I_q \iff \|A\mathbf{x}\| = \|\mathbf{x}\| \quad \text{for every } \mathbf{x} \in \mathbb{C}^q$

and

(16.13) $\qquad A^H A A^H = A^H \iff \|AA^H \mathbf{y}\| = \|A^H \mathbf{y}\| \quad \text{for every } \mathbf{y} \in \mathbb{C}^p.$

Proof. Since (16.12) is a special case of (16.13), it suffices to deal with the latter.

Suppose first that $\|AA^H \mathbf{y}\| = \|A^H \mathbf{y}\|$ for every $\mathbf{y} \in \mathbb{C}^p$ and let $\mathbf{x} \in \mathbb{C}^q$. Then $\mathbf{x} = \mathbf{u} + A^H \mathbf{y}$ for some choice of $\mathbf{u} \in \mathcal{N}_A$ and $\mathbf{y} \in \mathbb{C}^p$. Thus, as $\langle (I_q - A^H A) A^H \mathbf{y}, A^H \mathbf{y} \rangle = 0$, it is readily checked that

$$\langle (I_q - A^H A)\mathbf{x}, \mathbf{x} \rangle = \langle (I_q - A^H A)(\mathbf{u} + A^H \mathbf{y}), \mathbf{u} + A^H \mathbf{y} \rangle = \langle \mathbf{u}, \mathbf{u} \rangle \geq 0,$$

i.e., $(I_q - A^H A) \succeq O$. Consequently,

$$\langle (I_q - A^H A) A^H \mathbf{y}, A^H \mathbf{y} \rangle = 0 \implies \|(I_q - A^H A)^{1/2} A^H \mathbf{y}\| = 0$$
$$\implies (I_q - A^H A) A^H \mathbf{y} = \mathbf{0}.$$

Since these implications are valid for every $\mathbf{y} \in \mathbb{C}^p$, $(I_q - A^H A) A^H = O$.

The converse implication is easy and is left to the reader. $\square$

Lemma 16.10. *If $P \in \mathbb{C}^{n \times n}$ is a positive semidefinite matrix and $B \in \mathbb{C}^{n \times n}$ is a partial isometry with $\mathcal{N}_B = \mathcal{N}_P$, then*

(16.14) $\qquad\qquad BP = PB^H \implies B = B^H,$

(i.e., $BP = (BP)^H \implies B = B^H$).

Proof. Under the given assumptions,

$$BPB^H BPB^H = BP^2 B^H = PB^H BP = P^2 \implies BPB^H = P,$$

since the positive semidefinite matrix P^2 has exactly one positive semidefinite square root. But this in turn implies that

$$B^H P = B^H BPB^H = PB^H = BP.$$

Therefore, to complete the proof it suffices to show that $\mathcal{N}_{B^H} = \mathcal{N}_P$. But

$$B^H \mathbf{a} = \mathbf{0} \implies BPB^H \mathbf{a} = \mathbf{0} \implies P\mathbf{a} = \mathbf{0},$$

i.e., $\mathcal{N}_{B^H} \subseteq \mathcal{N}_P = \mathcal{N}_B$. Thus, as $\dim \mathcal{N}_B = n - \text{rank } B = n - \text{rank } B^H = \dim \mathcal{N}_{B^H}$, we see that $\mathcal{N}_{B^H} = \mathcal{N}_B = \mathcal{N}_P$. $\square$

16.5. Some useful formulas

It is useful to **keep in mind** that if $A = A^H$, then a number of formulas that were established earlier assume a more symmetric form:

Theorem 16.11. *If $A \in \mathbb{C}^{n \times n}$ and* $\operatorname{rank} A = r$, $r \geq 1$, *then:*

  (1) $A^H A = AA^H \implies \|A\| = \max\{|\langle A\mathbf{x}, \mathbf{x}\rangle| : \mathbf{x} \in \mathbb{C}^n \text{ and } \|\mathbf{x}\| = 1\}$.
  (2) $A \succeq O \implies \|A\| = \max\{\langle A\mathbf{x}, \mathbf{x}\rangle : \mathbf{x} \in \mathbb{C}^n \text{ and } \|\mathbf{x}\| = 1\}$.
  (3) $A \succeq O \implies$ *in the singular value decomposition* $A = V_1 S_1 U_1^H$ *in* (15.2)*, the two* $n \times r$ *isometric matrices coincide:* $V_1 = U_1$.
  (4) $A \succ O \implies \varphi(\mathbf{x}) = (\langle A\mathbf{x}, \mathbf{x}\rangle)^{1/2}$ *is a norm on* $\mathbb{C}^n$.
  (5) $A \succ O \implies \varphi(\mathbf{x}, \mathbf{y}) = \langle A\mathbf{x}, \mathbf{y}\rangle$ *is an inner product on* $\mathbb{C}^n$.

Proof. We shall verify (3) and leave the justification of the rest to the reader. Since $A \succeq O \implies A = A^H$, the corresponding singular value decompositions must coincide, i.e., $V_1 S_1 U_1^H = U_1 S_1 V_1^H$. Thus, in view of Corollary 16.8, $V_1 U_1^H = U_1 V_1^H$ and hence $V_1 = U_1 V_1^H U_1 = U_1 K$ with $K = V_1^H U_1$. Consequently,

$$I_r = V_1^H V_1 = K^H U_1^H U_1 K = K^H K, \quad \text{i.e., } K \text{ is unitary.}$$

Moreover, $KS_1 \succ O$, since $A = U_1 K S_1 U_1^H \succeq O$. Therefore, $KS_1 = S_1 K^H$ and hence

$$(KS_1)^2 = S_1 K^H K S_1 = S_1^2.$$

Thus, $KS_1 = S_1$, since they are both positive definite square roots of S_1^2. Therefore, $K = I_r$ and $V_1 = U_1$. □

Exercise 16.26. Verify items (1), (2), (4), and (5) in Theorem 16.11.

We remark that if $A, B \in \mathbb{C}^{n \times n}$, $A \succ B \succ O$, and $0 < t < 1$, then

$$(16.15) \qquad A^t - B^t = \frac{\sin \pi t}{\pi} \int_0^\infty x^t (xI_n + A)^{-1}(A - B)(xI_n + B)^{-1} dx.$$

Exercise 16.27. Use formula (16.15) to show that if $A, B \in \mathbb{C}^{n \times n}$, then

$$(16.16) \qquad A \succ B \succ O \implies A^t \succ B^t \quad \text{for} \quad 0 < t < 1.$$

Exercise 16.28. Let $A = \begin{bmatrix} x & 1 \\ 1 & 1 \end{bmatrix}$ and $B = \begin{bmatrix} 1 & 0 \\ 0 & 0 \end{bmatrix}$. Show that if $2 < x < 1 + \sqrt{2}$, then $A \succeq B \succeq O$, but $A^2 - B^2$ has one positive eigenvalue and one negative eigenvalue, i.e., $A \succeq B \succeq O$ does not imply that $A^2 \succeq B^2$.

16.6. Supplementary notes

This chapter is partially adapted from Chapter 12 in [**30**], which contains information on Toeplitz matrices, block Toeplitz matrices, and polynomial identities. Sections 16.4 and 16.5 are new, but (16.15) is discussed in [**30**].

Chapter 17

Determinants redux

This chapter deals with more advanced topics in the theory of determinants. The first three sections are devoted to developing the Binet-Cauchy formula for evaluating the determinants of matrix products of the form AB when $A, B^T \in \mathbb{C}^{n \times k}$ and $n > k$. Subsequently, some useful inequalities and the formulas of Jacobi and Sylvester are discussed.

17.1. Differentiating determinants

Theorem 17.1. *If*

$$\varphi(t) = \det \begin{bmatrix} f_{11}(t) & \cdots & f_{1n}(t) \\ \vdots & & \vdots \\ f_{n1}(t) & \cdots & f_{nn}(t) \end{bmatrix} = \det \begin{bmatrix} R_1(t) \\ \vdots \\ R_n(t) \end{bmatrix},$$

where the $f_{ij}(t)$ are smooth functions that can be differentiated freely with respect to t, $R_i(t) = \begin{bmatrix} f_{i1}(t) & \cdots & f_{in}(t) \end{bmatrix}$ for $i = 1, \ldots, n$, and the notation $|B|$ is used to denote the determinant of a matrix B, then

$$(17.1) \qquad \varphi'(t) = \begin{vmatrix} R_1'(t) \\ R_2(t) \\ \vdots \\ R_n(t) \end{vmatrix} + \begin{vmatrix} R_1(t) \\ R_2'(t) \\ \vdots \\ R_n(t) \end{vmatrix} + \cdots + \begin{vmatrix} R_1(t) \\ R_2(t) \\ \vdots \\ R_n'(t) \end{vmatrix}.$$

Discussion. Formula (17.1) follows from the fact that the determinant of a matrix is linear in each row of the matrix and is a continuous function of each entry in the matrix. Thus, if $n = 3$, then

$$\varphi(t+\varepsilon) = \begin{vmatrix} R_1(t+\varepsilon) - R_1(t) \\ R_2(t+\varepsilon) \\ R_3(t+\varepsilon) \end{vmatrix} + \begin{vmatrix} R_1(t) \\ R_2(t+\varepsilon) - R_2(t) \\ R_3(t+\varepsilon) \end{vmatrix} + \begin{vmatrix} R_1(t) \\ R_2(t) \\ R_3(t+\varepsilon) - R_3(t) \end{vmatrix} + \varphi(t)$$

and

$$\varphi'(t) = \lim_{\varepsilon \to 0} \frac{\varphi(t+\varepsilon) - \varphi(t)}{\varepsilon}$$

$$= \begin{vmatrix} f'_{11}(t) & f'_{12}(t) & f'_{13}(t) \\ f_{21}(t) & f_{22}(t) & f_{23}(t) \\ f_{31}(t) & f_{32}(t) & f_{33}(t) \end{vmatrix} + \begin{vmatrix} f_{11}(t) & f_{12}(t) & f_{13}(t) \\ f'_{21}(t) & f'_{22}(t) & f'_{23}(t) \\ f_{31}(t) & f_{32}(t) & f_{33}(t) \end{vmatrix} + \begin{vmatrix} f_{11}(t) & f_{12}(t) & f_{13}(t) \\ f_{21}(t) & f_{22}(t) & f_{23}(t) \\ f'_{31}(t) & f'_{32}(t) & f'_{33}(t) \end{vmatrix}.$$

The case of general n is treated in just the same way. □

Lemma 17.2. *If $A \in \mathbb{C}^{n \times n}$ and $\varphi(t) = \det(tI_n - A)$, then*

$$(17.2) \qquad \frac{\varphi'(t)}{\varphi(t)} = \mathrm{trace}(tI_n - A)^{-1} \quad \text{for } t \notin \sigma(A).$$

Discussion. If $n = 3$, $t \notin \sigma(A)$, and $I_3 = \begin{bmatrix} \mathbf{e}_1 & \mathbf{e}_2 & \mathbf{e}_3 \end{bmatrix}$, then (17.1) reduces to

$$\varphi'(t) = \begin{vmatrix} \mathbf{e}_1^T \\ R_2(t) \\ R_3(t) \end{vmatrix} + \begin{vmatrix} R_1(t) \\ \mathbf{e}_2^T \\ R_3(t) \end{vmatrix} + \begin{vmatrix} R_1(t) \\ R_2(t) \\ \mathbf{e}_3^T \end{vmatrix}$$

$$= (tI_n - A)_{\{1;1\}} + (tI_n - A)_{\{2;2\}} + (tI_n - A)_{\{3;3\}}$$

$$= \varphi(t)[((tI_n - A)^{-1})_{11} + ((tI_n - A)^{-1})_{22} + ((tI_n - A)^{-1})_{33}]$$

$$= \varphi(t) \, \mathrm{trace}(tI_n - A)^{-1}.$$

The general case is evaluated in exactly the same way. □

Lemma 17.3. *If $C \in \mathbb{C}^{n \times n}$, $t \in \mathbb{R}$, and $\varphi(t) = \det(I_n + tC)$, then*

$$(17.3) \qquad \varphi'(0) = \mathrm{trace}\, C.$$

Discussion. If $n = 3$, then

$$\varphi(t) = \det \begin{bmatrix} 1+tc_{11} & tc_{12} & tc_{13} \\ tc_{21} & 1+tc_{22} & tc_{23} \\ tc_{31} & tc_{32} & 1+tc_{33} \end{bmatrix} = \det \begin{bmatrix} R_1(t) \\ R_2(t) \\ R_3(t) \end{bmatrix}.$$

Thus,

$$\varphi'(t) = \begin{vmatrix} R'_1(t) \\ R_2(t) \\ R_3(t) \end{vmatrix} + \begin{vmatrix} R_1(t) \\ R'_2(t) \\ R_3(t) \end{vmatrix} + \begin{vmatrix} R_1(t) \\ R_2(t) \\ R'_3(t) \end{vmatrix} = \begin{vmatrix} \mathbf{e}_1^T C \\ R_2(t) \\ R_3(t) \end{vmatrix} + \begin{vmatrix} R_1(t) \\ \mathbf{e}_2^T C \\ R_3(t) \end{vmatrix} + \begin{vmatrix} R_1(t) \\ R_2(t) \\ \mathbf{e}_3^T C \end{vmatrix}$$

and
$$\varphi'(0) = \begin{vmatrix} c_{11} & c_{12} & c_{13} \\ 0 & 1 & 0 \\ 0 & 0 & 1 \end{vmatrix} + \begin{vmatrix} 1 & 0 & 0 \\ c_{21} & c_{22} & c_{23} \\ 0 & 0 & 1 \end{vmatrix} + \begin{vmatrix} 1 & 0 & 0 \\ 0 & 1 & 0 \\ c_{31} & c_{32} & c_{33} \end{vmatrix} = \operatorname{trace} C.$$

The computation for general n proceeds in the same way, just the bookkeeping is more elaborate. □

Lemma 17.4. *If $A, B \in \mathbb{C}^{n \times n}$, $A \succ O$, $B = B^H$, $h(t) = \ln \det(A + tB)$, and $t \in \mathbb{R}$, then*

(17.4) $$h'(0) = \operatorname{trace} A^{-1} B.$$

Proof. Let $C = A^{-1/2} B A^{-1/2}$ and $\varphi(t) = \det(I_n + tC)$. Then
$$h(t) = \ln[\det A \times \det(I_n + tC)] = h(0) + \ln \varphi(t)$$
and, in view of Lemma 17.3, $h'(0) = \varphi'(0)/\varphi(0) = \operatorname{trace} C$. □

Exercise 17.1. Give alternate proofs of formulas (17.2) and (17.3) using the Jordan decomposition for A.

Lemmas 17.2 and 17.3 are special cases of the following more general result:

Lemma 17.5. *Let*
$$F(t) = \begin{bmatrix} f_{11}(t) & \cdots & f_{1n}(t) \\ \vdots & & \vdots \\ f_{n1}(t) & \cdots & f_{nn}(t) \end{bmatrix} \quad \text{and} \quad \varphi(t) = \det F(t),$$

where the $f_{ij}(t)$ are smooth functions that can be differentiated freely with respect to t and $F(t)$ is invertible in the interval $a < t < b$. Then

(17.5) $$\frac{\varphi'(t)}{\varphi(t)} = \operatorname{trace}\{F'(t) F(t)^{-1}\} \quad \text{for} \quad a < t < b$$

and hence

(17.6) $$\varphi(t) = \varphi(c) \exp\left\{\int_c^t \operatorname{trace}\{F'(s) F(s)^{-1}\} ds\right\} \quad \text{for} \quad a < c, t < b.$$

Proof. Formula (17.1) exhibits $\varphi'(t)$ as a sum of n determinants, where in the j'th determinant the entries $f_{j1}(t), \ldots, f_{jn}(t)$ in the j'th row are replaced by $f'_{j1}(t), \ldots, f'_{jn}(t)$. Thus, upon restating this expression in terms of the minors of $F(t)$ and invoking the formula for the inverse of a matrix in terms of its minors ((1) of Theorem 7.5), we see that

(17.7) $$\varphi'(t) = \sum_{i,j=1}^n f'_{ij}(t) (-1)^{i+j} F(t)_{\{i;j\}} = \sum_{i,j=1}^n f'_{ij}(t) (F(t)^{-1})_{ji} \varphi(t),$$

which coincides with (17.5). Formula (17.6) is obtained by integrating (17.5) when $F(t)$ is real valued:

$$\int_c^t \operatorname{trace}\{F'(s)F(s)^{-1}\}ds = \int_c^t \frac{\varphi'(s)}{\varphi(s)}ds = \int_c^t \frac{d}{ds}\ln|\varphi(s)|ds$$
$$= \ln|\varphi(t)| - \ln|\varphi(c)| = \ln\frac{\varphi(t)}{\varphi(c)}.$$

The calculation uses the fact that $\varphi(t)$ does not change sign in the interval (a,b). It is then readily checked that (17.6) is also valid when $F(t)$ is complex valued. □

The first equality in (17.7) is valid even if $F(t)$ is not invertible and can be expressed as

(17.8) $$\varphi'(t) = \operatorname{trace}\{F'(t)G(t)\},$$

where the ij entry of $G(t)$ is equal to $(-1)^{i+j}F(t)_{\{j;i\}}$.

Exercise 17.2. Verify formula (17.5) in the special case that

$$\varphi(t) = \det(A + t\alpha\, \mathbf{e}_i \mathbf{e}_j^T), \quad \alpha \in \mathbb{C},$$

and A is invertible by evaluating

$$\lim_{\varepsilon\to 0}[\varphi(t+\varepsilon) - \varphi(t)]/\varepsilon.$$

17.2. The characteristic polynomial

The main objective of this section is to obtain explicit formulas for the coefficients $c_0, \ldots, c_{n-1}$ of the characteristic polynomial

$$\det(tI_n - A) = c_0 + c_1 t + \cdots + c_{n-1}t^{n-1} + t^n$$

of a matrix $A \in \mathbb{C}^{n\times n}$ in terms of the determinants of the principal submatrices of A. A $k \times k$ **principal submatrix** of $A \in \mathbb{C}^{n\times n}$ is a matrix that is obtained from A by deleting $n-k$ rows and $n-k$ columns with the same indices as the deleted rows.

Lemma 17.6. *If $A \in \mathbb{C}^{n\times n}$ and*

(17.9) $$\varphi(t) = \det(tI_n - A) = c_0 + c_1 t + \cdots + c_{n-1}t^{n-1} + t^n,$$

then $(-1)^k c_{n-k}$ is equal to the sum of the determinants of the $\binom{n}{k}$ principal $k \times k$ submatrices of A for $k = 1, \ldots, n$.

Proof. Clearly $c_j = \varphi^{(j)}(0)/j!$. The formula of interest is obtained by differentiating $\det(tI_n - A)$ with the help of formula (17.1). Notice that

$$R'_j(t) = \mathbf{e}_j^T \quad \text{and} \quad R''_j(t) = 0 \quad \text{for } j = 1, \ldots, n,$$

where $\mathbf{e}_j$ denotes the j'th column of I_n. For Sections 17.2 and 17.3 only, let $A(i_1,\ldots,i_k)$ denote the determinant of the $(n-k)\times(n-k)$ submatrix of A that is obtained by deleting rows $i_1,\ldots,i_k$ and columns $i_1,\ldots,i_k$ from A. Then

$$c_0 = \varphi(0) = (-1)^n \det A,$$
$$c_1 = \varphi^{(1)}(0) = (-1)^{n-1}\{A(1) + A(2) + \cdots + A(n)\},$$
$$c_2 = \frac{\varphi^{(2)}(0)}{2!} = (-1)^{n-2}\sum_{1\leq i_1 < i_2 \leq n} A(i_1, i_2),$$
$$\vdots$$
$$c_k = \frac{\varphi^{(k)}(0)}{k!} = (-1)^{n-k}\sum_{1\leq i_1 < \cdots < i_k \leq n} A(i_1,\ldots,i_k)$$

for $k = 0, \ldots, n-1$. $\square$

Example 17.1. If $A \in \mathbb{C}^{3\times 3}$, then in formula (17.9) $c_2 = -\{a_{11} + a_{22} + a_{33}\}$,

$$c_1 = \begin{vmatrix} a_{11} & a_{12} \\ a_{21} & a_{22} \end{vmatrix} + \begin{vmatrix} a_{22} & a_{23} \\ a_{32} & a_{33} \end{vmatrix} + \begin{vmatrix} a_{11} & a_{13} \\ a_{31} & a_{33} \end{vmatrix}, \quad \text{and} \quad c_0 = -\det A.$$

17.3. The Binet-Cauchy formula

The Binet-Cauchy formula expresses the determinant of the product AB of a matrix $A \in \mathbb{C}^{k\times n}$ and a matrix $B \in \mathbb{C}^{n\times k}$ in terms of the determinants of certain square subblocks of A and B when $n \geq k$.

Theorem 17.7. If $A = \begin{bmatrix} \mathbf{a}_1 & \cdots & \mathbf{a}_n \end{bmatrix} \in \mathbb{C}^{k\times n}$, $B^T = \begin{bmatrix} \mathbf{b}_1 & \cdots & \mathbf{b}_n \end{bmatrix} \in \mathbb{C}^{k\times n}$, and $k \leq n$, then

(17.10) $\det AB = \sum_{1\leq j_1 < \cdots < j_k \leq n} \det \begin{bmatrix} \mathbf{a}_{j_1} & \cdots & \mathbf{a}_{j_k} \end{bmatrix} \det \begin{bmatrix} \mathbf{b}_{j_1} & \cdots & \mathbf{b}_{j_k} \end{bmatrix}.$

Proof. Let

$$\psi(t) = \det(tI_k - AB) = c_0 + c_1 t + \cdots + c_k t^k$$

and

$$\varphi(t) = \det(tI_n - BA) = d_0 + d_1 t + \cdots + d_n t^n$$

denote the characteristic polynomials of AB and BA, respectively. We wish to calculate $\det AB = (-1)^k c_0$. Since

$$t^{n-k}(c_0 + c_1 t + \cdots + c_k t^k) = t^{n-k}\det(tI_k - AB) = \det(tI_n - BA)$$
$$= d_0 + d_1 t + \cdots + d_n t^n$$

for every point $t \in \mathbb{C}$, we see (with the help of Lemma 17.6) that

$$(-1)^k c_0 = (-1)^k d_{n-k} = (-1)^k \frac{\varphi^{(n-k)}(0)}{(n-k)!} = \sum (BA)(i_1, \ldots, i_{n-k}),$$

where the sum is over all tuples $1 \leq i_1 < \cdots < i_{n-k} \leq n$. But if $1 \leq j_1 < \cdots < j_k \leq n$ is the complementary set of indices to $i_1, \ldots, i_{n-k}$ (i.e., $\{i_1, \ldots, i_{n-k}\} \cup \{j_1, \ldots, j_k\} = \{1, \ldots, n\}$) and $\begin{bmatrix} \mathbf{e}_1 & \cdots & \mathbf{e}_n \end{bmatrix} = I_n$, then

$$\begin{aligned}
(BA)(i_1, \ldots, i_{n-k}) &= \det \left\{ \begin{bmatrix} \mathbf{e}_{j_1} & \cdots & \mathbf{e}_{j_k} \end{bmatrix}^T BA \begin{bmatrix} \mathbf{e}_{j_1} & \cdots & \mathbf{e}_{j_k} \end{bmatrix} \right\} \\
&= \det \{ \begin{bmatrix} \mathbf{e}_{j_1} & \cdots & \mathbf{e}_{j_k} \end{bmatrix}^T B \} \times \det \{ A \begin{bmatrix} \mathbf{e}_{j_1} & \cdots & \mathbf{e}_{j_k} \end{bmatrix} \} \\
&= \det \{ B^T \begin{bmatrix} \mathbf{e}_{j_1} & \cdots & \mathbf{e}_{j_k} \end{bmatrix} \} \times \det \{ A \begin{bmatrix} \mathbf{e}_{j_1} & \cdots & \mathbf{e}_{j_k} \end{bmatrix} \} \\
&= \det \begin{bmatrix} \mathbf{b}_{j_1} & \cdots & \mathbf{b}_{j_k} \end{bmatrix} \times \det \begin{bmatrix} \mathbf{a}_{j_1} & \cdots & \mathbf{a}_{j_k} \end{bmatrix}
\end{aligned}$$

and the sum of all such terms is equal to the right-hand side of (17.10). □

Exercise 17.3. Write the Binet-Cauchy formula for $\det AB$ for $A \in \mathbb{C}^{2 \times 3}$ and $B \in \mathbb{C}^{2 \times 3}$ in terms of the columns $\mathbf{a}_1, \mathbf{a}_2, \mathbf{a}_3$ of A and $\mathbf{c}_1, \mathbf{c}_2, \mathbf{c}_3$ of B^T. **Do not compute the relevant determinants.**

17.4. Inequalities for determinants

We begin with a pair of easy inequalities:

Theorem 17.8. *If $A, B \in \mathbb{C}^{n \times n}$ and $A \succeq B \succ O$, then $\det A \geq \det B$, with equality if and only if $A = B$.*

Proof. Under the given assumptions, the matrix $C = A^{-1/2} B A^{-1/2}$ is subject to the constraints $I_n \succeq C \succ O$ and hence the singular values $s_1, \ldots, s_n$ of C are all subject to the bounds $0 < s_j \leq 1$. Therefore $\det C = s_1 \cdots s_n \leq 1$, with equality if and only if $s_1 = \cdots = s_n = 1$, i.e., if and only if $C = I_n$. But this is equivalent to the assertion of the theorem. □

Theorem 17.9. *If $A = \begin{bmatrix} A_{11} & A_{12} \end{bmatrix}$ with components $A_{11} \in \mathbb{C}^{n \times p}$ and $A_{12} \in \mathbb{C}^{n \times q}$ and $p + q = n$, then*

(17.11) $$\det A^H A \leq \det A_{11}^H A_{11} \times \det A_{12}^H A_{12}.$$

Proof. We may assume that A is invertible, because otherwise the inequality is self-evident. But then $A_{11}^H A_{11}$ is invertible and the Schur complement formula applied to $A^H A = \begin{bmatrix} A_{11}^H A_{11} & A_{11}^H A_{12} \\ A_{12}^H A_{11} & A_{12}^H A_{12} \end{bmatrix}$ yields the identity

$$\det A^H A = \det A_{11}^H A_{11} \times \det [A_{12}^H A_{12} - A_{12}^H A_{11} (A_{11}^H A_{11})^{-1} A_{11}^H A_{12}]$$
$$\leq \det A_{11}^H A_{11} \times \det A_{12}^H A_{12},$$

thanks to Theorem 17.8. □

17.4. Inequalities for determinants

Corollary 17.10 (Hadamard's inequality). If $A = \begin{bmatrix} \mathbf{a}_1 & \cdots & \mathbf{a}_n \end{bmatrix} \in \mathbb{C}^{n \times n}$, then

$$|\det A| \leq \prod_{j=1}^{n} \|\mathbf{a}_j\|. \tag{17.12}$$

Proof. In view of Theorem 17.9 (applied $n-1$ times),

$$|\det A|^2 = \det A^H A \leq \prod_{j=1}^{n} \det \mathbf{a}_j^H \mathbf{a}_j = \prod_{j=1}^{n} \|\mathbf{a}_j\|^2. \qquad \square$$

In the rest of this section we establish some inequalities for the determinants of positive definite matrices that rest upon the fact (established in Example 9.1) that if $a_1, \ldots, a_n$ are positive numbers, then

$$(a_1 \cdots a_n)^{1/n} \leq \frac{a_1 + \cdots + a_n}{n}, \quad \text{with equality} \iff a_1 = \cdots = a_n. \tag{17.13}$$

Lemma 17.11. If $A \in \mathbb{C}^{n \times n}$ and $A \succ O$, then
(17.14)
$$(\det A)^{1/n} = \min \left\{ \frac{\operatorname{trace} AC}{n} : C \in \mathbb{C}^{n \times n},\ C \succ O,\ \text{and}\ \det C = 1 \right\}.$$

Proof. Since $A \succ O$, it admits a singular value decomposition of the form $A = VSV^H$ with $V \in \mathbb{C}^{n \times n}$ unitary and $S = \operatorname{diag}\{s_1, \ldots, s_n\}$. Consequently, the inequality (17.13) ensures that

$$(\det A)^{1/n} = (s_1 \cdots s_n)^{1/n} \leq \frac{s_1 + \cdots + s_n}{n} = \frac{\operatorname{trace} A}{n}.$$

If $C \in \mathbb{C}^{n \times n}$ is positive definite and $\det C = 1$, the same argument applied to $C^{1/2} A C^{1/2}$ yields the inequality

$$\begin{aligned}(\det A)^{1/n} &= \left(\det C^{1/2} A C^{1/2}\right)^{1/n} \\ &\leq \frac{\operatorname{trace} C^{1/2} A C^{1/2}}{n} = \frac{\operatorname{trace} AC}{n}.\end{aligned} \tag{17.15}$$

If $C = \delta A^{-1}$ with $\delta > 0$, then $C \succ O$ and $\det C = \delta^n / \det A$. Thus, $\det C = 1$ if and only if $\delta = (\det A)^{1/n}$, and for this choice of δ

$$\operatorname{trace} AC = n\delta = n(\det A)^{1/n}.$$

Therefore, (17.14) holds. $\qquad \square$

Lemma 17.12. If $A, B \in \mathbb{C}^{n \times n}$, $A \succ O$, and $B \succ O$, then

$$(\det(A+B))^{1/n} \geq (\det A)^{1/n} + (\det B)^{1/n} \tag{17.16}$$

with equality if and only if $A = \mu B$ for some $\mu > 0$.

Proof. In view of Lemma 17.11, there exists a positive definite matrix $C \in \mathbb{C}^{n \times n}$ with $\det C = 1$ such that
$$(\det (A + B))^{1/n} = \frac{\text{trace}\{(A + B)C\}}{n}.$$
The inequality (17.16) then follows easily from the fact that
$$\frac{\text{trace}\{(A + B)C\}}{n} = \frac{\text{trace}\{AC\}}{n} + \frac{\text{trace}\{BC\}}{n}$$
and
$$\frac{\text{trace}\{AC\}}{n} \geq (\det A)^{1/n} \quad \text{and} \quad \frac{\text{trace}\{BC\}}{n} \geq (\det B)^{1/n}.$$
Next, let $s_1 \geq \cdots \geq s_n$ denote the singular values of the matrix $D = A^{-1/2} B A^{-1/2}$. It is then readily checked that we have
$$\text{equality in (17.16)} \iff (\det (I_n + D))^{1/n} = 1 + (\det D)^{1/n}$$
$$\iff \{(1 + s_1) \cdots (1 + s_n)\}^{1/n} = 1 + (s_1 \cdots s_n)^{1/n}.$$
Let $f(x) = \{(x + s_1) \cdots (x + s_n)\}^{1/n}$ for $x > 0$. Then, $f(1) - f(0) = f'(x)$ for some point $x \in (0, 1)$ by the mean value theorem, and, in view of (17.13),
$$f'(x) = \frac{1}{n} f(x) \left\{ \frac{1}{x + s_1} + \cdots + \frac{1}{x + s_n} \right\}$$
$$\geq f(x) \left\{ \frac{1}{x + s_1} \cdots \frac{1}{x + s_n} \right\}^{1/n} = 1,$$
with equality if and only if $s_1 = \cdots = s_n$. Thus, it follows that
$$f(1) - f(0) = 1 \iff s_1 = \cdots = s_n = \mu \iff D = \mu I_n \iff A = \mu B,$$
as claimed. □

Exercise 17.4. Show that if $A, B \in \mathbb{C}^{n \times n}$, $A \succ O$, and $B \succ O$, then

(17.17) $\quad (\det (A + B))^{k/n} \geq (\det A)^{k/n} + (\det B)^{k/n} \quad$ for $k = 1, 2, \ldots$.

[HINT: Let $x = \{\det (A + B)\}^{-1} \det A$ and $y = \{\det (A + B)\}^{-1} \det B$. Then, $1 \geq x^{1/n} + y^{1/n} \implies 1 \geq x^{1/n} \implies x^{1/n} \geq x^{k/n}$, and similarly for y.]

Theorem 17.13. If $A, B \in \mathbb{C}^{n \times n}$, $A \succ O$, $B \succ O$, $f(A) = (\det A)^{1/n}$, and $g(A) = -\ln \det A$, then

(17.18) $\quad f(tA + (1 - t)B) \geq tf(A) + (1 - t)f(B) \quad$ for every $t \in [0, 1]$

and

(17.19) $\quad g(tA + (1 - t)B) \leq tg(A) + (1 - t)g(B) \quad$ for every $t \in [0, 1]$.

Moreover, equality holds in (17.19) for some $t \in (0, 1)$ if and only if $A = B$.

Proof. In view of Lemma 17.12,

$$f(tA + (1-t)B) \geq f(tA) + f((1-t)B) = \{\det tA\}^{1/n} + \{\det (1-t)B\}^{1/n}$$
$$= \{t^n \det A\}^{1/n} + \{(1-t)^n \det B\}^{1/n}$$
$$= tf(A) + (1-t)f(B).$$

Therefore, (17.18) holds.

To verify (17.19), recall that the function $h(x) = -\ln x$ is strictly convex and strictly monotone on $(0, \infty)$. Therefore, in view of (17.18),

$$h(f(tA + (1-t)B)) \leq h(tf(A) + (1-t)f(B))$$
$$\leq th(f(A)) + (1-t)h(f(B)),$$

which is equivalent to (17.19). Moreover, if equality holds in (17.19) for some $t \in (0,1)$, then

$$f(tA + (1-t)B) = tf(A) + (1-t)f(B) \quad \text{and} \quad f(A) = f(B).$$

The first equality implies that $A = \mu B$ for some $\mu > 0$ (by Lemma 17.12) and hence, the second equality implies that $\mu = 1$. □

Thus, $-f(A) = -(\det A)^{1/n}$ and $g(A) = -\ln \det A$ are both **convex functions on the convex set of positive definite matrices in** $\mathbb{C}^{n \times n}$; in fact, $g(A)$ is **strictly convex**.

Exercise 17.5. Show that if $C \in \mathbb{C}^{n \times n}$ and $C \succ O$, then

$$(17.20) \qquad \det \frac{I_n + C}{2} \geq (\det C)^{1/2}$$

with equality if and only if $C = I_n$. [HINT: The singular values of C satisfy $s_j^{1/2} = (1 \, s_j)^{1/2} \leq (1 + s_j)/2$ with equality if and only if $s_j = 1$.]

Exercise 17.6. Show that if $A, B \in \mathbb{C}^{n \times n}$, $A \succ O$, and $B \succ O$, then

$$(17.21) \qquad \det \frac{A + B}{2} \geq (\det A \det B)^{1/2},$$

with equality if and only if $A = B$. [HINT: Exploit Exercise 17.5.]

17.5. Some determinant identities

In this section we exploit the Schur complement formulas (3.18) and (3.19) to establish some useful identities.

Theorem 17.14. *If $A, B \in \mathbb{C}^{n \times n}$ are expressed in compatible four-block form with $A_{11}, B_{11} \in \mathbb{C}^{p \times p}$, $A_{22}, B_{22} \in \mathbb{C}^{q \times q}$, and $n = p + q$, then:*

(1) *If $AB = I_n$ and A_{11} is invertible, then $B_{22}(A_{22} - A_{21}A_{11}^{-1}A_{12}) = I_q$.*

(2) *If $AB = I_n$ and A_{22} is invertible, then $B_{11}(A_{11} - A_{12}A_{22}^{-1}A_{21}) = I_p$.*

Moreover, if $AB = I_n$, then

(17.22)
$$\det A_{11} = \det A \times \det B_{22} \quad \text{even if } A_{11} \text{ is not invertible, and}$$
$$\det A_{22} = \det A \times \det B_{11} \quad \text{even if } A_{22} \text{ is not invertible.}$$

Proof. If $AB = I_n$, then (1) and (2) follow by direct computation from formulas (3.18) and (3.19), respectively. (The strategy is to first express $B = A^{-1}$ as a product of the inverses of the three matrices that appear in the factorization of A.) It is then straightforward to verify the first (resp., second) formula in (17.22) under the assumptions imposed in (1) (resp., (2)).

Next, we wish to show that the first formula in (17.22) holds when $AB = I_n$ even if A_{11} is not invertible; the second follows from the first by interchanging A and B.

If A_{11} is not invertible, then there exist a matrix $C_{11} \in \mathbb{C}^{p \times p}$ and a $\delta > 0$ such that the matrices

$$A_{11} + \varepsilon C_{11} \quad \text{and} \quad A + \varepsilon \begin{bmatrix} C_{11} & O_{p \times q} \\ O_{q \times p} & O_{q \times q} \end{bmatrix} = A + \varepsilon C$$

are invertible when $0 < \varepsilon < \delta$. Therefore, there exists a matrix $D \in \mathbb{C}^{n \times n}$ that depends upon ε such that

$$(A + \varepsilon C)(B + D) = I_n \quad \text{and} \quad \det(A_{11} + \varepsilon C_{11}) = \det(A + \varepsilon C) \times \det(B_{22} + D_{22})$$

for every $\varepsilon \in (0, \delta)$. The first formula in the last display implies that

$$D = -\varepsilon (A + \varepsilon C)^{-1} C B = -\varepsilon (I_n + \varepsilon BC)^{-1} BCB$$

tends to 0 as $\varepsilon \downarrow 0$. Thus the second formula in that display tends to the first formula in (17.22) when $\varepsilon \downarrow 0$. $\square$

17.6. Jacobi's determinant identity

The formulas in (17.22) are the simplest forms of Jacobi's determinant identity; the general form is obtained by permuting rows and columns:

If $I \subset \{1, \ldots, n\}$ and $J \subset \{1, \ldots, n\}$ both contain q distinct integers and the complementary sets

$$I^c = \{1, \ldots, n\} \setminus I = \{i_1, \ldots, i_p\} \quad \text{and} \quad J^c = \{1, \ldots, n\} \setminus J = \{j_1, \ldots, j_p\}$$

and if $A_{\{I;J\}}$ denotes the matrix that is obtained from $A \in \mathbb{C}^{n \times n}$ by deleting the q rows indexed in I and the q columns indexed in J, then, since the remaining rows and columns are left in place,

$$A_{\{I;J\}} = \begin{bmatrix} a_{i_1 j_1} & a_{i_1 j_2} & \cdots & a_{i_1 j_p} \\ \vdots & & & \vdots \\ a_{i_p j_1} & a_{i_p j_2} & \cdots & a_{i_p j_p} \end{bmatrix},$$

with $i_1 < \cdots < i_p$ and $j_1 < \cdots < j_p$.

Theorem 17.15 (Jacobi's identity). *If $A, B \in \mathbb{C}^{n \times n}$ and $AB = I_n$, then, in terms of the preceding notation,*

(17.23) $\qquad \det A_{\{I;J\}} = (-1)^{(i_1 + \cdots + i_p + j_1 + \cdots + j_p)} \det A \times \det B_{\{J^c;I^c\}}.$

Proof. To obtain (17.23), choose a pair of permutation matrices $P, Q \in \mathbb{R}^{n \times n}$ such that the upper left-hand $p \times p$ corner $(PAQ)_{11}$ of PAQ is equal to $A_{\{I;J\}}$; then the lower right-hand $q \times q$ corner $(Q^T B P^T)_{22}$ of $(Q^T B P^T)$ is equal to $B_{\{J^c;I^c\}}$. Thus, as

$$AB = I_n \iff (PAQ)(Q^T B P^T) = I_n,$$

formula (17.22) ensures that

(17.24) $\qquad \det(PAQ) \times \det(Q^T B P^T)_{22} = \det(PAQ)_{11}.$

To complete the verification of Jacobi's identity (17.23), it remains only to check that $\det PQ = (-1)^{(i_1 + \cdots + i_p + j_1 + \cdots + j_p)}$. This stems from the fact that, as $i_k \geq k$, the transition from row i_k to the k'th position requires $i_k - k$ row interchanges and, similarly, as $j_k \geq k$, the transition of column j_k to the k'th position requires $j_k - k$ column interchanges. □

Example 17.2. If $A, B \in \mathbb{C}^{5 \times 5}$, $AB = I_5$, $I = \{2, 4\}$, $J = \{2, 5\}$, and $\mathbf{e}_j$ denotes the j'th column of I_5, let

$$P = \begin{bmatrix} \mathbf{e}_1 & \mathbf{e}_3 & \mathbf{e}_5 & \mathbf{e}_2 & \mathbf{e}_4 \end{bmatrix}^T \quad \text{and} \quad Q = \begin{bmatrix} \mathbf{e}_1 & \mathbf{e}_3 & \mathbf{e}_4 & \mathbf{e}_2 & \mathbf{e}_5 \end{bmatrix}.$$

Then $I^c = \{1, 3, 5\}$, $J^c = \{1, 3, 4\}$, $\det P = (-1)^3$, $\det Q = (-1)^2$,

$$(PAQ)_{11} = A_{\{I;J\}} = \begin{bmatrix} a_{11} & a_{13} & a_{14} \\ a_{31} & a_{33} & a_{34} \\ a_{51} & a_{53} & a_{54} \end{bmatrix}$$

and

$$(Q^T B P^T)_{22} = B_{\{J^c;I^c\}} = \begin{bmatrix} b_{22} & b_{24} \\ b_{52} & b_{54} \end{bmatrix}.$$

Therefore,

$$\begin{vmatrix} a_{11} & a_{13} & a_{14} \\ a_{31} & a_{33} & a_{34} \\ a_{51} & a_{53} & a_{54} \end{vmatrix} = -\det A \times \begin{vmatrix} b_{22} & b_{24} \\ b_{52} & b_{54} \end{vmatrix}.$$

17.7. Sylvester's determinant identity

If $A \in \mathbb{C}^{n \times n}$ is written in standard four-block form with $A_{11} \in \mathbb{C}^{p \times p}$, $A_{22} \in \mathbb{C}^{q \times q}$, and $p + q = n$, then the formulas

(17.25) $\qquad \det A = \det A_{11} \det[A_{22} - A_{21} A_{11}^{-1} A_{12}] \quad \text{if } A_{11} \text{ is invertible}$

and

(17.26) $\qquad \det A = \det A_{22} \det[A_{11} - A_{12} A_{22}^{-1} A_{21}] \quad \text{if } A_{22} \text{ is invertible}$

are obtained easily from the Schur complement formulas (3.18) and (3.19). Sylvester's determinant identity is an extension of formula (17.25) (resp., (17.26)) that is valid even if A_{11} (resp., A_{22}) is not invertible and A is not invertible.

Theorem 17.16 (Sylvester's identity). *If $A \in \mathbb{C}^{n \times n}$ is written as*

$$A = \begin{bmatrix} A_{11} & \mathbf{v}_1 & \cdots & \mathbf{v}_q \\ \mathbf{u}_1^T & b_{11} & \cdots & b_{1q} \\ \vdots & & & \vdots \\ \mathbf{u}_q^T & b_{q1} & \cdots & b_{qq} \end{bmatrix},$$

with $A_{11} \in \mathbb{C}^{p \times p}$, $\mathbf{u}_i, \mathbf{v}_i \in \mathbb{C}^p$ for $i = 1, \ldots, q$ and $b_{ij} \in \mathbb{C}$ for $i, j = 1, \ldots, q$ and $D \in \mathbb{C}^{q \times q}$ with entries $d_{ij} = \det \begin{bmatrix} A_{11} & \mathbf{v}_j \\ \mathbf{u}_i^T & b_{ij} \end{bmatrix}$ for $i, j = 1, \ldots, q$, then

(17.27) $\quad\quad\quad (\det A_{11})^{(q-1)} \det A = \det D$.

Proof. Suppose first that A_{11} is invertible. Then (17.25) is applicable and the entries c_{ij} in the matrix $C = A_{22} - A_{21} A_{11}^{-1} A_{12}$ are equal to

$$c_{ij} = b_{ij} - \mathbf{u}_i^T A_{11}^{-1} \mathbf{v}_j \quad \text{for } i, j = 1, \ldots, q,$$

and, by another application of (17.25),

$$d_{ij} = [b_{ij} - \mathbf{u}_i^T A_{11}^{-1} \mathbf{v}_j] \det A_{11} = c_{ij} \det A_{11}.$$

Thus,

$$\det A = \det A_{11} \det C = \det A_{11} \frac{\det D}{(\det A_{11})^q},$$

which serves to justify (17.27) when A_{11} is invertible.

If A_{11} is not invertible, then there exist a matrix $E_{11} \in \mathbb{C}^{p \times p}$ and a $\delta > 0$ such that the matrix $A_{11} + \varepsilon E_{11}$ is invertible when $0 < \varepsilon < \delta$. Correspondingly, let

$$A_\varepsilon = A + \varepsilon \begin{bmatrix} E_{11} & O_{p \times q} \\ O_{q \times p} & O_{q \times q} \end{bmatrix} = A + \varepsilon E$$

and let D_ε be the matrix with entries $d_{ij}(\varepsilon) = \det \begin{bmatrix} A_{11} + \varepsilon E_{11} & \mathbf{v}_j \\ \mathbf{u}_i^T & b_{ij} \end{bmatrix}$. Then

$$(\det [A_{11} + \varepsilon E_{11}])^{q-1} \det A_\varepsilon = \det D_\varepsilon \quad \text{for } 0 < \varepsilon < \delta$$

and (17.27) is obtained by letting $\varepsilon \downarrow 0$ in the last display. $\square$

Formula (17.27) is the simplest form of Sylvester's determinant identity; other forms are obtained by permuting rows and columns and reexpressing the entries in D in terms of minors that extract $q-1$ rows and $q-1$ columns.

17.7. Sylvester's determinant identity

Example 17.3. In keeping with the notation in Section 17.6, let $A_{\{i,j,k;r,s,t\}}$ denote the determinant of the $(n-3) \times (n-3)$ matrix that is obtained from $A \in \mathbb{C}^{n \times n}$ by deleting rows i, j, k and columns r, s, t. If $A \in \mathbb{C}^{n \times n}$, $i < j < k$, and $r < s < t$, then

$$(17.28) \qquad \det A \times (A_{\{i,j,k;r,s,t\}})^2 = \begin{vmatrix} A_{\{j,k;s,t\}} & A_{\{j,k;r,t\}} & A_{\{j,k;r,s\}} \\ A_{\{i,k;s,t\}} & A_{\{i,k;r,t\}} & A_{\{i,k;r,s\}} \\ A_{\{i,j;s,t\}} & A_{\{i,j;r,t\}} & A_{\{i,j;r,s\}} \end{vmatrix}.$$

Discussion. Let $I_n = \begin{bmatrix} \mathbf{e}_1 & \cdots & \mathbf{e}_n \end{bmatrix}$ and let $P, Q \in \mathbb{R}^{n \times n}$ be permutation matrices of the form $P = \begin{bmatrix} M & \mathbf{e}_i & \mathbf{e}_j & \mathbf{e}_k \end{bmatrix}^T$ and $Q = \begin{bmatrix} N & \mathbf{e}_r & \mathbf{e}_s & \mathbf{e}_t \end{bmatrix}$, respectively, where $M = \begin{bmatrix} \mathbf{e}_{j_1} & \cdots & \mathbf{e}_{j_{n-3}} \end{bmatrix}$ and $N = \begin{bmatrix} \mathbf{e}_{i_1} & \cdots & \mathbf{e}_{i_{n-3}} \end{bmatrix}$ with $j_1 \leq \cdots \leq j_{n-3}$ and $i_1 \leq \cdots \leq i_{n-3}$. Then $A_{\{i,j,k;r,s,t\}} = \det(M^T A N)$,

$$PAQ = \begin{bmatrix} M^T A N & M^T A \mathbf{e}_r & M^T A \mathbf{e}_s & M^T A \mathbf{e}_t \\ \mathbf{e}_i^T A N & a_{ir} & a_{is} & a_{it} \\ \mathbf{e}_j^T A N & a_{jr} & a_{js} & a_{jt} \\ \mathbf{e}_k^T A N & a_{kr} & a_{ks} & a_{kt} \end{bmatrix},$$

and, with D as in Theorem 17.16,

$$\det D = \begin{vmatrix} (-1)^{i+r} A_{\{j,k;s,t\}} & (-1)^{i+s} A_{\{j,k;r,t\}} & (-1)^{i+t} A_{\{j,k;r,s\}} \\ (-1)^{j+r} A_{\{i,k;s,t\}} & (-1)^{j+s} A_{\{i,k;r,t\}} & (-1)^{j+t} A_{\{i,k;r,s\}} \\ (-1)^{k+r} A_{\{i,j;s,t\}} & (-1)^{k+s} A_{\{i,j;r,t\}} & (-1)^{k+t} A_{\{i,j;r,s\}} \end{vmatrix} = (-1)^\nu \kappa,$$

where $\nu = r+s+t+i+j+k$ and κ is equal to the right-hand side of (17.28). Therefore, since $(-1)^\nu \det PQ = 1$, (17.28) is obtained from (17.27) with $q = 3$ by combining the preceding evaluations. $\diamond$

Exercise 17.7. Show that if $A_{\{i,j;r,s\}}$ denotes the determinant of the $(n-2) \times (n-2)$ matrix that is obtained from $A \in \mathbb{C}^{n \times n}$ by deleting rows i and j and columns r and s with $i < j$ and $r < s$, then

$$(17.29) \qquad \det A \times A_{\{i,j;r,s\}} = \begin{vmatrix} A_{\{j;s\}} & A_{\{j;r\}} \\ A_{\{i;s\}} & A_{\{i;r\}} \end{vmatrix}.$$

Exercise 17.8. Show that if $A \in \mathbb{C}^{n \times n}$ is expressed in standard four-block form with $A_{11} \in \mathbb{C}^{p \times p}$, $A_{22} \in \mathbb{C}^{q \times q}$, and $p + q = n$, then
(17.30)
$$(\det A_{22})^{p-1} \det A = \det[d_{ij}], \quad \text{where } d_{ij} = \det \begin{bmatrix} \mathbf{e}_i^T A_{11} \mathbf{e}_j & \mathbf{e}_i^T \\ A_{21} \mathbf{e}_j & A_{22} \end{bmatrix}$$

for $i, j = 1, \ldots, p$ and $\begin{bmatrix} \mathbf{e}_1 & \cdots & \mathbf{e}_p \end{bmatrix} = I_p$. [HINT: First obtain (17.30) when A_{22} is invertible.]

17.8. Supplementary notes

The idea of using the identity (7.4) to obtain the Binet-Cauchy formula is taken from an observation on page 253 of Tao [**71**]; a different strategy was followed in Chapter 5 of [**30**]. We have presented simple versions of the formulas of Jacobi and Sylvester in order not to get lost in notation, which can be a little overwhelming. A number of other proofs for Sylvester's formula are presented by Akritas, Akritas, and Malaschonok in [**2**]; see also the expository article by Brualdi and Schneider [**14**].

Formula (17.5) is attributed to Jacobi.

Chapter 18

Applications

In this chapter we consider a number of applications of the theory developed to this point, including Schur complements for semidefinite matrices in Section 18.5. En route, we discuss strictly convex functions and strictly convex normed linear spaces in Section 18.2.

18.1. A minimization problem

Let $A \in \mathbb{C}^{n \times n}$ with singular values $s_1 \geq \cdots \geq s_n$ and let

$$(18.1) \qquad f(\mathbf{x}) = \left\langle A \begin{bmatrix} \mathbf{u} \\ \mathbf{x} \end{bmatrix}, \begin{bmatrix} \mathbf{u} \\ \mathbf{x} \end{bmatrix} \right\rangle \quad \text{for } A \succ O, \ \mathbf{u} \in \mathbb{C}^p, \text{ and } \mathbf{x} \in \mathbb{C}^q,$$

where $p \geq 1$, $q \geq 1$, and $p + q = n$. Then, since $f(\mathbf{x})$ is continuous on $\mathbb{C}^q$ and

$$(18.2) \qquad 0 \leq s_n \|\mathbf{u}\|^2 \leq f(\mathbf{x}) \leq s_1\{\|\mathbf{u}\|^2 + \|\mathbf{x}\|^2\} \quad \text{for every } \mathbf{x} \in \mathbb{C}^q,$$

there exists at least one vector $\mathbf{x}_0 \in \mathbb{C}^q$ such that $f(\mathbf{x}_0) \leq f(\mathbf{x})$ for every vector $\mathbf{x} \in \mathbb{C}^q$. In the next section we shall show that $f(\mathbf{x})$ is strictly convex and hence there is exactly one such vector $\mathbf{x}_0$.

In this section we shall discuss a number of different ways of finding $\mathbf{x}_0$ and $f(\mathbf{x}_0)$. The first of these is a very nice application of factorization. Extensive use will be made of the block decompositions

$$(18.3) \qquad A = \begin{bmatrix} A_{11} & A_{12} \\ A_{21} & A_{22} \end{bmatrix} \quad \text{and} \quad A^{-1} = \begin{bmatrix} B_{11} & B_{12} \\ B_{21} & B_{22} \end{bmatrix},$$

where $A_{11}, B_{11} \in \mathbb{C}^{p \times p}$, $A_{21}, B_{21} \in \mathbb{C}^{q \times p}$, and $A_{22}, B_{22} \in \mathbb{C}^{q \times q}$.

Theorem 18.1. *If $A \succ O$, then*

(18.4) $$f(\mathbf{x}) \geq \langle (A_{11} - A_{12}A_{22}^{-1}A_{21})\mathbf{u}, \mathbf{u} \rangle = \mathbf{u}^H B_{11}^{-1} \mathbf{u}$$

for every $\mathbf{x} \in \mathbb{C}^q$, with equality if and only if $\mathbf{x} = -A_{22}^{-1}A_{21}\mathbf{u}$.

Proof (by factorization). In view of Theorem 16.5, $A = L^H L$, where $L \in \mathbb{C}^{n \times n}$ is an invertible lower triangular matrix. Therefore,

$$\left\langle A \begin{bmatrix} \mathbf{u} \\ \mathbf{x} \end{bmatrix}, \begin{bmatrix} \mathbf{u} \\ \mathbf{x} \end{bmatrix} \right\rangle = \left\langle \begin{bmatrix} L_{11} & O \\ L_{21} & L_{22} \end{bmatrix} \begin{bmatrix} \mathbf{u} \\ \mathbf{x} \end{bmatrix}, \begin{bmatrix} L_{11} & O \\ L_{21} & L_{22} \end{bmatrix} \begin{bmatrix} \mathbf{u} \\ \mathbf{x} \end{bmatrix} \right\rangle$$

$$= \|L_{11}\mathbf{u}\|^2 + \|L_{21}\mathbf{u} + L_{22}\mathbf{x}\|^2.$$

Thus, the minimum of interest is equal to $\|L_{11}\mathbf{u}\|^2$; it is achieved by choosing $\mathbf{x} = -L_{22}^{-1}L_{21}$.

It remains to express these conclusions in terms of the entries in the original matrix A by taking advantage of the formulas

$$\begin{bmatrix} A_{11} & A_{12} \\ A_{21} & A_{22} \end{bmatrix} = \begin{bmatrix} L_{11}^H & L_{21}^H \\ O & L_{22}^H \end{bmatrix} \begin{bmatrix} L_{11} & O \\ L_{21} & L_{22} \end{bmatrix}.$$

In particular, $A_{21} = L_{22}^H L_{21} = A_{12}^H$, $A_{22} = L_{22}^H L_{22}$, and hence

$$L_{11}^H L_{11} = A_{11} - L_{21}^H L_{21} = A_{11} - A_{12}A_{22}^{-1}A_{21} \quad \text{and} \quad L_{22}^{-1}L_{21} = A_{22}^{-1}A_{21}.$$

The equality in (18.4) follows from (3.20) but also drops out easily by noticing that $B = L^{-1}(L^H)^{-1}$ and $B_{11} = L_{11}^{-1}(L_{11}^H)^{-1}$.

Proof (by Schur complements). In view of formula (3.19),

$$f(\mathbf{x}) = \left\langle \begin{bmatrix} A_{11} - A_{12}A_{22}^{-1}A_{21} & O \\ O & A_{22} \end{bmatrix} \begin{bmatrix} \mathbf{u} \\ A_{22}^{-1}A_{21}\mathbf{u} + \mathbf{x} \end{bmatrix}, \begin{bmatrix} \mathbf{u} \\ A_{22}^{-1}A_{21}\mathbf{u} + \mathbf{x} \end{bmatrix} \right\rangle$$

$$= \langle (A_{11} - A_{12}A_{22}^{-1}A_{21})\mathbf{u}, \mathbf{u} \rangle + \langle A_{22}(A_{22}^{-1}A_{21}\mathbf{u} + \mathbf{x}), (A_{22}^{-1}A_{21}\mathbf{u} + \mathbf{x}) \rangle,$$

which leads easily to the conclusions of the theorem.

Proof (by completing the square). The three terms involving $\mathbf{x}$ in

$$f(\mathbf{x}) = \langle A_{11}\mathbf{u}, \mathbf{u} \rangle + \langle A_{12}\mathbf{x}, \mathbf{u} \rangle + \langle A_{21}\mathbf{u}, \mathbf{x} \rangle + \langle A_{22}\mathbf{x}, \mathbf{x} \rangle$$

can be written as

$$\langle A_{22}\mathbf{x}, \mathbf{x} \rangle + \langle A_{12}\mathbf{x}, \mathbf{u} \rangle + \langle A_{21}\mathbf{u}, \mathbf{x} \rangle + \|(A_{22})^{-1/2}A_{21}\mathbf{u}\|^2 - \|(A_{22})^{-1/2}A_{21}\mathbf{u}\|^2,$$

i.e., as a **square plus change**:

$$\|(A_{22})^{1/2}\mathbf{x} + (A_{22})^{-1/2}A_{21}\mathbf{u}\|^2 - \|(A_{22})^{-1/2}A_{21}\mathbf{u}\|^2.$$

Thus,

$$f(\mathbf{x}) = \|(A_{22})^{1/2}\mathbf{x} + (A_{22})^{-1/2}A_{21}\mathbf{u}\|^2 + \langle (A_{11} - A_{12}A_{22}^{-1}A_{21})\mathbf{u}, \mathbf{u} \rangle,$$

which also leads easily to the conclusions of the theorem.

Proof (by projection). Let $C = A^{1/2}$. Then upon expressing C in standard four-block form as

$$C = \begin{bmatrix} C_{11} & C_{12} \\ C_{21} & C_{22} \end{bmatrix} \quad \text{with } C_{11} \in \mathbb{C}^{p \times p} \text{ and } C_{22} \in \mathbb{C}^{q \times q},$$

we see that the orthogonal projection $\Pi_\mathcal{X}$ from $\mathbb{C}^n$ onto the subspace

$$\mathcal{X} = \left\{ \begin{bmatrix} C_{11} & C_{12} \\ C_{21} & C_{22} \end{bmatrix} \begin{bmatrix} \mathbf{0} \\ \mathbf{x} \end{bmatrix} : \mathbf{x} \in \mathbb{C}^q \right\} = \left\{ \begin{bmatrix} C_{12} \\ C_{22} \end{bmatrix} \mathbf{x} : \mathbf{x} \in \mathbb{C}^q \right\} = \mathcal{R} \begin{bmatrix} C_{12} \\ C_{22} \end{bmatrix}$$

of the vector $C \begin{bmatrix} \mathbf{u} \\ \mathbf{0} \end{bmatrix}$ is equal to

$$\Pi_\mathcal{X} C \begin{bmatrix} \mathbf{u} \\ \mathbf{0} \end{bmatrix} = \begin{bmatrix} C_{12} \\ C_{22} \end{bmatrix} \left(\begin{bmatrix} C_{12}^H & C_{22}^H \end{bmatrix} \begin{bmatrix} C_{12} \\ C_{22} \end{bmatrix} \right)^{-1} \begin{bmatrix} C_{12}^H & C_{22}^H \end{bmatrix} \begin{bmatrix} C_{11} \\ C_{21} \end{bmatrix} \mathbf{u}$$

$$= \begin{bmatrix} C_{12} \\ C_{22} \end{bmatrix} A_{22}^{-1} A_{21} \mathbf{u},$$

since $C^2 = A$. Consequently,

(18.5)
$$f(\mathbf{x}) = \left\| \begin{bmatrix} C_{12} \\ C_{22} \end{bmatrix} [A_{22}^{-1} A_{21} \mathbf{u} + \mathbf{x}] + (I - \Pi_\mathcal{X}) \begin{bmatrix} C_{11} \\ C_{21} \end{bmatrix} \mathbf{u} \right\|^2$$

$$= \left\| \begin{bmatrix} C_{12} \\ C_{22} \end{bmatrix} [A_{22}^{-1} A_{21} \mathbf{u} + \mathbf{x}] \right\|^2 + \left\| (I - \Pi_\mathcal{X}) \begin{bmatrix} C_{11} \\ C_{21} \end{bmatrix} \mathbf{u} \right\|^2,$$

which again leads easily to the conclusions of the theorem. □

Exercise 18.1. Complete the proof of Theorem 18.1 starting from (18.5).

18.2. Strictly convex functions and spaces

Recall that a function $f(\mathbf{x})$ that is defined on a convex set Q is said to be **strictly convex** if

$$f(t\mathbf{x} + (1-t)\mathbf{y}) < tf(\mathbf{x}) + (1-t)f(\mathbf{y}) \quad \text{when} \quad \mathbf{x} \neq \mathbf{y} \quad \text{and} \quad 0 < t < 1.$$

A **normed linear space** $\mathcal{X}$ with norm $\|\mathbf{x}\| = \|\mathbf{x}\|_\mathcal{X}$ is **strictly convex** if $\|t\mathbf{x}+(1-t)\mathbf{y}\| < t\|\mathbf{x}\|+(1-t)\|\mathbf{y}\|$ when $\mathbf{x} \neq \mathbf{y}$, $0 < t < 1$, and $\|\mathbf{x}\| = \|\mathbf{y}\| = 1$.

• **Warning:** The last condition is more restrictive than just requiring the function $f(\mathbf{x}) = \|\mathbf{x}\|$ to be strictly convex. Thus, for example, if $1 < p < \infty$, then the space $\mathbb{C}^n$ equipped with the norm $\|\mathbf{x}\|_p$ is strictly convex, even though the function $f(\mathbf{x}) = \|\mathbf{x}\|_p$ is not strictly convex: If $\mathbf{x} = \mu \mathbf{y}$, $\mu \geq 0$, and $t \in (0,1)$, then $\|t\mathbf{x} + (1-t)\mathbf{y}\|_p = t\|\mathbf{x}\|_p + (1-t)\|\mathbf{y}\|_p$ yet $\mathbf{y} \neq \mathbf{x}$ when $\mu \neq 1$; see Lemma 18.3 below for a more complete story.

Our main interest in convex functions and strictly convex functions stems from the central role that they play in extremal problems. We have already touched upon this in Chapter 9 and shall return to this theme for functions of many variables (after developing the necessary tools) in Chapter 21. At this point, we can state the following result:

Theorem 18.2. *If $f(\mathbf{x})$ is a strictly convex function that is defined on a convex set Q and if $\mathbf{x}_1, \mathbf{x}_2 \in Q$ and $f(\mathbf{x}_1) \leq f(\mathbf{x})$ and $f(\mathbf{x}_2) \leq f(\mathbf{x})$ for every point $\mathbf{x} \in Q$, then $\mathbf{x}_1 = \mathbf{x}_2$ and $f(\mathbf{x}_1) < f(\mathbf{x})$ if $\mathbf{x} \in Q$ and $\mathbf{x} \neq \mathbf{x}_1$.*

Proof. If $\mathbf{x}, \mathbf{x}_1 \in Q$, $\mathbf{x} \neq \mathbf{x}_1$, and $0 < t < 1$, then

$$f(\mathbf{x}_1) \leq f(\mathbf{x}_1 + t(\mathbf{x} - \mathbf{x}_1)) = f((1-t)\mathbf{x}_1 + t\mathbf{x}) < (1-t)f(\mathbf{x}_1) + tf(\mathbf{x}).$$

Consequently, $0 < t(f(\mathbf{x}) - f(\mathbf{x}_1))$ if $\mathbf{x} \in Q$ and $\mathbf{x} \neq \mathbf{x}_1$, which justifies the second assertion. But this in turn justifies the first assertion: If $\mathbf{x}_2 \neq \mathbf{x}_1$, then $f(\mathbf{x}_1) < f(\mathbf{x}_2) \leq f(\mathbf{x}_1)$, which is impossible. $\square$

Lemma 18.3. *If $\mathbf{x}, \mathbf{y} \in \mathcal{U}$, a finite-dimensional inner product space, and $0 < t < 1$, then*

(18.6) $\quad \|t\mathbf{x} + (1-t)\mathbf{y}\|_\mathcal{U} = t\|\mathbf{x}\|_\mathcal{U} + (1-t)\|\mathbf{y}\|_\mathcal{U} \iff$ *either* $\mathbf{y} = \mathbf{0}$
or $\mathbf{y} \neq \mathbf{0}$ *and* $\mathbf{x} = \mu\mathbf{y}$ *for some* $\mu \geq 0$.

However, if $0 < t < 1$ and $1 < s < \infty$, then

(18.7) $\quad \|t\mathbf{x} + (1-t)\mathbf{y}\|_\mathcal{U}^s = t\|\mathbf{x}\|_\mathcal{U}^s + (1-t)\|\mathbf{y}\|_\mathcal{U}^s \iff \mathbf{x} = \mathbf{y}$.

Proof. To minimize clutter, we shall drop the subscript $\mathcal{U}$ in the notation of the proof. Suppose first that $\|t\mathbf{x} + (1-t)\mathbf{y}\| = t\|\mathbf{x}\| + (1-t)\|\mathbf{y}\|$. Then

$$\begin{aligned}(t\|\mathbf{x}\| + (1-t)\|\mathbf{y}\|)^2 &= \|t\mathbf{x} + (1-t)\mathbf{y}\|^2 \\ &= t^2\|\mathbf{x}\|^2 + t(1-t)\{\langle \mathbf{x}, \mathbf{y}\rangle + \langle \mathbf{y}, \mathbf{x}\rangle\} + (1-t)^2\|\mathbf{y}\|^2 \\ &\leq t^2\|\mathbf{x}\|^2 + t(1-t)\{|\langle \mathbf{x}, \mathbf{y}\rangle| + |\langle \mathbf{y}, \mathbf{x}\rangle|\} + (1-t)^2\|\mathbf{y}\|^2 \\ &\leq t^2\|\mathbf{x}\|^2 + 2t(1-t)\|\mathbf{x}\|\,\|\mathbf{y}\| + (1-t)^2\|\mathbf{y}\|^2 \\ &= (t\|\mathbf{x}\| + (1-t)\|\mathbf{y}\|)^2.\end{aligned}$$

Thus, the two inequalities in the preceding display must both be equalities. Consequently,

(18.8) $\quad \langle \mathbf{x}, \mathbf{y}\rangle + \langle \mathbf{y}, \mathbf{x}\rangle = 2|\langle \mathbf{x}, \mathbf{y}\rangle| = 2\|\mathbf{x}\|\,\|\mathbf{y}\|.$

The second equality in (18.8) and the conditions for equality in the Cauchy-Schwarz inequality ensure that either $\mathbf{y} = \mathbf{0}$, or $\mathbf{y} \neq \mathbf{0}$ and $\mathbf{x} = \mu\mathbf{y}$ for some $\mu \in \mathbb{C}$. If $\mathbf{y} \neq \mathbf{0}$, then, in view of the first equality in (18.8), $\mu + \overline{\mu} = 2|\mu|$ and hence $\mu \geq 0$. This completes the proof of the implication $\implies$ in (18.6). The opposite implication is self-evident.

18.2. Strictly convex functions and spaces

To verify the implication $\Longrightarrow$ in (18.7), let $g(a) = a^s$ on $(0, \infty)$. Then, since g is monotone increasing and strictly convex, the equality

(18.9) $$\|t\mathbf{x} + (1-t)\mathbf{y}\|^s = t\|\mathbf{x}\|^s + (1-t)\|\mathbf{y}\|^s$$

implies that

$$\begin{aligned}tg(\|\mathbf{x}\|) + (1-t)g(\|\mathbf{y}\|) &= g(\|t\mathbf{x} + (1-t)\mathbf{y}\|) \\ &\leq g(t\|\mathbf{x}\| + (1-t)\|\mathbf{y}\|) \quad \text{(as } g(a) \leq g(b) \text{ if } a < b) \\ &\leq tg(\|\mathbf{x}\|) + (1-t)g(\|\mathbf{y}\|) \quad \text{(since } g \text{ is convex)}.\end{aligned}$$

Therefore, equality prevails, and thus, as g is strictly convex, $\|\mathbf{x}\| = \|\mathbf{y}\|$. Thus, (18.9) implies that $\|t\mathbf{x} + (1-t)\mathbf{y}\|^s = \|\mathbf{x}\|^s$ and hence that

$$\|t\mathbf{x} + (1-t)\mathbf{y}\| = \|\mathbf{x}\| = t\|\mathbf{x}\| + (1-t)\|\mathbf{y}\|.$$

Consequently, in view of (18.6), either $\mathbf{y} = \mathbf{0}$, or $\mathbf{y} \neq \mathbf{0}$ and $\mathbf{x} = \mu\mathbf{y}$ for some $\mu \geq 0$. But this together with the added fact that $\|\mathbf{x}\| = \|\mathbf{y}\|$ ensures that in both cases $\mathbf{x} = \mathbf{y}$. This completes the proof of the implication $\Longrightarrow$. The converse implication is self-evident. $\square$

Exercise 18.2. Show that if $f(\mathbf{x}) = g(h(\mathbf{x}))$, where $h(\mathbf{x})$ is convex (but not necessarily strictly convex) on $\mathbb{C}^n$ and $g(y)$ is increasing and strictly convex on $[0, \infty)$, then f is strictly convex. [HINT: Let the verification of (18.7) be your guide.]

Theorem 18.4. *If $\mathbf{x} \in \mathbb{C}^n$ and $1 < p < \infty$, then:*

(1) *The function $f(\mathbf{x}) = \|\mathbf{x}\|_p$ is convex, but not strictly convex.*

(2) *The function $f(\mathbf{x}) = \|\mathbf{x}\|_p^s$, with $1 < s < \infty$, is strictly convex.*

Proof. If $p = 2$, then this is immediate from Lemma 18.3. If $1 < p < \infty$, then a similar argument based on the conditions for equality in Minkowski's inequality works. $\square$

Exercise 18.3. Show that the function $f(\mathbf{x})$ defined in (18.1) is strictly convex.

Exercise 18.4. Show that $\mathbb{C}^n$ with the norm $\|\mathbf{x}\|_p$ is a strictly convex normed linear space if $1 < p < \infty$ but not if $p = 1$ or $p = \infty$. [HINT: Minkowski's inequality helps; for $p = 1$ consider $\begin{bmatrix} 1 & 0 \end{bmatrix}^T$ and $\begin{bmatrix} 0 & 1 \end{bmatrix}^T$; for $p = \infty$ consider $\begin{bmatrix} 1 & a \end{bmatrix}^T$ and $\begin{bmatrix} 1 & b \end{bmatrix}^T$ with $|a| < 1$ and $|b| < 1$.]

Exercise 18.5. Show that if $A \in \mathbb{C}^{n \times n}$ is positive definite and $B = A^{-1}$ is expressed in blocks as in (18.3), then

(18.10) $$\min_{\mathbf{x} \in \mathbb{C}^p} \left\langle A \begin{bmatrix} \mathbf{x} \\ \mathbf{v} \end{bmatrix}, \begin{bmatrix} \mathbf{x} \\ \mathbf{v} \end{bmatrix} \right\rangle = \langle (A_{22} - A_{21}A_{11}^{-1}A_{12})\mathbf{v}, \mathbf{v} \rangle = \mathbf{v}^H B_{22}^{-1} \mathbf{v}$$

and that this minimum is achieved by $\mathbf{x} = -(A_{11})^{-1}A_{12}\mathbf{v}$.

18.3. Fitting a line in $\mathbb{R}^2$

Let $\mathbf{x}_j = \begin{bmatrix} a_j & b_j \end{bmatrix}^T$, $j = 1, \ldots, k$, be a set of points in $\mathbb{R}^2$ with $a_j > 0$ and $b_j > 0$ for $j = 1, \ldots, k$. Let $\mathbf{u} = \begin{bmatrix} \cos\theta & \sin\theta \end{bmatrix}^T$ be a vector in $\mathbb{R}^2$ with $0 < \theta < \pi/2$ and let $\Pi_\mathcal{U}\mathbf{x} = \mathbf{u}(\mathbf{u}^H\mathbf{u})^{-1}\mathbf{u}^H\mathbf{x} = \mathbf{u}\mathbf{u}^H\mathbf{x}$ denote the orthogonal projection of $\mathbf{x} \in \mathbb{R}^2$ onto the subspace $\mathcal{U} = \mathrm{span}\{\mathbf{u}\} = \{t\mathbf{u} : t \in \mathbb{R}\}$. Then

$$\mathbf{x}_j = \Pi_\mathcal{U}\mathbf{x}_j + (I - \Pi_\mathcal{U})\mathbf{x}_j$$

and the square of the distance from $\mathbf{x}_j$ to $\mathcal{U}$ is equal to

$$\begin{aligned}\|(I - \Pi_\mathcal{U})\mathbf{x}_j\|^2 &= \langle (I - \Pi_\mathcal{U})\mathbf{x}_j, (I - \Pi_\mathcal{U})\mathbf{x}_j \rangle = \langle (I - \Pi_\mathcal{U})\mathbf{x}_j, \mathbf{x}_j \rangle \\ &= \|\mathbf{x}_j\|^2 - \|\Pi_\mathcal{U}\mathbf{x}_j\|^2 = a_j^2 + b_j^2 - (a_j\cos\theta + b_j\sin\theta)^2\,.\end{aligned}$$

We wish to choose $\theta \in (0, \pi/2)$ to minimize $\sum_{j=1}^k \|(I - \Pi_\mathcal{U})\mathbf{x}_j\|^2$.

In terms of the notation

(18.11) $\quad \mathbf{a} = \begin{bmatrix} a_1 & \cdots & a_k \end{bmatrix}^T, \quad \mathbf{b} = \begin{bmatrix} b_1 & \cdots & b_k \end{bmatrix}^T, \quad \text{and} \quad A = \begin{bmatrix} \mathbf{a} & \mathbf{b} \end{bmatrix},$

(18.12) $\quad \displaystyle\sum_{j=1}^k \|(I - \Pi_\mathcal{U})\mathbf{x}_j\|^2 = \sum_{j=1}^k \|\mathbf{x}_j\|^2 - \sum_{j=1}^k \|\Pi_\mathcal{U}\mathbf{x}_j\|^2 = \|\mathbf{a}\|^2 + \|\mathbf{b}\|^2 - f(\theta)\,,$

where

$$\begin{aligned}f(\theta) &= \sum_{j=1}^k \|\Pi_\mathcal{U}\mathbf{x}_j\|^2 = \|\mathbf{a}\|^2\cos^2\theta + \|\mathbf{b}\|^2\sin^2\theta + 2\langle \mathbf{a}, \mathbf{b}\rangle\cos\theta\sin\theta \\ &= \left\langle A\begin{bmatrix}\cos\theta\\ \sin\theta\end{bmatrix}, A\begin{bmatrix}\cos\theta\\ \sin\theta\end{bmatrix}\right\rangle.\end{aligned}$$

Since $f(\theta) \geq 0$, $\|\mathbf{a}\|^2 + \|\mathbf{b}\|^2 - f(\theta)$ will be minimized if we choose $\theta \in (0, \pi/2)$ to maximize $f(\theta)$.

Let $\alpha = \|\mathbf{a}\|^2 - \|\mathbf{b}\|^2$ and $\beta = 2\langle \mathbf{a}, \mathbf{b}\rangle$ for short. Then $\beta > 0$ and, upon invoking the trigonometric identities

$$\cos^2\theta = \frac{1 + \cos 2\theta}{2}, \quad \sin^2\theta = \frac{1 - \cos 2\theta}{2}, \quad \text{and} \quad \sin 2\theta = 2\cos\theta\sin\theta\,,$$

we see that

(18.13) $\qquad f(\theta) = \left(\|\mathbf{a}\|^2 + \|\mathbf{b}\|^2 + \alpha\cos 2\theta + \beta\sin 2\theta\right)/2\,.$

Moreover, by the Cauchy-Schwarz inequality,

$$|\alpha\cos 2\theta + \beta\sin 2\theta| \leq \sqrt{\alpha^2 + \beta^2}\sqrt{\cos^2 2\theta + \sin^2 2\theta} = \sqrt{\alpha^2 + \beta^2}\,,$$

with equality if and only if

$$\begin{bmatrix}\cos 2\theta\\ \sin 2\theta\end{bmatrix} = \gamma\begin{bmatrix}\alpha\\ \beta\end{bmatrix} \quad \text{for some } \gamma \in \mathbb{R}\,.$$

Therefore, $\gamma^2(\alpha^2+\beta^2) = \cos^2 2\theta + \sin^2 \theta = 1$, i.e., $\gamma = \pm(\alpha^2+\beta^2)^{-1/2}$. Since $\beta > 0$ and $\sin 2\theta > 0$ when $\theta \in (0, \pi/2)$, we must choose $\gamma = (\alpha^2+\beta^2)^{-1/2}$. Consequently,

$$\cos 2\theta = \frac{\alpha}{(\alpha^2+\beta^2)^{1/2}}, \quad \sin 2\theta = \frac{\beta}{(\alpha^2+\beta^2)^{1/2}}, \quad \text{and} \quad \cot 2\theta = \frac{\alpha}{\beta}.$$

Thus, as $\cot 2\theta$ decreases strictly monotonically from $+\infty$ to $-\infty$ as θ increases from 0 to $\pi/2$, there exists exactly one angle $\theta_1 \in (0, \pi/2)$ such that $\cot 2\theta_1 = \alpha/\beta$. If $\alpha > 0$, then $\theta_1 \in (0, \pi/4)$; if $\alpha = 0$, then $\theta_1 = \pi/4$; if $\alpha < 0$, then $\theta_1 \in (\pi/4, \pi/2)$.

Exercise 18.6. Show that the function $f(\theta)$ defined in (18.13) is subject to the bounds $s_2^2 \leq f(\theta) \leq s_1^2$, where

$$s_1^2 = \{\|\mathbf{a}\|^2 + \|\mathbf{b}\|^2 + \sqrt{\alpha^2+\beta^2}\}/2, \quad s_2^2 = \{\|\mathbf{a}\|^2 + \|\mathbf{b}\|^2 - \sqrt{\alpha^2+\beta^2}\}/2,$$

and s_1 and s_2 are the singular values of the matrix A in (18.11).

Exercise 18.7. Continuing Exercise 18.6, show that $f(\theta) > s_2$ if $0 < \theta \leq \pi/2$, but there exists exactly one angle $\theta_2 \in (\pi/2, \pi)$ such that $\cos 2\theta_2 = -\cos 2\theta_1$ and $\sin 2\theta_2 = -\sin 2\theta_1$, where $\theta_1 \in (0, \pi/2)$ and $f(\theta_1) = s_1^2$. Then check that $\theta_2 = \theta_1 + \pi/2$ and $f(\theta_2) = s_2^2$.

Exercise 18.8. Continuing Exercise 18.7, show that if A is the matrix defined in (18.11), then

$$A^T A = U S^2 U^{-1}, \quad \text{where } S^2 = \begin{bmatrix} s_1^2 & 0 \\ 0 & s_2^2 \end{bmatrix}, \quad U = \begin{bmatrix} \cos\theta_1 & \cos\theta_2 \\ \sin\theta_1 & \sin\theta_2 \end{bmatrix},$$

and $UU^T = I_2$, and then express U in terms of θ_1 only.

Exercise 18.9. Find the angle θ_1 for the best fitting line in the sense that it minimizes the total mean square distance for the points $\mathbf{x}_1 = \begin{bmatrix} 1 & 2 \end{bmatrix}^T$, $\mathbf{x}_2 = \begin{bmatrix} 2 & 1 \end{bmatrix}^T$, and $\mathbf{x}_3 = \begin{bmatrix} 2 & 3 \end{bmatrix}^T$.

18.4. Fitting a line in $\mathbb{R}^p$

If $\mathbf{x}, \mathbf{u}, \mathbf{v} \in \mathbb{R}^p$ and $\|\mathbf{v}\| = 1$, then the distance of $\mathbf{x}$ from the line $\{\mathbf{u} + t\mathbf{v} : t \in \mathbb{R}\}$ is equal to

$$\min_{t \in \mathbb{R}} \|\mathbf{x} - \mathbf{u} - t\mathbf{v}\| = \|\mathbf{x} - \mathbf{u} - \Pi_\mathcal{V}(\mathbf{x} - \mathbf{u})\|,$$

where $\Pi_\mathcal{V}$ denotes the orthogonal projection of $\mathbb{R}^p$ onto the vector space $\mathcal{V} = \text{span}\{\mathbf{v}\}$.

We wish to choose the line, i.e., the vectors $\mathbf{u}$ and $\mathbf{v}$, to minimize

$$\sum_{j=1}^{k} \|\mathbf{x}_j - \mathbf{u} - \Pi_\mathcal{V}(\mathbf{x}_j - \mathbf{u})\|^2$$

for a given set of k points $\mathbf{x}_1, \ldots, \mathbf{x}_k \in \mathbb{R}^p$.

Let $\mathbf{a} = k^{-1} \sum_{j=1}^{k} \mathbf{x}_j$. Then

$$\sum_{j=1}^{k} (\mathbf{x}_j - \mathbf{a}) = \mathbf{0}$$

and the sum of interest is equal to

(18.14)
$$\sum_{j=1}^{k} \|\mathbf{x}_j - \mathbf{a} - \Pi_\mathcal{V}(\mathbf{x}_j - \mathbf{a}) + \mathbf{a} - \mathbf{u} - \Pi_\mathcal{V}(\mathbf{a} - \mathbf{u})\|^2$$
$$= \sum_{j=1}^{k} \left\{ \|\mathbf{x}_j - \mathbf{a} - \Pi_\mathcal{V}(\mathbf{x}_j - \mathbf{a})\|^2 + \|\mathbf{a} - \mathbf{u} - \Pi_\mathcal{V}(\mathbf{a} - \mathbf{u})\|^2 \right\},$$

since

$$-2 \sum_{j=1}^{k} \langle \mathbf{x}_j - \mathbf{a} - \Pi_\mathcal{V}(\mathbf{x}_j - \mathbf{a}), \mathbf{a} - \mathbf{u} - \Pi_\mathcal{V}(\mathbf{a} - \mathbf{u}) \rangle = 0.$$

Thus, as both of the terms inside the curly brackets in the second line of (18.14) are nonnegative, the sum is minimized by choosing $\mathbf{u} = \mathbf{a}$ and then choosing the unit vector $\mathbf{v}$ to minimize

$$\sum_{j=1}^{k} \|\mathbf{x}_j - \mathbf{a} - \Pi_\mathcal{V}(\mathbf{x}_j - \mathbf{a})\|^2 = \sum_{j=1}^{k} \{\|\mathbf{x}_j - \mathbf{a}\|^2 - \|\Pi_\mathcal{V}(\mathbf{x}_j - \mathbf{a})\|^2\}$$
$$= \sum_{j=1}^{k} \{\|\mathbf{x}_j - \mathbf{a}\|^2 - \langle \mathbf{x}_j - \mathbf{a}, \mathbf{v}\rangle^2\}.$$

But this is the same as choosing $\mathbf{v}$ to maximize

$$\sum_{j=1}^{k} \langle \mathbf{x}_j - \mathbf{a}, \mathbf{v}\rangle^2 = \sum_{j=1}^{k} \mathbf{v}^T (\mathbf{x}_j - \mathbf{a})(\mathbf{x}_j - \mathbf{a})^T \mathbf{v} = \mathbf{v}^T Y Y^T \mathbf{v},$$

where $Y = \begin{bmatrix} \mathbf{x}_1 - \mathbf{a} & \cdots & \mathbf{x}_k - \mathbf{a} \end{bmatrix}$. Thus, if Y is expressed in terms of its singular value decomposition $Y = V_1 S_1 U_1^T$, then $YY^T = V_1 S_1^2 V_1^T$ and the sum is maximized by choosing $\mathbf{v}$ equal to the first column of V_1.

18.5. Schur complements for semidefinite matrices

In this section we shall show that if $A \succeq O$, then analogues of the Schur complement formulas considered in (3.18) and (3.19) hold even if neither of the block diagonal entries are invertible. (Similar formulas hold if $A \preceq O$.)

Recall that $B^\dagger$ denotes the Moore-Penrose inverse of B and that $BB^\dagger = (BB^\dagger)^H$ is the orthogonal projection onto the range of B.

Lemma 18.5. *If a positive semidefinite matrix $A \in \mathbb{C}^{n \times n}$ is written in standard four-block form as*

$$A = \begin{bmatrix} A_{11} & A_{12} \\ A_{21} & A_{22} \end{bmatrix}$$

with $A_{11} \in \mathbb{C}^{p \times p}$, $A_{22} \in \mathbb{C}^{q \times q}$, and $n = p + q$, then:

(1) $\mathcal{N}_{A_{11}} \subseteq \mathcal{N}_{A_{21}}$ *and* $\mathcal{N}_{A_{22}} \subseteq \mathcal{N}_{A_{12}}$.

(2) $\mathcal{R}_{A_{12}} \subseteq \mathcal{R}_{A_{11}}$ *and* $\mathcal{R}_{A_{21}} \subseteq \mathcal{R}_{A_{22}}$.

(3) $A_{11} A_{11}^\dagger A_{12} = A_{12}$, $A_{11} A_{11}^\dagger = A_{11}^\dagger A_{11}$, $A_{22} A_{22}^\dagger A_{21} = A_{21}$, *and* $A_{22} A_{22}^\dagger = A_{22}^\dagger A_{22}$.

(4) *The matrix A admits the (lower-upper) factorization*

(18.15) $$A = \begin{bmatrix} I_p & O \\ A_{21} A_{11}^\dagger & I_q \end{bmatrix} \begin{bmatrix} A_{11} & O \\ O & A_{22} - A_{21} A_{11}^\dagger A_{12} \end{bmatrix} \begin{bmatrix} I_p & A_{11}^\dagger A_{12} \\ O & I_q \end{bmatrix}.$$

(5) *The matrix A admits the (upper-lower) factorization*

(18.16) $$A = \begin{bmatrix} I_p & A_{12} A_{22}^\dagger \\ O & I_q \end{bmatrix} \begin{bmatrix} A_{11} - A_{12} A_{22}^\dagger A_{21} & O \\ O & A_{22} \end{bmatrix} \begin{bmatrix} I_p & O \\ A_{22}^\dagger A_{21} & I_q \end{bmatrix}.$$

Proof. Since $A \succeq O$, the inequality

$$\mathbf{x}^H (A_{11} \mathbf{x} + A_{12} \mathbf{y}) + \mathbf{y}^H (A_{21} \mathbf{x} + A_{22} \mathbf{y}) \geq 0$$

must be in force for every choice of $\mathbf{x} \in \mathbb{C}^p$ and $\mathbf{y} \in \mathbb{C}^q$. If $\mathbf{x} \in \mathcal{N}_{A_{11}}$, then this reduces to

$$\mathbf{x}^H A_{12} \mathbf{y} + \mathbf{y}^H (A_{21} \mathbf{x} + A_{22} \mathbf{y}) \geq 0$$

for every choice of $\mathbf{y} \in \mathbb{C}^q$ and hence, upon replacing $\mathbf{y}$ by $\varepsilon \mathbf{y}$, to

$$\varepsilon \mathbf{x}^H A_{12} \mathbf{y} + \varepsilon \mathbf{y}^H A_{21} \mathbf{x} + \varepsilon^2 \mathbf{y}^H A_{22} \mathbf{y} \geq 0$$

for every choice of $\varepsilon > 0$ as well. Consequently, upon dividing through by ε and then letting $\varepsilon \downarrow 0$, it follows that

$$\mathbf{x}^H A_{12} \mathbf{y} + \mathbf{y}^H A_{21} \mathbf{x} \geq 0$$

for every choice of $\mathbf{y} \in \mathbb{C}^q$. But if $\mathbf{y} = -A_{21} \mathbf{x}$, then, as $A_{12} = A_{21}^H$, the last inequality implies that

$$-2 \|A_{21} \mathbf{x}\|^2 = -2 \mathbf{x}^H A_{12} A_{21} \mathbf{x} \geq 0.$$

Therefore,
$$A_{21}\mathbf{x} = \mathbf{0},$$
i.e., $\mathcal{N}_{A_{11}} \subseteq \mathcal{N}_{A_{21}}$ and, since the orthogonal complements of these two sets satisfy the opposite inclusion,
$$\mathcal{R}_{A_{12}} = (\mathcal{N}_{A_{21}})^\perp \subseteq (\mathcal{N}_{A_{11}})^\perp = \mathcal{R}_{A_{11}},$$
as $A_{12} = A_{21}^H$ and $A_{11} = A_{11}^H$. This completes the verification of the first assertions in (1) and (2); the second assertions in (1) and (2) may be verified in much the same way.

Next, the formulas in (2) imply that there exists a pair of matrices $X \in \mathbb{C}^{p \times q}$ and $Y \in \mathbb{C}^{q \times p}$ such that $A_{12} = A_{11}X$ and $A_{21} = A_{22}Y$. Therefore,
$$A_{11}A_{11}^\dagger A_{12} = A_{11}A_{11}^\dagger A_{11}X = A_{11}X = A_{12}$$
and
$$A_{22}A_{22}^\dagger A_{21} = A_{22}A_{22}^\dagger A_{22}Y = A_{22}Y = A_{21}.$$
This justifies two of the formulas in (3); the other two follow from the fact that $A_{11} \succeq O$ and $A_{22} \succeq O$.

Items (4) and (5) are straightforward computations based on the formulas in (3) and their Hermitian transposes. They are left to the reader. □

Exercise 18.10. Show that if $A \succeq O$, then, in the notation of Lemma 18.5, $A_{11}A_{11}^\dagger = A_{11}^\dagger A_{11}$ and $A_{22}^\dagger A_{22} = A_{22}A_{22}^\dagger$.

Exercise 18.11. Verify the identities in (4) and (5) of Lemma 18.5.

18.6. von Neumann's inequality for contractive matrices

A matrix $A \in \mathbb{C}^{p \times q}$ is said to be **contractive** if $\|A\| \leq 1$.

Exercise 18.12. Show that if $A \in \mathbb{C}^{p \times q}$, then
$$(18.17) \qquad \|A\| \leq 1 \iff I_q - A^H A \succeq O \iff I_p - AA^H \succeq O.$$

Exercise 18.13. Show that if $\|A\| \leq 1$, then
$$(18.18) \qquad A(I_q - A^H A)^{1/2} = (I_p - AA^H)^{1/2} A.$$
[HINT: In the usual notation for svd's, $(I_q - A^H A)^{1/2} = U_1(I_r - S_1^2)^{1/2}U_1^H + U_2 U_2^H$; also keep in mind that diagonal matrices commute.]

Lemma 18.6. *If $A \in \mathbb{C}^{p \times p}$ and $\|A\| \leq 1$, then there exists an $np \times np$ unitary matrix B_n with the special property that*
$$(18.19) \quad E^T B_n^k E = A^k \quad \text{for} \quad E^T = \begin{bmatrix} I_p & O & \cdots & O \end{bmatrix}, \quad k = 0, \ldots, n-1,$$
when $n \geq 2$.

Proof. Let $D_A = (I_p - A^H A)^{1/2}$ and $D_{A^H} = (I_p - AA^H)^{1/2}$, for short. Then, since $AD_A = D_{A^H} A$, it is readily checked by direct calculation that the matrices

$$B_2 = \begin{bmatrix} A & D_{A^H} \\ D_A & -A^H \end{bmatrix}, \ B_3 = \begin{bmatrix} A & O & D_{A^H} \\ D_A & O & -A^H \\ O & I_p & O \end{bmatrix}, \ B_4 = \begin{bmatrix} A & O & O & D_{A^H} \\ D_A & O & O & -A^H \\ O & I_p & O & O \\ O & O & I_p & O \end{bmatrix},$$

$$\ldots, B_n = \begin{bmatrix} A & \Theta_{n-2}^T & D_{A^H} \\ D_A & \Theta_{n-2}^T & -A^H \\ \Theta_{n-2} & I_{(n-2)p} & \Theta_{n-2} \end{bmatrix} \quad \text{with} \quad \Theta_{n-2} = \begin{bmatrix} O_{p \times p} \\ \vdots \\ O_{p \times p} \end{bmatrix} \in \mathbb{C}^{(n-2)p}$$

possess the requisite properties for $n = 2, 3, 4$ and $n \geq 5$, respectively. □

Theorem 18.7. *If $A \in \mathbb{C}^{p \times p}$ and $\|A\| \leq 1$, then*

(18.20) $\qquad \|f(A)\| \leq \max\{|f(\lambda)| : \lambda \in \mathbb{C} \quad \text{and} \quad |\lambda| = 1\}$

for every polynomial $f(\lambda)$.

Proof. Let $f(\lambda)$ be a polynomial of degree at most n with $n \geq 2$ and let $B = B_{n+1}$ be a unitary matrix of the form indicated in Lemma 18.6. Then (18.19) holds and, as $B = UDU^H$ with U unitary and $D = \mathrm{diag}\{\lambda_1, \ldots, \lambda_{(n+1)p}\}$ with $|\lambda_j| = 1$ for $j = 1, \ldots, (n+1)p$, it is then easily checked by direct calculation that

$$\|f(A)\| = \|E^T f(B) E\| \leq \|f(B)\| = \|f(D)\| = \max\{|f(\lambda_j)| : j = 1, \ldots, np\},$$

which is clearly less than or equal to the right side of (18.20). □

18.7. Supplementary notes

Sections 18.3–18.6 are taken from [30]. The discussion in Sections 18.3 and 18.4 was adapted from Shuchat [69]. The discussion in Section 18.6 was adapted from Levy and Shalit [56]; see also [60] for further developments and the references in both for extensions to more general settings. Another approach to the minimization problem considered in Theorem 18.1 will be furnished in Example 21.1 for matrices $A \in \mathbb{R}^{n \times n}$.

Chapter 19

Discrete dynamical systems

A **discrete dynamical system** is a sequence of vectors $\mathbf{x}_0, \mathbf{x}_1, \mathbf{x}_2, \ldots$ in a set E that are generated by some rule. An intriguing example with a deceptively simple formulation is the $3x+1$ problem, wherein $E = \{1, 2, 3, \ldots\}$ is the set of positive integers and the sequence is $\{x, Tx, T^2x, \ldots\}$, in which $x \in E$, $Tx = 3x+1$ if x is odd, and $Tx = x/2$ if x is even. Thus, for example, the initial choice $x = 13$, leads to the sequence 13, 40, 20, 10, 5, 16, 8, 4, 2, 1 (if one stops at 1). It is conjectured that for every initial state $x \in E$, there exists an integer k such that $T^k x = 1$. This has been verified by computer for initial states up to 10^{20}, but a definitive answer is not known.

In this chapter we shall restrict our attention to sequences in $\mathbb{C}^p$ of the form

(19.1) $$\mathbf{x}_{k+1} = A\mathbf{x}_k + \mathbf{f}_k, \quad k = 0, 1, \ldots,$$

in which $A \in \mathbb{C}^{p \times p}$, $\mathbf{x}_0 \in \mathbb{C}^p$, and $\mathbf{f}_0, \mathbf{f}_1, \mathbf{f}_2, \ldots$ are specified vectors in $\mathbb{C}^p$; our objective is to understand the behavior of the solution $\mathbf{x}_n$ as n gets large. We shall subsequently use this understanding to study difference equations of the form (19.9).

19.1. Homogeneous systems

The system (19.1) is said to be **homogeneous** if $\mathbf{f}_k = \mathbf{0}$ for $k = 0, 1, \ldots$. The solution of the homogeneous system is

$$\mathbf{x}_n = A^n \mathbf{x}_0 \quad \text{for } n = 0, 1, \ldots.$$

This is a nice formula. However, it does not provide much insight into the behavior of $\mathbf{x}_n$. This is where the fact that A is similar to a Jordan matrix J comes into play:

(19.2) $\qquad A = VJV^{-1} \implies \mathbf{x}_n = VJ^nV^{-1}\mathbf{x}_0 \quad \text{for} \quad n = 0, 1, \ldots.$

The advantage of this new formulation is that J^n is relatively easy to compute: If A is diagonalizable, then

$$J = \mathrm{diag}\{\lambda_1, \ldots, \lambda_p\}, \quad J^n = \mathrm{diag}\{\lambda_1^n, \ldots, \lambda_p^n\},$$

and

$$\mathbf{x}_n = \sum_{j=1}^{p} d_j \lambda_j^n \mathbf{v}_j$$

is a linear combination of the eigenvectors $\mathbf{v}_j$ of A, alias the columns of V, with coefficients that are proportional to λ_j^n. If A is not diagonalizable, then

$$J = \mathrm{diag}\{J_1, \ldots, J_r\},$$

where each block entry J_i is a Jordan cell, and

$$J^n = \mathrm{diag}\{J_1^n, \ldots, J_r^n\}.$$

Consequently, the key issue reduces to understanding the behavior of the n'th power $(C_\lambda^{(m)})^n$ of the $m \times m$ Jordan cell $C_\lambda^{(m)}$ as n tends to ∞. Fortunately, this is still relatively easy:

Lemma 19.1. *If $N = C_\lambda^{(m)} - \lambda I_m = C_0^{(m)}$, then*

(19.3) $\qquad (C_\lambda^{(m)})^n = \sum_{j=0}^{m-1} \binom{n}{j} \lambda^{n-j} N^j \quad \text{when} \quad n \geq m.$

Proof. Since N commutes with λI_m, the binomial theorem is applicable and supplies the formula

$$(C_\lambda^{(m)})^n = (\lambda I_m + N)^n = \sum_{j=0}^{n} \binom{n}{j} \lambda^{n-j} N^j.$$

But this is the same as formula (19.3), since $N^j = 0$ for $j \geq m$. $\square$

The matrix $(C_\lambda^{(m)})^n$ is an upper triangular **Toeplitz matrix**, i.e., it is constant on diagonals. Thus, it is completely specified by its top row:

$$\left[\binom{n}{0}\lambda^n \quad \binom{n}{1}\lambda^{n-1} \quad \binom{n}{2}\lambda^{n-2} \quad \cdots \quad \binom{n}{m-1}\lambda^{n-m+1} \right].$$

19.1. Homogeneous systems

If $m = 3$ and $n > 3$ for example, then

$$(C_\lambda^{(3)})^n = \begin{bmatrix} \binom{n}{0}\lambda^n & \binom{n}{1}\lambda^{n-1} & \binom{n}{2}\lambda^{n-2} \\ 0 & \binom{n}{0}\lambda^n & \binom{n}{1}\lambda^{n-1} \\ 0 & 0 & \binom{n}{0}\lambda^n \end{bmatrix}.$$

Exercise 19.1. Show that if $J = \mathrm{diag}\{\lambda_1, \ldots, \lambda_p\}$, $V = \begin{bmatrix} \mathbf{v}_1 & \cdots & \mathbf{v}_p \end{bmatrix}$, and $(V^{-1})^T = \begin{bmatrix} \mathbf{w}_1 & \cdots & \mathbf{w}_p \end{bmatrix}$, then the solution (19.2) of the homogeneous system can be expressed in the form

$$\mathbf{x}_n = \sum_{j=1}^p \lambda_j^n \mathbf{v}_j \mathbf{w}_j^T \mathbf{x}_0.$$

Exercise 19.2. Show that if, in the setting of Exercise 19.1, $|\lambda_1| > |\lambda_j|$ for $j = 2, \ldots, p$, then

$$\lim_{n\uparrow\infty} \frac{1}{\lambda_1^n} \mathbf{x}_n = \mathbf{v}_1 \mathbf{w}_1^T \mathbf{x}_0.$$

Exercise 19.3. The output $\mathbf{u}_n$ of a chemical plant at time n, $n = 0, 1, \ldots$, is modeled by a system of the form $\mathbf{u}_n = A^n \mathbf{u}_0$. Show that if

$$A = \begin{bmatrix} 1 & -3/2 & 0 \\ 0 & 1/2 & 0 \\ 0 & 0 & 1/4 \end{bmatrix} \quad \text{and} \quad \mathbf{u}_0 = \begin{bmatrix} a \\ b \\ c \end{bmatrix}, \quad \text{then} \quad \lim_{n \to \infty} \mathbf{u}_n = \begin{bmatrix} a - 3b \\ 0 \\ 0 \end{bmatrix}.$$

Exercise 19.4. Find an explicit formula for the solution $\mathbf{u}_n$ of the system $\mathbf{u}_n = A^n \mathbf{u}_0$ when

$$A = \begin{bmatrix} 1 & 2 & 0 \\ 0 & 1 & 3 \\ 0 & 0 & 1 \end{bmatrix} \quad \text{and} \quad \mathbf{u}_0 = \begin{bmatrix} 2 \\ 3 \\ 0 \end{bmatrix}.$$

It is **not necessary to compute** V^{-1} in the formula for the solution in (19.2). It is enough to compute $V^{-1}\mathbf{x}_0$, which is often much less work (since only some of the minors of V may be needed): Set

$$\mathbf{y}_0 = V^{-1}\mathbf{x}_0 \quad \text{and solve the equation} \quad V\mathbf{y}_0 = \mathbf{x}_0.$$

Exercise 19.5. Calculate $V^{-1}\mathbf{x}_0$ when $V = \begin{bmatrix} 6 & 2 & 2 \\ 0 & 3 & 1 \\ 0 & 0 & 1 \end{bmatrix}$ and $\mathbf{x}_0 = \begin{bmatrix} 6 \\ 0 \\ 0 \end{bmatrix}$,

both directly (i.e., by first calculating V^{-1} and then calculating the product $V^{-1}\mathbf{x}_0$) and indirectly by solving the equation $V\mathbf{y}_0 = \mathbf{x}_0$, and compare the effort.

19.2. Nonhomogeneous systems

In this section we look briefly at the nonhomogeneous system (19.1). It is readily checked that

$$\begin{aligned}
\mathbf{x}_1 &= A\mathbf{x}_0 + \mathbf{f}_0, \\
\mathbf{x}_2 &= A\mathbf{x}_1 + \mathbf{f}_1 = A^2\mathbf{x}_0 + A\mathbf{f}_0 + \mathbf{f}_1, \\
\mathbf{x}_3 &= A\mathbf{x}_2 + \mathbf{f}_2 = A^3\mathbf{x}_0 + A^2\mathbf{f}_0 + A\mathbf{f}_1 + \mathbf{f}_2,
\end{aligned}$$

and, in general,

$$(19.4) \qquad \mathbf{x}_n = A^n \mathbf{x}_0 + \sum_{k=0}^{n-1} A^{n-1-k} \mathbf{f}_k.$$

Exercise 19.6. Let $\mathbf{u}_n = A^n \mathbf{x}_0$ for $n = 0, 1, \ldots$, $\mathbf{v}_n = \sum_{k=0}^{n-1} A^{n-1-k} \mathbf{f}_k$ for $n = 1, 2\ldots$, and $\mathbf{v}_0 = \mathbf{0}$. Show that $\mathbf{u}_n$ is a solution of the **homogeneous system** $\mathbf{u}_{n+1} = A\mathbf{u}_n$ for $n = 0, 1, \ldots$ with $\mathbf{u}_0 = \mathbf{x}_0$ and that $\mathbf{v}_n$ is a solution of the system $\mathbf{v}_{n+1} = A\mathbf{v}_n + \mathbf{f}_n$ for $n = 0, 1, \ldots$ with $\mathbf{v}_0 = \mathbf{0}$.

19.3. Second-order difference equations

A second-order homogeneous difference equation is an equation of the form

$$(19.5) \qquad a_0 x_n + a_1 x_{n+1} + a_2 x_{n+2} = 0, \quad n = 0, 1, \ldots, \text{ with } a_0 a_2 \neq 0,$$

where a_0, a_1, a_2 are fixed and x_0 and x_1 are given. The objective is to obtain a formula for x_n and, if possible, to understand how x_n behaves as $n \uparrow \infty$.

We shall solve this second-order difference equation by embedding it into a first-order vector equation by setting

$$\mathbf{x}_n = \begin{bmatrix} x_n \\ x_{n+1} \end{bmatrix} \quad \text{so that} \quad \mathbf{x}_{n+1} = \begin{bmatrix} 0 & 1 \\ -a_0/a_2 & -a_1/a_2 \end{bmatrix} \mathbf{x}_n \quad \text{for } n = 0, 1, \ldots.$$

Thus,

$$\mathbf{x}_n = A^n \mathbf{x}_0 \quad \text{with} \quad A = \begin{bmatrix} 0 & 1 \\ -a_0/a_2 & -a_1/a_2 \end{bmatrix} \quad \text{for } n = 0, 1, \ldots.$$

Since A is a companion matrix, Theorem 8.1 implies that

$$\det(\lambda I_2 - A) = (a_0 + a_1 \lambda + a_2 \lambda^2)/a_2 = (\lambda - \lambda_1)(\lambda - \lambda_2)$$

and that there are only two possible Jordan forms:

$$(19.6) \quad J = \begin{bmatrix} \lambda_1 & 0 \\ 0 & \lambda_2 \end{bmatrix} \quad \text{if} \quad \lambda_1 \neq \lambda_2 \quad \text{and} \quad J = \begin{bmatrix} \lambda_1 & 1 \\ 0 & \lambda_1 \end{bmatrix} \quad \text{if} \quad \lambda_1 = \lambda_2.$$

19.3. Second-order difference equations

Therefore, $A = UJU^{-1}$, where U is a Vandermonde matrix or a generalized Vandermonde matrix:

$$U = \begin{bmatrix} 1 & 1 \\ \lambda_1 & \lambda_2 \end{bmatrix} \quad \text{if} \quad \lambda_1 \neq \lambda_2 \quad \text{and} \quad U = \begin{bmatrix} 1 & 0 \\ \lambda_1 & 1 \end{bmatrix} \quad \text{if} \quad \lambda_1 = \lambda_2.$$

Moreover, since $a_0 \neq 0$ by assumption, $\lambda_1 \lambda_2 \neq 0$.

Case 1 ($\lambda_1 \neq \lambda_2$):

(19.7) $\quad \mathbf{x}_n = A^n \mathbf{u}_0 = UJ^n U^{-1} \mathbf{x}_0 = \begin{bmatrix} 1 & 1 \\ \lambda_1 & \lambda_2 \end{bmatrix} \begin{bmatrix} \lambda_1^n & 0 \\ 0 & \lambda_2^n \end{bmatrix} U^{-1} \mathbf{x}_0.$

Consequently,

(19.8) $\quad x_n = \begin{bmatrix} 1 & 0 \end{bmatrix} \begin{bmatrix} 1 & 1 \\ \lambda_1 & \lambda_2 \end{bmatrix} \begin{bmatrix} \lambda_1^n & 0 \\ 0 & \lambda_2^n \end{bmatrix} U^{-1} \mathbf{x}_0 = \begin{bmatrix} \lambda_1^n & \lambda_2^n \end{bmatrix} U^{-1} \mathbf{x}_0.$

However, it is not necessary to calculate U^{-1}. It suffices to note that formula (19.8) guarantees that x_n must be of the form

$$x_n = \alpha \lambda_1^n + \beta \lambda_2^n \quad (\lambda_1 \neq \lambda_2)$$

and then to solve for α and β from the given initial conditions: $x_0 = \alpha + \beta$ and $x_1 = \alpha \lambda_1 + \beta \lambda_2$.

Case 2 ($\lambda_1 = \lambda_2$):

$$\mathbf{x}_n = A^n \mathbf{x}_0 = U \begin{bmatrix} \lambda_1 & 1 \\ 0 & \lambda_1 \end{bmatrix}^n U^{-1} \mathbf{x}_0 = \begin{bmatrix} 1 & 0 \\ \lambda_1 & 1 \end{bmatrix} \begin{bmatrix} \lambda_1^n & n\lambda_1^{n-1} \\ 0 & \lambda_1^n \end{bmatrix} U^{-1} \mathbf{x}_0.$$

Consequently,

$$x_n = \begin{bmatrix} 1 & 0 \end{bmatrix} \begin{bmatrix} 1 & 0 \\ \lambda_1 & 1 \end{bmatrix} \begin{bmatrix} \lambda_1^n & n\lambda_1^{n-1} \\ 0 & \lambda_1^n \end{bmatrix} U^{-1} \mathbf{x}_0 = \begin{bmatrix} \lambda_1^n & n\lambda_1^{n-1} \end{bmatrix} U^{-1} \mathbf{x}_0$$

must be of the form

$$x_n = \alpha \lambda_1^n + \beta n \lambda_1^n.$$

The coefficients α and β are obtained from the initial conditions: $x_0 = \alpha$ and $x_1 = \alpha \lambda_1 + \beta \lambda_1$.

Notice that the second term in the solution was written as $\beta n \lambda_1^n$ and not as $\beta n \lambda_1^{n-1}$. This is possible because $\lambda_1 \neq 0$, and hence a (positive or negative) power of λ_1 can be absorbed into the constant β.

The preceding analysis leads to the following **recipe** for obtaining the solutions of the second-order (homogeneous) difference equation

$$a_2 x_{n+2} + a_1 x_{n+1} + a_0 x_n = 0 \quad \text{for } n = 0, 1, \ldots, \quad \text{with} \quad a_2 a_0 \neq 0$$

and initial conditions
$$x_0 = c \quad \text{and} \quad x_1 = d:$$

(1) Solve for the roots λ_1, λ_2 of the polynomial $a_2\lambda^2 + a_1\lambda + a_0$ and note that the factorization
$$a_2(\lambda - \lambda_1)(\lambda - \lambda_2) = a_2\lambda^2 + a_1\lambda + a_0$$
implies that $\lambda_1 \lambda_2 = a_0/a_2 \neq 0$.

(2) Express the solution as
$$x_n = \begin{cases} \alpha\lambda_1^n + \beta\lambda_2^n & \text{if } \lambda_1 \neq \lambda_2, \\ \alpha\lambda_1^n + \beta n \lambda_1^n & \text{if } \lambda_1 = \lambda_2 \end{cases}$$
for some choice of α and β.

(3) Solve for α and β by invoking the initial conditions:
$$\begin{cases} c = x_0 = \alpha + \beta & \text{and} \quad d = x_1 = \alpha\lambda_1 + \beta\lambda_2 \quad \text{if } \lambda_1 \neq \lambda_2, \\ c = x_0 = \alpha & \text{and} \quad d = x_1 = \alpha\lambda_1 + \beta\lambda_1 \quad \text{if } \lambda_1 = \lambda_2. \end{cases}$$

Example 19.1. Let $x_{n+2} - 3x_{n+1} - 4x_n = 0$ for $n = 0, 1, \ldots$, with initial conditions $x_0 = 5$ and $x_1 = 0$.

Discussion. The roots of the equation $\lambda^2 - 3\lambda - 4$ are $\lambda_1 = 4$ and $\lambda_2 = -1$. Therefore, the solution x_n must be of the form
$$x_n = \alpha 4^n + \beta(-1)^n \quad \text{for } n = 0, 1, \ldots.$$
The initial conditions $x_0 = 5$ and $x_1 = 0$ imply that
$$\alpha + \beta = 5 \quad \text{and} \quad \alpha 4 - \beta = 0.$$
Thus, $\alpha = 1$, $\beta = 4$, and
$$x_n = 4^n + 4(-1)^n \quad \text{for} \quad n = 0, 1, \ldots.$$

Example 19.2. Let $x_{n+2} - 2x_{n+1} + x_n = 0$ for $n = 0, 1, \ldots$ with initial conditions $x_0 = 3$ and $x_1 = 5$.

Discussion. The equation $\lambda^2 - 2\lambda + 1 = 0$ has two equal roots:
$$\lambda_1 = \lambda_2 = 1.$$
Therefore,
$$x_n = \alpha 1^n + \beta n 1^n = \alpha + \beta n.$$
Substituting the initial conditions
$$x_0 = \alpha = 3 \quad \text{and} \quad x_1 = \alpha + \beta = 5,$$
we see that $\alpha = 3$ and $\beta = 2$ and hence that
$$x_n = 3 + 2n \quad \text{for} \quad n = 0, 1, \ldots.$$

Exercise 19.7. Find an explicit formula for x_n, for $n = 0, 1, \ldots$, given that $x_0 = -1$, $x_1 = 2$, and $x_{k+1} = 3x_k - 2x_{k-1}$ for $k = 1, 2, \ldots$.

Exercise 19.8. The **Fibonacci sequence** x_n, $n = 0, 1, \ldots$, is prescribed by the initial conditions $x_0 = 1$, $x_1 = 1$ and the difference equation $x_{n+1} = x_n + x_{n-1}$ for $n = 1, 2, \ldots$. Find an explicit formula for x_n and use it to calculate the **golden mean**, $\lim_{n \uparrow \infty} x_{n+1}/x_n$.

Exercise 19.9. Let $x_n = tx_{n-1} + (1-t)x_{n+1}$ for $n = 1, 2, \ldots$. Evaluate the $\lim_{n \uparrow \infty} x_n$ as a function of t for $0 < t < 1$ when $x_0 = 0$ and $x_1 = 1$.

19.4. Higher-order difference equations

The solution of the p'th-order equation

(19.9) $\quad a_0 x_n + a_1 x_{n+1} + \cdots + a_p x_{n+p} = 0, \quad n = 0, 1, \ldots, \quad \text{with} \quad a_0 a_p \neq 0$

and given initial conditions $x_0, x_1, \ldots, x_{p-1}$ can be obtained from the solution of the first-order vector equation

$$\mathbf{x}_n = A\mathbf{x}_{n-1} \quad \text{for} \quad n = 0, 1, \ldots,$$

where

(19.10)

$$\mathbf{x}_n = \begin{bmatrix} x_n \\ x_{n+1} \\ \vdots \\ x_{n+p-1} \end{bmatrix}, \quad A = \begin{bmatrix} 0 & 1 & 0 & \cdots & 0 \\ 0 & 0 & 1 & & 0 \\ \vdots & & & & \vdots \\ 0 & 0 & 0 & & 1 \\ -a_0/a_p & -a_1/a_p & -a_2/a_p & \cdots & -a_{p-1}/a_p \end{bmatrix}.$$

The solution

$$x_n = \begin{bmatrix} 1 & 0 & \cdots & 0 \end{bmatrix} \mathbf{x}_n$$

is obtained from the top entry in $\mathbf{x}_n$. To see how this works, consider the case of a 6×6 companion matrix A with three distinct eigenvalues λ_1, λ_2, and λ_3 and $\det(\lambda I_6 - A) = (\lambda - \lambda_1)^3 (\lambda - \lambda_2)^2 (\lambda - \lambda_1)$ and let $\mathbf{v}(\lambda) = \begin{bmatrix} 1 & \lambda & \cdots & \lambda^5 \end{bmatrix}^T$. Then $A = VJV^{-1}$ with

$$V = \begin{bmatrix} \mathbf{v}(\lambda_1) & \mathbf{v}'(\lambda_1) & \dfrac{\mathbf{v}''(\lambda_1)}{2!} & \mathbf{v}(\lambda_2) & \mathbf{v}'(\lambda_2) & \mathbf{v}(\lambda_3) \end{bmatrix},$$

and $J = \mathrm{diag}\{C^{(3)}_{\lambda_1}, C^{(2)}_{\lambda_2}, C^{(1)}_{\lambda_3}\}$. Therefore,

$$x_n = \begin{bmatrix} 1 & 0 & \cdots & 0 \end{bmatrix} A^n \mathbf{x}_0 = \begin{bmatrix} 1 & 0 & \cdots & 0 \end{bmatrix} V \begin{bmatrix} C^{(3)}_{\lambda_1} & O & O \\ O & C^{(2)}_{\lambda_2} & O \\ O & O & \lambda_3 \end{bmatrix}^n \mathbf{c},$$

with $\mathbf{c} = V^{-1}\mathbf{x}_0$ and, as

$$\begin{bmatrix} 1 & 0 & 0 & 0 & 0 & 0 \end{bmatrix} V = \begin{bmatrix} 1 & 0 & 0 & 1 & 0 & 1 \end{bmatrix},$$

$$x_n = \left[\binom{n}{0}\lambda_1^n \quad \binom{n}{1}\lambda_1^{n-1} \quad \binom{n}{2}\lambda_1^{n-2} \quad \binom{n}{0}\lambda_2^n \quad \binom{n}{1}\lambda_2^{n-1} \quad \binom{n}{0}\lambda_3^n\right]\mathbf{c}$$

$$= \left[\lambda_1^n \quad \frac{n}{\lambda_1}\lambda_1^n \quad \frac{n(n-1)}{2\lambda_1^2}\lambda_1^n \quad \lambda_2^n \quad \frac{n}{\lambda_2}\lambda_2^n \quad \lambda_3^n\right]\mathbf{c}.$$

Thus, x_n is a linear combination of the top rows of $(C_{\lambda_1}^{(3)})^n$, $(C_{\lambda_2}^{(2)})^n$, and $(C_{\lambda_3}^{(1)})^n$. But, this is the same as saying that x_n is a linear combination of λ_1^n, $n\lambda_1^n$, $n^2\lambda_1^n$, λ_2^n, $n\lambda_2^n$, and λ_3^n; e.g.,

$$\begin{bmatrix} 1 & \frac{n}{\mu} & \frac{n(n-1)}{2\mu^2} \end{bmatrix} = \begin{bmatrix} 1 & n & n^2 \end{bmatrix} \begin{bmatrix} 1 & 0 & 0 \\ 0 & 1/\mu & -1/(2\mu^2) \\ 0 & 0 & 1/(2\mu^2) \end{bmatrix} \quad \text{when } \mu \neq 0$$

and the lower triangular matrix on the right is invertible.

This serves to motivate the following **recipe** for obtaining the solution of equation (19.9):

(1) Find the roots of the polynomial $a_0 + a_1\lambda + \cdots + a_p\lambda^p$.
(2) If $a_0 + a_1\lambda + \cdots + a_p\lambda^p = a_p(\lambda - \lambda_1)^{\alpha_1} \cdots (\lambda - \lambda_k)^{\alpha_k}$ with distinct roots $\lambda_1, \ldots, \lambda_k$, then the solution must be of the form

$$x_n = \sum_{j=1}^{k} p_j(n)\lambda_j^n, \quad \text{where} \quad p_j \text{ is a polynomial of degree } \alpha_j - 1.$$

(3) Invoke the initial conditions to solve for the coefficients of the polynomials p_j.

Discussion. The algorithm works because A is a companion matrix. Thus,

$$\det(\lambda I_p - A) = (a_0 + a_1\lambda + \cdots + a_p\lambda^p)/a_p$$

and hence, if

$$\det(\lambda I_p - A) = (\lambda - \lambda_1)^{\alpha_1} \cdots (\lambda - \lambda_k)^{\alpha_k}$$

with distinct roots $\lambda_1, \ldots, \lambda_k$, then A is similar to the Jordan matrix

$$J = \text{diag}\{C_{\lambda_1}^{(\alpha_1)}, \ldots, C_{\lambda_k}^{(\alpha_k)}\},$$

with one Jordan cell for each distinct eigenvalue, and the matrix U in the Jordan decomposition $A = UJU^{-1}$ is a generalized Vandermonde matrix. Therefore, the solution must be of the form indicated in (2).

Remark 19.2. The equation $a_0 + a_1\lambda + \cdots + a_p\lambda^p = 0$ for the eigenvalues of A may be obtained with minimum thought by letting $x_j = \lambda^j$ in equation (19.9) and then factoring out the highest common power of λ. The assumption $a_0 \neq 0$ ensures that the eigenvalues are all nonzero.

Exercise 19.10. Find the solution of the third-order difference equation
$$x_{n+3} - 3x_{n+2} + 3x_{n+1} - x_n = 0 \ , \ n = 0, 1, \ldots,$$
subject to the initial conditions $x_0 = 1$, $x_1 = 2$, and $x_2 = 8$. [HINT: $(x-1)^3 = x^3 - 3x^2 + 3x - 1$.]

19.5. Nonhomogeneous equations

The solution x_n of the p'th-order equation
(19.11)
$$a_0 x_n + a_1 x_{n+1} + \cdots + a_p x_{n+p} = f_n , \quad n = 0, 1, \ldots, \quad \text{with} \quad a_0 a_p \neq 0$$
and given initial conditions $x_j = c_j$ for $j = 0, \ldots, p-1$, can be obtained from the solution of the first-order vector equation
$$\mathbf{x}_n = A\mathbf{x}_{n-1} + \mathbf{f}_n \quad \text{for} \quad n = 0, 1, \ldots,$$
where $\mathbf{x}_n$ and A are as in (19.10) and $\mathbf{f}_n^T = \begin{bmatrix} 0 & \cdots & 0 & f_n/a_p \end{bmatrix}$. Let $\mathbf{e}_j$ denote the j'th column of I_p for $j = 1, \ldots, p$. Then, by (19.4), which is valid for any $A \in \mathbb{C}^{p \times p}$,
$$x_n = \mathbf{e}_1^T \mathbf{x}_n = \mathbf{e}_1^T A^n \mathbf{x}_0 + \mathbf{e}_1^T \sum_{j=0}^{n-1} A^{n-1-j} \mathbf{f}_j .$$

Exercise 19.11. Show that if $\mathbf{f}_n^T = \begin{bmatrix} 0 & \cdots & 0 & f_n/a_p \end{bmatrix}$ and $\mathbf{x}_n$ and A are as in (19.10), then $u_n = \mathbf{e}_1^T A^n \mathbf{x}_0$ for $n = 0, 1, \ldots$ is a solution of the homogeneous equation (19.9) with $u_j = c_j$ for $j = 0, \ldots, p-1$, whereas $v_n = \mathbf{e}_1^T \sum_{j=0}^{n-1} A^{n-1-j} \mathbf{f}_j$ for $n = 1, 2, \ldots$ and $v_0 = 0$ is a solution of (19.11) with $v_0 = \cdots = v_{p-1} = 0$.

19.6. Supplementary notes

This chapter was adapted from Chapter 13 of [**30**]. The $3x+1$ problem was formulated by L. Collatz in 1937. It is currently regarded as unsolvable.

A perhaps surprising application of the methods introduced in this chapter is to the computation of determinants of some special classes of matrices:

Exercise 19.12. Let $A_n = 5I_n + 2C_0^{(n)} + 2(C_0^{(n)})^T$. Compute $\det A_n$. [HINT: $\det A_n$ is a solution of the difference equation $x_n - 5x_{n-1} + 4x_{n-2} = 0$ for $n = 3, 4, \ldots$.]

Exercise 19.13. Let $A_n = bI_n + cC_0^{(n)} + c(C_0^{(n)})^T$ with $b, c \in \mathbb{R}$. Compute $\det A_n$.

Chapter 20

Continuous dynamical systems

A continuous dynamical system is a curve $\mathbf{x}(t)$ in a set E that evolves according to some rule as t runs over an interval $I \subseteq \mathbb{R}$. In this chapter we will focus initially on the special case in which $\mathbf{x}(t)$ is a solution of the first-order vector differential equation $\mathbf{x}'(t) = A\mathbf{x}(t) + \mathbf{f}(t)$, $\alpha \leq t < \beta$, based on a matrix $A \in \mathbb{C}^{p \times p}$ and a $p \times 1$ vector-valued function $\mathbf{f}(t)$. We shall then use the theory established for systems to develop an algorithm for solving differential equations of the form (20.17). We begin with some prerequisites.

20.1. Preliminaries on matrix-valued functions

Let
$$F(t) = \begin{bmatrix} f_{11}(t) & \cdots & f_{1q}(t) \\ \vdots & & \vdots \\ f_{p1}(t) & \cdots & f_{pq}(t) \end{bmatrix}$$
be a $p \times q$ matrix-valued functions with entries $f_{ij}(t)$ that are smooth functions. Then, the rules of matrix addition imply that
$$F'(t) = \lim_{\varepsilon \to 0} \frac{F(t + \varepsilon) - F(t)}{\varepsilon} = \begin{bmatrix} f'_{11}(t) & \cdots & f'_{1q}(t) \\ \vdots & & \vdots \\ f'_{p1}(t) & \cdots & f'_{pq}(t) \end{bmatrix}.$$
Consequently, the formula
$$\int_a^t F'(s)ds = F(t) - F(a)$$

will hold if and only if

$$\int_a^t F(s)ds = \begin{bmatrix} \int_a^t f_{11}(s)ds & \cdots & \int_a^t f_{1q}(s)ds \\ \vdots & & \vdots \\ \int_a^t f_{p1}(s)ds & \cdots & \int_a^t f_{pq}(s)ds \end{bmatrix},$$

i.e., **differentiation and integration of a matrix-valued function are carried out on each entry in the matrix separately:**

$$F'(t) = [f'_{ij}(t)] \quad \text{and} \quad \int_a^b F(s)ds = \left[\int_a^b f_{ij}(s)ds \right].$$

Moreover, if $B \in \mathbb{C}^{k \times p}$ and $C \in \mathbb{C}^{q \times r}$, then

$$(20.1) \quad B \int_a^b F(s)ds = \int_a^b BF(s)ds \quad \text{and} \quad \int_a^b F(s)ds\, C = \int_a^b F(s)C ds,$$

i.e., multiplication by constant matrices can be brought inside the integral. To verify this, notice, for example, that the ij entry of the first term on the left in (20.1) is equal to

$$\sum_{k=1}^p b_{ik} \left(\int_a^b F(s)ds \right)_{kj} = \sum_{k=1}^p b_{ik} \left(\int_a^b f_{kj}(s)ds \right) = \int_a^b \sum_{k=1}^p b_{ik} f_{kj}(s)ds$$

$$= \int_a^b (BF(s))_{ij} ds.$$

Similar considerations serve to justify the second formula in (20.1) and the rule for differentiating the product of two matrix-valued functions:

$$(F(s)G(s))' = F'(s)G(s) + F(s)G'(s).$$

Thus, if $F(s)$ and $G(s)$ are $p \times p$ matrix-valued functions such that $F(s)G(s) = I_p$, then

$$O = F'(s)G(s) + F(s)G'(s), \quad \text{i.e.,} \quad G'(s) = -F(s)^{-1}F'(s)F(s)^{-1}.$$

20.2. The exponential of a matrix

The exponential e^A of a matrix $A \in \mathbb{C}^{p \times p}$ is defined by the formula

$$(20.2) \quad e^A = \sum_{j=0}^\infty \frac{A^j}{j!} = I_p + A + \frac{A^2}{2!} + \cdots.$$

The two main facts that we shall use are:
(1) If $A, B \in \mathbb{C}^{p \times p}$ and $AB = BA$, then $e^{A+B} = e^A e^B$.
(2) If $F(t) = e^{tA}$, then $F'(t) = AF(t) = F(t)A$, since
$$\frac{F(t+h) - F(t)}{h} = \frac{e^{(t+h)A} - e^{tA}}{h} = e^{tA}\left(\frac{e^{hA} - I_p}{h}\right) \to e^{tA} A = A e^{tA}$$
as h tends to zero, and $F(0) = I_p$.

Exercise 20.1. Calculate e^A when $A = \begin{bmatrix} 0 & b^2 \\ c^2 & 0 \end{bmatrix}$.

Exercise 20.2. Calculate e^A when $A = \begin{bmatrix} a & b^2 \\ c^2 & a \end{bmatrix}$. [HINT: aI_2 and $A - aI_2$ commute.]

Exercise 20.3. Show that if $A, B \in \mathbb{C}^{p \times p}$, then
$$\lim_{t \to 0} \frac{e^{tA} e^{tB} e^{-tA} e^{-tB} - I_p}{t^2} = AB - BA.$$
[HINT: $e^{tA} \approx I_p + tA + (t^2/2)A^2$ when t is close to zero.]

Exercise 20.4. Show that if $A, B \in \mathbb{C}^{p \times p}$ and $e^{tA} e^{tB} = e^{tA+tB}$ for $t \in \mathbb{R}$, then $AB = BA$. [HINT: Exploit the formula in Exercise 20.3.]

20.3. Systems of differential equations

The set of solutions of the **homogeneous** system
$$(20.3) \qquad \mathbf{x}'(t) - A\mathbf{x}(t) = \mathbf{0}$$
is a vector space. In view of item (2) in the preceding section,
$$(20.4) \qquad \mathbf{x}(t) = e^{(t-a)A} \mathbf{c}$$
is a solution of (20.3) for $t \geq a$ with initial condition $\mathbf{x}(a) = \mathbf{c}$. If
$$A = VJV^{-1} \quad \text{for some Jordan matrix } J, \text{ then} \quad e^{tA} = Ve^{tJ}V^{-1}$$
and
$$(20.5) \qquad \mathbf{x}(t) = V e^{(t-a)J} \mathbf{d}, \quad \text{where} \quad \mathbf{d} = V^{-1} \mathbf{x}(a).$$
Note that **it is not necessary to calculate** V^{-1}, since only $\mathbf{d}$ is needed. The advantage of this new formula is that it is easy to calculate e^{tJ}: If
$$J = \mathrm{diag}\{\lambda_1, \ldots, \lambda_p\}, \quad \text{then} \quad e^{tJ} = \mathrm{diag}\{e^{t\lambda_1}, \ldots, e^{t\lambda_p}\}$$
and hence, upon writing $V = \begin{bmatrix} \mathbf{v}_1 & \cdots & \mathbf{v}_p \end{bmatrix}$ and $\mathbf{d}^T = \begin{bmatrix} d_1 & \cdots & d_p \end{bmatrix}$,
$$(20.6) \qquad \mathbf{x}(t) = \sum_{j=1}^{p} d_j e^{(t-a)\lambda_j} \mathbf{v}_j,$$

which exhibits the set $\{e^{t\lambda_1}\mathbf{v}_1, \ldots, e^{t\lambda_p}\mathbf{v}_p\}$ of vector-valued functions of t as a **basis** for the set of solutions of the homogeneous system (20.3), i.e., of the null space of the linear transformation that maps $\mathbf{x}(t)$ into $\mathbf{x}'(t) - A\mathbf{x}(t)$. If A is not diagonalizable, then

$$J = \text{diag}\{J_1, \ldots, J_r\} \quad \text{and} \quad e^{tJ} = \text{diag}\{e^{tJ_1}, \ldots, e^{tJ_r}\},$$

where each block entry J_i is a Jordan cell and the set of columns of Ve^{tJ} is a basis for the set of solutions of (20.3).

Lemma 20.1. *If $N = C_\lambda^{(m)} - \lambda I_m = C_0^{(m)}$, then*

(20.7)
$$e^{tC_\lambda^{(m)}} = e^{t(\lambda I_m + N)} = e^{t\lambda}e^{tN} = e^{t\lambda}\sum_{j=0}^{m-1}\frac{(tN)^j}{j!}.$$

Proof. Since $(\lambda I_m)N = N(\lambda I_m)$ and $N^m = O$,

$$e^{tC_\lambda^{(m)}} = e^{t(\lambda I_m + N)} = e^{t\lambda I_m}e^{tN} = e^{t\lambda}e^{tN}$$
$$= e^{t\lambda}\left\{I_m + tN + \cdots + \frac{t^{m-1}N^{m-1}}{(m-1)!}\right\}. \qquad \square$$

Formula (20.7) exhibits $e^{tC_\lambda^{(m)}}$ as an upper triangular Toeplitz matrix. Thus, for example, if $m = 3$, then $e^{tC_\lambda^{(3)}} = \begin{bmatrix} e^{t\lambda} & te^{t\lambda} & \frac{t^2}{2!}e^{t\lambda} \\ 0 & e^{t\lambda} & te^{t\lambda} \\ 0 & 0 & e^{t\lambda} \end{bmatrix}$. The same pattern propagates for every Jordan cell.

Exercise 20.5. Show that if $J = \text{diag}\{\lambda_1, \ldots, \lambda_p\}$, $V = \begin{bmatrix} \mathbf{v}_1 & \cdots & \mathbf{v}_p \end{bmatrix}$, and $(V^{-1})^T = \begin{bmatrix} \mathbf{w}_1 & \cdots & \mathbf{w}_p \end{bmatrix}$, then the solution (20.4) of the system (20.3) can be expressed in the form

$$\mathbf{x}(t) = \sum_{j=1}^p e^{(t-a)\lambda_j}\mathbf{v}_j\mathbf{w}_j^T\mathbf{x}(a).$$

Exercise 20.6. Show that if, in the setting of Exercise 20.5, $\lambda_1 > |\lambda_j|$ for $j = 2, \ldots, p$, then

$$\lim_{t\uparrow\infty} e^{-t\lambda_1}\mathbf{x}(t) = e^{-a\lambda_1}\mathbf{v}_1\mathbf{w}_1^T\mathbf{x}(a).$$

Exercise 20.7. Find an explicit formula for e^{tA} when

$$A = \begin{bmatrix} 0 & 1 & 0 \\ -1 & 0 & 1 \\ 0 & -1 & 0 \end{bmatrix}.$$

[HINT: You may use the fact that the eigenvalues of A are equal to 0, $i\sqrt{2}$, and $-i\sqrt{2}$.]

Exercise 20.8. Let $A = VJV^{-1}$, where $J = \begin{bmatrix} 2 & 1 & 0 \\ 0 & 2 & 0 \\ 0 & 0 & 3 \end{bmatrix}$, $V = [\mathbf{v}_1 \ \mathbf{v}_2 \ \mathbf{v}_3]$, and $(V^T)^{-1} = [\mathbf{w}_1 \ \mathbf{w}_2 \ \mathbf{w}_3]$. Evaluate the limit of the matrix-valued function $e^{-3t}e^{tA}$ as $t \uparrow \infty$.

20.4. Uniqueness

Formula (20.4) provides a (smooth) solution to the first-order vector differential equation (20.3). However, it remains to check that there are no others.

Lemma 20.2. *The differential equation (20.3) has only one solution $\mathbf{x}(t)$ with continuous derivative $\mathbf{x}'(t)$ on the interval $a \leq t \leq b$ that meets the specified initial condition at $t = a$.*

Proof. Suppose to the contrary that there are two solutions $\mathbf{x}(t)$ and $\mathbf{y}(t)$ and let $\mathbf{u}(t) = \mathbf{x}(t) - \mathbf{y}(t)$. Then $\mathbf{u}(a) = 0$ and

$$\mathbf{u}(t) = \int_a^t \mathbf{u}'(s)ds = \int_a^t A\mathbf{u}(s)ds\,.$$

Therefore, upon iterating the last equality, we obtain the formulas

$$\mathbf{u}(t) = A^n \int_a^t \int_a^{s_1} \cdots \int_a^{s_{n-1}} \mathbf{u}(s_n)ds_n \cdots ds_1 \quad \text{for } n = 2, 3, \ldots,$$

which, upon setting $M = \max\{\|\mathbf{u}(t)\| : a \leq t \leq b\}$, leads to the inequality

$$M \leq M\|A^n\|\frac{(b-a)^n}{n!} \leq M\|A\|^n\frac{(b-a)^n}{n!}\,.$$

If n is large enough, then $\|A\|^n(b-a)^n/n! < 1$ and hence

$$0 \leq M\left(1 - \|A\|^n\frac{(b-a)^n}{n!}\right) \leq 0\,.$$

Therefore, $M = 0$; i.e., there is only one smooth solution of the differential equation (20.3) that meets the given initial conditions. $\square$

Much the same sort of analysis leads to Gronwall's inequality:

Exercise 20.9. Let $h(t)$ be a continuous real-valued function on the interval $a \leq t \leq b$. Show that

$$\int_a^t h(s_2)\left\{\int_a^{s_2} h(s_1)ds_1\right\} ds_2 = \left(\int_a^t h(s)ds\right)^2/2!\,,$$

$$\int_a^t h(s_3)\left[\int_a^{s_3} h(s_2)\left\{\int_a^{s_2} h(s_1)ds_1\right\} ds_2\right] ds_3 = \left(\int_a^t h(s)ds\right)^3/3!\,,$$

etc.

Exercise 20.10 (Gronwall's inequality). Let $\alpha > 0$ and let $u(t)$ and $h(t)$ be continuous real-valued functions on the interval $a \leq t \leq b$ such that

$$u(t) \leq \alpha + \int_a^t h(s)u(s)ds \quad \text{and} \quad h(t) \geq 0 \quad \text{for} \quad a \leq t \leq b\,.$$

Show that

$$u(t) \leq \alpha \exp\left(\int_a^t h(s)ds\right) \quad \text{for} \quad a \leq t \leq b\,.$$

[HINT: Iterate the inequality and exploit Exercise 20.9.]

20.5. Isometric and isospectral flows

A matrix $B \in \mathbb{R}^{p \times p}$ is said to be **skew-symmetric** if $B = -B^T$. Analogously, $B \in \mathbb{C}^{p \times p}$ is said to be **skew-Hermitian** if $B = -B^H$.

Exercise 20.11. Let $B \in \mathbb{C}^{p \times p}$. Show that if B is skew-Hermitian, then B is normal and e^B is unitary.

Exercise 20.12. Let $F(t) = e^{tB}$, where $B \in \mathbb{R}^{p \times p}$. Show that $F(t)$ is an orthogonal matrix for every $t \in \mathbb{R}$ if and only if B is skew-symmetric. [HINT: If $F(t)$ is orthogonal, then the derivative $\{F(t)F(t)^T\}' = 0$.]

Exercise 20.13. Let $B \in \mathbb{R}^{p \times p}$ and let $\mathbf{x}(t)$, $t \geq 0$, denote the solution of the differential equation $\mathbf{x}'(t) = B\mathbf{x}(t)$ for $t \geq 0$ that meets the initial condition $\mathbf{x}(0) = \mathbf{c} \in \mathbb{R}^p$.

(a) Show that $\frac{d}{dt}\|\mathbf{x}(t)\|^2 = \mathbf{x}(t)^T(B + B^T)\mathbf{x}(t)$ for every $t \geq 0$.

(b) Show that if B is skew-symmetric, then $\|\mathbf{x}(t)\| = \|\mathbf{x}(0)\|$ for every $t \geq 0$.

Exercise 20.14. Let $A \in \mathbb{R}^{p \times p}$ and let $U(t)$ and $B(t)$, $t \geq 0$, be one-parameter families of real $p \times p$ matrices such that $U'(t) = B(t)U(t)$ for $t > 0$ and $U(0) = I_p$. Show that $F(t) = U(t)AU(t)^{-1}$ is a solution of the differential equation

(20.8) $\qquad F'(t) = B(t)F(t) - F(t)B(t) \qquad \text{for } t \geq 0\,.$

Exercise 20.15. Show that if $F(t)$ is the only smooth solution of a differential equation of the form (20.8) with suitably smooth $B(t)$, then $F(t) = U(t)F(0)U(t)^{-1}$ for $t \geq 0$. [HINT: Consider $U(t)F(0)U(t)^{-1}$ when $U(t)$ is a solution of $U'(t) = B(t)U(t)$ with $U(0) = I_p$.]

A pair of matrix-valued functions $F(t)$ and $B(t)$ that are related by equation (20.8) is said to be a **Lax pair**, and the solution $F(t) = U(t)F(0)U(t)^{-1}$ is said to be **isospectral** because its eigenvalues are independent of t.

20.6. Nonhomogeneous differential systems

In this section we shall consider the nonhomogeneous differential system

(20.9) $\mathbf{x}'(t) - A\mathbf{x}(t) = \mathbf{f}(t)$, $a \leq t < b$, with initial condition $\mathbf{x}(a) = \mathbf{c}$,

where $A \in \mathbb{R}^{n \times n}$ and $\mathbf{f}(t)$ is a continuous $n \times 1$ real vector-valued function on the interval $a \leq t < b$. Then, since

$$\mathbf{x}'(t) - A\mathbf{x}(t) = e^{tA} \left(e^{-tA} \mathbf{x}(t) \right)',$$

it is readily seen that the given system can be expressed as

$$\left(e^{-sA} \mathbf{x}(s) \right)' = e^{-sA} \mathbf{f}(s)$$

and hence, upon integrating both sides from a to a point $t \in (a,b)$, that

$$e^{-tA} \mathbf{x}(t) - e^{-aA} \mathbf{x}(a) = \int_a^t \left(e^{-sA} \mathbf{x}(s) \right)' ds = \int_a^t e^{-sA} \mathbf{f}(s) ds$$

or, equivalently, that

(20.10) $\quad \mathbf{x}(t) = e^{(t-a)A} \mathbf{c} + \int_a^t e^{(t-s)A} \mathbf{f}(s) ds \quad \text{for} \quad a \leq t < b$.

To explore formula (20.10) further, let

(20.11) $\quad \mathbf{u}(t) = e^{(t-a)A} \mathbf{c} \quad \text{and} \quad \mathbf{y}(t) = \int_a^t e^{(t-s)A} \mathbf{f}(s) ds$

and note that $\mathbf{u}(t)$ is a solution of the homogeneous equation

(20.12) $\quad \mathbf{u}'(t) = A\mathbf{u}(t) \quad \text{for } a \leq t \text{ with initial condition } \mathbf{u}(a) = \mathbf{c}$,

whereas $\mathbf{y}(t)$ is a solution of the equation

(20.13) $\quad \mathbf{y}'(t) = A\mathbf{y}(t) + \mathbf{f}(t) \quad \text{for } a \leq t \text{ with initial condition } \mathbf{y}(a) = \mathbf{0}$.

The key to this calculation is the general formula (for suitably smooth functions)

(20.14) $\quad \dfrac{d}{dt} \int_0^t g(t,s) f(s) ds = g(t,t) f(t) + \int_0^t \dfrac{\partial}{\partial t} g(t,s) f(s) ds$

applied to each entry of $\mathbf{y}(t)$. Thus,

$$\mathbf{u}'(t) + \mathbf{y}'(t) = A(\mathbf{u}(t) + \mathbf{y}(t)) + \mathbf{f}(t) \quad \text{and} \quad \mathbf{u}(a) + \mathbf{y}(a) = \mathbf{c},$$

as needed.

Exercise 20.16. Show that if $\mathbf{x}$ and $\mathbf{w}$ are solutions of (20.9) with $\mathbf{x}(a) = \mathbf{w}(a)$, then $\mathbf{x}(t) = \mathbf{w}(t)$ for $a \leq t < b$. [HINT: Lemma 20.2 is applicable.]

20.7. Second-order differential equations

Ordinary differential equations with constant coefficients can be solved by imbedding them in first-order vector differential equations and exploiting the theory developed in Sections 20.3 and 20.6. Thus, for example, to solve the second-order differential equation

(20.15) $\qquad a_0 x(t) + a_1 x'(t) + a_2 x''(t) = f(t) \quad \text{for } t \geq 0$

with $a_2 \neq 0$ and initial conditions $x(0) = c_1$ and $x'(0) = c_2$, let

$$\mathbf{x}(t) = \begin{bmatrix} x(t) \\ x'(t) \end{bmatrix}.$$

Then, since $x''(t) = (-a_0 x(t) - a_1 x'(t) + f(t))/a_2$,

$$\mathbf{x}'(t) = \begin{bmatrix} x'(t) \\ x''(t) \end{bmatrix} = \begin{bmatrix} 0 & 1 \\ -a_0/a_2 & -a_1/a_2 \end{bmatrix} \begin{bmatrix} x(t) \\ x'(t) \end{bmatrix} + \frac{1}{a_2} \begin{bmatrix} 0 \\ f(t) \end{bmatrix} = A\mathbf{x}(t) + \mathbf{f}(t)$$

for $t \geq 0$, with

(20.16) $A = \begin{bmatrix} 0 & 1 \\ -a_0/a_2 & -a_1/a_2 \end{bmatrix}$, $\mathbf{f}(t) = \frac{1}{a_2} \begin{bmatrix} 0 \\ f(t) \end{bmatrix}$, and $\mathbf{x}(0) = \begin{bmatrix} c_1 \\ c_2 \end{bmatrix} = \mathbf{c}.$

Thus,

$$\mathbf{x}(t) = e^{tA} \mathbf{c} + \int_0^t e^{(t-s)A} \mathbf{f}(s) ds = e^{tA} \mathbf{c} + \mathbf{y}(t)$$

and

$$x(t) = \begin{bmatrix} 1 & 0 \end{bmatrix} \mathbf{x}(t) = \begin{bmatrix} 1 & 0 \end{bmatrix} e^{tA} \mathbf{c} + \begin{bmatrix} 1 & 0 \end{bmatrix} \int_0^t e^{(t-s)A} \mathbf{f}(s) ds.$$

Let λ_1, λ_2 denote the roots of $a_0 + a_1\lambda + a_2\lambda^2$. Since A is a companion matrix, there are only two possible cases to consider. They correspond to the two Jordan forms described in (19.6).

Case 1 ($\lambda_1 \neq \lambda_2$):

$$e^{tA} = Ve^{tJ}V^{-1}, \quad \text{with} \quad V = \begin{bmatrix} 1 & 1 \\ \lambda_1 & \lambda_2 \end{bmatrix} \quad \text{and} \quad e^{tJ} = \begin{bmatrix} e^{\lambda_1 t} & 0 \\ 0 & e^{\lambda_2 t} \end{bmatrix},$$

and hence the solution $x(t)$ of equation (20.15) must be of the form

$$x(t) = \gamma e^{\lambda_1 t} + \delta e^{\lambda_2 t} + \begin{bmatrix} 1 & 0 \end{bmatrix} \int_0^t e^{(t-s)A} \mathbf{f}(s) ds$$

for some choice of the constants γ and δ.

Case 2 ($\lambda_1 = \lambda_2$):

$$e^{tA} = Ve^{tJ}V^{-1}, \quad \text{with} \quad V = \begin{bmatrix} 1 & 0 \\ \lambda_1 & 1 \end{bmatrix} \quad \text{and} \quad e^{tJ} = \begin{bmatrix} e^{\lambda_1 t} & te^{\lambda_1 t} \\ 0 & e^{\lambda_1 t} \end{bmatrix},$$

and hence the solution $x(t)$ of the equation must be of the form

$$x(t) = \gamma e^{\lambda_1 t} + \delta t e^{\lambda_1 t} + \begin{bmatrix} 1 & 0 \end{bmatrix} \int_0^t e^{(t-s)A} \mathbf{f}(s) ds$$

for some choice of the constants γ and δ.

In both cases, the constants γ and δ are determined by the initial conditions $x(0)$ and $x'(0)$. The **particular solution**

$$y(t) = \begin{bmatrix} 1 & 0 \end{bmatrix} \int_0^t e^{(t-s)A} \mathbf{f}(s) ds$$

does not influence the choice of these constants, since $y(0) = 0$ and

$$y'(t) = \begin{bmatrix} 1 & 0 \end{bmatrix} \left\{ \mathbf{f}(t) + \int_0^t A e^{(t-s)A} \mathbf{f}(s) ds \right\} \implies v'(0) = 0.$$

Exercise 20.17. Show that if A, $\mathbf{f}$, and $\mathbf{c}$ are as in (20.16) and $\mathbf{x}(t) = \begin{bmatrix} x_1(t) & x_2(t) \end{bmatrix}^T$ is a solution of the equation $\mathbf{x}'(t) = A\mathbf{x}(t) + \mathbf{f}(t)$ for $t \geq 0$ and $\mathbf{x}(0) = \mathbf{c}$, then $x_1(t)$ is a solution of (20.15) with $x_1(0) = c_1$ and $x_1'(0) = c_2$. [REMARK: The preceding discussion justified the passage from scalar equations to vector equations; the exercise runs the other way.]

20.8. Higher-order differential equations

The strategy for solving the p'th-order differential equation

(20.17) $\quad a_p x^{(p)}(t) + a_{p-1} x^{(p-1)}(t) + \cdots a_0 x(t) = f(t) \quad \text{for} \quad t \geq 0$

with constant coefficients that is subject to the constraint $a_p \neq 0$ and to the initial conditions

(20.18) $\quad\quad\quad\quad x(0) = c_1, \ldots, x^{(p-1)}(0) = c_p$

is the same as for the case $p = 2$ considered in Section 20.7. Again a recipe for the solution is obtained by identifying $x(t)$ as the first entry of the solution $\mathbf{x}(t) = \begin{bmatrix} x_1(t) & \cdots & x_p(t) \end{bmatrix}^T$ of the vector equation

(20.19) $\quad\quad\quad \mathbf{x}'(t) = A\mathbf{x}(t) + \mathbf{f}(t) \quad \text{for} \quad t \geq 0 \quad \text{with } \mathbf{x}(0) = \mathbf{c}$

based on $\mathbf{c} = \begin{bmatrix} c_1 & \cdots & c_p \end{bmatrix}^T$,

$$A = \begin{bmatrix} 0 & 1 & 0 & \cdots & 0 \\ 0 & 0 & 1 & \cdots & 0 \\ \vdots & & & \ddots & \vdots \\ 0 & 0 & 0 & \cdots & 1 \\ -a_0/a_p & -a_1/a_p & -a_2/a_p & \cdots & -a_{p-1}/a_p \end{bmatrix}, \text{ and } \mathbf{f}(t) = \frac{1}{a_p} \begin{bmatrix} 0 \\ 0 \\ \vdots \\ 0 \\ f(t) \end{bmatrix}$$

and then invoking formulas (20.10)–(20.13).

Let $\mathbf{e}_j$ denote the j'th column of I_p for $j = 1, \ldots, p$. Then, since

$$\mathbf{x}(t) = \mathbf{u}(t) + \mathbf{y}(t), \quad \text{with} \quad \mathbf{u}(t) = e^{tA}\mathbf{c} \quad \text{and} \quad \mathbf{y}(t) = \int_0^t e^{(t-s)A}\mathbf{f}(s)ds,$$

it follows that

$$x(t) = \mathbf{e}_1^T \mathbf{x}(t) = u(t) + y(t) \quad \text{with} \quad u(t) = \mathbf{e}_1^T \mathbf{u}(t) \text{ and } y(t) = \mathbf{e}_1^T \mathbf{y}(t)$$

for $t \geq 0$ and hence, by straightforward computations, that:

(1) $u(t)$ is a solution of the homogeneous equation

(20.20) $$a_0 u(t) + a_1 u^{(1)}(t) + \cdots + a_p u^{(p)}(t) = 0:$$

$$a_0 u(t) + a_1 u^{(1)}(t) + \cdots + a_p u^{(p)}(t)$$
$$= \mathbf{e}_1^T (a_0 I_p + a_1 A + \cdots + a_p A^p) \mathbf{u}(t) = 0,$$

since $a_0 I_p + a_1 A + \cdots + a_p A^p = O$ by the Cayley-Hamilton theorem.

(2) $u(t)$ meets the initial conditions $u^{(j-1)}(0) = c_j$ for $j = 1, \ldots, p$: $u^{(j-1)}(t) = \mathbf{e}_1^T A^{j-1} e^{tA} \mathbf{c}$ and hence $u^{(j-1)}(0) = \mathbf{e}_1^T A^{j-1} \mathbf{c} = c_j$ for $j = 1, \ldots, p$. (The last evaluation exploits the fact that $\mathbf{e}_1^T A = \mathbf{e}_2^T$, $\mathbf{e}_1^T A^2 = \mathbf{e}_2^T A = \mathbf{e}_3^T$, $\ldots$, $\mathbf{e}_1^T A^{p-1} = \mathbf{e}_p^T$.)

(3) $y(t)$ is a solution of (20.17) with $y^{(j)}(0) = 0$ for $j = 0, \ldots, p-1$:

(20.21) $$y(t) = \mathbf{e}_1^T \mathbf{y}(t) = \int_0^t h(t-s) f(s) ds \quad \text{with} \quad h(s) = \frac{\mathbf{e}_1^T e^{sA} \mathbf{e}_p}{a_p},$$

and, by repeated use of formula (20.14) (with $g(t,s) = \dfrac{\partial^k}{\partial t^k} h(t-s)$ for $k = 0, \ldots, p$), we obtain the evaluations

(20.22)
$$y^{(j)}(t) = h^{(j-1)}(0) f(t) + \int_0^t h^{(j)}(t-s) f(s) ds$$
$$= \frac{1}{a_p} \mathbf{e}_1^T \left(A^{j-1} \mathbf{e}_p f(t) + A^j \int_0^t e^{(t-s)A} \mathbf{e}_p f(s) ds \right),$$

for $j = 1, \ldots, p$. Thus, as $\mathbf{e}_1^T A^{j-1} = \mathbf{e}_j^T$ for $j = 1, \ldots, p$ and $\sum_{j=0}^p a_j A^j = O$,

$$a_0 y(t) + a_1 y_1(t) + \cdots + a_p y^{(p)}(t) = f(t)$$

and

$$y^{(j)}(0) = 0 \quad \text{for } j = 0, \ldots, p-1. \qquad \square$$

20.8. Higher-order differential equations

The **algorithm** for computing the solution $u(t) = \mathbf{e}_1^T V e^{tJ} V^{-1} \mathbf{c}$ of the homogeneous equation (20.20) is:

(1) Find the roots of the polynomial $a_p \lambda^p + a_{p-1} \lambda^{p-1} + \cdots + a_0$.

(2) If $a_p \lambda^p + a_{p-1} \lambda^{p-1} + \cdots + a_0 = a_p(\lambda - \lambda_1)^{\alpha_1} \cdots (\lambda - \lambda_k)^{\alpha_k}$ with k distinct roots $\lambda_1, \ldots, \lambda_k$, then the solution $u(t)$ of the homogeneous equation is **a linear combination of the top rows of** $e^{tC_{\lambda_j}^{(\alpha_j)}}$, $j = 1, \ldots, k$, which can be reexpressed in the form

$$u(t) = e^{t\lambda_1} p_1(t) + \cdots + e^{t\lambda_k} p_k(t),$$

where $p_j(t)$ is a polynomial of degree $\alpha_j - 1$ for $j = 1, \ldots, k$. In other words, the set of these entries is a **basis** for the vector space of solutions to the homogeneous differential equation (20.20).

(3) Find the coefficients of the polynomials $p_j(t)$ by imposing the initial conditions.

The same three steps serve to solve for $h(t) = a_p^{-1} \mathbf{e}_1^T e^{tA} \mathbf{e}_p$, except that now the initial conditions that are imposed in the third step are $h^{(j)}(0) = 0$ for $j = 0, \ldots, p-2$ and $h^{(p-1)}(0) = 1/a_p$; $y(t)$ is then obtained from (20.21).

Example 20.1. If $J = \mathrm{diag}\{C_{\lambda_1}^{(3)}, C_{\lambda_2}^{(2)}\}$, then

$$\mathbf{e}_1^T V e^{tJ} = \begin{bmatrix} 1 & 0 & 0 & 1 & 0 \end{bmatrix} e^{tJ} = \begin{bmatrix} e^{\lambda_1 t} & te^{\lambda_1 t} & \frac{t^2}{2} e^{\lambda_1 t} & e^{\lambda_2 t} & te^{\lambda_2 t} \end{bmatrix}.$$

The **key observation** is that $\mathbf{e}_1^T V$ selects the top rows of each block in e^{tJ}, i.e.,

$$\begin{bmatrix} 1 & 0 & 0 & 1 & 0 \end{bmatrix} \begin{bmatrix} a & b & c & 0 & 0 \\ * & * & * & 0 & 0 \\ * & * & * & 0 & 0 \\ 0 & 0 & 0 & d & e \\ 0 & 0 & 0 & * & * \end{bmatrix} = \begin{bmatrix} a & b & c & d & e \end{bmatrix}.$$

Example 20.2. The recipe for solving the third-order differential equation

$$a_3 x^{(3)}(t) + a_2 x^{(2)}(t) + a_1 x^{(1)}(t) + a_0 x(t) = f(t), \quad t \geq 0 \text{ and } a_3 \neq 0,$$

is:

(1) Solve for the roots $\lambda_1, \lambda_2, \lambda_3$ of the polynomial $a_3 \lambda^3 + a_2 \lambda^2 + a_1 \lambda + a_0$.

(2) The solution $u(t)$ to the homogeneous equation is

$$u(t) = \alpha e^{\lambda_1 t} + \beta e^{\lambda_2 t} + \gamma e^{\lambda_3 t} \text{ if } \lambda_1, \lambda_2, \lambda_3 \text{ are all different},$$
$$u(t) = \alpha e^{\lambda_1 t} + \beta t e^{\lambda_1 t} + \gamma e^{\lambda_3 t} \text{ if } \lambda_1 = \lambda_2 \neq \lambda_3,$$
$$u(t) = \alpha e^{\lambda_1 t} + \beta t e^{\lambda_1 t} + \gamma t^2 e^{\lambda_1 t} \text{ if } \lambda_1 = \lambda_2 = \lambda_3.$$

(3) Determine the constants α, β, γ from the initial conditions $x(0)$, $x'(0)$, and $x''(0)$.

The function $h(t)$ in (20.21) is of the same form as $u(t)$, but subject to the initial conditions $h(0) = h^{(1)}(0) = 0$ and $h^{(2)}(0) = 1/a_3$.

Exercise 20.18. Find the solution of the third-order differential equation
$$x^{(3)}(t) - 3x^{(2)}(t) + 3x^{(1)}(t) - x(t) = e^t, \ t \geq 0,$$
subject to the initial conditions $x(0) = 1$, $x^{(1)}(0) = 2$, $x^{(2)}(0) = 8$. [ANSWER: $x(t) = u(t) + y(t)$, where $u(t) = (1 + t + (5/2)t^2)e^t$ and $y(t) = t^3 e^t/6$.]

Exercise 20.19. Let $\mathbf{u}'(t) = \begin{bmatrix} 0 & \alpha \\ \alpha & 0 \end{bmatrix} \mathbf{u}(t)$ for $t \geq 0$. Show in two different ways that $\|\mathbf{u}(t)\|_2 = \|\mathbf{u}(0)\|_2$ if $\alpha + \overline{\alpha} = 0$: first by showing that the derivative of $\|\mathbf{u}(t)\|_2$ with respect to t is equal to zero and then by invoking Exercise 20.11.

Exercise 20.20. In the setting of Exercise 20.19, describe $\|\mathbf{u}(t)\|_2$ as $t \uparrow \infty$ if $\alpha + \overline{\alpha} \neq 0$.

Exercise 20.21. Evaluate $\lim_{t \uparrow \infty} t^{-2} e^{-2t} \mathbf{x}(t)$ for the solution $\mathbf{x}(t)$ of the equation
$$\mathbf{x}'(t) = \begin{bmatrix} 0 & 1 & 0 \\ 0 & 0 & 1 \\ 8 & -12 & 6 \end{bmatrix} \mathbf{x}(t), \ t \geq 0, \text{ when } \mathbf{x}(0) = \begin{bmatrix} 8 \\ 8 \\ 8 \end{bmatrix}.$$

20.9. Wronskians

Let $u_1(t), \ldots, u_p(t)$ be solutions of the homogeneous differential equation
$$x^{(p)}(t) + a_{p-1} x^{(p-1)}(t) + \cdots + a_1 x^{(1)}(t) + a_0 x(t) = 0, \ c \leq t \leq d,$$
and let $\mathbf{u}_j(t)^T = \begin{bmatrix} u_j(t) & u_j^{(1)}(t) & \cdots & u_j^{(p-1)}(t) \end{bmatrix}$. Then the function

(20.23) $\varphi(t) = \det \begin{bmatrix} \mathbf{u}_1(t) & \cdots & \mathbf{u}_p(t) \end{bmatrix} = \det \begin{bmatrix} u_1(t) & \cdots & u_p(t) \\ u_1^{(1)}(t) & \cdots & u_p^{(1)}(t) \\ \vdots & & \vdots \\ u_1^{(p-1)}(t) & \cdots & u_p^{(p-1)}(t) \end{bmatrix}$

is called the **Wronskian** of the functions $u_1(t), \ldots, u_p(t)$.

Exercise 20.22. Show that

(20.24) $\qquad \varphi(t) = \exp\{-(t-c)a_{p-1}\} \varphi(c).$

[HINT: Exploit formula (17.1) and item $4°$ of Theorem 7.2.]

Exercise 20.23. Show that the vectors $\mathbf{u}_j(t)$, $j = 1,\ldots,p$, in formula (20.23) are linearly independent at one point in the interval $c \leq t \leq d$ if and only if they are linearly independent at every point in the interval.

20.10. Supplementary notes

This chapter is partially adapted from Chapter 13 of [**30**].

Chapter 21

Vector-valued functions

In this chapter we shall discuss vector-valued functions of one and many variables and some of their applications. We begin with some **notation** for classes of functions with different degrees of smoothness.

Let Q be an open subset of $\mathbb{R}^n$. A function f that maps Q into $\mathbb{R}$ is said to belong to the class

$\mathcal{C}(Q)$ if f is continuous on Q,

$\mathcal{C}^k(Q)$ for some positive integer k if f and all its partial derivatives of order up to and including k are continuous on Q.

A vector-valued function $\mathbf{f}$ from Q into $\mathbb{R}^m$ is said to belong to one of the two classes listed above if all its components belong to that class. Moreover, on occasion, $\mathbf{f}$ is said to be **smooth** if it belongs to $\mathcal{C}^k(Q)$ for k large enough for the application at hand.

21.1. Mean value theorems

We begin with the classical mean value theorem for real-valued functions $f(x)$ of one variable x:

Theorem 21.1. *If Q is an open subset of $\mathbb{R}$, $f \in \mathcal{C}^1(Q)$, and Q contains the finite closed interval $[a, b]$, then*

(21.1) $$f(b) - f(a) = f'(c)(b-a)$$

for some point c in the open interval $(a, b) = \{x \in \mathbb{R} : a < x < b\}$.

Proof. See, e.g., [5]. □

The mean value theorem is a powerful tool for verifying inequalities.

Exercise 21.1. Show that if $0 < a < b$, then $\sqrt{ab} - a \leq (b-a)/2$.

Exercise 21.2. Show that if $a > 0$, $b > 0$, and $0 < t < 1$, then $(a+b)^t \leq a^t + b^t$. [HINT: If $a < b$, then $(a+b)^t - b^t = t(a/(c+b))^{1-t}a^t$ for some point $c \in (0, a)$.]

Exercise 21.3. Show that if $b > 1$ and $0 < t < 1$, then $|b - b^t| \leq (1-t)b \ln b$.

Exercise 21.4. Show that if $a > 0$, $b > 0$, and $1 < t < \infty$, then $(a+b)^t \geq a^t + b^t$. [HINT: See the hint to Exercise 21.2.]

Exercise 21.5. If Q is an open subset of $\mathbb{R}$, $f \in \mathcal{C}(Q)$, and Q contains the finite closed interval $[a, b]$, then

$$(21.2) \qquad \int_a^b f(x)dx = (b-a)f(c) \quad \text{for some point } c \in (a, b).$$

[HINT: Apply the mean value theorem to the function $h(y) = \int_a^y f(x)dx$.]

21.2. Taylor's formula with remainder

Theorem 21.2. *If Q is an open subset of $\mathbb{R}$, $f \in \mathcal{C}^n(Q)$, and Q contains the finite closed interval $[a, b]$, then*

$$(21.3) \qquad f(b) = f(a) + \sum_{k=1}^{n-1} f^{(k)}(a)\frac{(b-a)^k}{k!} + f^{(n)}(c)\frac{(b-a)^n}{n!}$$

for some point $c \in (a, b)$.

Proof. See, e.g., [5]. $\square$

Formula (21.3) may be used to approximate $f(b)$ by a sum that is expressed in terms of $f(a)$ and its derivatives $f^{(1)}(a), f^{(2)}(a), \ldots$, when b is close to a: If $b = a + h$, then (21.3) can be rewritten as

$$(21.4) \qquad f(a+h) - \left\{ f(a) + \sum_{k=1}^{n-1} f^{(k)}(a)\frac{h^k}{k!} \right\} = f^{(n)}(c)\frac{h^n}{n!}$$

and the right-hand side may be used to estimate the difference between the true value of $f(a+h)$ and the approximant

$$f(a) + \sum_{k=1}^{n-1} f^{(k)}(a)\frac{h^k}{k!}.$$

Thus, for example, in order to calculate $(27.1)^{5/3}$ to an accuracy of $1/100$, let

$$f(x) = x^{5/3}, \quad a = 27, \quad \text{and} \quad b = 27.1.$$

Then, as
$$f'(x) = \frac{5}{3}x^{2/3} \quad \text{and} \quad f''(x) = \frac{10}{9}x^{-1/3},$$
the formula
$$f(b) = f(a) + f'(a)(b-a) + f''(c)\frac{(b-a)^2}{2!}$$
translates to
$$(27.1)^{5/3} - \left\{(27)^{5/3} + \frac{5}{3}(27)^{2/3}\frac{1}{10}\right\} = \frac{10}{9}c^{-1/3}\frac{1}{200};$$
i.e.,
$$\left|(27.1)^{5/3} - \left\{3^5 + \frac{3}{2}\right\}\right| = \frac{c^{-1/3}}{180},$$
for some number c that lies between 27 and 27.1. In particular, this constraint implies that $c > 27$ and hence that $c^{-1/3} < \frac{1}{3}$. Consequently
$$\left|(27.1)^{5/3} - \left\{3^5 + \frac{3}{2}\right\}\right| < \frac{(1/3)}{180} = \frac{1}{540}.$$
Thus, the error in approximating $(27.1)^{5/3}$ by $3^5 + \frac{3}{2}$ is less than $1/540$.

Exercise 21.6. Show that $|(27.1)^{5/3} - (27)^{5/3}| > 3/2$.

Exercise 21.7. Show that

(21.5) $$e^t - \left(1 + t + \cdots + \frac{t^k}{k!}\right) = \int_0^t \frac{(t-s)^k}{k!}e^s\,ds \quad \text{for } t \geq 0.$$

[HINT: Integrate the right-hand side once by parts to reveal the key.]

21.3. Mean value theorem for functions of several variables

Let $f(\mathbf{x}) = f(x_1, \ldots, x_n)$ be a real-valued function of the vector $\mathbf{x}$ with components $x_1, \ldots, x_n$ and suppose that $f \in \mathcal{C}^1(Q)$ in some open subset Q of $\mathbb{R}^n$. Then the vector

(21.6) $$(\nabla f)(\mathbf{x}) = \begin{bmatrix} \frac{\partial f}{\partial x_1}(\mathbf{x}) & \cdots & \frac{\partial f}{\partial x_n}(\mathbf{x}) \end{bmatrix}$$

is called the **gradient** of f; it may also be written as a column vector or even as a matrix if f is a function of n^2 entries x_{ij} for $i, j = 1, \ldots, n$.

If $f \in \mathcal{C}^2(Q)$, then the matrix-valued function

(21.7) $$H_f(\mathbf{x}) = \begin{bmatrix} \frac{\partial^2 f}{\partial x_1 \partial x_1}(\mathbf{x}) & \cdots & \frac{\partial^2 f}{\partial x_1 \partial x_n}(\mathbf{x}) \\ \vdots & & \vdots \\ \frac{\partial^2 f}{\partial x_n \partial x_1}(\mathbf{x}) & \cdots & \frac{\partial^2 f}{\partial x_n \partial x_n}(\mathbf{x}) \end{bmatrix}$$

is called the **Hessian** of f.

Theorem 21.3. *If Q is an open convex subset of $\mathbb{R}^q$, f is a real-valued function in $\mathcal{C}^1(Q)$, and $\mathbf{a}, \mathbf{b} \in Q$, then*

(21.8) $$f(\mathbf{b}) - f(\mathbf{a}) = (\nabla f)(\mathbf{c})(\mathbf{b} - \mathbf{a})$$

for some point $\mathbf{c} = \mathbf{a} + t_0(\mathbf{b} - \mathbf{a})$, $0 < t_0 < 1$, in the open line segment between $\mathbf{a}$ and $\mathbf{b}$. If $f \in \mathcal{C}^2(Q)$, then

(21.9) $$f(\mathbf{b}) = f(\mathbf{a}) + (\nabla f)(\mathbf{a})(\mathbf{b} - \mathbf{a}) + \frac{1}{2}\langle H_f(\mathbf{c})(\mathbf{b} - \mathbf{a}), (\mathbf{b} - \mathbf{a})\rangle$$

for some point $\mathbf{c} = \mathbf{a} + t_0(\mathbf{b} - \mathbf{a})$, $0 < t_0 < 1$, in the open line segment between $\mathbf{a}$ and $\mathbf{b}$.

Proof. Let

$$\begin{aligned} h(t) &= f(\mathbf{a} + t(\mathbf{b} - \mathbf{a})) \\ &= f(x_1(t), \ldots, x_q(t)), \quad \text{where} \quad x_j(t) = a_j + t(b_j - a_j) \end{aligned}$$

for $0 \leq t \leq 1$. Then clearly $h \in \mathcal{C}(I)$ for some open interval I that contains $[0, 1]$, and

$$\begin{aligned} h'(t) &= \sum_{j=1}^{q} \frac{\partial f}{\partial x_j}(\mathbf{a} + t(\mathbf{b} - \mathbf{a}))(b_j - a_j) \\ &= (\nabla f)(\mathbf{a} + t(\mathbf{b} - \mathbf{a}))(\mathbf{b} - \mathbf{a}) \end{aligned}$$

exists for each point t in the open interval $(0, 1)$. Therefore, by Theorem 21.1, there exists a point $t_0 \in (0, 1)$ such that $h(1) - h(0) = h'(t_0)$. But, in view of the preceding calculation, this is easily seen to be the same as formula (21.8) with $\mathbf{c} = \mathbf{a} + t_0(\mathbf{b} - \mathbf{a})$.

If $f \in \mathcal{C}^2(Q)$, then Taylor's formula with remainder applied to $h(t)$ implies that

$$h(1) = h(0) + h'(0) \cdot 1 + h''(t_0) \cdot \frac{1^2}{2!}$$

for some point $t_0 \in (0, 1)$. But this is the same as saying that

$$f(\mathbf{b}) = f(\mathbf{a}) + \sum_{j=1}^{q} \frac{\partial f}{\partial x_j}(\mathbf{a})(b_j - a_j) + \frac{1}{2}\sum_{i,j=1}^{q}(b_i - a_i)\frac{\partial^2 f}{\partial x_i \partial x_j}(\mathbf{c})(b_j - a_j),$$

which coincides with (21.9). $\square$

Theorem 21.4. *If Q is an open convex subset of $\mathbb{R}^q$, f is a real-valued function in $\mathcal{C}^1(Q)$, and $\mathbf{a}, \mathbf{a} + \mathbf{u} \in Q$, then*

(21.10) $$\lim_{t \downarrow 0} \frac{f(\mathbf{a} + t\mathbf{u}) - f(\mathbf{a})}{t} = (\nabla f)(\mathbf{a})\mathbf{u}.$$

21.4. MVT for vector-valued functions

In particular, if $\mathbf{u} = \mathbf{e}_j$, the j'th column of I_q,

(21.11) $$\left(\frac{\partial f}{\partial x_j}\right)(\mathbf{a}) = \lim_{t \downarrow 0} \frac{f(\mathbf{a} + t\mathbf{e}_j) - f(\mathbf{a})}{t} = (\nabla f)(\mathbf{a})\mathbf{e}_j.$$

Proof. In view of formula (21.8),
$$\frac{f(\mathbf{a} + t\mathbf{u}) - f(\mathbf{a})}{t} = (\nabla f)(\mathbf{c})\mathbf{u}$$
where $\mathbf{c} = \mathbf{a} + t_0 t\mathbf{u}$ and $0 < t_0 < 1$. Therefore, $\|\mathbf{c} - \mathbf{a}\| \le t\|\mathbf{u}\|$, which tends to 0 as $t \downarrow 0$. Thus, as $(\nabla f)(\mathbf{x})$ is continuous, $(\nabla f)(\mathbf{c})$ tends to $(\nabla f)(\mathbf{a})$ as $t \downarrow 0$. This completes the justification of (21.10); (21.11) is a special case of (21.10). $\square$

Exercise 21.8. Show that if $A \in \mathbb{R}^{n \times n}$, $\mathbf{x} \in \mathbb{R}^n$, and $f(\mathbf{x}) = \langle A\mathbf{x}, \mathbf{x} \rangle$, then

(21.12) $\quad (\nabla f)(\mathbf{x})^T = (A + A^T)\mathbf{x}$ and $(H_f)(\mathbf{x}) = A + A^T$.

Exercise 21.9. Show that if $\mathbf{x} \in \mathbb{R}^n$, $\mathbf{b} \in \mathbb{R}^n$, and $f(\mathbf{x}) = \langle \mathbf{x}, \mathbf{b} \rangle$, then $(\nabla f)(\mathbf{x})^T = \mathbf{b}$ and $H_f(\mathbf{x}) = O$.

Exercise 21.10. Show that if $g \in \mathcal{C}^1(\mathbb{R}^q)$, $A \in \mathbb{R}^{q \times q}$, $\mathbf{a} \in \mathbb{R}^q$, and $f(\mathbf{x}) = g(\mathbf{a} + A\mathbf{x})$, then $(\nabla f)(\mathbf{x}) = (\nabla g)(\mathbf{a} + A\mathbf{x})A$.

21.4. Mean value theorems for vector-valued functions of several variables

We turn now to vector-valued functions of several variables. We assume that each of the components $f_i(\mathbf{x}) = f_i(x_1, \ldots, x_q)$, $i = 1, \ldots, p$, of $\mathbf{f}(\mathbf{x})$ is real valued. Thus, $\mathbf{f}(\mathbf{x})$ defines a mapping from some subset of $\mathbb{R}^q$ into $\mathbb{R}^p$. If Q is an open subset of $\mathbb{R}^q$ and $\mathbf{f} \in \mathcal{C}^1(Q)$, then the **Jacobian matrix**

$$J_\mathbf{f}(\mathbf{x}) = \begin{bmatrix} \frac{\partial f_1}{\partial x_1}(\mathbf{x}) & \cdots & \frac{\partial f_1}{\partial x_q}(\mathbf{x}) \\ \vdots & & \vdots \\ \frac{\partial f_p}{\partial x_1}(\mathbf{x}) & \cdots & \frac{\partial f_p}{\partial x_q}(\mathbf{x}) \end{bmatrix} \quad \text{of} \quad \mathbf{f}(\mathbf{x}) = \begin{bmatrix} f_1(x_1, \ldots, x_q) \\ \vdots \\ f_p(x_1, \ldots, x_q) \end{bmatrix}$$

is a continuous $p \times q$ matrix-valued function on Q.

Theorem 21.5. *If Q is an open convex subset of $\mathbb{R}^q$, $\mathbf{a}, \mathbf{b} \in Q$, and each of the components $f_i(\mathbf{x})$, $i = 1, \ldots, p$, of $\mathbf{f}(\mathbf{x})$ is a real-valued function in the class $\mathcal{C}^1(Q)$, then*

(21.13) $$\mathbf{f}(\mathbf{b}) - \mathbf{f}(\mathbf{a}) = \begin{bmatrix} (\nabla f_1)(\mathbf{c}_1) \\ \vdots \\ (\nabla f_p)(\mathbf{c}_p) \end{bmatrix} (\mathbf{b} - \mathbf{a})$$

for some set of points
$$\mathbf{c}_i = \mathbf{a} + t_i(\mathbf{b} - \mathbf{a}), \; 0 < t_i < 1, \; i = 1, \ldots, p.$$

Proof. This is an immediate consequence of the mean value theorem for functions of many variables (Theorem 21.3), applied to each component $f_i(\mathbf{x})$ of $\mathbf{f}(\mathbf{x})$ separately. □

The equality (21.13) is expressed in terms of p unknown points $\mathbf{c}_1, \ldots, \mathbf{c}_p$. The next theorem is formulated in terms of only one unknown point, but there is a price to pay: The equality is replaced by an inequality.

Theorem 21.6. *If Q is an open convex subset of $\mathbb{R}^q$, $\mathbf{a}, \mathbf{b} \in Q$, and each of the components $f_i(\mathbf{x})$, $i = 1, \ldots, p$, of $\mathbf{f}(\mathbf{x})$ is a real-valued function in the class $C^1(Q)$, then*

(21.14) $$\|\mathbf{f}(\mathbf{b}) - \mathbf{f}(\mathbf{a})\| \leq \|J_{\mathbf{f}}(\mathbf{c})\| \|\mathbf{b} - \mathbf{a}\|$$

for some point $\mathbf{c}$ in the open line segment between $\mathbf{a}$ and $\mathbf{b}$.

Proof. Let
$$\mathbf{u} = \mathbf{f}(\mathbf{b}) - \mathbf{f}(\mathbf{a})$$
and let
$$h(t) = \mathbf{u}^T \mathbf{f}(\mathbf{a} + t(\mathbf{b} - \mathbf{a})).$$
Then, by the classical mean value theorem,
$$h(1) - h(0) = h'(t_0)(1 - 0)$$
for some point $t_0 \in (0, 1)$. But now as
$$h(1) - h(0) = \mathbf{u}^T \mathbf{f}(\mathbf{b}) - \mathbf{u}^T \mathbf{f}(\mathbf{a}) = \|\mathbf{u}\|^2$$
and
$$h'(t) = \frac{d}{dt} \sum_{j=1}^{p} u_j f_j(\mathbf{a} + t(\mathbf{b} - \mathbf{a}))$$
$$= \sum_{j=1}^{p} u_j \sum_{k=1}^{q} \frac{\partial f_j}{\partial x_k}(\mathbf{a} + t(\mathbf{b} - \mathbf{a}))(b_k - a_k)$$
$$= \langle J_{\mathbf{f}}(\mathbf{a} + t(\mathbf{b} - \mathbf{a}))(\mathbf{b} - \mathbf{a}), \mathbf{u} \rangle,$$
the mean value theorem yields the formula
$$\|\mathbf{u}\|^2 = \langle J_{\mathbf{f}}(\mathbf{c})(\mathbf{b} - \mathbf{a}), \mathbf{u} \rangle$$
for some point
$$\mathbf{c} = \mathbf{a} + t_0(\mathbf{b} - \mathbf{a})$$
in the open line segment between $\mathbf{a}$ and $\mathbf{b}$. Thus, by the Cauchy-Schwarz inequality,
$$\|\mathbf{u}\|^2 \leq \|J_{\mathbf{f}}(\mathbf{c})(\mathbf{b} - \mathbf{a})\| \|\mathbf{u}\|,$$
which clearly implies the validity of (21.14) when $\mathbf{u} \neq \mathbf{0}$. Since (21.14) is also valid when $\mathbf{u} = \mathbf{0}$, the proof is complete. □

Exercise 21.11. Show that if $\mathbf{h}(\mathbf{x}) = \mathbf{f}(\mathbf{g}(\mathbf{x}))$, where $\mathbf{g}(\mathbf{x})$ is a smooth map of $\mathbb{R}^r$ into $\mathbb{R}^q$ and $\mathbf{f}$ is a smooth map of $\mathbb{R}^q$ into $\mathbb{R}^p$, then $J_\mathbf{h}(\mathbf{x}) = J_\mathbf{f}(\mathbf{g}(\mathbf{x}))J_\mathbf{g}(\mathbf{x})$.

21.5. Convex minimization problems

Convex functions have important extremal properties.

Theorem 21.7. *If Q is an open convex subset of $\mathbb{R}^n$ and $f \in \mathcal{C}^1(Q)$ is a real-valued convex function on Q, then the following implications are in force:*

(1) *If $\mathbf{a} \in Q$ and $(\nabla f)(\mathbf{a}) = \mathbf{0}$, then $f(\mathbf{a}) \leq f(\mathbf{b})$ for every point $\mathbf{b} \in Q$, i.e., f achieves its minimal value at the point $\mathbf{a}$.*

(2) *If $\mathbf{a} \in Q$, $(\nabla f)(\mathbf{a}) = \mathbf{0}$, and f is strictly convex, then $f(\mathbf{a}) < f(\mathbf{b})$ for every point $\mathbf{b} \in Q \setminus \{\mathbf{a}\}$, i.e., $f(\mathbf{c}) = f(\mathbf{a}) \iff \mathbf{c} = \mathbf{a}$.*

Proof. Suppose that $\mathbf{a}, \mathbf{b} \in Q$. Then the inequality (9.1) for convex functions implies that $f(\mathbf{a} + t(\mathbf{b} - \mathbf{a})) \leq (1-t)f(\mathbf{a}) + tf(\mathbf{b})$ and hence that

$$\frac{f(\mathbf{a} + t(\mathbf{b} - \mathbf{a})) - f(\mathbf{a})}{t} \leq f(\mathbf{b}) - f(\mathbf{a}) \quad \text{if } 0 < t < 1.$$

But this leads easily to the inequality in (1), since the term on the left tends to $(\nabla f)(\mathbf{a})(\mathbf{b} - \mathbf{a}) = 0$ as $t \downarrow 0$.

Suppose next that f is strictly convex and there exists a point $\mathbf{a} \in Q$ such that $f(\mathbf{a}) \leq f(\mathbf{b})$ for every point $\mathbf{b} \in Q$. Then, if $\mathbf{b} \neq \mathbf{a}$ and $0 < t < 1$,

$$f(\mathbf{a}) \leq f(t\mathbf{a} + (1-t)\mathbf{b}) < tf(\mathbf{a}) + (1-t)f(\mathbf{b}) \leq tf(\mathbf{b}) + (1-t)f(\mathbf{b}) = f(\mathbf{b}),$$

i.e., $f(\mathbf{a}) < f(\mathbf{b})$ when $\mathbf{a} \neq \mathbf{b}$, as claimed. $\square$

Exercise 21.12. Show that if Q is an open nonempty convex subset of $\mathbb{R}^n$, then a real-valued function $f \in \mathcal{C}^2(Q)$ is convex if and only if the Hessian $H_f(\mathbf{x}) \succeq 0$ on Q; and that f is strictly convex if $H_f(\mathbf{x}) \succ 0$ on Q.

Example 21.1. If $A \in \mathbb{R}^{n \times n}$, $A \succ O$ with block decomposition

$$A = \begin{bmatrix} A_{11} & A_{12} \\ A_{21} & A_{22} \end{bmatrix} \quad \text{and} \quad f(\mathbf{x}) = \left\langle A \begin{bmatrix} \mathbf{a} \\ \mathbf{x} \end{bmatrix}, \begin{bmatrix} \mathbf{a} \\ \mathbf{x} \end{bmatrix} \right\rangle$$

with $A_{11} \in \mathbb{R}^{p \times p}$, $\mathbf{a} \in \mathbb{R}^p$, $A_{22} \in \mathbb{R}^{q \times q}$, and $\mathbf{x} \in \mathbb{R}^q$, then

$$\frac{f(\mathbf{x} + \varepsilon\mathbf{u}) - f(\mathbf{x})}{\varepsilon} = 2\left\langle A \begin{bmatrix} \mathbf{a} \\ \mathbf{x} \end{bmatrix}, \begin{bmatrix} \mathbf{0} \\ \mathbf{u} \end{bmatrix} \right\rangle + \varepsilon \left\langle A \begin{bmatrix} \mathbf{0} \\ \mathbf{u} \end{bmatrix}, \begin{bmatrix} \mathbf{0} \\ \mathbf{u} \end{bmatrix} \right\rangle$$

which tends to

$$(\nabla f)(\mathbf{x})\mathbf{u} = 2\left\langle A \begin{bmatrix} \mathbf{a} \\ \mathbf{x} \end{bmatrix}, \begin{bmatrix} \mathbf{0} \\ \mathbf{u} \end{bmatrix} \right\rangle$$

as $\varepsilon \downarrow 0$. Thus,
$$(\nabla f)(\mathbf{x})^T = 2A_{21}\mathbf{a} + 2A_{22}\mathbf{x} \quad \text{for every } \mathbf{x} \in \mathbb{R}^q,$$
and, as A_{22} is invertible,
$$(\nabla f)(\mathbf{x})^T = \mathbf{0} \iff \mathbf{x} = \mathbf{x}_0, \quad \text{where } \mathbf{x}_0 = -A_{22}^{-1}A_{21}\mathbf{a}.$$
Moreover,
$$f(\mathbf{x}_0) = \langle (A_{11} - A_{12}A_{22}^{-1}A_{21})\mathbf{a}, \mathbf{a} \rangle,$$
and, since the Hessian $H_f(\mathbf{x}) = 2A_{22}$ is positive definite, $f(\mathbf{x}) \geq f(\mathbf{x}_0)$ for every $\mathbf{x} \in \mathbb{R}^q$ with strict inequality if $\mathbf{x} \neq \mathbf{x}_0$, thanks to Exercise 21.12. This gives a geometric interpretation to the Schur complement $A_{11} - A_{12}A_{22}^{-1}A_{21}$ for positive definite matrices A and is yet another way of treating the problem considered in Theorem 18.1 when $A \in \mathbb{R}^{n \times n}$ and $\mathbf{a} \in \mathbb{R}^p$. $\diamond$

Exercise 21.13. Show that if the positions and sizes of the vectors $\mathbf{a}$ and $\mathbf{x}$ in Example 21.1 are interchanged, then the minimum value of the new $f(\mathbf{x})$ will be $\langle (A_{22} - A_{21}A_{11}^{-1}A_{12})\mathbf{a}, \mathbf{a} \rangle$.

Example 21.2. Consider Example 21.1 but with $A = A^T$ only assumed to be positive semidefinite and rank $A_{22} = r$ with $1 \leq r < q$. Then $A_{22} = V_1 S_1 V_1^T$, where $V_1 \in \mathbb{R}^{q \times r}$, $V = \begin{bmatrix} V_1 & V_2 \end{bmatrix}$ is a real $q \times q$ unitary matrix, and $S_1 = \text{diag}\{s_1, \ldots, s_r\}$ with $s_1 \geq \cdots \geq s_r > 0$. Then, upon setting $\mathbf{x} = V_1 \mathbf{v} + V_2 \mathbf{w}$,
$$(\nabla f)(\mathbf{x})^T = 2A_{21}\mathbf{a} + 2A_{22}\mathbf{x} = 2A_{21}\mathbf{a} + 2V_1 S_1 V_1^T (V_1 \mathbf{v} + V_2 \mathbf{w})$$
$$= 2A_{21}\mathbf{a} + 2V_1 S_1 \mathbf{v},$$
and hence
$$(\nabla f)(\mathbf{x})^T = \mathbf{0} \iff \mathbf{v} = -S_1^{-1} V_1^T A_{21}\mathbf{a},$$
i.e., if and only if $\mathbf{x} = \mathbf{x}_0$, where
$$\mathbf{x}_0 = -V_1 S_1^{-1} V_1^T A_{21}\mathbf{a} + V_2 \mathbf{w} = -A_{22}^\dagger A_{21}\mathbf{a} + V_2 \mathbf{w}.$$
Lemma 18.5 ensures that $\mathcal{N}_{A_{22}} \subseteq \mathcal{N}_{A_{12}}$ and $A_{22} A_{22}^\dagger A_{21} = A_{21}$. Thus, as $V_2 \mathbf{w} \in \mathcal{N}_{A_{22}}$, $A_{12} V_2 \mathbf{w} = \mathbf{0}$ and $f(\mathbf{x}_0) = \mathbf{a}^T (A_{11} - A_{12} A_{22}^\dagger A_{21})\mathbf{a}$. $\diamond$

Exercise 21.14. Show that in the setting of Example 21.1, $f(\mathbf{x})^{1/2}$ is convex. [HINT: Identify $\langle A\mathbf{u}, \mathbf{u} \rangle^{1/2}$ as a norm on $\mathbb{R}^n$ and note that
$$f(t\mathbf{x} + (1-t)\mathbf{y}) = \left\langle A \begin{bmatrix} t\mathbf{a} + (1-t)\mathbf{a} \\ t\mathbf{x} + (1-t)\mathbf{y} \end{bmatrix}, \begin{bmatrix} t\mathbf{a} + (1-t)\mathbf{a} \\ t\mathbf{x} + (1-t)\mathbf{y} \end{bmatrix} \right\rangle.]$$

Exercise 21.15. Show that if $f(\mathbf{x})$ is a convex function on a convex subset Q of $\mathbb{R}^n$ and $f(\mathbf{x}) \geq 0$ when $\mathbf{x} \in Q$, then $f(\mathbf{x})^2$ is also convex on Q.

21.6. Supplementary notes

This chapter is partially adapted from Chapter 14 of [30].

Chapter 22

Fixed point theorems

A vector **x** is said to be a **fixed point** of a vector-valued function **f(x)** that maps a subset E of a vector space $\mathcal{V}$ into itself if $\mathbf{x} \in E$ and $\mathbf{f}(\mathbf{x}) = \mathbf{x}$.

22.1. A contractive fixed point theorem

Theorem 22.1. *Let* $\mathbf{f}(\mathbf{x})$ *be a continuous map of a closed subset* E *of a finite-dimensional normed linear space* $\mathcal{X}$ *into itself such that*

$$(22.1) \qquad \|\mathbf{f}(\mathbf{b}) - \mathbf{f}(\mathbf{a})\|_{\mathcal{X}} \leq K \|\mathbf{b} - \mathbf{a}\|_{\mathcal{X}}$$

for some constant $K, 0 < K < 1$, *and every pair of points* $\mathbf{a}, \mathbf{b}$ *in the set* E. *Then:*

(1) *There is exactly one point* $\mathbf{x}_* \in E$ *such that* $\mathbf{f}(\mathbf{x}_*) = \mathbf{x}_*$.
(2) *If* $\mathbf{x}_0 \in E$ *and* $\mathbf{x}_{n+1} = \mathbf{f}(\mathbf{x}_n)$ *for* $n = 0, 1, \ldots$, *then* $\mathbf{x}_1, \mathbf{x}_2, \ldots \in E$ *and*

$$\mathbf{x}_* = \lim_{n \uparrow \infty} \mathbf{x}_n \, ;$$

i.e., the limit exists and is independent of how the initial point $\mathbf{x}_0$ *is chosen.*
(3) $\|\mathbf{x}_* - \mathbf{x}_n\|_{\mathcal{X}} \leq \dfrac{K^n}{1-K} \|\mathbf{x}_1 - \mathbf{x}_0\|_{\mathcal{X}}$.

Proof. Choose any point $\mathbf{x}_0 \in E$ and then define the sequence of points $\mathbf{x}_1, \mathbf{x}_2, \ldots$ in E by the rule

$$\mathbf{x}_{n+1} = f(\mathbf{x}_n).$$

Then, setting $\|\mathbf{x}\| = \|\mathbf{x}\|_{\mathcal{X}}$ for short,

$$\|\mathbf{x}_2 - \mathbf{x}_1\| = \|f(\mathbf{x}_1) - f(\mathbf{x}_0)\| \leq K\|\mathbf{x}_1 - \mathbf{x}_0\|,$$
$$\|\mathbf{x}_3 - \mathbf{x}_2\| = \|f(\mathbf{x}_2) - f(\mathbf{x}_1)\| \leq K\|\mathbf{x}_2 - \mathbf{x}_1\| \leq K^2\|\mathbf{x}_1 - \mathbf{x}_0\|,$$
$$\vdots$$
$$\|\mathbf{x}_{n+1} - \mathbf{x}_n\| \leq K^n\|\mathbf{x}_1 - \mathbf{x}_0\|,$$

and hence

$$\|\mathbf{x}_{n+k} - \mathbf{x}_n\| \leq \|\mathbf{x}_{n+k} - \mathbf{x}_{n+k-1}\| + \cdots + \|\mathbf{x}_{n+1} - \mathbf{x}_n\|$$
$$\leq (K^{n+k-1} + \cdots + K^n)\|\mathbf{x}_1 - \mathbf{x}_0\|$$
$$\leq \frac{K^n}{1-K}\|\mathbf{x}_1 - \mathbf{x}_0\|.$$

Therefore, since K^n tends to 0 as $n \uparrow \infty$, this last bound guarantees that the sequence $\{\mathbf{x}_n\}$ is a Cauchy sequence in the closed subset E of $\mathcal{X}$. Thus, $\mathbf{x}_n$ converges to a limit $\mathbf{u} \in E$ as $n \uparrow \infty$.

The next step is to show that this limit, **which, as far as we know at this moment, may depend upon $\mathbf{x}_0$**, is a fixed point of $\mathbf{f}$:

$$\|\mathbf{f}(\mathbf{u}) - \mathbf{u}\| = \|\mathbf{f}(\mathbf{u}) - \mathbf{f}(\mathbf{x}_n) + \mathbf{x}_{n+1} - \mathbf{u}\|$$
$$\leq \|\mathbf{f}(\mathbf{u}) - \mathbf{f}(\mathbf{x}_n)\| + \|\mathbf{x}_{n+1} - \mathbf{u}\|$$
$$\leq K\|\mathbf{u} - \mathbf{x}_n\| + \|\mathbf{x}_{n+1} - \mathbf{u}\|.$$

Since the right-hand side tends to zero as $n \uparrow \infty$ and the left-hand side is independent of n, this implies that $\|\mathbf{f}(\mathbf{u}) - \mathbf{u}\| \leq \varepsilon$ for every $\varepsilon > 0$ and hence that $\mathbf{f}(\mathbf{u}) = \mathbf{u}$, i.e., $\mathbf{u}$ is a fixed point of $\mathbf{f}$.

Suppose next that $\mathbf{v}$ is also a fixed point of $\mathbf{f}$ in E. Then

$$0 \leq \|\mathbf{u} - \mathbf{v}\| = \|f(\mathbf{u}) - f(\mathbf{v})\| \leq K\|\mathbf{u} - \mathbf{v}\|.$$

Therefore,

$$0 \leq (1-K)\|\mathbf{u} - \mathbf{v}\| \leq 0.$$

This proves that $\mathbf{u} = \mathbf{v}$ and hence, upon denoting this one and only fixed point by $\mathbf{x}_*$, that (1) and (2) hold.

The upper bound in (3) is obtained from the inequality

$$\|\mathbf{x}_* - \mathbf{x}_n\| \leq \|\mathbf{x}_* - \mathbf{x}_{n+k}\| + \|\mathbf{x}_{n+k} - \mathbf{x}_n\| \leq \|\mathbf{x}_* - \mathbf{x}_{n+k}\| + \frac{K^n}{1-K}\|\mathbf{x}_1 - \mathbf{x}_0\|$$

by letting $k \uparrow \infty$. $\square$

22.2. A refined contractive fixed point theorem

The next theorem relaxes the constraint (22.1) that was imposed in the formulation of Theorem 22.1.

Theorem 22.2. *Let* $\mathbf{f}(\mathbf{x})$ *be a continuous map of a closed subset* E *of a finite-dimensional normed linear space* $\mathcal{X}$ *into itself such that the* j*'th iterate*

$$\mathbf{f}^{[j]}(\mathbf{x}) = \mathbf{f}(\mathbf{f}^{[j-1]}(\mathbf{x})) \quad \text{for } j = 2, 3, \ldots$$

of $\mathbf{f} = \mathbf{f}^{[1]}$ *satisfies the constraint*

$$\left\|\mathbf{f}^{[j]}(\mathbf{x}) - \mathbf{f}^{[j]}(\mathbf{y})\right\|_{\mathcal{X}} \leq K\|\mathbf{x} - \mathbf{y}\|_{\mathcal{X}}$$

for some constant K, $0 < K < 1$, *and some positive integer* j. *Then* $\mathbf{f}$ *has exactly one fixed point* $\mathbf{x}_*$ *in* E.

Proof. If $\mathbf{g}(\mathbf{x}) = \mathbf{f}^{[j]}(\mathbf{x})$ meets the indicated constraint, then, by Theorem 22.1, $\mathbf{g}$ has exactly one fixed point $\mathbf{x}_*$ in E. Moreover,

$$\mathbf{f}(\mathbf{x}_*) = \mathbf{f}(\mathbf{g}(\mathbf{x}_*)) = \mathbf{g}(\mathbf{f}(\mathbf{x}_*)).$$

But this exhibits $\mathbf{f}(\mathbf{x}_*)$ as a fixed point of $\mathbf{g}$. Thus, as $\mathbf{g}$ has only one fixed point in E, we must have

$$\mathbf{f}(\mathbf{x}_*) = \mathbf{x}_*;$$

i.e., $\mathbf{x}_*$ is a fixed point of $\mathbf{f}$. □

Example 22.1. Let $A = \begin{bmatrix} \alpha & \beta \\ 0 & \alpha \end{bmatrix}$ with $|\alpha| < 1$ and $|\beta| > 1$. Then, since $\|A\| \geq \{|\alpha|^2 + |\beta|^2\}^{1/2}$, the contractive fixed point theorem is not applicable to the vector-valued function $\mathbf{f}(\mathbf{x}) = A\mathbf{x}$. However, since

$$\|A^n\| = \left\|\begin{bmatrix} \alpha^n & n\alpha^{n-1}\beta \\ 0 & \alpha^n \end{bmatrix}\right\| \leq |\alpha^n| + n|\alpha^{n-1}||\beta|,$$

it will be applicable to $\mathbf{f}^{[n]}$ if n is large enough. ◇

Exercise 22.1. Show that if $A, C \in \mathbb{C}^{n \times n}$ and $N = C_0^{(n)}$, the $n \times n$ Jordan cell with zeros on the diagonal, then $A = C + NCN^H + \cdots + N^{n-1}C(N^H)^{n-1}$ is the one and only solution of the equation $A - NAN^H = C$. [HINT: Invoke Theorem 22.2 with $\mathbf{f}(A) = C + NAN^H$, and keep in mind that $\|N^k\| = 1$ if $k = 1, \ldots, n-1$.]

22.3. Other fixed point theorems

To add perspective, we list a number of fixed point theorems with less restrictive assumptions but (as is only fair) weaker conclusions.

We begin with the **Brouwer fixed point theorem**:

Theorem 22.3. *If $f(x)$ is a continuous mapping of the closed unit ball*
$$\overline{B} = \{\mathbf{x} \in \mathbb{R}^n : \|\mathbf{x}\| \leq 1\}$$
into itself, then there is at least one point $\mathbf{x} \in \overline{B}$ such that $f(\mathbf{x}) = \mathbf{x}$.

Proof. See, e.g., Chapter 14 of [**30**]. □

The Brouwer fixed point theorem can be strengthened to:

Theorem 22.4. *If $f(x)$ is a continuous mapping of a closed bounded convex subset of $\mathbb{R}^n$ into itself, then there is at least one point $\mathbf{x}$ in this set such that $f(\mathbf{x}) = \mathbf{x}$.*

Proof. See, e.g., Chapter 22 of [**30**]. □

Exercise 22.2. Invoke Theorem 22.4 to show that if $0 < a < b$,
$$E = \{\mathbf{x} \in \mathbb{R}^2 : a \leq x_1, x_2 \leq b\}, \quad \text{and} \quad \mathbf{f}\left(\begin{bmatrix} x_1 \\ x_2 \end{bmatrix}\right) = \begin{bmatrix} \sqrt{x_1 x_2} \\ (x_1 + x_2)/2 \end{bmatrix},$$
then $\mathbf{f}$ has a fixed point in E. [REMARK: Every vector $\mathbf{x} \in E$ with equal components is a fixed point.]

There are also more general fixed point theorems that are applicable in infinite-dimensional spaces:

The Leray-Schauder theorem: Every compact convex subset in a Banach space has the fixed point property; see, e.g., [**67**], for a start.

22.4. Applications of fixed point theorems

Example 22.2. Let $A \in \mathbb{C}^{p \times p}$, $B \in \mathbb{C}^{q \times q}$, and $C \in \mathbb{C}^{p \times q}$. Then the equation

(22.2) $$X - AXB = C$$

has a unique solution $X \in \mathbb{C}^{p \times q}$ if $\|A\|\|B\| < 1$. (Much stronger results will be obtained in Chapter 37.)

Discussion. Let $f(X) = C + AXB$. Then clearly f maps $\mathbb{C}^{p \times q}$ into $\mathbb{C}^{p \times q}$ and X is a solution of (22.2) if and only if $f(X) = X$. Thus, as
$$\|f(X) - f(Y)\| = \|A(X - Y)B\| \leq \|A\|\|X - Y\|\|B\|,$$

22.4. Applications of fixed point theorems

Theorem 22.1 (with $E = \mathbb{C}^{p \times q}$) guarantees that (22.2) has exactly one solution X_* in $\mathbb{C}^{p \times q}$ if $\|A\| \|B\| < 1$. Moreover, it is readily checked that

$$X_* \in \left\{ X \in \mathbb{C}^{p \times q} : \|X\| \leq \frac{\|C\|}{1 - \|A\| \|B\|} \right\},$$

since f maps the set in the last display into itself. $\diamond$

Example 22.3. Let $Q = (\alpha, \beta) \subset \mathbb{R}$ and suppose that $f \in \mathcal{C}^2(Q)$ maps Q into $\mathbb{R}$ and that there exists a point $b \in Q$ such that $f(b) = 0$ and $f'(b) \neq 0$, and let $E_\delta = \{x \in \mathbb{R} : |x - b| \leq \delta\}$. Suppose further that although $\delta > 0$, it is small enough so that $E_\delta \subset Q$ and $f'(x) \neq 0$ for $x \in E_\delta$. Then, by the mean value theorem,

$$f(b) = f(x) + (b-x)f'(c) \quad \text{with } c = tx + (1-t)b \text{ for some point } t \in (0,1).$$

Thus, as $f(b) = 0$ and $f'(c) \neq 0$,

$$b = x - \frac{f(x)}{f'(c)} \approx x - \frac{f(x)}{f'(x)}.$$

The function $N(x) = x - f(x)/f'(x)$ is called the **Newton step**. Since $N(x) = x$ if and only $f(x) = 0$, the problem of locating the zeros of $f(x)$ in E_δ is equivalent to finding the fixed points of $N(x)$ in E_δ. The **Newton method** sets up a sequence of points $x_{k+1} = N(x_k)$, $k = 0, 1, \ldots$, in E_δ that converge to b if the initial point x_0 was chosen close enough to b. This will now be justified by the contractive fixed point theorem; the speed of convergence and Newton's method for vector-valued functions of many variables will be discussed in Chapter 25.

Discussion. Let

$$\alpha_\delta = \max\{|f''(x)| : x \in E_\delta\}, \quad \beta_\delta = \max\left\{\frac{1}{|f'(x)|} : x \in E_\delta\right\},$$

and

$$\kappa_\delta = \max\{|f'(x)| : x \in E_\delta\},$$

and observe that if $0 < \delta_1 < \delta_2$, then $\alpha_{\delta_1} \leq \alpha_{\delta_2}$, $\beta_{\delta_1} \leq \beta_{\delta_2}$, and $\kappa_{\delta_1} \leq \kappa_{\delta_2}$. Then, since E_δ is a convex set and

$$N'(x) = 1 - \frac{f'(x)f'(x) - f(x)f''(x)}{f'(x)^2} = \frac{f(x)f''(x)}{f'(x)^2},$$

the mean value theorem implies that for every pair of points $x, y \in E_\delta$

$$N(x) - N(y) = (x-y)\frac{f(c)f''(c)}{f'(c)^2} \quad \text{for some point } c \in E_\delta.$$

Therefore,

$$|N(x) - N(y)| \leq |x-y|\, \alpha_\delta \beta_\delta^2 |f(c)| = |x-y|\, \alpha_\delta \beta_\delta^2 |f(c) - f(b)|$$
$$\leq |x-y|\, \alpha_\delta \beta_\delta^2 |c-b| \kappa_\delta \leq |x-y|\, \gamma_\delta \quad \text{with } \gamma_\delta = \delta \alpha_\delta \beta_\delta^2 \kappa_\delta.$$

If $\delta > 0$ is small enough, then $\gamma_\delta < 1$. Moreover, for such δ,
$$|N(x) - b| = |N(x) - N(b)| \leq |x - b|\gamma_\delta,$$
which ensures that $N(x)$ maps points $x \in E_\delta$ into E_δ. Therefore, the contractive fixed point theorem is applicable to ensure that $N(x)$ has exactly one fixed point in E_δ. Thus, as $b \in E_\delta$ and $N(b) = b$, that unique fixed point is b, and hence the sequence of points $x_{k+1} = N(x_k)$ will tend to b as $k \uparrow \infty$ for every choice of $x_0 \in E_\delta$. ◇

Example 22.4. Let $A \in \mathbb{C}^{p \times p}$ and $\mathbf{b} \in \mathbb{C}^p$. Then $A\mathbf{x} = \mathbf{b}$ if and only if $\mathbf{x}$ is a fixed point of the function $\mathbf{f}(\mathbf{x}) = \mathbf{b} - A\mathbf{x} + \mathbf{x}$. Let $B = I_p - A$ and suppose further that $\|B\| < 1$. Then A is invertible and hence $\mathbf{x}$ is a fixed point of $\mathbf{f}$ if and only if $\mathbf{x} = A^{-1}\mathbf{b}$. Thus, as $\mathbf{f}$ maps $\mathbb{C}^p$ into $\mathbb{C}^p$ and
$$\|\mathbf{f}(\mathbf{x}) - \mathbf{f}(\mathbf{y})\| \leq \|B\|\,\|\mathbf{x} - \mathbf{y}\| \quad \text{for every pair of vectors } \mathbf{x}, \mathbf{y} \in \mathbb{C}^p,$$
the contractive fixed point theorem (with $E = \mathbb{C}^p$) guarantees that if $\mathbf{x}_k = \mathbf{f}(\mathbf{x}_{k-1})$ for $k = 1, 2, \ldots$, then $\mathbf{x}_k = \mathbf{b} + B\mathbf{b} + \cdots + B^{k-1}\mathbf{b} + B^k \mathbf{x}_0$ tends to $A^{-1}\mathbf{b}$ as $k \uparrow \infty$ for every choice of the initial point $\mathbf{x}_0$. Moreover, as $\mathbf{f}$ maps the set
$$E = \left\{ \mathbf{x} \in \mathbb{C}^p : \|\mathbf{x}\| \leq \frac{\|\mathbf{b}\|}{1 - \|B\|} \right\}$$
into itself, $A^{-1}\mathbf{b} \in E$. ◇

Exercise 22.3. Show by direct computation that in the setting of Example 22.4, $\|\mathbf{x}_k - A^{-1}\mathbf{b}\| = \|B^k(\mathbf{x}_0 - A^{-1}\mathbf{b})\|$. [HINT: First verify the identity $I_n + B + \cdots + B^{k-1} = A^{-1}(I - B^k)$.]

Example 22.5. Let $E = \{(x,y) \in \mathbb{R}^2 : 0 \leq x \leq y \leq 1\}$ (i.e., E is the triangle with vertices $(0,0)$, $(0,1)$, and $(1,1)$ that lies above the line $y = x$) and let
$$\mathbf{f}(x,y) = \begin{bmatrix} f_1(x,y) \\ f_2(x,y) \end{bmatrix} = \begin{bmatrix} \sqrt{xy} \\ (x+y)/2 \end{bmatrix}.$$
Then, $\mathbf{f}$ maps E into E, since $(x,y) \in E \implies 0 \leq \sqrt{xy} \leq (x+y)/2 \leq 1$. Moreover, a point $(x,y) \in E$ is a fixed point of $\mathbf{f}$ if and only if
$$\begin{bmatrix} x \\ y \end{bmatrix} = \begin{bmatrix} \sqrt{xy} \\ (x+y)/2 \end{bmatrix} \iff x = y.$$
Thus, as $\mathbf{f}$ has infinitely many fixed points in E, there does not exist a positive constant $\gamma < 1$ such that $\|\mathbf{f}(\mathbf{u}) - \mathbf{f}(\mathbf{v})\| \leq \gamma \|\mathbf{u} - \mathbf{v}\|$ for all points $\mathbf{u}, \mathbf{v} \in E$. Nevertheless, the limit in the recursive scheme discussed in the proof of the contractive fixed point theorem exists and is a fixed point of $\mathbf{f}$, but now it depends upon the initial point in the recursion: If

(22.3) $\quad \begin{bmatrix} a_0 \\ b_0 \end{bmatrix} \in E \quad \text{and} \quad \begin{bmatrix} a_{k+1} \\ b_{k+1} \end{bmatrix} = \begin{bmatrix} \sqrt{a_k b_k} \\ (a_k + b_k)/2 \end{bmatrix} \quad \text{for } k = 0, 1, 2, \ldots,$

then, since $a_k \leq b_k$ for all nonnegative integers k, $a_{k+1} = \sqrt{a_k b_k} \geq a_k$, $b_{k+1} = (a_k + b_k)/2 \leq b_k$, and

$$b_{k+1} - a_{k+1} = \frac{a_k + b_k}{2} - \sqrt{a_k b_k} \leq \frac{a_k + b_k}{2} - a_k = \frac{b_k - a_k}{2} \leq \frac{b_0 - a_0}{2^{k+1}}.$$

Therefore, a_k tends to a limit α as $k \uparrow \infty$, b_k tends to a limit β as $k \uparrow \infty$, and

$$|\alpha - \beta| = |\alpha - a_k + (a_k - b_k) + b_k - \beta| \leq |\alpha - a_k| + |a_k - b_k| + |b_k - \beta|,$$

which implies that $\alpha = \beta$ and hence that it is a fixed point of $\mathbf{f}$. Notice that $a_k \leq \alpha = \beta \leq b_k$ for $k = 0, 1, 2, \ldots$. ◇

Exercise 22.4. Let $E = \{x \in \mathbb{R} : 0 \leq x \leq 1\}$ and let $f(x) = (1 + x^2)/2$. Show that:

(a) f maps E into E.

(b) There does not exist a positive constant $\gamma < 1$ such that $|f(b) - f(a)| \leq \gamma |b - a|$ for every choice of $a, b \in E$.

(c) f has exactly one fixed point $x_* \in E$.

Exercise 22.5. Show that the polynomial $p(x) = 1 - 4x + x^2 - x^3$ has exactly one root in the interval $0 \leq x \leq 1$. [HINT: Use the contractive fixed point theorem.]

Exercise 22.6. Show that the function $\mathbf{f}(x, y) = \begin{bmatrix} (1 - y)/2 \\ (1 + x^2)/3 \end{bmatrix}$ has a fixed point inside the set $\{(x, y) \in \mathbb{R}^2 : x^2 + y^2 \leq 1\}$.

22.5. Supplementary notes

This chapter is partially adapted from Chapters 14 and 22 in [**30**]. A perhaps surprising application of the sequence of vectors considered in (22.3) is to the evaluation of integrals of the form

$$I(a, b) = \int_0^{\pi/2} \frac{1}{\sqrt{a^2 \sin^2 \theta + b^2 \cos^2 \theta}} d\theta \quad \text{for } 0 < a < b < 1.$$

Gauss discovered that if $a = a_0$ and $b = b_0$, then $I(a, b) = I(a_k, b_k)$ for $k = 1, 2, \ldots$ and hence that if $a_k \to \alpha$, then $I(a, b) = I(\alpha, \alpha) = \pi/(2\alpha)$; see Duren [**26**] for a detailed account.

Chapter 23

The implicit function theorem

This chapter is devoted primarily to the implicit function theorem and a few of its applications. We shall, however, begin with the inverse function theorem and then use it to establish the implicit function theorem. (The two are really equivalent, but this direction makes for easier reading.) The **notation**

$$B_\alpha(\mathbf{x}_0) = \{\mathbf{x} \in \mathbb{R}^q : \|\mathbf{x} - \mathbf{x}_0\| < \alpha\}$$

for the ball with center $\mathbf{x}_0 \in \mathbb{R}^q$ and radius $\alpha > 0$ will be used frequently.

23.1. The inverse function theorem

If $A \in \mathbb{R}^{p \times p}$, then the function $\mathbf{f}(\mathbf{x}) = A\mathbf{x}$ maps $\mathbb{R}^p$ onto $\mathbb{R}^p$ if and only if $A = J_\mathbf{f}(\mathbf{x})$ is invertible. The inverse function theorem is an extension of this fact to a wider class of functions $\mathbf{f}(\mathbf{x})$. It makes extensive use of the observation that if $\mathbf{f} \in \mathcal{C}^1(Q)$ in some open set $Q \subseteq \mathbb{R}^p$ and $\mathbf{x}_0 \in Q$, then

$$\mathbf{f}(\mathbf{x}) \approx \mathbf{f}(\mathbf{x}_0) + J_\mathbf{f}(\mathbf{x}_0)(\mathbf{x} - \mathbf{x}_0) \quad \text{if } \mathbf{x} \in Q \text{ is close to } \mathbf{x}_0.$$

Theorem 23.1. *Suppose that the $p \times 1$ vector-valued function*

$$\mathbf{f}(\mathbf{x}) = \begin{bmatrix} f_1(x_1, \ldots, x_p) \\ \vdots \\ f_p(x_1, \ldots, x_p) \end{bmatrix}$$

maps an open set $Q \subseteq \mathbb{R}^p$ into $\mathbb{R}^p$, that $\mathbf{f} \in \mathcal{C}^1(Q)$, and that the Jacobian matrix

$$J_{\mathbf{f}}(\mathbf{x}) = \begin{bmatrix} \frac{\partial f_1}{\partial x_1}(\mathbf{x}) & \cdots & \frac{\partial f_1}{\partial x_p}(\mathbf{x}) \\ \vdots & & \vdots \\ \frac{\partial f_p}{\partial x_1}(\mathbf{x}) & \cdots & \frac{\partial f_p}{\partial x_p}(\mathbf{x}) \end{bmatrix}$$

is invertible at a point $\mathbf{x}_0 \in Q$ *and* $\mathbf{y}_0 = \mathbf{f}(\mathbf{x}_0)$. *Then there exists a pair of numbers* $\alpha > 0$ *and* $\beta > 0$ *such that* $B_\alpha(\mathbf{x}_0) \subset Q$ *and for each vector* $\mathbf{y} \in B_\beta(\mathbf{y}_0)$ *there exists exactly one vector* $\mathbf{x} = \mathbf{g}(\mathbf{y})$ *in the ball* $B_\alpha(\mathbf{x}_0)$ *such that* $\mathbf{y} = \mathbf{f}(\mathbf{x})$. *Moreover, the function* $\mathbf{g} \in \mathcal{C}^1(B_\beta(\mathbf{y}_0))$.

Proof. Let $A = J_{\mathbf{f}}(\mathbf{x}_0)$ and $\mathbf{w}(\mathbf{x}) = \mathbf{x} - A^{-1}\mathbf{f}(\mathbf{x})$ for $\mathbf{x} \in Q$. Then

(23.1) $\qquad J_{\mathbf{w}}(\mathbf{x}) = I_p - A^{-1}J_{\mathbf{f}}(\mathbf{x}) = A^{-1}[J_{\mathbf{f}}(\mathbf{x}_0) - J_{\mathbf{f}}(\mathbf{x})]\,.$

Now choose $\alpha > 0$ such that $B_\alpha(\mathbf{x}_0) \subseteq Q$ and $\|J_{\mathbf{w}}(\mathbf{x})\| \leq 1/2$ for $\mathbf{x} \in B_\alpha(\mathbf{x}_0)$. Then, as $A^{-1}J_{\mathbf{f}}(\mathbf{x}) = I_p - J_{\mathbf{w}}(\mathbf{x})$, this ensures that

(23.2) $\qquad J_{\mathbf{f}}(\mathbf{x}) \;$ is invertible for every point $\;\mathbf{x} \in B_\alpha(\mathbf{x}_0)\,.$

The strategy of the proof is to invoke the contractive fixed point theorem to show that if $\mathbf{y}$ is close enough to $\mathbf{y}_0$, then the vector-valued function

$$\mathbf{h}_{\mathbf{y}}(\mathbf{x}) = \mathbf{x} + A^{-1}[\mathbf{y} - \mathbf{f}(\mathbf{x})] = \mathbf{w}(\mathbf{x}) + A^{-1}\mathbf{y}$$

has exactly one fixed point $\mathbf{x}_* \in B_\alpha(\mathbf{x}_0)$. This then forces $\mathbf{f}(\mathbf{x}_*) = \mathbf{y}$. There are some additional fine points, since $B_\alpha(\mathbf{x}_0)$ is an open set and the contractive fixed point theorem is formulated for closed subsets of Q. To overcome this difficulty, the initial analysis is carried out in a closed nonempty subset $\overline{B_\delta(\mathbf{x}_0)}$ of $B_\alpha(\mathbf{x}_0)$. The details are presented in steps.

1. *If* $0 < \delta < \alpha$ *and* $\beta \leq \delta/(2\|A^{-1}\|)$, *then* $\mathbf{h}_{\mathbf{y}}(\mathbf{x})$ *has exactly one fixed point* $\mathbf{g}(\mathbf{y})$ *in* $\overline{B_\delta(\mathbf{x}_0)}$ *for each choice of* $\mathbf{y} \in B_\beta(\mathbf{y}_0)$.

In view of the bound (21.14), the condition $\|J_{\mathbf{w}}(\mathbf{x})\| \leq 1/2$ ensures that

$$\|\mathbf{w}(\mathbf{x}) - \mathbf{w}(\mathbf{v})\| \leq \frac{1}{2}\|\mathbf{x} - \mathbf{v}\| \quad \text{for every pair of points } \mathbf{x}, \mathbf{v} \in B_\alpha(\mathbf{x}_0)\,.$$

Consequently,

$$\|\mathbf{h}_{\mathbf{y}}(\mathbf{x}) - \mathbf{x}_0\| = \|\mathbf{w}(\mathbf{x}) - \mathbf{w}(\mathbf{x}_0) + A^{-1}[\mathbf{y} - \mathbf{f}(\mathbf{x}_0)]\|$$

$$\leq \frac{1}{2}\|\mathbf{x} - \mathbf{x}_0\| + \|A^{-1}\|\beta \leq \delta$$

for every point $\mathbf{x} \in \overline{B_\delta(\mathbf{x}_0)}$, i.e., $\mathbf{h}_{\mathbf{y}}$ maps $\overline{B_\delta(\mathbf{x}_0)}$ into itself. Moreover,

(23.3) $\qquad \|\mathbf{h}_{\mathbf{y}}(\mathbf{x}) - \mathbf{h}_{\mathbf{y}}(\mathbf{v})\| = \|\mathbf{w}(\mathbf{x}) - \mathbf{w}(\mathbf{v})\| \leq \frac{1}{2}\|\mathbf{x} - \mathbf{v}\|$

23.1. The inverse function theorem

for every pair of points $\mathbf{x}, \mathbf{v} \in \overline{B_\delta(\mathbf{x}_0)}$. Therefore, the contractive fixed point theorem ensures that $\mathbf{h_y}(\mathbf{x})$ has exactly one fixed point $\mathbf{x}_* \in \overline{B_\delta(\mathbf{x}_0)}$ and hence that $\mathbf{y} = \mathbf{f}(\mathbf{x}_*)$.

2. $\mathbf{h_y}(\mathbf{x})$ *has exactly one fixed point* $\mathbf{g}(\mathbf{y})$ *in* $B_\alpha(\mathbf{x}_0)$ *for each choice of* $\mathbf{y} \in B_\beta(\mathbf{y}_0)$.

If $\mathbf{v}_*$ and $\mathbf{x}_*$ are fixed points of $\mathbf{h_y}$ in $B_\alpha(\mathbf{x}_0)$, then

$$\|\mathbf{x}_* - \mathbf{v}_*\| = \|\mathbf{h_y}(\mathbf{x}_*) - \mathbf{h_y}(\mathbf{v}_*)\| = \|\mathbf{w}(\mathbf{x}_*) - \mathbf{w}(\mathbf{v}_*)\| \leq \frac{1}{2}\|\mathbf{x}_* - \mathbf{v}_*\|,$$

which is only viable if $\mathbf{x}_* = \mathbf{v}_*$.

3. $\mathbf{g} \in \mathcal{C}^1(B_\beta(\mathbf{y}_0))$.

Let $\mathbf{a}, \mathbf{b} \in B_\beta(\mathbf{y}_0)$. Then $\mathbf{u} = \mathbf{g}(\mathbf{a})$ and $\mathbf{v} = \mathbf{g}(\mathbf{b})$ belong to $B_\alpha(\mathbf{x}_0)$ and, in view of (23.3),

$$\frac{1}{2}\|\mathbf{v} - \mathbf{u}\| \geq \|\mathbf{w}(\mathbf{v}) - \mathbf{w}(\mathbf{u})\| = \|\mathbf{v} - \mathbf{u} - A^{-1}(\mathbf{f}(\mathbf{v}) - \mathbf{f}(\mathbf{u}))\|$$
$$= \|\mathbf{v} - \mathbf{u} - A^{-1}(\mathbf{b} - \mathbf{a})\| \geq \|\mathbf{v} - \mathbf{u}\| - \|A^{-1}(\mathbf{b} - \mathbf{a})\|.$$

Therefore,

(23.4) $$\|\mathbf{g}(\mathbf{b}) - \mathbf{g}(\mathbf{a})\| = \|\mathbf{v} - \mathbf{u}\| \leq 2\|A^{-1}\|\|\mathbf{b} - \mathbf{a}\|,$$

which clearly implies that $\mathbf{g}$ is continuous on $B_\beta(\mathbf{y}_0)$. Moreover,

$$\mathbf{b} - \mathbf{a} = \mathbf{f}(\mathbf{v}) - \mathbf{f}(\mathbf{u}) = A(t_1, \ldots, t_p)(\mathbf{v} - \mathbf{u}) = A(t_1, \ldots, t_p)(\mathbf{g}(\mathbf{b}) - \mathbf{g}(\mathbf{a})),$$

where $A(t_1, \ldots, t_p)$ is the $p \times p$ matrix-valued function with i'th row equal to

$$\mathbf{e}_i^T A(t_1, \ldots, t_p) = \nabla f_i(\mathbf{u} + t_i(\mathbf{v} - \mathbf{u}))$$

for some choice of $t_i \in (0, 1)$. Consequently, $\mathbf{v} = \mathbf{g}(\mathbf{b})$ tends to $\mathbf{u} = \mathbf{g}(\mathbf{a})$ as $\mathbf{b}$ tends to $\mathbf{a}$ and hence $A(t_1, \ldots, t_p)$ tends to $J_\mathbf{f}(\mathbf{u})$ as $\mathbf{b}$ tends to $\mathbf{a}$. Thus, as $J_\mathbf{f}(\mathbf{u})$ is invertible, $A(t_1, \ldots, t_p)$ will be invertible if $\mathbf{v}$ is close to $\mathbf{u}$. Thus, if $\mathbf{b} = \mathbf{a} + \varepsilon \mathbf{c}$ for some unit vector $\mathbf{c} \in \mathbb{R}^p$, then

$$J_\mathbf{f}(\mathbf{a})^{-1}\mathbf{c} = \lim_{\varepsilon \to 0} A(t_1, \ldots, t_p)^{-1}\mathbf{c} = \lim_{\varepsilon \to 0} \frac{\mathbf{g}(\mathbf{a} + \varepsilon\mathbf{c}) - \mathbf{g}(\mathbf{a})}{\varepsilon}.$$

Therefore $\mathbf{g} \in \mathcal{C}^1(Q)$, as claimed. $\square$

Corollary 23.2 (The open mapping theorem). *Let Q be an open nonempty subset of $\mathbb{R}^p$ and suppose that $\mathbf{f} \in \mathcal{C}^1(Q)$ maps Q into $\mathbb{R}^p$ and that $J_\mathbf{f}(\mathbf{x})$ is invertible for every point $\mathbf{x} \in Q$. Then $\mathbf{f}(Q)$ is an open subset of $\mathbb{R}^p$.*

Proof. Let $\mathbf{x}_0 \in Q$ and $\mathbf{y}_0 = \mathbf{f}(\mathbf{x}_0)$. Then there exists a pair of numbers $\alpha > 0$ and $\beta > 0$ such that $B_\alpha(\mathbf{x}_0) \subseteq Q$ and for each vector $\mathbf{y} \in B_\beta(\mathbf{y}_0)$ there exists exactly one vector $\mathbf{x} = \mathbf{g}(\mathbf{y})$ in $B_\alpha(\mathbf{x}_0)$ such that $\mathbf{f}(\mathbf{x}) = \mathbf{y}$. Thus,
$$B_\beta(\mathbf{y}_0) = \mathbf{f}(\mathbf{g}(B_\beta(\mathbf{y}_0))) \subseteq \mathbf{f}(B_\alpha(\mathbf{x}_0)) \subseteq \mathbf{f}(Q). \qquad \square$$

Exercise 23.1. Let
$$\mathbf{f}(\mathbf{x}) = \begin{bmatrix} x_1^2 - x_2 \\ x_2 + x_3 \\ x_3^2 - 2x_3 + 1 \end{bmatrix}, \quad \mathbf{x}_0 = \begin{bmatrix} 1 \\ -1 \\ -1 \end{bmatrix}, \quad \text{and} \quad \mathbf{y}_0 = \mathbf{f}(\mathbf{x}_0).$$

(a) Calculate $J_\mathbf{f}(\mathbf{x})$, $A = J_\mathbf{f}(\mathbf{x}_0)$, $J_\mathbf{f}(\mathbf{x})^{-1}$, and A^{-1}.

(b) Show that $\|A^{-1}\|^2 < 5/3$.

(c) Show that the equation $\mathbf{f}(\mathbf{x}) = \mathbf{y}_0$ has exactly two real solutions, but only one in the ball $B_2(\mathbf{x}_0)$.

Exercise 23.2. Let $\mathbf{u}_0 \in \mathbb{R}^2$, $\mathbf{f} \in \mathcal{C}^2(B_r(\mathbf{u}_0))$ and suppose that $\begin{bmatrix} (\nabla f_1)(\mathbf{u}) \\ (\nabla f_2)(\mathbf{v}) \end{bmatrix}$ is invertible for every pair of vectors $\mathbf{u}, \mathbf{v} \in B_r(\mathbf{u}_0)$. Show that if $\mathbf{a}, \mathbf{b} \in B_r(\mathbf{u}_0)$, then $\mathbf{f}(\mathbf{a}) = \mathbf{f}(\mathbf{b}) \iff \mathbf{a} = \mathbf{b}$.

Exercise 23.3. Show that the condition in Exercise 23.2 cannot be weakened to $\begin{bmatrix} (\nabla f_1)(\mathbf{u}) \\ (\nabla f_2)(\mathbf{u}) \end{bmatrix}$ is invertible for every vector $\mathbf{u} \in B_r(\mathbf{u}_0)$. [HINT: Consider the function $\mathbf{f}(\mathbf{x})$ with components $f_1(\mathbf{x}) = x_1 \cos x_2$ and $f_2(\mathbf{x}) = x_1 \sin x_2$ in a ball of radius 2π centered at the point $(3\pi, 2\pi)$.]

Exercise 23.4. Calculate the Jacobian matrix $J_\mathbf{f}(\mathbf{x})$ of the function $\mathbf{f}(\mathbf{x})$ with components $f_i(x_1, x_2, x_3) = x_i/(1 + x_1 + x_2 + x_3)$ for $i = 1, 2, 3$ that are defined at all points $\mathbf{x} \in \mathbb{R}^3$ with $x_1 + x_2 + x_3 \neq -1$.

Exercise 23.5. Show that the vector-valued function that is defined in Exercise 23.4 defines a one-to-one map from its domain of definition in $\mathbb{R}^3$ and find the inverse mapping.

23.2. The implicit function theorem

To warm up, consider first the problem of describing the set of solutions $\mathbf{u} \in \mathbb{R}^n$ to the equation $A\mathbf{u} = \mathbf{b}$, when $A \in \mathbb{R}^{p \times n}$, $n > p$, $\mathbf{b} \in \mathbb{R}^p$, and rank $A = p$. The rank condition implies that there exists an $n \times n$ permutation matrix P such that the last p columns of the matrix AP are linearly independent. Thus, upon writing

$$AP = \begin{bmatrix} A_{11} & A_{12} \end{bmatrix} \quad \text{and} \quad \begin{bmatrix} \mathbf{x} \\ \mathbf{y} \end{bmatrix} = P^T \mathbf{u}$$

23.2. The implicit function theorem

with $A_{12} \in \mathbb{R}^{p \times p}$ invertible, $\mathbf{x} \in \mathbb{R}^q$, $\mathbf{y} \in \mathbb{R}^p$, and $n = p + q$, the original equation can be rewritten as

$$\begin{aligned}\mathbf{0} &= \mathbf{b} - A\mathbf{u} = \mathbf{b} - APP^T\mathbf{u}\\ &= \mathbf{b} - \begin{bmatrix} A_{11} & A_{12} \end{bmatrix} \begin{bmatrix} \mathbf{x} \\ \mathbf{y} \end{bmatrix} = \mathbf{b} - A_{11}\mathbf{x} - A_{12}\mathbf{y}\,.\end{aligned}$$

Therefore,

(23.5) $$A\mathbf{u} - \mathbf{b} = \mathbf{0} \iff \mathbf{y} = A_{12}^{-1}(\mathbf{b} - A_{11}\mathbf{x})\,;$$

i.e., the constraint $\mathbf{g}(\mathbf{u}) = \mathbf{0}$ for the function $\mathbf{g}(\mathbf{u}) = A\mathbf{u} - \mathbf{b}$ implicitly prescribes some of the entries in $\mathbf{u}$ in terms of the others. The calculation rests on the fact that rank $J_{\mathbf{g}}(\mathbf{u}) = $ rank $A = p$. The implicit function theorem yields similar conclusions for a more general class of functions $\mathbf{g}(\mathbf{u})$.

Theorem 23.3. *Let Q be a nonempty open subset of $\mathbb{R}^n$ and let*

$$\mathbf{g}(\mathbf{u}) = \begin{bmatrix} g_1(u_1, \ldots, u_n) \\ \vdots \\ g_p(u_1, \ldots, u_n) \end{bmatrix} \quad \textit{belong to } \mathcal{C}^1(Q)$$

and suppose that $q = n - p \geq 1$ and that there exists a point $\mathbf{u}_0 \in Q$ such that

$$\mathbf{g}(\mathbf{u}_0) = \mathbf{0} \quad \textit{and} \quad \textit{rank}\, J_{\mathbf{g}}(\mathbf{u}_0) = p\,.$$

Suppose further (to ease the exposition) that the last p columns of $J_{\mathbf{g}}(\mathbf{u}_0)$ are linearly independent, and let

$$\mathbf{x} = \begin{bmatrix} u_1 \\ \vdots \\ u_q \end{bmatrix}, \quad \mathbf{x}_0 = \mathbf{a}\,, \quad A_{11}(\mathbf{u}) = \begin{bmatrix} \frac{\partial g_1}{\partial x_1}(\mathbf{u}) & \cdots & \frac{\partial g_1}{\partial x_q}(\mathbf{u}) \\ \vdots & & \vdots \\ \frac{\partial g_p}{\partial x_1}(\mathbf{u}) & \cdots & \frac{\partial g_p}{\partial x_q}(\mathbf{u}) \end{bmatrix},$$

$$\mathbf{y} = \begin{bmatrix} u_{q+1} \\ \vdots \\ u_{q+p} \end{bmatrix}, \quad \mathbf{y}_0 = \mathbf{b}\,, \quad \textit{and} \quad A_{12}(\mathbf{u}) = \begin{bmatrix} \frac{\partial g_1}{\partial y_1}(\mathbf{u}) & \cdots & \frac{\partial g_1}{\partial y_p}(\mathbf{u}) \\ \vdots & & \vdots \\ \frac{\partial g_p}{\partial y_1}(\mathbf{u}) & \cdots & \frac{\partial g_p}{\partial y_p}(\mathbf{u}) \end{bmatrix}.$$

Then there exists a pair of positive numbers α and β such that for every $\mathbf{x}$ in the ball

$$B_\beta(\mathbf{x}_0) = \{\mathbf{x} \in \mathbb{R}^q : \|\mathbf{x} - \mathbf{x}_0\| < \beta\}$$

there exists exactly one vector $\mathbf{y} = \varphi(\mathbf{x})$ in the ball

$$B_\alpha(\mathbf{y}_0) = \{\mathbf{y} \in \mathbb{R}^p : \|\mathbf{y} - \mathbf{y}_0\| < \alpha\}$$

such that $\mathbf{g}(\mathbf{x}, \varphi(\mathbf{x})) = \mathbf{0}$. Moreover, $\varphi \in \mathcal{C}^1(B_\beta(\mathbf{x}_0))$ and

(23.6) $\begin{bmatrix} \frac{\partial \varphi_1}{\partial x_1}(\mathbf{x}) & \cdots & \frac{\partial \varphi_1}{\partial x_q}(\mathbf{x}) \\ \vdots & & \vdots \\ \frac{\partial \varphi_p}{\partial x_1}(\mathbf{x}) & \cdots & \frac{\partial \varphi_p}{\partial x_q}(\mathbf{x}) \end{bmatrix} = -A_{12}(\mathbf{x}, \varphi(\mathbf{x}))^{-1} A_{11}(\mathbf{x}, \varphi(\mathbf{x})) \,.$

Proof. In order to apply the inverse function theorem, we embed $\mathbf{g}(\mathbf{u})$ in the function

$$\mathbf{f}(\mathbf{u}) = \begin{bmatrix} u_1 \\ \vdots \\ u_q \\ \mathbf{g}(\mathbf{u}) \end{bmatrix}. \quad \text{Then} \quad J_\mathbf{f}(\mathbf{u}) = \begin{bmatrix} I_q & O \\ A_{11}(\mathbf{u}) & A_{12}(\mathbf{u}) \end{bmatrix}$$

and, as $A_{12}(\mathbf{u}_0)$ is invertible by assumption, $J_\mathbf{f}(\mathbf{u}_0)$ is also invertible. Therefore, by the inverse function theorem, there exists a pair of numbers $\alpha > 0$ and $\beta > 0$ such that for every vector $\mathbf{v}$ in the ball $B_\beta(\mathbf{f}(\mathbf{u}_0))$, there exists exactly one vector $\mathbf{u}$ in the ball $B_\alpha(\mathbf{u}_0)$ such that $\mathbf{f}(\mathbf{u}) = \mathbf{v}$. Thus, if

$$\mathbf{v} = \begin{bmatrix} \mathbf{w} \\ \mathbf{0} \end{bmatrix} \quad \text{with} \quad \mathbf{w} \in \mathbb{C}^q,\, \mathbf{0} \in \mathbb{C}^p, \text{ and } \|\mathbf{w} - \mathbf{a}\| < \beta \,,$$

then

$$\|\mathbf{v} - \mathbf{f}(\mathbf{u}_0)\| = \left\| \begin{bmatrix} \mathbf{w} \\ \mathbf{0} \end{bmatrix} - \begin{bmatrix} \mathbf{a} \\ \mathbf{0} \end{bmatrix} \right\| = \|\mathbf{w} - \mathbf{a}\| < \beta \,.$$

Therefore, the vector $\mathbf{u} \in B_\alpha(\mathbf{u}_0)$ that meets the condition

$$\mathbf{f}(\mathbf{u}) = \begin{bmatrix} \mathbf{w} \\ \mathbf{0} \end{bmatrix} \quad \text{must be of the form} \quad \mathbf{u} = \begin{bmatrix} \mathbf{w} \\ \varphi(\mathbf{w}) \end{bmatrix} \,.$$

The inequality

$$\left\| \begin{bmatrix} \mathbf{w} \\ \varphi(\mathbf{w}) \end{bmatrix} - \begin{bmatrix} \mathbf{a} \\ \mathbf{b} \end{bmatrix} \right\| < \alpha$$

ensures that $\|\varphi(\mathbf{w}) - \mathbf{b}\| < \alpha$ and hence that for each $\mathbf{w} \in B_\beta(\mathbf{a})$ there exists at least one function $\varphi \in \mathcal{C}^{(1)}(B_\beta(\mathbf{a}))$ such that $\mathbf{g}(\mathbf{w}, \varphi(\mathbf{w})) = 0$ and $\varphi(\mathbf{w}) \in B_\alpha(\mathbf{b})$.

It remains to check that $\varphi(\mathbf{w})$ is uniquely specified at every point $\mathbf{w} \in B_\beta(\mathbf{a})$. To see this, fix $\mathbf{x} \in B_\beta(\mathbf{a})$, let $\mathbf{h}(\mathbf{y}) = \mathbf{y} - A_{12}(\mathbf{a}, \mathbf{b})^{-1} \mathbf{g}(\mathbf{x}, \mathbf{y})$, and observe that

$$\mathbf{h}(\mathbf{y}) = \mathbf{y} \iff \mathbf{g}(\mathbf{x}, \mathbf{y}) = O \quad \text{and} \quad J_\mathbf{h} = I_p - A_{12}(\mathbf{a}, \mathbf{b})^{-1} A_{12}(\mathbf{x}, \mathbf{y}) \,.$$

Therefore, if α and β are small enough, then $\|J_\mathbf{h}(\mathbf{y})\| \leq K < 1$ for all points $\mathbf{x} \in B_\beta(\mathbf{a})$ and $\mathbf{y} \in B_\alpha(\mathbf{b})$. Thus, if $\mathbf{c}, \mathbf{d} \in B_\alpha(\mathbf{b})$ and $\mathbf{g}(\mathbf{x}, \mathbf{c}) = \mathbf{g}(\mathbf{x}, \mathbf{d}) = \mathbf{0}$,

then
$$\|\mathbf{c}-\mathbf{d}\| = \|\mathbf{h}(\mathbf{c})-\mathbf{h}(\mathbf{d})\| \le \|J_\mathbf{h}(\mathbf{e})\|\,\|\mathbf{c}-\mathbf{d}\| \le K\,\|\mathbf{c}-\mathbf{d}\|$$
for a point $\mathbf{e} = \mathbf{c} + t(\mathbf{d}-\mathbf{c})$, $0 < t < 1$. But this in turn implies that $\mathbf{c} = \mathbf{d}$. $\square$

Exercise 23.6. Suppose that $g(x,y) = x^2 + y^2 - 1$ and
$$g(x_0, y_0) = 0 \quad \text{and} \quad \frac{\partial g}{\partial y}(x_0, y_0) \ne 0\,.$$
Find a pair of positive numbers α and β and a function $\varphi \in \mathcal{C}^1(B_\beta(\mathbf{x}_0))$ such that $\varphi(\mathbf{x}) \in B_\alpha(y_0)$ and $\mathbf{g}(\mathbf{x}, \varphi(\mathbf{x})) = 0$.

Exercise 23.7. Let $g_1(x, y_1, y_2) = x^2(y_1^2 + y_2^2) - 5$ and $g_2(x, y_1, y_2) = (x - y_2)^2 + y_1^2 - 2$. Show that in a neighborhood of the point $x = 1$, $y_1 = -1$, $y_2 = 2$, the curve of intersection of the two surfaces $g_1(x, y_1, y_2) = 0$ and $g_2(x, y_1, y_2) = 0$ can be described by a pair of functions $y_1 = \varphi_1(x)$ and $y_2 = \varphi_2(x)$.

Exercise 23.8. Let
$$S = \left\{ \begin{bmatrix} x_1 \\ x_2 \\ x_3 \\ x_4 \end{bmatrix} \in \mathbb{R}^4 : \begin{array}{c} x_1 - 2x_2 + 2x_3^2 - x_4 = 0 \\ \text{and} \\ x_1 - 2x_2 + 2x_3 - x_4 = 0 \end{array} \right\}, \mathbf{u} = \begin{bmatrix} 1 \\ 1 \\ 1 \\ 1 \end{bmatrix}, \text{ and } \mathbf{v} = \begin{bmatrix} 1 \\ 1 \\ 0 \\ -1 \end{bmatrix}.$$

Use the implicit function theorem to show that it is possible to solve for x_3 and x_4 as functions of x_1 and x_2 for points in S that are close to $\mathbf{u}$ and to points in S that are close to $\mathbf{v}$ and write down formulas for these functions for each of these two cases.

23.3. Continuous dependence of solutions

The implicit function theorem is often a useful tool to check the continuous dependence of the solution of an equation on the coefficients appearing in the equation. Suppose, for example, that $Y \in \mathbb{R}^{2 \times 2}$ is a solution of the matrix equation
$$A^T Y + YA = B$$
for some fixed choice of the matrices $A, B \in \mathbb{R}^{2 \times 2}$. To try to show that if A changes only a little, then Y will also change only a little, consider the matrix-valued function
$$F(A, Y) = A^T Y + YA - B$$
with entries
$$f_{ij}(A, Y) = \mathbf{e}_i^T(A^T Y + YA - B)\mathbf{e}_j\,, \quad i, j = 1, 2,$$

where $\mathbf{e}_1, \mathbf{e}_2$ denote the standard basis vectors in $\mathbb{R}^2$. Then, upon writing

$$Y = \begin{bmatrix} y_{11} & y_{12} \\ y_{21} & y_{22} \end{bmatrix},$$

it is readily checked that

$$\frac{\partial f_{ij}}{\partial y_{st}} = \mathbf{e}_i^T(A^T\mathbf{e}_s\mathbf{e}_t^T + \mathbf{e}_s\mathbf{e}_t^T A)\mathbf{e}_j = a_{si}\mathbf{e}_t^T\mathbf{e}_j + a_{tj}\mathbf{e}_i^T\mathbf{e}_s$$

and hence that

$$\begin{aligned} \frac{\partial f_{ij}}{\partial y_{11}} &= a_{1i}\mathbf{e}_1^T\mathbf{e}_j + a_{1j}\mathbf{e}_i^T\mathbf{e}_1\,, \\ \frac{\partial f_{ij}}{\partial y_{12}} &= a_{1i}\mathbf{e}_2^T\mathbf{e}_j + a_{2j}\mathbf{e}_i^T\mathbf{e}_1\,, \\ \frac{\partial f_{ij}}{\partial y_{21}} &= a_{2i}\mathbf{e}_1^T\mathbf{e}_j + a_{1j}\mathbf{e}_i^T\mathbf{e}_2\,, \\ \frac{\partial f_{ij}}{\partial y_{22}} &= a_{2i}\mathbf{e}_2^T\mathbf{e}_j + a_{2j}\mathbf{e}_i^T\mathbf{e}_2\,. \end{aligned}$$

Consequently,

$$\begin{bmatrix} \frac{\partial f_{11}}{\partial y_{11}} & \frac{\partial f_{11}}{\partial y_{12}} & \frac{\partial f_{11}}{\partial y_{21}} & \frac{\partial f_{11}}{\partial y_{22}} \\ \frac{\partial f_{12}}{\partial y_{11}} & \frac{\partial f_{12}}{\partial y_{12}} & \frac{\partial f_{12}}{\partial y_{21}} & \frac{\partial f_{12}}{\partial y_{22}} \\ \frac{\partial f_{21}}{\partial y_{11}} & \frac{\partial f_{21}}{\partial y_{12}} & \frac{\partial f_{21}}{\partial y_{21}} & \frac{\partial f_{21}}{\partial y_{22}} \\ \frac{\partial f_{22}}{\partial y_{11}} & \frac{\partial f_{22}}{\partial y_{12}} & \frac{\partial f_{22}}{\partial y_{21}} & \frac{\partial f_{22}}{\partial y_{22}} \end{bmatrix} = \begin{bmatrix} 2a_{11} & a_{21} & a_{21} & 0 \\ a_{12} & a_{11}+a_{22} & 0 & a_{21} \\ a_{12} & 0 & a_{11}+a_{22} & a_{21} \\ 0 & a_{12} & a_{12} & 2a_{22} \end{bmatrix}.$$

Now suppose that $F(A_0, Y_0) = 0$ and that the matrix on the right in the last identity is invertible when the terms a_{ij} are taken from A_0. Then the implicit function theorem guarantees the existence of a pair of numbers $\alpha > 0$ and $\beta > 0$ such that for every matrix A in the ball $B_\beta(A_0) = \{A \in \mathbb{R}^{2\times 2} : \|A - A_0\| < \beta\}$, there exists exactly one matrix $Y = \varphi(A)$ in the ball $B_\alpha(Y_0) = \{Y \in \mathbb{R}^{2\times 2} : \|Y - Y_0\| < \alpha\}$ such that $F(A, \varphi(A)) = 0$ and hence that $Y = \varphi(A)$ is a continuous function of A in the ball $B_\beta(A_0)$; in fact $\varphi \in \mathcal{C}^1(B_\beta(A_0))$.

23.4. Roots of polynomials

The next theorem is a special case of a general theorem, which guarantees that a small change in the coefficients of a polynomial causes only a small change in the roots of the polynomial. The general theorem is usually proved by Rouché's theorem from the theory of complex variables; see, e.g., pp. 153–154 of [9]. In this section we shall establish this result for polynomials with distinct roots via the implicit function theorem. The full result will be established later by invoking a different circle of ideas in Chapter 35. Another approach is considered in Exercise 35.7.

Theorem 23.4. *The roots of a polynomial*
$$f(\lambda) = \lambda^n + a_1 \lambda^{n-1} + \cdots + a_n$$
with distinct roots vary continuously with the coefficients $a_1, \ldots, a_n$.

Discussion. To illustrate the basic idea, consider the polynomial
$$f(\lambda) = \lambda^3 + \lambda^2 - 4\lambda + 6.$$
It has three distinct roots: $\lambda_1 = 1+i$, $\lambda_2 = 1-i$, and $\lambda_3 = -3$. Consequently, the equation

(23.7) $$(\mu + i\nu)^3 + a(\mu + i\nu)^2 + b(\mu + i\nu) + c = 0$$

in terms of the 5 real variables μ, ν, a, b, c is satisfied by the following choices:

$$\mu = 1, \quad \nu = 1, \quad a = 1, \, b = -4, \, c = 6,$$
$$\mu = 1, \quad \nu = -1, \, a = 1, \, b = -4, \, c = 6,$$
$$\mu = -3, \, \nu = 0, \quad a = 1, \, b = -4, \, c = 6.$$

To put this into the setting of the implicit function theorem, express
$$\begin{aligned} f(a, b, c, \mu + i\nu) &= (\mu + i\nu)^3 + a(\mu + i\nu)^2 + b(\mu + i\nu) + c \\ &= \mu^3 + 3\mu^2 i\nu + 3\mu(i\nu)^2 + (i\nu)^3 \\ &\quad + a(\mu^2 + 2\mu i\nu + (i\nu)^2) + b(\mu + i\nu) + c \end{aligned}$$
in terms of its real and imaginary parts as
$$f(a, b, c, \mu + i\nu) = f_1(a, b, c, \mu, \nu) + i f_2(a, b, c, \mu, \nu),$$
where
$$f_1(a, b, c, \mu, \nu) = \mu^3 - 3\mu\nu^2 + a\mu^2 - a\nu^2 + b\mu + c$$
and
$$f_2(a, b, c, \mu, \nu) = 3\mu^2\nu - \nu^3 + 2a\mu\nu + b\nu.$$

Thus, the study of the roots of the equation
$$\lambda^3 + a\lambda^2 + b\lambda + c = 0$$
with real coefficients a, b, c has been converted into the study of the solutions of the system
$$f_1(a, b, c, \mu, \nu) = 0,$$
$$f_2(a, b, c, \mu, \nu) = 0.$$

The implicit function theorem guarantees the continuous dependence of the pair (μ, ν) on (a, b, c) in the vicinity of a solution provided that the matrix

$$\triangle(a, b, c, \mu, \nu) = \begin{bmatrix} \frac{\partial f_1}{\partial \mu} & \frac{\partial f_1}{\partial \nu} \\ \frac{\partial f_2}{\partial \mu} & \frac{\partial f_2}{\partial \nu} \end{bmatrix}$$

is invertible. To explore this condition, observe first that

(23.8) $$\frac{\partial f_1}{\partial \mu} = 3\mu^2 - 3\nu^2 + 2a\mu + b = \frac{\partial f_2}{\partial \nu}$$

and

(23.9) $$\frac{\partial f_1}{\partial \nu} = -6\mu\nu - 2a\nu = -\frac{\partial f_2}{\partial \mu}.$$

Therefore,

(23.10) $$\det \begin{bmatrix} \frac{\partial f_1}{\partial \mu} & \frac{\partial f_1}{\partial \nu} \\ \frac{\partial f_2}{\partial \mu} & \frac{\partial f_2}{\partial \nu} \end{bmatrix} = \left|\frac{\partial f_1}{\partial \mu}\right|^2 + \left|\frac{\partial f_2}{\partial \mu}\right|^2 = \left|\frac{\partial f}{\partial \mu}\right|^2.$$

In the case at hand,

$$\frac{\partial f_1}{\partial \mu}(1, -4, 6, 1, 1) = -2 = \frac{\partial f_2}{\partial \nu}(1, -4, 6, 1, 1)$$

and

$$\frac{\partial f_1}{\partial \nu}(1, -4, 6, 1, 1) = -8 = -\frac{\partial f_2}{\partial \mu},$$

and hence

(23.11) $$\det \triangle(1, -4, 6, 1, 1) = \det \begin{bmatrix} -2 & -8 \\ 8 & -2 \end{bmatrix} = 2^2 + 8^2 = 68.$$

Thus, the implicit function theorem guarantees that if the coefficients a, b, c of the polynomial $\lambda^3 + a\lambda^2 + b\lambda + c$ change a little bit from $1, -4, 6$, then the root $1 + i$ will only change a little bit. Similar considerations apply to the other two roots in this example.

The same analysis applies to the simple roots of any polynomial $f(\lambda)$, since the identities $\partial f_1/\partial \mu = \partial f_2/\partial \nu$ and $\partial f_1/\partial \nu = -\partial f_2/\partial \mu$ (the **Cauchy-Riemann equations**) exhibited in (23.8) and (23.9) that connect the real and imaginary parts of f continue to hold and hence, if a polynomial $f(\lambda)$ of degree n has n distinct roots $\mu_1, \ldots, \mu_n$, then $|(\partial f/\partial \mu)(\mu_j)| \neq 0$. □

Exercise 23.9. Show that there exists a pair of numbers $\alpha > 0$ and $\beta > 0$ such that the polynomial $\lambda^3 + a\lambda^2 + b\lambda + c$ with real coefficients has exactly one root $\lambda = \mu + i\nu$ in the ball $\mu^2 + (\nu - 2)^2 < \beta$ if $(a-1)^2 + (b-4)^2 + (c-4)^2 < \alpha$. [HINT: The roots of the polynomial $\lambda^3 + \lambda^2 + 4\lambda + 4$ are $2i, -2i$, and -1.]

23.5. Supplementary notes

This chapter is partially adapted from Chapter 15 of [30]. However, there (in contrast to the present treatment) the proof of the inverse function theorem is based on the implicit function theorem, which was proved first. The monograph [52] by Krantz and Parks is a good source of additional information on the implicit function theorem and its applications.

Chapter 24

Extremal problems

This chapter is devoted primarily to classical extremal problems and extremal problems with constraints, which are resolved by the method of Lagrange multipliers.

24.1. Classical extremal problems

Let $f(\mathbf{x}) = f(x_1, \ldots, x_n)$ be a real-valued function of the variables $x_1, \ldots, x_n$ that is defined in some open set $\Omega \subset \mathbb{R}^n$ and suppose that $f \in \mathcal{C}^1(\Omega)$ and $\mathbf{a} \in \Omega$. Then Theorem 21.4 guarantees that the **directional derivative**

$$(24.1) \qquad (D_{\mathbf{u}}f)(\mathbf{a}) = \lim_{\varepsilon \downarrow 0} \frac{f(\mathbf{a} + \varepsilon \mathbf{u}) - f(\mathbf{a})}{\varepsilon}$$

exists for every choice of $\mathbf{u} \in \mathbb{R}^n$ with $\|\mathbf{u}\| = 1$ and supplies the formula

$$(24.2) \qquad (D_{\mathbf{u}}f)(\mathbf{a}) = (\nabla f)(\mathbf{a})\mathbf{u}.$$

If $\mathbf{a}$ is a local maximum, then

$$\mathbf{f}(\mathbf{a}) \geq f(\mathbf{a} + \varepsilon \mathbf{u})$$

for all unit vectors $\mathbf{u}$ and all sufficiently small positive numbers ε. Thus,

$$\varepsilon > 0 \implies \frac{f(\mathbf{a} + \varepsilon \mathbf{u}) - f(\mathbf{a})}{\varepsilon} \leq 0 \implies (D_{\mathbf{u}}f)(\mathbf{a}) = (\nabla f)(\mathbf{a})\mathbf{u} \leq 0$$

for all unit vectors $\mathbf{u} \in \mathbb{R}^n$. However, since the same inequality holds when $\mathbf{u}$ is replaced by $-\mathbf{u}$, it follows that the last inequality must in fact be an equality: If $\mathbf{a}$ is a local maximum, then

$$(\nabla f)(\mathbf{a})\mathbf{u} = (D_{\mathbf{u}}f)(\mathbf{a}) = 0$$

for all directions $\mathbf{u}$. Therefore, as similar arguments lead to the same conclusion when $\mathbf{a}$ is a local minimum point for $f(\mathbf{x})$, we obtain the following result:

Theorem 24.1. *Let Q be an open subset of $\mathbb{R}^n$ and let $f \in \mathcal{C}^1(Q)$. If a vector $\mathbf{a} \in Q$ is a local maximum or a local minimum for $f(\mathbf{x})$, then*

(24.3) $$(\nabla f)(\mathbf{a}) = O_{1 \times n}.$$

- **Warning:** The condition (24.3) is necessary but not sufficient for $\mathbf{a}$ to be a local extremum (i.e., a local maximum or a local minimum). Thus, for example, the point $(0,0)$ is not a local extremum for the function $f(x_1, x_2) = x_1^2 - x_2^2$, even though $(\nabla f)(0,0) = \begin{bmatrix} 0 & 0 \end{bmatrix}$. **Convex functions that are defined on convex sets** are a happy exception, as we have already seen in Section 21.5.

If $f \in \mathcal{C}^2(Q)$, then additional insight may be obtained from the Hessian $H_f(\mathbf{a})$: If $B_\varepsilon(\mathbf{a}) = \{\mathbf{x} \in \mathbb{R}^q : \|\mathbf{a} - \mathbf{x}\| < \varepsilon\} \subset Q$ for some $\varepsilon > 0$, then, in view of formula (21.9), which is applicable as $B_\varepsilon(\mathbf{a})$ is convex,

(24.4) $$(\nabla f)(\mathbf{a}) = 0_{1 \times n} \implies f(\mathbf{a} + \varepsilon \mathbf{u}) - f(\mathbf{a}) = \frac{1}{2}\varepsilon^2 \langle H_f(\mathbf{c})\mathbf{u}, \mathbf{u} \rangle$$

for some point $\mathbf{c}$ on the open line segment between $\mathbf{a} + \varepsilon \mathbf{u}$ and $\mathbf{a}$. Thus, if $0 < \delta < \varepsilon$ and

(24.5) $H_f(\mathbf{x}) \succeq O$ for every $\mathbf{x} \in B_\delta(\mathbf{a})$, then $\mathbf{a}$ is a local minimum for f,

whereas if

(24.6) $H_f(\mathbf{x}) \preceq O$ for every $\mathbf{x} \in B_\delta(\mathbf{a})$, then $\mathbf{a}$ is a local maximum for f.

This leads easily to the following conclusions (consistent with Theorem 21.7):

Theorem 24.2. *If $Q \subseteq \mathbb{R}^n$ is an open set that contains the point $\mathbf{a}$, $f(\mathbf{x}) = f(x_1, \ldots, x_n)$ belongs to $\mathcal{C}^2(Q)$, and $(\nabla f)(\mathbf{a}) = 0_{1 \times n}$, then*

(24.7) $\quad\quad H_f(\mathbf{a}) \succ O \implies \mathbf{a}$ *is a local minimum for* $f(\mathbf{x})$,

(24.8) $\quad\quad H_f(\mathbf{a}) \prec O \implies \mathbf{a}$ *is a local maximum for* $f(\mathbf{x})$.

Proof. Assertion (24.7) follows from (24.5) and the observation that if $H_\mathbf{f}(\mathbf{a}) \succ O$, then there exists a $\delta > 0$ such that $H_f(\mathbf{x}) \succeq O$ for every $\mathbf{x} \in B_\delta(\mathbf{a})$. The verification that (24.6) implies (24.8) is similar. $\square$

Exercise 24.1. Let $A, B \in \mathbb{R}^{n \times n}$ and suppose that $A \succ O$, $B = B^T$, and $\|A - B\| < \lambda_{\min}$, where $\lambda_{\min}$ denotes the smallest eigenvalue of A. Show that $B \succ O$. [HINT: $\langle B\mathbf{u}, \mathbf{u} \rangle = \langle A\mathbf{u}, \mathbf{u} \rangle + \langle (B - A)\mathbf{u}, \mathbf{u} \rangle$.]

Exercise 24.2. Let $A, B \in \mathbb{R}^{n \times n}$ and suppose that $A \succ O$ and $B = B^T$. Show that $A + \varepsilon B \succ O$ if $|\varepsilon|$ is sufficiently small.

24.1. Classical extremal problems

Theorem 24.2 implies that the behavior of a smooth function $f(\mathbf{x})$ in the vicinity of a point $\mathbf{a}$ at which $(\nabla f)(\mathbf{a}) = 0_{1 \times n}$ depends critically on the eigenvalues of the real symmetric matrix $H_f(\mathbf{a})$.

Example 24.1. Let $f(u,v) = \alpha(u-1)^2 + \beta(v-2)^3$ with nonzero coefficients $\alpha \in \mathbb{R}$ and $\beta \in \mathbb{R}$. Then

$$\frac{\partial f}{\partial u}(u,v) = 2\alpha(u-1) \quad \text{and} \quad \frac{\partial f}{\partial v}(u,v) = 3\beta(v-2)^2.$$

Hence,

$$(\nabla f)(u,v) = [0 \quad 0] \text{ if } u = 1 \text{ and } v = 2.$$

However, the point $(1,2)$ is not a local maximum point or a local minimum point for the function $f(u,v)$, since

$$f(1+\varepsilon_1, 2+\varepsilon_2) - f(1,2) = \alpha\varepsilon_1^2 + \beta\varepsilon_2^3,$$

which changes sign as ε_2 passes through zero when $\varepsilon_1 = 0$. This is consistent with the fact that the Hessian

$$H_f(u,v) = \begin{bmatrix} \frac{\partial^2 f}{\partial u^2} & \frac{\partial^2 f}{\partial u \partial v} \\ \frac{\partial^2 f}{\partial v \partial u} & \frac{\partial^2 f}{\partial v^2} \end{bmatrix} = \begin{bmatrix} 2\alpha & 0 \\ 0 & 6\beta(v-2) \end{bmatrix},$$

and

$$H_f(1,2) = \begin{bmatrix} 2\alpha & 0 \\ 0 & 0 \end{bmatrix}$$

is neither positive definite nor negative definite. $\diamond$

Exercise 24.3. Show that the Hessian $H_g(1,2)$ of the function

$$g(u,v) = \alpha(u-1)^2 + \beta(v-2)^4$$

is the same as the Hessian $H_f(1,2)$ of the function considered in the preceding example.

• **Warning:** A local minimum or maximum point is not necessarily an absolute minimum or maximum point:

Exercise 24.4. Show that the point $(1,2)$ is an absolute minimum for the function

$$g(u,v) = (u-1)^2 + (v-2)^4,$$

but it is not even a local minimum point for the function

$$f(u,v) = (u-1)^2 + (v-2)^3.$$

Exercise 24.5. Let $f \in \mathcal{C}^2(\mathbb{R}^2)$ and suppose that $(\nabla f)(a,b) = \begin{bmatrix} 0 & 0 \end{bmatrix}$, and let λ_1 and λ_2 denote the eigenvalues of $H_f(a,b)$. Show that the point (a,b) is: (i) a local minimum for f if $\lambda_1 > 0$ and $\lambda_2 > 0$, (ii) a local maximum for f if $\lambda_1 < 0$ and $\lambda_2 < 0$, (iii) neither a local maximum nor a local minimum if $|\lambda_1 \lambda_2| > 0$, but $\lambda_1 \lambda_2 < 0$.

Exercise 24.6. In many textbooks on calculus the conclusions formulated in Exercise 24.5 are given in terms of the second-order partial derivatives $\alpha = (\partial^2 f/\partial x^2)(a,b)$, $\beta = (\partial^2 f/\partial y^2)(a,b)$, and $\gamma = (\partial^2 f/\partial x \partial y)(a,b)$ by the conditions (i) $\alpha > 0$, $\beta > 0$, and $\alpha\beta - \gamma^2 > 0$; (ii) $\alpha < 0$, $\beta < 0$, and $\alpha\beta - \gamma^2 > 0$; (iii) $\alpha\beta - \gamma^2 < 0$, respectively. Show that the two formulations are equivalent.

24.2. Extremal problems with constraints

In this section we shall consider extremal problems with constraints, using the method of **Lagrange multipliers**.

Let $\mathbf{a}$ be a local extremum of the function $f(x_1, \ldots, x_n)$ when the variables $(x_1, \ldots, x_n)$ belong to the surface S that is defined by the constraint

$$g(x_1, \ldots, x_n) = 0.$$

If $\mathbf{x}(t)$, $-1 \leq t \leq 1$, is any smooth curve on this surface with $\mathbf{x}(0) = \mathbf{a}$, then $g(\mathbf{x}(t)) = g(x_1(t), \ldots, x_n(t)) = 0$ for $-1 < t < 1$ and hence, **upon writing gradients as column vectors**,

$$\frac{d}{dt} g(\mathbf{x}(t)) = \langle (\nabla g)(\mathbf{x}(t)), \mathbf{x}'(t) \rangle = 0 \quad \text{for all} \quad t \in (-1, 1).$$

In particular,

(24.9) $$\langle (\nabla g)(\mathbf{a}), \mathbf{x}'(0) \rangle = \langle (\nabla g)(\mathbf{x}(0)), \mathbf{x}'(0) \rangle = 0$$

for all such curves $\mathbf{x}(t)$. At the same time, since $\mathbf{a}$ is a local extremum for $f(\mathbf{x})$ on S, $t = 0$ is also a local extremum for the function $f(\mathbf{x}(t))$. Therefore,

(24.10) $$0 = \frac{d}{dt} f(\mathbf{x}(t))|_{t=0} = \langle (\nabla f)(\mathbf{a}), \mathbf{x}'(0) \rangle.$$

Thus, if $\text{rank}\, (\nabla g)(\mathbf{a}) = 1$ and the span of the set of possible vectors $\mathbf{x}'(0)$ fills out an $(n-1)$-dimensional space, then

(24.11) $$(\nabla f)(\mathbf{a}) = \lambda (\nabla g)(\mathbf{a}), \quad \text{i.e.,} \quad \frac{\partial f}{\partial x_i}(\mathbf{a}) = \lambda \frac{\partial g}{\partial x_i}(\mathbf{a})$$

for some constant $\lambda \in \mathbb{R}$ and $i = 1, \ldots, n$.

24.2. Extremal problems with constraints

Example 24.2. Let $g(\mathbf{x}) = \|\mathbf{x}\|^2 - 1$ and let $\mathbf{a} \in \mathbb{R}^n$ be a local extremum for a smooth function $f(\mathbf{x}) = f(x_1, \ldots, x_n)$ on the unit sphere $S = \{\mathbf{x} \in \mathbb{R}^n : g(\mathbf{x}) = 0\}$. If $\mathbf{u} \in S$ and

$$\varphi(t) = \|\mathbf{a} + t\mathbf{u}\|, \quad \text{then the curve} \quad \mathbf{x}(t) = \frac{\mathbf{a} + t\mathbf{u}}{\varphi(t)}$$

belongs to S for $-1 < t < 1$, $\mathbf{x}(0) = \mathbf{a}$, and

$$\mathbf{x}'(t) = \frac{\varphi(t)\mathbf{u} - \varphi'(t)(\mathbf{a} + t\mathbf{u})}{\varphi(t)^2}.$$

Thus, as $\varphi(0) = 1$ and $\varphi(t)^2 = \langle \mathbf{a} + t\mathbf{u}, \mathbf{a} + t\mathbf{u}\rangle$,

$$2\varphi(t)\varphi'(t) = 2\langle \mathbf{u}, \mathbf{a} + t\mathbf{u}\rangle.$$

Therefore,

$$\varphi'(0) = \langle \mathbf{u}, \mathbf{a}\rangle \quad \text{and} \quad \mathbf{x}'(0) = \mathbf{u} - \langle \mathbf{u}, \mathbf{a}\rangle \mathbf{a} = (I_n - \Pi_{\mathcal{A}})\mathbf{u},$$

where $\Pi_{\mathcal{A}}$ denotes the orthogonal projection of $\mathbb{R}^n$ onto $\mathcal{A} = \{c\mathbf{a} : c \in \mathbb{R}\}$. Since $\mathbf{u}$ is an arbitrary vector in $\mathbb{R}^n$, the condition

$$0 = \langle \mathbf{v}, \mathbf{x}'(0)\rangle = \langle \mathbf{v}, (I_n - \Pi_{\mathcal{A}})\mathbf{u}\rangle = \langle (I_n - \Pi_{\mathcal{A}})\mathbf{v}, \mathbf{u}\rangle$$

implies that $(I_n - \Pi_{\mathcal{A}})\mathbf{v} = \mathbf{0}$, i.e., $\mathbf{v} = \langle \mathbf{v}, \mathbf{a}\rangle \mathbf{a}$ is a constant multiple of the vector $\mathbf{a}$. Therefore, in view of (24.9) and (24.10), $\nabla g(\mathbf{a})$ and $\nabla f(\mathbf{a})$ are both constant multiples of $\mathbf{a}$. But, this in turn implies that (24.11) holds, since $(\nabla g)(\mathbf{a}) = 2\mathbf{a}$ and $\|\mathbf{a}\| = 1$. ◇

Analogously, if $\mathbf{u}_0$ is a local extremum for $f(\mathbf{x})$ subject to p constraints $g_1(\mathbf{x}) = 0, \ldots, g_p(\mathbf{x}) = 0$ and

$$\operatorname{rank} J_{\mathbf{g}}(\mathbf{u}_0) = \operatorname{rank} \begin{bmatrix} (\nabla g_1)(\mathbf{u}_0) \\ \vdots \\ (\nabla g_p)(\mathbf{u}_0) \end{bmatrix} = p,$$

then

(24.12) $\quad (\nabla f)(\mathbf{u}_0) = \lambda_1 (\nabla g_1)(\mathbf{u}_0) + \cdots + \lambda_p (\nabla g_p)(\mathbf{u}_0).$

Our next objective is to justify this conclusion with the help of the implicit function theorem.

Theorem 24.3. Let $f(\mathbf{u}) = f(u_1, \ldots, u_n)$, $g_j(\mathbf{u}) = g_j(u_1, \ldots, u_n)$, $j = 1, \ldots, p$, be real-valued functions in $\mathcal{C}^1(Q)$ for some open set Q in $\mathbb{R}^n$, where $p < n$. Let

$$S = \{(u_1, \ldots, u_n) \in Q : g_j(u_1, \ldots, u_n) = 0 \text{ for } j = 1, \ldots, p\}$$

and assume that there exist a point $\mathbf{u}_0 \in S$ and a number $\delta > 0$ such that:

(1) $B_\delta(\mathbf{u}_0) \subseteq Q$ and either $f(\mathbf{u}) \geq f(\mathbf{u}_0)$ for all $\mathbf{u} \in S \cap B_\delta(\mathbf{u}_0)$ or $f(\mathbf{u}) \leq f(\mathbf{u}_0)$ for all $\mathbf{u} \in S \cap B_\delta(\mathbf{u}_0)$.

$$\text{(2) rank} \begin{bmatrix} (\nabla g_1)(\mathbf{u}) \\ \vdots \\ (\nabla g_p)(\mathbf{u}) \end{bmatrix} = p \quad \text{for all points } \mathbf{u} \text{ in the ball } B_\delta(\mathbf{u}_0).$$

Then there exists a set of p constants $\lambda_1, \ldots, \lambda_p$ such that (24.12) holds.

Proof. To ease the exposition, we shall assume that the last p columns of $J_\mathbf{g}(\mathbf{u}_0)$ are linearly independent. Then the implicit function theorem guarantees the existence of a pair of constants $\alpha > 0$ and $\beta > 0$ such that if

$$\mathbf{u} = \begin{bmatrix} \mathbf{x} \\ \mathbf{y} \end{bmatrix} \quad \text{with} \quad \mathbf{x} \in \mathbb{R}^q, \quad \mathbf{y} \in \mathbb{R}^p, \quad p+q = n, \quad \text{and} \quad \mathbf{u}_0 = \begin{bmatrix} \mathbf{x}_0 \\ \mathbf{y}_0 \end{bmatrix},$$

then for each point $\mathbf{x}$ in the ball $B_\beta(\mathbf{x}_0)$ there exists exactly one point $\mathbf{y} = \varphi(\mathbf{x})$ in the ball $B_\alpha(\mathbf{y}_0)$ such that $g_i(\mathbf{x}, \varphi(\mathbf{x})) = 0$ for $i = 1, \ldots, p$. Moreover, $\varphi \in \mathcal{C}^1(B_\beta(\mathbf{x}_0))$. Thus, if $\mathbf{w} \in \mathbb{R}^q$ and $\|\mathbf{w}\| < \beta$, then

$$\mathbf{x}(t) = \mathbf{x}_0 + t\mathbf{w}, \quad -1 \leq t \leq 1,$$

belongs to the ball $B_\beta(\mathbf{x}_0)$. If $\alpha^2 + \beta^2 \leq \delta^2$, then

$$\mathbf{u}(t) = \begin{bmatrix} \mathbf{x}(t) \\ \varphi(\mathbf{x}(t)) \end{bmatrix} \in B_\delta(\mathbf{u}_0) \quad \text{and} \quad \mathbf{u}'(0) = \begin{bmatrix} \mathbf{x}'(0) \\ \mathbf{y}'(0) \end{bmatrix} = \begin{bmatrix} \mathbf{w} \\ C\mathbf{w} \end{bmatrix},$$

where

$$C = \begin{bmatrix} \frac{\partial \varphi_1}{\partial x_1}(\mathbf{x}_0) & \cdots & \frac{\partial \varphi_1}{\partial x_q}(\mathbf{x}_0) \\ \vdots & & \vdots \\ \frac{\partial \varphi_p}{\partial x_1}(\mathbf{x}_0) & \cdots & \frac{\partial \varphi_p}{\partial x_q}(\mathbf{x}_0) \end{bmatrix}.$$

Thus, the set of $\mathbf{u}'(0)$ that is obtained this way spans a q-dimensional subspace of $\mathbb{R}^n$. Moreover,

$$(24.13) \qquad \frac{d}{dt} f(\mathbf{u}(t))|_{t=0} = (\nabla f)(\mathbf{u}_0)\mathbf{u}'(0) = 0$$

and

$$g_i(\mathbf{u}(t)) = 0 \quad \text{for} \quad -1 < t < 1 \quad \text{and} \quad i = 1, \ldots, p.$$

Therefore,

$$\frac{d}{dt} g_i(\mathbf{u}(t)) = (\nabla g_i)(\mathbf{u}(t))\mathbf{u}'(t) = 0 \quad \text{for} \quad -1 < t < 1 \quad \text{and} \quad i = 1, \ldots, p.$$

In particular,
$$(\nabla g_i)(\mathbf{u}_0)\mathbf{u}'(0) = (\nabla g_i)(\mathbf{u}(0))\mathbf{u}'(0) = 0$$
for $i = 1, \ldots, p$. Consequently, the vector $\mathbf{u}'(0)$ belongs to the null space $\mathcal{N}_A$ of the $p \times n$ matrix
$$A = J_{\mathbf{g}}(\mathbf{u}_0) = \begin{bmatrix} (\nabla g_1)(\mathbf{u}_0) \\ \vdots \\ (\nabla g_p)(\mathbf{u}_0) \end{bmatrix}.$$
Since rank $A = p$, by assumption, $\dim \mathcal{N}_A = q = n - p$. Therefore, the considered set of vectors $\mathbf{u}'(0)$ spans $\mathcal{N}_A$. Thus, if $\mathbf{v} \in \mathbb{R}^n$ and
$$\mathbf{v}^T \mathbf{u}'(0) = 0$$
for all vectors $\mathbf{u}'(0)$ of the indicated form, then $\mathbf{v} \in \mathcal{R}_{A^T}$. Therefore, $(\nabla f)(\mathbf{u}_0)^T \in \mathcal{R}_{A^T}$, which is equivalent to the asserted conclusion. □

24.3. Examples

Example 24.3. Let $A \in \mathbb{R}^{p \times q}$ and let
$$f(\mathbf{x}) = \langle A\mathbf{x}, A\mathbf{x} \rangle \quad \text{and} \quad g(\mathbf{x}) = \langle \mathbf{x}, \mathbf{x} \rangle - 1.$$
The problem is to find the local extrema of $f(\mathbf{x})$ subject to the constraint $g(\mathbf{x}) = 0$.

Discussion. The first order of business is to verify the following formulas for the gradients, **written as column vectors**:

(24.14) $$(\nabla f)(\mathbf{x}) = 2A^T A\mathbf{x},$$
(24.15) $$(\nabla g)(\mathbf{x}) = 2\mathbf{x}.$$

If $\mathbf{a}$ is a local extremum for this problem, then rank $\nabla g(\mathbf{a}) = 1$ and hence, in view of the preceding discussion, there exists a constant $\alpha \in \mathbb{R}$ such that
$$(\nabla f)(\mathbf{a}) = \alpha (\nabla g)(\mathbf{a}).$$
But this in turn implies that
$$2A^T A \mathbf{a} = 2\alpha \mathbf{a},$$
and hence, since $\mathbf{a} \neq \mathbf{0}$, that α is an eigenvalue of $A^T A$. Thus, the local extrema, if any, must be eigenvalues of $A^T A$. Since
$$A^T A \mathbf{u}_j = s_j^2 \mathbf{u}_j \quad \text{for } j = 1, \ldots, q,$$
where $s_1 \geq \cdots \geq s_q \geq 0$ and $\mathbf{u}_1, \ldots, \mathbf{u}_q$ are orthonormal vectors in $\mathbb{R}^q$, it is readily checked that $f(\mathbf{x}) \leq f(\mathbf{u}_1) = s_1^2$ for every $\mathbf{x} \in \mathbb{R}^q$ for which $g(\mathbf{x}) = 0$ and hence that $\mathbf{u}_1$ is an extreme point for this problem.

Suppose further, for the sake of definiteness, that $s_1 > s_2 \geq \cdots \geq s_{q-1} > s_q$. Then, since
$$f(a\mathbf{u}_j + b\mathbf{u}_k) = a^2 s_j^2 + b^2 s_k^2 \quad \text{and} \quad g(a\mathbf{u}_j + b\mathbf{u}_k) = a^2 + b^2 - 1 \quad \text{for} \quad j \neq k,$$
we see that when $b \neq 0$,
$$f(a\mathbf{u}_j + b\mathbf{u}_1) = a^2 s_j^2 + b^2 s_1^2 > a^2 s_j^2 + b^2 s_j^2 = s_j^2 \quad \text{if} \quad j = 2, \ldots, q-1$$
and
$$f(a\mathbf{u}_j + b\mathbf{u}_q) = a^2 s_j^2 + b^2 s_q^2 < a^2 s_j^2 + b^2 s_j^2 = s_j^2 \quad \text{if} \quad j = 2, \ldots, q-1.$$
Thus, $\mathbf{u}_2, \ldots, \mathbf{u}_{q-1}$ are not local maxima or minima of f under the given constraints. $\diamond$

Example 24.4. Let A be a real $p \times q$ matrix of rank $r > 1$ with $A^T A = US^2 U^T$, where $U = \begin{bmatrix} \mathbf{u}_1 & \cdots & \mathbf{u}_q \end{bmatrix}$ is a real $q \times q$ unitary matrix and $S = \text{diag}\{s_1^2, \ldots, s_q^2\}$ with $s_1 \geq \cdots \geq s_q \geq 0$ and let
$$f(\mathbf{x}) = \langle A\mathbf{x}, A\mathbf{x} \rangle, \quad g_1(\mathbf{x}) = \langle \mathbf{x}, \mathbf{x} \rangle - 1, \quad \text{and} \quad g_2(\mathbf{x}) = \langle \mathbf{x}, \mathbf{u}_1 \rangle.$$
The problem is to find
$$\max f(\mathbf{x}) \quad \text{subject to the constraints} \quad g_1(\mathbf{x}) = 0 \quad \text{and} \quad g_2(\mathbf{x}) = 0.$$

Discussion. In view of (24.12), there exists a pair of real constants α and β such that
$$(\nabla f)(\mathbf{a}) = \alpha (\nabla g_1)(\mathbf{a}) + \beta (\nabla g_2)(\mathbf{a})$$
at each local extremum $\mathbf{a}$ of the given problem. Therefore,

(24.16) $$2 A^T A \mathbf{a} = 2\alpha \mathbf{a} + \beta \mathbf{u}_1,$$

and the constraints supply the supplementary information that
$$\langle \mathbf{a}, \mathbf{a} \rangle = 1 \quad \text{and} \quad \langle \mathbf{a}, \mathbf{u}_1 \rangle = 0.$$
Thus, as $\mathbf{u}_1, \ldots, \mathbf{u}_q$ is an orthonormal basis for $\mathbb{R}^q$,
$$\mathbf{a} = \sum_{j=1}^q c_j \mathbf{u}_j, \quad \text{where} \quad c_j = \langle \mathbf{a}, \mathbf{u}_j \rangle \quad \text{for} \quad j = 1, \ldots, q.$$
This last formula exhibits once again the advantage of working with an orthonormal basis: It is easy to calculate the coefficients. In particular, the constraint $g_2(\mathbf{a}) = 0$ forces $c_1 = 0$ and hence that
$$\mathbf{a} = \sum_{j=2}^q c_j \mathbf{u}_j.$$
Substituting this into formula (24.16), we obtain
$$2\alpha \sum_{j=2}^q c_j \mathbf{u}_j + \beta \mathbf{u}_1 = 2 A^T A \sum_{j=2}^q c_j \mathbf{u}_j = 2 \sum_{j=2}^q c_j s_j^2 \mathbf{u}_j.$$

24.3. Examples

Therefore, $\beta = 0$ and $c_j s_j^2 = \alpha c_j$ for $j = 2, \ldots, q$. Moreover, since the constraint
$$g_1(\mathbf{a}) = 0 \implies \sum_{j=2}^{q} c_j^2 = 1,$$
it is easily seen that
$$\langle A\mathbf{a}, A\mathbf{a} \rangle = \sum_{j=2}^{q} c_j^2 s_j^2 \leq \left(\sum_{j=2}^{q} c_j^2 \right) s_2^2 = s_2^2$$
and hence that the maximum value of $\langle A\mathbf{a}, A\mathbf{a} \rangle$ subject to the two given constraints is obtained by choosing $\alpha = s_2^2, c_2 = 1$, and $c_j = 0$ for $j = 3, \ldots, q$. This gives a geometric interpretation to s_2. Analogous interpretations hold for the other singular values of A. ◇

Example 24.5. The problem is to find

$$\max f(\mathbf{x}) \quad \text{subject to the constraints} \quad g_1(\mathbf{x}) = 0 \quad \text{and} \quad g_2(\mathbf{x}) = 0$$

when $f(\mathbf{x})$ and $g_1(\mathbf{x})$ are the same as in Example 24.4, but now $g_2(\mathbf{x}) = \langle \mathbf{x}, 3\mathbf{u}_1 + 4\mathbf{u}_2 \rangle$, $q \geq 3$, and $s_1 > s_2$.

Discussion. In view of (24.12), there exists a pair of real constants α and β such that
$$(\nabla f)(\mathbf{a}) = \alpha(\nabla g_1)(\mathbf{a}) + \beta(\nabla g_2)(\mathbf{a})$$
at each local extremum $\mathbf{a}$ of the given problem. Therefore,

(24.17) $$2A^T A \mathbf{a} = 2\alpha \mathbf{a} + \beta(3\mathbf{u}_1 + 4\mathbf{u}_2),$$

and the constraints supply the supplementary information that
$$\langle \mathbf{a}, \mathbf{a} \rangle = 1 \quad \text{and} \quad \langle \mathbf{a}, 3\mathbf{u}_1 + 4\mathbf{u}_2 \rangle = 0.$$
Thus, as $\mathbf{u}_1, \ldots, \mathbf{u}_q$ is an orthonormal basis for $\mathbb{R}^q$,
$$\mathbf{a} = \sum_{j=1}^{q} c_j \mathbf{u}_j, \quad \text{where} \quad c_j = \langle \mathbf{a}, \mathbf{u}_j \rangle \quad \text{for} \quad j = 1, \ldots, q.$$
Consequently,

(24.18) $$2 \sum_{j=1}^{q} c_j s_j^2 \mathbf{u}_j = 2\alpha \sum_{j=1}^{q} c_j \mathbf{u}_j + \beta(3\mathbf{u}_1 + 4\mathbf{u}_2),$$

which, upon matching coefficients, implies that
$$2c_1 s_1^2 = 2\alpha c_1 + 3\beta, \quad 2c_2 s_2^2 = 2\alpha c_2 + 4\beta \quad \text{and} \quad 2c_j s_j^2 = 2\alpha c_j \text{ for } j \geq 3.$$
Moreover, the constraints $g_1(\mathbf{a}) = 0$ and $g_2(\mathbf{a}) = 0$ yield the supplementary conditions $\sum_{j=1}^{q} c_j^2 = 1$ and $3c_1 + 4c_2 = 0$, respectively. Thus, as $8c_1 s_1^2 = 8\alpha c_1 + 12\beta$ and $6c_2 s_2^2 = 6\alpha c_2 + 12\beta$, it follows that $8c_1 s_1^2 - 6c_2 s_2^2 = 8\alpha c_1 - 6\alpha c_2$

and hence, as $c_2 = -3c_1/4$, that $c_1(16s_1^2 + 9s_2^2) = 25c_1\alpha$. If $c_1 \neq 0$, then $\alpha = (16s_1^2 + 9s_2^2)/25 > s_2^2 \geq s_j^2$ when $j \geq 3$. Therefore, $c_j = 0$ for $j \geq 3$ and consequently $c_1^2 + c_2^2 = 1$. Therefore, either $c_1 = 4/5$ and $c_2 = -3/5$, or $c_1 = -4/5$ and $c_2 = 3/5$. In both cases $f(c_1\mathbf{u}_1 + c_2\mathbf{u}_2) = (16s_1^2 + 9s_2^2)/25$.

On the other hand, if $c_1 = 0$, then also $c_2 = 0$ and $\mathbf{f}(\sum_{j=3}^{q} c_j \mathbf{u}_j) \leq s_3^2 < (16s_1^2 + 9s_2^2)/25$, so this option does not yield a maximum. ◇

Exercise 24.7. In the setting of Example 24.5, find the minimum value of $f(\mathbf{x})$ over all vectors $\mathbf{x} \in \mathbb{R}^4$ that are subject to the same two constraints that are considered there.

Example 24.6. Let $X \in \mathbb{R}^{p \times p}$, $\begin{bmatrix} \mathbf{e}_1 & \cdots & \mathbf{e}_p \end{bmatrix} = I_p$, $f(X) = (\det X)^2$, and $g(X) = \operatorname{trace} X^T X - 1$. The problem is to compute $\max\{f(X) : X \in \mathbb{R}^{p \times p} \text{ and } g(X) = 0\}$. If A is a local extremum and $E_{ij} = \mathbf{e}_i \mathbf{e}_j^T$ for $i, j = 1, \ldots, p$, then, in view of (24.11), there exists a number $\mu \in \mathbb{R}$ such that

$$(24.19) \qquad \lim_{\varepsilon \to 0} \frac{f(A + \varepsilon E_{ij}) - f(A)}{\varepsilon} = \mu \lim_{\varepsilon \to 0} \frac{g(A + \varepsilon E_{ij}) - g(A)}{\varepsilon}$$

for $i, j = 1, \ldots, p$, i.e.,

$$\frac{\partial f}{\partial x_{ij}}(X) = \lambda \frac{\partial g}{\partial x_{ij}}(X) \quad \text{for } i, j = 1, \ldots, p \text{ when } X = A.$$

We shall carry out the implications of this condition in a series of exercises.

Exercise 24.8. Show that (24.19) holds if and only if

$$2(-1)^{i+j} A_{\{i;j\}} \det A = \mu \, 2 \mathbf{e}_i^T A \mathbf{e}_j = \mu \, 2 a_{ij} \quad \text{for } i, j = 1, \ldots, p.$$

Exercise 24.9. Show that the condition in Exercise 24.8 holds if and only if $f(A) I_p = \mu A A^T$ and hence that if this condition is met, then $\mu = p f(A)$, which in turn implies that $A A^T = p^{-1} I_p$. Consequently, $(\det A)^2 = \det A A^T = p^{-p}$.

Exercise 24.10. Confirm that p^{-p} is indeed the maximum value of f under the given constraints. [HINT: It is enough to show that $\det(p^{-1} I_p + X) \leq p^{-p}$ for every symmetric matrix $X \in \mathbb{R}^{p \times p}$ with trace $X = 0$.]

Exercise 24.11. Show that the problem considered in Example 24.6 is equivalent to finding the maximum value of the product $s_1^2 \cdots s_p^2$ for a set of real numbers $s_1, \ldots, s_p$ that are subject to the constraint $s_1^2 + \cdots + s_p^2 = 1$.

Exercise 24.12. Let $B \in \mathbb{R}^{n \times n}$ be positive definite with entries b_{ij}, let

$$Q_m = \{A \in \mathbb{R}^{n \times n} : A \succ O \text{ and } a_{ij} = b_{ij} \text{ for } |i-j| \leq m\}, \quad 0 \leq m < n,$$

and let $f(X) = \ln \det X$. Show that if $\begin{bmatrix} \mathbf{e}_1 & \cdots & \mathbf{e}_n \end{bmatrix} = I_n$, $\widetilde{A} \in Q_m$, and $f(\widetilde{A}) \geq f(A)$ for every matrix $A \in Q_m$, then $\mathbf{e}_i^T (\widetilde{A})^{-1} \mathbf{e}_j = 0$ for $|i-j| > m$.

24.4. Supplementary notes

This chapter is adapted from Chapter 16 in [**30**], which contains additional exercises. Exercise 24.12 is connected with the maximum entropy completion problem that will be considered in Chapter 38.

Chapter 25

Newton's method

In Example 22.3 we used the contractive fixed point theorem to show that if $Q = (\alpha, \beta) \subset \mathbb{R}$, $f \in \mathcal{C}^2(Q)$ maps Q into $\mathbb{R}$, and there exists a point $b \in Q$ such that $f(b) = 0$ and $f'(b) \neq 0$, then there exists a $\delta > 0$ such that:

(1) The set $E_\delta = \{x \in \mathbb{R} : |x - b| \leq \delta\}$ is a subset of Q and $f'(x) \neq 0$ for $x \in E_\delta$.

(2) The **Newton step** $N(x) = x - f(x)/f'(x)$ maps E_δ into itself.

(3) The sequence of points $x_{k+1} = N(x_k)$ tends to b as $k \uparrow \infty$ for every choice of $x_0 \in E_\delta$.

In this chapter we will find estimates for the speed of this convergence, first for scalar-valued functions and then, after developing the Newton method for vector-valued functions, for vector-valued functions too.

25.1. Newton's method for scalar functions

Let $b \in \mathbb{R}$,
$$\overline{B_\delta(b)} = \{y \in \mathbb{R} : |y - b| \leq \delta\}, \quad \text{and} \quad N(x) = x - f(x)/f'(x).$$

Theorem 25.1. *Let Q be a nonempty open subset of $\mathbb{R}$, let $f \in \mathcal{C}^2(Q)$ map Q into $\mathbb{R}$, and suppose that there exists a point $b \in Q$ such that $f(b) = 0$ and $f'(b) \neq 0$. Then there exists a pair of constants $\delta > 0$ and $\gamma_\delta > 0$ with $\delta \gamma_\delta < 1$ such that $\overline{B_\delta(b)} \subset Q$, $f'(y)$ is invertible for every point $y \in \overline{B_\delta(b)}$, and:*

(1) $|N(x) - b| \leq |x - b|^2 \gamma_\delta < |x - b|$ *for every point* $x \in \overline{B_\delta(b)}$.

275

(2) If $x_0 \in \overline{B_\delta(b)}$ and $x_j = N(x_{j-1})$ for $j = 1, 2, \ldots$, then $x_j \in \overline{B_\delta(b)}$ and

(25.1) $\qquad |x_j - b| \leq \gamma_\delta^{-1}\{|x_0 - b|\gamma_\delta\}^{2^j} \quad for\ j = 1, 2, \ldots$.

Proof. Choose $\delta > 0$ (but small enough) so that $\overline{B_\delta(b)} \subset Q$ and $|f'(y)| > 0$ for $y \in \overline{B_\delta(b)}$; and let

$$\alpha_\delta = \max\{|f''(y)| : y \in \overline{B_\delta(b)}\} \quad \text{and} \quad \beta_\delta = \max\{|f'(y)|^{-1} : y \in \overline{B_\delta(b)}\}.$$

Then, since $f(b) = 0$, the Taylor theorem with remainder implies that

$$0 = f(x) + (b-x)f'(x) + \frac{(b-x)^2}{2!}f''(c)$$

for $x \in \overline{B_\delta(b)}$ and some point c between x and b. Thus, as $c \in \overline{B_\delta(b)}$,

$$|N(x) - b| = \frac{|(b-x)^2 f''(c)|}{2|f'(x)|} \leq \frac{(b-x)^2 \alpha_\delta \beta_\delta}{2} \leq |b-x|\delta\gamma_\delta$$

with $\gamma_\delta = \alpha_\delta \beta_\delta / 2$. Since $\gamma_{\delta_1} \leq \gamma_{\delta_2}$ when $\delta_1 \leq \delta_2$, we can, by shrinking δ if need be, assume that $\delta\gamma_\delta < 1$. Consequently, (1) holds; (2) follows easily from (1):

If $|x_k - b| \leq (\gamma_\delta)^{-1}(|x_0 - b|\gamma_\delta)^{2^k}$ for some positive integer k, then, in view of (1),

$$|x_{k+1} - b| = |N(x_k) - b| \leq |x_k - b|^2 \gamma_\delta \leq \left(\frac{(|x_0 - b|\gamma_\delta)^{2^k}}{\gamma_\delta}\right)^2 \gamma_\delta,$$

which coincides with the right-hand side of (25.1) if $j = k+1$. $\square$

Example 25.1. Let $f(x) = x^2 - 3x + 2 = (x-2)(x-1)$. Then clearly $f \in C^2(\mathbb{R})$, $f(2) = f(1) = 0$, $f'(x) = 2x - 3$, $f''(x) = 2$, and

$$N(x) = x - (x^2 - 3x + 2)/(2x - 3) = (x^2 - 2)/(2x - 3).$$

Thus, in terms of the notation introduced in the proof of Theorem 25.1, $\alpha_\delta = 2$ for every $\delta > 0$ and (focusing on the root $b = 2$)

$$|f'(x)^{-1}| = \frac{1}{|2x-3|} = \frac{1}{|2(x-2)+1|} \leq \frac{1}{1-2\delta}$$

if $|x - 2| \leq \delta$ and $\delta < 1/2$. Consequently, $\beta_\delta = 1/(1-2\delta)$ and

$$\frac{\alpha_\delta \beta_\delta}{2}\delta = \frac{\delta}{1-2\delta} < 1 \iff \delta < 1/3.$$

This constraint is met by the initial choice $\mathbf{x}_0 = 15/8 = 2 - 1/8$. Then

$$\mathbf{x}_1 = 97/48 = 2 + \frac{1}{48}, \quad \mathbf{x}_2 = 2 + \frac{1}{2400}, \quad \ldots$$

Exercise 25.1. Show that the **Newton recursion** $x_{n+1} = N(x_n)$, $n = 0, 1, \ldots$, for solving the equation $x^2 - a = 0$ to find the square roots of $a > 0$ is
$$x_{n+1} = \frac{1}{2}\left(x_n + \frac{a}{x_n}\right) \quad \text{if} \quad x_n \neq 0\,,$$
and calculate x_1, x_2, x_3 when $a = 4$ and $x_0 = \pm 1$.

Exercise 25.2. Show that in the setting of Exercise 25.1
$$|x_{n+1} - x_n| \leq \frac{1}{2}|x_n^{-1}|\,|x_n - x_{n-1}|^2\,.$$

Exercise 25.3. Show that the Newton recursion $x_{n+1} = N(x_n)$, $n = 0, 1, \ldots$, for solving the equation $x^3 - a = 0$ to find the cube roots of a is
$$x_{n+1} = \frac{1}{3}\left(2x_n + \frac{a}{x_n^2}\right) \quad \text{if} \quad x_n \neq 0\,,$$
and calculate x_1, x_2, x_3 when $a = 8$ and $x_0 = \pm 1$.

25.2. Newton's method for vector-valued functions

Newton's method for vector-valued functions is an iterative scheme for solving equations of the form $\mathbf{f}(\mathbf{x}) = \mathbf{0}$ for vector-valued functions $\mathbf{f} \in \mathcal{C}^2(Q)$ that map an open subset Q of $\mathbb{R}^p$ into $\mathbb{R}^p$. The underlying idea is that if $\mathbf{f}(\mathbf{b}) = \mathbf{0}$ and the Jacobian matrix

$$J_{\mathbf{f}}(\mathbf{x}) = \begin{bmatrix} \frac{\partial f_1}{\partial x_1}(\mathbf{x}) & \cdots & \frac{\partial f_1}{\partial x_p}(\mathbf{x}) \\ \vdots & & \vdots \\ \frac{\partial f_p}{\partial x_1}(\mathbf{x}) & \cdots & \frac{\partial f_p}{\partial x_p}(\mathbf{x}) \end{bmatrix}$$

is invertible at the point $\mathbf{x} = \mathbf{b}$, then, since
$$\mathbf{0} = \mathbf{f}(\mathbf{b}) \approx \mathbf{f}(\mathbf{x}) + J_{\mathbf{f}}(\mathbf{x})(\mathbf{b} - \mathbf{x})$$
and $J_{\mathbf{f}}(\mathbf{x})$ is invertible if $\mathbf{x}$ is close enough to $\mathbf{b}$,
$$\mathbf{b} \approx \mathbf{x} - J_{\mathbf{f}}(\mathbf{x})^{-1}\mathbf{f}(\mathbf{x})\,.$$
The vector-valued function
(25.2) $$N(\mathbf{x}) = \mathbf{x} - J_{\mathbf{f}}(\mathbf{x})^{-1}\mathbf{f}(\mathbf{x})$$
is called the **Newton step**. Clearly, $N(\mathbf{b}) = \mathbf{b}$, and it turns out that if $\|\mathbf{x} - \mathbf{b}\|$ is small enough, then $\|N(\mathbf{x}) - \mathbf{b}\| < \|\mathbf{x} - \mathbf{b}\|$. Thus, a reasonable strategy is to define a sequence of points $\mathbf{x}_{j+1} = N(\mathbf{x}_j)$, $j = 0, 1, \ldots$, and then to try to show that the sequence of points $\mathbf{x}_{j+1} = N(\mathbf{x}_j)$, $j = 0, 1, \ldots$, converges to $\mathbf{b}$ as $j \uparrow \infty$.

Theorem 25.2. *Let Q be a nonempty open subset of $\mathbb{R}^p$, let $\mathbf{f} \in \mathcal{C}^2(Q)$ map Q into $\mathbb{R}^p$, and suppose that there exists a point $\mathbf{b} \in Q$ such that*

$$\mathbf{f}(\mathbf{b}) = \mathbf{0} \quad \text{and the Jacobian matrix } J_{\mathbf{f}}(\mathbf{x}) \text{ is invertible at } \mathbf{b}.$$

Then there exists a pair of numbers $\delta > 0$ and $\gamma_\delta > 0$ with $\delta\gamma_\delta < 1$ such that $\overline{B_\delta(\mathbf{b})} \subset Q$ and the following implications hold:

(1) $\mathbf{x} \in \overline{B_\delta(\mathbf{b})} \implies N(\mathbf{x}) \in \overline{B_\delta(\mathbf{b})}$ *and*

(25.3) $$\|N(\mathbf{x}) - \mathbf{b}\| \leq \gamma_\delta \|\mathbf{x} - \mathbf{b}\|^2 \leq \delta\gamma_\delta \|\mathbf{x} - \mathbf{b}\|.$$

(2) *If $\mathbf{x}_0 \in \overline{B_\delta(\mathbf{b})}$ and $\mathbf{x}_j = N(\mathbf{x}_{j-1})$ for $j = 1, 2, \ldots$, then $\mathbf{x}_j \in \overline{B_\delta(\mathbf{b})}$ and*

(25.4) $$\|\mathbf{x}_j - \mathbf{b}\| \leq \gamma_\delta^{-1} \{\gamma_\delta \|\mathbf{x}_0 - \mathbf{b}\|\}^{2^j} \quad for \ j = 1, 2, \ldots.$$

Discussion. The proof is similar in spirit to the proof for the scalar case; however, the estimates are a little more complicated.

If $\mathbf{f} \in \mathcal{C}^2(Q)$, $\mathbf{u} \in \mathbb{C}^p$, and $\mathbf{x}(t) = \mathbf{a} + t(\mathbf{b} - \mathbf{a})$ for $0 \leq t \leq 1$, then the function $h(t) = \mathbf{u}^T \mathbf{f}(\mathbf{x}(t))$ can be differentiated twice to obtain a formula that is expressed in terms of the Hessian H_{f_j} of each component f_j of $\mathbf{f}$ and the components u_j of $\mathbf{u}$:

$$h''(t) = \sum_{j=1}^{p} u_j \langle H_{f_j}(\mathbf{x}(t))(\mathbf{b} - \mathbf{a}), (\mathbf{b} - \mathbf{a}) \rangle.$$

Thus, as $h(1) = h(0) + h'(0) + \frac{1}{2}h''(t_0)$ for some point $t_0 \in (0, 1)$,

$$\mathbf{u}^T[\mathbf{f}(\mathbf{b}) - \mathbf{f}(\mathbf{a})] = \mathbf{u}^T J_{\mathbf{f}}(\mathbf{a})(\mathbf{b} - \mathbf{a}) + \frac{1}{2} \sum_{j=1}^{p} u_j \langle H_{f_j}(\mathbf{c})(\mathbf{b} - \mathbf{a}), (\mathbf{b} - \mathbf{a}) \rangle$$

with $\mathbf{c} = \mathbf{a} + t_0(\mathbf{b} - \mathbf{b})$. If $\mathbf{f}(\mathbf{b}) = \mathbf{0}$, this reduces to

$$\langle J_{\mathbf{f}}(\mathbf{a})\{N(\mathbf{a}) - \mathbf{b}\}, \mathbf{u} \rangle = \frac{1}{2} \sum_{j=1}^{p} u_j \langle H_{f_j}(\mathbf{c})(\mathbf{b} - \mathbf{a}), (\mathbf{b} - \mathbf{a}) \rangle = \frac{1}{2} \langle \mathbf{v}, \mathbf{u} \rangle,$$

for every choice of $\mathbf{u} \in \mathbb{R}^p$, where $\mathbf{v} \in \mathbb{R}^p$ is the vector with entries

$$v_j = \langle H_{f_j}(\mathbf{c})(\mathbf{b} - \mathbf{a}), (\mathbf{b} - \mathbf{a}) \rangle.$$

Therefore,

$$|\langle J_{\mathbf{f}}(\mathbf{a})\{N(\mathbf{a}) - \mathbf{b}\}, \mathbf{u} \rangle| \leq \frac{1}{2} \|\mathbf{v}\| \|\mathbf{u}\|,$$

which implies that

$$\|J_{\mathbf{f}}(\mathbf{a})\{N(\mathbf{a}) - \mathbf{b}\}\| \leq \frac{1}{2} \|\mathbf{v}\|$$

and hence that

$$\|N(\mathbf{a}) - \mathbf{b}\| = \|J_{\mathbf{f}}(\mathbf{a})^{-1} J_{\mathbf{f}}(\mathbf{a})\{N(\mathbf{a}) - \mathbf{b}\}\| \leq \frac{1}{2} \|J_{\mathbf{f}}(\mathbf{a})^{-1}\| \|\mathbf{v}\|.$$

Consequently, upon setting

(25.5) $\alpha_\delta = \max\{\|H_{f_j}(\mathbf{x})\| : \mathbf{x} \in \overline{B_\delta(b)} \text{ and } j = 1, \ldots, p\}$,

$$\beta_\delta = \max\{\|J_\mathbf{f}(\mathbf{x})^{-1}\| : \mathbf{x} \in \overline{B_\delta(b)}\}, \quad \text{and} \quad \gamma_\delta = \beta_\delta \frac{1}{2}\sqrt{p}\, \alpha_\delta,$$

it is readily checked that

$$\|\mathbf{v}\| \leq \sqrt{p}\, \alpha_\delta \|\mathbf{a} - \mathbf{b}\|^2$$

and hence that

$$\|N(\mathbf{a}) - \mathbf{b}\| \leq \beta_\delta \frac{1}{2}\sqrt{p}\, \alpha_\delta \|\mathbf{a} - \mathbf{b}\|^2 = \gamma_\delta \|\mathbf{a} - \mathbf{b}\|^2.$$

Moreover, as $\alpha_{\delta_1} \leq \alpha_{\delta_2}$ and $\beta_{\delta_1} \leq \beta_{\delta_2}$ if $\delta_1 \leq \delta_2$, we can, by shrinking δ if need be, ensure that $\delta\gamma_\delta < 1$ and hence that (25.3) holds; (25.4) is an easy consequence of (25.3) (see the tail end of the proof of Theorem 25.1 for the strategy). □

Example 25.2. Let $\mathbf{f}(\mathbf{x}) = \begin{bmatrix} f_1(\mathbf{x}) \\ f_2(\mathbf{x}) \end{bmatrix} = \begin{bmatrix} x_1^2 + x_2^2 - 1 \\ x_1^2 - x_2 \end{bmatrix}$. Then $\mathbf{f} \in \mathcal{C}^2(\mathbb{R}^2)$ and, since

$$\frac{\partial f_1}{\partial x_1} = 2x_1, \quad \frac{\partial f_1}{\partial x_2} = 2x_2, \quad \frac{\partial f_2}{\partial x_1} = 2x_1, \quad \text{and} \quad \frac{\partial f_2}{\partial x_2} = -1,$$

it is readily checked that

$$J_\mathbf{f}(\mathbf{x}) = \begin{bmatrix} 2x_1 & 2x_2 \\ 2x_1 & -1 \end{bmatrix}, \quad H_{f_1}(\mathbf{x}) = \begin{bmatrix} 2 & 0 \\ 0 & 2 \end{bmatrix}, \quad \text{and} \quad H_{f_2}(\mathbf{x}) = \begin{bmatrix} 2 & 0 \\ 0 & 0 \end{bmatrix}.$$

Thus, the upper bound α_δ defined in (25.5) is equal to 2 for every $\delta > 0$. Moreover, $\mathbf{f}(\mathbf{x}) = \mathbf{0}$ if and only if $x_1 = \pm x_2^{1/2}$ and $x_2 = (-1 + \sqrt{5})/2$, and as $J_\mathbf{f}(\mathbf{x})$ is invertible at these two points (which are the points at which the parabola intersects the circle) that are approximately equal to $(\pm 0.786, 0.618)$, there clearly exists a $\delta > 0$ and a $\beta_\delta > 0$ such that if $\mathbf{b}$ is one of these points and $\|\mathbf{x} - \mathbf{b}\| \leq \delta$, then

$$\|J_\mathbf{f}(\mathbf{x})^{-1}\| = \left\| \{-2x_1(1 + 2x_2)\}^{-1} \begin{bmatrix} -1 & -2x_2 \\ -2x_1 & 2x_1 \end{bmatrix} \right\| \leq \beta_\delta.$$

Therefore, the Newton recursion will converge if the initial point $\mathbf{x}_0$ is chosen close enough to $\mathbf{b}$. ◇

Exercise 25.4. Show that if Q is an open subset of $\mathbb{R}^p$ and $\mathbf{f} \in \mathcal{C}^2(Q)$ maps Q into $\mathbb{R}^p$ and $\overline{B_\delta(\mathbf{b})} \subset Q$ for some $\delta > 0$, then there exists a constant $\alpha_\delta > 0$ such that $\|J_\mathbf{f}(\mathbf{x}) - J_\mathbf{f}(\mathbf{y})\| \leq \alpha_\delta \|\mathbf{x} - \mathbf{y}\|$ for $\mathbf{x}, \mathbf{y} \in \overline{B_\delta(\mathbf{b})}$.

Exercise 25.5. Show that if, in the setting of Exercise 25.4, $\mathbf{f}(\mathbf{b}) = \mathbf{0}$ and $\beta_\delta = \max\{\|J_\mathbf{f}(\mathbf{x})^{-1}\| : \mathbf{x} \in \overline{B_\delta(\mathbf{b})}\} < \infty$, then the Newton step $N(\mathbf{x})$ satisfies the equality

$$(25.6) \qquad N(\mathbf{x}) - \mathbf{b} = J_\mathbf{f}(\mathbf{x})^{-1} \int_0^1 \{J_\mathbf{f}(\mathbf{x} + s(\mathbf{b} - \mathbf{x})) - J_\mathbf{f}(\mathbf{x})\} ds (\mathbf{b} - \mathbf{x})$$

and hence $\|N(\mathbf{x}) - \mathbf{b}\| \leq \alpha_\delta \beta_\delta \int_0^1 s\, ds \|\mathbf{b} - \mathbf{x}\|^2$ for every point $\mathbf{x} \in \overline{B_\delta(\mathbf{b})}$. [HINT: If $\mathbf{h}(s) = \mathbf{f}(\mathbf{x} + s(\mathbf{b} - \mathbf{x}))$, then $\mathbf{h}(1) - \mathbf{h}(0) = \int_0^1 \mathbf{h}'(s) ds$.]

25.3. Supplementary notes

This chapter is an expanded and revised version of Section 14.7 in [**30**]. The discussion of Newton's method was originally adapted from [**65**]. A more sophisticated version due to Kantorovich may be found in the book [**67**] by Saaty and Bram.

Chapter 26

Matrices with nonnegative entries

We shall say that a matrix $A \in \mathbb{R}^{p \times q}$ with entries a_{ij} belongs to the class

- $\mathbb{R}_{\geq}^{p \times q}$ if $a_{ij} \geq 0$ for $i = 1, \ldots, p$ and $j = 1, \ldots, q$;
- $\mathbb{R}_{>}^{p \times q}$ if $a_{ij} > 0$ for $i = 1, \ldots, p$ and $j = 1, \ldots, q$.

The class of $\mathbb{R}_{\geq}^{n \times n}$ (resp., $\mathbb{R}_{>}^{n \times n}$) of matrices with nonnegative (resp., positive) entries **is not the same** as the class of $n \times n$ positive semidefinite (resp., positive definite) matrices:

$$\text{if } A = \begin{bmatrix} 1 & -1 \\ -1 & 2 \end{bmatrix}, \text{ then } A \succ O, \text{ but } A \notin \mathbb{R}_{\geq}^{n \times n};$$

$$\text{if } A = \begin{bmatrix} 1 & 2 \\ 2 & 2 \end{bmatrix}, \text{ then } A \in \mathbb{R}_{>}^{n \times n}, \text{ but } A \text{ is not } \succeq O.$$

A matrix $A \in \mathbb{R}_{\geq}^{n \times n}$ is said to be **irreducible** if for every pair of indices $i, j \in \{1, \ldots, n\}$ there exists an integer $k \geq 1$ such that the ij entry of A^k is positive, i.e., in terms of the standard basis $\mathbf{e}_i$, $i = 1, \ldots, n$, of $\mathbb{R}^n$, if

$$\langle A^k \mathbf{e}_j, \mathbf{e}_i \rangle > 0 \quad \text{for some positive integer } k \text{ that may depend upon } i, j.$$

This is less restrictive than assuming that there exists an integer $k \geq 1$ such that $A^k \in \mathbb{R}_{>}^{n \times n}$.

Example 26.1. The matrix $A = \begin{bmatrix} 0 & 1 \\ 1 & 0 \end{bmatrix}$ is irreducible, but $A^k \notin \mathbb{R}_{>}^{2 \times 2}$ for any positive integer k. ◇

The main objective of this chapter is to show that if $A \in \mathbb{R}_{\geq}^{n \times n}$ is irreducible, then $r_\sigma(A)$, the spectral radius of A, is an eigenvalue of A with algebraic multiplicity equal to one and there exists a pair of vectors $\mathbf{u}, \mathbf{v} \in \mathbb{R}_{\geq}^n$ such that $A\mathbf{u} = r_\sigma(A)\mathbf{u}$, $A^T\mathbf{v} = r_\sigma(A)\mathbf{v}$, and $\mathbf{v}^T\mathbf{u} = 1$. These facts are part and parcel of the Perron-Frobenius theorem. To warm up we first consider a special case, wherein these conclusions are obtained easily under more restrictive conditions on A. This case will also serve as a useful review of a number of concepts that were introduced earlier.

26.1. A warm-up theorem

We begin with a general observation that is formulated as a lemma for ease of future reference.

Lemma 26.1. *If $A \in \mathbb{R}_{\geq}^{n \times n}$, $A\mathbf{x} = \mu\mathbf{x}$ for $\mu \in \mathbb{C}$ and some nonzero vector $\mathbf{x} \in \mathbb{C}^n$, and $\mathbf{v}$ is the vector with components $v_i = |x_i|$ for $i = 1, \ldots, n$, then*

$$(26.1) \qquad |\mu|\langle \mathbf{v}, \mathbf{v}\rangle \leq \langle A\mathbf{v}, \mathbf{v}\rangle.$$

Proof. Since

$$|\mu|v_i = |\mu x_i| = \left|\sum_{j=1}^n a_{ij}x_j\right| \leq \sum_{j=1}^n a_{ij}v_j \quad \text{for } i = 1, \ldots, n,$$

it follows that

$$|\mu|\sum_{i=1}^n |v_i|^2 \leq \sum_{i,j=1}^n v_i a_{ij} v_j,$$

which is equivalent to (26.1). □

Theorem 26.2. *If $A \in \mathbb{R}_{\geq}^{n \times n}$ is symmetric with eigenvalues $\lambda_1 \geq \lambda_2 \geq \cdots \geq \lambda_n$, repeated according to their algebraic multiplicity, then:*

(1) $\lambda_1 > 0$.
(2) $\langle A\mathbf{x}, \mathbf{x}\rangle \leq \lambda_1 \langle \mathbf{x}, \mathbf{x}\rangle$ *for every vector* $\mathbf{x} \in \mathbb{R}^n$.
(3) $\lambda_1 I_n - A \succeq O$.
(4) *If $A\mathbf{x} = \lambda_1\mathbf{x}$ for some nonzero vector $\mathbf{x} \in \mathbb{C}^n$ and $\mathbf{v}$ is the vector with entries $v_j = |x_j|$ for $j = 1, \ldots, n$, then $A\mathbf{v} = \lambda_1\mathbf{v}$ and $\mathbf{v} \in \mathbb{R}_{>}^n$.*
(5) *The algebraic multiplicity of λ_1 is equal to 1.*
(6) *If $\mu \in \sigma(A)$ and $\mu \neq \lambda_1$, then $|\mu| < \lambda_1$ (and hence $\lambda_1 = r_\sigma(A)$).*

Proof. Under the given assumptions, $A = UDU^T$, where $U \in \mathbb{R}^{n \times n}$ is unitary, $D = \mathrm{diag}\{\lambda_1, \ldots, \lambda_n\}$ with $\lambda_1 \geq \cdots \geq \lambda_n$. Consequently, (1) holds, since

$$\lambda_1 + \cdots + \lambda_n = \mathrm{trace}\, A = \sum_{i=1}^n a_{ii} > 0.$$

26.1. A warm-up theorem

To verify (2), observe that if $\mathbf{x} \in \mathbb{R}^n$ and $\mathbf{y} = U^T\mathbf{x}$, then

$$\langle A\mathbf{x}, \mathbf{x}\rangle = \langle UDU^T\mathbf{x}, \mathbf{x}\rangle = \langle DU^T\mathbf{x}, U^T\mathbf{x}\rangle = \sum_{j=1}^{n} \lambda_j y_j^2$$

$$\leq \lambda_1 \sum_{j=1}^{n} y_j^2 = \lambda_1 \langle \mathbf{y}, \mathbf{y}\rangle = \lambda_1 \langle U^T\mathbf{x}, U^T\mathbf{x}\rangle = \lambda_1 \langle \mathbf{x}, \mathbf{x}\rangle.$$

Assertion (3) is then immediate from (2), since $A = A^T$.

To verify (4), let $\mathbf{x} \in \mathcal{N}_{(A-\lambda_1 I_n)}$ and let $\mathbf{v}$ be the vector with $v_i = |x_i|$ for $i = 1, \ldots, n$. Then, in view of Lemma 26.1, $\lambda_1 \langle \mathbf{v}, \mathbf{v}\rangle \leq \langle A\mathbf{v}, \mathbf{v}\rangle$, whereas (3) implies that $\lambda_1 \langle \mathbf{v}, \mathbf{v}\rangle \geq \langle A\mathbf{v}, \mathbf{v}\rangle$. Therefore,

$$0 = \langle (\lambda_1 I_n - A)\mathbf{v}, \mathbf{v}\rangle = \|(\lambda_1 I_n - A)^{1/2}\mathbf{v}\|^2,$$

which implies that $(\lambda_1 I_n - A)^{1/2}\mathbf{v} = \mathbf{0}$ and hence that $\lambda_1 \mathbf{v} = A\mathbf{v}$. Consequently, $\lambda_1 v_i = \sum_{j=1}^{n} a_{ij} v_j > 0$ for $i = 1, \ldots, n$. This completes the proof of (4).

To verify (5), suppose that $\lambda_1 \mathbf{x} = A\mathbf{x}$ and $\lambda_1 \mathbf{y} = A\mathbf{y}$ for a pair of nonzero vectors $\mathbf{x}, \mathbf{y} \in \mathbb{R}^n$ with entries $x_1, \ldots, x_n$ and $y_1, \ldots, y_n$, respectively. Then the vector $\mathbf{u} = y_1 \mathbf{x} - x_1 \mathbf{y}$ also belongs to $\mathcal{N}_{(\lambda_1 I_n - A)}$. Therefore, either $\mathbf{u} = \mathbf{0}$ or $u_j \neq 0$ for $j = 1, \ldots, n$. Since $u_1 = 0$, $\mathbf{u} = \mathbf{0}$. Thus, the geometric multiplicity of λ_1 is equal to one. But as $A \in \mathbb{R}^{n\times n}$ and $A = A^T$, the algebraic multiplicity of each eigenvalue of A is equal to its geometric multiplicity.

The first step in the proof of (6) is to observe that if $A\mathbf{x} = \mu \mathbf{x}$ for some nonzero vector $\mathbf{x} \in \mathbb{C}^n$ with components $x_1, \ldots, x_n$ and $\mathbf{v}$ is the vector with $v_i = |x_i|$ for $i = 1, \ldots, n$, then in view of Lemma 26.1 and (3),

$$|\mu| \langle \mathbf{v}, \mathbf{v}\rangle \leq \langle A\mathbf{v}, \mathbf{v}\rangle \leq \lambda_1 \langle \mathbf{v}, \mathbf{v}\rangle.$$

Therefore, $|\mu| \leq \lambda_1$. Thus, to complete the proof of (6), it remains only to show that $-\lambda_1 \notin \sigma(A)$. However, if $-\lambda_1 x_i = \sum_{j=1}^{n} a_{ij} x_j$ and $\mathbf{w}$ is the vector with $w_i = |x_i|$ for $i = 1, \ldots, n$, then, by another application of (3) and Lemma 26.1, $A\mathbf{w} = \lambda_1 \mathbf{w}$. Consequently,

$$\lambda_1(|x_i| - x_i) = \sum_{j=1}^{n} a_{ij}[|x_j| + x_j].$$

Thus, if we say $x_m > 0$ for any index m, then

$$0 = \lambda_1(|x_m| - x_m) = \sum_{j=1}^{n} a_{mj}[|x_j| + x_j] > 0,$$

which is clearly not viable. Consequently, $x_j = -|x_j|$ for $j = 1, \ldots, n$. But then

$$2\lambda_1 |x_i| = \lambda_1(|x_i| - x_i) = \sum_{j=1}^n a_{ij}[|x_j| - |x_j|] = 0 \quad \text{for } i = 1, \ldots, n.$$

Therefore, $-\lambda_1$ is not an eigenvalue of A. $\square$

Exercise 26.1. Show that the matrix
$A = \begin{bmatrix} 0 & 1 \\ 1 & 1 \end{bmatrix}$ is irreducible, but the matrix $B = \begin{bmatrix} 1 & 1 \\ 0 & 1 \end{bmatrix}$ is not irreducible
and, more generally, that every triangular matrix $B \in \mathbb{R}_{\geq}^{n \times n}$ is not irreducible.

Exercise 26.2. Let $A \in \mathbb{R}_{\geq}^{n \times n}$ and let $D \in \mathbb{R}_{>}^{n \times n}$ be a diagonal matrix. Show that A is irreducible $\iff$ AD is irreducible $\iff$ DA is irreducible.

26.2. The Perron-Frobenius theorem

We begin with a preliminary lemma. The **notation**

(26.2) $$\mathbf{a} \geq \mathbf{b} \quad (\text{resp., } \mathbf{a} > \mathbf{b})$$

for a pair of vectors $\mathbf{a}, \mathbf{b} \in \mathbb{R}^n$ means that $\mathbf{a} - \mathbf{b} \in \mathbb{R}_{\geq}^n$ (resp., $\mathbf{a} - \mathbf{b} \in \mathbb{R}_{>}^n$).

Lemma 26.3. If $A \in \mathbb{R}_{\geq}^{n \times n}$, $\mathbf{x} \in \mathbb{R}_{\geq}^n$, and there exists a matrix $P \in \mathbb{R}_{>}^{n \times n}$ such that $AP = PA$, then

(26.3) $$A\mathbf{x} - r_\sigma(A)\mathbf{x} \in \mathbb{R}_{\geq}^n \iff A\mathbf{x} = r_\sigma(A)\mathbf{x}.$$

Moreover, if $A\mathbf{x} - r_\sigma(A)\mathbf{x} \in \mathbb{R}_{\geq}^{n \times n}$ and $\varphi(t)$ is a polynomial such that

(26.4) $$\varphi(A) \in \mathbb{R}_{>}^{n \times n} \quad \text{and} \quad \varphi(r_\sigma(A)) > 0,$$

then

(26.5) $$\varphi(A)\mathbf{x} \in \mathbb{R}_{>}^n \iff \mathbf{x} \in \mathbb{R}_{>}^n.$$

Proof. The equivalence (26.3) clearly holds when $A\mathbf{x} - r_\sigma(A)\mathbf{x} = \mathbf{0}$. Thus, to verify (26.3) under the given assumptions, it suffices to focus on the case when $A\mathbf{x} - r_\sigma(A)\mathbf{x} \neq \mathbf{0}$. But, if $A\mathbf{x} - r_\sigma(A)\mathbf{x} \in \mathbb{R}_{\geq}^n$ and $A\mathbf{x} - r_\sigma(A)\mathbf{x} \neq \mathbf{0}$, then $\mathbf{x} \neq \mathbf{0}$, $A \neq O$, $P\mathbf{x} \in \mathbb{R}_{>}^n$, and

$$AP\mathbf{x} - r_\sigma(A)P\mathbf{x} = P(A\mathbf{x} - r_\sigma(A)\mathbf{x}) \in \mathbb{R}_{>}^n.$$

Therefore,

(26.6) $$\varepsilon = \min_{i=1,\ldots,n} \frac{(AP\mathbf{x} - r_\sigma(A)P\mathbf{x})_i}{(P\mathbf{x})_i} > 0$$

and

$$AP\mathbf{x} \geq [r_\sigma(A) + \varepsilon]P\mathbf{x}.$$

Thus,
$$A^k P\mathbf{x} \geq [r_\sigma(A) + \varepsilon]^k P\mathbf{x} > \mathbf{0}$$
for every positive integer k and
$$\|A^k\| \geq \frac{\|A^k P\mathbf{x}\|}{\|P\mathbf{x}\|} \geq [r_\sigma(A) + \varepsilon]^k.$$
But this implies that
$$r_\sigma(A) = \lim_{k\uparrow\infty} \|A^k\|^{1/k} \geq r_\sigma(A) + \varepsilon$$
and hence that $\varepsilon = 0$, which contradicts (26.6). Consequently, the initial assumption that $A\mathbf{x} - r_\sigma(A)\mathbf{x} \neq \mathbf{0}$ is not viable. Therefore, $A\mathbf{x} = r_\sigma(A)\mathbf{x}$, i.e., the implication $\Longrightarrow$ in (26.3) holds. Thus, as the opposite implication is self-evident, the proof of (26.3) is complete.

Finally, if $A\mathbf{x} - r_\sigma(A)\mathbf{x} \in \mathbb{R}^{n\times n}_{\geq}$ and φ is a polynomial for which (26.4) holds, then $A\mathbf{x} = r_\sigma(A)\mathbf{x}$ by (26.3), since $\varphi(A) \in \mathbb{R}^{n\times n}_{\geq}$ and $\varphi(A)$ commutes with A. Thus, $\varphi(A)\mathbf{x} = \varphi(r_\sigma(A))\mathbf{x}$, and hence, as $\varphi(r_\sigma(A)) > 0$, (26.5) must hold. $\square$

In future applications of Lemma 26.3, P will usually be taken equal to A^k for some positive integer k, or $(I_n + A)^{n-1}$, depending upon the constraints imposed on A.

Exercise 26.3. Show that if $A \in \mathbb{R}^{n\times n}_{\geq}$ and if A is irreducible, then $(I_n + A)^{n-1} \in \mathbb{R}^{n\times n}_{>}$. [HINT: If $\mathbf{e}_i^T (I_n + A)^{n-1} \mathbf{e}_j = 0$, then $\mathbf{e}_i^T A^k \mathbf{e}_j = 0$ for every positive integer k.]

Theorem 26.4 (Perron-Frobenius). *If $A \in \mathbb{R}^{n\times n}_{\geq}$ is irreducible, then $A \neq O_{n\times n}$ and:*

(1) $r_\sigma(A) \in \sigma(A)$.

(2) *There exists a pair of vectors $\mathbf{u}, \mathbf{v} \in \mathbb{R}^n_{>}$ such that*

(26.7) $\qquad A\mathbf{u} = r_\sigma(A)\mathbf{u}, \quad A^T\mathbf{v} = r_\sigma(A)\mathbf{v}, \quad \text{and} \quad \mathbf{v}^T\mathbf{u} = 1$.

(3) *The algebraic multiplicity of $r_\sigma(A)$ as an eigenvalue of A is equal to one.*

(4) *If $\mathbf{x}$ is an eigenvector of A with nonnegative entries, then $A\mathbf{x} = r_\sigma(A)\mathbf{x}$.*

If also $A^k \in \mathbb{R}^{n\times n}_{>}$ for some positive integer k, then the following supplementary assertions are valid:

(5) *If $\mu \in \sigma(A)$ and $\mu \neq r_\sigma(A)$, then $|\mu| < r_\sigma(A)$.*

(6) $\lim_{k\uparrow\infty} r_\sigma(A)^{-k} A^k = \mathbf{u}\mathbf{v}^T$.

Proof. The proof is divided into steps.

1. *If $\mu \in \mathbb{C}$ and $\mathbf{x} \in \mathbb{C}^n$ is a nonzero vector such that $A\mathbf{x} = \mu\mathbf{x}$ and $|\mu| = r_\sigma(A)$, then $x_i \neq 0$ for $i = 1, \ldots, n$ and the vector $\mathbf{w}$ with entries $w_i = |x_i|$ for $i = 1, \ldots, n$ is in the nullspace of $A - r_\sigma(A)I_n$, i.e., $A\mathbf{w} = r_\sigma(A)\mathbf{w}$.*

Under the given assumptions,

$$r_\sigma(A)w_i = |\mu x_i| = \left|\sum_{j=1}^n a_{ij}x_j\right| \leq \sum_{j=1}^n a_{ij}|x_j| = \sum_{j=1}^n a_{ij}w_i,$$

i.e., $A\mathbf{w} - r_\sigma(A)\mathbf{w} \in \mathbb{R}^n_\geq$. Therefore, by Lemma 26.3 with $P = (I_n + A)^{n-1}$, $A\mathbf{w} = r_\sigma(A)\mathbf{w}$ and $\mathbf{w} \in \mathbb{R}^n_>$.

2. *Verification of (1) and (2).*

Step 1, applied first to A and then to A^T ensures that there exists a pair of vectors $\mathbf{u}, \mathbf{v} \in \mathbb{R}^n_>$ such that $A\mathbf{u} = r_\sigma(A)\mathbf{u}$ and $A^T\mathbf{v} = r_\sigma(A)\mathbf{v}$. Moreover, $\mathbf{u}$ and $\mathbf{v}$ can be normalized to achieve the condition $\mathbf{v}^T\mathbf{u} = 1$. Thus, (1) and (2) hold.

3. *The geometric multiplicity of $r_\sigma(A)$ as an eigenvalue of A is equal to one.*

Let $\mathbf{x} = \begin{bmatrix} x_1 \cdots x_n \end{bmatrix}^T$ and $\mathbf{y} = \begin{bmatrix} y_1 \cdots y_n \end{bmatrix}^T$ be any two nonzero vectors in $\mathbb{C}^n$ such that $A\mathbf{x} = r_\sigma(A)\mathbf{x}$ and $A\mathbf{y} = r_\sigma(A)\mathbf{y}$. Then $y_1\mathbf{x} - x_1\mathbf{y}$ is also in the null space of the matrix $A - r_\sigma(A)I_n$. Thus, by step 2, $x_1 \neq 0$, $y_1 \neq 0$, and either $y_1\mathbf{x} - x_1\mathbf{y} = \mathbf{0}$ or $|y_1 x_j - x_1 y_j| > 0$ for $j = 1, \ldots, n$. However, the second alternative is clearly impossible, since the first entry in the vector $y_1\mathbf{x} - x_1\mathbf{y}$ is equal to zero. Thus, $\mathbf{x}$ and $\mathbf{y}$ are linearly dependent.

4. *Verification of (3).*

Let $\lambda_1, \ldots, \lambda_n$ denote the eigenvalues of A repeated in accordance with their algebraic multiplicity, let $\mathbf{u}, \mathbf{v} \in \mathbb{R}^n_>$ satisfy (26.7), and let $\lambda_1 = r_\sigma(A)$,

$$\varphi(\lambda) = \det(\lambda I_n - A), \quad \text{and} \quad \psi(\lambda) = \det(\lambda I_n - [A - \lambda_1 \mathbf{u}\mathbf{v}^T]).$$

Then, since $(\lambda I_n - A)^{-1}\lambda_1\mathbf{u} = (\lambda - \lambda_1)^{-1}\lambda_1\mathbf{u}$ for $\lambda \notin \sigma(A)$,

$$\psi(\lambda) = \det(\lambda I_n - A + \lambda_1 \mathbf{u}\mathbf{v}^T)$$
$$= \det(\lambda I_n - A)\det(I_n + (\lambda I_n - A)^{-1}\lambda_1\mathbf{u}\mathbf{v}^T)$$
$$= \varphi(\lambda)\det(I_n + (\lambda - \lambda_1)^{-1}\lambda_1\mathbf{u}\mathbf{v}^T)$$
$$= \varphi(\lambda)\left(1 + \frac{\lambda_1\mathbf{v}^T\mathbf{u}}{\lambda - \lambda_1}\right) = \lambda\frac{\varphi(\lambda)}{\lambda - \lambda_1} = \lambda\frac{\varphi(\lambda) - \varphi(\lambda_1)}{\lambda - \lambda_1} \quad \text{for } \lambda \notin \sigma(A).$$

26.2. The Perron-Frobenius theorem

Thus, as $\lambda_1 \varphi'(\lambda_1) = \psi(\lambda_1)$, λ_1 will be a simple root of $\varphi(\lambda)$ if and only if $\psi(\lambda_1) \neq 0$. But, if $(A - \lambda_1 \mathbf{u}\mathbf{v}^T)\mathbf{x} = \lambda_1 \mathbf{x}$ for some vector $\mathbf{x} \in \mathbb{C}^n$, then

$$\lambda_1 \mathbf{v}^T \mathbf{x} = \mathbf{v}^T (A - \lambda_1 \mathbf{u}\mathbf{v}^T)\mathbf{x} = \lambda_1 \mathbf{v}^T \mathbf{x} - \lambda_1 \mathbf{v}^T \mathbf{x} = \mathbf{0}.$$

Therefore, $A\mathbf{x} = \lambda_1 \mathbf{x}$ and hence, in view of step 3, $\mathbf{x} = \alpha \mathbf{u}$ for some $\alpha \in \mathbb{C}$. Thus,

$$\alpha = \alpha \mathbf{v}^T \mathbf{u} = \mathbf{v}^T \mathbf{x} = 0.$$

But this means that the null space of $A - \lambda_1 \mathbf{u}\mathbf{v}^T$ is equal to zero and, consequently, $\psi(\lambda_1) \neq 0$, as needed to complete the proof of this step.

5. *Verification of* (4).

If $A\mathbf{x} = \mu\mathbf{x}$ for some nonzero vector $\mathbf{x} \in \mathbb{R}^n_{\geq}$, then

$$\mu \langle \mathbf{x}, \mathbf{v} \rangle = \langle A\mathbf{x}, \mathbf{v} \rangle = \langle \mathbf{x}, A^T \mathbf{v} \rangle = r_\sigma(A) \langle \mathbf{x}, \mathbf{v} \rangle.$$

Therefore, $\mu = r_\sigma(A)$ and hence as this eigenvalue has geometric multiplicity equal to one, $\mathbf{x} = \alpha \mathbf{u}$ for some $\alpha > 0$.

6. *If* $c_1, \ldots, c_n \in \mathbb{C} \setminus \{0\}$ *and* $|c_1 + \cdots + c_n| = |c_1| + \cdots + |c_n|$, *then* $c_j = e^{i\theta}|c_j|$ *for some* $\theta \in [0, 2\pi)$ *and* $j = 1, \ldots, n$.

To minimize the bookkeeping, let $n = 3$. Then, under the given assumptions,

$$|c_1| + |c_2| + |c_3| = |c_1 + c_2 + c_3| \leq |c_1| + |c_2 + c_3| \leq |c_1| + |c_2| + |c_3|.$$

Therefore, equality prevails throughout. In particular, $|c_2 + c_3| = |c_2| + |c_3|$, and hence, upon setting $c_3/c_2 = re^{i\alpha}$ with $0 \leq \alpha < 2\pi$ and $r > 0$, $|1 + re^{i\alpha}| = 1 + r$. But this is only possible if $1 + 2r\cos\alpha + r^2 = 1 + 2r + r^2$, which forces $\alpha = 0$. Thus, $c_3 = rc_2$ for some $r > 0$. Since also $|c_1 + c_2| = |c_1| + |c_2|$, the same reasoning yields the equality $c_2 = \delta c_1$ for some $\delta > 0$. Therefore, $c_j = \gamma_j c_1 = \gamma_j e^{i\theta}|c_1|$ with $\gamma_j > 0$ for $j = 1, 2, 3$. Consequently, $c_j = e^{i\theta}|c_j|$ for $j = 1, 2, 3$ in the case at hand and, in general, for $j = 1, \ldots, n$.

7. *Verification of* (5).

Suppose that $A\mathbf{x} = \mu\mathbf{x}$ for some nonzero vector $\mathbf{x} \in \mathbb{C}^n$ and that $|\mu| = r_\sigma(A)$ and $B = A^k \in \mathbb{R}^{n \times n}_{\geq}$. Then $B\mathbf{x} = \mu^k \mathbf{x}$ and $|\mu^k| = r_\sigma(B)$. Thus, if $\mathbf{w}$ is the vector with entries $w_i = |x_i|$ for $i = 1, \ldots, n$, then

$$r_\sigma(B) w_i = |\mu|^k |x_i| = \left| \sum_{j=1}^n b_{ij} x_j \right| \leq \sum_{j=1}^n b_{ij} |x_j| = \sum_{j=1}^n b_{ij} w_j$$

for $i = 1, \ldots, n$, i.e., $B\mathbf{w} - r_\sigma(B)\mathbf{w} \in \mathbb{R}^n_{\geq}$. Therefore, $B\mathbf{w} - r_\sigma(B)\mathbf{w} = \mathbf{0}$, thanks to Lemma 26.3, and hence $\mathbf{w} \in \mathbb{R}^n_{>}$. Consequently,

$$\sum_{j=1}^n b_{1j}|x_j| = \left|\sum_{j=1}^n b_{1j}x_j\right|,$$

which, in view of step 6, implies that $x_j = e^{i\theta}|x_j| = e^{i\theta}w_j$ for $j = 1, \ldots, n$, i.e., $\mathbf{x} = e^{i\theta}\mathbf{w}$. Consequently, $|\mu| = r_\sigma(A) \implies \mu = r_\sigma(A)$, or, to put it another way, $\mu \neq r_\sigma(A) \implies |\mu| < r_\sigma(A) \implies$ (5) holds.

8. *Verification of* (6).

Let $A = UJU^{-1}$ with $J = \begin{bmatrix} r_\sigma(A) & O_{1 \times (n-1)} \\ O_{(n-1) \times 1} & J_1 \end{bmatrix}$ and let $\mathbf{u}_1$ (resp,. $\mathbf{v}_1$) denote the first column of U (resp., $V = (U^{-1})^T$). Then

$$\lim_{k \uparrow \infty} \frac{A^k}{r_\sigma(A)^k} = U \begin{bmatrix} 1 & O \\ O & O \end{bmatrix} V^T = \mathbf{u}_1 \mathbf{v}_1^T.$$

But, as

$$AU = UJ \implies A\mathbf{u}_1 = r_\sigma(A)\mathbf{u}_1 \implies \mathbf{u}_1 = \alpha \mathbf{u},$$

$$V^T A = JV^T \implies \mathbf{v}_1^T A = r_\sigma(A)\mathbf{v}_1^T \implies \mathbf{v}_1 = \beta \mathbf{v},$$

and

$$V^T U = I_n \implies \mathbf{v}_1^T \mathbf{u}_1 = 1 \implies \alpha\beta \mathbf{v}^T \mathbf{u} = 1 \implies \alpha\beta = 1,$$

it follows that

$$\mathbf{u}_1 \mathbf{v}_1^T = \alpha\beta \mathbf{u} \mathbf{v}^T = \mathbf{u} \mathbf{v}^T,$$

as claimed. □

Exercise 26.4. Show that if, in the setting of Theorem 26.4, $B = A - r_\sigma(A)\mathbf{u}\mathbf{v}^T$, then $r_\sigma(B) < r_\sigma(A)$. [HINT: See the verification of step 8 in the proof of Theorem 26.4.]

Exercise 26.5. Show by explicit calculation that the six assertions of Theorem 26.4 hold for the matrix $A = \begin{bmatrix} 1 & 4 \\ 1 & 1 \end{bmatrix}$.

Exercise 26.6. Show that the matrix $A = \begin{bmatrix} 0 & 1 & 0 \\ 0 & 0 & 1 \\ 1 & 0 & 0 \end{bmatrix}$ is irreducible and then check that the first four assertions of Theorem 26.4 hold, but the last two do not.

26.2. The Perron-Frobenius theorem

Exercise 26.7. Show that if $\mathbf{u}$ and $\mathbf{v}$ are vectors in $\mathbb{R}^3_{\geq}$ that meet the conditions in (26.7) for the matrix A specified in Exercise 26.6, then

$$\lim_{n\uparrow\infty} \frac{1}{n} \sum_{k=1}^n r_\sigma(A)^{-k} A^k = \mathbf{u}\mathbf{v}^T.$$

Exercise 26.8. Let $A \in \mathbb{R}^{n\times n}_{\geq}$ be an irreducible matrix with spectral radius $r_\sigma(A) = 1$ and let $B = A - r_\sigma(A)\mathbf{u}\mathbf{v}^T$, where $\mathbf{u}, \mathbf{v} \in \mathbb{R}^n_{>}$ meet the conditions in (26.7). Show that:

(a) $\sigma(B) \subset \sigma(A) \cup \{0\}$, but $1 \notin \sigma(B)$.
(b) $\lim_{N\to\infty} \frac{1}{N} \sum_{k=1}^N B^k = 0$.
(c) $B^k = A^k - \mathbf{u}\mathbf{v}^T$ for $k = 1, 2, \ldots$.
(d) $\lim_{N\to\infty} \frac{1}{N} \sum_{k=1}^N A^k = \mathbf{u}\mathbf{v}^T$.

[HINT: If $r_\sigma(B) < 1$, then it is readily checked that $B^k \to O$ as $k \to \infty$. However, if $r_\sigma(B) = 1$, then B may have complex eigenvalues of the form $e^{i\theta}$ and a more careful analysis is required that exploits the fact that $\lim_{N\to\infty} \frac{1}{N} \sum_{k=1}^N e^{ik\theta} = 0$ if $e^{i\theta} \neq 1$.]

Exercise 26.9. Show that if $A = C^{(3)}_\mu$, then there does not exist a pair of vectors $\mathbf{u}$ and $\mathbf{v}$ in $\mathbb{R}^3_{\geq}$ that meet the three conditions in (26.7).

Lemma 26.5. If $A \in \mathbb{R}^{n\times n}_{\geq}$ is irreducible, $B \in \mathbb{R}^{n\times n}_{\geq}$, and $A - B \in \mathbb{R}^{n\times n}_{\geq}$, then:

(1) $r_\sigma(A) \geq r_\sigma(B)$.
(2) $r_\sigma(A) = r_\sigma(B) \iff A = B$.

Proof. Let $\beta \in \sigma(B)$ with $|\beta| = r_\sigma(B)$, let $B\mathbf{y} = \beta\mathbf{y}$ for some nonzero vector $\mathbf{y} \in \mathbb{C}^n$, and let $\mathbf{w} \in \mathbb{R}^n$ be the vector with components $w_i = |y_i|$ for $i = 1, \ldots, n$. Then

(26.8)
$$r_\sigma(B)w_i = |\beta||y_i| = \left|\sum_{j=1}^n b_{ij}y_j\right|$$
$$\leq \sum_{j=1}^n b_{ij}|y_j| \leq \sum_{j=1}^n a_{ij}w_j \quad \text{for } i = 1,\ldots,n.$$

By Theorem 26.4, there exists a vector $\mathbf{v} \in \mathbb{R}^n_{>}$ such that $A^T\mathbf{v} = r_\sigma(A)\mathbf{v}$. Consequently

$$r_\sigma(B)\langle \mathbf{w}, \mathbf{v}\rangle \leq \langle A\mathbf{w}, \mathbf{v}\rangle = \langle \mathbf{w}, A^T\mathbf{v}\rangle = r_\sigma(A)\langle \mathbf{w}, \mathbf{v}\rangle.$$

Therefore, since $\langle \mathbf{w}, \mathbf{v}\rangle > 0$, (1) holds.

Suppose next that $r_\sigma(A) = r_\sigma(B)$. Then the inequality (26.8) implies that
$$A\mathbf{w} - r_\sigma(A)\mathbf{w} \geq \mathbf{0}$$
and hence, by Lemma 26.3 with $P = (I_n + A)^{n-1}$, that $A\mathbf{w} - r_\sigma(A)\mathbf{w} = \mathbf{0}$ and $\mathbf{w} > \mathbf{0}$. But this in turn implies that
$$\sum_{j=1}^n (a_{ij} - b_{ij})w_j = 0 \quad \text{for} \quad i = 1, \ldots, n$$
and thus, as $a_{ij} - b_{ij} \geq 0$ and $w_j > 0$, we must have $a_{ij} = b_{ij}$ for every choice of $i, j \in \{1, \ldots, n\}$, i.e., $r_\sigma(A) = r_\sigma(B) \implies A = B$. The other direction is self-evident. □

Exercise 26.10. Show that if $\mathbf{e} = \begin{bmatrix} 1 & 1 & \cdots & 1 \end{bmatrix}^T$ and $A \in \mathbb{R}_{\geq}^{n \times n}$ is irreducible then:

(a) $\mathcal{C} = \{\mathbf{x} \in \mathbb{R}_{\geq}^n : \langle \mathbf{x}, \mathbf{e} \rangle = 1\}$ is a closed convex set.

(b) The function $\mathbf{f}(\mathbf{x}) = (\langle A\mathbf{x}, \mathbf{e} \rangle)^{-1} A\mathbf{x}$ has a fixed point in $\mathcal{C}$.

(c) If $\mathbf{f}(\mathbf{x}) = \mathbf{x}$ for some $\mathbf{x} \in \mathcal{C}$, then $\mathbf{x} \in \mathbb{R}_{>}^n$ and $r_\sigma(A) = \langle A\mathbf{x}, \mathbf{e} \rangle$.

Exercise 26.11. Show that if $A \in \mathbb{R}_{\geq}^{n \times n}$ is irreducible, then
$$\min_i \left\{ \sum_{j=1}^n a_{ij} \right\} \leq r_\sigma(A) \leq \max_i \left\{ \sum_{j=1}^n a_{ij} \right\}.$$

Exercise 26.12. Show that if $A \in \mathbb{R}^{n \times n}$ with entries $a_{ij} \geq 0$ when $i \neq j$, then $e^{tA} \in \mathbb{R}_{\geq}^{n \times n}$ for every $t \geq 0$. [HINT: There exists a $\delta > 0$ such that $A + \delta I_n \in \mathbb{R}_{\geq}^{n \times n}$.]

Exercise 26.13. Show that the implication in Exercise 26.12 is really an equivalence, i.e., if $A \in \mathbb{R}^{n \times n}$ and $e^{tA} \in \mathbb{R}_{\geq}^{n \times n}$ for every $t \geq 0$, then $a_{ij} \geq 0$ when $i \neq j$.

26.3. Supplementary notes

The idea for a warm-up theorem came from the paper by Ninio [63]. The treatment of the Perron-Frobenius theorem here differs considerably from the treatment in [30]. It contains more information and the exposition is simpler; it is partially adapted from an internet source that unfortunately I cannot recover and so cannot reference. Exercises 26.12 and 26.13 are adapted from an article by Glück [40].

Chapter 27

Applications of matrices with nonnegative entries

In this chapter a number of applications of matrices with nonnegative entries are considered.

27.1. Stochastic matrices

A matrix $A \in \mathbb{R}_{\geq}^{p \times q}$ with entries a_{ij}, $i = 1, \ldots, p$, $j = 1, \ldots, q$, is said to be a **stochastic matrix** if

$$\sum_{j=1}^{q} a_{ij} = 1 \quad \text{for } i = 1, \ldots, p. \tag{27.1}$$

Lemma 27.1. *If $A \in \mathbb{R}_{\geq}^{n \times n}$ is a stochastic matrix and $\mathbf{a} = \begin{bmatrix} 1 & 1 & \cdots & 1 \end{bmatrix}^T$, then*

$$A\mathbf{a} = \mathbf{a} \quad \text{and} \quad r_\sigma(A) = 1. \tag{27.2}$$

Proof. The first assertion in (27.2) is immediate from the definition of a stochastic matrix. To verify the second, assume that $\mu \mathbf{x} = A\mathbf{x}$ for some nonzero vector $\mathbf{x} \in \mathbb{C}^n$ and let $\delta = \max\{|x_i| : i = 1, \ldots, n\}$. Then, since

$$|\mu x_i| = \left| \sum_{j=1}^{n} a_{ij} x_j \right| \leq \sum_{j=1}^{n} a_{ij} |x_j| \leq \sum_{j=1}^{n} a_{ij} \delta = \delta \quad \text{for } i = 1, \ldots, n,$$

it follows that $|\mu|\delta \leq \delta$ and hence that $|\mu| \leq 1$ for every $\mu \in \sigma(A)$. Thus, as $1 \in \sigma(A)$, $r_\sigma(A) = 1$. □

Exercise 27.1. Show that $A \in \mathbb{R}^{n \times n}$ is a stochastic matrix if and only if A^k is a stochastic matrix for every positive integer k.

Exercise 27.2. Show that if $A \in \mathbb{R}^{n \times n}$ is an irreducible stochastic matrix with entries a_{ij} for $i, j = 1, \ldots, n$, then there exists a positive vector $\mathbf{u} \in \mathbb{R}^n$ with entries u_i for $i = 1, \ldots, n$ such that $u_j = \sum_{i=1}^{n} u_i a_{ij}$ for $j = 1, \ldots, n$. [HINT: Exploit Theorem 26.4.]

Exercise 27.3. Show that the matrix $A = \begin{bmatrix} 1/2 & 0 & 1/2 \\ 1/4 & 1/2 & 1/4 \\ 1/8 & 3/8 & 1/2 \end{bmatrix}$ is an irreducible stochastic matrix and find a positive vector $\mathbf{u} \in \mathbb{R}^3$ that meets the conditions discussed in Exercise 27.2.

27.2. Behind Google

In a library of n documents, the Google search engine associates a vector in $\mathbb{R}^n_{\geq}$ to each document. The entries of each such vector are based on a weighted average of the number of links of this document to other documents with overlapping sets of keywords; the vectors are normalized so that their entries sum to one. Thus, if $G = \begin{bmatrix} \mathbf{g}_1 & \cdots & \mathbf{g}_n \end{bmatrix}$ is the array of the n vectors corresponding to the n documents, then G^T is a stochastic matrix. Let

$$\mathbf{a}^T = \begin{bmatrix} 1 & \cdots & 1 \end{bmatrix}, \quad A = \frac{1}{n}\mathbf{a}\mathbf{a}^T, \quad \text{and} \quad B = tG + (1-t)A$$

for some fixed choice of $t \in (0,1)$; B is the **Google matrix**.

Theorem 27.2. *If $0 < t < 1$ and the matrix $B = tG + (1-t)A$ has k distinct eigenvalues $\lambda_1, \ldots, \lambda_k$ with $|\lambda_1| \geq \cdots \geq |\lambda_k|$, then*

(27.3) $\qquad \lambda_1 = 1 \quad \text{and} \quad |\lambda_j| \leq t \quad \text{for } j = 2, \ldots, k.$

Moreover, if $B = UJU^{-1}$, $U = \begin{bmatrix} \mathbf{u}_1 & \cdots & \mathbf{u}_n \end{bmatrix}$, J is in Jordan form, and $B\mathbf{u}_1 = \mathbf{u}_1$, then

(27.4) $\qquad \lim_{m \uparrow \infty} \left\| B^m \left(\sum_{j=1}^{n} c_j \mathbf{u}_j \right) - c_1 \mathbf{u}_1 \right\| = 0$

for every choice of $c_1, \ldots, c_n \in \mathbb{R}$.

Proof. Since B^T is a stochastic matrix, $B^T \mathbf{a} = \mathbf{a}$, and $r_\sigma(B^T) = 1$. Thus, as $\sigma(B^T) = \sigma(B)$, $1 \in \sigma(B)$ and $r_\sigma(B) = 1$. Moreover, as $B \in \mathbb{R}^{n \times n}_{>}$, Theorem 26.4 ensures that if $\lambda \in \sigma(B)$ and $\lambda \neq 1$, then $|\lambda| < 1$. Therefore, $\lambda_1 = 1$ and $|\lambda_j| < 1$ for $j = 2, \ldots, k$. Consequently, $J = \begin{bmatrix} 1 & O_{1 \times n-1} \\ O_{n-1 \times 1} & J' \end{bmatrix}$.

27.2. Behind Google

Suppose next, for example, that $B \begin{bmatrix} \mathbf{u}_2 & \mathbf{u}_3 & \mathbf{u}_4 \end{bmatrix} = \begin{bmatrix} \mathbf{u}_2 & \mathbf{u}_3 & \mathbf{u}_4 \end{bmatrix} C^{(3)}_{\lambda_2}$. Then

$$\langle \mathbf{u}_2, \mathbf{a} \rangle = \langle \mathbf{u}_2, B^T \mathbf{a} \rangle = \langle B\mathbf{u}_2, \mathbf{a} \rangle = \lambda_2 \langle \mathbf{u}_2, \mathbf{a} \rangle \implies \langle \mathbf{u}_2, \mathbf{a} \rangle = 0,$$

$$\langle \mathbf{u}_3, \mathbf{a} \rangle = \langle \mathbf{u}_3, B^T \mathbf{a} \rangle = \langle B\mathbf{u}_3, \mathbf{a} \rangle = \lambda_2 \langle \mathbf{u}_3, \mathbf{a} \rangle + \langle \mathbf{u}_2, \mathbf{a} \rangle = \lambda_2 \langle \mathbf{u}_3, \mathbf{a} \rangle$$
$$\implies \langle \mathbf{u}_3, \mathbf{a} \rangle = 0,$$

$$\langle \mathbf{u}_4, \mathbf{a} \rangle = \langle \mathbf{u}_4, B^T \mathbf{a} \rangle = \langle B\mathbf{u}_4, \mathbf{a} \rangle = \lambda_2 \langle \mathbf{u}_4, \mathbf{a} \rangle + \langle \mathbf{u}_3, \mathbf{a} \rangle = \lambda_2 \langle \mathbf{u}_4, \mathbf{a} \rangle$$
$$\implies \langle \mathbf{u}_4, \mathbf{a} \rangle = 0.$$

The same argument is applicable to each of the Jordan chains associated with $\mathbf{u}_2, \ldots, \mathbf{u}_n$. Thus,

$$\langle \mathbf{u}_j, \mathbf{a} \rangle = 0 \quad \text{for } j = 2, \ldots, n$$

and hence $A \begin{bmatrix} \mathbf{u}_2 & \cdots & \mathbf{u}_n \end{bmatrix} = O_{n,n-1}$ and

$$tG \begin{bmatrix} \mathbf{u}_2 & \cdots & \mathbf{u}_n \end{bmatrix} = B \begin{bmatrix} \mathbf{u}_2 & \cdots & \mathbf{u}_n \end{bmatrix} = \begin{bmatrix} \mathbf{u}_1 & \cdots & \mathbf{u}_n \end{bmatrix} J \begin{bmatrix} \mathbf{e}_2 & \cdots & \mathbf{e}_n \end{bmatrix}$$
$$= \begin{bmatrix} \mathbf{u}_2 & \cdots & \mathbf{u}_n \end{bmatrix} J'.$$

Consequently, $\lambda_2, \ldots, \lambda_k \in \sigma(tG)$ and thus, as $r_\sigma(tG) = t\, r_\sigma(G) = t$, $|\lambda_j| \leq t$ for $j = 1, \ldots, k$. This completes the proof of (27.3).

Next, as G^T is a stochastic matrix,

$$nAG^m \mathbf{u}_j = \langle G^m \mathbf{u}_j, \mathbf{a} \rangle \mathbf{a} = \langle \mathbf{u}_j, (G^T)^m \mathbf{a} \rangle \mathbf{a} = \langle \mathbf{u}_j, \mathbf{a} \rangle \mathbf{a} = 0$$

for $j = 2, \ldots, n$. Therefore,

$$B^m \mathbf{u}_j = t^m G^m \mathbf{u}_j \quad \text{for } j = 2, \ldots, n.$$

Thus, if $\mathbf{x} = \sum_{j=1}^n c_j \mathbf{u}_j$, then, as $B\mathbf{u}_1 = \mathbf{u}_1$,

$$\|B^m \mathbf{x} - c_1 \mathbf{u}_1\| = \left\| B^m \left(\sum_{j=1}^n c_j \mathbf{u}_j - c_1 \mathbf{u}_1 \right) \right\| = \left\| B^m \sum_{j=2}^n c_j \mathbf{u}_j \right\|$$
$$= t^m \left\| G^m \sum_{j=2}^n c_j \mathbf{u}_j \right\| \leq t^m \|G^m\| \left\| \sum_{j=2}^n c_j \mathbf{u}_j \right\|,$$

which tends to 0 as $m \uparrow \infty$, since $\lim_{m \uparrow \infty} \|G^m\|^{1/m} = r_\sigma(G)$ and $r_\sigma(G) = r_\sigma(G^T) = 1$, because G^T is a stochastic matrix. $\square$

The ranking of documents is based on the entries in a good approximation to $\mathbf{u}_1$ (normalized so that $\|\mathbf{u}_1\|_1 = 1$) that is obtained by computing $B^m \mathbf{x}$ for large enough m. The numbers involved are reportedly on the order of $n = 25{,}000{,}000{,}000$, $k = 100$ with $t = .85$. Presumably, these computations are tractable because most of the entries in the vectors $\mathbf{g}_j$ are equal to zero.

27.3. Leslie matrices

Matrices of the form

$$A = \begin{bmatrix} f_1 & f_2 & f_3 & \cdots & f_{n-1} & f_n \\ s_1 & 0 & 0 & \cdots & 0 & 0 \\ 0 & s_2 & 0 & \cdots & 0 & 0 \\ \vdots & & & & & \vdots \\ 0 & 0 & 0 & \cdots & s_{n-1} & 0 \end{bmatrix}$$

in which $f_j \geq 0$ for $j = 1, \ldots, n$ and $0 \leq s_j \leq 1$ for $j = 1, \ldots, n-1$ serve to model population growth for many different species; they are called **Leslie matrices**. In order to simplify the exposition we shall restrict our attention to the case in which $f_j > 0$ for $j = 1, \ldots, n$, $0 < s_j \leq 1$ for $j = 1, \ldots, n-1$, and $n \geq 2$. Then $A^n \in \mathbb{R}_>^{n \times n}$ and hence (all six assertions of) Theorem 26.4 are applicable.

To get some feeling for the properties of this class of matrices observe that if $n = 4$ and

$$Z_4 = \begin{bmatrix} 0 & 0 & 0 & 1 \\ 0 & 0 & 1 & 0 \\ 0 & 1 & 0 & 0 \\ 1 & 0 & 0 & 0 \end{bmatrix}, \quad \text{then} \quad Z_4 = Z_4^T, \; Z_4 Z_4^T = I_4,$$

and

$$Z_4 \begin{bmatrix} f_1 & f_2 & f_3 & f_4 \\ s_1 & 0 & 0 & 0 \\ 0 & s_2 & 0 & 0 \\ 0 & 0 & s_3 & 0 \end{bmatrix} Z_4 = \begin{bmatrix} 0 & s_3 & 0 & 0 \\ 0 & 0 & s_2 & 0 \\ 0 & 0 & 0 & s_1 \\ f_4 & f_3 & f_2 & f_1 \end{bmatrix},$$

which has the same structure as the companion matrices that we studied earlier. In particular, if $s_1 = s_2 = s_3 = 1$, then

(27.5) $\det(\lambda I_4 - A) = \det(\lambda I_4 - Z_4 A Z_4) = -f_4 - f_3\lambda - f_2\lambda^2 - f_1\lambda^3 + \lambda^4$.

Theorem 26.4 ensures that $r_\sigma(A)$ is an eigenvalue of A. Since $f_j > 0$ for $j = 1, \ldots, 4$, this polynomial has exactly one positive root. Consequently, this root must be equal to $r_\sigma(A)$.

Analogously, the spectral radius of an $n \times n$ Leslie matrix is equal to the positive root of the polynomial

$\det(\lambda I_n - A) = \lambda^n - f_1\lambda^{n-1} - s_1 f_2 \lambda^{n-2} - s_2 s_1 f_3 \lambda^{n-3} - \cdots - s_{n-1} \cdots s_1 f_n$.

This polynomial is of the form

$$g(x) = x^n - \sum_{j=0}^{n-1} a_j x^j \quad \text{with} \quad a_j > 0 \text{ for } j = 0, \ldots, n-1.$$

Thus, $g(x) = 0$ if and only if $x^n = \sum_{j=0}^{n-1} a_j x^j$, which is very special in that both the left and right sides of the last equality are polynomials with positive coefficients (and hence both sides are convex functions of x on $[0, \infty)$). Moreover, as $g(0) = -a_0 < 0$ and $g(b) > 0$ if $b > \max\{\sum_{j=0}^{n-1} a_j, 1\}$, there must exist a point $\alpha \in (0, b)$ at which $g(\alpha) = 0$, i.e., $\alpha^n = \sum_{j=0}^{n-1} a_j \alpha^j$. Therefore,

$$\alpha g'(\alpha) = n\alpha^n - \sum_{j=1}^{n-1} j a_j \alpha^j = na_0 + \sum_{j=1}^{n-1} (n-j) a_j \alpha^j > 0.$$

Consequently, as $g(\alpha) = 0$ and $g'(\alpha) \neq 0$, this root may be computed by Newton's method.

Example 27.1. Let

$$A = \begin{bmatrix} 1 & 1 & 3 \\ 1 & 0 & 0 \\ 0 & 1 & 0 \end{bmatrix}. \quad \text{Then } f(x) = \det(xI_3 - A) = x^3 - x^2 - x - 3,$$

and, since $f(0) = -3$, $f(1) = -4$, $f(2) = -1$, and $f(3) = 12$, there exists a point μ with $2 < \mu < 3$ at which $f(\mu) = 0$. Moreover, since $f'(x) = 3x^2 - 2x - 1$ and $\mu^3 = \mu^2 + \mu + 3$,

$$\mu f'(\mu) = 3\mu^3 - 2\mu^2 - \mu = 3(\mu^2 + \mu + 3) - 2\mu^2 - \mu > 0,$$

as it should be in view of the preceding analysis. Thus, as $f(\mu) = 0$ and $f'(\mu) \neq 0$, the sequence of points $x_k = N(x_{k-1})$, $k = 1, 2, \ldots$, in the Newton recursion will converge to μ if x_0 is chosen close enough to μ. Since $2 < \mu < 3$, it is reasonable to choose $x_0 = 2$ and to hope that this is close enough to μ to ensure that the sequence converges. Since

$$N(x) = x - \frac{f(x)}{f'(x)} = \frac{2x^3 - x^2 + 3}{3x^2 - 2x - 1},$$

$$N(2) = 15/7 \approx 2.14, \quad \text{and} \quad N(15/7) = \frac{6204}{2912} \approx 2.13,$$

there are grounds for optimism. $\diamond$

27.4. Minimum matrices

The entries of the **minimum matrix** $A_{\min} \in \mathbb{R}^{n \times n}$ are $a_{ij} = \min\{i, j\}$. Thus, for example, if $n = 4$, then

$$A_{\min} = \begin{bmatrix} 1 & 1 & 1 & 1 \\ 1 & 2 & 2 & 2 \\ 1 & 2 & 3 & 3 \\ 1 & 2 & 3 & 4 \end{bmatrix},$$

which can be expressed as

$$A_{\min} = \begin{bmatrix} 1 & 1 & 1 & 1 \\ 1 & 1 & 1 & 1 \\ 1 & 1 & 1 & 1 \\ 1 & 1 & 1 & 1 \end{bmatrix} + \begin{bmatrix} 0 & 0 & 0 & 0 \\ 0 & 1 & 1 & 1 \\ 0 & 1 & 1 & 1 \\ 0 & 1 & 1 & 1 \end{bmatrix} + \begin{bmatrix} 0 & 0 & 0 & 0 \\ 0 & 0 & 0 & 0 \\ 0 & 0 & 1 & 1 \\ 0 & 0 & 1 & 1 \end{bmatrix} + \begin{bmatrix} 0 & 0 & 0 & 0 \\ 0 & 0 & 0 & 0 \\ 0 & 0 & 0 & 0 \\ 0 & 0 & 0 & 1 \end{bmatrix}$$

$$= \mathbf{a}_1 \mathbf{a}_1^T + \mathbf{a}_2 \mathbf{a}_2^T + \mathbf{a}_3 \mathbf{a}_3^T + \mathbf{a}_4 \mathbf{a}_4^T = L L^T,$$

where

$$L = \begin{bmatrix} \mathbf{a}_1 & \mathbf{a}_2 & \mathbf{a}_3 & \mathbf{a}_4 \end{bmatrix} = \begin{bmatrix} 1 & 0 & 0 & 0 \\ 1 & 1 & 0 & 0 \\ 1 & 1 & 1 & 0 \\ 1 & 1 & 1 & 1 \end{bmatrix} \quad \text{and} \quad L^{-1} = \begin{bmatrix} 1 & 0 & 0 & 0 \\ -1 & 1 & 0 & 0 \\ 0 & -1 & 1 & 0 \\ 0 & 0 & -1 & 1 \end{bmatrix}$$

(see (3.11) for L^{-1}). Therefore, $A_{\min}$ is positive definite and

$$(A_{\min})^{-1} = (L^T)^{-1} L^{-1} = \begin{bmatrix} 2 & -1 & 0 & 0 \\ -1 & 2 & -1 & 0 \\ 0 & -1 & 2 & -1 \\ 0 & 0 & -1 & 1 \end{bmatrix}.$$

If λ is an eigenvalue of $(A_{\min})^{-1}$, then $\lambda > 0$ and, as $A_{\min} \in \mathbb{R}^{n \times n}$ and $A_{\min} = A_{\min}^H$, there exists a nonzero vector $\mathbf{x} \in \mathbb{R}^4$ such that

$$\begin{bmatrix} 2 & -1 & 0 & 0 \\ -1 & 2 & -1 & 0 \\ 0 & -1 & 2 & -1 \\ 0 & 0 & -1 & 1 \end{bmatrix} \begin{bmatrix} x_1 \\ x_2 \\ x_3 \\ x_4 \end{bmatrix} = \lambda \begin{bmatrix} x_1 \\ x_2 \\ x_3 \\ x_4 \end{bmatrix}$$

and hence that

$$-x_{j-1} + 2x_j - x_{j+1} = \lambda x_j \quad \text{for } j = 1, \ldots, 4 \quad \text{with } x_0 = 0 \text{ and } x_5 = x_4.$$

Let $\delta = \max\{|x_1|, \ldots, |x_4|\}$. Then,

$$(2 - \lambda) x_j = x_{j-1} + x_{j+1} \implies |2 - \lambda|\, |x_j| = |x_{j-1} + x_{j+1}| \leq 2\delta$$
$$\implies |2 - \lambda|\, \delta \leq 2\delta \implies |2 - \lambda| \leq 2.$$

(This is a special case of Geršgorin's theorem that will be discussed in Chapter 36.) Therefore, $2 - \lambda = 2\cos\theta$ for exactly one choice of $\theta \in [0, \pi]$ and the general solution of an equation of the form $-x_{j-1} + 2\cos\theta\, x_j - x_{j+1} = 0$ is of the form $x_j = \alpha \omega_1^j + \beta \omega_2^j$, where ω_1, ω_2 are the roots of $1 - 2\cos\theta\, \omega + \omega^2 = 0$, i.e.,

$$\omega = \frac{2\cos\theta \pm \sqrt{4\cos^2\theta - 4}}{2} = \cos\theta \pm i\sin\theta.$$

Consequently, $x_k = \alpha e^{ik\theta} + \beta e^{-ik\theta}$ and the condition $x_0 = 0$ implies that $\alpha = -\beta$ and hence that we may choose $x_k = \sin k\theta$ for $k = 0, \ldots, 4$. This in turn imposes the restriction that $0 < \theta < \pi$, and so in particular that $e^{i\theta} - 1 \neq 0$. It remains only to choose θ so that $x_5 = x_4$. But

$$\sin(n+1)\theta = \sin n\theta \iff e^{i(n+1)\theta} - e^{in\theta} = e^{-i(n+1)\theta} - e^{-in\theta}$$
$$\iff e^{in\theta}(e^{i\theta} - 1) = e^{-i(n+1)\theta}(1 - e^{i\theta}) \iff e^{i(2n+1)\theta} = -1.$$

Consequently, if $n = 4$, the permissible values of θ are $\theta_j = (2j+1)\pi/9$ for $j = 0, \ldots, 3$. Correspondingly, the eigenvalues of $(A_{\min})^{-1}$ are $\lambda_j = 2(1 - \cos\theta_j)$ for $j = 0, \ldots, 3$ and hence $r_\sigma(A_{\min}) = \{2(1 - \cos(\pi/9))\}^{-1}$ and

$$A_{\min}\mathbf{x} = r_\sigma(A_{\min})\mathbf{x} \quad \text{for} \quad \mathbf{x} = \left[\sin(\pi/9), \sin(3\pi/9), \sin(5\pi/9), \sin(7\pi/9)\right]^T.$$

Notice that $\mathbf{x} \in \mathbb{R}^4_>$, in keeping with the Perron-Frobenius theorem.

27.5. Doubly stochastic matrices

A matrix $A \in \mathbb{R}^{n \times n}_{\geq}$ is said to be a **doubly stochastic matrix** if both A and A^T are stochastic matrices, i.e., if

$$(27.6) \quad \sum_{j=1}^n a_{ij} = 1 \quad \text{for } i = 1, \ldots, n \quad \text{and} \quad \sum_{i=1}^n a_{ij} = 1 \quad \text{for } j = 1, \ldots, n.$$

Exercise 27.4. Find the eigenvalues of the (doubly) stochastic matrix $A = \begin{bmatrix} \mathbf{e}_1 & \mathbf{e}_3 & \mathbf{e}_2 & \mathbf{e}_4 \end{bmatrix}$ based on the columns $\mathbf{e}_j$, $j = 1, \ldots, 4$, of I_4.

Exercise 27.5. Show that the set of doubly stochastic $n \times n$ matrices is a convex set and that a matrix $A \in \mathbb{R}^{n \times n}$ is an extreme point of this set if and only if A is a permutation matrix. [HINT: $0 \leq a_{ij} \leq 1$.]

Theorem 27.3 (Birkhoff-von Neumann). *A matrix $A \in \mathbb{R}^{n \times n}$ is doubly stochastic if and only if it is a convex combination of permutation matrices.*

Proof. The set of $n \times n$ doubly stochastic matrices is a closed bounded convex subset of $\mathbb{R}^{n \times n}$. The extreme points of this set are the $n \times n$ permutation matrices. Thus, the fact that every doubly stochastic matrix is a convex combination of permutation matrices follows from the Krein-Milman theorem. The converse implication is self-evident. $\square$

Example 27.2. To illustrate Theorem 27.3, let

$$A = \frac{1}{9}\begin{bmatrix} 3 & 5 & 1 \\ 4 & 4 & 1 \\ 2 & 0 & 7 \end{bmatrix} \quad \text{and} \quad P_1 = \begin{bmatrix} 0 & 1 & 0 \\ 1 & 0 & 0 \\ 0 & 0 & 1 \end{bmatrix}.$$

Then A is doubly stochastic and

$$A_1 = (1 - 4/9)^{-1}\left(A - \frac{4}{9}P_1\right) = \frac{1}{5}\begin{bmatrix} 3 & 1 & 1 \\ 0 & 4 & 1 \\ 2 & 0 & 3 \end{bmatrix}$$

is a doubly stochastic matrix with one more 0 than the matrix A. Similarly,

$$A_2 = (1 - 1/5)^{-1}\left(A_1 - \frac{1}{5}\begin{bmatrix} 0 & 0 & 1 \\ 0 & 1 & 0 \\ 1 & 0 & 0 \end{bmatrix}\right) = \frac{1}{4}\begin{bmatrix} 3 & 1 & 0 \\ 0 & 3 & 1 \\ 1 & 0 & 3 \end{bmatrix}$$

is a doubly stochastic matrix with one more zero than A_1. The procedure terminates at the k'th step if A_k is a permutation matrix. ◇

Exercise 27.6. Show that the doubly stochastic matrix A considered in Example 27.2 is a convex combination of permutation matrices.

Exercise 27.7. Show that if $A \in \mathbb{R}^{n \times n}$ is a doubly stochastic matrix that is not a permutation matrix, σ is a permutation of the integers $\{1, \ldots, n\}$ such that $t_1 = \min\{a_{1\sigma(1)}, \ldots, a_{n\sigma(n)}\}$ belongs to the open interval $(0, 1)$, and P_1 is the permutation matrix with 1's in the $i\sigma(i)$ position for $i = 1, \ldots, n$, then $A_1 = (1 - t_1)^{-1}(A - t_1 P_1)$ is a doubly stochastic matrix with at least one more zero entry than A and $A = t_1 P_1 + (1 - t_1)A_1$. [REMARK: If A_1 is not a permutation matrix, then the preceding argument can be repeated.]

Exercise 27.8. Show that the representation of a doubly stochastic matrix $A \in \mathbb{R}^{n \times n}$ as a convex combination of permutation matrices is unique if $n = 1, 2, 3$, but not if $n \geq 4$. [HINT: If $n \geq 4$, then $n! > n^2$.]

27.6. Inequalities of Ky Fan and von Neuman

In this section we shall use the Birkhoff-von Neumann theorem (Theorem 27.3) and the Hardy-Littlewood-Polya rearrangement lemma (Lemma 27.4) to obtain bounds on the trace of the product AB of a pair of matrices in terms of the eigenvalues and the singular values of A and of B.

Lemma 27.4. *Let $\mathbf{a}$ and $\mathbf{b}$ be vectors in $\mathbb{R}^n$ with entries $a_1 \geq a_2 \geq \cdots \geq a_n$ and $b_1 \geq b_2 \geq \cdots \geq b_n$, respectively. Then*

$$(27.7) \qquad \mathbf{a}^T P \mathbf{b} \leq \mathbf{a}^T \mathbf{b}$$

for every $n \times n$ permutation matrix P.

Proof. Let $P = \sum_{j=1}^n \mathbf{e}_j \mathbf{e}_{\sigma(j)}^T$ for some one-to-one mapping σ of the integers $\{1, \ldots, n\}$ onto themselves, and suppose that $P \neq I_n$. Then $\mathbf{a}^T P \mathbf{b} = \sum_{j=1}^n a_j b_{\sigma(j)}$ and there exists a smallest positive integer k such that $\sigma(k) \neq k$, i.e.,

$$\sum_{j=1}^n a_j b_{\sigma(j)} = a_1 b_1 + \cdots + a_{k-1} b_{k-1} + \sum_{j=k}^n a_j b_{\sigma(j)}.$$

Thus, $\sigma(k) > k$ and $k = \sigma(\ell)$, for some integer $\ell > k$. Therefore,

$$(a_k - a_\ell)(b_{\sigma(\ell)} - b_{\sigma(k)}) = (a_k - a_\ell)(b_k - b_{\sigma(k)}) \geq 0,$$

and hence
$$a_k b_{\sigma(k)} + a_\ell b_{\sigma(\ell)} \le a_k b_{\sigma(\ell)} + a_\ell b_{\sigma(k)} = a_k b_k + a_\ell b_{\sigma(k)}\,.$$

In the same way, one can rearrange the remaining terms to obtain the inequality (27.7). □

Example 27.3. Suppose that in the setting of Lemma 27.4, $n = 5$, $\sigma(1) = 1$, $\sigma(2) = 4$, $\sigma(3) = 5$, $\sigma(4) = 3$, and $\sigma(5) = 2$. Then
$$\mathbf{a}^T P_\sigma \mathbf{b} = a_1 b_1 + a_2 b_4 + a_3 b_5 + a_4 b_3 + a_5 b_2$$
$$\le a_1 b_1 + a_2 b_2 + a_3 b_5 + a_4 b_3 + a_5 b_4\,,$$
since
$$(a_2 b_2 + a_5 b_4) - (a_2 b_4 + a_5 b_2) = (a_2 - a_5)(b_2 - b_4) \ge 0\,.$$
Thus, the sum is "increased" by performing the interchange that pairs b_2 with a_2. Next, to pair b_3 with a_3, we consider the terms in the sum in which a_3 and b_3 appear and observe that
$$a_3 b_5 + a_4 b_3 \le a_3 b_3 + a_4 b_5 \quad (\text{since } (a_3 - a_4)(b_3 - b_5) \ge 0)\,.$$
Finally, to pair b_4 with a_4, we consider the terms in which a_4 and b_4 appear and observe that
$$a_4 b_5 + a_5 b_4 \le a_4 b_4 + a_5 b_5\,.$$

Exercise 27.9. Show by explicit calculation that if, in the setting of Lemma 27.4, $n = 4$, $\sigma(1) = 1$, $\sigma(2) = 4$, $\sigma(3) = 2$, and $\sigma(4) = 3$, then
$$\sum_{j=1}^{4} a_j b_{\sigma(j)} = a_1 b_1 + a_2 b_4 + a_3 b_2 + a_4 b_3 \le a_1 b_1 + a_2 b_2 + a_3 b_4 + a_4 b_3\,.$$
[HINT: The inequality $(a_2 - a_3)(b_2 - b_4) \ge 0$ is the key.]

Lemma 27.5. *If $U \in \mathbb{C}^{n \times n}$ is a unitary matrix with entries u_{ij}, then the matrix $W \in \mathbb{R}^{n \times n}$ with entries $w_{ij} = |u_{ij}|^2$ is doubly stochastic.*

Proof. Let v_{ij} denote the ij entry of U^H. Then $w_{ij} = u_{ij}\overline{u_{ij}} = u_{ij} v_{ji}$ for $i, j = 1, \ldots, n$. Therefore, in self-evident notation,
$$\sum_{j=1}^{n} w_{ij} = \sum_{j=1}^{n} u_{ij} v_{ji} = (UU^H)_{ii} = (I_n)_{ii} = 1 \quad \text{for } i = 1, \ldots, n$$
and
$$\sum_{i=1}^{n} w_{ij} = \sum_{i=1}^{n} u_{ij} v_{ji} = (U^H U)_{jj} = (I_n)_{jj} = 1 \quad \text{for } j = 1, \ldots, n\,,$$
as claimed. □

Doubly stochastic matrices of the special form considered in Lemma 27.5 are often referred to as **orthostochastic** matrices. Not every doubly stochastic matrix is an orthostochastic matrix; see, e.g., Exercise 27.10.

Exercise 27.10. Show that the doubly stochastic matrix $A = \frac{1}{6}\begin{bmatrix} 0 & 3 & 3 \\ 3 & 1 & 2 \\ 3 & 2 & 1 \end{bmatrix}$ is not an orthostochastic matrix. [HINT: Show that a matrix U with $|u_{ij}| = \sqrt{a_{ij}}$ is not unitary.]

Theorem 27.6 (Ky Fan). *Let $A, B \in \mathbb{C}^{n \times n}$ be Hermitian matrices with eigenvalues*
$$\mu_1 \geq \mu_2 \geq \cdots \geq \mu_n \quad \text{and} \quad \nu_1 \geq \nu_2 \geq \cdots \geq \nu_n,$$
respectively (counting algebraic multiplicities). Then

(27.8) $\qquad \mu_1 \nu_n + \cdots + \mu_n \nu_1 \leq \operatorname{trace} AB \leq \mu_1 \nu_1 + \cdots + \mu_n \nu_n.$

Proof. Under the given assumptions there exists a pair of $n \times n$ unitary matrices U and V such that
$$A = U D_A U^H \quad \text{and} \quad B = V D_B V^H,$$
where
$$D_A = \operatorname{diag}\{\mu_1, \ldots, \mu_n\} \quad \text{and} \quad D_B = \operatorname{diag}\{\nu_1, \ldots, \nu_n\}.$$
Thus
$$\begin{aligned} \operatorname{trace} AB &= \operatorname{trace}\{U D_A U^H V D_B V^H\} \\ &= \operatorname{trace}\{D_A X D_B X^H\} \\ &= \sum_{i,j=1}^n \mu_i |x_{ij}|^2 \nu_j = \sum_{i,j=1}^n \mu_i w_{ij} \nu_j, \end{aligned}$$
where x_{ij} denotes the ij entry of the unitary matrix $X = U^H V$ and $w_{ij} = |x_{ij}|^2$, $i,j = 1, \ldots, n$, is the ij entry of the doubly stochastic matrix $W \in \mathbb{R}^{n \times n}$. Consequently, by the Birkhoff-von Neumann theorem,
$$W = \sum_{j=1}^k t_j P_j \quad \text{with } t_j > 0 \text{ and } \sum_{j=1}^k t_j = 1$$
is a convex combination of permutation matrices P_j. Thus, upon setting $\mathbf{u}^T = [\mu_1, \ldots, \mu_n]$ and $\mathbf{v}^T = [\nu_1, \ldots, \nu_n]$ and invoking Lemma 27.4, it is readily seen that
$$\operatorname{trace} AB = \mathbf{u}^T W \mathbf{v} = \sum_{j=1}^k t_j \mathbf{u}^T P_j \mathbf{v} \leq \sum_{j=1}^k t_j \mathbf{u}^T \mathbf{v} = \mathbf{u}^T \mathbf{v},$$

which completes the proof of the upper bound. To obtain the lower bound, apply the upper bound to the pair A and $-B$. $\square$

Remark 27.7. The upper (resp., lower) bound in (27.8) is achieved if and only if there exists a unitary matrix that diagonalizes both A and B and preserves the order of the eigenvalues of both matrices (resp., reverses the order of the eigenvalues of one of the matrices): If the upper bound in (27.8) is achieved for a pair of Hermitian matrices $A, B \in \mathbb{C}^{n \times n}$ with eigenvalues $\mu_1 \geq \cdots \geq \mu_n$ and $\nu_1 \geq \cdots \geq \nu_n$, respectively, and $U(t)$ is a family of unitary matrices for $t \in \mathbb{R}$ with $U(0) = I_n$, then $\sigma(U(t)BU(t)^H) = \sigma(B)$ and the function

$$f(t) = \text{trace}\{AU(t)BU(t)^H\}$$

achieves its maximum value at $t = 0$. Thus, as $U(t) = e^{tK}$ is unitary if $t \in \mathbb{R}$, $K \in \mathbb{C}^{n \times n}$, and $K^H = -K$, and

$$f'(t) = \text{trace}\{AU'(t)BU(t)^H\} + \text{trace}\{AU(t)BU'(t)^H\},$$

we see that

$$f'(0) = \text{trace}\{AKB - ABK\} = \text{trace}\{K(BA - AB)\} = 0$$

for every skew Hermitian matrix $K \in \mathbb{C}^{n \times n}$. Since $C = BA - AB = -C^H$, we can choose $K = C^H$ to obtain trace $C^H C = 0$ and hence that $C = O$, i.e., $AB = BA$. But this in turn implies that there exists a single unitary matrix W that serves to diagonalize both A and B, i.e., $A = W\Delta_A W^H$ and $B = W\Delta_B W^H$. Consequently, trace $AB = \text{trace } \Delta_A \Delta_B = \sum_{i=1}^{n} \mu_i \nu_{\sigma(i)}$ for some permutation σ. In view of Lemma 27.4, this sum will only equal the maximum if $\nu_{\sigma(i)} = \nu_i$.

Exercise 27.11. Let $A, B \in \mathbb{C}^{n \times n}$ be normal matrices with eigenvalues (counting algebraic multiplicities) $\mu_1, \ldots, \mu_n$ and $\nu_1, \ldots, \nu_n$, respectively, that are indexed so that $|\mu_1| \geq \cdots \geq |\mu_n|$ and $|\nu_1| \geq \cdots \geq |\nu_n|$. Show that $|\text{trace } AB| \leq |\mu_1||\nu_1| + \cdots + |\mu_n||\nu_n|$. [HINT: The proof of Theorem 27.6 is a helpful guide.]

Exercise 27.12. Show that if $A, B \in \mathbb{C}^{n \times n}$ with singular values $s_1 \geq s_2 \geq \cdots \geq s_n$ and $t_1 \geq t_2 \geq \cdots \geq t_n$, respectively, then

$$|\text{trace } B^H A| \leq \left\{\sum_{j=1}^{n} s_j^2\right\}^{1/2} \left\{\sum_{j=1}^{n} t_j^2\right\}^{1/2}$$

and then identify this inequality as the Cauchy-Schwarz inequality in an appropriately defined inner product space.

Theorem 27.8 (von Neumann). *If $A \in \mathbb{C}^{p \times q}$ and $B \in \mathbb{C}^{q \times p}$ with singular values $s_1 \geq s_2 \geq \cdots \geq s_q$ and $t_1 \geq t_2 \geq \cdots \geq t_p$, respectively, then*

$$(27.9) \qquad |\operatorname{trace} AB| \leq \sum_{i=1}^{m} s_i t_i, \quad \text{where } m = \min\{\operatorname{rank} A, \operatorname{rank} B\}.$$

Proof. The singular value decompositions $A = V_1 S_1 U_1^H$ and $B = X_1 T_1 Y_1^H$, wherein $V_1 \in \mathbb{C}^{p \times r_1}$, $U_1 \in \mathbb{C}^{q \times r_1}$, $X_1 \in \mathbb{C}^{q \times r_2}$, and $Y_1 \in \mathbb{C}^{p \times r_2}$ are isometric matrices, $r_1 = \operatorname{rank} A$, and $r_2 = \operatorname{rank} B$, ensure that

$$|\operatorname{trace} AB| = |\operatorname{trace} V_1 S_1 U_1^H X_1 T_1 Y_1^H| = |\operatorname{trace} S_1 U_1^H X_1 T_1 Y_1^H V_1|$$

$$= |\operatorname{trace} S_1 M_1 T_1 N_1| = \left|\sum_{i=1}^{r_1} \sum_{j=1}^{r_2} s_i m_{ij} t_j n_{ji}\right|$$

$$\leq \frac{1}{2} \sum_{i=1}^{r_1} \sum_{j=1}^{r_2} s_i t_j [|m_{ij}|^2 + |n_{ji}|^2],$$

where m_{ij} is the ij entry of the $r_1 \times r_2$ matrix $M_1 = U_1^H X_1$ and n_{ji} is the ji entry of the $r_2 \times r_1$ matrix $N_1 = Y_1^H V_1$.

The matrices M_1 and N_1 are not unitary matrices, unless $p = q = r_1 = r_2$. However, they can be embedded in unitary matrices. Thus, for example, if U_2 and X_2 are chosen so that $U = \begin{bmatrix} U_1 & U_2 \end{bmatrix}$ and $X = \begin{bmatrix} X_1 & X_2 \end{bmatrix}$ are both $q \times q$ unitary matrices, then M_1 is the upper left $r_1 \times r_2$ corner of the unitary matrix

$$M = \begin{bmatrix} U_1^H \\ U_2^H \end{bmatrix} \begin{bmatrix} X_1 & X_2 \end{bmatrix} \in \mathbb{C}^{q \times q} \quad \text{with entries } m_{ij} \text{ for } i, j = 1, \ldots, q.$$

Consequently, the matrix $W \in \mathbb{C}^{q \times q}$ with $w_{ij} = |m_{ij}|^2$ for $i, j = 1, \ldots, q$ is doubly stochastic and hence, upon expressing $W = \sum_{k=1}^{r} \mu_k P_k$ as a convex combination of permutation matrices and setting

$$\mathbf{a} = \begin{bmatrix} s_1 & \cdots & s_{r_1} & 0 & \cdots & 0 \end{bmatrix}^T \quad \text{and} \quad \mathbf{b} = \begin{bmatrix} t_1 & \cdots & t_{r_2} & 0 & \cdots & 0 \end{bmatrix}^T,$$

both in $\mathbb{R}^q$ with entries $a_1 \geq \cdots \geq a_q$ and $b_1 \geq \cdots \geq b_q$, it is readily seen that

$$\sum_{i=1}^{r_1} \sum_{j=1}^{r_2} s_i t_j |m_{ij}|^2 = \sum_{i,j=1}^{q} a_i b_j w_{ij} = \langle W\mathbf{b}, \mathbf{a} \rangle = \sum_{k=1}^{r} \mu_k \langle P_k \mathbf{b}, \mathbf{a} \rangle$$

$$\leq \sum_{k=1}^{r} \mu_k \langle \mathbf{b}, \mathbf{a} \rangle = \langle \mathbf{b}, \mathbf{a} \rangle = \sum_{i=1}^{r_1 \wedge r_2} s_i t_i,$$

where $r_1 \wedge r_2 = \min\{r_1, r_2\}$. Since $\sum_{i=1}^{r_1} \sum_{j=1}^{r_2} s_i t_j |n_{ji}|^2$ is subject to the same bound, this serves to complete the proof. $\square$

Exercise 27.13. Show that the set of $p \times q$ stochastic matrices is a convex set with q^p extreme points.

27.7. Supplementary notes

The discussion of minimum matrices is partially adapted from the article [11] by R. Bhatia. Notice that $A_{\min} \succ O$ and its inverse is restricted to a band. Thus, if $A(a,b,c) = \begin{bmatrix} 1 & 1 & a & b \\ 1 & 2 & 2 & c \\ \overline{a} & 2 & 3 & 3 \\ \overline{b} & \overline{c} & 3 & 4 \end{bmatrix}$ is positive definite, then, as will follow from Theorem 38.3, $\det A(a,b,c) \leq \det A_{\min}$.

The Birkhoff-von Neumann theorem can be proved directly by relatively elementary arguments instead of referring to the Krein-Milman theorem, much as in Example 27.2. The missing ingredient is a proof that if $A \in \mathbb{R}^{n \times n}$ is a doubly stochastic matrix that is not a permutation matrix, then there exists a permutation σ such that $a_{j\sigma(j)} > 0$ for $j = 1, \ldots, n$. This is usually justified by invoking the theory of **permanents**; see, e.g., Chapter 23 of [30]. Remark 27.7 is adapted from Section 28 of Bhatia [10].

Chapter 28

Eigenvalues of Hermitian matrices

This chapter is devoted to a number of classical results on the eigenvalues of Hermitian matrices.

28.1. The Courant-Fischer theorem

The **notation**

(28.1) $\quad (A;\mathcal{X})_{\min} = \min\{\langle A\mathbf{x},\mathbf{x}\rangle : \mathbf{x} \in \mathcal{X} \text{ and } \|\mathbf{x}\| = 1\}$

and

(28.2) $\quad (A;\mathcal{X})_{\max} = \max\{\langle A\mathbf{x},\mathbf{x}\rangle : \mathbf{x} \in \mathcal{X} \text{ and } \|\mathbf{x}\| = 1\}$

for Hermitian matrices $A \in \mathbb{C}^{n \times n}$ and subspaces $\mathcal{X}$ of $\mathbb{C}^n$ will be convenient. These definitions are meaningful because $\langle A\mathbf{x}, \mathbf{x}\rangle$ is a real number when $A = A^H$.

Theorem 28.1 (Courant-Fischer). *Let $A \in \mathbb{C}^{n \times n}$ be a Hermitian matrix with eigenvalues $\lambda_1 \geq \cdots \geq \lambda_n$ and let $\mu_j = \lambda_{n+1-j}$ for $j = 1, \ldots, n$ (so that the numbers $\mu_1, \ldots, \mu_n$ run through the same set of eigenvalues but are now indexed so that $\mu_1 \leq \cdots \leq \mu_n$). Then*

(28.3) $\quad \lambda_j = \mu_{n+1-j} = \max\{(A;\mathcal{X})_{\min} : \dim \mathcal{X} = j\} \quad \text{for } j = 1, \ldots, n$

and

(28.4) $\quad \mu_j = \lambda_{n+1-j} = \min\{(A;\mathcal{X})_{\max} : \dim \mathcal{X} = j\} \quad \text{for } j = 1, \ldots, n.$

Proof. Let $\mathbf{u}_1, \ldots, \mathbf{u}_n$ be an orthonormal set of eigenvectors of A corresponding to the eigenvalues $\lambda_1, \ldots, \lambda_n$. Then, to verify (28.3), let $\mathcal{X}$ be any

j-dimensional subspace of $\mathbb{C}^n$ and let $\mathcal{U}_j = \text{span}\{\mathbf{u}_j, \ldots, \mathbf{u}_n\}$. Then, since $\dim \mathcal{U}_j = n + 1 - j$ and $\dim(\mathcal{X} + \mathcal{U}_j) \leq n$,

$$\dim(\mathcal{X} \cap \mathcal{U}_j) = \dim \mathcal{X} + \dim \mathcal{U}_j - \dim(\mathcal{X} + \mathcal{U}_j)$$
$$\geq j + n + 1 - j - n = 1.$$

Thus, $\mathcal{X} \cap \mathcal{U}_j \neq \{0\}$, and if $\mathbf{v} \in \mathcal{X} \cap \mathcal{U}_j$ with $\|\mathbf{v}\| = 1$, then

$$\mathbf{v} = \sum_{i=j}^{n} c_i \mathbf{u}_i,$$

and hence

$$\langle A\mathbf{v}, \mathbf{v} \rangle = \sum_{i=j}^{n} \lambda_i |c_i|^2 \leq \lambda_j \sum_{i=j}^{n} |c_i|^2 = \lambda_j.$$

Therefore,

$$(A; \mathcal{X})_{\min} \leq \lambda_j \quad \text{for every } j\text{-dimensional subspace } \mathcal{X} \text{ of } \mathbb{C}^n.$$

Consequently,

$$\max\{(A; \mathcal{X})_{\min} : \dim \mathcal{X} = j\} \leq \lambda_j.$$

Since the upper bound is attained when $\mathcal{X} = \text{span}\{\mathbf{u}_1, \ldots, \mathbf{u}_j\}$, the verification of (28.3) is complete.

Next, (28.4) may be verified by applying (28.3) to $-A$ and noting that $\lambda_j(-A) = -\lambda_{n+1-j}(A)$, $(-A; \mathcal{X})_{\min} = -(A; \mathcal{X})_{\max}$, and

$$\max\{-(A; \mathcal{X})_{\max} : \dim \mathcal{X} = j\} = -\min\{(A; \mathcal{X})_{\max} : \dim \mathcal{X} = j\}.$$

Another option, which some may find to be more congenial, is to imitate the proof of (28.3) but with

$$\mathcal{U}_j = \text{span}\{\mathbf{u}_1, \ldots, \mathbf{u}_{n-j+1}\}.$$

Details are left to the reader. $\square$

Exercise 28.1. Show that if $A, B \in \mathbb{C}^{n \times n}$ are Hermitian matrices such that $\lambda_1(A) \geq \cdots \geq \lambda_n(A)$, $\lambda_1(B) \geq \cdots \geq \lambda_n(B)$, and $\langle A\mathbf{x}, \mathbf{x} \rangle \leq \langle B\mathbf{x}, \mathbf{x} \rangle$ for every vector $\mathbf{x} \in \mathbb{C}^n$, then $\lambda_j(A) \leq \lambda_j(B)$ for $j = 1, \ldots, n$.

Exercise 28.2. Show that if $A \in \mathbb{C}^{n \times n}$ is a Hermitian matrix with eigenvalues $\mu_1 \leq \cdots \leq \mu_n$ and $\mathcal{X}^\perp$ denotes the orthogonal complement of $\mathcal{X}$ in $\mathbb{C}^n$, then

$$\mu_{n-j+1} = \min_{\mathcal{X} \in \mathcal{S}_j} \max\left\{\langle A\mathbf{x}, \mathbf{x} \rangle : \mathbf{x} \in \mathcal{X}^\perp \text{ and } \|\mathbf{x}\| = 1\right\} \text{ for } j = 1, \ldots, n.$$

Exercise 28.3. Show that if $A \in \mathbb{C}^{n \times n}$ is a Hermitian matrix with eigenvalues $\mu_1 \leq \cdots \leq \mu_n$ and $\mathcal{X}^\perp$ denotes the orthogonal complement of $\mathcal{X}$ in $\mathbb{C}^n$, then

$$\mu_j = \max_{\mathcal{X} \in \mathcal{S}_j} \min\left\{\frac{\langle A\mathbf{x}, \mathbf{x} \rangle}{\langle \mathbf{x}, \mathbf{x} \rangle} : \mathbf{x} \in \mathcal{X}^\perp \text{ and } \mathbf{x} \neq \mathbf{0}\right\} \text{ for } j = 1, \ldots, n.$$

Exercise 28.4. Let $A \in \mathbb{C}^{n \times n}$ be a Hermitian matrix with eigenvalues $\lambda_1 \geq \cdots \geq \lambda_n$. Show that $\lambda_n \leq \min a_{ii} \leq \max a_{ii} \leq \lambda_1$.

Exercise 28.5. Use the Courant-Fischer theorem to give another proof of item (2) of Theorem 15.6. [HINT: Compare the eigenvalues of $A^H A$ and $A^H B^H BA$.]

Exercise 28.6. Use the Courant-Fischer theorem to give another proof of item (3) of Theorem 15.6. [HINT: Compare the eigenvalues of $A^H A$ and $C^H A^H A C$.]

Exercise 28.7. Let $A \in \mathbb{C}^{n \times n}$, and let $\beta_1 \geq \cdots \geq \beta_n$ and $\delta_1 \geq \cdots \geq \delta_n$ denote the eigenvalues of the Hermitian matrices $B = (A + A^H)/2$ and $C = (A - A^H)/(2i)$, respectively. Show that
$$\beta_n \leq \frac{\lambda + \overline{\lambda}}{2} \leq \beta_1 \quad \text{and} \quad \delta_n \leq \frac{\lambda - \overline{\lambda}}{2i} \leq \delta_1 \quad \text{for every point } \lambda \in \sigma(A).$$
[HINT: If $A\mathbf{x} = \lambda \mathbf{x}$ and $\|\mathbf{x}\| = 1$, then $\lambda + \overline{\lambda} = \langle A\mathbf{x}, \mathbf{x} \rangle + \langle \mathbf{x}, A\mathbf{x} \rangle$.]

28.2. Applications of the Courant-Fischer theorem

If a Hermitian matrix undergoes a small Hermitian perturbation, then its eigenvalues will only change a little (but multiplicities may change):

Theorem 28.2. *If* $A, B \in \mathbb{C}^{n \times n}$ *are Hermitian matrices with eigenvalues* $\lambda_1(A) \geq \cdots \geq \lambda_n(A)$ *and* $\lambda_1(B) \geq \cdots \geq \lambda_n(B)$, *respectively, then*
$$(28.5) \qquad |\lambda_j(A) - \lambda_j(B)| \leq \|A - B\| \quad \text{for } j = 1, \ldots, n\,.$$

Proof. Let $\mathcal{X}$ be any j-dimensional subspace of $\mathbb{C}^n$. Then
$$\langle A\mathbf{x}, \mathbf{x} \rangle = \langle B\mathbf{x}, \mathbf{x} \rangle + \langle (A-B)\mathbf{x}, \mathbf{x} \rangle$$
$$\leq \langle B\mathbf{x}, \mathbf{x} \rangle + \|A - B\|$$
for every vector $\mathbf{x} \in \mathcal{X}$ with $\|\mathbf{x}\| = 1$. Therefore,
$$(A; \mathcal{X})_{\min} \leq \langle B\mathbf{x}, \mathbf{x} \rangle + \|A - B\| \quad \text{for each vector } \mathbf{x} \in \mathcal{X} \text{ with } \|\mathbf{x}\| = 1$$
and hence
$$(A; \mathcal{X})_{\min} \leq (B; \mathcal{X})_{\min} + \|A - B\|$$
$$\leq \max\{(B; \mathcal{X})_{\min} : \dim \mathcal{X} = j\} + \|A - B\|$$
$$= \lambda_j(B) + \|A - B\|\,.$$
Thus,
$$\lambda_j(A) - \lambda_j(B) \leq \|A - B\|$$
and, as follows by interchanging A and B,
$$\lambda_j(B) - \lambda_j(A) \leq \|B - A\| = \|A - B\|\,.$$
But these two inequalities serve to justify (28.5). $\square$

We turn next to the **Cauchy interlacing theorem**.

Theorem 28.3. *Let*
$$A = \begin{bmatrix} A_{11} & A_{12} \\ A_{21} & A_{22} \end{bmatrix} = A^H \quad \text{with } A_{11} \in \mathbb{C}^{p \times p}, \ A_{22} \in \mathbb{C}^{q \times q}, \ n = p+q,$$
and suppose that $\lambda_1(A) \geq \cdots \geq \lambda_n(A)$ *and* $\lambda_1(A_{11}) \geq \cdots \geq \lambda_p(A_{11})$. *Then*
$$(28.6) \qquad \lambda_{j+q}(A) \leq \lambda_j(A_{11}) \leq \lambda_j(A) \quad \text{for } j = 1, \ldots, p.$$

Proof. For each subspace $\mathcal{Y}$ of $\mathbb{C}^p$, let
$$\widehat{\mathcal{Y}} = \left\{ \begin{bmatrix} \mathbf{y} \\ \mathbf{0} \end{bmatrix} \in \mathbb{C}^n : \mathbf{y} \in \mathcal{Y} \right\}.$$
Then $\widehat{\mathcal{Y}}$ is a subspace of $\mathbb{C}^n$ and $(A_{11}; \mathcal{Y})_{\min} = (A; \widehat{\mathcal{Y}})_{\min}$, since
$$\langle A_{11}\mathbf{y}, \mathbf{y} \rangle = \left\langle A \begin{bmatrix} \mathbf{y} \\ \mathbf{0} \end{bmatrix}, \begin{bmatrix} \mathbf{y} \\ \mathbf{0} \end{bmatrix} \right\rangle \quad \text{when } \begin{bmatrix} \mathbf{y} \\ \mathbf{0} \end{bmatrix} \in \widehat{\mathcal{Y}}.$$
Therefore,
$$\begin{aligned} \lambda_j(A_{11}) &= \max\{(A_{11}; \mathcal{Y})_{\min} : \dim \mathcal{Y} = j\} \\ &= \max\{(A; \widehat{\mathcal{Y}})_{\min} : \dim \mathcal{Y} = j\} \\ &\leq \max\{(A; \mathcal{X})_{\min} : \mathcal{X} \text{ is a subspace of } \mathbb{C}^n \text{ and } \dim \mathcal{X} = j\} \\ &= \lambda_j(A) \quad \text{for } j = 1, \ldots, p. \end{aligned}$$
This completes the justification of the upper bound in (28.6).

The lower bound in (28.6) is obtained by applying the upper bound to $-A$ to get
$$\lambda_k(-A_{11}) \leq \lambda_k(-A) \quad \text{for } k = 1, \ldots, p$$
and then noting that
$$\lambda_k(-A_{11}) = -\lambda_{p+1-k}(A_{11}) \quad \text{and} \quad \lambda_k(-A) = -\lambda_{n+1-k}(A).$$
The final inequality emerges upon setting $j = p+1-k$, which in turn implies that $n + 1 - k = n + 1 - (p+1-j) = j + q$. $\square$

A **tridiagonal** Hermitian matrix $A_n \in \mathbb{R}^{n \times n}$ of the form
$$A_n = \sum_{j=1}^{n} a_j \mathbf{e}_j \mathbf{e}_j^T + \sum_{j=1}^{n-1} b_j (\mathbf{e}_j \mathbf{e}_{j+1}^T + \mathbf{e}_{j+1} \mathbf{e}_j^T) = \begin{bmatrix} a_1 & b_1 & 0 & \cdots & 0 & 0 \\ b_1 & a_2 & b_2 & \cdots & 0 & 0 \\ 0 & b_2 & a_3 & \cdots & 0 & 0 \\ \vdots & & & & & \vdots \\ 0 & 0 & 0 & \cdots & b_{n-1} & a_n \end{bmatrix}$$

with $b_j > 0$ and $a_j \in \mathbb{R}$ is termed a **Jacobi matrix**.

Exercise 28.8. Show that a Jacobi matrix A_{n+1} has $n+1$ distinct eigenvalues $\lambda_1 < \cdots < \lambda_{n+1}$ and that if $\mu_1 < \cdots < \mu_n$ denote the eigenvalues of the Jacobi matrix A_n, then $\lambda_j < \mu_j < \lambda_{j+1}$ for $j = 1, \ldots, n$.

28.3. Ky Fan's maximum principle

Theorem 28.4. *If* $A = A^H \in \mathbb{C}^{n \times n}$ *with eigenvalues* $\lambda_1 \geq \cdots \geq \lambda_n$, *then*

$$(28.7) \qquad \sum_{j=1}^{k} \lambda_j = \max\{\operatorname{trace} X^H A X \,:\, X \in \mathbb{C}^{n \times k} \text{ and } X^H X = I_k\}$$

and

$$(28.8) \qquad \sum_{j=n-k+1}^{n} \lambda_j = \min\{\operatorname{trace} X^H A X \,:\, X \in \mathbb{C}^{n \times k} \text{ and } X^H X = I_k\}$$

for $k = 1, \ldots, n$.

Proof. Let $\begin{bmatrix} X & Y \end{bmatrix}$ be a unitary matrix with isometric blocks $X \in \mathbb{C}^{n \times k}$ and $Y \in \mathbb{C}^{n \times (n-k)}$, where $1 \leq k \leq n-1$, and let

$$B = \begin{bmatrix} X^H \\ Y^H \end{bmatrix} A \begin{bmatrix} X & Y \end{bmatrix} = \begin{bmatrix} X^H A X & X^H A Y \\ Y^H A X & Y^H A Y \end{bmatrix} = \begin{bmatrix} B_{11} & B_{12} \\ B_{21} & B_{22} \end{bmatrix}.$$

Then, by Cauchy's interlacing theorem,

$$\lambda_j(A) = \lambda_j(B) \geq \lambda_j(B_{11}) \quad \text{for } j = 1, \ldots, k\,.$$

Therefore,

$$(28.9) \qquad \sum_{j=1}^{k} \lambda_j(A) \geq \sum_{j=1}^{k} \lambda_j(B_{11}) = \operatorname{trace} B_{11} = \operatorname{trace} X^H A X\,.$$

On the other hand, if $A = UDU^H$ with $D = \operatorname{diag}\{\lambda_1, \ldots, \lambda_n\}$, U unitary, and $X^H = \begin{bmatrix} I_k & O \end{bmatrix} U^H$, then equality is achieved in (28.9). This completes the verification of (28.7) for $k = 1, \ldots, n-1$. Since the case $k = n$ is self-evident, the proof of (28.7) is complete. The proof of (28.8) is left to the reader. □

Exercise 28.9. Verify (28.8).

28.4. The sum of two Hermitian matrices

In this section we shall survey a number of classical inequalities for the eigenvalues of the sum $A + B$ of two Hermitian matrices.

Theorem 28.5. *If* $A \in \mathbb{C}^{n \times n}$ *and* $B \in \mathbb{C}^{n \times n}$ *are Hermitian matrices and* $\lambda_i(C)$, $i = 1, \ldots, n$, *denotes the i'th eigenvalue of* $C = C^H$ *indexed in descending order, i.e.,* $\lambda_1(C) \geq \cdots \geq \lambda_n(C)$, *then the following inequalities* are in force:

(1) ***(Weyl's inequalities)***

(28.10) $\quad \lambda_{i+j-1}(A+B) \leq \lambda_i(A) + \lambda_j(B) \quad for\ i,j \geq 1\ and\ i+j-1 \leq n$.

(28.11) $\quad \lambda_i(A) + \lambda_n(B) \leq \lambda_i(A+B) \leq \lambda_i(A) + \lambda_1(B) \quad for\ i=1,\ldots,n$.

(28.12) $\quad |\lambda_i(A+B) - \lambda_i(A)| \leq \|B\| \quad for\ i=1,\ldots,n$.

(2) ***(Ky Fan's inequality)***

(28.13) $\quad \displaystyle\sum_{i=1}^{k} \lambda_i(A+B) - \sum_{i=1}^{k} \lambda_i(A) \leq \sum_{i=1}^{k} \lambda_i(B) \quad for\ k=1,\ldots,n$,

with equality when $k=n$.

(3) ***(Lidskii's inequality)***

(28.14) $\quad \displaystyle\sum_{j=1}^{k} \lambda_{i_j}(A+B) - \sum_{j=1}^{k} \lambda_{i_j}(A) \leq \sum_{i=1}^{k} \lambda_i(B)$

for $1 \leq i_1 < \cdots < i_k \leq n$ and $k=1,\ldots,n$, with equality when $k=n$.

Proof. Since A and B are Hermitian, there exist three orthonormal sets of vectors $A\mathbf{u}_j = a_j\mathbf{u}_j$, $B\mathbf{v}_j = b_j\mathbf{v}_j$, and $(A+B)\mathbf{w}_j = c_j\mathbf{w}_j$, for $j=1,\ldots,n$, where $a_j = \lambda_j(A)$, $b_j = \lambda_j(B)$, and $c_j = \lambda_j(A+B)$. The rest of the proof is divided into steps.

1. *Verification of* (28.10)–(28.12).

If
$$\mathcal{U} = \mathrm{span}\{\mathbf{u}_i,\ldots,\mathbf{u}_n\},\ \mathcal{V} = \mathrm{span}\{\mathbf{v}_j,\ldots,\mathbf{v}_n\},\ \mathcal{W} = \mathrm{span}\{\mathbf{w}_1,\ldots,\mathbf{w}_k\},$$
with $k = i+j-1$, and $\mathcal{Y} = \mathcal{U} \cap \mathcal{W}$, then
$$\begin{aligned}\dim \mathcal{Y} &= \dim \mathcal{U} + \dim \mathcal{W} - \dim(\mathcal{U}+\mathcal{W}) \\ &\geq (n-i+1) + (i+j-1) - n = j\end{aligned}$$
and
$$\begin{aligned}\dim \mathcal{Y} \cap \mathcal{V} &= \dim \mathcal{Y} + \dim \mathcal{V} - \dim(\mathcal{Y}+\mathcal{V}) \\ &\geq j + (n-j+1) - n = 1\,.\end{aligned}$$

Therefore, there exists a unit vector $\mathbf{x} \in \mathcal{U} \cap \mathcal{V} \cap \mathcal{W}$. Thus,
$$c_{i+j-1} \leq \langle (A+B)\mathbf{x},\mathbf{x}\rangle = \langle A\mathbf{x},\mathbf{x}\rangle + \langle B\mathbf{x},\mathbf{x}\rangle \leq a_i + b_j\,,$$
i.e., (28.10) holds.

The upper bound in (28.11) is obtained by choosing $j = 1$ in (28.10). The lower bound follows by applying the upper bound to the matrices $-A$ and $-B$.

The inequality (28.12) is an easy consequence of (28.11),
$$\lambda_n(B) \leq \lambda_i(A+B) - \lambda_i(A) \leq \lambda_1(B) \quad \text{for } i = 1, \ldots, n,$$
and the fact that $\|B\| = \max\{|b_1|, |b_n|\}$.

2. *Verification of* (28.13) *and* (28.14).

It suffices to verify (28.14), since (28.13) is a special case of (28.14). Towards this end, fix $1 \leq k \leq n-1$ and note that (28.14) holds if and only if
$$\sum_{j=1}^{k}[\lambda_{i_j}(A + B - b_k I_n) - \lambda_{i_j}(A)] \leq \sum_{i=1}^{k} \lambda_i(B - b_k I_n)$$
and that
$$B - b_k I_n = \sum_{i=1}^{n}(b_i - b_k)\mathbf{v}_i \mathbf{v}_i^H \leq B_+, \quad \text{where} \quad B_+ = \sum_{i=1}^{k}(b_i - b_k)\mathbf{v}_i \mathbf{v}_i^H.$$
Therefore, by Exercise 28.1,
$$\lambda_i(A + B - b_k I_n) \leq \lambda_i(A + B_+) \quad \text{and} \quad \lambda_i(A) \leq \lambda_i(A + B_+)$$
for $i = 1, \ldots, n$. Consequently,
$$\sum_{j=1}^{k}[\lambda_{i_j}(A + B - b_k I_n) - \lambda_{i_j}(A)] \leq \sum_{j=1}^{k}[\lambda_{i_j}(A + B_+) - \lambda_{i_j}(A)]$$
$$\leq \sum_{i=1}^{n}[\lambda_i(A + B_+) - \lambda_i(A)]$$
$$= \text{trace}(A + B_+) - \text{trace}(A)$$
$$= \text{trace } B_+ = \sum_{i=1}^{k}(b_i - b_k),$$
which is equivalent to (28.14). □

28.5. On the right-differentiability of eigenvalues

The next theorem gives more refined information on the behavior of the eigenvalues of a Hermitian matrix A under small (Hermitian) perturbations of A than the inequality (28.12). It is a key step in understanding the behavior of the singular values of a matrix A under small perturbations of A.

Theorem 28.6. If $A, B \in \mathbb{C}^{n \times n}$, $A = A^H$ with eigenvalues $\lambda_1(A) \geq \cdots \geq \lambda_n(A)$, k of which are distinct with geometric multiplicities $\gamma_1, \ldots, \gamma_k$, respectively, and $B = B^H$, then

$$\lim_{\mu \downarrow 0} \frac{\lambda_j(A + \mu B) - \lambda_j(A)}{\mu} = \nu_j, \qquad (28.15)$$

where the ν_j are the eigenvalues of a block diagonal matrix

$$\mathrm{diag}\{C_{11}, \ldots, C_{kk}\} \quad \text{(based on blocks } C_{jj} \text{ of size } \gamma_j \times \gamma_j\text{)}$$

that is extracted from a matrix $C \in \mathbb{C}^{n \times n}$ that is similar to B and $\sum_{j=1}^n \nu_j = \mathrm{trace}\, B$. (Here, too, $\lambda_1(A + \mu B) \geq \cdots \geq \lambda_n(A + \mu B)$.)

Discussion. Since $A = A^H$, it admits a representation of the form $A = UDU^H$ with $U \in \mathbb{C}^{n \times n}$ unitary and $D = \mathrm{diag}\{\lambda_1, \ldots, \lambda_n\}$. Therefore,

$$\lambda_j(A + \mu B) = \lambda_j(U[D + \mu U^H B U]U^H) = \lambda_j(D + \mu C) \quad \text{with } C = U^H B U.$$

Moreover,

$$\lambda_j(D + \mu C) = \lambda_j(e^{\mu M}(D + \mu C)e^{-\mu M})$$

for every choice of $M \in \mathbb{C}^{n \times n}$ and $\mu \in \mathbb{R}$. But,

$$e^{\mu M}(D + \mu C)e^{-\mu M}$$
$$= \left(I_n + \mu M + \frac{\mu^2}{2!}M^2 + \cdots\right)(D + \mu C)\left(I_n - \mu M + \frac{\mu^2}{2!}M^2 + \cdots\right)$$
$$= D + \mu(MD - DM + C) + \cdots,$$

which for small μ behaves essentially like $D + \mu(MD - DM + C)$. To make the picture as transparent as possible, suppose that

$$D = \begin{bmatrix} \kappa_1 I_3 & O & O \\ O & \kappa_2 I_4 & O \\ O & O & \kappa_3 I_2 \end{bmatrix} \quad \text{with } \kappa_1 > \kappa_2 > \kappa_3$$

so that $\lambda_1 = \lambda_2 = \lambda_3 = \kappa_1$, $\lambda_4 = \lambda_5 = \lambda_6 = \lambda_7 = \kappa_2$, and $\lambda_8 = \lambda_9 = \kappa_3$. Suppose further that $C = [C_{ij}]$ and $M = [M_{ij}]$ for $i, j = 1, 2, 3$ with blocks C_{ii} and M_{ii} of size 3×3 for $i = 1$, 4×4 for $i = 2$, and 2×2 for $i = 3$. Then, since $(MD - DM + C)_{ij} = M_{ij}\kappa_j - \kappa_i M_{ij} + C_{ij}$, we can choose

$$M_{ij} = \frac{C_{ij}}{\kappa_i - \kappa_j} \quad \text{if } i \neq j$$

so that $(MD - DM + C) = \mathrm{diag}\{C_{11}, C_{22}, C_{33}\}$. Consequently,

$$\lambda_i(D + \mu C) = \lambda_i(\lambda_1 I_3 + \mu C_{11}) = \lambda_i + \mu \nu_i \quad \text{for } i = 1, 2, 3,$$
$$\lambda_i(D + \mu C) = \lambda_i(\lambda_4 I_4 + \mu C_{22}) = \lambda_i + \mu \nu_i \quad \text{for } i = 4, \ldots, 7,$$
$$\lambda_i(D + \mu C) = \lambda_i(\lambda_8 I_2 + \mu C_{33}) = \lambda_i + \mu \nu_i \quad \text{for } i = 8, 9,$$

when μ is small, where $\nu_1 \geq \cdots \geq \nu_3$ are the eigenvalues of C_{11}, $\nu_4 \geq \cdots \geq \nu_7$ are the eigenvalues of C_{22}, and $\nu_8 \geq \nu_9$ are the eigenvalues of C_{33}. Therefore, the limit (28.15) clearly holds for this example, and

$$\sum_{j=1}^{9} \nu_j = \sum_{i=1}^{3} \operatorname{trace} C_{ii} = \operatorname{trace} C = \operatorname{trace} B\,.$$

The general case is treated in exactly the same way; the only difference is that the bookkeeping is more elaborate. $\square$

Exercise 28.10. Show that in the setting and notation of Theorem 28.6, $\sum_{j=1}^{n} \lambda_j \nu_j = \operatorname{trace} DC$. [HINT: Look first at the example considered in the proof of the theorem.]

28.6. Sylvester's law of inertia

In this section we shall use the **notation**

- $\mathcal{E}_+(A) =$ the number of positive eigenvalues of A,
- $\mathcal{E}_-(A) =$ the number of negative eigenvalues of A,
- $\mathcal{E}_0(A) =$ the number of zero eigenvalues of A

(**counting multiplicities in all three**) to track the location of the eigenvalues of Hermitian matrices $A \in \mathbb{C}^{n \times n}$. Thus,

$$\mathcal{E}_+(A) + \mathcal{E}_-(A) = \operatorname{rank} A \quad \text{and} \quad \mathcal{E}_0(A) = \dim \mathcal{N}_A\,.$$

Theorem 28.7. *If $A \in \mathbb{C}^{n \times n}$ is Hermitian and $C \in \mathbb{C}^{m \times n}$, then*

$$\mathcal{E}_+(CAC^H) \leq \mathcal{E}_+(A) \quad \text{and} \quad \mathcal{E}_-(CAC^H) \leq \mathcal{E}_-(A)\,,$$

with equality in both if and only if $\operatorname{rank} A = \operatorname{rank} CAC^H$.

Proof. Since A and $B = CAC^H$ are Hermitian matrices, there exists a pair of invertible matrices $X \in \mathbb{C}^{n \times n}$ and $Y \in \mathbb{C}^{m \times m}$ such that

$$A = X \begin{bmatrix} I_{s_1} & O & O \\ O & -I_{t_1} & O \\ O & O & O \end{bmatrix} X^H \quad \text{and} \quad B = Y \begin{bmatrix} I_{s_2} & O & O \\ O & -I_{t_2} & O \\ O & O & O \end{bmatrix} Y^H,$$

where $s_1 = \mathcal{E}_+(A)$, $t_1 = \mathcal{E}_-(A)$, $s_2 = \mathcal{E}_+(B)$, and $t_2 = \mathcal{E}_-(B)$. Therefore,

$$\begin{bmatrix} I_{s_2} & O & O \\ O & -I_{t_2} & O \\ O & O & O \end{bmatrix} = Q \begin{bmatrix} I_{s_1} & O & O \\ O & -I_{t_1} & O \\ O & O & O \end{bmatrix} Q^H,$$

where $Q = Y^{-1}CX$. Thus, upon expressing the $m \times n$ matrix Q in block form as
$$Q = \begin{bmatrix} Q_{11} & Q_{12} & Q_{13} \\ Q_{21} & Q_{22} & Q_{23} \\ Q_{31} & Q_{32} & Q_{33} \end{bmatrix},$$
where the heights of the block rows are s_2, t_2, and $m - s_2 - t_2$ and the widths of the block columns are s_1, t_1, and $n - s_1 - t_1$, respectively, it is readily seen that
$$I_{s_2} = Q_{11}Q_{11}^H - Q_{12}Q_{12}^H \quad \text{and} \quad I_{t_2} = -Q_{21}Q_{21}^H + Q_{22}Q_{22}^H.$$
Therefore,
$$Q_{11}Q_{11}^H = I_{s_2} + Q_{12}Q_{12}^H \quad \text{and} \quad Q_{22}Q_{22}^H = I_{t_2} + Q_{21}Q_{21}^H.$$
Thus,
$$s_2 = \operatorname{rank} Q_{11}Q_{11}^H \leq \operatorname{rank} Q_{11} \leq s_1 \quad \text{and}$$
$$t_2 = \operatorname{rank} Q_{22}Q_{22}^H \leq \operatorname{rank} Q_{22} \leq t_1.$$
Consequently,
$$\operatorname{rank} A = s_1 + t_1 \geq s_2 + t_1 \geq s_2 + t_2 = \operatorname{rank} B,$$
which clearly displays the fact that
$$\operatorname{rank} A = \operatorname{rank} B \iff \mathcal{E}_+(A) = \mathcal{E}_+(B) \text{ and } \mathcal{E}_-(A) = \mathcal{E}_-(B). \qquad \square$$

Corollary 28.8 (Sylvester's law of inertia). *If $A \in \mathbb{C}^{n \times n}$ is Hermitian and $C \in \mathbb{C}^{n \times n}$ is invertible, then*
$$\mathcal{E}_+(CAC^H) = \mathcal{E}_+(A), \quad \mathcal{E}_-(CAC^H) = \mathcal{E}_-(A), \quad \text{and} \quad \mathcal{E}_0(CAC^H) = \mathcal{E}_0(A).$$

Exercise 28.11. Show that if $A, B \in \mathbb{C}^{n \times n}$, $A \succeq O$, and $B = B^H$, then $\mathcal{E}_-(A - B) \leq \mathcal{E}_+(B)$. [HINT: $A - B = \begin{bmatrix} I_n & I_n \end{bmatrix} \begin{bmatrix} A & O \\ O & -B \end{bmatrix} \begin{bmatrix} I_n \\ I_n \end{bmatrix}$.]

28.7. Supplementary notes

Sections 28.1 and 28.6 are adapted from [30]. Sections 28.4 and 28.3 are adapted from Bhatia [9]. Theorem 28.6 is adapted from Theorem A4 in the paper [58] by R. Lippert, which has stronger results than are presented here. The fact that the limit (28.15) does not change if $\lambda_j(A + \mu B)$ is replaced by $\lambda_j(D + \mu(MD - DM + C))$ can be justified by invoking (28.5). There is an analogue of Theorem 28.6 in which the limit as $\mu \downarrow 0$ is replaced by the limit as $\mu \uparrow 0$. These two limits are not always the same. They will be if A has n distinct eigenvalues.

Chapter 29

Singular values redux I

In this chapter we shall establish extremal characterizations of the partial sums $s_1(A) + \cdots + s_k(A)$ and partial products $s_1(A) \cdots s_k(A)$, $k = 1, \ldots, q$, of the singular values $s_j(A)$ of a matrix $A \in \mathbb{C}^{p \times q}$. These characterizations will then be used to develop a new class of norms on the space $\mathbb{C}^{p \times q}$ and to establish upper bounds on the eigenvalues of a square matrix in terms of its singular values.

29.1. Sums of singular values

The next theorem supplies the extension of formula (15.14) that was used in Chapter 15; the proof rests on von Neumann's trace inequality (27.9).

Theorem 29.1. *If $A \in \mathbb{C}^{p \times q}$ and $k \in \{1, \ldots, q\}$, then*

$$(29.1) \quad \sum_{j=1}^{k} s_j(A) = \max\{|\operatorname{trace}(Y^H A X)| : X \in \mathbb{C}^{q \times k}, Y \in \mathbb{C}^{p \times k},$$

$$\text{and } X^H X = Y^H Y = I_k\}.$$

Proof. Let $m = \min\{\operatorname{rank} A, \operatorname{rank} XY^H\}$. In view of (27.9),

$$|\operatorname{trace}(Y^H A X)| = |\operatorname{trace}(A X Y^H)| \leq \sum_{j=1}^{m} s_j(A) s_j(XY^H) = \sum_{j=1}^{k} s_j(A),$$

since $\operatorname{rank} XY^H = k$ and $s_j(XY^H) = 1$ for $j = 1, \ldots, k$.

Equality is obtained by choosing X and Y appropriately: If rank $A = r$ and the singular value decomposition of A is $A = V_1 S_1 U_1^H$ with isometric factors $V_1 \in \mathbb{C}^{p \times r}$ and $U_1 \in \mathbb{C}^{q \times r}$ and $1 \leq k < r$, then $V_1 = \begin{bmatrix} V_{11} & V_{12} \end{bmatrix}$ and $U_1 = \begin{bmatrix} U_{11} & U_{12} \end{bmatrix}$ with $V_{11} \in \mathbb{C}^{p \times k}$ and $U_{11} \in \mathbb{C}^{q \times k}$. Thus, if $X = U_{11}$ and $Y = V_{11}$, then

$$Y^H A X = V_{11}^H \begin{bmatrix} V_{11} & V_{12} \end{bmatrix} S_1 \begin{bmatrix} U_{11}^H \\ U_{12}^H \end{bmatrix} U_{11} = \begin{bmatrix} I_k & O \end{bmatrix} S_1 \begin{bmatrix} I_k \\ O \end{bmatrix}$$

and hence $\mathrm{trace}(Y^H A X) = s_1 + \cdots + s_k$.

If $k = r$, then the equality $\mathrm{trace}(Y^H A X) = s_1 + \cdots + s_r$ is obtained by choosing $X = U_1$ and $Y = V_1$. $\square$

Corollary 29.2. *If* $A, B \in \mathbb{C}^{p \times q}$, *then*

$$(29.2) \quad \sum_{j=1}^{k} s_j(A+B) \leq \sum_{j=1}^{k} s_j(A) + \sum_{j=1}^{k} s_j(B) \quad \text{for } k = 1, \ldots, q.$$

Proof. In view of Theorem 29.1,

$$|\langle (A+B)X, Y \rangle| = |\langle AX, Y \rangle + \langle BX, Y \rangle| \leq |\langle AX, Y \rangle| + |\langle BX, Y \rangle|$$

$$\leq \sum_{j=1}^{k} s_j(A) + \sum_{j=1}^{k} s_j(B)$$

for every pair of isometric matrices $X \in \mathbb{C}^{q \times k}$ and $Y \in \mathbb{C}^{p \times k}$. The inequality (29.2) is obtained by choosing X and Y to maximize the first term in the last display. $\square$

Our next objective is to extend the inequality (29.2) to

$$\sum_{j=1}^{k} s_j(A+B)^t \leq \sum_{j=1}^{k} (s_j(A) + s_j(B))^t \quad \text{for } k = 1, \ldots, q \text{ and } 1 \leq t < \infty$$

by a method called **majorization**.

29.2. Majorization

The main result of this section is Theorem 29.4. It is convenient to begin with a lemma.

Lemma 29.3. *Let* $\{a_1, \ldots, a_n\}$ *and* $\{b_1, \ldots, b_n\}$ *be two sequences of real numbers such that*

$$a_1 \geq a_2 \geq \cdots \geq a_n, \quad b_1 \geq b_2 \cdots \geq b_n,$$

$$(29.3) \quad \text{and} \quad \sum_{j=1}^{k} a_j \leq \sum_{j=1}^{k} b_j \quad \text{for } k = 1, \ldots, n,$$

and let
$$(x-s)_+ = \begin{cases} x-s & \text{for } x-s>0, \\ 0 & \text{for } x-s \le 0. \end{cases}$$
Then
$$\sum_{j=1}^{k}(a_j-s)_+ \le \sum_{j=1}^{k}(b_j-s)_+$$
for every $s \in \mathbb{R}$.

Proof. Let
$$\alpha(s) = (a_1-s)_+ + \cdots + (a_k-s)_+ \quad \text{and} \quad \beta(s) = (b_1-s)_+ + \cdots + (b_k-s)_+$$
and consider the following cases:

(1) If $s < a_k$, then
$$\begin{aligned}\alpha(s) &= (a_1-s) + \cdots + (a_k-s) \\ &\le (b_1-s) + \cdots + (b_k-s) \\ &\le (b_1-s)_+ + \cdots + (b_k-s)_+ = \beta(s).\end{aligned}$$

(2) If $a_j \le s \le a_{j-1}$ for $j = 2, \ldots, k$, then
$$\begin{aligned}\alpha(s) &= (a_1-s)_+ + \cdots + (a_j-s)_+ \\ &= (a_1-s) + \cdots + (a_{j-1}-s) \\ &\le (b_1-s) + \cdots + (b_{j-1}-s) \\ &\le (b_1-s)_+ + \cdots + (b_k-s)_+ = \beta(s).\end{aligned}$$

(3) If $s \ge a_1$, then $\alpha(s) = 0$ and so $\beta(s) \ge \alpha(s)$, since $\beta(s) \ge 0$. □

Theorem 29.4. If $a_1 \ge \cdots \ge a_n \ge a$ and $b_1 \ge \cdots \ge b_n \ge a$, then

(29.4)
$$\sum_{j=1}^{k} a_j \le \sum_{j=1}^{k} b_j \quad \text{for } k=1,\ldots,n$$
$$\implies \sum_{j=1}^{k} \varphi(a_j) \le \sum_{j=1}^{k} \varphi(b_j) \quad \text{for } k=1,\ldots,n$$

and every convex function $\varphi \in \mathcal{C}^2(Q)$ on the interval $Q = (a, \infty)$ for which $\lim_{b \downarrow a} \varphi(b) = 0$, $\lim_{b \downarrow a} \varphi'(b) = 0$, and $\lim_{b \downarrow a} b\varphi'(b) = 0$.

Proof. The key to the proof is the integral representation

(29.5)
$$\varphi(x) = \int_a^\infty (x-u)_+ \varphi''(u)\,du$$

for functions $\varphi(x)$ that meet the given assumptions. Then, since $\varphi(x)$ is convex, $\varphi''(u) \geq 0$ for $u \in (a, \infty)$ and hence, in view of Lemma 29.3,

$$\sum_{j=1}^{k} \varphi(a_j) = \sum_{j=1}^{k} \int_{a}^{\infty} (a_j - u)_{+} \varphi''(u) du \leq \sum_{j=1}^{k} \int_{a}^{\infty} (b_j - u)_{+} \varphi''(u) du$$

$$= \sum_{j=1}^{k} \varphi(b_j).$$

It remains to verify (29.5). Towards this end, choose $x > b > a$. Then

$$\varphi(x) - \varphi(b) = \int_{b}^{x} \varphi'(s) ds = \int_{b}^{x} \left\{ \int_{b}^{s} \varphi''(u) du + \varphi'(b) \right\} ds$$

$$= \int_{b}^{x} \left(\int_{u}^{x} ds \right) \varphi''(u) du + (x - b) \varphi'(b)$$

$$= \int_{b}^{x} (x - u) \varphi''(u) du + (x - b) \varphi'(b).$$

Formula (29.5) is obtained by letting $b \downarrow a$ and replacing $(x - u)$ by $(x - u)_{+}$. □

Corollary 29.5. *Let $\{a_1, \ldots, a_n\}$ and $\{b_1, \ldots, b_n\}$ be two sequences of real numbers such that $a_1 \geq a_2 \geq \cdots \geq a_n$, $b_1 \geq b_2 \geq \cdots \geq b_n$, and*

$$\sum_{j=1}^{k} a_j \leq \sum_{j=1}^{k} b_j \quad \text{for} \quad k = 1, \ldots, n.$$

Then

$$\sum_{j=1}^{k} e^{a_j} \leq \sum_{j=1}^{k} e^{b_j} \quad \text{for} \quad k = 1, \ldots, n.$$

Proof. This is an immediate consequence of Theorem 29.4 and the formula

$$e^x = \int_{-\infty}^{x} (x - s) e^s ds = \int_{-\infty}^{\infty} (x - s)_{+} e^s ds,$$

which corresponds to the choice $\varphi(x) = e^x$ and $a = -\infty$. □

Exercise 29.1. Show that if $a_1 \geq \cdots \geq a_n > 0$, $b_1 \geq \cdots \geq b_n > 0$, and $1 > t > 0$, then

$$\sum_{j=1}^{k} a_j \leq \sum_{j=1}^{k} b_j \text{ for } k = 1, \ldots, n \implies \sum_{j=1}^{k} a_j^t \geq \sum_{j=1}^{k} b_j^t \text{ for } k = 1, \ldots, n.$$

29.3. Norms based on sums of singular values

Theorem 29.6. *If $A, B \in \mathbb{C}^{p\times q}$ and $1 \le t < \infty$, then*

$$(29.6) \qquad \sum_{j=1}^{k} s_j(A+B)^t \le \sum_{j=1}^{k}(s_j(A) + s_j(B))^t \quad \text{for } k = 1, \ldots, q$$

and the function

$$(29.7) \qquad \varphi_{k,t}(A) = \left(\sum_{j=1}^{k} s_j(A)^t\right)^{1/t}$$

defines a norm on $\mathbb{C}^{p\times q}$ for each choice of $k \in \{1, \ldots, q\}$ and $t \in [1, \infty)$.

Proof. Let $a_j = s_j(A+B)$ and $b_j = s_j(A) + s_j(B)$ for $j = 1, \ldots, q$. Then $a_1 \ge \cdots \ge a_q \ge 0$, $b_1 \ge \cdots \ge b_q \ge 0$, and, in view of (29.2), $\sum_{j=1}^{k} a_j \le \sum_{j=1}^{k} b_j$ for $k = 1, \ldots, q$. Thus, as the function $\varphi(x) = x^t$ is convex on $(0, \infty)$ and $\varphi(x)$, $\varphi'(x)$, and $x\varphi'(x)$ all tend to zero as $x \downarrow 0$ when $t > 1$, Theorem 29.4 ensures that

$$\sum_{j=1}^{k} s_j(A+B)^t = \sum_{j=1}^{k} a_j^t \le \sum_{j=1}^{k} b_j^t = \sum_{j=1}^{k}(s_j(A) + s_j(B))^t \quad \text{for } 1 < t < \infty.$$

Consequently, as

$$\left(\sum_{j=1}^{k}(s_j(A) + s_j(B))^t\right)^{1/t} \le \left(\sum_{j=1}^{k}(s_j(A))^t\right)^{1/t} + \left(\sum_{j=1}^{k}(s_j(B))^t\right)^{1/t}$$

for $1 \le t < \infty$ by Minkowski's inequality (9.24),

$$\varphi_{k,t}(A+B) \le \varphi_{k,t}(A) + \varphi_{k,t}(B),$$

i.e., the triangle inequality holds. The remaining properties of a norm are easily verified; the details are left to the reader. $\square$

29.4. Unitarily invariant norms

A norm $\varphi(A)$ on $\mathbb{C}^{p\times q}$ is said to be **unitarily invariant** if $\varphi(UAV) = \varphi(A)$ for every $A \in \mathbb{C}^{p\times q}$ and every pair of unitary matrices $U \in \mathbb{C}^{p\times p}$ and $V \in \mathbb{C}^{q\times q}$. The norm $\varphi_{k,t}(A)$ considered in (29.7) is an example of a unitarily invariant norm.

A norm $g(\mathbf{x})$ on $\mathbb{R}^n$ is said to be a **symmetric gauge function** if:

(1) $g(P\mathbf{x}) = g(\mathbf{x})$ for every $\mathbf{x} \in \mathbb{R}^n$ and every permutation matrix $P \in \mathbb{R}^{n \times n}$.

(2) $g(D\mathbf{x}) = g(\mathbf{x})$ for every $\mathbf{x} \in \mathbb{R}^n$ and every orthogonal diagonal matrix $D \in \mathbb{R}^{n \times n}$.

The classical norms $\|\mathbf{x}\|_t$, $1 \leq t \leq \infty$, are all symmetric gauge functions.

A theorem of **von Neumann** establishes a connection between these two classes of norms. To formulate it, the notation

$$\mathbf{s}_A = \begin{bmatrix} s_1(A) & \cdots & s_n(A) \end{bmatrix}^T \quad \text{for } A \in \mathbb{C}^{n \times n}$$

and

$$D_{\mathbf{x}} = \operatorname{diag}\{x_1, \ldots, x_n\} \quad \text{for } \mathbf{x} \in \mathbb{R}^n$$

will be convenient.

Theorem 29.7. *If $A \in \mathbb{C}^{n \times n}$ and g is a symmetric gauge function on $\mathbb{R}^n$, then the function $\varphi(A) = g(\mathbf{s}_A)$ is a unitarily invariant norm on $\mathbb{C}^{n \times n}$.*

Conversely, if $A \in \mathbb{C}^{n \times n}$ and $\varphi(A)$ is a unitarily invariant norm on $\mathbb{C}^{n \times n}$, then the function $g(\mathbf{x}) = \varphi(D_{\mathbf{x}})$ is a symmetric gauge function.

Proof. See, e.g., Theorem IV.2.1 on page 91 in Bhatia [9]. □

29.5. Products of singular values

The main result in this section is Theorem 29.9. We begin with a preliminary lemma.

Lemma 29.8. *Let $A \in \mathbb{C}^{p \times q}$, let $s_1 \geq \cdots \geq s_q$ denote the singular values of A, and let $1 \leq k \leq q$. Then*

$$\det(W^H A^H A W) \leq s_1^2 \cdots s_k^2 \det(W^H W) \quad \text{for every choice of } W \in \mathbb{C}^{q \times k}.$$

Proof. Let $A \in \mathbb{C}^{p \times q}$ and $A^H A = US^2 U^H$ with $U \in \mathbb{C}^{q \times q}$ unitary, $S = \operatorname{diag}\{s_1, \ldots, s_q\}$, and $W \in \mathbb{C}^{q \times k}$. Then

$$\det\{W^H A^H A W\} = \det\{W^H U S^2 U^H W\} = \det X X^H$$

with $X = W^H US \in \mathbb{C}^{k \times q}$. Thus, upon writing $X \in \mathbb{C}^{k \times q}$ as an array $X = \begin{bmatrix} \mathbf{x}_1 & \cdots & \mathbf{x}_q \end{bmatrix}$ of q columns of height k and setting

$$Y = W^H U = \begin{bmatrix} \mathbf{y}_1 & \cdots & \mathbf{y}_q \end{bmatrix}$$

so that $X = YS$, the Binet-Cauchy formula implies that

$$\det XX^H = \sum \det \begin{bmatrix} \mathbf{x}_{i_1} & \cdots & \mathbf{x}_{i_k} \end{bmatrix} \det \begin{bmatrix} \mathbf{x}_{i_1}^H \\ \vdots \\ \mathbf{x}_{i_k}^H \end{bmatrix}$$

$$= \sum |\det \begin{bmatrix} \mathbf{x}_{i_1} & \cdots & \mathbf{x}_{i_k} \end{bmatrix}|^2$$

$$= \sum |\det \begin{bmatrix} \mathbf{y}_{i_1} & \cdots & \mathbf{y}_{i_k} \end{bmatrix}|^2 s_{i_1}^2 \cdots s_{i_k}^2$$

$$\leq \sum |\det \begin{bmatrix} \mathbf{y}_{i_1} & \cdots & \mathbf{y}_{i_k} \end{bmatrix}|^2 s_1^2 \cdots s_k^2,$$

where each sum is over all k-tuples $1 \leq i_1 < \cdots < i_k \leq q$. Thus,

$$\det\{W^H A^H A W\} = \det XX^H \leq s_1^2 \cdots s_k^2 \det YY^H = s_1^2 \cdots s_k^2 \det W^H W,$$

as claimed. □

Theorem 29.9. *If* $A \in \mathbb{C}^{p \times q}$ *with singular values* $s_1 \geq \cdots \geq s_q$ *and* $k \leq \min\{p, q\}$, *then*

(29.8) $\quad s_1^2 \cdots s_k^2 = \max\left\{\det\{W^H A^H A W\}; \; W \in \mathbb{C}^{q \times k} \text{ and } W^H W = I_k\right\}.$

Proof. Lemma 29.8 ensures that

$$\det\{W^H A^H A W\} \leq s_1^2 \cdots s_k^2 \quad \text{for every } W \in \mathbb{C}^{q \times k} \text{ such that } W^H W = I_k.$$

If $A^H A = U S^2 U^H$ with $U \in \mathbb{C}^{q \times q}$ unitary and $S = \text{diag}\{s_1, \ldots, s_q\}$, then equality in (29.8) is achieved by choosing

$$W = U \begin{bmatrix} I_k \\ O_{(q-k) \times k} \end{bmatrix}. \qquad \square$$

Exercise 29.2. Show that if $A \in \mathbb{C}^{p \times q}$ with singular values $s_1 \geq \cdots \geq s_q$ and $1 \leq k \leq \text{rank } A$, then
(29.9)
$$\max\left\{\frac{\det\{W^H A^H A W\}}{\det W^H W} : W \in \mathbb{C}^{q \times k} \text{ and } \text{rank } W = k\right\} = s_1^2 \cdots s_k^2.$$

Exercise 29.3. Let $A, B \in \mathbb{C}^{n \times n}$ with singular values $s_1 \geq \cdots \geq s_n$ and $t_1 \geq \cdots \geq t_n$, respectively. Show that

$$\max\left\{|\text{trace}(UAVB)| : U, V \in \mathbb{C}^{n \times n} \text{ and } U^H U = V^H V = I_n\right\} = \sum_{j=1}^n s_j t_j.$$

[HINT: Theorem 27.8.]

29.6. Eigenvalues versus singular values

Lemma 29.10. *Let $A \in \mathbb{C}^{n \times n}$, let $s_1 \geq \cdots \geq s_n$ denote the singular values of A, and suppose that the eigenvalues $\lambda_1, \ldots, \lambda_n$ of A, repeated according to their algebraic multiplicity, are indexed so that $|\lambda_1| \geq |\lambda_2| \geq \cdots \geq |\lambda_n|$. Then*
$$|\lambda_1| \cdots |\lambda_k| \leq s_1 \cdots s_k \quad \text{for} \quad k = 1, \ldots, n \,.$$

Proof. By Schur's theorem, there exists a unitary matrix $U \in \mathbb{C}^{n \times n}$ such that $T = U^H A U$ is upper triangular and $t_{jj} = \lambda_j$ for $j = 1, \ldots, n$. Thus, if $k < n$ and $U = \begin{bmatrix} U_1 & U_2 \end{bmatrix}$ with $U_1 \in \mathbb{C}^{n \times k}$ and $W = U_1$, then

$$W^H A^H A W = U_1^H U T^H T U^H U_1 = \begin{bmatrix} I_k & O \end{bmatrix} T^H T \begin{bmatrix} I_k \\ O \end{bmatrix},$$

which, upon writing the upper triangular matrix T in compatible block form, can be expressed as

$$\begin{bmatrix} I_k & O \end{bmatrix} \begin{bmatrix} T_{11}^H & O \\ T_{12}^H & T_{22}^H \end{bmatrix} \begin{bmatrix} T_{11} & T_{12} \\ O & T_{22} \end{bmatrix} \begin{bmatrix} I_k \\ O \end{bmatrix} = T_{11}^H T_{11},$$

where T_{11} denotes the upper left-hand $k \times k$ corner of T. Therefore,

$$\begin{aligned}
|\lambda_1 \cdots \lambda_k|^2 &= |\det T_{11}|^2 = \det\{T_{11}^H T_{11}\} \\
&= \det\{W^H U T^H T U^H W\} = \det\{W^H U T^H U^H U T U^H W\} \\
&= \det\{W^H A^H A W\} \\
&\leq s_1^2 \cdots s_k^2 \det\{W^H W\} = s_1^2 \cdots s_k^2,
\end{aligned}$$

by Lemma 29.8. $\square$

Corollary 29.11. *Let $A \in \mathbb{C}^{n \times n}$ with singular values $s_1 \geq \cdots \geq s_n$ and eigenvalues $\lambda_1, \ldots, \lambda_n$, repeated according to their algebraic multiplicity and indexed so that $|\lambda_1| \geq \cdots \geq |\lambda_n|$. Then $s_k = 0 \implies \lambda_k = 0$ (i.e., $|\lambda_k| > 0 \implies s_k > 0$).*

Theorem 29.12. *Let $A \in \mathbb{C}^{n \times n}$, let $s_1, \ldots, s_n$ denote the singular values of A, and let $\lambda_1, \ldots, \lambda_n$ denote the eigenvalues of A, repeated according to their algebraic multiplicity and indexed so that $|\lambda_1| \geq \cdots \geq |\lambda_n|$. Then*

(29.10) $$\sum_{j=1}^{k} |\lambda_j|^t \leq \sum_{j=1}^{k} s_j^t \quad \text{for } t > 0 \text{ and } k = 1, \ldots, n\,.$$

Proof. Lemma 29.10 guarantees that

$$|\lambda_1| \cdots |\lambda_k| \leq s_1 \cdots s_k \quad \text{for } k = 1, \ldots, n\,.$$

Suppose that $|\lambda_k| > 0$. Then

$$\ln |\lambda_1| + \cdots + \ln |\lambda_k| \leq \ln s_1 + \cdots + \ln s_k$$

and hence, if $t > 0$,
$$t(\ln|\lambda_1| + \cdots + \ln|\lambda_k|) \leq t(\ln s_1 + \cdots + \ln s_k)$$
or, equivalently,
$$\ln|\lambda_1|^t + \cdots + \ln|\lambda_k|^t \leq \ln s_1^t + \cdots + \ln s_k^t.$$
Consequently, Corollary 29.5 is applicable to the numbers $a_j = \ln|\lambda_j|^t$ and $b_j = \ln s_j^t$ for $j = 1, \ldots, k$ and yields the inequality
$$e^{\ln|\lambda_1|^t} + \cdots + e^{\ln|\lambda_k|^t} \leq e^{\ln s_1^t} + \cdots + e^{\ln s_k^t},$$
which is equivalent to
$$|\lambda_1|^t + \cdots + |\lambda_k|^t \leq s_1^t + \cdots + s_k^t.$$
Thus we have established the inequality for every integer $k \in \{1, \ldots, n\}$ for which $|\lambda_k| > 0$. However, this is really enough, because if $\lambda_\ell = 0$, then $|\lambda_j| \leq s_j$ for $j = \ell, \ldots, n$. Thus, for example, if $n = 5$ and $|\lambda_3| > 0$ but $\lambda_4 = 0$, then the inequality (29.10) holds for $k = 1, 2, 3$ by the preceding analysis. However, it must also hold for $k = 4$ and $k = 5$, since $\lambda_4 = 0 \implies \lambda_5 = 0$ and thus $|\lambda_4| \leq s_4$ and $|\lambda_5| \leq s_5$. □

29.7. Supplementary notes

The proof of Theorem 29.4 is adapted from the proof of Lemma 3.4 in Gohberg-Krein [43], which is formulated with less restrictive smoothness conditions on the convex function $\varphi(x)$. They credit Herman Weyl and Hardy, Littlewood, and Polya [47]. They also present a number of inequalities for the real and imaginary parts of eigenvalues. There is an extensive literature on inequalities for eigenvalues and singular values of matrices; see, e.g., Ando [3] and Bhatia [9] and the references cited therein.

Chapter 30

Singular values redux II

In this chapter we shall develop a number of inequalities for singular values, partially for future use and partially because a number of them play a significant role in assorted algorithms in numerical analysis.

Recall that if $A \in \mathbb{C}^{p \times q}$ and rank $A = r \geq 1$, then $A^H A$ can be expressed in the form $A^H A = US^2 U^H$, where $U \in \mathbb{C}^{q \times q}$ is unitary and $S = \text{diag}\{s_1, \ldots, s_q\}$, where $s_\geq \cdots \geq s_r > 0$ and the remaining singular values of A, if any, are all equal to zero (i.e., $s_j = 0$ for $j = r+1, \ldots, q$ if $r < q$).

30.1. Sums of powers of singular values

In this section we develop two extremal characterizations for sums of powers of the singular values of a matrix A and then use these characterizations to establish some inequalities in terms of the entries in A.

Lemma 30.1. *If* $D \in \mathbb{C}^{n \times n}$ *is a positive semidefinite diagonal matrix and* $\mathbf{x} \in \mathbb{C}^n$, *then*

(30.1) $\quad (\langle D\mathbf{x}, \mathbf{x} \rangle)^t \leq \langle D^t \mathbf{x}, \mathbf{x} \rangle \quad \text{for } \|\mathbf{x}\| = 1 \text{ and } 1 \leq t < \infty$

and

(30.2) $\quad (\langle D\mathbf{x}, \mathbf{x} \rangle)^t \geq \langle D^t \mathbf{x}, \mathbf{x} \rangle \quad \text{for } \|\mathbf{x}\| = 1 \text{ and } 0 < t \leq 1.$

Proof. The asserted inequalities are self-evident if $t = 1$. Therefore, it suffices to verify (30.1) for $1 < t < \infty$ and (30.2) for $0 < t < 1$.

If $t > 1$, $1/s = 1 - 1/t$, $d_{jj} = \delta_j$, and $\|\mathbf{x}\| = 1$, then, by Hölder's inequality,

$$\langle D\mathbf{x}, \mathbf{x}\rangle = \sum_{j=1}^n \delta_j |x_j|^2 = \sum_{j=1}^n \delta_j |x_j|^{2/t} |x_j|^{2/s}$$

$$\leq \left(\sum_{j=1}^n \delta_j^t |x_j|^2\right)^{1/t} \left(\sum_{j=1}^n |x_j|^2\right)^{1/s} = \left(\sum_{j=1}^n \delta_j^t |x_j|^2\right)^{1/t},$$

which is equivalent to (30.1), since $\sum_{j=1}^n \delta_j^t |x_j|^2 = \langle D^t \mathbf{x}, \mathbf{x}\rangle$.

Suppose next that $0 < t < 1$. Then $1 < t^{-1} < \infty$. Thus, if $C \in \mathbb{C}^{n \times n}$ is a positive semidefinite diagonal matrix, then

$$(\langle C\mathbf{x}, \mathbf{x}\rangle)^{1/t} \leq \langle C^{1/t}\mathbf{x}, \mathbf{x}\rangle \, \|\mathbf{x}\|^{2(1-t)/t}$$

by (30.1). The inequality (30.2) is obtained by setting $C = D^t$ and raising both sides to the t'th power. $\square$

Exercise 30.1. Show that if $V \in \mathbb{C}^{n \times k}$ is isometric and $D \in \mathbb{C}^{k \times k}$ is a positive semidefinite diagonal matrix, then

(30.3) $\quad \left(\langle VDV^H \mathbf{x}, \mathbf{x}\rangle\right)^t \leq \langle VD^t V^H \mathbf{x}, \mathbf{x}\rangle \quad$ for $\|\mathbf{x}\| = 1$ and $1 \leq t < \infty$

and

(30.4) $\quad \left(\langle VDV^H \mathbf{x}, \mathbf{x}\rangle\right)^t \geq \langle VD^t V^H \mathbf{x}, \mathbf{x}\rangle \quad$ for $\|\mathbf{x}\| = 1$ and $0 < t \leq 1$.

Theorem 30.2. If $A \in \mathbb{C}^{p \times q}$, $\operatorname{rank} A = r \geq 1$, $k \in \{1, \ldots, q\}$, and $\mathbf{e}_j$ denotes the j'th column of I_k, then

(30.5)
$$\max\left\{\sum_{j=1}^k \|AW\mathbf{e}_j\|^t : W \in \mathbb{C}^{q \times k} \text{ is isometric}\right\}$$
$$= \sum_{j=1}^k s_j(A)^t \quad \text{if } 2 \leq t < \infty$$

and

(30.6)
$$\min\left\{\sum_{j=1}^k \|AW\mathbf{e}_j\|^t : W \in \mathbb{C}^{q \times k} \text{ is isometric}\right\}$$
$$= \sum_{j=q-k+1}^q s_j(A)^t \quad \text{if } 0 < t \leq 2.$$

30.1. Sums of powers of singular values

Proof. Since $A^H A = U S^2 U^H$ with $S = \text{diag}\{s_1(A), \ldots, s_q(A)\}$ and $U \in \mathbb{C}^{q \times q}$ unitary,

$$\|AW\mathbf{e}_j\|^2 = \langle A^H A W \mathbf{e}_j, W \mathbf{e}_j \rangle = \langle U S^2 U^H W \mathbf{e}_j, W \mathbf{e}_j \rangle = \langle S^2 X \mathbf{e}_j, X \mathbf{e}_j \rangle$$

with $X = U^H W$. Therefore,

$$\sum_{j=1}^{k} \|AW\mathbf{e}_j\|^t = \sum_{j=1}^{k} \left(\langle S^2 X \mathbf{e}_j, X \mathbf{e}_j \rangle \right)^{t/2}.$$

Moreover, since $U \in \mathbb{C}^{q \times q}$ is unitary and $W \in \mathbb{C}^{q \times k}$ is isometric, X is also isometric and hence $\|X\mathbf{e}_j\| = 1$. Thus, Lemma 30.1 is applicable. There are two cases to consider:

1. If $2 \leq t < \infty$, then $1 \leq t/2 < \infty$ and hence, in view of (30.1),

$$\sum_{j=1}^{k} \|AW\mathbf{e}_j\|^t = \sum_{j=1}^{k} \left(\langle S^2 X \mathbf{e}_j, X \mathbf{e}_j \rangle \right)^{t/2} \leq \sum_{j=1}^{k} \langle S^t X \mathbf{e}_j, X \mathbf{e}_j \rangle$$

$$= \text{trace}(X^H S^t X) = \text{trace}(S^t X X^H) \leq \sum_{j=1}^{k} s_j(A)^t,$$

by the upper bound in the Ky Fan inequality (27.8), since the eigenvalues of S^t are $s_1(A)^t, \ldots, s_q(A)^t$ and the eigenvalues of XX^H are equal to 1 with multiplicity k and 0 with multiplicity $q - k$. This upper bound is valid for every choice of the isometric matrix W. Equality holds if W is chosen equal to U_1 in the block decomposition $U = \begin{bmatrix} U_1 & U_2 \end{bmatrix}$ of the unitary matrix U with $U_1 \in \mathbb{C}^{q \times k}$, because then $X = U^H W = \begin{bmatrix} I_k \\ O \end{bmatrix}$ and hence

$$\sum_{j=1}^{k} (\langle S^2 X \mathbf{e}_j, X \mathbf{e}_j \rangle)^{t/2} = \sum_{j=1}^{k} s_j(A)^t.$$

2. If $0 < t \leq 2$, then $0 < t/2 \leq 1$ and, in view of (30.2),

$$\sum_{j=1}^{k} \|AW\mathbf{e}_j\|^t = \sum_{j=1}^{k} \left(\langle S^2 X \mathbf{e}_j, X \mathbf{e}_j \rangle \right)^{t/2} \geq \sum_{j=1}^{k} \langle S^t X \mathbf{e}_j, X \mathbf{e}_j \rangle$$

$$= \text{trace } X^H S^t X = \text{trace } S^t X X^H \geq \sum_{j=q-k+1}^{q} s_j(A)^t,$$

by the lower bound in the Ky Fan inequality (27.8). This lower bound is valid for every choice of the isometric matrix W. Equality holds if $W = U_2$

in the block decomposition $U = \begin{bmatrix} U_1 & U_2 \end{bmatrix}$ of the unitary matrix U with $U_2 \in \mathbb{C}^{q \times k}$, because then $X = U^H W = \begin{bmatrix} O \\ I_k \end{bmatrix}$. $\square$

Corollary 30.3. *If $A \in \mathbb{C}^{n \times n}$ and* rank $A = r \geq 1$, *then*

$$\text{(30.7)} \qquad \sum_{j=1}^{n} |\langle A\mathbf{u}_j, \mathbf{u}_j \rangle|^t \leq \sum_{j=1}^{r} s_j(A)^t \quad \text{for } 1 \leq t < \infty$$

and every orthonormal basis $\{\mathbf{u}_1, \ldots, \mathbf{u}_n\}$ *of* $\mathbb{C}^n$.

Proof. Since rank $A = r$, A admits a singular value decomposition of the form $A = V_1 S_1 U_1^H$, with isometric factors $V_1, U_1 \in \mathbb{C}^{n \times r}$. Thus, if $\mathbf{x} \in \mathbb{C}^n$ and $1 \leq t < \infty$, then

$$|\langle A\mathbf{x}, \mathbf{x} \rangle|^t = |\langle V_1 S_1 U_1^H \mathbf{x}, \mathbf{x} \rangle|^t = |\langle S_1^{1/2} U_1^H \mathbf{x}, S_1^{1/2} V_1^H \mathbf{x} \rangle|^t$$
$$\leq \|S_1^{1/2} U_1^H \mathbf{x}\|^t \|S_1^{1/2} V_1^H \mathbf{x}\|^t \leq \frac{\|S_1^{1/2} U_1^H \mathbf{x}\|^{2t} + \|S_1^{1/2} V_1^H \mathbf{x}\|^{2t}}{2}.$$

Consequently, if $\{\mathbf{u}_1, \ldots, \mathbf{u}_n\}$ is any orthonormal basis of $\mathbb{C}^n$, then, in view of Theorem 30.2,

$$\sum_{j=1}^{n} |\langle A\mathbf{u}_j, \mathbf{u}_j \rangle|^t \leq \frac{1}{2} \left\{ \sum_{j=1}^{n} (s_j(S_1^{1/2} U_1^H))^{2t} + \sum_{j=1}^{n} (s_j(S_1^{1/2} V_1^H))^{2t} \right\}$$
$$\leq \frac{1}{2} \left\{ \sum_{j=1}^{n} s_j(A)^t + \sum_{j=1}^{n} s_j(A)^t \right\},$$

as claimed. $\square$

30.2. Inequalities for singular values in terms of A

Theorem 30.4. *If $A \in \mathbb{C}^{p \times q}$ with* rank $A = r \geq 1$ *and positive singular values $s_1 \geq \cdots \geq s_r$, then*

$$\text{(30.8)} \qquad \sum_{i=1}^{p} \sum_{j=1}^{q} |a_{ij}|^t \leq \sum_{j=1}^{r} s_j(A)^t \quad \text{if } 2 \leq t < \infty$$

and

$$\text{(30.9)} \qquad \sum_{i=1}^{p} \sum_{j=1}^{q} |a_{ij}|^t \geq \sum_{j=1}^{r} s_j(A)^t \quad \text{if } 0 < t \leq 2.$$

Proof. Let $\mathbf{a}_j$ denote the j'th column of A, $j = 1, \ldots, q$. If $t \geq 2$, $\|\mathbf{a}_j\|_t \leq \|\mathbf{a}_j\|_2$, i.e.,

$$\sum_{i=1}^{p} |a_{ij}|^t \leq \|\mathbf{a}_j\|_2^t.$$

Consequently,

$$\sum_{j=1}^{q}\sum_{i=1}^{p} |a_{ij}|^t \leq \sum_{j=1}^{q} \|\mathbf{a}_j\|^t \leq \sum_{j=1}^{q} s_j(A)^t,$$

where the last inequality follows from (30.5) upon choosing $k = q$ and $W = I_q$. This completes the proof of (30.8).

The proof of (30.9) rests upon the inequalities

(30.10) $$\sum_{i=1}^{p} |a_{ij}|^t \geq \|\mathbf{a}_j\|_2^t \quad \text{if } 0 < t \leq 2$$

and

(30.11) $$\sum_{j=1}^{q} \|\mathbf{a}_j\|_2^t \geq \sum_{j=1}^{q} s_j(A)^t \quad \text{if } 0 < t \leq 2.$$

The verification of (30.10) is relegated to Exercise 30.2; the inequality (30.11) is immediate from (30.6). The details are left to the reader. □

Exercise 30.2. Verify the inequality (30.10). [HINT: Use Exercise 21.2.]

30.3. Perturbation of singular values

In this section we shall investigate the properties of $s_j(A)$ under small changes in the matrix A. The next lemma shows that $s_j(A)$ is a continuous function of A.

Lemma 30.5. *If* $A, B \in \mathbb{C}^{p \times q}$, *then*

(30.12) $$|s_j(A) - s_j(B)| \leq \|A - B\| \quad \text{for } j = 1, \ldots, q.$$

Proof. If $j = 1$, then (30.12) is just another way of expressing the well-known inequality $|\|A\| - \|B\|| \leq \|A - B\|$.

If $j > 1$, then

$$s_j(A) = \min\{\|A - C\| : C \in \mathbb{C}^{p \times q} \quad \text{and} \quad \operatorname{rank} C \leq j - 1\}.$$

Therefore,

$$s_j(A) \leq \|A - C\| = \|B - C + A - B\| \leq \|B - C\| + \|A - B\|$$

for every $C \in \mathbb{C}^{p \times q}$ with $\operatorname{rank} C \leq j - 1$. Thus, upon minimizing the term

on the right over all admissible C, we obtain the inequality
$$s_j(A) \leq s_j(B) + \|A - B\|.$$
But this justifies (30.12), since A and B may be interchanged. □

Lemma 30.6. *If $A, B \in \mathbb{C}^{p \times q}$ and $t \geq 1$, then*

(30.13) $|s_j(A)^t - s_j(B)^t| \leq t \|A - B\| (s_j(A) + s_j(B))^{t-1}$ *for $j = 1, \ldots, q$.*

Proof. Let $f(x) = x^t$. If $a \geq 0$, $b \geq 0$, and $t > 1$, then, by the mean value theorem, there exists a point $c = \tau a + (1 - \tau)b$, $0 < \tau < 1$, such that
$$a^t - b^t = t(a - b) c^{t-1} = t(a - b)(\tau a + (1 - \tau)b)^{t-1}.$$
Consequently,
$$|s_j(A)^t - s_j(B)^t| = t|s_j(A) - s_j(B)|(\tau s_j(A) + (1 - \tau)s_j(B))^{t-1}$$
$$\leq t\|A - B\|(s_j(A) + s_j(B))^{t-1},$$
as claimed. □

Our next objective is to obtain information on the singular values of A from information on the eigenvalues of the matrix $\mathbb{A}$ constructed in Exercise 30.3 (below) and Theorem 28.6.

Exercise 30.3. Show that if $A \in \mathbb{C}^{p \times q}$ with rank $A = r \geq 1$ and $p + q = n$, then the nonzero eigenvalues $\delta_1(\mathbb{A}) \geq \cdots \geq \delta_r(\mathbb{A})$ and $\delta_{n+r-1}(\mathbb{A}) \geq \cdots \geq \delta_n(\mathbb{A})$ of the Hermitian matrix $\mathbb{A} = \begin{bmatrix} O & A \\ A^H & O \end{bmatrix}$ can be expressed in terms of the nonzero singular values $s_j(A)$ by the formulas
$$\delta_j(\mathbb{A}) = s_j(A) \quad \text{and} \quad \delta_{n+1-j}(\mathbb{A}) = -s_j(A) \quad \text{for } j = 1, \ldots, r.$$

Theorem 30.7. *If $A, B \in \mathbb{C}^{n \times n}$ and A has the singular value decomposition $A = VSU^H$ with unitary factors $V, U \in \mathbb{C}^{n \times n}$, and $C = V^H BU$, then*

(30.14) $\displaystyle\lim_{\mu \downarrow 0} \frac{s_j(A + \mu B)^t - s_j(A)^t}{\mu} = t s_j(A)^{t-1} \nu_j$ *for $j = 1, \ldots, n$*

and

(30.15) $\displaystyle\lim_{\mu \downarrow 0} \sum_{j=1}^{n} \frac{s_j(A + \mu B)^t - s_j(A)^t}{\mu} = t \Re \text{ trace}\{US^{t-1}V^H B\}$

for every choice of $t > 1$. In (30.14), $\nu_1, \ldots, \nu_n$ are the eigenvalues of a block diagonal submatrix of $(C + C^H)/2$ and $\sum_{j=1}^{n} \nu_j = \text{trace}(C + C^H)/2$.

If $s_j(A) > 0$, then (30.14) holds for every $t > 0$; if A is invertible, then (30.15) holds for every $t > 0$.

30.3. Perturbation of singular values

Proof. Since the matrices

$$\mathbb{A} = \begin{bmatrix} O & A \\ A^H & O \end{bmatrix} \quad \text{and} \quad \mathbb{B} = \begin{bmatrix} O & B \\ B^H & O \end{bmatrix}$$

are Hermitian, Theorem 28.6 guarantees that

$$\lim_{\mu \downarrow 0} \frac{\lambda_j(\mathbb{A} + \mu \mathbb{B}) - \lambda_j(\mathbb{A})}{\mu} = \nu_j$$

exists for $j = 1, \ldots, 2n$. Moreover, if $A = VSU^H$ is the singular value decomposition of A and Z_n is defined in terms of the standard basis $\{\mathbf{e}_1, \ldots, \mathbf{e}_n\}$ by the formula $Z_n = \mathbf{e}_1 \mathbf{e}_n^H + \cdots + \mathbf{e}_n \mathbf{e}_1^H$, then the matrix

$$W = \frac{1}{\sqrt{2}} \begin{bmatrix} V & O \\ O & U \end{bmatrix} \begin{bmatrix} I_n & Z_n \\ I_n & -Z_n \end{bmatrix} = \frac{1}{\sqrt{2}} \begin{bmatrix} V & VZ_n \\ U & -UZ_n \end{bmatrix}$$

is unitary and

$$W^H(\mathbb{A} + \mu \mathbb{B})W = \Sigma + \mu F,$$

where

$$\Sigma = \begin{bmatrix} S & O \\ O & -Z_n S Z_n \end{bmatrix}, \quad F = \frac{1}{2} \begin{bmatrix} C + C^H & (C^H - C)Z_n \\ Z_n(C - C^H) & -Z_n(C + C^H)Z_n \end{bmatrix},$$

and $C = V^H B U$. Therefore,

$$\nu_j = \lim_{\mu \downarrow 0} \frac{\lambda_j(\mathbb{A} + \mu \mathbb{B}) - \lambda_j(\mathbb{A})}{\mu} = \lim_{\mu \downarrow 0} \frac{\lambda_j(\Sigma + \mu F) - \lambda_j(\Sigma)}{\mu}$$

$$= \lim_{\mu \downarrow 0} \frac{\lambda_j(\Sigma + \mu F) - s_j(A)}{\mu}$$

for $j = 1, \ldots, n$ and the numbers $\nu_1, \ldots, \nu_n$ are the eigenvalues of a certain block diagonal submatrix of $(C + C^H)/2$. Consequently, as

$$s_j(A + \mu B) = \lambda_j(\mathbb{A} + \mu \mathbb{B}) \quad \text{for } j = 1, \ldots, n \text{ and } \mu \in \mathbb{R},$$

it follows that

(30.16) $$\lim_{\mu \downarrow 0} \frac{s_j(A + \mu B) - s_j(A)}{\mu} = \nu_j \quad \text{for } j = 1, \ldots, n$$

and hence that

(30.17) $$s_j(A + \mu B) = s_j(A) + \mu[\nu_j + \varepsilon(\mu)], \quad \text{where} \quad \lim_{\mu \downarrow 0} \varepsilon(\mu) = 0.$$

Thus,

$$s_j(A + \mu B)^t = (s_j(A) + \mu[\nu_j + \varepsilon(\mu)])^t,$$

which is of the form $f(x) = (s_j(A) + x)^t$ with $x = \mu[\nu_j + \varepsilon(\mu)]$. By the mean value theorem,

$$f(x) - f(0) = xf'(\xi) = xt(s_j(A) + \xi)^{t-1} \quad \text{for some point } \xi \text{ with } |\xi| < |x|.$$

Therefore,

(30.18) $$\frac{s_j(A+\mu B)^t - s_j(A)^t}{\mu} = t[\nu_j + \varepsilon(\mu)](s_j(A) + \xi)^{t-1},$$

which tends to the right-hand side of (30.14) as $\mu \downarrow 0$ when $t \geq 1$ even if $s_j(A) = 0$. If $s_j(A) > 0$, then the same conclusion holds for every $t > 0$. Then, since $\sum_{j=1}^n \nu_j = \operatorname{trace}(C + C^H)/2$ and S is a diagonal matrix,

$$\sum_{j=1}^n s_j(A)^{t-1} \nu_j = \frac{1}{2} \operatorname{trace} S^{t-1}(C + C^H) = \Re \operatorname{trace} S^{t-1} C$$

$$= \Re \operatorname{trace}\{S^{t-1} V^H B U\} = \Re \operatorname{trace}\{U S^{t-1} V^H B\}.$$

This completes the justification of formula (30.15) for every $A \in \mathbb{C}^{n \times n}$ when $t > 1$ and for invertible matrices $A \in \mathbb{C}^{n \times n}$ when $t > 0$. (If rank $A = r < n$, then S^{t-1} is only defined for $t > 1$. The constraint $t > 1$ ensures that the limit in (30.14) is equal to zero for $j > r$.) □

Exercise 30.4. Verify the first equality in the last display. [HINT: Look at Exercise 28.10.]

Notice that if $A, B \in \mathbb{R}^{n \times n}$, then the limit in (30.15) is a linear function of B. The extra complication of taking the real part enters because we are allowing complex matrices.

It is tempting to conclude that if $t > 1$, then the limit in (30.14) is also equal to the real part of a function $f(A; B)$ that is linear in B. However, the next example shows that this is not the case.

Example 30.1. If

$$A = \begin{bmatrix} 1 & 0 \\ 0 & 1 \end{bmatrix}, \quad b \in \mathbb{C}, \quad \text{and} \quad B = \begin{bmatrix} 0 & b \\ 0 & 0 \end{bmatrix},$$

then the eigenvalues of the matrix

$$(A + \mu B)^H (A + \mu B) = \begin{bmatrix} 1 & \mu b \\ \mu \bar{b} & 1 + \mu^2 |b|^2 \end{bmatrix}$$

are equal to

$$\frac{2 + \mu^2 |b|^2 \pm \sqrt{4\mu^2 |b|^2 + \mu^4 |b|^4}}{2}.$$

Thus,

$$\lim_{\mu \downarrow 0} \frac{s_1(A + \mu B)^2 - s_1(A)^2}{\mu} = \lim_{\mu \downarrow 0} \frac{\mu |b|^2 + \sqrt{4|b|^2 + \mu^2 |b|^4}}{2} = |b|,$$

whereas

$$\lim_{\mu \downarrow 0} \frac{s_2(A + \mu B)^2 - s_2(A)^2}{\mu} = \lim_{\mu \downarrow 0} \frac{\mu |b|^2 - \sqrt{4|b|^2 + \mu^2 |b|^4}}{2} = -|b|.$$

30.3. Perturbation of singular values

Neither of these two limits is equal to the real part of a linear function of B. However,

$$(30.19) \qquad \lim_{\mu\downarrow 0} \sum_{j=1}^{2} \frac{s_j(A+\mu B)^2 - s_j(A)^2}{\mu} = 0,$$

which is a linear function of B. ◇

Example 30.2. Let

$$A = \begin{bmatrix} a_{11} & 0 \\ 0 & a_{22} \end{bmatrix} \quad \text{and} \quad B = \begin{bmatrix} b_{11} & 0 \\ 0 & b_{22} \end{bmatrix} \quad \text{with } |a_{11}| > |a_{22}|.$$

Then, since $|a_{11} + \mu b_{11}| > |a_{22} + \mu b_{22}|$ if $\mu > 0$ is sufficiently small,

$$s_1(A+\mu B) = |a_{11} + \mu b_{11}| \quad \text{and} \quad s_2(A+\mu B) = |a_{22} + \mu b_{22}| \quad \text{for such } \mu.$$

Consequently,

$$s_1(A+\mu B)^t - s_1(A)^t = |a_{11} + \mu b_{11}|^t - |a_{11}|^t = |a_{11}|^t \left\{ \left|1 + \mu \frac{b_{11}}{a_{11}}\right|^t - 1 \right\},$$

which, upon setting $\alpha + i\beta = b_{11}/a_{11}$ (with $\alpha, \beta \in \mathbb{R}$), can be expressed as

$$s_1(A+\mu B)^t - s_1(A)^t = |a_{11}|^t \{|1 + \mu(\alpha + i\beta)|^t - 1\}$$
$$= |a_{11}|^t \{|(1 + \mu\alpha)^2 + (\mu\beta)^2|^{t/2} - 1\}.$$

Therefore,

$$(30.20) \qquad \lim_{\mu\downarrow 0} \frac{s_1(A+\mu B)^t - s_1(A)^t}{\mu} = t|a_{11}|^t \alpha = t|a_{11}|^t \Re \frac{b_{11}}{a_{11}},$$

and hence the limit is of the form $t\Re f_1(A; B)$, where f_1 is linear in B. A similar formula holds when s_1 is replaced by s_2 if $|a_{22}| > 0$.

However, if $a_{22} = 0$, then

$$\lim_{\mu\downarrow 0} \frac{s_2(A+\mu B)^t - s_2(A)^t}{\mu} = \lim_{\mu\downarrow 0} \frac{|\mu b_{22}|^t}{\mu} = \begin{cases} 0 & \text{if } t > 1, \\ |b_{22}| & \text{if } t = 1. \end{cases}$$

Thus, the limit exists for $t \geq 1$; however, it can only be expressed as the real part of a function $f_2(A; B)$ that is linear in B if $t > 1$. ◇

Exercise 30.5. Let

$$A = \begin{bmatrix} 1 & 0 & 0 & 0 \\ 0 & 1 & 0 & 0 \\ 0 & 0 & 0 & 0 \\ 0 & 0 & 0 & 0 \end{bmatrix}, \quad B = \begin{bmatrix} B_{11} & B_{12} \\ B_{21} & B_{22} \end{bmatrix} = \begin{bmatrix} 1 & 0 & 0 & 1 \\ 0 & 2 & 1 & 0 \\ 0 & 1 & 3 & 0 \\ 1 & 0 & 0 & 4 \end{bmatrix}, \quad \widehat{B} = \begin{bmatrix} B_{11} & O \\ O & B_{22} \end{bmatrix},$$

where $B_{ij} \in \mathbb{C}^{2\times 2}$ for $i,j = 1,2$. Compute

$$\lim_{\mu\downarrow 0} \frac{\lambda_j(A+\mu\widehat{B}) - \lambda_j(A)}{\mu} \quad \text{for } j=1,\ldots,4,$$

where $\lambda_1(C) \geq \cdots \geq \lambda_n(C)$ for all matrices C that intervene.

Exercise 30.6. Show by direct computation that $\sigma(B) \cap \sigma(\widehat{B}) = \emptyset$ in Exercise 30.5, but nevertheless $\sum_{j=1}^{4} \lambda_j(\widehat{B}) = \sum_{j=1}^{4} \lambda_j(B)$. Explain why this is not surprising.

The basic facts for $A, B \in \mathbb{C}^{n\times n}$ are:

(1) The limit in (30.14) exists for each singular value $s_j(A)$, but it is not necessarily equal to the real part of a linear function of B.

(2) The limit of the full sum is

(30.21) $$\lim_{\mu\downarrow 0} \sum_{j=1}^{n} \frac{s_j(A+\mu B)^t - s_j(A)^t}{\mu} = t\Re f_t(A;B),$$

where $f_t(A;B)$ is linear in B when $1 < t < \infty$.

Exercise 30.7. Show that if $\mathbf{a}$ and $\mathbf{b}$ belong to an inner product space $\mathcal{U}$, then

$$\lim_{\mu\downarrow 0} \frac{\langle \mathbf{a}+\mu\mathbf{b}, \mathbf{a}+\mu\mathbf{b}\rangle_{\mathcal{U}} - \langle \mathbf{a},\mathbf{a}\rangle_{\mathcal{U}}}{\mu} = 2\Re\langle \mathbf{b},\mathbf{a}\rangle_{\mathcal{U}},$$

i.e., the limit is the real part of a complex-valued function that is linear in $\mathbf{b}$.

Exercise 30.8. Compute the limit referred to in the previous exercise explicitly when $\mathcal{U} = \mathbb{C}^{p\times q}$ equipped with the inner product $\langle A,B\rangle_{\mathcal{U}} = \text{trace } B^H A$, and express it as the real part of a complex-valued function that is linear in B.

Exercise 30.9. Show that if $A \in \mathbb{C}^{p\times q}$, then $f(A) = \sum_{j=1}^{k} s_j(A)^t$ is convex on $\mathbb{C}^{p\times q}$ for each choice of $k \in \{1,\ldots,q\}$ and $1 \leq t < \infty$, and it is strictly convex if $1 < t < \infty$. [HINT: $f(A)^{1/t}$ is a norm when $1 \leq t < \infty$.]

Exercise 30.10. A real-valued function φ is said to be Fréchet differentiable on $\mathbb{C}^{p\times q}$ if

(30.22) $$\lim_{\mu\downarrow 0} \frac{\varphi(A+\mu B) - \varphi(A)}{\mu} = \Re f(A;B)$$

exists and $f(A;B)$ is linear in B for every choice of $A, B \in \mathbb{C}^{p \times q}$. Show that if a Fréchet differentiable function φ on $\mathbb{C}^{p \times q}$ has a local maximum or local minimum at A, then the limit in (30.22) is equal to zero for every $B \in \mathbb{C}^{p \times q}$.

30.4. Supplementary notes

The last section was developed to provide backing for the next chapter.

Chapter 31

Approximation by unitary matrices

Given a basis $\{\mathbf{a}_1, \ldots, \mathbf{a}_n\}$ of $\mathbb{C}^n$, we wish to find an orthonormal basis $\{\mathbf{w}_1, \ldots, \mathbf{w}_n\}$ of $\mathbb{C}^n$ such that the error incurred by replacing $\sum_{j=1}^n c_j \mathbf{a}_j$ by $\sum_{j=1}^n c_j \mathbf{w}_j$ is small for a reasonable set of coefficients, say for all c_j with $\sum_{j=1}^n |c_j|^2 \leq 1$. Upon setting $A = \begin{bmatrix} \mathbf{a}_1 & \cdots & \mathbf{a}_n \end{bmatrix}$, $W = \begin{bmatrix} \mathbf{w}_1 & \cdots & \mathbf{w}_n \end{bmatrix}$, and $\mathbf{c} = \begin{bmatrix} c_1 & \cdots & c_n \end{bmatrix}^T$, the identity

$$\sum_{j=1}^n c_j \mathbf{a}_j - \sum_{j=1}^n c_j \mathbf{w}_j = (A - W)\mathbf{c}$$

enables us to reformulate the problem to that of approximating an invertible matrix $A \in \mathbb{C}^{n \times n}$ by a unitary matrix $W \in \mathbb{C}^{n \times n}$. This chapter is devoted to evaluating this approximation in terms of the function

$$(31.1) \qquad \varphi_t(X) = \sum_{j=1}^n s_j(X)^t \quad \text{for } X \in \mathbb{C}^{n \times n} \text{ and } t > 0,$$

which is equal to the t'th power of a norm when $t \geq 1$. The main conclusion is:

Theorem 31.1. *If $A \in \mathbb{C}^{n \times n}$ is invertible and $t > 1$, then*

$$(31.2) \quad \min\left\{\varphi_t(A - W) : WW^H = W^H W = I_n\right\} = \sum_{j=1}^n |s_j(A) - 1|^t.$$

Moreover, in terms of the unitary factors V and U in the singular value decomposition $A = VSU^H$:

(1) *The minimum in* (31.2) *is attained by exactly one unitary matrix* $W = VU^H$, *which can be characterized as the one and only unitary matrix for which* $W^H A$ *is positive definite* (*as in* (6) *of Theorem* 15.1).

(2) *If* $A \succ O$, *then the minimum in* (31.2) *is attained by* $W = I_n$.

31.1. Approximation in the Frobenius norm

To warm up, we shall first evaluate (31.2) for $t = 2$, the square of the **Frobenius norm**, because this computation is easy, thanks to the identities

$$\varphi_2(A - W) = \sum_{j=1}^{n} s_j(A - W)^2 = \text{trace}(A - W)^H(A - W)$$

$$= \sum_{j=1}^{n} \|(A - W)\mathbf{e}_j\|^2 = \sum_{i,j=1}^{n} |a_{ij} - w_{ij}|^2,$$

where $\mathbf{e}_j$ denotes the j'th column of I_n. Since A is presumed to be invertible, it admits a singular value decomposition $A = VSU^H$ in which $V, U \in \mathbb{C}^{n \times n}$ are unitary and S is the diagonal $n \times n$ matrix with entries $s_1(A) \geq \cdots \geq s_n(A) > 0$. Thus, in view of (2) and (3) of Theorem 15.6,

$$s_j(A - W) = s_j(VSU^H - W) = s_j(V[S - V^H WU]U^H) = s_j(S - Z),$$

wherein $Z = V^H WU$ is unitary. Consequently,

$$\varphi_2(A - W) = \varphi_2(S - Z) = \sum_{i=1}^{n} |s_i - z_{ii}|^2 + \sum_{i,j=1, i \neq j}^{n} |z_{ij}|^2 \geq \sum_{i=1}^{n} |s_i - z_{ii}|^2,$$

with equality if and only $z_{ij} = 0$ for $i \neq j$, i.e., if and only if Z is a diagonal matrix. Since Z is also unitary, the constraint $|z_{ii}| = 1$ must also be met. Thus, as $s_i > 0$, the sum $\sum_{i=1}^{n} |s_i - z_{ii}|^2$ will attain its minimum value when $z_{ii} = 1$ for $i = 1, \ldots, n$, i.e., when $V^H WU = Z = I_n$, i.e., when $W = VU^H$. Thus, $\varphi_2(A - W)$ is minimized by choosing $W = VU^H$, the outside factors in the singular value decomposition $A = VSU^H$. But then (31.2) holds for $t = 2$ and $W^H A = UV^H VSU^H = USU^H$, which is clearly positive definite.

It remains to show that there is exactly one unitary matrix W with the property that $W^H A$ is positive definite. But, if W_1 and W_2 are unitary matrices such that both $W_1^H A$ and $W_2^H A$ are positive definite, then

$$W_1^H A = (W_1^H A)^H = A^H W_1 \quad \text{and} \quad W_2^H A = (W_2^H A)^H = A^H W_2.$$

31.2. Approximation in other norms

Therefore,

$$(W_1^H A)^2 = (W_1^H A)^H (W_1^H A) = A^H A = (W_2^H A)^H (W_2^H A) = (W_2^H A)^2 \,.$$

But this serves to identify both $W_1^H A$ and $W_2^H A$ as positive definite square roots of the positive definite matrix $A^H A$. Therefore, they must coincide: $W_1^H A = W_2^H A$. Thus, as A is invertible, $W_1 = W_2$. This completes the proof of (1).

Finally, to verify (2), observe that if $A \succ O$, then $V = U$ and the minimum is achieved by $W = VV^H = I_n$. □

31.2. Approximation in other norms

In this section we shall establish Theorem 31.1. We begin, however, with a general principle that will be useful in the rest of this section.

Lemma 31.2. *If* $P, Z \in \mathbb{C}^{n \times n}$, $P \succeq 0$, *and* $Z^H Z = Z Z^H = I_n$, *then*

(31.3) $\quad PZ = Z^H P \quad \text{if and only if} \quad PZ = ZP \text{ and } \sigma(PZ) \subset \mathbb{R}.$

Proof. If $PZ = Z^H P$, then

$$(Z^H P Z)^2 = Z^H P Z Z^H P Z = Z^H P P Z = P Z Z^H P = P^2 \,.$$

Therefore, $Z^H P Z = P$, since they are both positive semi-definite square roots of P^2. Consequently $PZ = ZP$, and, as $PZ = (PZ)^H$, $\sigma(PZ) \subset \mathbb{R}$.

Conversely, if $PZ = ZP$, then, since P and Z are both normal matrices, Theorem 13.5 ensures that there exists an orthonormal basis $\{\mathbf{u}_1, \ldots, \mathbf{u}_n\}$ such that $P\mathbf{u}_j = \rho_j \mathbf{u}_j$ with $\rho_j \geq 0$ and $Z\mathbf{u}_j = \omega_j \mathbf{u}_j$ with $|\omega_j| = 1$ for $j = 1, \ldots, n$. Consequently, $PZ\mathbf{u}_j = \rho_j \omega_j \mathbf{u}_j$ for $j = 1, \ldots, n$, and hence, in view of the condition $\sigma(PZ) \subset \mathbb{R}$, $\rho_j \omega_j = \rho_j \overline{\omega_j}$. Therefore, $Z^H \mathbf{u}_j = \overline{\omega_j} \mathbf{u}_j = \omega_j \mathbf{u}_j$ and $PZ\mathbf{u}_j = Z^H P \mathbf{u}_j$ for $j = 1, \ldots, n$. □

It is readily checked that if $A = VSU^H$ with unitary factors V and U, then

$$\varphi_t(A - VU^H) = \varphi_t(V(S - I_n)U^H) = \varphi_t(S - I_n)$$

and hence that the minimum in (31.2) (which exists since φ_t is a continuous function acting on a closed bounded subset of $\mathbb{C}^{n \times n}$) is $\leq \varphi_t(S - I_n)$. Theorem 31.1 guarantees that the last inequality is in fact an equality when $t > 1$. The verification is carried out next in a number of small steps, many of which are of independent interest.

1. If $X, Y \in \mathbb{C}^{n \times n}$, $\alpha, \beta \in \mathbb{C}$, and $t \geq 1$, then
$$|\varphi_t(X + \alpha Y) - \varphi_t(X + \beta Y)| \leq tn|\alpha - \beta|\|Y\|(2\|X\| + (|\alpha| + |\beta|)\|Y\|)^{t-1}.$$

In view of Lemma 30.6,
$$|\varphi_t(X + \alpha Y) - \varphi_t(X + \beta Y)|$$
$$\leq \sum_{j=1}^{n} |s_j(X + \alpha Y)^t - s_j(X + \beta Y)^t|$$
$$\leq \sum_{j=1}^{n} t|\alpha - \beta|\|Y\|\{s_j(X + \alpha Y) + s_j(X + \beta Y)\}^{t-1}$$
$$\leq tn|\alpha - \beta|\|Y\|\{s_1(X + \alpha Y) + s_1(X + \beta Y)\}^{t-1},$$
which leads easily to the asserted inequality.

2. If W is a unitary matrix such that $\varphi_t(A - W) \leq \varphi_t(A - B)$, B unitary, and $\widehat{V}\widehat{S}\widehat{U}^H$ is the singular value decomposition of $A - W$, then

(31.4) $\quad \langle \widehat{U}\widehat{S}^{t-1}\widehat{V}^H W\mathbf{x}, \mathbf{x}\rangle \in \mathbb{R} \quad$ for $t \geq 1$ and every vector $\mathbf{x} \in \mathbb{C}^n$.

Let $\{\mathbf{u}_1, \ldots, \mathbf{u}_n\}$ be an orthonormal basis for $\mathbb{C}^n$ and let
$$M(\theta) = e^{i\theta}\mathbf{u}_1\mathbf{u}_1^H + \sum_{j=2}^{n} \mathbf{u}_j\mathbf{u}_j^H \quad \text{for } -\pi < \theta < \pi.$$

Then $M(\theta)$ is a unitary matrix with $M(0) = I_n$. Thus, if $W \in \mathbb{C}^{n \times n}$ is a unitary matrix such that $\varphi_t(A - W) \leq \varphi_t(A - B)$ as B runs through the set of unitary matrices in $\mathbb{C}^{n \times n}$, then
$$\varphi_t(A - WM(\theta)) = \varphi_t(A - W - W(M(\theta) - I_n))$$
$$= \varphi_t(A - W - (e^{i\theta} - 1)W\mathbf{u}_1\mathbf{u}_1^H)$$
$$\geq \varphi_t(A - W)$$

for small $|\theta|$, since $|e^{i\theta} - 1| \leq |\theta|$ when $\theta \in \mathbb{R}$. To ease the reading, let $X = A - W$, $Y = W\mathbf{u}_1\mathbf{u}_1^H$, $f(\theta) = \varphi_t(X - (e^{i\theta} - 1)Y)$, and $g(\theta) = \varphi_t(X - i\theta Y)$. Then the last inequality implies that

(31.5) $\quad \dfrac{f(\theta) - f(0)}{\theta} = \dfrac{f(\theta) - g(\theta) + g(\theta) - g(0)}{\theta} \geq 0$

when $\theta > 0$ and is sufficiently small. Since $|e^{i\theta} - 1 - i\theta| \leq |\theta|^2/2!$, the inequality in step 1 ensures that $\lim_{\theta \downarrow 0}[f(\theta) - g(\theta)]/\theta = 0$. Therefore, in

31.2. Approximation in other norms

view of formula (30.15),

$$\lim_{\theta\downarrow 0}\frac{f(\theta)-f(0)}{\theta}=\lim_{\theta\downarrow 0}\frac{g(\theta)-g(0)}{\theta}=t\Re\,\text{trace}\,\{-i\widehat{U}\widehat{S}^{t-1}\widehat{V}^H Y\}\geq 0\,.$$

Since the opposite inequality is obtained by setting $M(\theta)=e^{-i\theta}\mathbf{u}_1\mathbf{u}_1^H+\sum_{j=2}^n\mathbf{u}_j\mathbf{u}_j^H$ with $\theta>0$, we see that

$$0=t\Re\,\text{trace}\,\{-i\widehat{U}\widehat{S}^{t-1}\widehat{V}^H Y\}=t\Re\,\{-i\langle\widehat{U}\widehat{S}^{t-1}\widehat{V}^H W\mathbf{u}_1,\mathbf{u}_1\rangle\}\,,$$

which yields (31.4), since $\mathbf{u}_1$ is any unit vector in $\mathbb{C}^n$.

3. *If W is as in* **2** *and $t\geq 1$, then* $\widehat{U}\widehat{S}^{t-1}\widehat{V}^H W = W^H\widehat{V}\widehat{S}^{t-1}\widehat{U}^H$.

This is an immediate consequence of step 2 and (11.25):
If $C\in\mathbb{C}^{n\times n}$ and $\langle C\mathbf{x},\mathbf{x}\rangle\in\mathbb{R}$ for every vector $\mathbf{x}\in\mathbb{C}^n$, then

$$\langle(C-C^H)\mathbf{x},\mathbf{x}\rangle=\langle C\mathbf{x},\mathbf{x}\rangle-\langle\mathbf{x},C\mathbf{x}\rangle=0$$

for every $\mathbf{x}\in\mathbb{C}^n$, and hence $C=C^H$.

4. *If W is as in* **2** *and $t\geq 1$, then* $\widehat{S}^{t-1}\widehat{V}^H W\widehat{U}=\widehat{V}^H W\widehat{U}\widehat{S}^{t-1}$ *and* $\sigma(\widehat{S}^{t-1}\widehat{V}^H W\widehat{U})\subset\mathbb{R}$.

Let $Z=\widehat{V}^H W\widehat{U}$. Then, $Z\in\mathbb{C}^{n\times n}$ is unitary, $\widehat{S}^{t-1}\succeq O$, and the preceding step implies that $\widehat{S}^{t-1}Z=Z^H\widehat{S}^{t-1}$. Thus, $\widehat{S}^{t-1}Z=Z\widehat{S}^{t-1}$ and $\sigma(\widehat{S}^{t-1}Z)\subset\mathbb{R}$ by Lemma 31.2.

5. *If W is as in* **2** *and $t>1$, then* $\widehat{S}\widehat{V}^H W\widehat{U}=\widehat{V}^H W\widehat{U}\widehat{S}=\widehat{U}^H W^H\widehat{V}\widehat{S}$.

To this point we know that $\widehat{S}^{t-1}Z=Z\widehat{S}^{t-1}=Z^H\widehat{S}^{t-1}$ and $\sigma(\widehat{S}^{t-1}Z)\subset\mathbb{R}$ when $Z=\widehat{V}^H W\widehat{U}$ and $t>1$. But as $\widehat{S}^{t-1}$ and Z are both normal operators, Theorem 13.5 ensures that there exists an orthonormal basis $\{\mathbf{u}_1,\ldots,\mathbf{u}_n\}$ of $\mathbb{C}^n$ such that $Z\mathbf{u}_j=\omega_j\mathbf{u}_j$ with $|\omega_j|=1$ and $\widehat{S}^{t-1}\mathbf{u}_j=\rho_j\mathbf{u}_j$ with $\rho_j\geq 0$ for $j=1,\ldots,n$. Moreover, $\omega_j=\pm 1$ if $\rho_j>0$. Consequently, $\widehat{S}\mathbf{u}_j=\rho_j^{1/(t-1)}\mathbf{u}_j$ and

$$\widehat{S}Z\mathbf{u}_j=\rho_j^{1/(t-1)}\omega_j\mathbf{u}_j=Z\widehat{S}\mathbf{u}_j=\rho_j^{1/(t-1)}\overline{\omega_j}\mathbf{u}_j=Z^H\widehat{S}\mathbf{u}_j\quad\text{for }j=1,\ldots,n\,.$$

Therefore, **5** holds.

6. *If W is as in* **2**, $A \succ O$, *and* $t > 1$, *then* $AW = WA$ *and* $W = W^H$.

Recall that $\widehat{V}\widehat{S}\widehat{U}^H$ is the singular value decomposition of $A - W$. Then, as $A = A^H$ and W is unitary,

$$AW - I_n = (A - W)^H W = \widehat{U}\{\widehat{S}\widehat{V}^H W \widehat{U}\}\widehat{U}^H = \widehat{U}\{\widehat{U}^H W^H \widehat{V}\widehat{S}\}\widehat{U}^H$$
$$= W^H \widehat{V}\widehat{S}\widehat{U}^H = W^H(A - W) = W^H A - I_n.$$

Therefore, $AW = W^H A$ and hence $AW = WA$ by Lemma 31.2. Thus, $W^H A = WA$ and, as A is invertible, $W = W^H$.

7. *If W is as in* **2**, $A \succ O$, *and* $t > 1$, *then* $\varphi_t(A - W) \geq \sum_{j=1}^{n} |s_j(A) - 1|^t$ *with equality if and only if* $W = I_n$.

Since A and W are normal matrices that commute, Theorem 13.5 ensures that there exists an orthonormal basis $\{\mathbf{u}_1, \ldots, \mathbf{u}_n\}$ of $\mathbb{C}^n$ such that

$$A\mathbf{u}_j = \mu_j \mathbf{u}_j \quad \text{and} \quad W\mathbf{u}_j = \omega_j \mathbf{u}_j \quad \text{for } j = 1, \ldots, n.$$

Consequently, $(A - W)\mathbf{u}_j = (\mu_j - \omega_j)\mathbf{u}_j$ for $j = 1, \ldots, n$. Thus, if the eigenvectors are indexed so that

$$|\mu_1 - \omega_1| \geq |\mu_2 - \omega_2| \geq \cdots \geq |\mu_n - \omega_n|,$$

then Theorem 29.12 ensures that

$$\sum_{j=1}^{n} s_j(A - W)^t \geq \sum_{j=1}^{n} |\lambda_j(A - W)|^t = \sum_{j=1}^{n} |\mu_j - \omega_j|^t.$$

However, $\mu_j > 0$, since $A \succ O$, and $\omega_j = \pm 1$, since $W = W^H$ and $W^H W = I_n$. Therefore, the sum on the far right is clearly minimized when $\omega_j = 1$ for $j = 1, \ldots, n$, i.e., if and only if $W = I_n$. Moreover, since $A \succ O$, $\sum_{j=1}^{n} |\mu_j - 1|^t = \sum_{j=1}^{n} |s_j(A) - 1|^t$.

8. *Verification of* (1) *and* (2) *of Theorem* 31.1.

If $A = VSU^H$ with $S = \text{diag}\{s_1(A), \ldots, s_n(A)\}$ and $V, U \in \mathbb{C}^{n \times n}$ unitary, then

$$s_j(A - W) = s_j(VSU^H - W) = s_j(V[S - V^H W U]U^H) = s_j(S - V^H W U)$$

for every unitary matrix $W \in \mathbb{C}^{n \times n}$. Thus, as $S \succ O$ and $V^H W U$ is unitary, step 7 ensures that

$$\varphi_t(A - W) = \varphi_t(S - V^H W U) \geq \varphi_t(S - I_n) = \sum_{j=1}^{n} |s_j(A) - 1|^t,$$

with equality if and only if $W = VU^H$. Thus (1) holds; (2) then follows from (1) because $VU^H = I_n$ when $A \succeq O$. $\square$

31.2. Approximation in other norms

Remark 31.3. The passage from $S^{t-1}Z = ZS^{t-1}$ to $SZ = ZS$ in the verification of step 5 exploited the fact that S^{t-1} and Z are both normal matrices. It is also of interest to show that *if $B \in \mathbb{C}^{n \times n}$ and $S^\alpha B = BS^\alpha$ for some $\alpha > 0$, then $SB = BS$.*

Towards this end, we first show that

$$(31.6) \qquad S^\alpha B = BS^\alpha \implies S^{\alpha/2} B = BS^{\alpha/2}.$$

Under the given assumptions

$$S^\alpha B - S^{\alpha/2} B S^{\alpha/2} = BS^\alpha - S^{\alpha/2} B S^{\alpha/2}$$

and hence

$$S^{\alpha/2}(S^{\alpha/2} B - BS^{\alpha/2}) + (S^{\alpha/2} B - BS^{\alpha/2})S^{\alpha/2} = O.$$

Thus, if q_{ij} denotes the ij entry of the matrix $Q = S^{\alpha/2} B - BS^{\alpha/2}$, then

$$s_i^{\alpha/2} q_{ij} + q_{ij} s_j^{\alpha/2} = 0,$$

which implies that $q_{ij} = 0$ if $s_i^{\alpha/2} + s_j^{\alpha/2} > 0$. If rank $S = r$ and $r < n$, then this ensures that $q_{ij} = 0$ for all the indices ij except possibly if $i > r$ and $j > r$. However, if $i > r$ and $j > r$, then $q_{ij} = 0$ by definition. Thus, (31.6) holds and serves to ensure that if $S^\alpha B = BS^\alpha$ for some $\alpha > 0$, then the same equality holds for an $\alpha \in (0, \delta)$ for any choice of $\delta > 0$, no matter how small. Consequently, there exists a positive integer m such that $m\alpha < 1 \leq (m+1)\alpha$ and $0 < \alpha < \delta$. Let $\tau = m\alpha$. Then $S^\tau B = BS^\tau$ and $0 < 1 - \tau \leq \delta$. Therefore,

$$(31.7) \qquad \|SB - BS\| = \|(S - S^\tau)B - B(S - S^\tau)\| \leq \|S - S^\tau\| 2\|B\| \leq \kappa \delta,$$

where κ is a positive constant that depends only upon S. Thus, $SB = BS$.

Exercise 31.1. Verify the upper bound in (31.7). [HINT: Exercise 21.3 does part of the job.]

Exercise 31.2. Show that if $A, U \in \mathbb{C}^{n \times n}$, $A \succeq O$, and U is unitary, then $\|I_n - A\| \leq \|U - A\| \leq \|I_n + A\|$.

Exercise 31.3. Show that if $A, U \in \mathbb{C}^{n \times n}$, $A \succeq O$, U is unitary, and $t \geq 1$, then $\varphi_t(I_n - A) \leq \varphi_t(U - A) \leq \varphi_t(I_n + A)$.

Exercise 31.4. Show that if $A, U \in \mathbb{C}^{n \times n}$, $A \succ O$, U is unitary, and $t > 0$, then $\varphi_t(I_n - A) \leq \varphi_t(U - A) \leq \varphi_t(I_n + A)$.

Exercise 31.5. Show that if $0 < a < 1$,

$$A = \begin{bmatrix} 1+a & 0 \\ 0 & 1-a \end{bmatrix}, \quad \text{and} \quad W_\theta = \begin{bmatrix} \cos\theta & -\sin\theta \\ \sin\theta & \cos\theta \end{bmatrix},$$

then $\varphi_1(A - W_\theta) = 2a\varphi_1(A - I_2)$ for all points θ for which $\cos\theta > (2-a^2)/2$. [HINT: Exploit the formula $(s_1 + s_2)^2 = \text{trace}\, B^H B + 2\{\det B^H B\}^{1/2}$ for the singular values $s_1 \geq s_2$ of $B \in \mathbb{C}^{2 \times 2}$ to minimize the calculations.]

31.3. Supplementary notes

This chapter was adapted from a 1980 article [1] by Aiken, Erdos, and Goldstein that was motivated by a problem in quantum chemistry. In [1], the analysis was carried out in a Hilbert space setting. Exercises 31.2–31.5 are adapted from [1].

Chapter 32

Linear functionals

This chapter develops a number of basic properties of linear functionals and related applications, including the Hahn-Banach extension theorem and the Hahn-Banach separation theorem in finite-dimensional normed linear spaces.

32.1. Linear functionals

A linear transformation from a complex (resp., real) vector space $\mathcal{X}$ into $\mathbb{C}$ (resp., $\mathbb{R}$) is called a **linear functional**. Thus, if f belongs to the set $\mathcal{X}'$ of linear functionals on $\mathcal{X}$, then $f(\alpha \mathbf{x} + \beta \mathbf{y}) = \alpha f(\mathbf{x}) + \beta f(\mathbf{y})$ for every choice of $\mathbf{x}, \mathbf{y} \in \mathcal{X}$ and every pair of scalars α, β.

The set $\mathcal{X}'$ is a vector space with respect to the natural rules of addition and scalar multiplication: if $f, g \in \mathcal{X}'$ and α is a scalar, then

$$(f+g)(\mathbf{x}) = f(\mathbf{x}) + g(\mathbf{x}) \quad \text{and} \quad (\alpha f)(\mathbf{x}) = \alpha f(\mathbf{x}) \quad \text{for every } \mathbf{x} \in \mathcal{X}.$$

The main facts to keep in mind are:

Theorem 32.1. *If $\mathcal{X}$ is a finite-dimensional vector space, then:*

(1) *$\mathcal{X}'$ is a finite-dimensional vector space and $\dim \mathcal{X} = \dim \mathcal{X}'$.*

(2) *If $\mathcal{X}$ is a normed linear space with norm $\|\mathbf{x}\|_{\mathcal{X}}$, then $\mathcal{X}'$ is a normed linear space with norm*

$$\begin{aligned} \|f\|_{\mathcal{X}'} &= \max\{|f(\mathbf{x})| : \mathbf{x} \in \mathcal{X} \text{ and } \|\mathbf{x}\|_{\mathcal{X}} = 1\} \\ &= \max\{|f(\mathbf{x})| : \mathbf{x} \in \mathcal{X} \text{ and } \|\mathbf{x}\|_{\mathcal{X}} \leq 1\} \\ &= \max\left\{\frac{|f(\mathbf{x})|}{\|\mathbf{x}\|_{\mathcal{X}}} : \mathbf{x} \in \mathcal{X} \text{ and } \mathbf{x} \neq \mathbf{0}\right\}. \end{aligned}$$

The numerical value of $\|f\|_{\mathcal{X}'}$ depends upon the choice of norm in $\mathcal{X}$; see Exercise 32.2.

(3) $(\mathcal{X}')' = \mathcal{X}$.

Proof. To verify (1), suppose that $\{\mathbf{x}_1, \ldots, \mathbf{x}_n\}$ is a basis for $\mathcal{X}$ and define the linear functionals f_j, $j = 1, \ldots, n$, by the rule

$$f_j(\mathbf{x}_i) = \begin{cases} 1 & \text{if } j = i, \\ 0 & \text{if } j \neq i. \end{cases}$$

Then, if $f \in \mathcal{X}'$, $\mathbf{x} = \sum_{j=1}^n c_j \mathbf{x}_j$, and $f(\mathbf{x}_j) = \alpha_j$,

$$f(\mathbf{x}) = f\left(\sum_{j=1}^n c_j \mathbf{x}_j\right) = \sum_{j=1}^n c_j f(\mathbf{x}_j) = \sum_{j=1}^n c_j \alpha_j = \sum_{j=1}^n c_j \alpha_j f_j(\mathbf{x}_j)$$

$$= \sum_{j=1}^n \alpha_j f_j \left(\sum_{i=1}^n c_i \mathbf{x}_i\right) = \sum_{j=1}^n \alpha_j f_j(\mathbf{x}) = \sum_{j=1}^n (\alpha_j f_j)(\mathbf{x}),$$

i.e., $f = \sum_{j=1}^n \alpha_j f_j$. This proves that $\text{span}\{f_1, \ldots, f_n\} = \mathcal{X}'$.

To complete the verification of (1), it remains to show that the linear functionals $f_1, \ldots, f_n$ are linearly independent. But if $\sum_{j=1}^n a_j f_j(\mathbf{x}) = 0$ for some set of scalars $a_1, \ldots, a_n$ and every vector $\mathbf{x} \in \mathcal{X}$, then

$$a_i = \sum_{j=1}^n a_j f_j(\mathbf{x}_i) = 0 \quad \text{for } i = 1, \ldots, n.$$

Thus, $\{f_1, \ldots, f_n\}$ is a basis for $\mathcal{X}'$ and hence (1) holds.

The verification of (2) and (3) is left to the reader. □

Exercise 32.1. Let $f(\mathbf{x})$ be a linear functional on $\mathbb{C}^n$ and let $f(\mathbf{e}_j) = a_j$ for $j = 1, \ldots, n$, where $\mathbf{e}_j$ denotes the j'th column of I_n. Show that if $A = \begin{bmatrix} a_1 & \cdots & a_n \end{bmatrix}$ and $1 \leq s \leq \infty$, then

$$\max\{|f(\mathbf{x})| : \|\mathbf{x}\|_s = 1\} = \|A\|_{s,\infty}.$$

Exercise 32.2. Let $f(\mathbf{x})$ be a linear functional on $\mathbb{C}^n$ and let $f(\mathbf{e}_j) = a_j$ for $j = 1, \ldots, n$, where $\mathbf{e}_j$ denotes the j'th column of I_n. Show that if $\mathbf{a} = \begin{bmatrix} a_1 & \cdots & a_n \end{bmatrix}^T$, then

$$\max\{|f(\mathbf{x})| : \|\mathbf{x}\|_s = 1\} = \|\mathbf{a}\|_{s'},$$

where $s' = s/(s-1)$ if $1 < s < \infty$; $s' = 1$ if $s = \infty$; and $s' = \infty$ if $s = 1$. [HINT: The inequality $|f(\mathbf{x})| \leq \|\mathbf{a}\|_{s'} \|\mathbf{x}\|_s$ is covered by Theorem 10.4. Equality is obtained by making an appropriate choice for $\mathbf{x}$; if $1 < s < \infty$, try the vector $\mathbf{x}$ with entries $x_j = \overline{a_j} |a_j|^{s'-2}$ when $a_j \neq 0$.]

32.1. Linear functionals

Theorem 32.2. *If f is a linear functional from a complex (resp., real) vector space $\mathcal{X}$ into $\mathbb{C}$ (resp., $\mathbb{R}$), then:*

(1) *The set*
$$\mathcal{N}_f = \{\mathbf{x} \in \mathcal{X} : f(\mathbf{x}) = 0\}$$
is a subspace of $\mathcal{X}$.

(2) *If $\mathcal{N}_f \neq \mathcal{X}$, then there exists a vector $\mathbf{u} \in \mathcal{X}$ such that*
$$\mathcal{X} = \begin{cases} \mathcal{N}_f \dotplus \{\alpha \mathbf{u} : \alpha \in \mathbb{R}\} & \text{if } \mathcal{X} \text{ is a real vector space}, \\ \mathcal{N}_f \dotplus \{\alpha \mathbf{u} : \alpha \in \mathbb{C}\} & \text{if } \mathcal{X} \text{ is a complex vector space}. \end{cases}$$

(3) *If $\mathcal{X}$ is a finite-dimensional inner product space, then there exists exactly one vector $\mathbf{v} \in \mathcal{X}$ such that*
$$(32.1) \qquad f(\mathbf{x}) = \langle \mathbf{x}, \mathbf{v} \rangle_{\mathcal{X}} \quad \text{for every vector } \mathbf{x} \in \mathcal{X}.$$

Proof. The proof of (1) is left to the reader.

To verify (2), suppose that $\mathcal{N}_f \neq \mathcal{X}$. Then there exists a vector $\mathbf{u} \in \mathcal{X}$ such that $f(\mathbf{u}) \neq 0$. Thus, as

$$(32.2) \qquad \mathbf{x} = \left(\mathbf{x} - \frac{f(\mathbf{x})}{f(\mathbf{u})}\mathbf{u}\right) + \left(\frac{f(\mathbf{x})}{f(\mathbf{u})}\mathbf{u}\right) \quad \text{and} \quad \left(\mathbf{x} - \frac{f(\mathbf{x})}{f(\mathbf{u})}\mathbf{u}\right) \in \mathcal{N}_f,$$

it is clear that
$$\mathcal{X} = \mathcal{N}_f + \text{span}\{\mathbf{u}\}.$$
Moreover, this sum is direct, because if $\mathbf{w} = \alpha \mathbf{u}$ belongs to $\mathcal{N}_f$, then $f(\mathbf{w}) = \alpha f(\mathbf{u}) = 0$, which forces $\alpha = 0$ and hence $\mathbf{w} = \mathbf{0}$. Therefore, (2) holds.

If $\mathcal{X}$ is a finite-dimensional inner product space and $\mathcal{N}_f \neq \mathcal{X}$, then there exists a nonzero vector $\mathbf{u} \in \mathcal{X}$ that is orthogonal to $\mathcal{N}_f$. Thus, as $f(\mathbf{u}) \neq 0$, the formulas in (32.2) are in force. Moreover, as $\mathbf{u}$ is orthogonal to $\mathcal{N}_f$,

$$\langle \mathbf{x}, \mathbf{u} \rangle_{\mathcal{X}} = \frac{f(\mathbf{x})}{f(\mathbf{u})} \langle \mathbf{u}, \mathbf{u} \rangle_{\mathcal{X}},$$

i.e., (32.1) holds with $\mathbf{v} = \overline{f(\mathbf{u})}\, \mathbf{u}/\langle \mathbf{u}, \mathbf{u} \rangle_{\mathcal{X}}$ when $\mathcal{N}_f \neq \mathcal{X}$. Since (32.1) holds with $\mathbf{v} = \mathbf{0}$ when $\mathcal{N}_f = \mathcal{X}$, the proof of the existence of at least one vector $\mathbf{v} \in \mathcal{X}$ for which (32.1) holds is complete. It remains to show that there is only one such vector $\mathbf{v}$. This is left to the reader. $\square$

Exercise 32.3. Show that there is only one vector $\mathbf{v} \in \mathcal{X}$ for which (32.1) holds.

The next two exercises are formulated in terms of the **notation** $\mathbf{x} + \mathcal{V} = \{\mathbf{x} + \mathbf{v} : \mathbf{v} \in \mathcal{V}\}$ for a vector $\mathbf{x}$ and a vector space $\mathcal{V}$.

Exercise 32.4. Show that if $\mathcal{V}$ and $\mathcal{W}$ are subspaces of a vector space $\mathcal{X}$ and $\mathbf{x}, \mathbf{y} \in \mathcal{X}$, then

$$(32.3) \qquad \mathbf{x} + \mathcal{V} = \mathbf{y} + \mathcal{W} \iff \mathcal{V} = \mathcal{W} \text{ and } \mathbf{x} - \mathbf{y} \in \mathcal{V} \cap \mathcal{W}.$$

A subset $\mathcal{Q}$ of an n-dimensional vector space $\mathcal{X}$ is a **hyperplane** if there exists a vector $\mathbf{x} \in \mathcal{X}$ and an $(n-1)$-dimensional subspace $\mathcal{V}$ of $\mathcal{X}$ such that $\mathcal{Q} = \mathbf{x} + \mathcal{V}$.

Exercise 32.5. Show that a subset $\mathcal{Q}$ of an n-dimensional vector space $\mathcal{X}$ is a hyperplane if and only if there exists a nonzero linear functional $f \in \mathcal{X}'$ and a scalar α such that $\mathcal{Q} = \{\mathbf{x} \in \mathcal{X} : f(\mathbf{x}) = \alpha\}$.

32.2. Extensions of linear functionals

If f is a linear functional on a vector space $\mathcal{X}$, then the function $\varphi(\mathbf{x}) = |f(\mathbf{x})|$ meets the following two conditions:

(32.4) $\qquad \varphi(\mathbf{x} + \mathbf{y}) \leq \varphi(\mathbf{x}) + \varphi(\mathbf{y}) \quad \text{and} \quad \varphi(\alpha \mathbf{x}) = \alpha \varphi(\mathbf{x})$

for every choice of $\mathbf{x}, \mathbf{y} \in \mathcal{X}$ and $\alpha \geq 0$. A real-valued function $\varphi(\mathbf{x})$ that meets the two conditions in (32.4) is said to be **sublinear**.

Lemma 32.3. *If f is a linear functional on a proper subspace $\mathcal{V}$ of a finite-dimensional real vector space $\mathcal{X}$ and p is a sublinear functional on all of $\mathcal{X}$ such that $f(\mathbf{v}) \leq p(\mathbf{v})$ for every $\mathbf{v} \in \mathcal{V}$, then there exists a linear functional g on $\mathcal{X}$ such that $g(\mathbf{v}) = f(\mathbf{v})$ for every $\mathbf{v} \in \mathcal{V}$ and $g(\mathbf{x}) \leq p(\mathbf{x})$ for every $\mathbf{x} \in \mathcal{X}$.*

Proof. Since $\mathcal{X}$ is finite dimensional, it suffices to show that if $\mathbf{w} \in \mathcal{X} \setminus \mathcal{V}$, then there exists a linear functional g on the vector space $\mathcal{V} + \{\alpha \mathbf{w} : \alpha \in \mathbb{R}\}$ such that

$$g(\mathbf{v} + \alpha \mathbf{w}) = f(\mathbf{v}) + \alpha g(\mathbf{w}) \quad \text{and} \quad g(\mathbf{v} + \alpha \mathbf{w}) \leq p(\mathbf{v} + \alpha \mathbf{w})$$

for every $\alpha \in \mathbb{R}$ and every $\mathbf{v} \in \mathcal{V}$. The last inequality will hold if and only if $g(\mathbf{v} \pm \alpha \mathbf{w}) \leq p(\mathbf{v} \pm \alpha \mathbf{w})$ for every $\alpha > 0$ and every $\mathbf{v} \in \mathcal{V}$. But this imposes two constraints on $g(\mathbf{w})$:

$$g(\mathbf{w}) \leq p(\alpha^{-1}\mathbf{v} + \mathbf{w}) - f(\alpha^{-1}\mathbf{v}) \quad \text{and} \quad -g(\mathbf{w}) \leq p(\alpha^{-1}\mathbf{v} - \mathbf{w}) - f(\alpha^{-1}\mathbf{v}).$$

In order to meet these constraints, we must choose $g(\mathbf{w})$ so that

$$f(\mathbf{x}) - p(\mathbf{x} - \mathbf{w}) \leq g(\mathbf{w}) \leq p(\mathbf{y} + \mathbf{w}) - f(\mathbf{y}) \quad \text{for every choice of } \mathbf{x}, \mathbf{y} \in \mathcal{V},$$

if possible. But the development

$$f(\mathbf{x}) + f(\mathbf{y}) = f(\mathbf{x} + \mathbf{y}) \leq p(\mathbf{x} + \mathbf{y}) = p(\mathbf{x} - \mathbf{w} + \mathbf{y} + \mathbf{w}) \leq p(\mathbf{x} - \mathbf{w}) + p(\mathbf{y} + \mathbf{w})$$

implies that

$$m = \max\{f(\mathbf{x}) - p(\mathbf{x} - \mathbf{w}) : \mathbf{x} \in \mathcal{V}\} \leq \min\{p(\mathbf{y} + \mathbf{w}) - f(\mathbf{y}) : \mathbf{y} \in \mathcal{V}\} = M.$$

Therefore, if we set $g(\mathbf{w}) = \gamma$ for any point $\gamma \in [m, M]$ and take $\alpha > 0$, we see that

$$g(\mathbf{x} + \alpha \mathbf{w}) = f(\mathbf{x}) + \alpha \gamma \leq f(\mathbf{x}) + \alpha(p(\mathbf{y} + \mathbf{w}) - f(\mathbf{y})) \quad \text{for every } \mathbf{y} \in \mathcal{V}$$

and
$$g(\mathbf{x} - \alpha\mathbf{w}) = f(\mathbf{x}) - \alpha\gamma \geq f(\mathbf{x}) - \alpha(p(\mathbf{y} - \mathbf{w}) - f(\mathbf{y})) \quad \text{for every } \mathbf{y} \in \mathcal{V}.$$
Thus, if $\mathbf{y} = \mathbf{x}/\alpha$, then it is readily checked that
$$g(\mathbf{x} + \alpha\mathbf{w}) \leq f(\mathbf{x}) + \alpha(p(\alpha^{-1}\mathbf{x} + \mathbf{w}) - f(\alpha^{-1}\mathbf{x})) = p(\mathbf{x} + \alpha\mathbf{w})$$
and
$$g(\mathbf{x} - \alpha\mathbf{w}) \leq f(\mathbf{x}) + \alpha(p(\alpha^{-1}\mathbf{x} - \mathbf{w}) - f(\alpha^{-1}\mathbf{x})) = p(\mathbf{x} - \alpha\mathbf{w}),$$
as needed to define an extension on a space of dimension equal to $\dim \mathcal{V} + 1$. Since $\mathcal{X}$ is finite dimensional, we can obtain an extension on the full space $\mathcal{X}$ by repeating this procedure a finite number of times. □

The next theorem is a finite-dimensional version of the **Hahn-Banach extension theorem**.

Theorem 32.4. *If f is a linear functional on a proper subspace $\mathcal{V}$ of a finite-dimensional normed linear space $\mathcal{X}$, then there exists a linear functional g on the full space $\mathcal{X}$ such that:*

(1) $f(\mathbf{v}) = g(\mathbf{v})$ *for every vector* $\mathbf{v} \in \mathcal{V}$.
(2) $\max\{|f(\mathbf{v})| : \mathbf{v} \in \mathcal{V} \text{ and } \|\mathbf{v}\|_{\mathcal{X}} \leq 1\}$
$= \max\{|g(\mathbf{x})| : \mathbf{x} \in \mathcal{X} \text{ and } \|\mathbf{x}\|_{\mathcal{X}} \leq 1\}.$

Discussion. If $\mathcal{X}$ is a real normed linear space and $\kappa = \max\{|f(\mathbf{v})| : \mathbf{v} \in \mathcal{V} \text{ and } \|\mathbf{v}\|_{\mathcal{X}} \leq 1\}$, then $p(\mathbf{x}) = \kappa \|\mathbf{x}\|_{\mathcal{X}}$ is a sublinear function on $\mathcal{X}$ and $f(\mathbf{v}) \leq p(\mathbf{v})$ for every vector $\mathbf{v} \in \mathcal{V}$. Therefore, Lemma 32.3 guarantees that there exists a linear functional g on the full space $\mathcal{X}$ such that (1) holds and $g(\pm\mathbf{x}) \leq p(\pm\mathbf{x}) = \kappa \|\mathbf{x}\|_{\mathcal{X}}$. Consequently, $|g(\mathbf{x})| \leq \kappa \|\mathbf{x}\|_{\mathcal{X}}$, which leads easily to (2).

These conclusions can be extended to complex normed linear spaces; see, e.g., the proof of Theorem 7.30 in [**30**]. □

32.3. The Minkowski functional

Let $\mathcal{X}$ be a normed linear space and let $\mathcal{Q} \subseteq \mathcal{X}$. Then the functional
$$p_{\mathcal{Q}}(\mathbf{x}) = \inf\left\{t > 0 : \frac{\mathbf{x}}{t} \in \mathcal{Q}\right\}$$
is called the **Minkowski functional**. If the indicated set of t is empty, then $p_{\mathcal{Q}}(\mathbf{x}) = \infty$.

Recall that in a normed linear space $\mathcal{X}$, $B_r(\mathbf{0}) = \{\mathbf{x} \in \mathcal{X} : \|\mathbf{x}\| < r\}$, and a subset $\mathcal{B}$ of $\mathcal{X}$ is **bounded** if there exists an $R > 0$ such that $\mathcal{B} \subset B_R(\mathbf{0})$.

Lemma 32.5. *If Q is a convex subset of a normed linear space $\mathcal{X}$ such that*

$$Q \supseteq B_r(\mathbf{0}) \quad \text{for some} \quad r > 0,$$

then:

(1) $p_Q(\mathbf{x}+\mathbf{y}) \leq p_Q(\mathbf{x}) + p_Q(\mathbf{y})$ *for* $\mathbf{x}, \mathbf{y} \in \mathcal{X}$.

(2) $p_Q(\alpha \mathbf{x}) = \alpha p_Q(\mathbf{x})$ *for* $\alpha \geq 0$ *and* $\mathbf{x} \in \mathcal{X}$.

(3) $|p_Q(\mathbf{x}) - p_Q(\mathbf{y})| \leq (2/r)\|\mathbf{x}-\mathbf{y}\|_{\mathcal{X}}$ *for* $\mathbf{x}, \mathbf{y} \in \mathcal{X}$ *(and hence $p_Q(\mathbf{x})$ is a continuous function of $\mathbf{x}$ on $\mathcal{X}$).*

(4) *If Q is open, then* $Q = \{\mathbf{x} \in \mathcal{X} : p_Q(\mathbf{x}) < 1\}$.

(5) *If Q is closed, then* $Q = \{\mathbf{x} \in \mathcal{X} : p_Q(\mathbf{x}) \leq 1\}$.

(6) *If Q is bounded, then $p_Q(\mathbf{x}) = 0 \implies \mathbf{x} = \mathbf{0}$.*

Proof. Let $\mathbf{x}, \mathbf{y} \in \mathcal{X}$ and suppose that $\alpha^{-1}\mathbf{x} \in Q$ and $\beta^{-1}\mathbf{y} \in Q$ for some choice of $\alpha > 0$ and $\beta > 0$. Then, since Q is convex,

$$\frac{\mathbf{x}+\mathbf{y}}{\alpha+\beta} = \frac{\alpha}{\alpha+\beta}(\alpha^{-1}\mathbf{x}) + \frac{\beta}{\alpha+\beta}(\beta^{-1}\mathbf{y})$$

belongs to Q and hence,

$$p_Q(\mathbf{x}+\mathbf{y}) \leq \alpha + \beta .$$

Consequently, upon letting α run through a sequence of values $\alpha_1 \geq \alpha_2 \geq \cdots$ that tend to $p_Q(\mathbf{x})$ and letting β run through a sequence of values $\beta_1 \geq \beta_2 \geq \cdots$ that tend to $p_Q(\mathbf{y})$, it is readily seen that

$$p_Q(\mathbf{x}+\mathbf{y}) \leq p_Q(\mathbf{x}) + p_Q(\mathbf{y}) .$$

Suppose next that $\alpha > 0$ and $p_Q(\mathbf{x}) = a$. Then there exists a sequence of numbers $t_1, t_2, \ldots$ such that $t_j > 0$,

$$\frac{\mathbf{x}}{t_j} \in Q \quad \text{and} \quad \lim_{j \uparrow \infty} t_j = a .$$

Therefore, since

$$\frac{\alpha \mathbf{x}}{\alpha t_j} \in Q \quad \text{and} \quad \lim_{j \uparrow \infty} \alpha t_j = \alpha a ,$$

$$p_Q(\alpha \mathbf{x}) \leq \alpha p_Q(\mathbf{x}) .$$

However, the same argument yields the opposite inequality:

$$\alpha p_Q(\mathbf{x}) = \alpha p_Q(\alpha^{-1}\alpha \mathbf{x}) \leq \alpha \alpha^{-1} p_Q(\alpha \mathbf{x}) = p_Q(\alpha \mathbf{x}) .$$

Therefore, equality prevails. This completes the proof of (2) when $\alpha > 0$. But, (2) also holds when $\alpha = 0$, because $p_Q(\mathbf{0}) = 0$.

Next, to verify (3), we first observe that
$$p_Q(\mathbf{x}) \leq (2/r)\|\mathbf{x}\|_{\mathcal{X}} \quad \text{for every point } \mathbf{x} \in \mathcal{X},$$
since $p_Q(\mathbf{0}) = 0$ and $((2/r)\|\mathbf{x}\|)_{\mathcal{X}}^{-1}\mathbf{x} \in B_r(\mathbf{0})$ if $\mathbf{x} \neq \mathbf{0}$. Thus,
$$p_Q(\mathbf{x}) = p_Q(\mathbf{x} - \mathbf{y} + \mathbf{y}) \leq p_Q(\mathbf{x} - \mathbf{y}) + p_Q(\mathbf{y}) \leq (2/r)\|\mathbf{x} - \mathbf{y}\|_{\mathcal{X}} + p_Q(\mathbf{y})$$
and
$$p_Q(\mathbf{y}) = p_Q(\mathbf{y} - \mathbf{x} + \mathbf{x}) \leq p_Q(\mathbf{y} - \mathbf{x}) + p_Q(\mathbf{x}) \leq (2/r)\|\mathbf{y} - \mathbf{x}\|_{\mathcal{X}} + p_Q(\mathbf{x})$$
for every choice of $\mathbf{x}, \mathbf{y} \in \mathcal{X}$, which justifies (3).

Items (4) and (5) are left to the reader.

Finally, to verify (6), suppose that $p_Q(\mathbf{x}) = 0$. Then there exists a sequence of points $\alpha_1 \geq \alpha_2 \geq \cdots$ decreasing to 0 such that $\alpha_j^{-1}\mathbf{x} \in Q$. Therefore, since $Q \subseteq B_R(\mathbf{0})$ for some $R > 0$, the inequality $\|\alpha_j^{-1}\mathbf{x}\|_{\mathcal{X}} \leq R$ implies that $\|\mathbf{x}\|_{\mathcal{X}} \leq \alpha_j R$ for $j = 1, 2 \ldots$ and hence that $\mathbf{x} = \mathbf{0}$. $\square$

Exercise 32.6. Complete the proof of Lemma 32.5 by verifying items (4) and (5).

Exercise 32.7. Show that in the setting of Lemma 32.5, $p_Q(\mathbf{x}) < 1 \Longrightarrow \mathbf{x} \in Q$ and $\mathbf{x} \in Q \Longrightarrow p_Q(\mathbf{x}) \leq 1$.

32.4. Separation theorems

In this section we shall use the Hahn-Banach extension theorem to justify the key steps in the proof of Theorem 32.9, which is the main conclusion of this section.

Lemma 32.6. *If $\mathcal{A}$ is a nonempty open convex set in a real normed linear space $\mathcal{X}$ such that $\mathbf{0} \notin \mathcal{A}$, then there exists a linear functional $f \in \mathcal{X}'$ such that $f(\mathbf{a}) > 0$ for every vector $\mathbf{a} \in \mathcal{A}$.*

Proof. Choose a point $\mathbf{a}_0 \in \mathcal{A}$ and let $p(\mathbf{x}) = p_Q(\mathbf{x})$, the Minkowski functional for the set
$$Q = \mathbf{a}_0 - \mathcal{A} = \{\mathbf{a}_0 - \mathbf{a} : \mathbf{a} \in \mathcal{A}\}.$$
Then Q is an open convex subset of $\mathcal{X}$ that contains the point $\mathbf{0}$ and hence
$$Q = \{\mathbf{x} \in \mathcal{X} : p(\mathbf{x}) < 1\}.$$
Thus, as $\mathbf{a}_0 \notin Q$, $p(\mathbf{a}_0) \geq 1$. Let
$$\mathcal{Y} = \{\alpha \mathbf{a}_0 : \alpha \in \mathbb{R}\} \quad \text{and define} \quad h(\alpha \mathbf{a}_0) = \alpha p(\mathbf{a}_0) \quad \text{for all } \alpha \in \mathbb{R}.$$
Then h is a linear functional on $\mathcal{Y}$ such that $h(\alpha \mathbf{a}_0) = p(\alpha \mathbf{a}_0)$ if $\alpha \geq 0$, and
$$h(\alpha \mathbf{a}_0) = \alpha p(\mathbf{a}_0) < 0 \leq p(\alpha \mathbf{a}_0) \quad \text{if } \alpha < 0.$$

Therefore, $h(\mathbf{y}) \leq p(\mathbf{y})$ for every $\mathbf{y} \in \mathcal{Y}$. Consequently, Lemma 32.3 ensures that there exists a linear functional $f \in \mathcal{X}'$ such that $f(\mathbf{y}) = h(\mathbf{y})$ for every $\mathbf{y} \in \mathcal{Y}$ and $f(\mathbf{x}) \leq p(\mathbf{x})$ for every $\mathbf{x} \in \mathcal{X}$. Therefore,
$$1 > p(\mathbf{a}_0 - \mathbf{a}) \geq f(\mathbf{a}_0 - \mathbf{a}) = f(\mathbf{a}_0) - f(\mathbf{a}) = p(\mathbf{a}_0) - f(\mathbf{a}),$$
i.e., $f(\mathbf{a}) > p(\mathbf{a}_0) - 1 \geq 0$ for every $\mathbf{a} \in \mathcal{A}$, as needed. □

Lemma 32.7. *If $\mathcal{A}$ and $\mathcal{B}$ are nonempty convex sets in a finite-dimensional real normed linear space such that $\mathcal{A} \cap \mathcal{B} = \emptyset$ and $\mathcal{A}$ is open, then there exists a linear functional $f \in \mathcal{X}'$ and a point $c \in \mathbb{R}$ such that*

(32.5) $\qquad f(\mathbf{b}) \leq c < f(\mathbf{a}) \quad$ *for every $\mathbf{a} \in \mathcal{A}$ and $\mathbf{b} \in \mathcal{B}$.*

Proof. Let $\mathcal{Q} = \mathcal{A} - \mathcal{B}$. Then $\mathcal{Q}$ is an open convex subset of $\mathcal{X}$ that does not contain the point $\mathbf{0}$, since $\mathcal{A} \cap \mathcal{B} = \emptyset$. Therefore, by Lemma 32.6, there exists a linear functional $f \in \mathcal{X}'$ such that $f(\mathbf{q}) > 0$ for every vector $\mathbf{q} \in \mathcal{Q}$. Thus, $f(\mathbf{a} - \mathbf{b}) = f(\mathbf{a}) - f(\mathbf{b}) > 0$ for every $\mathbf{a} \in \mathcal{A}$ and every $\mathbf{b} \in \mathcal{B}$. But, as $f(\mathcal{A})$ is an open convex subset of $\mathbb{R}$, $f(\mathcal{A}) = (c, d)$, and hence
$$f(\mathbf{b}) \leq c < f(\mathbf{a}) \quad \text{for every } \mathbf{a} \in \mathcal{A} \text{ and } \mathbf{b} \in \mathcal{B}.$$
as claimed. □

Lemma 32.8. *If $\mathcal{B}$ is a nonempty closed convex set in a finite-dimensional real normed linear space $\mathcal{X}$ such that $\mathcal{B} \subset \mathcal{Q}$ for some open subset $\mathcal{Q}$ of $\mathcal{X}$ and $\mathcal{B}$ is bounded, then there exists an $r > 0$ such that $\mathcal{B} + B_r(\mathbf{0}) \subseteq \mathcal{Q}$ for some $r > 0$.*

Proof. If the assertion is false, then there exists a sequence of vectors $\mathbf{b}_j + \mathbf{x}_j$ such that $\mathbf{b}_j \in \mathcal{B}$ and $\|\mathbf{x}_j\| < 1/j$ for $j = 1, 2, \ldots$ such that $\mathbf{b}_j + \mathbf{x}_j \in \mathcal{X} \setminus \mathcal{Q}$. Since $\mathcal{B}$ is a closed bounded subset of $\mathcal{X}$, a subsequence of the $\mathbf{b}_j$ tends to a limit $\mathbf{b}$. But then $\mathbf{b} \in \mathcal{B} \cap (\mathcal{X} \setminus \mathcal{Q})$, since both of these sets are closed. But this contradicts the fact that $\mathcal{B} \subset \mathcal{Q}$. □

Theorem 32.9. *If $\mathcal{A}$ and $\mathcal{B}$ are nonempty closed convex sets in a finite-dimensional real normed linear space $\mathcal{X}$ such that $\mathcal{A} \cap \mathcal{B} = \emptyset$ and $\mathcal{B}$ is bounded, then there exist a linear functional $f \in \mathcal{X}'$ and a pair of numbers $c, d \in \mathbb{R}$ such that*

(32.6) $\qquad\qquad\qquad f(\mathbf{b}) < c < d < f(\mathbf{a})$

for every choice of $\mathbf{a} \in \mathcal{A}$ and $\mathbf{b} \in \mathcal{B}$ (i.e., $\mathcal{A}$ and $\mathcal{B}$ are strictly separated by the hyperplane $\mathcal{Q} = \{\mathbf{x} : f(\mathbf{x}) = c + 2^{-1}(d - c)\}$).

Proof. Since $\mathcal{B} \subseteq \mathcal{X} \setminus \mathcal{A}$ and $\mathcal{X} \setminus \mathcal{A}$ is an open subset of $\mathcal{X}$, Lemma 32.8 ensures that there exists an $r > 0$ such that $\mathcal{B} + B_r(\mathbf{0}) \subseteq \mathcal{X} \setminus \mathcal{A}$. Therefore,
$$(\mathcal{B} + B_{(r/2)}(\mathbf{0})) \bigcap (\mathcal{A} + B_{(r/2)}(\mathbf{0})) = \emptyset.$$

Thus, as the two sets on the left are disjoint convex open sets, Lemma 32.7 guarantees that there exists a linear functional $f \in \mathcal{X}'$ such that $f(\mathbf{b}+\mathbf{x}) \leq d < f(\mathbf{a}+\mathbf{y})$ for every $\mathbf{b} \in \mathcal{B}$, $\mathbf{a} \in \mathcal{A}$, and $\mathbf{x}, \mathbf{y} \in B_{(r/2)}(\mathbf{0})$. To finish, fix $\mathbf{x} \in B_{r/2}(\mathbf{0})$ such that $f(\mathbf{x}) > 0$. Then

$$f(\mathbf{b}) \leq d - f(\mathbf{x}) < d - 2^{-1} f(\mathbf{x}) < d < f(\mathbf{a})$$

and (32.6) follows upon setting $c = d - 2^{-1} f(\mathbf{x})$. $\square$

Theorem 32.9 supplies a key step in the proof of Farkas's lemma, which is an important result in linear programming.

Lemma 32.10 (Farkas). *If $A \in \mathbb{R}^{p \times q}$ and $\mathbf{b} \in \mathbb{R}^p$, then either*

(1) *there exists a vector $\mathbf{x} \in \mathbb{R}_{\geq}^q$ such that $A\mathbf{x} = \mathbf{b}$ or*

(2) *there exists a vector $\mathbf{y} \in \mathbb{R}^p$ such that $A^T \mathbf{y} \in \mathbb{R}_{\geq}^q$ and $\mathbf{b}^T \mathbf{y} < 0$,*

but not both.

Proof. Let $\mathcal{A} = \{A\mathbf{x} : \mathbf{x} \in \mathbb{R}_{\geq}^q\}$. If $\mathbf{b} \in \mathcal{A}$, then (1) is in force. If $\mathbf{b} \notin \mathcal{A}$, then $\mathcal{C} = \{\mathbf{a} - \mathbf{b} : \mathbf{a} \in \mathcal{A}\}$ is a closed convex subset of $\mathbb{R}^p$ and $\mathbf{0} \notin \mathcal{C}$. Therefore, Theorem 32.9 (with $\mathcal{B} = \{\mathbf{0}\}$ and $\mathcal{A}$ replaced by $\mathcal{C}$) guarantees that there exist a linear functional f on $\mathbb{R}^p$ and a number $c \in \mathbb{R}$ such that

$$f(\mathbf{a} - \mathbf{b}) > c > f(\mathbf{0}) = 0 \quad \text{for every } \mathbf{a} \in \mathcal{A}.$$

Thus, there exists a vector $\mathbf{y} \in \mathbb{R}^p$ such that $\mathbf{y}^T(A\mathbf{x} - \mathbf{b}) > 0$ for every $\mathbf{x} \in \mathbb{R}_{\geq}^q$. Consequently, $\mathbf{y}^T \mathbf{b} < 0$ and

(32.7) $$\mathbf{x}^T A^T \mathbf{y} > \mathbf{y}^T \mathbf{b} \quad \text{for every } \mathbf{x} \in \mathbb{R}_{\geq}^q.$$

But this implies that $A^T \mathbf{y} \in \mathbb{R}_{\geq}^q$, because if $(A^T \mathbf{y})_j < 0$ for some $j \in \{1, \ldots, q\}$, then there exists a $t > 0$ such that $t\mathbf{e}_j^T (A^T \mathbf{y}) < \mathbf{y}^T \mathbf{b}$, which violates the inequality (32.7).

Finally, if $\mathbf{x} \in \mathbb{R}_{\geq}^q$ is a vector for which (1) holds and $\mathbf{y} \in \mathbb{R}^p$ is a vector for which (2) holds, then $\mathbf{y}^T A\mathbf{x} = \mathbf{y}^T \mathbf{b}$, which is impossible, since $\mathbf{y}^T A\mathbf{x} \geq 0$ and $\mathbf{y}^T \mathbf{b} < 0$. Therefore, (1) and (2) cannot both hold. $\square$

32.5. Another path

In this section we shall present a direct proof of a special case of Theorem 32.9 that is formulated in a finite-dimensional real inner product space $\mathcal{X}$, because it is instructive.

Warning: To avoid clutter, we shall drop the subscripts $\mathcal{X}$ in $\|\mathbf{x}\|_\mathcal{X}$ and $\langle \mathbf{x}, \mathbf{y} \rangle_\mathcal{X}$ throughout this section.

Lemma 32.11. *If $\mathcal{Q}$ is nonempty closed convex subset of a finite-dimensional real inner product space $\mathcal{X}$ and $\mathbf{a} \in \mathcal{X}$, then there exists exactly one vector $\mathbf{q_a} \in \mathcal{Q}$ such that*

(32.8) $$\|\mathbf{a} - \mathbf{q_a}\| \leq \|\mathbf{a} - \mathbf{q}\| \quad \text{for every } \mathbf{q} \in \mathcal{Q}.$$

Proof. Let $d = \inf\{\|\mathbf{a} - \mathbf{q}\| : \mathbf{q} \in \mathcal{Q}\}$. If $\mathbf{a} \in \mathcal{Q}$, then $d = 0$ and $\mathbf{q_a} = \mathbf{a}$. If $\mathbf{a} \notin \mathcal{Q}$, Then there exists a sequence of vectors $\mathbf{q}_1, \mathbf{q}_2, \ldots$ such that

$$\|\mathbf{a} - \mathbf{q}_j\| \leq d + 1/j.$$

Since this sequence,

$$\|\mathbf{q}_j\| = \|\mathbf{q}_j - \mathbf{a} + \mathbf{a}\| \leq \|\mathbf{q}_j - \mathbf{a}\| + \|\mathbf{a}\| \leq d + 1/j + \|\mathbf{a}\|,$$

is bounded, a subsequence $\mathbf{q}_{j_1}, \mathbf{q}_{j_2}, \ldots$ will tend to a limit $\mathbf{q}'$, and the bounds

$$d \leq \|\mathbf{a} - \mathbf{q}'\| \leq \|\mathbf{a} - \mathbf{q}_{j_k}\| + \|\mathbf{q}_{j_k} - \mathbf{q}'\| \leq d + 1/j_k + \|\mathbf{q}_{j_k} - \mathbf{q}'\|$$

clearly imply that

(32.9) $$\|\mathbf{a} - \mathbf{q}'\| = d \leq \|\mathbf{a} - \mathbf{q}\| \quad \text{for every } \mathbf{q} \in \mathcal{Q}.$$

If also $\mathbf{q}'' \in \mathcal{Q}$ and

$$\|\mathbf{a} - \mathbf{q}''\| \leq \|\mathbf{a} - \mathbf{q}\| \quad \text{for every } \mathbf{q} \in \mathcal{Q},$$

then, by the parallelogram law,

$$4d^2 = 2\|\mathbf{q}' - \mathbf{a}\|^2 + 2\|\mathbf{q}'' - \mathbf{a}\|^2 = \|\mathbf{q}' - \mathbf{a} + \mathbf{q}'' - \mathbf{a}\|^2 + \|\mathbf{q}' - \mathbf{q}''\|^2$$
$$= 4\|\mathbf{a} - (\mathbf{q}' + \mathbf{q}'')/2\|^2 + \|\mathbf{q}' - \mathbf{q}''\|^2$$
$$\geq 4d^2 + \|\mathbf{q}' - \mathbf{q}''\|^2.$$

Therefore, $\mathbf{q}' = \mathbf{q}''$. $\square$

Lemma 32.12. *Let $\mathcal{Q}$ be a closed nonempty convex subset of $\mathcal{X}$, let $\mathbf{a} \in \mathcal{X}$, and let $\mathbf{q_a}$ be the unique element in $\mathcal{Q}$ that is closest to $\mathbf{a}$. Then*

(32.10) $$\langle \mathbf{q} - \mathbf{q_a}, \mathbf{a} - \mathbf{q_a} \rangle \leq 0 \quad \text{for every } \mathbf{q} \in \mathcal{Q}.$$

Proof. Let $\mathbf{q} \in \mathcal{Q}$. Then clearly $(1 - t)\mathbf{q_a} + t\mathbf{q} \in \mathcal{Q}$ for every number t in the interval $0 \leq t \leq 1$. Therefore,

$$\begin{aligned}\|\mathbf{a} - \mathbf{q_a}\|^2 &\leq \|\mathbf{a} - (1-t)\mathbf{q_a} - t\mathbf{q}\|^2 \\ &= \|\mathbf{a} - \mathbf{q_a} - t(\mathbf{q} - \mathbf{q_a})\|^2 \\ &= \|\mathbf{a} - \mathbf{q_a}\|^2 - 2t\langle \mathbf{a} - \mathbf{q_a}, \mathbf{q} - \mathbf{q_a}\rangle + t^2\|\mathbf{q} - \mathbf{q_a}\|^2,\end{aligned}$$

since $t \in \mathbb{R}$ and $\mathcal{X}$ is a real inner product space. But this in turn implies that

$$2t\langle \mathbf{a} - \mathbf{q_a}, \mathbf{q} - \mathbf{q_a}\rangle \leq t^2\|\mathbf{q} - \mathbf{q_a}\|^2$$

32.5. Another path

and hence that
$$2\langle \mathbf{a} - \mathbf{q_a}, \mathbf{q} - \mathbf{q_a}\rangle \leq t\|\mathbf{q} - \mathbf{q_a}\|^2$$
for every t in the interval $0 < t \leq 1$. The inequality (32.10) now drops out easily upon letting $t \downarrow 0$. □

Exercise 32.8. Show that if $\mathcal{X} = \mathbb{R}^n$ in Lemma 32.12, then $2\langle \mathbf{a} - \mathbf{q_a}, \mathbf{q} - \mathbf{q_a}\rangle$ is equal to the directional derivative
$$(D_\mathbf{u} f)(\mathbf{x}_0) = \lim_{t \downarrow 0} \frac{f(\mathbf{x}_0 + t\mathbf{u}) - f(\mathbf{x})}{t} = \langle (\nabla f)(\mathbf{x}_0), \mathbf{u}\rangle$$
for $f(\mathbf{x}) = \|\mathbf{x}\|^2$ and appropriate choices of $\mathbf{x}_0$ and $\mathbf{u}$ (when $(\nabla f)(\mathbf{x}_0)$ is written as a column vector). When is $(\nabla f)(\mathbf{x}_0) = 0$?

Lemma 32.13. *Let $\mathcal{Q}$ be a closed nonempty convex subset of $\mathcal{X}$, let $\mathbf{a}, \mathbf{b} \in \mathcal{X}$, and let $\mathbf{q_a}$ and $\mathbf{q_b}$ denote the unique elements in $\mathcal{Q}$ that are closest to $\mathbf{a}$ and $\mathbf{b}$, respectively. Then*

(32.11) $$\|\mathbf{q_a} - \mathbf{q_b}\| \leq \|\mathbf{a} - \mathbf{b}\|.$$

Proof. Let
$$\alpha = 2\langle \mathbf{q_b} - \mathbf{q_a}, \mathbf{a} - \mathbf{q_a}\rangle \quad \text{and} \quad \beta = 2\langle \mathbf{q_a} - \mathbf{q_b}, \mathbf{b} - \mathbf{q_b}\rangle.$$
In view of Lemma 32.12, $\alpha \leq 0$ and $\beta \leq 0$. Therefore
$$\begin{aligned}\|\mathbf{a} - \mathbf{b}\|^2 &= \|(\mathbf{a} - \mathbf{q_a}) - (\mathbf{b} - \mathbf{q_b}) + (\mathbf{q_a} - \mathbf{q_b})\|^2 \\ &= \|(\mathbf{a} - \mathbf{q_a}) - (\mathbf{b} - \mathbf{q_b})\|^2 - \alpha - \beta + \|\mathbf{q_a} - \mathbf{q_b}\|^2 \\ &\geq \|\mathbf{q_a} - \mathbf{q_b}\|^2,\end{aligned}$$
as claimed. □

The inequality (32.11) implies that if $\mathcal{Q}$ is a closed nonempty convex subset of $\mathcal{X}$, then the mapping from $\mathbf{a} \in \mathcal{X} \to \mathbf{q_a} \in \mathcal{Q}$ is continuous. This fact will be used to advantage in the next proof:

Theorem 32.14. *If $\mathcal{A}$ and $\mathcal{B}$ are nonempty closed convex sets in a finite-dimensional real inner product space $\mathcal{X}$ such that $\mathcal{A} \cap \mathcal{B} = \emptyset$ and $\mathcal{B}$ is bounded, then there exist a linear functional $f \in \mathcal{X}'$ and a pair of numbers $c, d \in \mathbb{R}$ such that*

(32.12) $$f(\mathbf{b}) < c < d < f(\mathbf{a})$$

for every choice of $\mathbf{a} \in \mathcal{A}$ and $\mathbf{b} \in \mathcal{B}$.

Proof. Let $\mathbf{a_x}$ denote the unique point in $\mathcal{A}$ that is closest to $\mathbf{x} \in \mathcal{X}$. Then, by Lemma 32.12,

$$\langle \mathbf{x} - \mathbf{a_x}, \mathbf{a} - \mathbf{a_x} \rangle \leq 0 \quad \text{for every} \quad \mathbf{a} \in \mathcal{A}.$$

Moreover, in view of Lemma 32.13,

$$g(\mathbf{x}) = \|\mathbf{x} - \mathbf{a_x}\|$$

is a continuous function of $\mathbf{x} \in \mathcal{X}$. In particular, g is continuous on the closed bounded set $\mathcal{B}$, and hence there exists a vector $\mathbf{b}_0 \in \mathcal{B}$ such that

$$\|\mathbf{b}_0 - \mathbf{a}_{\mathbf{b}_0}\| \leq \|\mathbf{b} - \mathbf{a}_\mathbf{b}\| \leq \|\mathbf{b} - \mathbf{a}_{\mathbf{b}_0}\|$$

for every $\mathbf{b} \in \mathcal{B}$. Let $\mathbf{a}_0 = \mathbf{a}_{\mathbf{b}_0}$. Then $\|\mathbf{b}_0 - \mathbf{a}_0\| \leq \|\mathbf{b} - \mathbf{a}_0\|$ for every $\mathbf{b} \in \mathcal{B}$, and hence, as $\mathcal{B}$ is convex,

$$\begin{aligned}
\|\mathbf{b}_0 - \mathbf{a}_0\|^2 &\leq \|(1-t)\mathbf{b}_0 + t\mathbf{b} - \mathbf{a}_0\|^2 \\
&= \|t(\mathbf{b} - \mathbf{b}_0) - (\mathbf{a}_0 - \mathbf{b}_0)\|^2 \\
&= t^2 \|\mathbf{b} - \mathbf{b}_0\|^2 - 2t \langle \mathbf{b} - \mathbf{b}_0, \mathbf{a}_0 - \mathbf{b}_0 \rangle + \|\mathbf{a}_0 - \mathbf{b}_0\|^2
\end{aligned}$$

for $0 \leq t \leq 1$. But this implies that

$$2\langle \mathbf{b} - \mathbf{b}_0, \mathbf{a}_0 - \mathbf{b}_0 \rangle \leq t \|\mathbf{b} - \mathbf{b}_0\|^2$$

for every t in the interval $0 < t \leq 1$ and hence that

$$\langle \mathbf{b} - \mathbf{b}_0, \mathbf{a}_0 - \mathbf{b}_0 \rangle \leq 0 \quad \text{for every} \quad \mathbf{b} \in \mathcal{B},$$

which, upon setting $f(\mathbf{x}) = \langle \mathbf{x}, \mathbf{a}_0 - \mathbf{b}_0 \rangle$, yields the inequality

(32.13) $\quad f(\mathbf{b}_0) = \langle \mathbf{b}_0, \mathbf{a}_0 - \mathbf{b}_0 \rangle \geq \langle \mathbf{b}, \mathbf{a}_0 - \mathbf{b}_0 \rangle = f(\mathbf{b}) \quad \text{for every} \quad \mathbf{b} \in \mathcal{B}.$

Moreover, since $\mathbf{a}_0 = \mathbf{a}_{\mathbf{b}_0}$, Lemma 32.12 implies that

$$\langle \mathbf{a}_0 - \mathbf{a}, \mathbf{a}_0 - \mathbf{b}_0 \rangle \leq 0 \quad \text{for every} \quad \mathbf{a} \in \mathcal{A}$$

and hence that

(32.14) $\quad f(\mathbf{a}) = \langle \mathbf{a}, \mathbf{a}_0 - \mathbf{b}_0 \rangle \geq \langle \mathbf{a}_0, \mathbf{a}_0 - \mathbf{b}_0 \rangle = f(\mathbf{a}_0) \quad \text{for every} \quad \mathbf{a} \in \mathcal{A}.$

Consequently, in view of (32.14) and (32.13),

$$f(\mathbf{a}) \geq f(\mathbf{a}_0) = \|\mathbf{a}_0 - \mathbf{b}_0\|^2 + f(\mathbf{b}_0) \geq \|\mathbf{a}_0 - \mathbf{b}_0\|^2 + f(\mathbf{b}),$$

for every choice of $\mathbf{a} \in \mathcal{A}$ and $\mathbf{b} \in \mathcal{B}$. The inequality (32.6) is obtained by setting $d = f(\mathbf{a}_0) - \varepsilon$ and $c = f(\mathbf{b}_0) + \varepsilon$ with $0 < 2\varepsilon < \|\mathbf{a}_0 - \mathbf{b}_0\|^2$. □

32.6. Supplementary notes

This chapter is partially adapted from Chapters 7 and 22 in [**30**]. Section 32.4 is adapted from Chapter 4 of Conway [**19**], which establishes a separation theorem from a Hahn-Banach extension theorem in a much more general setting, whereas Section 32.5 is adapted from Section 2.4 in the monograph by Webster [**75**], which is an eminently readable source of supplementary information on convexity in $\mathbb{R}^n$. Almost any textbook on functional analysis will contain more general formulations of the finite-dimensional versions of the Hahn-Banach extension theorem (Theorem 32.4) and the Hahn-Banach separation theorem (Theorem 32.9) considered here; see, e.g., Bollobás [**12**] and Conway [**19**].

Chapter 33

A minimal norm problem

In this chapter we will consider the following problem:

Given $A \in \mathbb{C}^{p \times n}$ and a vector $\mathbf{b} \in \mathcal{R}_A$, find

(33.1) $$\min\{\|\mathbf{x}\|_1 : \mathbf{x} \in \mathcal{S}_\mathbf{b}\},$$

where

(33.2) $$\mathcal{S}_\mathbf{b} = \{\mathbf{x} \in \mathbb{C}^n : A\mathbf{x} = \mathbf{b}\}.$$

If the minimization in (33.1) is carried out with respect to $\|\mathbf{x}\|_2$ instead of $\|\mathbf{x}\|_1$, then this problem is similar to problems that were resolved in Chapter 15 by invoking the singular value decomposition $A = V_1 S_1 U_1^H$, in which $V_1 \in \mathbb{C}^{p \times r}$ and $U_1 \in \mathbb{C}^{n \times r}$ are isometric matrices, $S_1 = \mathrm{diag}\{s_1, \ldots, s_r\}$ is a positive definite matrix, and $r = \mathrm{rank}\, A$. In particular,

$$A\mathbf{x} = \mathbf{b} \iff V_1 S_1 U_1^H \mathbf{x} = \mathbf{b} \iff U_1^H \mathbf{x} = S_1^{-1} V_1^H \mathbf{b}.$$

If $r = n$, then U_1 is unitary and the problem is not interesting because the equation $A\mathbf{x} = \mathbf{b}$ will have only one solution:

$$\mathbf{x} = U_1 S_1^{-1} V_1^H \mathbf{b} = A^\dagger \mathbf{b} = (A^H A)^{-1} A^H \mathbf{b},$$

and hence $\mathcal{S}_\mathbf{b}$ is a set with only one vector.

If $r < n$, then U_1 can be embedded in a unitary matrix $U = \begin{bmatrix} U_1 & U_2 \end{bmatrix}$ and

(33.3) $$\mathcal{S}_\mathbf{b} = \{A^\dagger \mathbf{b} + U_2 \mathbf{y} : \mathbf{y} \in \mathbb{C}^{n-r}\}.$$

Thus, as $\|A^\dagger \mathbf{b} + U_2 \mathbf{y}\|_2^2 = \|A^\dagger \mathbf{b}\|_2^2 + \|U_2 \mathbf{y}\|_2^2$, it is clear that

(33.4) $$\min \{\|\mathbf{x}\|_2 : \mathbf{x} \in \mathcal{S}_\mathbf{b}\} = \|A^\dagger \mathbf{b}\|_2.$$

Minimization with respect to $\|\mathbf{x}\|_1$ requires a totally different approach that falls within the class of dual extremal problems that will be introduced in the next section.

Exercise 33.1. Verify the formulas in (33.3) and (33.4).

33.1. Dual extremal problems

Recall that if $\mathcal{X}$ is a finite-dimensional normed linear space and $f \in \mathcal{X}'$, the set of linear functionals on $\mathcal{X}$, then

$$\|f\|_{\mathcal{X}'} = \max\{|f(\mathbf{x})| : \mathbf{x} \in \mathcal{X} \text{ and } \|\mathbf{x}\|_\mathcal{X} = 1\}.$$

Moreover, if $\mathcal{X} = \mathbb{C}^n$, $\mathbf{x} = \begin{bmatrix} x_1 & \cdots & x_n \end{bmatrix}^T$, and $\mathbf{v} = \begin{bmatrix} v_1 & \cdots & v_n \end{bmatrix}^T$, then $f(\mathbf{x}) = \sum_{j=1}^n x_j v_j$ belongs to $\mathcal{X}'$ and, as we shall see shortly,

$$\|\mathbf{x}\|_\mathcal{X} = \|\mathbf{x}\|_1 \implies \|f\|_{\mathcal{X}'} = \|\mathbf{v}\|_\infty.$$

Theorem 33.1. *Let $\mathcal{X}$ be a finite-dimensional complex normed linear space, let $\mathcal{U}$ be a subspace of $\mathcal{X}$, and let $\mathcal{U}^\circ = \{f \in \mathcal{X}' : f(\mathbf{u}) = 0 \text{ for every } \mathbf{u} \in \mathcal{U}\}$. Then for each vector $\mathbf{x} \in \mathcal{X}$,*

(33.5) $$\min_{\mathbf{u} \in \mathcal{U}} \|\mathbf{x} - \mathbf{u}\|_\mathcal{X} = \max_{\substack{f \in \mathcal{U}^\circ \\ \|f\|_{\mathcal{X}'} \leq 1}} |f(\mathbf{x})|.$$

Proof. If $\mathbf{x} \in \mathcal{U}$, then (33.5) is self-evident, since both sides of the asserted equality are equal to zero. Suppose therefore that $\mathbf{x} \notin \mathcal{U}$. Then for any $f \in \mathcal{U}^\circ$ with $\|f\|_{\mathcal{X}'} \leq 1$ and any $\mathbf{u} \in \mathcal{U}$,

$$\begin{aligned} |f(\mathbf{x})| = |f(\mathbf{x} - \mathbf{u})| &\leq \|f\|_{\mathcal{X}'} \|\mathbf{x} - \mathbf{u}\|_\mathcal{X} \\ &\leq \|\mathbf{x} - \mathbf{u}\|_\mathcal{X}. \end{aligned}$$

Therefore,

(33.6) $$\max\{|f(\mathbf{x})| : f \in \mathcal{U}^\circ \text{ and } \|f\|_{\mathcal{X}'} \leq 1\} \leq \min\{\|\mathbf{x} - \mathbf{u}\|_\mathcal{X} : \mathbf{u} \in \mathcal{U}\}.$$

To obtain the opposite inequality for $\mathbf{x} \notin \mathcal{U}$, define the linear functional g on the subspace

$$\mathcal{W} = \{\alpha \mathbf{x} + \mathbf{u} : \alpha \in \mathbb{C} \text{ and } \mathbf{u} \in \mathcal{U}\}$$

by the formula

$$g(\alpha \mathbf{x} + \mathbf{u}) = \alpha d, \quad \text{with } d = \min\{\|\mathbf{x} - \mathbf{u}\|_\mathcal{X} : \mathbf{u} \in \mathcal{U}\}.$$

Then

$$\begin{aligned} \alpha = 0 &\implies \|\alpha \mathbf{x} + \mathbf{u}\|_\mathcal{X} = \|\mathbf{u}\|_\mathcal{X} \geq 0 = |g(\alpha \mathbf{x} + \mathbf{u})|, \\ \alpha \neq 0 &\implies \|\alpha \mathbf{x} + \mathbf{u}\|_\mathcal{X} = |\alpha| \|\mathbf{x} + \mathbf{u}/\alpha\|_\mathcal{X} \geq |\alpha| d = |g(\alpha \mathbf{x} + \mathbf{u})|. \end{aligned}$$

Thus,
$$|g(\mathbf{w})| \leq \|\mathbf{w}\|_{\mathcal{X}} \quad \text{for every } \mathbf{w} \in \mathcal{W}.$$
Theorem 32.4 guarantees the existence of a linear functional h on the full space $\mathcal{X}$ such that
$$h(\mathbf{w}) = g(\mathbf{w}) \quad \text{for every } \mathbf{w} \in \mathcal{W}$$
and
$$\|h\|_{\mathcal{X}'} = \max\left\{|g(\mathbf{w})| : \|\mathbf{w}\|_{\mathcal{X}} = 1\right\}.$$
Thus, $h \in \mathcal{U}^\circ$, $\|h\|_{\mathcal{X}'} \leq 1$, and $h(\mathbf{x}) = g(\mathbf{x}) = d$. Therefore,

(33.7)
$$\max\left\{|f(\mathbf{x})| : f \in \mathcal{U}^\circ \text{ and } \|f\|_{\mathcal{X}'} \leq 1\right\} \geq |h(\mathbf{x})| = |h(\mathbf{x} - \mathbf{u})|$$
$$= \min\{\|\mathbf{x} - \mathbf{u}\|_{\mathcal{X}} : \mathbf{u} \in \mathcal{U}\},$$

and the proof is complete. □

Exercise 33.2. Show that the conclusions of Theorem 33.1 are also valid when $\mathcal{X}$ is a finite-dimensional real normed linear space. [HINT: $f \in \mathcal{U}^\circ \iff -f \in \mathcal{U}^\circ$ and $\mathbf{u} \in \mathcal{U} \iff -\mathbf{u} \in \mathcal{U}$.]

Remark 33.2. Theorem 33.1 is also valid if $\mathcal{X}$ is a Banach space, but with inf in place of min and sup in place of max:

If $\mathbf{x} \in \mathcal{X}$, then
$$\inf\{\|\mathbf{x} - \mathbf{u}\| : \mathbf{u} \in \mathcal{U}\} = \max\{|f(\mathbf{x})| : f \in \mathcal{U}^\circ \text{ and } \|f\| \leq 1\},$$
whereas if $f \in \mathcal{X}'$, then
$$\min\{\|f - g\| : g \in \mathcal{U}^\circ\} = \sup\{|f(\mathbf{u})| : \mathbf{u} \in \mathcal{U} \text{ and } \|\mathbf{u}\| \leq 1\}.$$
The proof is much the same, except that the Hahn-Banach theorem is invoked in place of the finite-dimensional version considered here.

33.2. Preliminary calculations

If $\mathbf{u} = \begin{bmatrix} u_1 & \cdots & u_n \end{bmatrix}^T$ and $\mathbf{v} = \begin{bmatrix} v_1 & \cdots & v_n \end{bmatrix}^T$, it is readily seen that

(33.8)
$$|\langle \mathbf{u}, \mathbf{v} \rangle| = \left|\sum_{j=1}^n u_j \overline{v_j}\right| \leq \|\mathbf{u}\|_1 \|\mathbf{v}\|_\infty.$$

However, more is true:

Theorem 33.3. *If $\mathbf{u}, \mathbf{v} \in \mathbb{C}^n$, then*

(33.9)
$$\|\mathbf{v}\|_1 = \max\left\{|\langle \mathbf{u}, \mathbf{v} \rangle| : \|\mathbf{u}\|_\infty \leq 1\right\}$$

and

(33.10)
$$\|\mathbf{v}\|_\infty = \max\left\{|\langle \mathbf{u}, \mathbf{v} \rangle| : \|\mathbf{u}\|_1 \leq 1\right\}.$$

Moreover, if equality holds in (33.8), *then*

(33.11) $$u_j = 0 \quad if \quad |v_j| < \|\mathbf{v}\|_\infty.$$

Proof. To verify (33.9) when $\mathbf{v} \neq \mathbf{0}$, let $\mathbf{u}$ be the vector with entries

$$u_j = \begin{cases} v_j/|v_j| & \text{if } v_j \neq 0, \\ 0 & \text{if } v_j = 0. \end{cases}$$

Then $\|\mathbf{u}\|_\infty = 1$ and

$$\langle \mathbf{u}, \mathbf{v} \rangle = \sum_{j=1}^n u_j \overline{v_j} = \sum_{j=1}^n |v_j| = \|\mathbf{u}\|_\infty \|\mathbf{v}\|_1.$$

To verify (33.10) when $\mathbf{v} \neq \mathbf{0}$, let $\{1, \ldots, n\} = \Omega_1 \cup \Omega_2$, where

(33.12) $$\Omega_1 = \{j : |v_j| = \|\mathbf{v}\|_\infty\} \quad \text{and} \quad \Omega_2 = \{j : |v_j| < \|\mathbf{v}\|_\infty\}$$

and let $\mathbf{u}$ be the vector with entries

$$u_j = \begin{cases} t_j v_j/|v_j| & \text{if } j \in \Omega_1, \\ 0 & \text{if } j \in \Omega_2, \end{cases}$$

where $t_j > 0$ and $\sum_{j \in \Omega_1} t_j = 1$. Then $\|\mathbf{u}\|_1 = \sum_{j \in \Omega_1} t_j = 1$ and

$$\langle \mathbf{u}, \mathbf{v} \rangle = \sum_{j=1}^n u_j \overline{v_j} = \sum_{j \in \Omega_1} |u_j| \|\mathbf{v}\|_\infty = \|\mathbf{u}\|_1 \|\mathbf{v}\|_\infty.$$

To verify (33.11), suppose that

$$\left| \sum_{j=1}^n u_j \overline{v_j} \right| = \|\mathbf{u}\|_1 \|\mathbf{v}\|_\infty.$$

Then, in terms of the notation (33.12),

$$\sum_{j=1}^n |u_j| \|\mathbf{v}\|_\infty = \|\mathbf{u}\|_1 \|\mathbf{v}\|_\infty = \left| \sum_{j=1}^n u_j \overline{v_j} \right|$$

$$\leq \sum_{j=1}^n |u_j \overline{v_j}| = \sum_{j \in \Omega_1} |u_j| \|\mathbf{v}\|_\infty + \sum_{j \in \Omega_2} |u_j| |v_j|,$$

which implies that

$$0 \geq \sum_{j \in \Omega_2} |u_j| [\|\mathbf{v}\|_\infty - |v_j|] \geq \sum_{j \in \Omega_2} |u_j| \varepsilon$$

for some $\varepsilon > 0$ and hence that $u_j = 0$ if $j \in \Omega_2$. □

33.2. Preliminary calculations

Exercise 33.3. Show that if $\mathbf{u}, \mathbf{v} \in \mathbb{C}^n$, then

(33.13) $$\|\mathbf{u}\|_1 = \max\{|\langle \mathbf{u}, \mathbf{v} \rangle| : \|\mathbf{v}\|_\infty \leq 1\}$$

and

(33.14) $$\|\mathbf{u}\|_\infty = \max\{|\langle \mathbf{u}, \mathbf{v} \rangle| : \|\mathbf{v}\|_1 \leq 1\}.$$

In view of (33.11), solutions x of the minimization problem with respect to the norm $\|\mathbf{x}\|_1$ will typically have many zero entries.

The situation is markedly different if the minimization is carried out with respect to $\|\mathbf{x}\|_2$, or with respect to any of the other classical norms $\|\mathbf{x}\|_t$ for $1 < t < \infty$, as is spelled out in the next theorem.

Theorem 33.4. *If $\mathbf{u}, \mathbf{v} \in \mathbb{C}^n$, $1 < t < \infty$, and $t' = t/(t-1)$, then*

(33.15) $$\|\mathbf{u}\|_t = \max\{|\langle \mathbf{u}, \mathbf{v} \rangle| : \|\mathbf{v}\|_{t'} = 1\}$$

and

(33.16) $$\|\mathbf{v}\|_{t'} = \max\{|\langle \mathbf{u}, \mathbf{v} \rangle| : \|\mathbf{u}\|_t = 1\}.$$

Moreover, if $\mathbf{u} \neq \mathbf{0}$ and $\langle \mathbf{u}, \mathbf{v} \rangle = \|\mathbf{u}\|_t$ for some $\mathbf{v} \in \mathbb{C}^n$ with $\|\mathbf{v}\|_{t'} = 1$, then

(33.17) $v_j = 0$ *if* $u_j = 0$ *and* $v_j = \gamma \dfrac{u_j}{|u_j|^{2-t}}$ *for some $\gamma \in \mathbb{C}$ if $u_j \neq 0$.*

Proof. Let $\mathbf{u} = \begin{bmatrix} u_1 & \cdots & u_n \end{bmatrix}^T$ and $\mathbf{v} = \begin{bmatrix} v_1 & \cdots & v_n \end{bmatrix}^T$. Then Hölder's inequality ensures that

$$|\langle \mathbf{u}, \mathbf{v} \rangle| = \left|\sum_{j=1}^n u_j \overline{v_j}\right| \leq \|\mathbf{u}\|_t \|\mathbf{v}\|_{t'} \quad \text{with } 1 < t < \infty \text{ and } t' = \frac{t}{t-1}.$$

Therefore,

$$\max\{|\langle \mathbf{u}, \mathbf{v} \rangle| : \|\mathbf{v}\|_{t'} = 1\} \leq \|\mathbf{u}\|_t.$$

Moreover, if $u_j \neq 0$ and $\gamma > 0$, then

$$u_j \overline{v_j} = \gamma |u_j|^t \iff v_j = \gamma u_j |u_j|^{t-2} \implies \sum_{j=1}^n u_j \overline{v_j} = \gamma \sum_{j=1}^n |u_j|^t$$

and hence the equality

$$\sum_{j=1}^n u_j \overline{v_j} = \|\mathbf{u}\|_t$$

is attained by choosing

$$\gamma = \|\mathbf{u}\|_t \left\{\sum_{j=1}^n |u_j|^t\right\}^{-1} = \left\{\sum_{j=1}^n |u_j|^t\right\}^{-1/t'}.$$

It is also readily checked that $\|\mathbf{v}\|_{t'} = 1$ and hence that (33.15) holds; (33.16) is obtained from (33.15) by interchanging the roles of $\mathbf{u}$ and $\mathbf{v}$ and t and t'.

Finally, since

$$|\langle \mathbf{u}, \mathbf{v} \rangle| = \left| \sum_{j=1}^{n} u_j \overline{v_j} \right| \leq \sum_{j=1}^{n} |u_j| |v_j| \leq \|\mathbf{u}\|_t \|\mathbf{v}\|_{t'},$$

it is clear that $|\langle \mathbf{u}, \mathbf{v} \rangle| = \|\mathbf{u}\|_t \|\mathbf{v}\|_{t'}$ if and only if the two inequalities in the last display are both equalities, i.e., if and only if there exist a $\theta \in [0, 2\pi)$ and a $\delta \geq 0$ such that

$$u_j \overline{v_j} = e^{i\theta} |u_j v_j| \quad \text{and} \quad |v_j|^{t'} = \delta |u_j|^t \quad \text{for } j = 1, \ldots, n;$$

see Exercise 9.16 and Theorem 9.6. The second condition implies that

$$|v_j| = \delta^{1-t^{-1}} |u_j|^{t-1} \quad \text{for } j = 1, \ldots, n.$$

Thus, $v_j = 0$ if $u_j = 0$ and, in view of the first condition,

$$\overline{v_j} = e^{i\theta} \frac{|u_j|}{u_j} |v_j| = e^{i\theta} \frac{\overline{u_j}}{|u_j|} |v_j| = e^{i\theta} \delta^{1-t^{-1}} \overline{u_j} |u_j|^{t-2} \quad \text{if } u_j \neq 0.$$

Therefore, (33.17) holds. $\square$

The next theorem provides some explicit evaluations for $\|f\|_{\mathcal{X}'}$ when $f \in \mathcal{X}'$ and $\mathcal{X} = \mathbb{C}^n$.

Theorem 33.5. *If $\mathcal{X} = \mathbb{C}^n$ and $f(\mathbf{x}) = \langle \mathbf{x}, \mathbf{v} \rangle$ for some vector $\mathbf{v} \in \mathbb{C}^n$, then:*

(1) $\|\mathbf{x}\|_{\mathcal{X}} = \|\mathbf{x}\|_1 \implies \|f\|_{\mathcal{X}'} = \|\mathbf{v}\|_\infty$.
(2) $\|\mathbf{x}\|_{\mathcal{X}} = \|\mathbf{x}\|_\infty \implies \|f\|_{\mathcal{X}'} = \|\mathbf{v}\|_1$.
(3) $\|\mathbf{x}\|_{\mathcal{X}} = \|\mathbf{x}\|_t, 1 < t < \infty, \implies \|f\|_{\mathcal{X}'} = \|\mathbf{v}\|_{t'}$, *where $1/t' = 1 - 1/t$.*

Proof. Assertions (1)–(3) follow from the fact that

$$\|f\|_{\mathcal{X}'} = \max \{|f(\mathbf{x})| : \mathbf{x} \in \mathbb{C}^n \text{ and } \|\mathbf{x}\|_{\mathcal{X}} \leq 1\}$$
$$= \max \{|\langle \mathbf{x}, \mathbf{v} \rangle| : \mathbf{x} \in \mathbb{C}^n \text{ and } \|\mathbf{x}\|_{\mathcal{X}} \leq 1\}$$

and the formulas (33.10), (33.9), and (33.16), respectively. $\square$

33.3. Evaluation of (33.1)

Theorem 33.6. *If $A \in \mathbb{C}^{p \times n}$ with rank $A = p$, $p < n$, and $\mathbf{b} \in \mathcal{R}_A$, then*

(33.18)
$$\min \{\|\mathbf{x}\|_1 : \mathbf{x} \in \mathcal{S}_\mathbf{b}\}$$
$$= \max \{|\langle \mathbf{b}, \mathbf{y} \rangle| : \mathbf{y} \in \mathbb{C}^p \text{ and } \|A^H \mathbf{y}\|_\infty \leq 1\}.$$

Moreover, if $1 < s < \infty$ and $1/s + 1/t = 1$, then

(33.19) $\min \{\|\mathbf{x}\|_s : \mathbf{x} \in \mathcal{S}_\mathbf{b}\} = \max \{|\mathbf{y}^H \mathbf{b}| : \mathbf{y} \in \mathbb{C}^p \text{ and } \|A^H \mathbf{y}\|_t \leq 1\}.$

Proof. This theorem is a special case of Theorem 33.1, in which $\mathcal{X} = \mathbb{C}^n$ and $\mathcal{U} = \mathcal{N}_A$. Thus, a linear functional $f(\mathbf{x}) = \langle \mathbf{x}, \mathbf{v} \rangle$ belongs to $\mathcal{U}^\circ$ if and only if $\mathbf{v} = A^H \mathbf{y}$ for some vector $\mathbf{y} \in \mathbb{C}^p$. Moreover, since $\|\mathbf{x}\|_\mathcal{X} = \|\mathbf{x}\|_1$,

$$\|f\|_{\mathcal{X}'} = \|A^H \mathbf{y}\|_\infty \quad \text{and} \quad \langle \mathbf{x}, A^H \mathbf{y} \rangle = \langle A\mathbf{x}, \mathbf{y} \rangle = \langle \mathbf{b}, \mathbf{y} \rangle.$$

Consequently, (33.18) holds; (33.19) is left to the reader. □

Lemma 33.7. *If $\widehat{\mathbf{x}} \in \mathbb{C}^n$ is a solution of the minimization problem in (33.18) and $\widehat{\mathbf{y}} \in \mathbb{C}^p$ is a solution of the maximization problem in (33.18), then*

(33.20) $$|\langle \widehat{\mathbf{x}}, A^H \widehat{\mathbf{y}} \rangle| = \|\widehat{\mathbf{x}}\|_1 \|A^H \widehat{\mathbf{y}}\|_\infty$$

and (hence)

(33.21) $$\mathbf{e}_j^T \widehat{\mathbf{x}} = 0 \quad \textit{if} \quad |\mathbf{e}_j^T A^H \widehat{\mathbf{y}}| < \|A^H \widehat{\mathbf{y}}\|_\infty.$$

Proof. Under the given assumptions, it is readily seen with the help of (33.9) that

$$\|\widehat{\mathbf{x}}\|_1 = \min\{\|\widehat{\mathbf{x}} - \mathbf{u}\|_1 : \mathbf{u} \in \mathcal{N}_A\} = \max\{|\langle \widehat{\mathbf{x}}, A^H \mathbf{y} \rangle| : \|A^H \mathbf{y}\|_\infty = 1\}$$
$$= |\langle \widehat{\mathbf{x}}, A^H \widehat{\mathbf{y}} \rangle| \leq \|\widehat{\mathbf{x}}\|_1 \|A^H \widehat{\mathbf{y}}\|_\infty \leq \|\widehat{\mathbf{x}}\|_1$$

and hence that (33.20) holds. Formula (33.21) then follows from formula (33.11) in Theorem 33.3. □

Exercise 33.4. Verify the equality (33.19).

Exercise 33.5. Show that if $s = t = 2$ in (33.19), then

$$\min\{\|\mathbf{x}\|_2 : \mathbf{x} \in \mathcal{S}_\mathbf{b}\} = \max\{|\mathbf{y}^H \mathbf{b}| : \mathbf{y} \in \mathcal{N}_{A^H} \text{ and } \|\mathbf{y}\|_\infty \leq 1\} = \|A^\dagger \mathbf{b}\|.$$

Exercise 33.6. Show that if $A \in \mathbb{C}^{p \times q}$, $\mathbf{c} \in \mathbb{C}^q$, $1 < s < \infty$, and $1/s + 1/t = 1$, then

$$\min\{\|\mathbf{c} - \mathbf{y}\|_s : \mathbf{y} \in \mathcal{N}_A\} = \max\{|\mathbf{c}^H A^H \mathbf{x}| : \mathbf{x} \in \mathbb{C}^p \text{ and } \|A^H \mathbf{x}\|_t \leq 1\}.$$

33.4. A numerical example

Let

$$A = \begin{bmatrix} 1 & 1/2 & 1/3 & 1/4 & 1/5 \\ 1 & -1/2 & 1/4 & -1/8 & 1/16 \end{bmatrix} \quad \text{and} \quad \mathbf{b} = \begin{bmatrix} 3 \\ 4 \end{bmatrix}.$$

Then $\{\mathbf{y} \in \mathbb{R}^2 : \|A^T \mathbf{y}\|_\infty \leq 1\}$ is clearly a convex subset of $\mathbb{R}^2$, which is given explicitly by the set of $\mathbf{y} = \begin{bmatrix} c & d \end{bmatrix}^T$ that satisfy the following 5 sets of inequalities:

$-1 \leq c + d \leq 1, \quad -2 \leq c - d \leq 2, \quad -1 \leq c/3 + d/4 \leq 1, \quad -1 \leq c/4 - d/8 \leq 1,$

and

$$-1 \leq c/5 + d/16 \leq 1.$$

The first two sets of constraints define a closed parallelogram Q with vertices

$$\mathbf{v}_1 = \begin{bmatrix} 1/2 \\ -3/2 \end{bmatrix}, \quad \mathbf{v}_2 = \begin{bmatrix} 3/2 \\ -1/2 \end{bmatrix}, \quad \mathbf{v}_3 = \begin{bmatrix} -1/2 \\ 3/2 \end{bmatrix}, \quad \text{and} \quad \mathbf{v}_4 = \begin{bmatrix} -3/2 \\ 1/2 \end{bmatrix};$$

and it turns out that the last three constraints are automatically satisfied by the points in Q; see Exercise 33.7. Thus, to this point, we have reduced the problem to finding $\max\{|\langle \mathbf{b}, \mathbf{y} \rangle| : \mathbf{y} \in Q\}$. But, as $\mathbf{y} \in Q$ if and only if $-\mathbf{y} \in Q$, this maximum is equal to

$$\max\{\langle \mathbf{b}, \mathbf{y} \rangle : \mathbf{y} \in Q\}.$$

However, since the four vertices are exactly the extreme points of Q, the Krein-Milman theorem ensures that it suffices to evaluate $\langle \mathbf{b}, \mathbf{v}_j \rangle$ for these four points $\mathbf{v}_1, \ldots, \mathbf{v}_4$. Thus, as

$$\langle \mathbf{b}, \mathbf{v}_1 \rangle = -9/2, \quad \langle \mathbf{b}, \mathbf{v}_2 \rangle = 5/2, \quad \langle \mathbf{b}, \mathbf{v}_3 \rangle = 9/2, \quad \langle \mathbf{b}, \mathbf{v}_4 \rangle = -5/2,$$

it follows that

$$\max\{\langle \mathbf{b}, \mathbf{y} \rangle : \|A^T \mathbf{y}\|_\infty \leq 1\} = \max\{\langle \mathbf{b}, \mathbf{y} \rangle : \mathbf{y} \in \Omega\} = 9/2 = \langle \mathbf{b}, \mathbf{v}_3 \rangle,$$

i.e., $\widehat{\mathbf{y}} = \mathbf{v}_3$. Moreover, as

$$A^T \mathbf{v}_3 = \begin{bmatrix} 1 & 1 \\ 1/2 & -1/2 \\ 1/3 & 1/4 \\ 1/4 & -1/8 \\ 1/5 & 1/16 \end{bmatrix} \begin{bmatrix} -1/2 \\ 3/2 \end{bmatrix} = \begin{bmatrix} 1 \\ -1 \\ 5/24 \\ -5/16 \\ 67/80 \end{bmatrix} \quad \text{and} \quad \|A^T \mathbf{v}_3\|_\infty = 1,$$

Theorem 33.3 implies that the last three entries of $\widehat{\mathbf{x}}$ are equal to zero. Consequently,

$$A\widehat{\mathbf{x}} = \begin{bmatrix} 1 & 1/2 & 1/3 & 1/4 & 1/5 \\ 1 & -1/2 & 1/4 & -1/8 & 1/16 \end{bmatrix} \widehat{\mathbf{x}} = \begin{bmatrix} 1 & 1/2 \\ 1 & -1/2 \end{bmatrix} \begin{bmatrix} \widehat{x}_1 \\ \widehat{x}_2 \end{bmatrix} = \begin{bmatrix} 3 \\ 4 \end{bmatrix},$$

which implies that $\widehat{\mathbf{x}} = \begin{bmatrix} 7/2 & -1 & 0 & 0 & 0 \end{bmatrix}^T$ and hence that $\|\widehat{\mathbf{x}}\|_1 = 9/2$, as it should be in order that

$$\langle \widehat{\mathbf{x}}, A^T \widehat{\mathbf{y}} \rangle = \langle \mathbf{b}, \widehat{\mathbf{y}} \rangle = \langle \mathbf{b}, \mathbf{v}_3 \rangle = \frac{9}{2} = \|\widehat{\mathbf{x}}\|_1 \|A^T \mathbf{v}_3\|_\infty.$$

Exercise 33.7. Verify the claim that the last three constraints in the set of five constraints listed above are automatically met by every point in the closed parallelogram Q. [HINT: It is enough to check that every convex combination of the extreme points $\mathbf{v}_1, \ldots, \mathbf{v}_4$ meets these three constraints. Why?]

33.5. A review

To summarize, we solved the problem formulated in (33.1) by the following steps:

(1) Find the extreme points $\mathbf{v}_1, \ldots, \mathbf{v}_k$ of the closed convex bounded set $\{\mathbf{y} \in \mathbb{R}^p : \|A^T\mathbf{y}\|_\infty \leq 1\}$.
(2) Find the extreme points $\mathbf{v}_{i_1}, \ldots, \mathbf{v}_{i_r}$ at which $\max\{\langle \mathbf{b}, \mathbf{v}_j \rangle : j = 1, \ldots, k\}$ is attained and let $\widehat{\mathbf{y}}$ be any convex combination of these r extreme points. Then $\widehat{\mathbf{y}}$ is a solution of the maximization problem.
(3) Let $\Omega_2 = \{j : |(A^T\widehat{\mathbf{y}})_j| < \|A^T\widehat{\mathbf{y}}\|_\infty$. Then $\widehat{\mathbf{x}}$ is a solution of the equation $A\widehat{\mathbf{x}} = \mathbf{b}$ with $(\widehat{\mathbf{x}})_j = 0$ if $j \in \Omega_2$; it is a solution of the minimization problem.
(4) Check that $\langle \widehat{\mathbf{x}}, A^T\widehat{\mathbf{y}} \rangle = \|\widehat{\mathbf{x}}\|_1 \|A^T\widehat{\mathbf{y}}\|_\infty$.

33.6. Supplementary notes

This chapter was adapted from a recent article [18] by R. Cheng and Y. Xu, which treats an infinite-dimensional version of the problem considered here. The problem (33.1) is in some sense an approximation to the following open problem:

Given $A \in \mathbb{C}^{p \times n}$ with rank $A = p$ *and* $p < n$ *and a vector* $\mathbf{b} \in \mathbb{C}^p$, *find*

(33.22) $\qquad \min\{\nu(\mathbf{x}) : \mathbf{x} \in \mathbb{C}^n \quad \text{and} \quad A\mathbf{x} = \mathbf{b}\},$

where $\nu(\mathbf{x}) =$ *the number of nonzero entries in* $\mathbf{x}$.

Chapter 34

Conjugate gradients

The method of **conjugate gradients** is an iterative approach to solving equations of the form $A\mathbf{x} = \mathbf{b}$ for matrices $A \in \mathbb{R}^{n \times n}$ that are positive definite and vectors $\mathbf{b} \in \mathbb{R}^n$. It is based on the observation that if $A \succ O$, then the gradient $(\nabla \varphi)(\mathbf{x})$ of the function

(34.1) $$\varphi(\mathbf{x}) = \frac{1}{2}\langle A\mathbf{x}, \mathbf{x}\rangle - \langle \mathbf{b}, \mathbf{x}\rangle$$

(written as a column vector) is equal to

$$(\nabla \varphi)(\mathbf{x}) = \begin{bmatrix} \frac{\partial \varphi}{\partial x_1} \\ \vdots \\ \frac{\partial \varphi}{\partial x_n} \end{bmatrix} = A\mathbf{x} - \mathbf{b},$$

and the Hessian $H_\varphi(\mathbf{x}) = A \succ O$. Therefore, φ is strictly convex, and hence the solution $\mathbf{x}$ of the equation $A\mathbf{x} = \mathbf{b}$ is the unique point in $\mathbb{R}^n$ at which $\varphi(\mathbf{x})$ attains its minimum value, i.e.,

(34.2) $$\varphi(A^{-1}\mathbf{b}) < \varphi(\mathbf{x}) \quad \text{if} \quad \mathbf{x} \in \mathbb{R}^n \text{ and } \mathbf{x} \neq A^{-1}\mathbf{b}.$$

The method of conjugate gradients exploits this fact in order to find the solution of this equation recursively, as the limit of the solutions $\mathbf{x}_1, \mathbf{x}_2, \ldots$ of a sequence of minimization problems.

Lemma 34.1. *If $A \in \mathbb{R}^{n \times n}$ is positive definite and Q is a closed convex subset of $\mathbb{R}^n$, then there exists exactly one point $\mathbf{q} \in Q$ at which the function $\varphi(\mathbf{x})$ defined by formula (34.1) attains its minimum value, i.e., at which*

$$\varphi(\mathbf{q}) \leq \varphi(\mathbf{x}) \quad \text{for every} \quad \mathbf{x} \in Q.$$

Proof. Let $s_1 \geq \cdots \geq s_n$ denote the singular values of A. Then, $s_n > 0$ and

$$\begin{aligned}\varphi(\mathbf{x}) &= \frac{1}{2}\langle A\mathbf{x}, \mathbf{x}\rangle - \langle \mathbf{b}, \mathbf{x}\rangle \\ &\geq \frac{1}{2}s_n\|\mathbf{x}\|_2^2 - \|\mathbf{b}\|_2\|\mathbf{x}\|_2 \\ &= \|\mathbf{x}\|_2\left(\frac{1}{2}s_n\|\mathbf{x}\|_2 - \|\mathbf{b}\|_2\right),\end{aligned}$$

which is clearly positive if $\|\mathbf{x}\|_2 > 2\|\mathbf{b}\|_2/s_n$. Thus, as $\varphi(\mathbf{0}) = 0$, $\varphi(\mathbf{x})$ will achieve its lowest values in the set $Q \cap \{\mathbf{x} : \|\mathbf{x}\|_2 \leq 2\|\mathbf{b}\|_2/s_n\}$. Since this set is closed and bounded and $\varphi(\mathbf{x})$ is a continuous function of $\mathbf{x}$, $\varphi(\mathbf{x})$ will attain its minimum value in this set. Moreover, as the Hessian of φ is equal to A and $A \succ O$, $\varphi(\mathbf{x})$ is strictly convex (see also Exercise 34.2). Therefore, Theorem 18.2 ensures that $\varphi(\mathbf{x})$ attains its minimum at exactly one point in the subset Q. $\square$

Exercise 34.1. Show that if $\mathbf{u} \in \mathbb{R}^n$ and $\mathcal{V}$ is a subspace of $\mathbb{R}^n$, then the set $\mathbf{u} + \mathcal{V} = \{\mathbf{u} + \mathbf{v} : \mathbf{v} \in \mathcal{V}\}$ is a closed convex subset of $\mathbb{R}^n$.

Exercise 34.2. Verify directly that the function $\varphi(\mathbf{x})$ defined by formula (34.1) is strictly convex. [HINT: Check that the identity

$$(34.3) \quad \varphi(t\mathbf{u} + (1-t)\mathbf{v}) = t\varphi(\mathbf{u}) + (1-t)\varphi(\mathbf{v}) - \frac{1}{2}t(1-t)\langle A(\mathbf{u}-\mathbf{v}), \mathbf{u}-\mathbf{v}\rangle$$

is valid for every pair of vectors $\mathbf{u}$ and $\mathbf{v}$ in $\mathbb{R}^n$ and every number $t \in \mathbb{R}$.]

Let

$$(34.4) \qquad \mathcal{H}_j = \text{span}\{\mathbf{b}, A\mathbf{b}, \ldots, A^{j-1}\mathbf{b}\} \quad \text{for } j = 1, \ldots, n.$$

Then, in view of Lemma 34.1 and Exercise 34.1, there exists a sequence of points $\mathbf{x}_j \in \mathbb{R}^n$ such that

$$(34.5) \qquad \mathbf{x}_j \in \mathcal{H}_j \quad \text{and} \quad \varphi(\mathbf{x}_j) < \varphi(\mathbf{x}) \quad \text{if } \mathbf{x} \in \mathcal{H}_j \text{ and } \mathbf{x} \neq \mathbf{x}_j.$$

Lemma 34.2. *If $A \in \mathbb{R}^{n \times n}$ is positive definite, $\mathbf{b} \in \mathbb{R}^n$, and the spaces $\mathcal{H}_j$ are defined by (34.4), then:*

(1) $A^{-1}\mathbf{b} \in \mathcal{H}_n$.

(2) *If $\mathcal{H}_k = \mathcal{H}_{k+1}$ for some positive integer $k \leq n-2$, then $\mathcal{H}_{k+1} = \mathcal{H}_{k+2}$.*

(3) *If $A^{-1}\mathbf{b} \in \mathcal{H}_k$ for some positive integer $k < n$, then $A^{-1}\mathbf{b} = \mathbf{x}_k$.*

Proof. Let

$$\det(\lambda I_n - A) = a_0 + a_1\lambda + \cdots + a_{n-1}\lambda^{n-1} + \lambda^n.$$

34. Conjugate gradients

Then the Cayley-Hamilton theorem implies that $a_0 I_n + \cdots + a_{n-1} A^{n-1} + A^n = O$ and hence, as $a_0 = \det(-A^n) = (-1)^n \det A \neq 0$, that

$$I_n = -\frac{1}{a_0}[a_1 A + \cdots + a_{n-1} A^{n-1} + A^n].$$

Therefore,

$$A^{-1}\mathbf{b} = -\frac{1}{a_0}[a_1 I_n + \cdots + a_{n-1} A^{n-2} + A^{n-1}]\mathbf{b}$$

which belongs to $\mathcal{H}_n$. Thus, (1) holds.

To verify (2), suppose next that $\mathcal{H}_k = \mathcal{H}_{k+1}$ for some integer $k \leq n-2$. Then $A^k \mathbf{b} \in \mathcal{H}_k$. Therefore, $A^{k+1}\mathbf{b} \in \mathcal{H}_{k+1}$, which ensures that $\mathcal{H}_{k+1} = \mathcal{H}_{k+2}$, as claimed.

Finally, if $A^{-1}\mathbf{b} \in \mathcal{H}_k$ for some positive integer $k < n$, then

$$\varphi(A^{-1}\mathbf{b}) \leq \varphi(\mathbf{x}_k) \leq \varphi(A^{-1}\mathbf{b}),$$

since

$$\varphi(A^{-1}\mathbf{b}) \leq \varphi(\mathbf{x}) \quad \text{for every } \mathbf{x} \in \mathbb{R}^n$$

and

$$\varphi(\mathbf{x}_k) \leq \varphi(\mathbf{x}) \quad \text{for every } \mathbf{x} \in \mathcal{H}_k.$$

Thus, Lemma 34.1 ensures that $A^{-1}\mathbf{b} = \mathbf{x}_k$. □

The **notation**

$$\langle \mathbf{u}, \mathbf{v} \rangle_A = \langle A\mathbf{u}, \mathbf{v} \rangle_{\text{st}} = \mathbf{v}^H A \mathbf{u}$$

will be used in the sequel.

Lemma 34.3. *If $\mathcal{H}_j$ is a proper subspace of $\mathcal{H}_{j+1}$, then*

(34.6) $$\langle A\mathbf{x}_j - \mathbf{b}, \mathbf{x} \rangle = 0 \quad \text{for every } \mathbf{x} \in \mathcal{H}_j$$

and

(34.7) $$\langle \mathbf{x}_{j+1} - \mathbf{x}_j, \mathbf{x} \rangle_A = 0 \quad \text{for every } \mathbf{x} \in \mathcal{H}_j.$$

Proof. If $\mathbf{x} \in \mathcal{H}_j$, then

$$\varphi(\mathbf{x}_j + \varepsilon \mathbf{x}) - \varphi(\mathbf{x}_j) \geq 0 \quad \text{for every } \mathbf{x} \in \mathcal{H}_j \text{ and } \varepsilon > 0.$$

Thus, as the left-hand side is equal to $\varepsilon \langle (\nabla \varphi)(\mathbf{c}), \mathbf{x} \rangle$ for some point $\mathbf{c}$ between $\mathbf{x}_j + \varepsilon \mathbf{x}$ and $\mathbf{x}_j$, it follows that

$$\langle (\nabla \varphi)(\mathbf{x}_j), \mathbf{x} \rangle = \lim_{\varepsilon \downarrow 0} \frac{\varphi(\mathbf{x}_j + \varepsilon \mathbf{x}) - \varphi(\mathbf{x}_j)}{\varepsilon} \geq 0$$

for every vector $\mathbf{x} \in \mathcal{H}_j$. But, by the same reasoning,

$$\langle (\nabla \varphi)(\mathbf{x}_j), -\mathbf{x} \rangle \geq 0$$

for every vector $\mathbf{x} \in \mathcal{H}_j$. Thus, (34.6) holds.

To verify (34.7), it suffices to note that

$$\langle \mathbf{x}_{j+1} - \mathbf{x}_j, \mathbf{x} \rangle_A = \langle A(\mathbf{x}_{j+1} - \mathbf{x}_j), \mathbf{x} \rangle_{\text{st}}$$
$$= \langle A\mathbf{x}_{j+1} - \mathbf{b}, \mathbf{x} \rangle_{\text{st}} - \langle A\mathbf{x}_j - \mathbf{b}, \mathbf{x} \rangle_{\text{st}} = 0$$

for every $\mathbf{x} \in \mathcal{H}_j$ by (34.6). $\square$

34.1. The recursion

In this section we shall establish a recursive procedure for calculating the points $\mathbf{x}_1, \mathbf{x}_2, \ldots$ specified in (34.5) that avoids the need for solving minimization problems. It is convenient to express some formulas in terms of the orthogonal projection $\Pi_{\mathcal{X}_j}$ from $\mathbb{R}^n$ onto the subspace $\mathcal{X}_j = \{\alpha(\mathbf{x}_j - \mathbf{x}_{j-1}) : \alpha \in \mathbb{R}\}$ with respect to the inner product $\langle \cdot, \cdot \rangle_A$:

$$(34.8) \qquad \Pi_{\mathcal{X}_j} : \mathbf{u} \in \mathbb{R}^n \mapsto \frac{\langle \mathbf{u}, \mathbf{x}_j - \mathbf{x}_{j-1} \rangle_A}{\langle \mathbf{x}_j - \mathbf{x}_{j-1}, \mathbf{x}_j - \mathbf{x}_{j-1} \rangle_A} (\mathbf{x}_j - \mathbf{x}_{j-1})$$

for $j = 1, \ldots, n$.

Lemma 34.4. *If $\mathcal{H}_j$ is a proper subspace of $\mathcal{H}_{j+1}$ and $\mathbf{x}_0 = \mathbf{0}$, then:*

(1) *The set of vectors $\{A\mathbf{x}_0 - \mathbf{b}, A\mathbf{x}_1 - \mathbf{b}, \ldots, A\mathbf{x}_j - \mathbf{b}\}$ is a basis for $\mathcal{H}_{j+1}$ that is orthogonal with respect to the standard inner product.*

(2) *The set of vectors $\{\mathbf{x}_1 - \mathbf{x}_0, \mathbf{x}_2 - \mathbf{x}_1, \ldots, \mathbf{x}_{j+1} - \mathbf{x}_j\}$ is a basis for $\mathcal{H}_{j+1}$ that is orthogonal with respect to the inner product $\langle \cdot, \cdot \rangle_A$.*

(3) *There exists exactly one vector $\mathbf{v}_j \in \mathcal{H}_{j+1}$ such that $\langle \mathbf{v}_j, A\mathbf{y} \rangle = 0$ for every vector $\mathbf{y} \in \mathcal{H}_j$ and $A\mathbf{x}_j - \mathbf{b} - \mathbf{v}_j \in \mathcal{H}_j$; it may be specified in terms of the projection $\Pi_{\mathcal{X}_j}$ by the formulas*

$$(34.9) \qquad \mathbf{v}_j = A\mathbf{x}_j - \mathbf{b} - \Pi_{\mathcal{X}_j}(A\mathbf{x}_j - \mathbf{b}) = \Pi_{\mathcal{X}_{j+1}}(A\mathbf{x}_j - \mathbf{b}).$$

Proof. The first two assertions follow easily from (34.6) and (34.7), respectively.

Suppose next that $\mathbf{v}_j$ and $\mathbf{w}_j$ are vectors in $\mathcal{H}_{j+1}$ that meet the two constraints in (3). Then

$$\mathbf{v}_j - \mathbf{w}_j = (A\mathbf{x}_j - \mathbf{b} - \mathbf{w}_j) - (A\mathbf{x}_j - \mathbf{b} - \mathbf{v}_j) \in \mathcal{H}_j$$

and $\mathbf{v}_j - \mathbf{w}_j$ is orthogonal to $\mathcal{H}_j$ in the inner product $\langle \cdot, \cdot \rangle_A$. Therefore, $\mathbf{v}_j = \mathbf{w}_j$. Moreover, in view of (2),

$$A\mathbf{x}_j - \mathbf{b} = \sum_{i=1}^{j+1} c_i(\mathbf{x}_i - \mathbf{x}_{i-1}), \quad \text{with} \quad c_i = \frac{\langle A\mathbf{x}_j - \mathbf{b}, \mathbf{x}_i - \mathbf{x}_{i-1} \rangle_A}{\langle \mathbf{x}_i - \mathbf{x}_{i-1}, \mathbf{x}_i - \mathbf{x}_{i-1} \rangle_A}.$$

34.1. The recursion

The condition $\langle A\mathbf{x}_j - \mathbf{b}, \mathbf{x}\rangle = 0$ for $\mathbf{x} \in \mathcal{H}_j$ forces $c_i = 0$ for $i = 1, \ldots, j-1$. Thus,
$$A\mathbf{x}_j - \mathbf{b} = c_{j+1}(\mathbf{x}_{j+1} - \mathbf{x}_j) + c_j(\mathbf{x}_j - \mathbf{x}_{j-1}).$$
Therefore, the vector
$$\mathbf{v}_j = c_{j+1}(\mathbf{x}_{j+1} - \mathbf{x}_j) = A\mathbf{x}_j - \mathbf{b} - c_j(\mathbf{x}_j - \mathbf{x}_{j-1})$$
meets the two constraints in (3). Since $c_i(\mathbf{x}_i - \mathbf{x}_{i-1}) = \Pi_{\mathcal{X}_i}(A\mathbf{x}_j - \mathbf{b})$, the two formulas for $\mathbf{v}_j$ in the last display coincide with the formulas in (34.9). Thus, the proof is complete. □

Lemma 34.5. *The initial vectors in the recursion are*

(34.10) $$\mathbf{x}_1 = \frac{\langle \mathbf{b}, \mathbf{b}\rangle}{\langle A\mathbf{b}, \mathbf{b}\rangle}\mathbf{b}$$

and

(34.11) $$\mathbf{v}_1 = A\mathbf{x}_1 - \mathbf{b} - \frac{\langle A\mathbf{x}_1 - \mathbf{b}, A\mathbf{b}\rangle}{\langle A\mathbf{b}, \mathbf{b}\rangle}\mathbf{b}.$$

Proof. Since $\mathbf{x}_1 = t\mathbf{b}$ for some $t \in \mathbb{R}$,
$$\varphi(t\mathbf{b}) = \frac{t^2}{2}\langle A\mathbf{b}, \mathbf{b}\rangle - t\langle \mathbf{b}, \mathbf{b}\rangle,$$
which clearly achieves its minimum value when $t = t_1 = \langle \mathbf{b}, \mathbf{b}\rangle/\langle A\mathbf{b}, \mathbf{b}\rangle$. Thus, (34.10) holds; (34.11) follows from the first equality in (34.9), since $\mathcal{X}_1 = \{\alpha\mathbf{b} : \alpha \in \mathbb{R}\}$. □

The next three exercises are a warmup to the general recursion that will be presented in Theorem 34.6.

Exercise 34.3. Verify the identities $\langle A\mathbf{x}_1 - \mathbf{b}, \mathbf{x}_2 - \mathbf{x}_1\rangle_A = \langle \mathbf{v}_1, \mathbf{x}_2 - \mathbf{x}_1\rangle_A$ and $\langle A\mathbf{x}_1 - \mathbf{b}, \mathbf{x}_2 - \mathbf{x}_1\rangle_A = -\langle \mathbf{x}_2 - \mathbf{x}_1, \mathbf{x}_2 - \mathbf{x}_1\rangle_A$. [HINT: $A(\mathbf{x}_2 - \mathbf{x}_1) = (A\mathbf{x}_2 - \mathbf{b}) - (A\mathbf{x}_1 - \mathbf{b})$.]

Exercise 34.4. Show that
$$\mathbf{x}_2 = \mathbf{x}_1 - \gamma_1 \mathbf{v}_1 \quad \text{with } \gamma_1 = \frac{\langle A\mathbf{x}_1 - \mathbf{b}, \mathbf{v}_1\rangle}{\langle A\mathbf{v}_1, \mathbf{v}_1\rangle}.$$

[HINT: Use the second equality in (34.9) and the two identities that were referred to in Exercise 34.3.]

Exercise 34.5. Show that
$$\mathbf{v}_2 = A\mathbf{x}_2 - \mathbf{b} - \beta_1 \mathbf{v}_1 \quad \text{with } \beta_1 = \frac{\langle A\mathbf{x}_2 - \mathbf{b}, A\mathbf{v}_1\rangle}{\langle A\mathbf{v}_1, \mathbf{v}_1\rangle}.$$

Theorem 34.6. *The general recursion is given by the formulas*

(34.12) $$\mathbf{x}_{j+1} = \mathbf{x}_j - \gamma_j \mathbf{v}_j \quad \text{with } \gamma_j = \frac{\langle A\mathbf{x}_j - \mathbf{b}, \mathbf{v}_j \rangle}{\langle A\mathbf{v}_j, \mathbf{v}_j \rangle}$$

and

(34.13) $$\mathbf{v}_{j+1} = A\mathbf{x}_{j+1} - \mathbf{b} - \beta_j \mathbf{v}_j \quad \text{with } \beta_j = \frac{\langle A\mathbf{x}_{j+1} - \mathbf{b}, A\mathbf{v}_j \rangle}{\langle A\mathbf{v}_j, \mathbf{v}_j \rangle}$$

for $j = 1, \ldots, n-1$ and the initial conditions $\mathbf{x}_1$ and $\mathbf{v}_1$ given by (34.10) and (34.11), respectively.

Proof. Formula (34.12) is obtained by inverting the second formula for $\mathbf{v}_j$ in (34.9):

$$\mathbf{v}_j = \frac{\langle A\mathbf{x}_j - \mathbf{b}, \mathbf{x}_{j+1} - \mathbf{x}_j \rangle_A}{\langle \mathbf{x}_{j+1} - \mathbf{x}_j, \mathbf{x}_{j+1} - \mathbf{x}_j \rangle_A} (\mathbf{x}_{j+1} - \mathbf{x}_j),$$

in order to obtain an expression for $\mathbf{x}_{j+1} - \mathbf{x}_j$ in terms of $\mathbf{v}_j$. The verification rests on the following two extensions of the identities in Exercise 34.3:

$$\langle A\mathbf{x}_j - \mathbf{b}, \mathbf{x}_{j+1} - \mathbf{x}_j \rangle_A = \langle A\mathbf{x}_j - \mathbf{b} - \mathbf{v}_j, \mathbf{x}_{j+1} - \mathbf{x}_j \rangle_A + \langle \mathbf{v}_j, \mathbf{x}_{j+1} - \mathbf{x}_j \rangle_A$$
$$= \langle \mathbf{v}_j, \mathbf{x}_{j+1} - \mathbf{x}_j \rangle_A$$

and

$$\langle \mathbf{x}_{j+1} - \mathbf{x}_j, \mathbf{x}_{j+1} - \mathbf{x}_j \rangle_A = \langle (A\mathbf{x}_{j+1} - \mathbf{b}) - (A\mathbf{x}_j - \mathbf{b}), \mathbf{x}_{j+1} - \mathbf{x}_j \rangle$$
$$= -\langle A\mathbf{x}_j - \mathbf{b}, \mathbf{x}_{j+1} - \mathbf{x}_j \rangle.$$

These identities enable us to write

$$\mathbf{x}_{j+1} - \mathbf{x}_j = \frac{\langle \mathbf{x}_{j+1} - \mathbf{x}_j, \mathbf{x}_{j+1} - \mathbf{x}_j \rangle_A}{\langle A\mathbf{x}_j - \mathbf{b}, \mathbf{x}_{j+1} - \mathbf{x}_j \rangle_A} \mathbf{v}_j = -\frac{\langle A\mathbf{x}_j - \mathbf{b}, \mathbf{x}_{j+1} - \mathbf{x}_j \rangle}{\langle \mathbf{v}_j, \mathbf{x}_{j+1} - \mathbf{x}_j \rangle_A} \mathbf{v}_j,$$

which is equivalent to (34.12), since $\mathbf{x}_{j+1} - \mathbf{x}_j = \delta_j \mathbf{v}_j$.

Formula (34.13) is immediate from (34.9):

$$\mathbf{v}_{j+1} = A\mathbf{x}_{j+1} - \mathbf{b} - \Pi_{\mathcal{X}_{j+1}}(A\mathbf{x}_{j+1} - \mathbf{b}),$$

since

$$\Pi_{\mathcal{X}_{j+1}}(A\mathbf{x}_{j+1} - \mathbf{b}) = \frac{\langle A\mathbf{x}_{j+1} - \mathbf{b}, \mathbf{x}_{j+1} - \mathbf{x}_j \rangle_A}{\langle \mathbf{x}_{j+1} - \mathbf{x}_j, \mathbf{x}_{j+1} - \mathbf{x}_j \rangle_A} (\mathbf{x}_{j+1} - \mathbf{x}_j) = \beta_j \mathbf{v}_j. \quad \square$$

Exercise 34.6. Let $A = \begin{bmatrix} 2 & 1 \\ 1 & 1 \end{bmatrix}$ and $\mathbf{b} = -\begin{bmatrix} 1 \\ 1 \end{bmatrix}$. Show that $A \succ O$ and then use the formulas in Lemma 34.5 and Theorem 34.6 to solve the equation $A\mathbf{x} = \mathbf{b}$.

Exercise 34.7. Let

$$A = \begin{bmatrix} 3 & 1 & 0 \\ 1 & 1 & 1 \\ 0 & 1 & 2 \end{bmatrix} \quad \text{and} \quad \mathbf{b} = \begin{bmatrix} 0 \\ 0 \\ 1 \end{bmatrix}.$$

Show that $A \succ O$ and then use the formulas in Lemma 34.5 and Theorem 34.6 to solve the equation $A\mathbf{x} = \mathbf{b}$.

34.2. Convergence estimates

In this section we shall develop upper bounds on $\|\mathbf{x}_j - A^{-1}\mathbf{b}\|_A$.

Theorem 34.7. *If $A \in \mathbb{R}^{n \times n}$ is positive definite with singular values $s_1 \geq \cdots \geq s_n$ and singular value decomposition $A = VSV^T$ and $A^{-1}\mathbf{b} \notin \mathcal{H}_k$, then*

$$(34.14) \qquad \|\mathbf{x}_k - A^{-1}\mathbf{b}\|_A^2 = \min\left\{ \sum_{i=1}^n \frac{c_i^2}{s_i} q(s_i)^2 : q \in \mathcal{P}_k \text{ and } q(0) = 1 \right\},$$

where $\mathcal{P}_k$ denotes the set of polynomials of degree $\leq k$ and $c_1, \ldots, c_n$ are the entries in the vector $\mathbf{c} = V^T \mathbf{b}$.

Proof. The proof is divided into steps.

1. *Verification of the formula*

$$(34.15) \qquad \|\mathbf{x}_k - A^{-1}\mathbf{b}\|_A = \min\{\|\mathbf{x} - A^{-1}\mathbf{b}\|_A : \mathbf{x} \in \mathcal{H}_k\}.$$

If $\mathbf{x} \in \mathcal{H}_k$ and $k < m = \dim \mathcal{H}$, then $A^{-1}\mathbf{b} = \mathbf{x}_m$ and

$$\|\mathbf{x} - A^{-1}\mathbf{b}\|_A^2 = \|\mathbf{x} - \mathbf{x}_k + \mathbf{x}_k - \mathbf{x}_m\|_A^2 = \|\mathbf{x} - \mathbf{x}_k\|_A^2 + \|\mathbf{x}_k - \mathbf{x}_m\|_A^2,$$

since $\mathbf{x}_m - \mathbf{x}_k$ is orthogonal to $\mathcal{H}_k$ with respect to the inner product $\langle \cdot, \cdot \rangle_A$. Therefore, the assertion is now self-evident.

2. *Verification of the formula*

$$\|\mathbf{x}_k - A^{-1}\mathbf{b}\|_A = \min_{p \in \mathcal{P}_{k-1}} \{\|p(A)\mathbf{b} - A^{-1}\mathbf{b}\|_A\}.$$

This is immediate from the fact that $\mathbf{x} \in \mathcal{H}_k$ if and only if $\mathbf{x} = p(A)\mathbf{b}$ for some polynomial of degree $\leq k - 1$.

3. Verification of (34.14).

Since $A = VSV^T$ with $V \in \mathbb{R}^{n \times n}$ unitary,

$$\langle A[p(A)\mathbf{b} - A^{-1}\mathbf{b}], [p(A)\mathbf{b} - A^{-1}\mathbf{b}]\rangle$$
$$= \langle VSV^T[Vp(S)V^T - VS^{-1}V^T]\mathbf{b}, [Vp(S)V^T - VS^{-1}V^T]\mathbf{b}\rangle$$
$$= \langle VS[p(S)V^T - S^{-1}V^T]\mathbf{b}, V[p(S)V^T - S^{-1}V^T]\mathbf{b}\rangle$$
$$= \langle S^{-1}[Sp(S) - I_n]\mathbf{c}, [Sp(S) - I_n]\mathbf{c}\rangle,$$

which yields (34.14). □

We shall not attempt to evaluate the minimization indicated on the right-hand side of (34.14). Instead, we shall obtain an upper bound on the right-hand side of (34.14) by choosing a particular polynomial $q \in \mathcal{P}_k$ with $q(0) = 1$. Towards this end, we introduce the **Chebyshev polynomial**

$$T_k(x) = \frac{1}{2}\{(x + \sqrt{x^2 - 1})^k + (x - \sqrt{x^2 - 1})^k\}, \quad k = 0, 1, \ldots,$$

which really is a polynomial in x of degree k (since odd powers of the term $\sqrt{x^2 - 1}$ cancel out). Moreover,

$$T_k(x) > 2^{-1}(x + \sqrt{x^2 - 1})^k > 1/2 \quad \text{if } x > 1$$

and $|T_k(x)| \leq 1$ if $|x| \leq 1$, since

$$T_k(\cos\theta) = \frac{(\cos\theta + i\sin\theta)^k + (\cos\theta - i\sin\theta)^k}{2}$$
$$= \frac{e^{ik\theta} + e^{-ik\theta}}{2}$$
$$= \cos k\theta.$$

Now, assuming that $s_1 > s_n$, let

$$h_k(x) = T_k\left(\frac{s_1 + s_n - 2x}{s_1 - s_n}\right) \quad \text{and} \quad q_k(x) = \frac{h_k(x)}{h_k(0)}.$$

Then $h_k(x)$ is a polynomial of degree k in x and $|h_k(s_i)| \leq 1$ for $i = 1, \ldots, n$, since

$$-1 = \frac{s_n - s_1}{s_1 - s_n} \leq \frac{s_1 + s_n - 2s_i}{s_1 - s_n} \leq \frac{s_1 - s_n}{s_1 - s_n} = 1.$$

Consequently, $q_k(x)$ is a polynomial of degree k in x such that $q_k(0) = 1$ and
(34.16)
$$q_k(s_i) \leq \frac{1}{h_k(0)} = \frac{2}{\mu^k + \mu^{-k}}, \text{ where } \mu = \frac{\sqrt{\kappa}+1}{\sqrt{\kappa}-1} \text{ and } \kappa = s_1/s_n > 1.$$

The upper bounds in (34.16) depend essentially upon the observation that
$$\frac{\kappa+1}{\kappa-1} + \sqrt{\left(\frac{\kappa+1}{\kappa-1}\right)^2 - 1} = \frac{\kappa+1}{\kappa-1} + \frac{2\sqrt{\kappa}}{\kappa-1}$$
$$= \frac{(\sqrt{\kappa}+1)^2}{(\sqrt{\kappa}+1)(\sqrt{\kappa}-1)} = \frac{\sqrt{\kappa}+1}{\sqrt{\kappa}-1} = \mu$$

and, by a similar calculation,
$$\frac{\kappa+1}{\kappa-1} - \sqrt{\left(\frac{\kappa+1}{\kappa-1}\right)^2 - 1} = \mu^{-1}.$$

Upon inserting these bounds into (34.14), we see that if $A \succ O$ and $s_1 > s_n$, then

(34.17)
$$\|\mathbf{x}_k - A^{-1}\mathbf{b}\|_A^2 \leq \sum_{j=1}^n \frac{c_j^2}{s_j}\left(\frac{2}{\mu^k + \mu^{-k}}\right)^2$$
$$= \langle A^{-1}\mathbf{b}, \mathbf{b}\rangle_{st} \left(\frac{2}{\mu^k + \mu^{-k}}\right)^2.$$

The number $\kappa = s_1/s_n$ is called the **condition number** of the matrix A. If κ is close to 1, then μ is large and the upper bound in (34.17) tends to zero quickly.

34.3. Krylov subspaces

The spaces $\mathcal{H}_j$ generated by $\mathbf{b} \in \mathbb{R}^n$ and $A \in \mathbb{R}^{n \times n}$ that are defined in (34.4) are examples of **Krylov subspaces**. The k'th Krylov subspace of $\mathbb{R}^n$ generated by a nonzero vector $\mathbf{u} \in \mathbb{R}^n$ and a matrix $A \in \mathbb{R}^{n \times n}$ is defined by the formula

(34.18) $\qquad \mathcal{H}_k = \text{span}\{\mathbf{u}, A\mathbf{u}, \ldots, A^{k-1}\mathbf{u}\}$ for $k = 1, 2, \ldots$.

Clearly, $\dim \mathcal{H}_1 = 1$, $\dim \mathcal{H}_k \leq k$ for $k = 2, 3, \ldots$, and, if $\mathcal{H}_{k+1} = \mathcal{H}_k$ for some positive integer k, then $\mathcal{H}_j = \mathcal{H}_k$ for every integer $j \geq k$.

Exercise 34.8. Show that if $\mathcal{H}_{k+1} = \mathcal{H}_k$ for some positive integer k, then $\mathcal{H}_j = \mathcal{H}_k$ for every integer $j \geq k$. [HINT: If $\mathcal{H}_{k+1} = \mathcal{H}_k$ for some positive integer k, then $A^k\mathbf{u} = c_{k-1}A^{k-1}\mathbf{u} + \cdots + c_0\mathbf{u}$.]

Exercise 34.9. Let $A \in \mathbb{R}^{n \times n}$, let $\mathbf{u} \in \mathbb{R}^n$, and let $k \geq 1$ be a positive integer. Show that if $A \succ 0$, then the matrix

$$\begin{bmatrix} \langle A\mathbf{u}, \mathbf{u} \rangle & \langle A^2\mathbf{u}, \mathbf{u} \rangle & \cdots & \langle A^k\mathbf{u}, \mathbf{u} \rangle \\ \langle A\mathbf{u}, A\mathbf{u} \rangle & \langle A^2\mathbf{u}, A\mathbf{u} \rangle & \cdots & \langle A^k\mathbf{u}, A\mathbf{u} \rangle \\ \vdots & & \cdots & \vdots \\ \langle A\mathbf{u}, A^{k-1}\mathbf{u} \rangle & \langle A^2\mathbf{u}, A^{k-1}\mathbf{u} \rangle & \cdots & \langle A^k\mathbf{u}, A^{k-1}\mathbf{u} \rangle \end{bmatrix}$$

is invertible if and only if the vectors $\mathbf{u}, A\mathbf{u}, \ldots, A^{k-1}\mathbf{u}$ are linearly independent in $\mathbb{R}^n$.

34.4. The general conjugate gradient method

In this section we shall outline the conjugate gradient algorithm for computing a sequence of points $\mathbf{x}_0, \mathbf{x}_1, \ldots$ that tend to the solution of the equation $A\mathbf{x} = \mathbf{b}$ when $A \succ O$, starting from an arbitrary point $\mathbf{x}_0 \in \mathbb{R}^n$ for which $A\mathbf{x}_0 - \mathbf{b} \neq \mathbf{0}$ in a series of exercises. Correspondingly, the spaces

(34.19) $\quad \mathcal{H}_j = \mathrm{span}\{\mathbf{u}, A\mathbf{u}, \ldots, A^{j-1}\mathbf{u}\} \quad \text{with} \quad \mathbf{u} = A\mathbf{x}_0 - \mathbf{b} \neq \mathbf{0}$

for $j = 1, \ldots, n$.

Exercise 34.10. Show that if $\mathbf{x}_0 \in \mathbb{R}^n$ and $\mathcal{X}$ is a subspace of $\mathbb{R}^n$, then there exists exactly one point $\mathbf{u} \in \mathbf{x}_0 + \mathcal{X}$ such that $\varphi(\mathbf{u}) \leq \varphi(\mathbf{x})$ for every point $\mathbf{x} \in \mathbf{x}_0 + \mathcal{X}$.

Exercise 34.11. Let $\mathbf{x}_0 \in \mathbb{R}^n$ and assume that $\mathbf{u} = A\mathbf{x}_0 - \mathbf{b} \neq \mathbf{0}$ and let $\mathbf{x}_j \in \mathbf{x}_0 + \mathcal{H}_j$ be the unique point in that set such $\varphi(\mathbf{x}_j) \leq \varphi(\mathbf{x})$ for every point $\mathbf{x} \in \mathbf{x}_0 + \mathcal{H}_j$. Show that:

(a) $\mathbf{x}_j - \mathbf{x}_0 \in \mathcal{H}_j$ for $j = 1, 2, \ldots, \ell$.

(b) $\mathbf{x}_j - \mathbf{x}_{j-1} \in \mathcal{H}_j$ for $j = 1, 2, \ldots, \ell$.

(c) $A\mathbf{x}_{j-1} - \mathbf{b} \in \mathcal{H}_j$ for $j = 1, 2, \ldots, \ell$.

Exercise 34.12. Show that $\langle A\mathbf{x}_j - \mathbf{b}, \mathbf{x} \rangle = 0$ for every vector $\mathbf{x} \in \mathbf{x}_0 + \mathcal{H}_j$ and hence that:

(a) $A\mathbf{x}_j = \mathbf{b}$ if and only if $\mathcal{H}_{j+1} = \mathcal{H}_j$.

(b) If $1 \leq j \leq \ell$, then the vectors $A\mathbf{x}_0 - \mathbf{b}, A\mathbf{x}_1 - \mathbf{b}, \ldots, A\mathbf{x}_{j-1} - \mathbf{b}$ form an orthogonal basis for $\mathcal{H}_j$ with respect to the standard inner product.

(c) If $1 \leq j \leq \ell$, then the vectors $\mathbf{x}_1 - \mathbf{x}_0, \mathbf{x}_2 - \mathbf{x}_1, \ldots, \mathbf{x}_j - \mathbf{x}_{j-1}$ form an orthogonal basis for $\mathcal{H}_j$ with respect to the "A"-inner product, i.e., $\langle A(\mathbf{x}_s - \mathbf{x}_{s-1}), \mathbf{x}_t - \mathbf{x}_{t-1} \rangle = 0$ if $s, t = 1, \ldots, \ell$ and $s \neq t$.

Exercise 34.13. Show that if $1 \leq j \leq \ell - 1$, then:
(a) $A\mathbf{x}_j - \mathbf{b} \in \text{span}\{\mathbf{x}_{j+1} - \mathbf{x}_j, \mathbf{x}_j - \mathbf{x}_{j-1}\}$.
(b) There exists exactly one vector $\mathbf{v}_j \in \mathcal{H}_{j+1}$ that is of the form $\mathbf{v}_j = \alpha_j(\mathbf{x}_{j+1} - \mathbf{x}_j)$ with $\alpha_j \in \mathbb{R}$ such that $A\mathbf{x}_j - \mathbf{b} - \mathbf{v}_j \in \mathcal{H}_j$.

Exercise 34.14. Use the formulas
$$\mathbf{x}_{j+1} - \mathbf{x}_j = \gamma_j \mathbf{v}_j \quad \text{and} \quad A\mathbf{x}_{j+1} - \mathbf{b} = \mathbf{v}_{j+1} + \beta_j \mathbf{v}_j$$
to establish the recursions
$$\mathbf{x}_{j+1} = \mathbf{x}_j - \frac{\langle A\mathbf{x}_j - \mathbf{b}, \mathbf{v}_j \rangle}{\langle A\mathbf{v}_j, \mathbf{v}_j \rangle} \mathbf{v}_j,$$
$$\mathbf{v}_{j+1} = A\mathbf{x}_{j+1} - \mathbf{b} - \frac{\langle A(A\mathbf{x}_{j+1} - \mathbf{b}), \mathbf{v}_j \rangle}{\langle A\mathbf{v}_j, \mathbf{v}_j \rangle} \mathbf{v}_j,$$
for $j = 0, \ldots, \ell - 1$, with initial conditions $\mathbf{x}_0$ and $\mathbf{v}_0 = A\mathbf{x}_0 - \mathbf{b}$.

The recursions in Exercise 34.14 enable us to solve sequentially for $\mathbf{x}_1$, $\mathbf{v}_1$, $\mathbf{x}_2$, $\mathbf{v}_2$, ... and only involve simple computations; it is not necessary to compute A^{-1} or to solve minimization problems.

Exercise 34.15. Use the recursions in Exercise 34.14 to solve the equation $A\mathbf{x} = \mathbf{b}$ when $A = \begin{bmatrix} 2 & 1 \\ 1 & 1 \end{bmatrix}$, $\mathbf{b} = \begin{bmatrix} 3 \\ 2 \end{bmatrix}$, and $\mathbf{x}_0 = \begin{bmatrix} 1 \\ 0 \end{bmatrix}$.

34.5. Supplementary notes

The discussion of the general conjugate gradient method in Section 34.4 is based partially on the analysis in Section 16.6 of [30], which was adapted from the discussion in the monographs of Luenberger [59] and of Trefethen and Bau [72]. The analysis in Sections 34.1 and 34.2 does not appear in [30]. The convergence estimates in Section 34.2 were adapted from the online article [68] by Shewchuk.

Chapter 35

Continuity of eigenvalues

In this chapter we shall use complex function theory to show that if $A, B \in \mathbb{C}^{n \times n}$ and $\|A - B\|$ is small, then $\sigma(A)$ is close to $\sigma(B)$ in a sense that will be made precise in Theorem 35.7, which is the main result of this chapter. We shall assume that the reader is familiar with the elements of complex function theory. A quick introduction is presented in Chapter 17 of [**30**]. However, all you need to proceed is a willingness to accept the following facts:

(1) A complex-valued function $f(\lambda)$ of the complex variable λ that is defined in an open set $\Omega \subseteq \mathbb{C}$ is said to be **holomorphic** (or **analytic**) in Ω if the limit

(35.1) $$f'(\lambda) = \lim_{\xi \to 0} \frac{f(\lambda + \xi) - f(\lambda)}{\xi}$$

exists for every point $\lambda \in \Omega$. This is a very strong constraint because the variable ξ in this difference ratio is complex and the definition requires the limit to be the same regardless of how ξ tends to zero. In fact, it is so strong that

$$f \text{ analytic in } \Omega \implies f' \text{ analytic in } \Omega$$
$$\implies f'' \text{ analytic in } \Omega \implies \cdots,$$

and hence (see Exercise 35.1) the function f and its successive **derivatives** $f', f'' \ldots$ are all continuous in Ω.

(2) The **contour integral** $\int_\Gamma f(\lambda)d\lambda$ of a continuous complex-valued function that is defined on a smooth curve Γ that is parametrized by $\gamma(t)$, $a \le t \le b$, is defined by the formula

$$\text{(35.2)} \qquad \int_\Gamma f(\lambda)d\lambda = \int_a^b f(\gamma(t))\gamma'(t)dt\,.$$

The curve Γ is said to be **simple** if $a < t_1, t_2 < b$ and $t_1 \ne t_2$, then $\gamma(t_1) \ne \gamma(t_2)$.

(3) If $f(\lambda)$ is holomorphic in some open nonempty set Ω and Γ is a simple closed piecewise smooth curve in Ω (think of a rubber band) that is directed counterclockwise such that all the points enclosed by Γ also belong to Ω, then

$$\text{(35.3)} \qquad \frac{1}{2\pi i}\int_\Gamma f(\lambda)d\lambda = 0$$

and

$$\text{(35.4)} \qquad \frac{1}{2\pi i}\int_\Gamma \frac{f(\lambda)}{(\lambda-\omega)^{k+1}}d\lambda = \begin{cases} \frac{f^{(k)}(\omega)}{k!} & \text{if } \omega \text{ is inside } \Gamma, \\ 0 & \text{if } \omega \text{ is outside } \Gamma \end{cases}$$

for $k = 0, 1, \ldots$. (Here $f^{(0)} = f$ and $f^{(k)}$ denotes the k'th derivative of f for $k = 1, 2, \ldots$.)

The numerical value of the integral in (35.2) depends upon the curve Γ, but not upon the particular choice of the (one-to-one) function $\gamma(t)$ that is used to describe the curve.

Exercise 35.1. Show that if f is analytic in an open set $\Omega \subseteq \mathbb{C}$, then f is continuous in Ω. [HINT: $|f(\lambda+\xi) - f(\lambda)| = |\xi| |(f(\lambda+\xi) - f(\lambda))/\xi - f'(\lambda) + f'(\lambda)|$.]

Exercise 35.2. Show that if f and g are analytic in a nonempty open set $\Omega \subseteq \mathbb{C}$, then fg and $f+g$ are analytic in Ω, and that if $|f(\omega)| > 0$ for every point $\omega \in \Omega$, then $1/f$ is also analytic in Ω.

Exercise 35.3. Show that if $f(\lambda)$ is a polynomial, then f and e^f are analytic in $\mathbb{C}$; but if $f(\lambda) = (\lambda - \lambda_1)^{-1}(\lambda - \lambda_2)^{-1}$ with $\lambda_1 \ne \lambda_2$, then f is analytic in $\mathbb{C} \setminus \{\lambda_1, \lambda_2\}$.

Exercise 35.4. Show that if f is analytic in an open set $\Omega \subseteq \mathbb{C}$ and $\omega \in \Omega$, then the function

$$g(\lambda) = \begin{cases} \frac{f(\lambda) - f(\omega)}{\lambda - \omega} & \text{if } \lambda \ne \omega, \\ f'(\omega) & \text{if } \lambda = \omega \end{cases}$$

is analytic in Ω.

Remark 35.1. The formulas in (35.4) are all obtained from (35.3). In particular, the formula for $k = 0$,

$$(35.5) \qquad \frac{1}{2\pi i} \int_\Gamma \frac{f(\lambda)}{(\lambda - \omega)} d\lambda = f(\omega) \quad \text{for } \omega \text{ inside } \Gamma,$$

may be obtained by applying (35.3) to the function g considered in Exercise 35.4.

Formula (35.5) implies that

$$\frac{f(\omega + \xi) - f(\omega)}{\xi} = \frac{1}{2\pi i \xi} \int_\Gamma \left\{ \frac{1}{\lambda - \omega - \xi} - \frac{1}{\lambda - \omega} \right\} d\lambda$$

$$= \frac{1}{2\pi i} \int_\Gamma \left\{ \frac{1}{(\lambda - \omega - \xi)(\lambda - \omega)} \right\} d\lambda$$

when $|\xi| > 0$ and $\omega + \xi$ is inside Γ and hence, upon letting $\xi \to 0$, yields the formula in (35.4) for $k = 1$:

$$(35.6) \qquad f'(\omega) = \frac{1}{2\pi i} \int_\Gamma \frac{f(\lambda)}{(\lambda - \omega)^2} d\lambda \quad \text{for } \omega \text{ inside } \Gamma.$$

Analogously, the formula for $k = 2$ is obtained from the formula for $k = 1$ by writing

$$\frac{f'(\omega + \xi) - f'(\omega)}{\xi} = \frac{1}{2\pi i \xi} \int_\Gamma \left\{ \frac{1}{(\lambda - \omega - \xi)^2} - \frac{1}{(\lambda - \omega)^2} \right\} d\lambda$$

$$= \frac{1}{2\pi i} \int_\Gamma \left\{ \frac{2(\lambda - \omega) - \xi}{(\lambda - \omega - \xi)^2 (\lambda - \omega)^2} \right\} d\lambda$$

and then letting $\xi \to 0$. This justifies the claim that if f is analytic in Ω, then f' is also analytic in Ω.

In much the same way one can obtain the formula for $f^{(k)}(\omega)$ from the formula for $f^{(k-1)}(\omega)$; see Exercise 35.5.

Exercise 35.5. Show that if $\omega \in \Omega$, $|\xi| > 0$, and $\omega + \xi \in \Omega$, then

$$\lim_{\xi \to 0} \frac{f^{(k-1)}(\omega + \xi) - f^{(k-1)}(\omega)}{\xi (k-1)!} = k \frac{1}{2\pi i} \int_\Gamma f(\lambda) \left\{ \frac{1}{(\lambda - \omega)^{(k+1)}} \right\} d\lambda.$$

35.1. Contour integrals of matrix-valued functions

The contour integral

$$\int_\Gamma F(\lambda) d\lambda$$

of a $p \times q$ matrix-valued function

$$F(\lambda) = \begin{bmatrix} f_{11}(\lambda) & \cdots & f_{1q}(\lambda) \\ \vdots & & \vdots \\ f_{p1}(\lambda) & \cdots & f_{pq}(\lambda) \end{bmatrix}$$

is defined by the formula

$$\int_\Gamma F(\lambda)d\lambda = \begin{bmatrix} a_{11} & \cdots & a_{1q} \\ \vdots & & \vdots \\ a_{p1} & \cdots & a_{pq} \end{bmatrix},$$

where

$$a_{ij} = \int_\Gamma f_{ij}(\lambda)d\lambda \,, \ i=1,\ldots,p,\ j=1,\ldots,q\,;$$

i.e., each entry is integrated separately. At first glance, this may seem unnatural. However, it is a consequence of the fact that

$$F'(\lambda) = \begin{bmatrix} f'_{11}(\lambda) & \cdots & f'_{1q}(\lambda) \\ \vdots & & \vdots \\ f'_{p1}(\lambda) & \cdots & f'_{pq}(\lambda) \end{bmatrix}.$$

It is readily checked that

$$\int_\Gamma \{F(\lambda)+G(\lambda)\}d\lambda = \int_\Gamma F(\lambda)d\lambda + \int_\Gamma G(\lambda)d\lambda$$

and that if B and C are appropriately sized constant matrices, then

$$\int_\Gamma BF(\lambda)C d\lambda = B\left(\int_\Gamma F(\lambda)d\lambda\right)C\,.$$

Moreover, if $\varphi(\lambda)$ is a scalar-valued function and $C \in \mathbb{C}^{p\times q}$, then

$$\int_\Gamma \varphi(\lambda)C d\lambda = \left(\int_\Gamma \varphi(\lambda)d\lambda\right)C\,.$$

Lemma 35.2. *If $\Gamma = \{\gamma(t) : a \le t \le b\}$ is a simple smooth curve that is parametrized by a function $\gamma(t) \in \mathcal{C}^1(Q)$, where Q is an open subset of $\mathbb{R}$ that contains $[a,b]$ and $F(\lambda)$ is a continous $p\times q$ matrix-valued function on Γ, then*

$$\left\|\int_\Gamma F(\lambda)d\lambda\right\| \le \int_a^b \|F(\gamma(t))\|\,|\gamma'(t)|dt\,.$$

Proof. This is a straightforward consequence of the triangle inequality applied to the Riemann sums that are used to approximate the integral. □

Lemma 35.3. *Let $C_\mu^{(k)} = \mu I_k + N$ be a Jordan cell of size $k \times k$ and let Γ be a simple piecewise smooth counterclockwise directed closed curve in the complex plane $\mathbb{C}$ that does not intersect the point μ. Then*

(35.7) $$\frac{1}{2\pi i}\int_\Gamma (\lambda I_k - C_\mu^{(k)})^{-1}d\lambda = \begin{cases} I_k & \textit{if } \mu \textit{ is inside } \Gamma, \\ O_{k\times k} & \textit{if } \mu \textit{ is outside } \Gamma. \end{cases}$$

35.1. Contour integrals of matrix-valued functions

Proof. Clearly,
$$\lambda I_k - C_\mu^{(k)} = (\lambda - \mu)I_k - N$$
and, since $N^k = O_{k \times k}$,
$$(\lambda I_k - C_\mu^{(k)})^{-1} = (\lambda - \mu)^{-1}\left(I_k - \frac{N}{\lambda - \mu}\right)^{-1}$$
$$= \frac{I_k}{\lambda - \mu} + \frac{N}{(\lambda - \mu)^2} + \cdots + \frac{N^{k-1}}{(\lambda - \mu)^k},$$
when $\lambda \neq \mu$. Therefore,
$$\frac{1}{2\pi i}\int_\Gamma (\lambda I_k - C_\mu^{(k)})^{-1} d\lambda = \sum_{j=1}^k \left\{\frac{1}{2\pi i}\int_\Gamma \frac{1}{(\lambda - \mu)^j} d\lambda\right\} N^{j-1}.$$

But this yields the asserted formula, since
$$\frac{1}{2\pi i}\int_\Gamma \frac{1}{(\lambda - \mu)^j} d\lambda = 0 \quad \text{if} \quad j > 1$$
and
$$\frac{1}{2\pi i}\int_\Gamma \frac{1}{\lambda - \mu} d\lambda = \begin{cases} 1 & \text{if } \mu \text{ is inside } \Gamma, \\ 0 & \text{if } \mu \text{ is outside } \Gamma. \end{cases} \qquad \square$$

Let $A \in \mathbb{C}^{n \times n}$ admit a Jordan decomposition of the form

(35.8) $\quad A = UJU^{-1} = [U_1 \cdots U_\ell] \begin{bmatrix} J_1 & & \\ & \ddots & \\ & & J_\ell \end{bmatrix} \begin{bmatrix} V_1^T \\ \vdots \\ V_\ell^T \end{bmatrix},$

where $J_1, \ldots, J_\ell$ denote the Jordan cells of J, $U_1, \ldots, U_\ell$ denote the corresponding block columns of U, and $V_1^T, \ldots, V_\ell^T$ denote the corresponding block rows of U^{-1}. Consequently,

(35.9) $\quad A = \sum_{i=1}^\ell U_i J_i V_i^T$

and, if J_i is a Jordan cell of size $n_i \times n_i$, then

(35.10) $\quad (\lambda I_n - A)^{-1} = U(\lambda I_n - J)^{-1} U^{-1} = \sum_{i=1}^\ell U_i(\lambda I_{n_i} - J_i)^{-1} V_i^T.$

If A has k distinct eigenvalues with geometric multiplicities $\gamma_1, \ldots, \gamma_k$, then $\ell = \gamma_1 + \cdots + \gamma_k$ in formula (35.10).

Lemma 35.4. Let $A \in \mathbb{C}^{n \times n}$ admit a Jordan decomposition of the form (35.8), where the $n_i \times n_i$ Jordan cell $J_i = \beta_i I_{n_i} + N_{n_i}$, and let Γ be a simple piecewise smooth counterclockwise directed closed curve in $\mathbb{C}$ that does not intersect any of the eigenvalues of A. Then

$$(35.11) \qquad \frac{1}{2\pi i} \int_\Gamma (\lambda I_n - A)^{-1} d\lambda = \sum_{i=1}^{\ell} U_i X_i V_i^T,$$

where

$$X_i = \begin{cases} I_{n_i} & \text{if } \beta_i \text{ is inside } \Gamma, \\ 0_{n_i \times n_i} & \text{if } \beta_i \text{ is outside } \Gamma. \end{cases}$$

Proof. This is an easy consequence of (35.10) and Lemma 35.3. □

It is readily checked that the sum on the right-hand side of formula (35.11) is a projection:

$$\left(\sum_{i=1}^{\ell} U_i X_i V_i^T\right)^2 = \sum_{i=1}^{\ell} U_i X_i V_i^T.$$

Therefore, the integral on the left-hand side of (35.11),

$$(35.12) \qquad P_\Gamma^A = \frac{1}{2\pi i} \int_\Gamma (\lambda I_n - A)^{-1} d\lambda,$$

is also a projection. It is termed the **Riesz projection**.

Lemma 35.5. Let $A \in \mathbb{C}^{n \times n}$, let $\det(\lambda I_n - A) = (\lambda - \lambda_1)^{\alpha_1} \cdots (\lambda - \lambda_k)^{\alpha_k}$, where the points $\lambda_1, \ldots, \lambda_k$ are distinct, and let Γ be a simple piecewise smooth counterclockwise directed closed curve that does not intersect any of the eigenvalues of A. Then

$$(35.13) \qquad \operatorname{rank} P_\Gamma^A = \sum_{i \in G} \alpha_i \quad \text{where} \quad G = \{i : \lambda_i \text{ is inside } \Gamma\}.$$

Proof. The conclusion rests on the observation that

$$\operatorname{rank} \sum_{i=1}^{\ell} U_i X_i V_i^T = \operatorname{rank}\left(U \left(\operatorname{diag}\{X_1, \ldots, X_\ell\}\right) V^T\right)$$
$$= \operatorname{rank}(\operatorname{diag}\{X_1, \ldots, X_\ell\}),$$

since U and V^T are invertible. Therefore, the rank of the indicated sum is equal to the sum of the sizes of the nonzero X_i that intervene in the formula for P_Γ^A, which agrees with formula (35.13). □

35.2. Continuous dependence of the eigenvalues

In this section we shall use the projection formulas P_Γ^A to establish the continuous dependence of the eigenvalues of a matrix $A \in \mathbb{C}^{n \times n}$ on A. The strategy is to show that if $B \in \mathbb{C}^{n \times n}$ is sufficiently close to A, then $\|P_\Gamma^A - P_\Gamma^B\| < 1$, and hence, as will be shown in the next lemma, rank $P_\Gamma^A = $ rank P_Γ^B.

Lemma 35.6. *If $P, Q \in \mathbb{C}^{n \times n}$, $P = P^2$, $Q = Q^2$, and $\|P - Q\| < 1$, then*

(35.14)
$$\operatorname{rank} P = \operatorname{rank} Q.$$

Proof. If $\|P-Q\| < 1$, then the matrix $I_n \pm (P-Q)$ is invertible. Therefore,
$$\operatorname{rank} P = \operatorname{rank}\{P(I_n - (P-Q))\} = \operatorname{rank}\{PQ\}$$
$$= \operatorname{rank}\{(I_n - (Q-P))Q\} = \operatorname{rank} Q. \qquad \square$$

The **notation**
$$D_r(\mu) = \{\lambda \in \mathbb{C} : |\lambda - \mu| < r\}$$
will be used in the next theorem.

Theorem 35.7. *Let $A, B \in \mathbb{C}^{n \times n}$ and let*
$$\det(\lambda I_n - A) = (\lambda - \lambda_1)^{\alpha_1} \cdots (\lambda - \lambda_k)^{\alpha_k},$$
where the k points $\lambda_1, \ldots, \lambda_k$ are distinct. Then for every $r > 0$, there exists a $\delta > 0$ such that if $\|A - B\| < \delta$, then
$$\sigma(B) \subset \bigcup_{i=1}^k D_r(\lambda_i).$$
Moreover, if
$$r < r_0 = \frac{1}{2} \min\{|\lambda_i - \lambda_j| : i \neq j\},$$
then each disk $D_r(\lambda_i)$ will contain exactly α_i eigenvalues of B, counting algebraic multiplicities.

Proof. Let $r < r_0$ and let $\Gamma_j = \Gamma_j(r)$ denote a circle of radius r centered at λ_j and directed counterclockwise. This constraint on r ensures that $\Gamma_j \cap \Gamma_i = \emptyset$ if $j \neq i$ and that λ_j is the only eigenvalue of A inside the closed disc $\overline{D_r(\lambda_j)} = \{\lambda \in \mathbb{C} : |\lambda - \lambda_j| \leq r\}$. The remainder of the proof is divided into steps.

1. If $0 < r < r_0$ and $\lambda \in \bigcup_{j=1}^k \Gamma_j(r)$, then $\lambda I_n - A$ is invertible and there exists a constant δ_r such that

(35.15) $\qquad \|(\lambda I_n - A)^{-1}\| \leq \delta_r \quad \text{for every point } \lambda \in \bigcup_{j=1}^k \Gamma_j(r).$

It is clear that $\lambda I_n - A$ is invertible for every point $\lambda \in \bigcup_{j=1}^k \Gamma_j(r)$. Moreover, upon invoking the Jordan decomposition $A = UJU^{-1}$, we may write

$$\|(\lambda I_n - A)^{-1}\| = \|U(\lambda I_n - J)^{-1}U^{-1}\| \leq \|U\| \|(\lambda I_n - J)^{-1}\| \|U^{-1}\|.$$

Thus, as $J = \text{diag}\{J_{\lambda_1}, \ldots, J_{\lambda_k}\}$ with blocks $J_{\lambda_j} = \lambda_j I_{\alpha_j} + T_j$, where $T_j = \text{diag}\{C_{\lambda_j}^{(k_1)}, \ldots, C_{\lambda_j}^{(k_m)}\}$ (with $k_1 + \cdots + k_m = \alpha_j$ and $m = \gamma_j$) is a strictly upper triangular matrix of size $\alpha_j \times \alpha_j$ such that $\|T_j\| \leq 1$ and $T_j^{\alpha_j} = O$,

$$\|(\lambda I_{\alpha_j} - J_{\lambda_j})^{-1}\| = \left\|(\lambda - \lambda_j)^{-1}\left(I_{\alpha_j} - \frac{T_j}{\lambda - \lambda_j}\right)^{-1}\right\|$$

$$= \left\|\sum_{i=1}^{\alpha_j} \frac{(T_j)^{i-1}}{(\lambda - \lambda_j)^i}\right\| \leq \sum_{i=1}^{\alpha_j} \left(\frac{1}{|\lambda - \lambda_j|}\right)^i$$

$$\leq \sum_{i=1}^{\alpha_j} \frac{1}{r^i} \leq \sum_{i=1}^n \frac{1}{r^i}.$$

Therefore, (35.15) holds with $\delta_r = \|U\| \|U^{-1}\| \sum_{i=1}^n r^{-i}$.

2. If $\|A - B\| < 1/\delta_r$, then $\lambda I_n - B$ is invertible and

$$\|(\lambda I_n - A)^{-1} - (\lambda I_n - B)^{-1}\| \leq \frac{\delta_r^2 \|B - A\|}{1 - \delta_r \|B - A\|} \quad \text{for every } \lambda \in \bigcup_{j=1}^k \Gamma_j(r).$$

Let $X = (\lambda I_n - A)^{-1}(B - A)$. Then

$$\lambda I_n - B = (\lambda I_n - A - (B - A)) = (\lambda I_n - A)(I_n - X)$$

is invertible if $\lambda \in \bigcup_{j=1}^k \Gamma_j(r)$ and $\|A - B\| < \frac{1}{\delta_r}$, because then $I_n - X$ is invertible, since $\|X\| \leq \delta_r \|A - B\| < 1$. Moreover,

$$\|(\lambda I_n - A)^{-1} - (\lambda I_n - B)^{-1}\| = \|(I_n - (I_n - X)^{-1})(\lambda I_n - A)^{-1}\|$$

$$= \|X(I_n - X)^{-1}(\lambda I_n - A)^{-1}\|$$

$$\leq \delta_r \frac{\|X\|}{1 - \|X\|} \leq \frac{\delta_r^2 \|A - B\|}{1 - \delta_r \|A - B\|},$$

as claimed.

3. If $\|B - A\| < (\delta_r + \delta_r^2 r)^{-1}$, then each disk $D_r(\lambda_i)$ will contain exactly α_i points of spectrum of B, counting multiplicities.

In view of step 2 and Lemma 35.2,

$$\|P^A_{\Gamma_j} - P^B_{\Gamma_j}\| = \left\| \frac{1}{2\pi i} \int_{\Gamma_j} [(\lambda I_n - A)^{-1} - (\lambda I_n - B)^{-1}] d\lambda \right\|$$

$$\leq \frac{r\, \delta_r^2 \|B - A\|}{1 - \delta_r \|B - A\|} < 1.$$

Then, by Lemma 35.6,

$$\operatorname{rank} P^B_{\Gamma_j} = \operatorname{rank} P^A_{\Gamma_j} = \alpha_j$$

for $j = 1, \ldots, k$, and hence by Lemma 35.5,

$$\operatorname{rank} P^B_{\Gamma_j} = \text{the sum of the algebraic multiplicities of the eigenvalues}$$
$$\text{of } B \text{ inside } D_r(\lambda_j),$$

which yields the desired result. □

Exercise 35.6. Verify the claim that $\|T_j\| \leq 1$ and $T_j^{\alpha_j} = O$ that was made in step 1 in the proof of Theorem 35.7.

Exercise 35.7. Show that the roots $\lambda_1, \ldots, \lambda_n$ of the polynomial $p(\lambda) = a_0 + a_1 \lambda + \cdots + a_n \lambda^n$ with $a_n \neq 0$ depend continuously on the coefficients $a_0, \ldots, a_n$ of the polynomial. [HINT: Exploit companion matrices.]

35.3. Matrices with distinct eigenvalues

In this section we shall focus on matrices $A \in \mathbb{C}^{n \times n}$ with n distinct eigenvalues.

Lemma 35.8. *If $A \in \mathbb{C}^{n \times n}$ has n distinct eigenvalues, then there exists a number $\delta > 0$ such that every matrix $B \in \mathbb{C}^{n \times n}$ for which $\|A - B\| < \delta$ also has n distinct eigenvalues.*

Proof. This is an easy consequence of Theorem 35.7. □

The next result shows that any matrix $A \in \mathbb{C}^{n \times n}$ can be approximated arbitrarily well by a matrix $B \in \mathbb{C}^{n \times n}$ with n distinct eigenvalues; i.e., the class of $n \times n$ matrices with n distinct eigenvalues is **dense** in $\mathbb{C}^{n \times n}$.

Lemma 35.9. *If $A \in \mathbb{C}^{n \times n}$ and $\delta > 0$ are specified, then there exists a matrix $B \in \mathbb{C}^{n \times n}$ with n distinct eigenvalues such that $\|A - B\| < \delta$.*

Proof. Let $A = UJU^{-1}$ be a Jordan decomposition of the matrix A and let D be a diagonal matrix such that the diagonal entries of the matrix $D + J$ are all distinct and

$$|d_{ii}| \leq (\|U\| \|U^{-1}\|)^{-1} \delta$$

for $i = 1, \ldots, n$. Then the matrix $B = U(D + J)U^{-1}$ has n distinct eigenvalues and
$$\|A - B\| = \|UDU^{-1}\| \leq \|U\|\|D\|\|U^{-1}\| \leq \delta.$$
□

Note that this lemma does not say that every matrix B that meets the inequality $\|A - B\| < \delta$ has n distinct eigenvalues.

Since the class of $n \times n$ matrices with n distinct eigenvalues is a subclass of the set of diagonalizable matrices, it is reasonable to ask whether or not a diagonalizable matrix remains diagonalizable if some of its entries are changed just a little. The answer is: not always!

Exercise 35.8. Let $A = \begin{bmatrix} 1 & 0 \\ 0 & 1 \end{bmatrix}$ and $B = \begin{bmatrix} 1 & \beta \\ 0 & 1 \end{bmatrix}$, where $\beta \neq 0$. Show that $\|A - B\| = |\beta|$ but that A is diagonalizable, whereas B is not.

Exercise 35.9. Let $A \in \mathbb{C}^{p \times q}$ and suppose that $\operatorname{rank} A = k$ and $k < \min\{p, q\}$. Show that for every $\varepsilon > 0$ there exists a matrix $B \in \mathbb{C}^{p \times q}$ such that $\|A - B\| < \varepsilon$ and $\operatorname{rank} B = k + 1$.

Some conclusions

A subset $\mathcal{X}$ of the set of $p \times q$ matrices $\mathbb{C}^{p \times q}$ is said to be **open** if for every matrix $A \in \mathcal{X}$ there exists a number $\delta > 0$ such that the **open ball**
$$B_\delta(A) = \{C \in \mathbb{C}^{p \times q} : \|A - C\| < \delta\}$$
is also a subset of $\mathcal{X}$. The meaning of this condition is that if the entries in a matrix $A \in \mathcal{X}$ are changed only slightly, then the new perturbed matrix will also belong to the class $\mathcal{X}$. This is a significant property in applications and computations because the entries in any matrix that is obtained from data or from a numerical algorithm are only known approximately. The preceding analysis implies that:

- $\{A \in \mathbb{C}^{p \times q} : \operatorname{rank} A = \min\{p, q\}\}$ is an open subset of $\mathbb{C}^{p \times q}$.
- $\{A \in \mathbb{C}^{p \times q} : \operatorname{rank} A < \min\{p, q\}\}$ is not an open subset of $\mathbb{C}^{p \times q}$.
- $\{A \in \mathbb{C}^{n \times n} \text{ with } n \text{ distinct eigenvalues}\}$ is an open subset of $\mathbb{C}^{n \times n}$.
- $\{A \in \mathbb{C}^{n \times n} : A \text{ is diagonalizable}\}$ is not an open subset of $\mathbb{C}^{n \times n}$.

The set $\{A \in \mathbb{C}^{n \times n} : \text{ with } n \text{ distinct eigenvalues}\}$ is particularly useful because it is both open and **dense** in $\mathbb{C}^{n \times n}$, thanks to Lemma 35.9. Open dense sets are said to be **generic**.

Exercise 35.10. Show that $\{A \in \mathbb{C}^{n \times n} : A \text{ is invertible}\}$ is a generic set.

35.4. Supplementary notes

Some additional applications of complex function theory to linear algebra will be discussed in the next chapter. In particular, another verification of formula (11.20) for the spectral radius of a matrix $A \in \mathbb{C}^{n \times n}$ and formulas for fractional powers A^t when $A \succ O$ will be presented there. The continuous dependence of the eigenvalues of a matrix A that is established in Theorem 35.7 is (in the terminology of the article [57] by Li and Zhang) topological continuity, not functional continuity. In particular, they note the following result, which is based on Theorem 5.2 on page 109 of Kato's monograph [51]:

Theorem 35.10. *If $A(\mu) \in \mathbb{C}^{n \times n}$ is a continuous function of μ for all points μ in a connected domain $\Omega \subset \mathbb{C}$ such that either (1) Ω is a real interval or (2) the eigenvalues of $A(\lambda)$ are real, then there exist n eigenvalues (counted with algebraic multiplicities) of $A(\mu)$ that can be parametrized as continuous functions $\lambda_1(\mu), \ldots, \lambda_n(\mu)$ from Ω to $\mathbb{C}$. In the second case, one can choose $\lambda_1(\mu) \geq \cdots \geq \lambda_n(\mu)$.*

Chapter 36

Eigenvalue location problems

In this chapter we shall continue to use complex function theory to extract information about the eigenvalues of matrices.

36.1. Geršgorin disks

The **notation**
$$\Delta(\alpha; r) = \{\lambda \in \mathbb{C} : |\lambda - \alpha| \leq r\}$$
for the closed disk with center α and radius r will be used in this section. Let $A \in \mathbb{C}^{n \times n}$ with entries a_{ij} and let
$$\rho_i(A) = \sum_{j=1}^{n} |a_{ij}| - |a_{ii}| \quad \text{for} \quad i = 1, \ldots, n.$$

The set $\Delta(a_{ii}; \rho_i(A))$ is called the i'th **Geršgorin disk** of A.

Theorem 36.1. *If $A \in \mathbb{C}^{n \times n}$, then:*

(1) $\sigma(A) \subset \bigcup_{i=1}^{n} \Delta(a_{ii}; \rho_i(A))$.

(2) *A union Ω_1 of k Geršgorin disks that has no points in common with the union Ω_2 of the remaining $n - k$ Geršgorin disks contains exactly k eigenvalues of A, counting their algebraic multiplicities.*

Proof. If $\lambda \in \sigma(A)$, then there exists a nonzero vector $\mathbf{u} \in \mathbb{C}^n$ with components $u_1, \ldots, u_n$ such that $(\lambda I_n - A)\mathbf{u} = \mathbf{0}$. Suppose that
$$|u_k| = \max\{|u_j| : j = 1, \ldots, n\}.$$

393

Then the identity

$$(\lambda - a_{kk})u_k = \sum_{j=1}^{n} a_{kj}u_j - a_{kk}u_k$$

implies that

$$\begin{aligned}|(\lambda - a_{kk})u_k| &\leq \sum_{j=1}^{n} |a_{kj}||u_j| - |a_{kk}||u_k| \\ &\leq \rho_k(A)|u_k|.\end{aligned}$$

Therefore,

$$\lambda \in \Delta(a_{kk}; \rho_k(A)) \subset \bigcup_{i=1}^{n} \Delta(a_{ii}; \rho_i(A)).$$

This completes the proof of (1).

Next, to verify (2), let $D = \mathrm{diag}\{a_{11}, \ldots, a_{nn}\}$ and let

$$B(t) = D + t(A - D) \quad \text{for} \quad 0 \leq t \leq 1.$$

Then

$$b_{ij}(t) = d_{ij} + t(a_{ij} - d_{ij}) = \begin{cases} a_{ii} & \text{if } i = j, \\ ta_{ij} & \text{if } i \neq j, \end{cases}$$

and hence each eigenvalue λ of $B(t)$ belongs to one of the disks $\Delta(a_{ii}; t\rho_i(A))$, i.e.,

$$\sigma(B(t)) \subset \bigcup_{i=1}^{n} \Delta(a_{ii}; t\rho_i(A)).$$

Now, let Γ be a simple smooth closed curve that contains Ω_1 in its interior and Ω_2 in its exterior such that

$$\min\{|\omega - \lambda| : \omega \in \Omega_1 \cup \Omega_2 \quad \text{and} \quad \lambda \in \Gamma\} \geq \delta$$

for some $\delta > 0$. Then $\lambda I_n - B(t)$ is invertible for every choice of $\lambda \in \Gamma$ and $t \in [0, 1]$ and there exists a finite positive number β such that

$$\|(\lambda I_n - B(t))^{-1}\| \leq \beta \quad \text{for every choice of } \lambda \in \Gamma \text{ and } t \in [0, 1].$$

Thus,

$$\begin{aligned}\|(\lambda I_n &- B(t))^{-1} - (\lambda I_n - B(s))^{-1}\| \\ &= \|(\lambda I_n - B(t))^{-1}(B(t) - B(s))(\lambda I_n - B(s))^{-1}\| \\ &\leq \beta^2 \|A - D\| |t - s|\end{aligned}$$

and hence if the curve Γ is parameterized by the function $\gamma(u)$, $a \leq u \leq b$, the corresponding projectors meet the constraint

$$\|P_\Gamma^{B(t)} - P_\Gamma^{B(s)}\| = \left\|\frac{1}{2\pi i}\int_\Gamma [(\lambda I_n - B(t))^{-1} - (\lambda I_n - B(s))^{-1}]d\lambda\right\|$$
$$\leq \frac{1}{2\pi}\int_a^b |\gamma'(u)|du\, \beta^2 \|A - D\|\, |t - s|,$$

which is < 1 if $|t - s\|$ is small enough. Thus, by partitioning $[0,1]$ and proceeding from 0 to 1 in a sequence of small steps, we see that

$$k = \operatorname{rank} P_\Gamma^{B(0)} = \operatorname{rank} P_\Gamma^{B(1)} = \operatorname{rank} P_\Gamma^A$$

and hence that A has k eigenvalues inside Γ, counting algebraic multiplicities. But, as $\sigma(A) \subset \Omega_1 \cup \Omega_2$, these k eigenvalues must belong to Ω_1. $\square$

Exercise 36.1. Show that the spectral radius $r_\sigma(A)$ of a matrix $A \in \mathbb{C}^{n \times n}$ is subject to the bound

$$r_\sigma(A) \leq \max\left\{\sum_{j=1}^n |a_{ij}| : i = 1, \ldots, n\right\}.$$

Exercise 36.2. Show that the spectral radius $r_\sigma(A)$ of a matrix $A \in \mathbb{C}^{n \times n}$ is subject to the bound

$$r_\sigma(A) \leq \max\left\{\sum_{i=1}^n |a_{ij}| : j = 1, \ldots, n\right\}.$$

Exercise 36.3. Let $A \in \mathbb{C}^{n \times n}$. Show that if $|a_{ii}| > \rho_i(A)$ for $i = 1, \ldots, n$, then A is invertible.

Exercise 36.4. Let $A \in \mathbb{C}^{n \times n}$. Show that A is a diagonal matrix if and only if $\sigma(A) = \bigcup_{i=1}^n \Delta(a_{ii}; 0)$.

Exercise 36.5. Show that if $A \in \mathbb{C}^{n \times n}$ is a circulant, then all the Geršgorin disks $\Delta(a_{jj}; \rho_j(A))$, $j = 1, \ldots, n$, are the same and then check that $\sigma(A) \subseteq \Delta(a_{11}; \rho_1(A))$

36.2. Spectral radius redux

In this section we shall use the methods of complex analysis to obtain a simple proof of the formula

$$r_\sigma(A) = \lim_{k\uparrow\infty} \|A^k\|^{1/k}$$

for the spectral radius

$$r_\sigma(A) = \max\{|\lambda| : \lambda \in \sigma(A)\}$$

of a matrix $A \in \mathbb{C}^{n\times n}$. Since the inequality
$$r_\sigma(A) \leq \|A^k\|^{1/k} \quad \text{for } k = 1, 2, \ldots$$
is easily verified, it suffices to show that

(36.1) $$\lim_{k\uparrow\infty} \|A^k\|^{1/k} \leq r_\sigma(A).$$

Lemma 36.2. *If $A \in \mathbb{C}^{n\times n}$ and $\sigma(A)$ belongs to the set of points enclosed by a simple piecewise smooth counterclockwise directed closed curve Γ, then*

(36.2) $$A^k = \frac{1}{2\pi i} \int_\Gamma \lambda^k (\lambda I_n - A)^{-1} d\lambda.$$

Proof. Let Γ_r denote a circle of radius r centered at zero and directed counterclockwise, and suppose that $r > \|A\|$ and $g(\lambda) = \lambda^k$. Then

$$\frac{1}{2\pi i} \int_{\Gamma_r} \lambda^k (\lambda I_n - A)^{-1} d\lambda = \frac{1}{2\pi i} \int_{\Gamma_r} g(\lambda) \sum_{j=0}^{\infty} \frac{A^j}{\lambda^{j+1}} d\lambda$$

$$= \sum_{j=0}^{\infty} \left\{ \frac{1}{2\pi i} \int_{\Gamma_r} \frac{g(\lambda)}{\lambda^{j+1}} d\lambda \right\} A^j$$

$$= \sum_{j=0}^{\infty} \frac{g^{(j)}(0)}{j!} A^j = A^k.$$

The assumption $r > \|A\|$ is used to guarantee the uniform convergence of the sum $\sum_{j=0}^{\infty} \lambda^{-j} A^j$ on Γ_r and subsequently to justify the interchange in the order of summation and integration. Thus, to this point, we have established formula (36.2) for $\Gamma = \Gamma_r$ and $r > \|A\|$. However, since $\lambda^k (\lambda I_n - A)^{-1}$ is holomorphic in an open set that contains the points between and on the curves Γ and Γ_r, it follows that

$$\frac{1}{2\pi i} \int_\Gamma \lambda^k (\lambda I_n - A)^{-1} d\lambda = \frac{1}{2\pi i} \int_{\Gamma_r} \lambda^k (\lambda I_n - A)^{-1} d\lambda. \qquad \square$$

Corollary 36.3. *If $A \in \mathbb{C}^{n\times n}$ and $\sigma(A)$ belongs to the set of points enclosed by a simple piecewise smooth counterclockwise directed closed curve Γ, then*

(36.3) $$p(A) = \frac{1}{2\pi i} \int_\Gamma p(\lambda) (\lambda I_n - A)^{-1} d\lambda$$

for every polynomial $p(\lambda)$.

Theorem 36.4. *If $A \in \mathbb{C}^{n\times n}$, then*

(36.4) $$\lim_{k\uparrow\infty} \|A^k\|^{1/k} = r_\sigma(A);$$

i.e., the limit exists and is equal to the modulus of the maximum eigenvalue of A.

36.2. Spectral radius redux

Proof. Fix $\epsilon > 0$, let $r = r_\sigma(A) + \epsilon$, and let
$$\delta_r = \max\{\|(\lambda I_n - A)^{-1}\| : |\lambda| = r\}.$$
Then, by formula (36.2),
$$\begin{aligned}\|A^k\| &= \left\|\frac{1}{2\pi i}\int_0^{2\pi}(re^{i\theta})^k(re^{i\theta}I_n - A)^{-1}ire^{i\theta}d\theta\right\| \\ &\leq \frac{1}{2\pi}\int_0^{2\pi}r^k\|(re^{i\theta}I_n - A)^{-1}\|rd\theta \\ &\leq r^{k+1}\delta_r.\end{aligned}$$
Thus,
$$\|A^k\|^{1/k} \leq r(r\delta_r)^{1/k}$$
and, as $(r\delta_r)^{1/k} \to 1$ as $k \uparrow \infty$, there exists a positive integer N such that
$$\|A^k\|^{1/k} \leq r_\sigma(A) + 2\epsilon \quad \text{for every integer } k \geq N.$$
Therefore, since $r_\sigma(A) \leq \|A^k\|^{1/k}$ for $k = 1, 2, \ldots$, it follows that
$$0 \leq \|A^k\|^{1/k} - r_\sigma(A) \leq 2\varepsilon \quad \text{for every integer } k \geq N,$$
which serves to establish formula (36.4). □

Exercise 36.6. Show that if $A \in \mathbb{C}^{n\times n}$ and $\sigma(A)$ belongs to the set of points enclosed by a simple piecewise smooth counterclockwise directed closed curve Γ, then

(36.5) $$\frac{1}{2\pi i}\int_\Gamma e^\lambda(\lambda I_n - A)^{-1}d\lambda = \sum_{j=0}^\infty \frac{A^j}{j!}.$$

Let $A \in \mathbb{C}^{n\times n}$ and let $f(\lambda)$ be holomorphic in an open set Ω that contains $\sigma(A)$. Then, in view of formulas (36.3) and (36.5) it is reasonable to **define**

(36.6) $$f(A) = \frac{1}{2\pi i}\int_\Gamma f(\lambda)(\lambda I_n - A)^{-1}d\lambda,$$

where Γ is any simple piecewise smooth counterclockwise directed closed curve in Ω that encloses $\sigma(A)$ such that every point inside Γ also belongs to Ω. This definition is independent of the choice of Γ and is consistent with the definitions of $f(A)$ considered earlier.

Exercise 36.7. Show that if $A \in \mathbb{C}^{n\times n}$, $f(\lambda)$ and $g(\lambda)$ are holomorphic in an open set Ω that contains $\sigma(A)$, and Γ is any simple piecewise smooth counterclockwise directed closed curve in Ω that encloses $\sigma(A)$ such that every point inside Γ also belongs to Ω, then

(36.7)
$$f(A)\left\{\frac{1}{2\pi i}\int_\Gamma g(\zeta)(\zeta I_n - A)^{-1}d\zeta\right\} = \frac{1}{2\pi i}\int_\Gamma f(\lambda)g(\lambda)(\lambda I_n - A)^{-1}d\lambda.$$

[HINT: Use the identity
$$(\lambda I_n - A)^{-1}(\zeta I_n - A)^{-1} = (\zeta - \lambda)^{-1}\{(\lambda I_n - A)^{-1} - (\zeta I_n - A)^{-1}\}$$
to reexpress $2\pi i f(A) 2\pi i g(A)$ as
$$\int_\Gamma f(\lambda)(\lambda I_n - A)^{-1}\left\{\int_{\Gamma_1}\frac{g(\zeta)}{\zeta - \lambda}d\zeta\right\}d\lambda - \int_{\Gamma_1} g(\zeta)(\zeta I_n - A)^{-1}\left\{\int_\Gamma \cdots\right\}d\zeta,$$
where the contour Γ sits inside the contour Γ_1.]

Exercise 36.8. Show that if, in terms of the notation introduced in (35.9), $A = U_1 C_\alpha^{(p)} V_1^T + U_2 C_\beta^{(q)} V_2^T$, then

(36.8)
$$f(A) = U_1 \sum_{j=0}^{p-1} \frac{f^{(j)}(\alpha)}{j!}(C_0^{(p)})^j V_1^T + U_2 \sum_{j=0}^{q-1} \frac{f^{(j)}(\alpha)}{j!}(C_0^{(q)})^j V_2^T$$

for every function $f(\lambda)$ that is holomorphic in an open set that contains the points α and β.

Exercise 36.9. Show that in the setting of Exercise 36.8

(36.9)
$$\det(\lambda I_n - f(A)) = (\lambda - f(\alpha))^p (\lambda - f(\beta))^q.$$

Exercise 36.10. Show that if $A \in \mathbb{C}^{n\times n}$ and $f(\lambda)$ is holomorphic in an open set that contains $\sigma(A)$, then

(36.10)
$$\det(\lambda I_n - A) = (\lambda - \lambda_1)^{\alpha_1}\cdots(\lambda - \lambda_k)^{\alpha_k}$$
$$\implies \det(\lambda I_n - f(A)) = (\lambda - f(\lambda_1))^{\alpha_1}\cdots(\lambda - f(\lambda_k))^{\alpha_k}.$$

[HINT: The main ideas are contained in Exercises 36.8 and 36.9. The rest is just more elaborate bookkeeping.]

Theorem 36.5 (The spectral mapping theorem). Let $A \in \mathbb{C}^{n\times n}$ and let $f(\lambda)$ be holomorphic in an open set that contains $\sigma(A)$. Then

(36.11)
$$\mu \in \sigma(f(A)) \iff f(\mu) \in \sigma(f(A)).$$

Proof. This is immediate from formula (36.10). □

36.3. Shifting eigenvalues

In this section we extend the discussion in Section 8.4 by dropping the restriction that the geometric multiplicity of each eigenvalue of the given matrix A is equal to one. We will now work with a matrix

$$\mathcal{C} = \begin{bmatrix} B & AB & \cdots & A^{n-1}B \end{bmatrix},$$

in which $A \in \mathbb{C}^{n \times n}$ and $B \in \mathbb{C}^{n \times k}$ with $k \geq 1$. In the control theory literature, a pair of matrices $(A, B) \in \mathbb{C}^{n \times n} \times \mathbb{C}^{n \times k}$ with $k \geq 1$ is said to be a **controllable pair** if $\operatorname{rank} \mathcal{C} = n$.

Lemma 36.6. *If $(A, B) \in \mathbb{C}^{n \times n} \times \mathbb{C}^{n \times k}$ is a controllable pair and $k > 1$, then there exist a matrix $C \in \mathbb{C}^{k \times n}$ and a vector $\mathbf{b} \in \mathcal{R}_B$ such that $(A + BC, \mathbf{b})$ is a controllable pair.*

Discussion. Suppose for the sake of definiteness that $A \in \mathbb{C}^{10 \times 10}$, $B = [\mathbf{b}_1 \cdots \mathbf{b}_5]$, and then permute the columns in the matrix

$$\begin{bmatrix} B & AB & \cdots & A^{n-1}B \end{bmatrix}$$

to obtain the matrix

$$\begin{bmatrix} \mathbf{b}_1 & A\mathbf{b}_1 & \cdots & A^9\mathbf{b}_1 & \cdots & \mathbf{b}_5 & A\mathbf{b}_5 & \cdots & A^9\mathbf{b}_5 \end{bmatrix}.$$

Then, moving from left to right, discard vectors that may be expressed as linear combinations of vectors that sit to their left. Suppose further that

$$\begin{aligned}
A^3\mathbf{b}_1 &\in \operatorname{span}\{\mathbf{b}_1, A\mathbf{b}_1, A^2\mathbf{b}_1\}, \\
A^5\mathbf{b}_2 &\in \operatorname{span}\{\mathbf{b}_1, A\mathbf{b}_1, A^2\mathbf{b}_1, \mathbf{b}_2, \ldots, A^4\mathbf{b}_2\}, \\
A^2\mathbf{b}_3 &\in \operatorname{span}\{\mathbf{b}_1, A\mathbf{b}_1, A^2\mathbf{b}_1, \mathbf{b}_2, \ldots, A^4\mathbf{b}_2, \mathbf{b}_3, A\mathbf{b}_3\}
\end{aligned}$$

and the vectors in each of the sets on the right are linearly independent. Then the matrix

$$Q = [\mathbf{b}_1 \; A\mathbf{b}_1 \; A^2\mathbf{b}_1 \; \mathbf{b}_2 \; A\mathbf{b}_2 \; A^2\mathbf{b}_2 \; A^3\mathbf{b}_2 \; A^4\mathbf{b}_2 \; \mathbf{b}_3 \; A\mathbf{b}_3]$$

is invertible. Let $\mathbf{e}_j$ denote the j'th column of I_5 and let $\mathbf{f}_k$ denote the k'th column of I_{10} and set

$$G = [\mathbf{0} \; \mathbf{0} \; \mathbf{e}_2 \; \mathbf{0} \; \mathbf{0} \; \mathbf{0} \; \mathbf{0} \; \mathbf{e}_3 \; \mathbf{0} \; \mathbf{0}]$$

and $C = GQ^{-1}$. Then

$$\begin{aligned}
(A+BC)\mathbf{b}_1 &= A\mathbf{b}_1 + BGQ^{-1}Q\mathbf{f}_1 = A\mathbf{b}_1, \\
(A+BC)^2\mathbf{b}_1 &= (A+BC)A\mathbf{b}_1 = A^2\mathbf{b}_1 + BGQ^{-1}Q\mathbf{f}_2 \\
&= A^2\mathbf{b}_1, \\
(A+BC)^3\mathbf{b}_1 &= (A+BC)A^2\mathbf{b}_1 = A^3\mathbf{b}_1 + BGQ^{-1}Q\mathbf{f}_3 \\
&= A^3\mathbf{b}_1 + B\mathbf{e}_2 = A^3\mathbf{b}_1 + \mathbf{b}_2.
\end{aligned}$$

Thus,
$$\begin{bmatrix} \mathbf{b}_1 & (A+BC)\mathbf{b}_1 & (A+BC)^2\mathbf{b}_1 & (A+BC)^3\mathbf{b}_1 \end{bmatrix}$$
$$= Q \begin{bmatrix} \mathbf{f}_1 & \mathbf{f}_2 & \mathbf{f}_3 & \sum_{i=1}^{4} c_{4i}\mathbf{f}_i \end{bmatrix}$$

with $c_{44} = 1$. Similar considerations lead to the conclusion that
$$[\mathbf{b}_1 \ (A+BC)\mathbf{b}_1 \ \cdots \ (A+BC)^9\mathbf{b}_1] = QU,$$
where U is an upper triangular matrix with ones on the diagonal. Therefore, since QU is invertible, $(A + BC, \mathbf{b}_1)$ is a controllable pair. □

Theorem 36.7. *Let* $(A, B) \in \mathbb{C}^{n \times n} \times \mathbb{C}^{n \times k}$ *be a controllable pair and let* $\{\mu_1, \ldots, \mu_n\}$ *be any set of points in* $\mathbb{C}$ *(not necessarily distinct). Then there exists a matrix* $K \in \mathbb{C}^{k \times n}$ *such that*
$$\det(\lambda I_n - A - BK) = (\lambda - \mu_1) \cdots (\lambda - \mu_n).$$

Proof. If $k = 1$, then the conclusion is given in Theorem 8.10. If $k > 1$, then, by Lemma 36.6, there exist a matrix $C \in \mathbb{C}^{k \times n}$ and a vector $\mathbf{b} \in \mathcal{R}_B$ such that $(A + BC, \mathbf{b})$ is a controllable pair. Therefore, by Theorem 8.10, there exists a vector $\mathbf{u} \in \mathbb{C}^n$ such that
$$\det(\lambda I_n - A - BC - \mathbf{b}\mathbf{u}^H) = (\lambda - \mu_1) \cdots (\lambda - \mu_n).$$
However, $BC + \mathbf{b}\mathbf{u}^H = B(C + \mathbf{v}\mathbf{u}^H)$ for some $\mathbf{v} \in \mathbb{C}^k$, since $\mathbf{b} \in \mathcal{R}_B$. Thus, the proof may be completed by choosing $K = C + \mathbf{v}\mathbf{u}^H$. □

Exercise 36.11. Show that $(A, B) \in \mathbb{C}^{n \times n} \times \mathbb{C}^{n \times k}$ is a controllable pair if and only if rank $\begin{bmatrix} A - \lambda I_n & B \end{bmatrix} = n$ for every point $\lambda \in \mathbb{C}$.

36.4. The Hilbert matrix

The matrix $A_n \in \mathbb{C}^{(n+1) \times (n+1)}$ with jk entry

(36.12) $$a_{jk} = \frac{1}{j+k+1} \quad \text{for } j, k = 0, \ldots, n$$

is called the **Hilbert matrix**. It is readily checked that $A_n \succ O$ by setting $f(x) = \sum_{k=0}^{n} c_k x^k$ and noting that

$$\int_0^1 |f(x)|^2 dx = \int_0^1 \sum_{j,k=0}^{n} c_k x^k \overline{c_j} x^j dx$$

$$= \sum_{j,k=0}^{n} \frac{\overline{c_j} c_k}{j+k+1} = \mathbf{c}^H A_n \mathbf{c},$$

where $\mathbf{c}^H = \begin{bmatrix} \overline{c_0} & \cdots & \overline{c_n} \end{bmatrix}$.

Lemma 36.8. *The Hilbert matrix A_n defined by (36.12) is subject to the bound*

(36.13) $\qquad \|A_n\| \leq \pi \quad$ *for every choice of the positive integer n.*

Proof. Let Γ_1 denote the semicircle of radius 1 in the open upper half-plane with base $[-1, 1]$ directed counterclockwise, let Γ_2 denote the arc of that semicircle, and set $g(\lambda) = \sum_{k=0}^{n} |c_k| \lambda^k$. Then, by the preceding calculations,

$$\begin{aligned}
\mathbf{c}^H A_n \mathbf{c} &= \int_0^1 |f(x)|^2 dx \leq \int_0^1 g(x)^2 dx \leq \int_{-1}^1 g(x)^2 dx \\
&= \int_{\Gamma_1} g(\lambda)^2 d\lambda - \int_{\Gamma_2} g(\lambda)^2 d\lambda = -\int_{\Gamma_2} g(\lambda)^2 d\lambda \\
&= -i \int_0^\pi g(e^{i\theta})^2 e^{i\theta} d\theta \leq \int_0^\pi |g(e^{i\theta})|^2 d\theta \\
&= \frac{1}{2} \int_{-\pi}^\pi |g(e^{i\theta})|^2 d\theta = \pi \sum_{k=0}^n |c_k|^2 = \pi \|\mathbf{c}\|^2 .
\end{aligned}$$

Thus, the bound (36.13) holds, as claimed. $\square$

36.5. Fractional powers

Since the function $f(\lambda) = \lambda^t$ is holomorphic in $\mathbb{C} \setminus (-\infty, 0]$ when $t > 0$, formula (36.6) may be used to obtain useful bounds on fractional powers of positive definite matrices. In particular, if $A \in \mathbb{C}^{n \times n}$ and $A \succ O$, then

$$A^{1/2} = \frac{1}{2\pi i} \int_\Gamma \sqrt{\lambda} (\lambda I_n - A)^{-1} d\lambda$$

for any simple closed piecewise smooth curve Γ in the open right half-plane that includes the eigenvalues of A in its interior. Moreover, if $A, B \in \mathbb{C}^{n \times n}$, $A \succ B \succ O$, and $0 < t < 1$, then

(36.14) $\qquad A^t - B^t = \dfrac{1}{2\pi i} \int_\Gamma \lambda^t \left\{ (\lambda I_n - A)^{-1} - (\lambda I_n - B)^{-1} \right\} d\lambda$,

for an appropriately chosen contour Γ. The formula (16.15) is obtained by passing to appropriate limits; see, e.g., Section 17.10 in [**30**] for details.

36.6. Supplementary notes

The argument used to bound the Hilbert matrix that is presented in Section 36.4 is taken from [**47**]. The authors credit this approach to Fejér and F. Riesz and present several other ways of verifying this bound (including one based on Lagrange multipliers by J. W. S. Cassels that was presented as an exercise in the first edition of [**30**]); for yet another approach, see the discussion of the Hilbert matrix in Peller [**64**]

Section 36.3 is adapted from Section 20.9 in [30], which is partially adapted from Heymann [49].

If $C \in \mathbb{C}^{p \times n}$, $A \in \mathbb{C}^{n \times n}$, and $\mathfrak{O} = \begin{bmatrix} C \\ CA \\ \vdots \\ CA^{n-1} \end{bmatrix}$, then the pair (C, A) is said to be **observable** if $\mathcal{N}_{\mathfrak{O}} = \{\mathbf{0}\}$.

Exercise 36.12. Show that a pair of matrices $(C, A) \in \mathbb{C}^{p \times n} \times \mathbb{C}^{n \times n}$ is observable if and only if rank $\begin{bmatrix} \lambda I_n - A \\ C \end{bmatrix}$ for every point $\lambda \in \mathbb{C}$.

The conditions in Exercises 36.11 and 36.12 are often referred to as Hautus tests, or Popov-Belevich-Hautus tests. For additional equivalences for controllability and observability, see, e.g., Lemmas 19.2 and 19.3 in [30].

Chapter 37

Matrix equations

In this chapter we shall analyze the existence and uniqueness of solutions to a number of matrix equations that occur frequently in applications. The **notation**

(37.1) $\quad \mathbb{C}_R = \{\lambda \in \mathbb{C} : \lambda + \overline{\lambda} > 0\} \quad \text{and} \quad \mathbb{C}_L = \{\lambda \in \mathbb{C} : \lambda + \overline{\lambda} < 0\}$

for the open right and open left half-plane, respectively, will be useful.

37.1. The equation $X - AXB = C$

In this section we shall study the equation $X - AXB = C$ for appropriately sized matrices A, B, C, and X. If $A = \text{diag}\{\lambda_1, \ldots, \lambda_p\}$ and $B = \text{diag}\{\mu_1, \ldots, \mu_q\}$ are diagonal matrices, then it is easily seen that the equation $x_{ij} - \lambda_i x_{ij} \mu_j = c_{ij}$ for the ij entry x_{ij} of X has exactly one solution if and only if $1 - \lambda_i \mu_j \neq 0$ for $i = 1, \ldots, p$ and $j = 1, \ldots, q$. This condition on the eigenvalues of A and B holds for nondiagonal matrices too but requires a little more work to justify.

Lemma 37.1. *Let $A \in \mathbb{C}^{p \times p}$, $B \in \mathbb{C}^{q \times q}$ and let $\lambda_1, \ldots, \lambda_k$ and $\beta_1, \ldots, \beta_m$ denote the distinct eigenvalues of the matrices A and B, respectively; and let T denote the linear transformation from $\mathbb{C}^{p \times q}$ into $\mathbb{C}^{p \times q}$ that is defined by the rule*

(37.2) $\qquad T : X \in \mathbb{C}^{p \times q} \mapsto X - AXB \in \mathbb{C}^{p \times q}.$

Then $\mathcal{N}_T = \{O_{p \times q}\}$ if and only if

(37.3) $\qquad \lambda_i \beta_j \neq 1 \quad \text{for} \quad i = 1, \ldots, k \quad \text{and} \quad j = 1, \ldots, m.$

Proof. Let $A\mathbf{u}_i = \lambda_i \mathbf{u}_i$ and $B^T \mathbf{v}_j = \beta_j \mathbf{v}_j$ for some pair of nonzero vectors $\mathbf{u}_i \in \mathbb{C}^p$ and $\mathbf{v}_j \in \mathbb{C}^q$ and let $X = \mathbf{u}_i \mathbf{v}_j^T$. Then the formula

$$TX = \mathbf{u}_i \mathbf{v}_j^T - A\mathbf{u}_i \mathbf{v}_j^T B = (1 - \lambda_i \beta_j) \mathbf{u}_i \mathbf{v}_j^T$$

clearly implies that the condition (37.3) is necessary for $\mathcal{N}_T = \{O_{p \times q}\}$.

To prove the sufficiency of this condition, invoke the Jordan decomposition $B = UJU^{-1}$ with $U = \begin{bmatrix} U_1 & \cdots & U_m \end{bmatrix}$ and $J = \text{diag}\{J_{\beta_1}, \ldots, J_{\beta_m}\}$, in which $U_j \in \mathbb{C}^{q \times k_j}$ and $J_{\beta_j} \in \mathbb{C}^{k_j \times k_j}$ is upper triangular with β_j on the diagonal for $j = 1, \ldots, m$ and k_j is equal to the algebraic multiplicity of β_j for $j = 1, \ldots, m$. Then, since $J_{\beta_j} = \beta_j I_{k_j} + N_j$ and $(N_j)^{k_j} = O$,

$$X - AXB = O \iff X - AXUJU^{-1} = O \iff XU - AXUJ = O$$
$$\iff XU_j = AXU_j J_{\beta_j} \quad \text{for } j = 1, \ldots, m$$
$$\iff (I_p - \beta_j A)XU_j = AXU_j N_j \quad \text{for } j = 1, \ldots, m.$$

Therefore, since $(I_p - \beta_j A)$ is invertible when (37.3) is in force,

$$X - AXB = O \iff XU_j = M_j XU_j N_j \quad \text{for } j = 1, \ldots, m$$

with $M_j = (I_p - \beta_j A)^{-1} A$. But upon iterating the last displayed equality n times we obtain

$$XU_j = M_j^n XU_j N_j^n = O \quad \text{for } j = 1, \ldots, m \quad \text{if } n \geq q.$$

Therefore, $X = XUU^{-1} = \begin{bmatrix} XU_1 & \cdots & XU_m \end{bmatrix} U^{-1} = O$ is the only solution of the equation $X - AXB = O$ when $\lambda_i \beta_j \neq 1$. $\square$

Theorem 37.2. *Let $A \in \mathbb{C}^{p \times p}$, $B \in \mathbb{C}^{q \times q}$ and let $\lambda_1, \ldots, \lambda_k$ and $\beta_1, \ldots, \beta_m$ denote the distinct eigenvalues of the matrices A and B, respectively. Then, for each choice of $C \in \mathbb{C}^{p \times q}$, the equation*

(37.4) $$X - AXB = C$$

has exactly one solution $X \in \mathbb{C}^{p \times q}$ if and only if (37.3) holds.

Proof. This is immediate from Lemma 37.1 and the principle of conservation of dimension: If T is the linear transformation that is defined by the rule (37.2), then

$$pq = \dim \mathcal{N}_T + \dim \mathcal{R}_T.$$

Therefore, T maps onto $\mathbb{C}^{p \times q}$ if and only if $\mathcal{N}_T = \{O_{p \times q}\}$, i.e., if and only if $\lambda_i \beta_j \neq 1$ for every choice of i and j. $\square$

Corollary 37.3. *If $A \in \mathbb{C}^{p \times p}$, $B \in \mathbb{C}^{q \times q}$, $C \in \mathbb{C}^{p \times q}$, and $r_\sigma(A) r_\sigma(B) < 1$, then $X = \sum_{j=0}^{\infty} A^j C B^j$ is the only solution of equation (37.4).*

Exercise 37.1. Use Theorem 22.2, the refined fixed point theorem, to justify Corollary 37.3 a second way (that does not depend upon Theorem 37.2).

Corollary 37.4. Let $A \in \mathbb{C}^{p \times p}$, $C \in \mathbb{C}^{p \times p}$ and let $\lambda_1, \ldots, \lambda_p$ denote the eigenvalues of the matrix A. Then the **Stein equation**

(37.5) $$X - A^H X A = C$$

has exactly one solution $X \in \mathbb{C}^{p \times p}$ if and only if $1 - \overline{\lambda_i}\lambda_j \neq 0$ for every choice of i and j.

Exercise 37.2. Verify Corollary 37.4.

Exercise 37.3. Let $A = C_\alpha^{(2)}$ and $B = C_\beta^{(2)}$ and suppose that $\alpha\beta = 1$. Show that the equation $X - AXB = C$ has no solutions if either $c_{21} \neq 0$ or $\alpha c_{11} \neq \beta c_{22}$.

Exercise 37.4. Let $A = C_\alpha^{(2)}$ and $B = C_\beta^{(2)}$ and suppose that $\alpha\beta = 1$. Show that if $c_{21} = 0$ and $\alpha c_{11} = \beta c_{22}$, then the equation $X - AXB = C$ has infinitely many solutions.

Exercise 37.5. Find the unique solution $X \in \mathbb{C}^{p \times p}$ of equation (37.5) when $A = C_0^{(p)}$, $C = \mathbf{e}_1 \mathbf{u}^H + \mathbf{u}\mathbf{e}_1^H$, and $\mathbf{u}^H = \begin{bmatrix} \overline{t_0} & \overline{t_1} & \cdots & \overline{t_{p-1}} \end{bmatrix}$.

Exercise 37.6. Let $A = C_0^{(4)}$ and let T denote the linear transformation from $\mathbb{C}^{4 \times 4}$ into itself that is defined by the formula $TX = X - A^H X A$.

(a) Calculate $\dim \mathcal{N}_T$.

(b) Show that a matrix $X \in \mathbb{C}^{4 \times 4}$ is a solution of the matrix equation

$$X - A^H X A = \begin{bmatrix} a & b & c & d \\ e & 0 & 0 & 0 \\ f & 0 & 0 & 0 \\ g & 0 & 0 & 0 \end{bmatrix} \text{ if and only if } X \text{ is a \textbf{Toeplitz}}$$

matrix (i.e., $x_{ij} = x_{i+1,j+1}$) with $x_{1j} = c_{1j}$ and $x_{j1} = c_{j1}$ for $j = 1, \ldots, 4$.

37.2. The Sylvester equation $AX - XB = C$

The strategy for studying the equation $AX - XB = C$ is much the same as for the equation $X - AXB = C$. Again the special case in which $A = \operatorname{diag}\{\lambda_1, \ldots, \lambda_p\}$ and $B = \operatorname{diag}\{\beta_1, \ldots, \beta_q\}$ are diagonal matrices points the way: The equation $\lambda_i x_{ij} - x_{ij}\beta_j = c_{ij}$ for the ij entry x_{ij} of X has exactly one solution if and only if $\lambda_i - \beta_j \neq 0$ for $i = 1, \ldots, p$ and $j = 1, \ldots, q$. This condition on the eigenvalues of A and B holds for nondiagonal matrices too but requires a little more work to justify.

Lemma 37.5. Let $A \in \mathbb{C}^{p \times p}$, $B \in \mathbb{C}^{q \times q}$ and let $\lambda_1, \ldots, \lambda_k$ and $\beta_1, \ldots, \beta_m$ denote the distinct eigenvalues of the matrices A and B, respectively, and let T denote the linear transformation from $\mathbb{C}^{p \times q}$ into $\mathbb{C}^{p \times q}$ that is defined by the rule

(37.6) $$T: X \in \mathbb{C}^{p \times q} \mapsto AX - XB \in \mathbb{C}^{p \times q}.$$

Then $\mathcal{N}_T = \{O_{p \times q}\}$ if and only if

(37.7) $$\lambda_i - \beta_j \neq 0 \quad \text{for} \quad i = 1, \ldots, p \quad \text{and} \quad j = 1, \ldots, q.$$

Proof. Let $A\mathbf{u}_i = \lambda_i \mathbf{u}_i$ and $B^T \mathbf{v}_j = \beta_j \mathbf{v}_j$ for some pair of nonzero vectors $\mathbf{u}_i \in \mathbb{C}^p$ and $\mathbf{v}_j \in \mathbb{C}^q$ and let $X = \mathbf{u}_i \mathbf{v}_j^T$. Then the formula

$$TX = A\mathbf{u}_i \mathbf{v}_j^T - \mathbf{u}_i \mathbf{v}_j^T B = (\lambda_i - \beta_j) \mathbf{u}_i \mathbf{v}_j^T$$

clearly implies that the condition (37.7) is necessary for $\mathcal{N}_T = \{O_{p \times q}\}$.

To prove the sufficiency of this condition, invoke the Jordan decomposition $B = UJU^{-1}$ with $U = \begin{bmatrix} U_1 & \cdots & U_m \end{bmatrix}$ and $J = \text{diag}\{J_{\beta_1}, \ldots, J_{\beta_m}\}$, in which $U_j \in \mathbb{C}^{q \times k_j}$ and $J_{\beta_j} \in \mathbb{C}^{k_j \times k_j}$ for $j = 1, \ldots, m$ and k_j is equal to the algebraic multiplicity of β_j for $j = 1, \ldots, m$. Then, since $J_{\beta_j} = \beta_j I_{k_j} + N_j$ and $(N_j)^{k_j} = O$,

$$AX - XB = O \iff AX - XUJU^{-1} = O \iff AXU - XUJ = O$$
$$\iff AXU_j = XU_j J_{\beta_j} \quad \text{for } j = 1, \ldots, m$$
$$\iff (A - \beta_j I_p)XU_j = XU_j N_j \quad \text{for } j = 1, \ldots, m.$$

Therefore, since $(A - \beta_j I_p)$ is invertible when (37.7) is in force,

$$AX - XB = O \iff XU_j = M_j XU_j N_j \quad \text{for } j = 1, \ldots, m$$

with $M_j = (A - \beta_j I_p)^{-1} A$. But upon iterating the last displayed equality n times we obtain

$$XU_j = M_j^n XU_j N_j^n = O \quad \text{for } j = 1, \ldots, m \quad \text{if } n \geq q.$$

Therefore $X = XUU^{-1} = \begin{bmatrix} XU_1 & \cdots & XU_m \end{bmatrix} U^{-1} = O$ is the only solution of the equation $AX - XB = O$ when $\lambda_i \beta_j \neq 1$. $\square$

Theorem 37.6. *Let $A \in \mathbb{C}^{p \times p}$, $B \in \mathbb{C}^{q \times q}$ and let $\lambda_1, \ldots, \lambda_k$ and $\beta_1, \ldots, \beta_m$ denote the distinct eigenvalues of the matrices A and B, respectively. Then, for each choice of $C \in \mathbb{C}^{p \times q}$, the equation*

$$AX - XB = C$$

has exactly one solution $X \in \mathbb{C}^{p \times q}$ if and only if (37.7) holds, i.e., if and only if $\sigma(A) \cap \sigma(B) = \emptyset$.

Proof. This is an immediate corollary of Lemma 37.5 and the principle of conservation of dimension. $\square$

37.2. The Sylvester equation $AX - XB = C$

Exercise 37.7. Let $A \in \mathbb{C}^{n \times n}$. Show that the **Lyapunov equation**
$$A^H X + XA = Q \tag{37.8}$$
has exactly one solution for each choice of $Q \in \mathbb{C}^{n \times n}$ if and only if $\sigma(A) \cap \sigma(-A^H) = \emptyset$.

Lemma 37.7. *If $A, Q \in \mathbb{C}^{n \times n}$ and if $\sigma(A) \subset \mathbb{C}_L$ and $-Q \succeq O$, then the Lyapunov equation (37.8) has exactly one solution $X \in \mathbb{C}^{n \times n}$. Moreover, this solution is positive semidefinite.*

Proof. Since $\sigma(A) \subset \mathbb{C}_L$, the matrix
$$Z = -\int_0^\infty e^{tA^H} Q e^{tA} dt$$
is well-defined and is positive semidefinite. Moreover,
$$\begin{aligned}
A^H Z &= -\int_0^\infty A^H e^{tA^H} Q e^{tA} dt \\
&= -\int_0^\infty \left(\frac{d}{dt} e^{tA^H}\right) Q e^{tA} dt \\
&= -\left\{ e^{tA^H} Q e^{tA} \Big|_{t=0}^\infty - \int_0^\infty e^{tA^H} \frac{d}{dt}(Q e^{tA}) dt \right\} \\
&= Q + \int_0^\infty e^{tA^H} Q e^{tA} dt\, A \\
&= Q - ZA\,.
\end{aligned}$$

Thus, the matrix Z is a solution of the Lyapunov equation (37.8) and hence, as the assumption $\sigma(A) \subset \mathbb{C}_L$ implies that $\sigma(A) \cap \sigma(-A^H) = \emptyset$, Theorem 37.6 (as reformulated for (37.8) in Exercise 37.7) ensures that this equation has only one solution. Therefore, $X = Z$ is positive semidefinite. $\square$

Exercise 37.8. Show that if, in addition to the assumptions that $\sigma(A) \subset \mathbb{C}_L$ and $-Q \succeq O$, it is also assumed in Lemma 37.7 that $A, Q \in \mathbb{R}^{n \times n}$, then the solution X of (37.8) belongs to $\mathbb{R}^{n \times n}$.

Exercise 37.9. Let $A \in \mathbb{C}^{n \times n}$. Show that if $\sigma(A) \subset \mathbb{C}_R$, the open right half-plane, then the equation $A^H X + XA = Q$ has exactly one solution for every choice of $Q \in \mathbb{C}^{n \times n}$ and that this solution can be expressed as
$$X = \int_0^\infty e^{-tA^H} Q e^{-tA} dt$$
for every choice of $Q \in \mathbb{C}^{n \times n}$. [HINT: Integrate the formula
$$A^H \int_0^\infty e^{-tA^H} Q e^{-tA} dt = -\int_0^\infty \frac{d}{dt}\left(e^{-tA^H} Q\right) e^{-tA} dt$$
by parts.]

Exercise 37.10. Show that in the setting of Exercise 37.9, the solution X can also be expressed as

$$X = -\frac{1}{2\pi}\int_{-\infty}^{\infty}(i\mu I_n + A^H)^{-1}Q(i\mu I_n - A)^{-1}d\mu.$$

[HINT: Write $A^H X = -\lim_{R\uparrow\infty}\frac{1}{2\pi i}\int_{-R}^{R}(A^H + i\mu I_n - i\mu I_n)\{\cdots\}d\mu$ and evaluate the integral by adding a semicircle of radius R to complete the contour keeping (35.4) in mind.]

Exercise 37.11. Let $A = \text{diag}\{A_{11}, A_{22}\}$ be a block diagonal matrix in $\mathbb{C}^{n\times n}$ with $\sigma(A_{11}) \subset \mathbb{C}_R$ and $\sigma(A_{22}) \subset \mathbb{C}_L$, let $Q \in \mathbb{C}^{n\times n}$, and let $Y \in \mathbb{C}^{n\times n}$ and $Z \in \mathbb{C}^{n\times n}$ be solutions of the Lyapunov equation $A^H X + XA = Q$. Show that if Y and Z are written in block form consistent with the block decomposition of A, then $Y_{11} = Z_{11}$ and $Y_{22} = Z_{22}$.

Exercise 37.12. Let $A, Q \in \mathbb{C}^{n\times n}$. Show that if $\sigma(A) \cap i\mathbb{R} = \emptyset$ and if Y and Z are both solutions of the same Lyapunov equation $A^H X + XA = Q$ and $Y - Z \succeq O$, then $Y = Z$. [HINT: To warm up, suppose first that $A = \text{diag}\{A_{11}, A_{22}\}$, where $\sigma(A_{11}) \subset \mathbb{C}_R$ and $\sigma(A_{22}) \subset \mathbb{C}_L$ and consider Exercise 37.11.]

Exercise 37.13. Let $A = \sum_{j=1}^{3}e_j e_{j+1}^T = C_0^{(4)}$ and let T denote the linear transformation from $\mathbb{C}^{4\times 4}$ into itself that is defined by the formula $TX = A^H X - XA$.

(a) Calculate $\dim \mathcal{N}_T$.

(b) Show that a matrix $X \in \mathbb{C}^{4\times 4}$ with entries x_{ij} is a solution of the matrix equation $A^H X - XA = \begin{bmatrix} 0 & -a & -b & -c \\ a & 0 & 0 & 0 \\ b & 0 & 0 & 0 \\ c & 0 & 0 & 0 \end{bmatrix}$ if and only if X is a **Hankel matrix** (i.e., $x_{ij} = x_{i+1,j-1}$) with $x_{11} = a$, $x_{12} = b$, and $x_{13} = c$.

Exercise 37.14. Show that if $A, B, C \in \mathbb{C}^{n\times n}$, then

$$X(t) = e^{tA}Ce^{-tB}$$

is a solution of the differential equation

$$X'(t) = AX(t) - X(t)B$$

that meets the initial condition $X(0) = C$.

37.3. $AX = XB$

The next result complements Lemma 37.5. It deals with the case where the nullspace $\mathcal{N}_T$ of the linear transformation T introduced in (37.6) is not equal to zero.

Lemma 37.8. *Let $A, X, B \in \mathbb{C}^{n \times n}$ and suppose that $AX = XB$. Then there exists a matrix $C \in \mathbb{C}^{n \times n}$ such that $AX = X(B+C)$ and $\sigma(B+C) \subseteq \sigma(A)$.*

Proof. If X is invertible, then $\sigma(A) = \sigma(B)$; i.e., the matrix $C = O$ does the trick. Suppose therefore that X is not invertible and that $C_\beta^{(k)}$ is a $k \times k$ Jordan cell in the Jordan decomposition of $B = UJU^{-1}$ such that $\beta \notin \sigma(A)$. Then there exists a subblock $W \in \mathbb{C}^{n \times k}$ of U such that $BW = WC_\beta^{(k)}$. Therefore,
$$AXW - XWC_\beta^{(k)} = AXW - XBW = O$$
and hence, as $\beta \notin \sigma(A)$, Lemma 37.5 implies that $XW = O$. Thus, if
$$B_1 = B + W(C_\alpha^{(k)} - C_\beta^{(k)})V^H,$$
where $V^H \in \mathbb{C}^{k \times n}$ is the block of rows in U^{-1} corresponding to the columns in W (i.e., $UJU^{-1} = WC_\beta^{(k)}V^H + \cdots$) and $\alpha \in \sigma(A)$, then
$$XB_1 = XB = AX,$$
and the diagonal entry of the block under consideration in the Jordan decomposition of B_1 now belongs to $\sigma(A)$ and not to $\sigma(B)$. Moreover, none of the other Jordan blocks in the Jordan decomposition of B are affected by this change. The same procedure can now be applied to change the diagonal entry of any Jordan cell in the Jordan decomposition of B_1 from a point that is not in $\sigma(A)$ to a point that is in $\sigma(A)$. The proof is completed by repeating this procedure. $\square$

Exercise 37.15. Let $A, X, B \in \mathbb{C}^{n \times n}$. Show that if $AX = XB$ and the columns of $V \in \mathbb{C}^{n \times k}$ form a basis for $\mathcal{N}_X$, then there exists a matrix $L \in \mathbb{C}^{k \times n}$ such that $\sigma(B + VL) \subseteq \sigma(A)$.

37.4. Special classes of solutions

Let $A \in \mathbb{C}^{n \times n}$ and now let:

- $\mathcal{E}_+(A)$ = the number of zeros of $\det(\lambda I_n - A)$ in $\mathbb{C}_R$,
- $\mathcal{E}_-(A)$ = the number of zeros of $\det(\lambda I_n - A)$ in $\mathbb{C}_L$,
- $\mathcal{E}_0(A)$ = the number of zeros of $\det(\lambda I_n - A)$ in $i\mathbb{R}$,

counting multiplicities in all three. The triple $(\mathcal{E}_+(A), \mathcal{E}_-(A), \mathcal{E}_0(A))$ is called the **inertia** of A. Since multiplicities are counted,
$$\mathcal{E}_+(A) + \mathcal{E}_-(A) + \mathcal{E}_0(A) = n.$$

Theorem 37.9. *Let $A \in \mathbb{C}^{n \times n}$ and suppose that $\sigma(A) \cap i\mathbb{R} = \emptyset$. Then there exists a Hermitian matrix $G \in \mathbb{C}^{n \times n}$ such that:*

(1) $A^H G + GA \succ O$.
(2) $\mathcal{E}_+(G) = \mathcal{E}_+(A)$, $\mathcal{E}_-(G) = \mathcal{E}_-(A)$, and $\mathcal{E}_0(G) = \mathcal{E}_0(A) = 0$.

Proof. Suppose first that $\mathcal{E}_+(A) = p \geq 1$ and $\mathcal{E}_-(A) = q \geq 1$. Then the assumption $\sigma(A) \cap i\mathbb{R} = \emptyset$ guarantees that $p + q = n$ and hence that A admits a Jordan decomposition UJU^{-1} of the form
$$A = U \begin{bmatrix} J_1 & O \\ O & J_2 \end{bmatrix} U^{-1}$$
with $J_1 \in \mathbb{C}^{p \times p}$, $\sigma(J_1) \subset \mathbb{C}_R$, $J_2 \in \mathbb{C}^{q \times q}$, and $\sigma(J_2) \subset \mathbb{C}_L$.

Let $P_{11} \in \mathbb{C}^{p \times p}$ and $P_{22} \in \mathbb{C}^{q \times q}$ be positive definite matrices. Then it is readily checked, much as in the proof of Lemma 37.7, that
$$X_{11} = \int_0^\infty e^{-tJ_1^H} P_{11} e^{-tJ_1} dt$$
is a positive definite solution of the equation
$$J_1^H X_{11} + X_{11} J_1 = P_{11}$$
and that
$$X_{22} = -\int_0^\infty e^{tJ_2^H} P_{22} e^{tJ_2} dt$$
is a negative definite solution of the equation
$$J_2^H X_{22} + X_{22} J_2 = P_{22}.$$
(The two integrals are well-defined because $\sigma(J_1) \subset \mathbb{C}_R \implies \sigma(J_1^H) \subset \mathbb{C}_R$ and $\sigma(J_2) \subset \mathbb{C}_L \implies \sigma(J_2^H) \subset \mathbb{C}_L$.) Let
$$X = \mathrm{diag}\{X_{11}, X_{22}\} \quad \text{and} \quad P = \mathrm{diag}\{P_{11}, P_{22}\}.$$
Then $J^H X + XJ = P$ and hence
$$(U^H)^{-1} J^H U^H (U^H)^{-1} X U^{-1} + (U^H)^{-1} X U^{-1} U J U^{-1} = (U^H)^{-1} P U^{-1}.$$
Thus, the matrix $G = (U^H)^{-1} X U^{-1}$ is a solution of the equation
$$A^H G + GA = (U^H)^{-1} P U^{-1}$$
and hence, as $(U^H)^{-1} P U^{-1} \succ O$, (1) holds when $p > 0$ and $q > 0$. The cases $p = 0$, $q = n$ and $p = n$, $q = 0$ are left to the reader.

Sylvester's inertia theorem (which is discussed in Section 28.6) guarantees that $\mathcal{E}_\pm(G) = \mathcal{E}_\pm(X) = \mathcal{E}_\pm(A)$, which justifies (2). □

Exercise 37.16. Complete the proof of Theorem 37.9 by verifying the cases $p = 0$ and $p = n$.

Exercise 37.17. Find a Hermitian matrix $G \in \mathbb{C}^{2\times 2}$ that fulfills the conditions of Theorem 37.9 when $A = \mathrm{diag}\{1+i, 1-i\}$.

37.5. Supplementary notes

This chapter is adapted from Chapter 18 and Section 20.8 of [**30**]. A number of refinements may be found in Lancaster and Tismenetsky [**55**].

Chapter 38

A matrix completion problem

In this chapter we shall consider the problem of filling in the missing entries of a partially specified positive definite matrix $Z \in \mathbb{C}^{n \times n}$.

Towards this end, we shall suppose that the entries z_{ij} of Z are specified for those indices (i,j) that belong to a proper subset Ω of $\{(i,j) : i,j = 1,\ldots,n\}$ and shall let

$$\mathcal{Z}_\Omega = \{A \in \mathbb{C}^{n \times n} : A \succ O \quad \text{and} \quad a_{ij} = z_{ij} \text{ when } (i,j) \in \Omega\}.$$

Since $A \succ O \implies A = A^H$, we shall assume that Ω is symmetric, i.e., $(i,j) \in \Omega \iff (j,i) \in \Omega$.

The set $\mathcal{Z}_\Omega$ can also be conveniently described in terms of the standard basis vectors $\mathbf{e}_1,\ldots,\mathbf{e}_n$ for $\mathbb{C}^n$ and the transformation Π_Ω that acts on $A \in \mathbb{C}^{n \times n}$ by the rule

$$\Pi_\Omega A = \sum_{(i,j) \in \Omega} \langle A, \mathbf{e}_i \mathbf{e}_j^T \rangle \mathbf{e}_i \mathbf{e}_j^T = \sum_{(i,j) \in \Omega} \text{trace}\{\mathbf{e}_j \mathbf{e}_i^T A\} \mathbf{e}_i \mathbf{e}_j^T$$
$$= \sum_{(i,j) \in \Omega} \text{trace}\{\mathbf{e}_i^T A \mathbf{e}_j\} \mathbf{e}_i \mathbf{e}_j^T = \sum_{(i,j) \in \Omega} a_{ij}\, \mathbf{e}_i \mathbf{e}_j^T,$$

i.e.,

$$\mathcal{Z}_\Omega = \{A \in \mathbb{C}^{n \times n} : A \succ O \quad \text{and} \quad \Pi_\Omega A = \Pi_\Omega Z\}.$$

In these terms, a basic question is:

(38.1) For which sets Ω does there exist a matrix $A \in \mathbb{C}^{n \times n}$ such that $A \succ O$ and $A^{-1} - \Pi_\Omega A^{-1} = O$?

We shall see that a necessary condition is that no principal submatrix of A is in the complement of Ω (and hence that $a_{ii} \in \Omega$ for $i = 1, \ldots, n$). But that is not the whole story.

Exercise 38.1. Show that the transformation Π_Ω is an orthogonal projection in the inner product space $\mathbb{C}^{n \times n}$ equipped with the inner product $\langle A, B \rangle = \operatorname{trace} B^H A$.

38.1. Constraints on Ω

Let $g(X) = \ln \det X$ and recall that $-g(X)$ is a strictly convex function on the convex set of positive definite matrices $X \in \mathbb{C}^{n \times n}$ and so too on the set $\mathcal{Z}_\Omega$, since $\mathcal{Z}_\Omega$ is also a convex set.

The main result of this section is Theorem 38.3. We begin, however, with a number of preliminary lemmas.

Lemma 38.1. *If $A \in \mathcal{Z}_\Omega$, $B = B^H \in \mathbb{C}^{n \times n}$, and $\Pi_\Omega B = O$, then $A + \mu B \in \mathcal{Z}_\Omega$ for every choice of $\mu \in \mathbb{R}$ for which $|\mu| \, \|B\| < s_n$, the smallest singular value of A.*

Proof. Under the given assumptions, $A + \mu B \succ O$, since
$$\langle (A + \mu B)\mathbf{x}, \mathbf{x} \rangle = \langle A\mathbf{x}, \mathbf{x} \rangle + \langle \mu B \mathbf{x}, \mathbf{x} \rangle \geq s_n \langle \mathbf{x}, \mathbf{x} \rangle - |\mu| \, \|B\| \, \langle \mathbf{x}, \mathbf{x} \rangle > 0$$
for every nonzero vector $\mathbf{x} \in \mathbb{C}^n$. Moreover, $(A + \mu B) \in \mathcal{Z}_\Omega$, since $\Pi_\Omega(A + \mu B) = \Pi_\Omega A$. $\square$

Lemma 38.2. *If Ω is symmetric and the matrix $A \in \mathcal{Z}_\Omega$ is a local extremum for the function $g(X) = \ln \det X$ in $\mathcal{Z}_\Omega$, then:*

(1) $\mathbf{e}_i^T (A^{-1}) \mathbf{e}_j = (A^{-1})_{ij} = 0$ *for every pair of points $(i,j) \notin \Omega$.*

(2) *The indices $(i,i) \in \Omega$ for $i = 1, \ldots, n$.*

Proof. If A is a local extremum for $g(X)$ in $\mathcal{Z}_\Omega$ and if $B = B^H$ and $\Pi_\Omega B = O$, then $A + \mu B \in \mathcal{Z}_\Omega$ for sufficiently small $\mu \in \mathbb{R}$ and, in view of Lemma 17.4,
$$0 = \lim_{\mu \downarrow 0} \frac{g(A + \mu B) - g(A)}{\mu} = \operatorname{trace} A^{-1} B \,.$$
Thus, if $B = \alpha \mathbf{e}_j \mathbf{e}_i^T + \overline{\alpha} \mathbf{e}_i \mathbf{e}_j^T$ and $(i,j) \notin \Omega$, then
$$\operatorname{trace} A^{-1} B = \alpha \mathbf{e}_i^T (A^{-1}) \mathbf{e}_j + \overline{\alpha} \mathbf{e}_j^T (A^{-1}) \mathbf{e}_i = 0 \,.$$
Therefore,
$$(A^{-1})_{ij} + (A^{-1})_{ji} = 0 \text{ if } \alpha \in \mathbb{R} \quad \text{and} \quad (A^{-1})_{ij} - (A^{-1})_{ji} = 0 \text{ if } i\alpha \in \mathbb{R} \,.$$
Thus, (1) holds; and (2) follows from (1), since $(A^{-1})_{ii} > 0$. $\square$

There exist sets Ω for which $\mathcal{Z}_\Omega$ **does not have a local extremum**.

38.1. Constraints on Ω

Example 38.1. Let $\Omega = \{(1,2), (2,1)\}$ and $Z = \begin{bmatrix} ? & 1 \\ 1 & ? \end{bmatrix}$. Then

$$\mathcal{Z}_\Omega = \left\{ X \in \mathbb{C}^{2\times 2} : X = \begin{bmatrix} a & 1 \\ 1 & b \end{bmatrix} \text{ with } a > 0,\ b > 0,\ \text{and } ab > 1 \right\}.$$

The function $g(X)$ does not have a local extremum in $\mathcal{Z}_\Omega$: $g(X) = \ln(ab-1)$ and $\ln[(a+\alpha)(b+\beta) - 1] > \ln(ab - 1)$ if $\alpha b + \beta a + \alpha\beta > 0$, whereas $\ln[(a+\alpha)(b+\beta) - 1] < \ln(ab - 1)$ if $\alpha b + \beta a + \alpha\beta < 0$. ◇

Exercise 38.2. Show that $\mathcal{Z}_\Omega$ is a closed subset of $\mathbb{C}^{n\times n}$, i.e., if $X \in \mathbb{C}^{n\times n}$, $X_k \in \mathcal{Z}_\Omega$, and $\lim_{k\uparrow\infty} \|X - X_k\| = 0$, then $X \in \mathcal{Z}_\Omega$.

Theorem 38.3. *If $(i,i) \in \Omega$ for $i = 1, \ldots, n$ and $\mathcal{Z}_\Omega \neq \emptyset$, then there exists exactly one matrix $A \in \mathcal{Z}_\Omega$ such that*

(38.2) $\ln \det A > \ln \det X \quad \text{if} \quad X \in \mathcal{Z}_\Omega \text{ and } X \neq A$

(and hence $(A^{-1})_{ij} = 0$ for $(i,j) \notin \Omega$).

Proof. Let $\gamma = \max\{|z_{ii}| : i = 1, \ldots, n\}$ and proceed in steps.

1. *If $B \in \mathcal{Z}_\Omega$, then $|b_{ij}| \leq \gamma$, $\|B\| \leq n\gamma$, and $\det B \leq \gamma^n$.*

Since $B \succ O$,

$$\begin{bmatrix} b_{ii} & b_{ij} \\ b_{ji} & b_{jj} \end{bmatrix} \succ O \quad \text{for every choice of } 1 \leq i < j \leq n.$$

Therefore, $|b_{ij}|^2 \leq b_{ii} b_{jj} = z_{ii} z_{jj} \leq \gamma^2$, and hence $|b_{ij}| \leq \gamma$.

The second assertion follows by writing $B = \begin{bmatrix} \mathbf{b}_1 & \cdots & \mathbf{b}_n \end{bmatrix}$ as an array of its columns and noting that if $\mathbf{c} = \begin{bmatrix} c_1 & \cdots & c_n \end{bmatrix}^T$, then

$$\|B\mathbf{c}\| = \left\| \sum_{j=1}^n c_j \mathbf{b}_j \right\| \leq \sum_{j=1}^n |c_j| \|\mathbf{b}_j\| \leq \sum_{j=1}^n |c_j| \sqrt{n}\, \gamma \leq \|\mathbf{c}\|\, n\,\gamma,$$

since $\|\mathbf{b}_j\| \leq \sqrt{n}\gamma$ and $\sum_{j=1}^n |c_j| \leq \|\mathbf{c}\|\sqrt{n}$.

Finally, invoking the singular value decomposition $B = VSV^H$ with $V \in \mathbb{C}^{n\times n}$ unitary and $S = \mathrm{diag}\{s_1, \ldots, s_n\}$, the inequality between the geometric and arithmetic means ensures that

$$(\det B)^{1/n} = (s_1 \cdots s_n)^{1/n} \leq \frac{s_1 + \cdots + s_n}{n} = \frac{\mathrm{trace}\, B}{n} \leq \gamma,$$

which justifies the last assertion of this step.

2. *There exists exactly one matrix $A \in \mathcal{Z}_\Omega$ such that (38.2) holds.*

In view of step 1, $\{\det X : X \in \mathcal{Z}_\Omega\}$ is a bounded subset of $\mathbb{R}$. By assumption, $\mathcal{Z}_\Omega \neq \emptyset$. Therefore, there exists a matrix $C \in \mathcal{Z}_\Omega$. Thus, the set $\{X \in \mathcal{Z}_\Omega : \det X \geq \det C\}$ is a nonempty, closed, bounded, convex set and hence **2** follows from the fact that $-g$ is strictly convex. $\square$

The assumption that $\mathcal{Z}_\Omega \neq \emptyset$ in the formulation of Theorem 38.3 is essential; an example of a set Ω that contains the points (i,i) but $\mathcal{Z}_\Omega = \emptyset$ is developed in the next six exercises, which culminate in Exercise 38.8.

Exercise 38.3. Let $A = \begin{bmatrix} a & \mathbf{x}^H & \mu \\ \mathbf{x} & B & \mathbf{y} \\ \overline{\mu} & \mathbf{y}^H & d \end{bmatrix}$, where $B \in \mathbb{C}^{(n-2) \times (n-2)}$, $\mathbf{x}, \mathbf{y} \in \mathbb{C}^{n-2}$, and $a, b \in \mathbb{C}$ are known, but $\mu \in \mathbb{C}$ is not. Show that (in terms of the notation $|C|$ for $\det C$)

$$(38.3) \quad |A||B| = \begin{vmatrix} B & \mathbf{y} \\ \mathbf{y}^H & d \end{vmatrix} \begin{vmatrix} a & \mathbf{x}^H \\ \mathbf{x} & B \end{vmatrix} - \begin{vmatrix} \mathbf{x}^H & \mu \\ B & \mathbf{y} \end{vmatrix} \begin{vmatrix} \mathbf{x} & B \\ \overline{\mu} & \mathbf{y}^H \end{vmatrix}.$$

[HINT: Use Sylvester's formula.]

Exercise 38.4. Show that if $\begin{bmatrix} a & \mathbf{x}^H \\ \mathbf{x} & B \end{bmatrix} \succ O$ and $\begin{bmatrix} B & \mathbf{y} \\ \mathbf{y}^H & d \end{bmatrix} \succ O$ in the setting of Exercise 38.3, then

$$A \succ O \iff |\mu - \mathbf{x}^H B^{-1} \mathbf{y}|^2 < \frac{\det \begin{bmatrix} a & \mathbf{x}^H \\ \mathbf{x} & B \end{bmatrix} \det \begin{bmatrix} B & \mathbf{y} \\ \mathbf{y}^H & d \end{bmatrix}}{(\det B)^2}.$$

Exercise 38.5. Show that if A is a positive definite matrix of the form specified in Exercise 38.3, then

$$|A| \leq |B|(d - \mathbf{y}^H B^{-1} \mathbf{y})(a - \mathbf{x}^H B^{-1} \mathbf{x})$$

with equality if and only if $\mu = \mathbf{x}^H B^{-1} \mathbf{y}$.

Exercise 38.6. Show that if $\mu = \mathbf{x}^H B^{-1} \mathbf{y}$ in Exercise 38.3, then $(A^{-1})_{n1} = (A^{-1})_{1n} = 0$.

Exercise 38.7. Show that if $\beta > 1$, then $\begin{bmatrix} \beta & 1 & \mu \\ 1 & \beta & 1 \\ \overline{\mu} & 1 & \beta \end{bmatrix} \succ O$ if and only if $|\mu - \frac{1}{\beta}| < \beta - \frac{1}{\beta}$, whereas $\begin{bmatrix} \beta & 1 & -1 \\ 1 & \beta & \mu \\ -1 & \overline{\mu} & \beta \end{bmatrix} \succ O$ if and only if $|\mu + \frac{1}{\beta}| < \beta - \frac{1}{\beta}$.

38.2. The central diagonals are specified

Exercise 38.8. Show that if $1 < \beta < \sqrt{2}$, then there does not exist a choice of $\mu \in \mathbb{C}$ for which the matrix

$$A = \begin{bmatrix} \beta & 1 & 1 & -1 \\ 1 & \beta & 1 & \mu \\ 1 & 1 & \beta & 1 \\ -1 & \overline{\mu} & 1 & \beta \end{bmatrix}$$

is positive definite. [HINT: If $1 < \beta < \sqrt{2}$, then, in view of Exercise 38.7, there does not exist a choice of $\mu \in \mathbb{C}$ for which $A_{\{11\}}$ and $A_{\{33\}}$ are both positive definite.]

38.2. The central diagonals are specified

Our next objective is to formulate necessary and sufficient conditions on the specified entries z_{ij} that ensure that $\mathcal{Z}_\Omega \neq \emptyset$ when

(38.4) $\qquad \Omega = \Lambda_m \stackrel{\text{def}}{=} \{(i,j) : i, j = 1, \ldots, n \quad \text{and} \quad |i - j| \leq m\}$,

i.e., when the entries in the $2m + 1$ central diagonals of the matrix are specified.

It is tempting to set the unknown entries equal to zero. However, the matrix that is obtained this way is not necessarily positive definite; see Exercise 38.9 for a simple example, which also displays the fact that there may be many ways to fill in the missing entries to obtain a positive definite completion, even though there is only one way to fill in the missing entries so that the ij entries of the inverse of this completion are equal to zero if $(i,j) \notin \Lambda_m$.

We shall present an algorithm for obtaining this particular completion that is based on factorization and shall present another proof of the fact that it can be characterized as the completion which maximizes the determinant. Because of this property this particular completion is commonly referred to as the **maximum entropy completion**.

Exercise 38.9. Show that if $x \in \mathbb{R}$, then the matrix $\begin{bmatrix} 3 & 2 & x \\ 2 & 2 & 1 \\ x & 1 & 1 \end{bmatrix}$ is positive definite if and only $(x-1)^2 < 1/2$. However, there is only one choice of x for which $A \succ O$ and $(A^{-1})_{31} = (A^{-1})_{13} = 0$.

It is convenient to begin with a lemma:

Lemma 38.4. *If* $B = YY^H$, $Y \in \mathbb{C}^{n \times n}$ *is an invertible triangular matrix (upper or lower)*, $1 \leq i, j \leq n$, *and* $1 \leq k \leq n-1$, *then*

(38.5) $\qquad b_{ij} = 0 \quad \text{for} \quad |i - j| \geq k \iff y_{ij} = 0 \quad \text{for} \quad |i - j| \geq k$.

Discussion. The verification of (38.5) becomes transparent if the calculations are organized properly. The underlying ideas are best conveyed by example. Let $B \in \mathbb{C}^{4 \times 4}$, $k = 2$ and suppose that Y is lower triangular. Then, in terms of the standard basis $\mathbf{e}_1, \ldots, \mathbf{e}_4$ for $\mathbb{C}^4$,

$$b_{31} = \mathbf{e}_3^T B \mathbf{e}_1 = \mathbf{e}_3^T Y Y^H \mathbf{e}_1 = \begin{bmatrix} y_{31} & y_{32} & y_{33} & 0 \end{bmatrix} \begin{bmatrix} \overline{y_{11}} \\ 0 \\ 0 \\ 0 \end{bmatrix} = y_{31} \overline{y_{11}},$$

$$b_{41} = \mathbf{e}_4^T B \mathbf{e}_1 = \mathbf{e}_4^T Y Y^H \mathbf{e}_1 = \begin{bmatrix} y_{41} & y_{42} & y_{43} & y_{44} \end{bmatrix} \begin{bmatrix} \overline{y_{11}} \\ 0 \\ 0 \\ 0 \end{bmatrix} = y_{41} \overline{y_{11}},$$

and

$$b_{42} = \mathbf{e}_4^T B \mathbf{e}_2 = \mathbf{e}_4^T Y Y^H \mathbf{e}_2 = \begin{bmatrix} y_{41} & y_{42} & y_{43} & y_{44} \end{bmatrix} \begin{bmatrix} \overline{y_{21}} \\ \overline{y_{22}} \\ 0 \\ 0 \end{bmatrix} = y_{41} \overline{y_{21}} + y_{42} \overline{y_{22}}.$$

Since Y is presumed to be invertible, $y_{ii} \neq 0$ for $i = 1, \ldots, 4$, and hence the preceding three formulas clearly imply that

$$b_{31} = b_{41} = b_{42} = 0 \iff y_{31} = y_{41} = y_{42} = 0.$$

The verification of (38.5) in the general setting for lower triangular Y goes through in exactly the same way; the only difference is that the bookkeeping is a little more complicated.

The proof for upper triangular Y is left to the reader. □

Recall the **notation**

$$B_{[j,k]} = \begin{bmatrix} b_{jj} & \cdots & b_{jk} \\ \vdots & & \vdots \\ b_{kj} & \cdots & b_{kk} \end{bmatrix} \quad \text{for } 1 \leq j \leq k \leq n.$$

Theorem 38.5. *If $\Omega = \Lambda_m$ for some nonnegative integer $m \leq n - 1$, then*

(38.6) $$\mathcal{Z}_{\Lambda_m} \neq \emptyset \iff Z_{[j,j+m]} = \begin{bmatrix} z_{jj} & \cdots & z_{j,j+m} \\ \vdots & & \vdots \\ z_{j+m,j} & \cdots & z_{j+m,j+m} \end{bmatrix} \succ O$$

for $j = 1, \ldots, n - m$.

Moreover, if these conditions are met, then there is exactly one matrix

(38.7) $$A \in \mathcal{Z}_{\Lambda_m} \quad \text{with} \quad (A^{-1})_{ij} = 0 \text{ for } (i,j) \notin \Lambda_m.$$

38.2. The central diagonals are specified

(This unique matrix A is given by the formula $A = (X^H)^{-1}DX^{-1}$; X and D are defined in the proof.)

Proof. The proof of the implication $\Longrightarrow$ in (38.6) is easy and is left to the reader.

Conversely, if the matrices on the right-hand side of (38.6) are positive definite, then the following implications are in force:

1. *There exists exactly one lower triangular matrix $X \in \mathbb{C}^{n \times n}$ such that*

$$(38.8) \qquad Z_{[j,j+m]} \begin{bmatrix} x_{jj} \\ \vdots \\ x_{j+m,j} \end{bmatrix} = \begin{bmatrix} 1 \\ 0 \\ \vdots \\ 0 \end{bmatrix} \quad \text{for} \quad j = 1, \ldots, n-m,$$

$$(38.9) \qquad Z_{[j,n]} \begin{bmatrix} x_{jj} \\ \vdots \\ x_{n,j} \end{bmatrix} = \begin{bmatrix} 1 \\ 0 \\ \vdots \\ 0 \end{bmatrix} \quad \text{for} \quad j = n-m+1, \ldots, n,$$

and $x_{ij} = 0$ for $(i,j) \notin \Lambda_m$.

This is self-evident since the matrices $Z_{[j,j+m]}$ and $Z_{[j,n]}$ in (38.8) and (38.9) are positive definite.

2. *The diagonal matrix D with entries $d_{ii} = x_{ii}$ is positive definite.*

This is an easy computation.

3. *The matrix $A = (X^H)^{-1}DX^{-1}$ belongs to $\mathcal{Z}_{\Lambda_m}$ and $(A^{-1})_{ij} = 0$ for $(i,j) \notin \Lambda_m$.*

The fact that $(A^{-1})_{ij} = 0$ for $(i,j) \notin \Lambda_m$ follows from Lemma 38.4, since $x_{ij} = 0$ for $(i,j) \notin \Lambda_m$. It remains therefore only to check that $a_{ij} = z_{ij}$ for $(i,j) \in \Lambda_m$. Towards this end, let $U = (X^H)^{-1}D$. Then U is an upper triangular matrix with diagonal entries $u_{ii} = 1$ for $i = 1, \ldots, n$. Consequently, the formula $A = (X^H)^{-1}DX^{-1}$ can be rewritten as

$$AX = U,$$

which in turn implies that (38.8) and (38.9) hold with $A_{[j,j+m]}$ and $A_{[j,n]}$ in place of $Z_{[j,j+m]}$ and $Z_{[j,n]}$, respectively. Therefore, the numbers $c_{ij} = z_{ij} - a_{ij}$ are solutions of the equations (38.8) and (38.9), but with right-hand sides equal to zero. Thus, for example, if $n = 5$ and $m = 2$, then

$$\begin{bmatrix} c_{11} & c_{12} & c_{13} \\ c_{21} & c_{22} & c_{23} \\ c_{31} & c_{32} & c_{33} \end{bmatrix} \begin{bmatrix} x_{11} \\ x_{21} \\ x_{31} \end{bmatrix} = \begin{bmatrix} 0 \\ 0 \\ 0 \end{bmatrix}, \quad \begin{bmatrix} c_{22} & c_{23} & c_{24} \\ c_{32} & c_{33} & c_{34} \\ c_{42} & c_{43} & c_{44} \end{bmatrix} \begin{bmatrix} x_{22} \\ x_{32} \\ x_{42} \end{bmatrix} = \begin{bmatrix} 0 \\ 0 \\ 0 \end{bmatrix},$$

$$\begin{bmatrix} c_{33} & c_{34} & c_{35} \\ c_{43} & c_{44} & c_{45} \\ c_{53} & c_{54} & c_{55} \end{bmatrix} \begin{bmatrix} x_{33} \\ x_{43} \\ x_{53} \end{bmatrix} = \begin{bmatrix} 0 \\ 0 \\ 0 \end{bmatrix}, \quad \begin{bmatrix} c_{44} & c_{45} \\ c_{54} & c_{55} \end{bmatrix} \begin{bmatrix} x_{44} \\ x_{54} \end{bmatrix} = \begin{bmatrix} 0 \\ 0 \end{bmatrix}, \quad \text{and}$$

$$c_{55} x_{55} = 0.$$

But, since the x_{jj} are positive and each of the five submatrices are Hermitian, it is readily seen that $c_{ij} = 0$ for all the indicated entries; i.e., $a_{ij} = z_{ij}$ for $|i-j| \le 2$. (Start with $c_{55}x_{55} = 0$ and work your way back.) Thus, the matrix $A = (X^H)^{-1} D X^{-1}$ is a positive definite completion. □

A word on the intuition behind the proof of Theorem 38.5 is in order. It rests on the observation that if there exists a matrix A that meets the constraints in (38.7), then, since $A^{-1} \succ O$, there exists an invertible lower triangular matrix $Y \in \mathbb{C}^{n \times n}$ with positive diagonal entries y_{ii}, $i = 1, \ldots, n$, such that $A^{-1} = YY^H$. Lemma 38.4 then guarantees that $y_{ij} = 0$ for $(i,j) \notin \Lambda_m$. Thus, the matrix $U = (Y^H)^{-1} D$ with $D = \mathrm{diag}\{y_{11}, \ldots, y_{nn}\}$ is upper triangular with $u_{ii} = 1$ for $i = 1, \ldots, n$ and $A = U X^{-1}$ with $X = DY$. The system of equations considered in step 1 in the proof of the theorem is obtained by considering certain subblocks of the system $AX = U$. The next example should help to make this more transparent.

Example 38.2. If $A \in \mathcal{Z}_{\Lambda_m}$, then the matrix equation $AX = U$ for $n = 4$ and $m = 1$ is

$$\begin{bmatrix} z_{11} & z_{12} & ? & ? \\ z_{21} & z_{22} & z_{23} & ? \\ ? & z_{32} & z_{33} & z_{34} \\ ? & ? & z_{43} & z_{44} \end{bmatrix} \begin{bmatrix} x_{11} & 0 & 0 & 0 \\ x_{21} & x_{22} & 0 & 0 \\ 0 & x_{32} & x_{33} & 0 \\ 0 & 0 & x_{43} & x_{44} \end{bmatrix} = \begin{bmatrix} 1 & u_{12} & u_{13} & u_{14} \\ 0 & 1 & u_{23} & u_{24} \\ 0 & 0 & 1 & u_{34} \\ 0 & 0 & 0 & 1 \end{bmatrix}.$$

Therefore, the entries x_{ij} in X for $(i,j) \in \Lambda_m$ must satisfy the equations

(38.10) $$\begin{bmatrix} z_{11} & z_{12} \\ z_{21} & z_{22} \end{bmatrix} \begin{bmatrix} x_{11} \\ x_{21} \end{bmatrix} = \begin{bmatrix} 1 \\ 0 \end{bmatrix}, \quad \begin{bmatrix} z_{22} & z_{23} \\ z_{32} & z_{33} \end{bmatrix} \begin{bmatrix} x_{22} \\ x_{32} \end{bmatrix} = \begin{bmatrix} 1 \\ 0 \end{bmatrix},$$

(38.11) $$\begin{bmatrix} z_{33} & z_{34} \\ z_{43} & b_{44} \end{bmatrix} \begin{bmatrix} x_{33} \\ x_{43} \end{bmatrix} = \begin{bmatrix} 1 \\ 0 \end{bmatrix}, \quad \text{and} \quad z_{44} x_{44} = 1.$$

Next, since $x_{jj} > 0$ for $j = 1, \ldots, 4$, the matrices $D = \mathrm{diag}\{x_{11}, \ldots, x_{44}\}$ and

(38.12) $$A = (X^H)^{-1} D X^{-1}$$

38.2. The central diagonals are specified

are positive definite. Moreover, equations (38.10) and (38.11) are in force, but with a_{ij} in place of z_{ij}. Therefore, the numbers $c_{ij} = z_{ij} - a_{ij}$ are solutions of the equations

$$(38.13) \quad \begin{bmatrix} c_{11} & c_{12} \\ c_{21} & c_{22} \end{bmatrix} \begin{bmatrix} x_{11} \\ x_{21} \end{bmatrix} = \begin{bmatrix} 0 \\ 0 \end{bmatrix}, \quad \begin{bmatrix} c_{22} & c_{23} \\ c_{32} & c_{33} \end{bmatrix} \begin{bmatrix} x_{22} \\ x_{32} \end{bmatrix} = \begin{bmatrix} 0 \\ 0 \end{bmatrix},$$

$$(38.14) \quad \begin{bmatrix} c_{33} & c_{34} \\ c_{43} & c_{44} \end{bmatrix} \begin{bmatrix} x_{33} \\ x_{43} \end{bmatrix} = \begin{bmatrix} 0 \\ 0 \end{bmatrix}, \quad \text{and} \quad c_{44}x_{44} = 0.$$

But, since the x_{jj} are positive and each of the four submatrices are Hermitian, it is readily seen that $c_{ij} = 0$ for all the indicated entries; i.e., $a_{ij} = z_{ij}$ for $|i - j| \leq 1$. (Start with $c_{44}x_{44} = 0$ and work your way up.) Thus, $A \in \mathcal{Z}_{\Lambda_m}$. $\diamond$

We next give an independent proof (based on factorization) that the matrix A that satisfies the constraints in (38.7) maximizes the determinant.

Theorem 38.6. *If the conditions in (38.6) are in force and if A meets the constraints in (38.7) and C is any matrix in $\mathcal{Z}_{\Lambda_m}$, then $\det A \geq \det C$, with equality if and only if $C = A$.*

Proof. In view of Theorem 16.5,

$$(38.15) \quad A = (X^H)^{-1} D X^{-1} \quad \text{and} \quad C = (Y^H)^{-1} G Y^{-1},$$

where $X \in \mathbb{C}^{n \times n}$ and $Y \in \mathbb{C}^{n \times n}$ are lower triangular matrices with ones on the diagonal and $D \in \mathbb{C}^{n \times n}$ and $G \in \mathbb{C}^{n \times n}$ are positive definite diagonal matrices. Moreover, the formulas

$$C = A + (C - A) \quad \text{and} \quad W = Y^{-1} X$$

imply that

$$W^H G W = D + X^H (C - A) X.$$

Thus, as W is lower triangular with ones on the diagonal and $x_{ij} = 0$ for $i \geq m + j$, the diagonal entries of $X^H (C - A) X$ are all equal to zero. Consequently,

$$d_{jj} = \sum_{s=j}^{n} g_{ss} |w_{sj}|^2 = g_{jj} + \sum_{s>j} g_{ss} |w_{sj}|^2 \geq g_{jj},$$

with strict inequality unless $w_{sj} = 0$ for $s > j$, i.e., unless $W = I_n$ and hence $C = A$. $\square$

Exercise 38.10. Show that in terms of the notation in (38.15) $D \succeq G$, with equality if and only if $A = C$.

38.3. A moment problem

Positive definite matrices $A \in \mathbb{C}^{n \times n}$ with $(A^{-1})_{ij} = 0$ for $(i,j) \notin \Lambda_m$ occur naturally in certain classes of moment problems. (See also Section 27.4.)

Example 38.3. If $p(\lambda) = \lambda - \omega$, $|\omega| > 1$, and

$$t_j = \frac{1}{2\pi} \int_0^{2\pi} e^{-ij\theta} |p(e^{i\theta})|^{-2} d\theta \quad \text{for } j = 0, \pm 1, \ldots, \pm 3,$$

then the matrix

$$A = \begin{bmatrix} t_0 & t_{-1} & t_{-2} & t_{-3} \\ t_1 & t_0 & t_{-1} & t_{-2} \\ t_2 & t_1 & t_0 & t_{-1} \\ t_3 & t_2 & t_1 & t_0 \end{bmatrix}$$

is positive definite and $(A^{-1})_{ij} = 0$ for $|i - j| > 1$.

Discussion. It is readily checked that

$$t_j = \frac{\omega^{-j}}{|\omega|^2 - 1} = \overline{t_{-j}} \quad \text{for } j = 0, \ldots, 3,$$

and hence (since the ranks of the relevant subblocks are all less than 3) that the minors $A_{\{1;3\}} = A_{\{1;4\}} = A_{\{2;4\}}$ are all equal to zero. $\diamond$

38.4. Supplementary notes

The section on the maximum entropy completion problem is adapted from the paper [**32**]. The analysis therein exploits factorization. The pioneering work on this problem was done by J. P. Burg [**15**], [**16**], who used Lagrange multipliers to maximize the determinant of a class of partially specified positive semidefinite matrices, as in Exercise 24.12. A description of the set of all the completions to the problem considered in [**32**] may be found, e.g., in Chapter 10 of [**28**]. Reformulations of this completion problem as a convex optimization problem, which in turn led to significant generalizations, were presented in Grone, Johnson, de Sá, and Wolkowicz [**45**]; for additional perspective, see Glunt, Hayden, Johnson, and Tarazaga [**41**], the survey paper by Johnson [**50**], and Theorem 3.2 of the earlier paper by Deutsch and Schneider [**23**]. Exercises 38.3–38.8 are adapted from an example that is presented in [**50**]. Example 38.3 is a special case of Theorem 12.20 in [**30**]. The algebraic structure underlying Theorem 38.5 is clarified in [**33**]; see also Gohberg, Kaashoek, and Woerdeman [**42**] for further generalizations.

Chapter 39

Minimal norm completions

In its simplest form, the minimal norm completion problem is formulated in terms of a given set of three scalars a, b, and c and the objective is to choose a scalar z so that the norm of the 2×2 matrix $\begin{bmatrix} a & b \\ c & z \end{bmatrix}$ is as small as possible. It is easily checked that

(39.1) $$\left\| \begin{bmatrix} a & b \\ c & z \end{bmatrix} \right\| \geq \gamma, \quad \text{where } \gamma = \max\left\{ \left\| \begin{bmatrix} a \\ c \end{bmatrix} \right\|, \| \begin{bmatrix} a & b \end{bmatrix} \| \right\}.$$

Thus,
$$|c|^2 \leq \gamma^2 - |a|^2 \quad \text{and} \quad |b|^2 \leq \gamma^2 - |a|^2.$$

Therefore, there exists a pair of scalars $x, y \in \mathbb{C}$ with $|x| \leq 1$ and $|y| \leq 1$ such that
$$c = x(\gamma^2 - |a|^2)^{1/2} \quad \text{and} \quad b = (\gamma^2 - |a|^2)^{1/2} y.$$

Upon making these substitutions and setting $z = -x\bar{a}y$, the matrix of interest can be expressed in factored form as

$$\begin{bmatrix} a & b \\ c & z \end{bmatrix} = \begin{bmatrix} 1 & 0 \\ 0 & x \end{bmatrix} E \begin{bmatrix} 1 & 0 \\ 0 & y \end{bmatrix}, \quad \text{with} \quad E = \begin{bmatrix} a & (\gamma^2 - |a|^2)^{1/2} \\ (\gamma^2 - |a|^2)^{1/2} & -\bar{a} \end{bmatrix}.$$

Thus, as $E^H E = \gamma^2 I_2$, it is readily checked that the minimum norm is in fact equal to γ.

The matrix analogue of this conclusion is presented in Theorem 39.6. To obtain it we shall need a more subtle variant of the elementary observation that if A is invertible, then:

(1) $A^H A \succeq C^H C \implies C = KA$ for some K with $\|K\| \leq 1$.
(2) $AA^H \succeq BB^H \implies B = AL$ for some L with $\|L\| \leq 1$.

The next lemma serves to show that the preceding two implications are valid even if A is not invertible.

Lemma 39.1. *If $A \in \mathbb{C}^{p \times q}$, $B \in \mathbb{C}^{p \times r}$, and $C \in \mathbb{C}^{s \times q}$, then:*

(1) $A^H A \succeq C^H C \implies$ *there exists exactly one matrix $K \in \mathbb{C}^{s \times p}$ such that $KA = C$ and $K\mathbf{u} = \mathbf{0}$ for every vector $\mathbf{u} \in \mathcal{N}_{A^H}$.*

(2) $AA^H \succeq BB^H \implies$ *there exists exactly one matrix $L \in \mathbb{C}^{q \times r}$ such that $B = AL$ and $L^H \mathbf{u} = \mathbf{0}$ for every vector $\mathbf{u} \in \mathcal{N}_A$.*

Moreover, both of these uniquely specified matrices are contractions: $\|K\| \leq 1$ and $\|L\| \leq 1$.

Proof. The inequality $A^H A \succeq C^H C$ implies that $\|A\mathbf{x}\| \geq \|C\mathbf{x}\|$ for every vector $\mathbf{x} \in \mathbb{C}^q$ and hence that $\mathcal{N}_A \subseteq \mathcal{N}_C$. Thus, $\mathcal{R}_{A^H} \supseteq \mathcal{R}_{C^H}$, since

$$\mathbb{C}^q = \mathcal{N}_A \oplus \mathcal{R}_{A^H} = \mathcal{N}_C \oplus \mathcal{R}_{C^H}.$$

Consequently, there exists a matrix $G \in \mathbb{C}^{p \times s}$ such that $A^H G = C^H$. Thus, upon setting $K = G^H$, we see that $C\mathbf{u} = KA\mathbf{u}$ for every $\mathbf{u} \in \mathbb{C}^q$. We shall also choose K so that $K\mathbf{v} = \mathbf{0}$ for every vector $\mathbf{v} \in \mathcal{N}_{A^H}$. This serves to define K uniquely. Moreover, since every vector $\mathbf{x} \in \mathbb{C}^p$ can be written as $\mathbf{x} = A\mathbf{u} + \mathbf{v}$ with $\mathbf{u} \in \mathbb{C}^q$ and $\mathbf{v} \in \mathcal{N}_{A^H}$, it is easily seen that

$$\|K\mathbf{x}\| = \|KA\mathbf{u} + K\mathbf{v}\| = \|KA\mathbf{u}\| = \|C\mathbf{u}\| \leq \|A\mathbf{u}\| \leq \|\mathbf{x}\|$$

and hence that $\|K\| \leq 1$.

The proof of the second assertion and the fact that $\|L\| \leq 1$ goes through in much the same way. The details are left to the reader. $\square$

39.1. A minimal norm completion problem

This section treats a **minimal norm completion problem**. The main result is Theorem 39.5, which is usually referred to as **Parrott's lemma**. The proof rests essentially on three preliminary lemmas that will be presented first.

Lemma 39.2. *If $A \in \mathbb{C}^{p \times q}$, $B \in \mathbb{C}^{p \times r}$, and $C \in \mathbb{C}^{s \times q}$, then:*

(1) $\|A\| \leq \gamma \iff \gamma^2 I_q - A^H A \succeq O \iff \gamma^2 I_p - AA^H \succeq O$.

(2) $\gamma \geq \left\| \begin{bmatrix} A \\ C \end{bmatrix} \right\| \iff \gamma^2 I_q - A^H A \succeq C^H C$.

(3) $\gamma^2 I_q - A^H A \succeq C^H C \iff C = X(\gamma^2 I_q - A^H A)^{1/2}$ for some $X \in \mathbb{C}^{s \times q}$ with $\|X\| \leq 1$.

(4) $\gamma \geq \left\| \begin{bmatrix} A & B \end{bmatrix} \right\| \iff \gamma^2 I_p - AA^H \succeq BB^H$.

(5) $\gamma^2 I_p - AA^H \succeq BB^H \iff B = (\gamma^2 I_p - AA^H)^{1/2} Y$ for some $Y \in \mathbb{C}^{p \times r}$ with $\|Y\| \leq 1$.

39.1. A minimal norm completion problem

Proof. (1), (2), and (4) are easy consequences of the definitions. Thus, for example,

$$\gamma \geq \|A\| \iff \gamma^2 \langle \mathbf{x}, \mathbf{x} \rangle \geq \langle A\mathbf{x}, A\mathbf{x} \rangle \quad \text{for every vector } \mathbf{x} \in \mathbb{C}^q$$
$$\iff \langle (\gamma^2 I_q - A^H A)\mathbf{x}, \mathbf{x} \rangle \geq 0 \quad \text{for every vector } \mathbf{x} \in \mathbb{C}^q.$$

Lemma 39.1 then justifies the verification of (3) from (2) and (5) from (4). □

Lemma 39.3. *If $A \in \mathbb{C}^{p \times q}$ and $\|A\| \leq \gamma$, then:*

(39.2) $(\gamma^2 I_q - A^H A)^{1/2} A^H = A^H (\gamma^2 I_p - AA^H)^{1/2},$

(39.3) $(\gamma^2 I_p - AA^H)^{1/2} A = A(\gamma^2 I_q - A^H A)^{1/2}$

and the matrix

(39.4) $E = \begin{bmatrix} A & (\gamma^2 I_p - AA^H)^{1/2} \\ (\gamma^2 I_q - A^H A)^{1/2} & -A^H \end{bmatrix}$

satisfies the identity

$$EE^H = \begin{bmatrix} \gamma^2 I_p & O \\ O & \gamma^2 I_q \end{bmatrix} = \gamma^2 I_n.$$

Proof. The first two identities may be established with the aid of the singular value decomposition of A; the identity $EE^H = \gamma^2 I_n$ is an easy consequence of the first two. □

Lemma 39.4. *If $A \in \mathbb{C}^{p \times q}$, $B \in \mathbb{C}^{p \times r}$, $C \in \mathbb{C}^{s \times q}$, and*

$$\max \left\{ \left\| \begin{bmatrix} A \\ C \end{bmatrix} \right\|, \| [A \quad B] \| \right\} \leq \gamma,$$

then there exists a matrix $D \in \mathbb{C}^{s \times r}$ such that

$$\left\| \begin{bmatrix} A & B \\ C & D \end{bmatrix} \right\| \leq \gamma.$$

Proof. The given inequality implies that

$$\gamma^2 I_q - A^H A \succeq C^H C \quad \text{and} \quad \gamma^2 I_p - AA^H \succeq BB^H.$$

Therefore, by Lemma 39.1,

(39.5) $B = (\gamma^2 I_p - AA^H)^{1/2} X \quad \text{and} \quad C = Y(\gamma^2 I_q - A^H A)^{1/2}$

for some choice of $X \in \mathbb{C}^{p \times r}$ and $Y \in \mathbb{C}^{s \times q}$ with $\|X\| \leq 1$ and $\|Y\| \leq 1$. Thus, upon setting $D = -YA^H X$, it is readily seen that

$$\begin{bmatrix} A & B \\ C & D \end{bmatrix} = \begin{bmatrix} I_p & O \\ O & Y \end{bmatrix} E \begin{bmatrix} I_q & O \\ O & X \end{bmatrix},$$

where E is given by formula (39.4), and hence that

$$\left\| \begin{bmatrix} A & B \\ C & D \end{bmatrix} \right\| \leq \left\| \begin{bmatrix} I_p & O \\ O & Y \end{bmatrix} \right\| \|E\| \left\| \begin{bmatrix} I_q & O \\ O & X \end{bmatrix} \right\| = \|E\| = \gamma,$$

since $EE^H = \gamma^2 I_n$ by Lemma 39.3 and the norm of each of the two block diagonal matrices is equal to one. □

Theorem 39.5. *If* $A \in \mathbb{C}^{p \times q}$, $B \in \mathbb{C}^{p \times r}$, *and* $C \in \mathbb{C}^{s \times q}$, *then*

(39.6)
$$\min\left\{ \left\| \begin{bmatrix} A & B \\ C & D \end{bmatrix} \right\| : D \in \mathbb{C}^{s \times r} \right\} = \max\left\{ \left\| \begin{bmatrix} A \\ C \end{bmatrix} \right\|, \| [\, A \ B \,] \| \right\}.$$

Proof. Let γ be equal to the right-hand side of (39.6). Then Lemma 39.4 implies that there exists a matrix $D \in \mathbb{C}^{s \times r}$ such that

$$\left\| \begin{bmatrix} A & B \\ C & D \end{bmatrix} \right\| \leq \gamma.$$

Therefore, the left-hand side of (39.6) is less than or equal to the right-hand side of (39.6). Thus, as the left-hand side of (39.6) is $\geq \gamma$, equality must prevail. □

39.2. A description of all solutions to the minimal norm completion problem

Theorem 39.6. *A matrix* $D \in \mathbb{C}^{s \times r}$ *achieves the minimum in* (39.6) *if and only if it can be expressed in the form*

(39.7) $\quad D = -Y A^H X + (I_s - Y Y^H)^{1/2} Z (I_r - X^H X)^{1/2},$

where

(39.8) $\quad X = \left\{ (\gamma^2 I_p - A A^H)^{1/2} \right\}^\dagger B, \quad Y = C \left\{ (\gamma^2 I_q - A^H A)^{1/2} \right\}^\dagger,$

and

(39.9) $\quad Z$ *is any matrix in* $\mathbb{C}^{s \times r}$ *such that* $Z^H Z \leq \gamma^2 I_r$.

Discussion. The proof is outlined in a series of exercises with hints.

Exercise 39.1. Show that the inequality

$$\begin{bmatrix} A^H & C^H \\ B^H & D^H \end{bmatrix} \begin{bmatrix} A & B \\ C & D \end{bmatrix} \preceq \gamma^2 I_{q+r}$$

holds if and only if

(39.10) $\qquad \begin{bmatrix} C^H \\ D^H \end{bmatrix} [\, C \ \ D \,] \preceq \gamma^2 I_{q+r} - \begin{bmatrix} A^H \\ B^H \end{bmatrix} [\, A \ \ B \,].$

39.2. A description of all solutions to the completion problem

Exercise 39.2. Show that

$$\gamma^2 I_{q+r} - \begin{bmatrix} A^H \\ B^H \end{bmatrix} \begin{bmatrix} A & B \end{bmatrix} = M^H M,$$

where

$$M = \begin{bmatrix} (\gamma^2 I_q - A^H A)^{1/2} & -A^H X \\ O & \gamma(I_r - X^H X)^{1/2} \end{bmatrix}.$$

[HINT: Use Lemma 39.3 and the formulas in (39.5).]

Exercise 39.3. Show that there exists exactly one matrix $\begin{bmatrix} K_1 & K_2 \end{bmatrix}$ with components $K_1 \in \mathbb{C}^{s \times q}$ and $K_2 \in \mathbb{C}^{s \times r}$ such that

$$\begin{aligned} \begin{bmatrix} C & D \end{bmatrix} &= \begin{bmatrix} K_1 & K_2 \end{bmatrix} M \\ &= \begin{bmatrix} K_1(\gamma^2 I_q - A^H A)^{1/2} & -K_1 A^H X + K_2 \gamma (I_r - X^H X)^{1/2} \end{bmatrix} \end{aligned}$$

and

$$K_1 \mathbf{u}_1 + K_2 \mathbf{u}_2 = \mathbf{0} \quad \text{if} \quad M^H \begin{bmatrix} \mathbf{u}_1 \\ \mathbf{u}_2 \end{bmatrix} = \mathbf{0}.$$

[HINT: Use (39.10), the identity in Exercise 39.2, and Lemma 39.1.]

Exercise 39.4. Show that $K_1 = Y$. [HINT: First check that

$$M^H \begin{bmatrix} \mathbf{u}_1 \\ \mathbf{u}_2 \end{bmatrix} = \mathbf{0} \iff (\gamma^2 I_q - A^H A)^{1/2} \mathbf{u}_1 = \mathbf{0} \quad \text{and}$$

$$-X^H A \mathbf{u}_1 + \gamma (I_r - X^H X)^{1/2} \mathbf{u}_2 = \mathbf{0}$$

$$\iff (\gamma^2 I_q - A^H A)^{1/2} \mathbf{u}_1 = \mathbf{0} \text{ and } (I_r - X^H X)^{1/2} \mathbf{u}_2 = \mathbf{0},$$

because

$$X^H A = B^H \left((\gamma^2 I_q - A^H A)^{1/2}\right)^\dagger A = B^H A \left((\gamma^2 I_p - A A^H)^{1/2}\right)^\dagger$$

and $\mathcal{N}_{W^H} = \mathcal{N}_{W^\dagger}$ for any matrix $W \in \mathbb{C}^{k \times k}$.]

Exercise 39.5. Complete the proof. [HINT: Extract the formula

$$D = -K_1 A^H X + \gamma K_2 (I_q - X X^H)^{1/2}$$

from Exercise 39.3 and then, taking note of the fact that $K_1 K_1^H + K_2 K_2^H \preceq I_s$, replace K_1 by Y and γK_2 by $(I_s - Y Y^H)^{1/2} Z$.]

□

39.3. Supplementary notes

This chapter is based largely on Section 12.13 in [**30**]. The minimal norm completion problem is adapted from Feintuch [**36**] and Zhou, Glover, and Doyle [**77**], both of which cite Davis, Kahan, and Weinberger [**17**] as a basic reference for this problem. Lemma 39.1 is a special case of (part of) Douglas's lemma [**25**], which is applicable to bounded linear operators in Hilbert space.

Chapter 40

The numerical range

Let $A \in \mathbb{C}^{n \times n}$. The set
$$W(A) = \{\langle A\mathbf{x}, \mathbf{x}\rangle : \mathbf{x} \in \mathbb{C}^n \text{ and } \|\mathbf{x}\| = 1\}$$
is called the **numerical range** of A.

40.1. The numerical range is convex

The objective of this section is to show that $W(A)$ is a convex subset of $\mathbb{C}$. We begin with a special case.

Lemma 40.1. *If $B \in \mathbb{C}^{n \times n}$ and if $0 \in W(B)$ and $1 \in W(B)$, then every point $t \in [0,1]$ also belongs to $W(B)$.*

Proof. Let $\mathbf{x}, \mathbf{y} \in \mathbb{C}^n$ be vectors such that $\|\mathbf{x}\| = \|\mathbf{y}\| = 1$, $\langle B\mathbf{x}, \mathbf{x}\rangle = 1$, and $\langle B\mathbf{y}, \mathbf{y}\rangle = 0$ and let
$$\mathbf{u}_t = t\gamma\mathbf{x} + (1-t)\mathbf{y},$$
where $|\gamma| = 1$ and $0 \le t \le 1$. Then
$$\begin{aligned}\langle B\mathbf{u}_t, \mathbf{u}_t\rangle &= t^2\langle B\mathbf{x}, \mathbf{x}\rangle + t(1-t)\{\gamma\langle B\mathbf{x},\mathbf{y}\rangle + \overline{\gamma}\langle B\mathbf{y},\mathbf{x}\rangle\}\\ &\quad + (1-t)^2\langle B\mathbf{y},\mathbf{y}\rangle\\ &= t^2 + t(1-t)\{\gamma\langle B\mathbf{x},\mathbf{y}\rangle + \overline{\gamma}\langle B\mathbf{y},\mathbf{x}\rangle\}.\end{aligned}$$
The next step is to show that there exists a choice of γ such that
$$\gamma\langle B\mathbf{x},\mathbf{y}\rangle + \overline{\gamma}\langle B\mathbf{y},\mathbf{x}\rangle$$
is a real number. To this end it is convenient to write
$$B = C + iD$$

in terms of its **real** and **imaginary** parts

$$C = \frac{B + B^H}{2} \quad \text{and} \quad D = \frac{B - B^H}{2i}.$$

Then, since C and D are both Hermitian matrices,

$$\begin{aligned}
\gamma \langle B\mathbf{x}, \mathbf{y}\rangle + \overline{\gamma}\langle B\mathbf{y}, \mathbf{x}\rangle &= \gamma\langle C\mathbf{x},\mathbf{y}\rangle + i\gamma\langle D\mathbf{x},\mathbf{y}\rangle + \overline{\gamma}\langle C\mathbf{y},\mathbf{x}\rangle + i\overline{\gamma}\langle D\mathbf{y},\mathbf{x}\rangle \\
&= \gamma c + \overline{\gamma c} + i\{\gamma d + \overline{\gamma d}\},
\end{aligned}$$

where

$$c = \langle C\mathbf{x},\mathbf{y}\rangle \quad \text{and} \quad d = \langle D\mathbf{x},\mathbf{y}\rangle$$

are both independent of t. Now, in order to eliminate the imaginary component, set

$$\gamma = \begin{cases} 1 & \text{if } d = 0, \\ i|d|^{-1}\overline{d} & \text{if } d \neq 0. \end{cases}$$

Then, for this choice of γ,

$$\langle B\mathbf{u}_t, \mathbf{u}_t\rangle = t^2 + t(1-t)(\gamma c + \overline{\gamma c}).$$

Moreover, since $\langle B\mathbf{x}, \mathbf{x}\rangle = 1$ and $\langle B\mathbf{y}, \mathbf{y}\rangle = 0$, the vectors $\mathbf{x}$ and $\mathbf{y}$ are linearly independent. Thus,

$$\begin{aligned}
\|u_t\|^2 &= \|t\gamma\mathbf{x} + (1-t)\mathbf{y}\|^2 \\
&= t^2 + (1-t)t\{\langle \gamma\mathbf{x},\mathbf{y}\rangle + \langle \mathbf{y},\gamma\mathbf{x}\rangle\} + (1-t)^2 > 0
\end{aligned}$$

for every choice of t in the interval $0 \leq t \leq 1$. Therefore,

$$\mathbf{v}_t = \frac{\mathbf{u}_t}{\|\mathbf{u}_t\|}$$

is a well-defined unit vector and

$$\langle B\mathbf{v}_t, \mathbf{v}_t\rangle = \frac{t^2 + t(1-t)\{\gamma c + \overline{\gamma c}\}}{t^2 + t(1-t)\{\langle \gamma\mathbf{x},\mathbf{y}\rangle + \langle \mathbf{y},\gamma\mathbf{x}\rangle\} + (1-t)^2}$$

is a continuous real-valued function of t on the interval $0 \leq t \leq 1$ such that

$$\langle B\mathbf{v}_0,\mathbf{v}_0\rangle = 0 \quad \text{and} \quad \langle B\mathbf{v}_1,\mathbf{v}_1\rangle = 1.$$

Therefore, the equation

$$\langle B\mathbf{v}_t,\mathbf{v}_t\rangle = \mu$$

has at least one solution $t \in [0,1]$ for every choice of $\mu \in [0,1]$. $\square$

Theorem 40.2 (Toeplitz-Hausdorff). *The numerical range $W(A)$ of a matrix $A \in \mathbb{C}^{n\times n}$ is a convex subset of $\mathbb{C}$.*

Proof. The objective is to show that if $\|\mathbf{x}\| = \|\mathbf{y}\| = 1$ and if

$$\langle A\mathbf{x},\mathbf{x}\rangle = a \quad \text{and} \quad \langle A\mathbf{y},\mathbf{y}\rangle = b,$$

then for each choice of the number t in the interval $0 \le t \le 1$, there exists a vector $\mathbf{u}_t$ such that
$$\|\mathbf{u}_t\| = 1 \quad \text{and} \quad \langle A\mathbf{u}_t, \mathbf{u}_t \rangle = ta + (1-t)b \,.$$
If $a = b$, then $ta + (1-t)b = a = b$, and hence we can choose $\mathbf{u}_t = \mathbf{x}$ or $\mathbf{u}_t = \mathbf{y}$. Suppose therefore that $a \ne b$ and let
$$B = \alpha A + \beta I_n \,,$$
where α, β are solutions of the system of equations
$$\begin{aligned} a\alpha + \beta &= 1 \\ b\alpha + \beta &= 0 \,. \end{aligned}$$
Then
$$\begin{aligned} \langle B\mathbf{x}, \mathbf{x} \rangle &= \langle \alpha A\mathbf{x}, \mathbf{x} \rangle + \beta \langle \mathbf{x}, \mathbf{x} \rangle \\ &= \alpha a + \beta = 1 \end{aligned}$$
and
$$\begin{aligned} \langle B\mathbf{y}, \mathbf{y} \rangle &= \langle \alpha A\mathbf{y}, \mathbf{y} \rangle + \beta \langle \mathbf{y}, \mathbf{y} \rangle \\ &= \alpha b + \beta = 0 \,. \end{aligned}$$
Therefore, by Lemma 40.1, for each choice of t in the interval $0 \le t \le 1$, there exists a vector $\mathbf{w}_t$ such that
$$\|\mathbf{w}_t\| = 1 \quad \text{and} \quad \langle B\mathbf{w}_t, \mathbf{w}_t \rangle = t \,.$$
But this in turn is the same as to say that
$$\begin{aligned} \alpha \langle A\mathbf{w}_t, \mathbf{w}_t \rangle + \beta \langle \mathbf{w}_t, \mathbf{w}_t \rangle &= t + (1-t)0 \\ &= t(\alpha a + \beta) + (1-t)(b\alpha + \beta) \\ &= \alpha\{ta + (1-t)b\} + \beta \,. \end{aligned}$$
Thus, as $\langle \mathbf{w}_t, \mathbf{w}_t \rangle = 1$ and $\alpha \ne 0$,
$$\langle A\mathbf{w}_t, \mathbf{w}_t \rangle = ta + (1-t)b \,,$$
as claimed. □

40.2. Eigenvalues versus numerical range

Let $\lambda_1, \ldots, \lambda_k$ denote the distinct eigenvalues of a matrix $A \in \mathbb{C}^{n \times n}$. The set
$$\operatorname{conv} \sigma(A) = \left\{ \sum_{j=1}^{k} t_j \lambda_j : t_j \ge 0 \quad \text{and} \quad \sum_{j=1}^{k} t_j = 1 \right\}$$

of convex combinations of $\lambda_1, \ldots, \lambda_k$ is called the **convex hull** of $\sigma(A)$. Since the numerical range $W(A)$ is convex and $\lambda_j \in W(A)$ for $j = 1, \ldots, k$, it is clear that

(40.1) $\qquad \operatorname{conv} \sigma(A) \subseteq W(A) \quad \text{for every} \quad A \in \mathbb{C}^{n \times n}$.

In general, however, these two sets can be quite different. If

$$A = \begin{bmatrix} 0 & 0 \\ 1 & 0 \end{bmatrix},$$

for example, then

$$\sigma(A) = \{0\} \quad \text{and} \quad W(A) = \{a\overline{b} : a, b \in \mathbb{C} \text{ and } |a|^2 + |b|^2 = 1\}.$$

The situation for normal matrices is markedly different:

Theorem 40.3. *Let $A \in \mathbb{C}^{n \times n}$ be a normal matrix, i.e., $AA^H = A^H A$. Then the convex hull of $\sigma(A)$ is equal to the numerical range of A, i.e.,*

$$\operatorname{conv} \sigma(A) = W(A).$$

Proof. Since A is normal, it is unitarily equivalent to a diagonal matrix, i.e., there exists a unitary matrix $U \in \mathbb{C}^{n \times n}$ such that

$$U^H A U = \operatorname{diag}\{\lambda_1, \ldots, \lambda_n\}.$$

The columns $\mathbf{u}_1, \ldots, \mathbf{u}_n$ of U form an orthonormal basis for $\mathbb{C}^n$. Thus, if $\mathbf{x} \in \mathbb{C}^n$ and $\|\mathbf{x}\| = 1$, then

$$\mathbf{x} = \sum_{i=1}^{n} c_i \mathbf{u}_i$$

is a linear combination of $\mathbf{u}_1, \ldots, \mathbf{u}_n$,

$$\langle A\mathbf{x}, \mathbf{x} \rangle = \left\langle A \sum_{i=1}^{n} c_i \mathbf{u}_i, \sum_{j=1}^{n} c_j \mathbf{u}_j \right\rangle = \sum_{i,j=1}^{n} \lambda_i c_i \overline{c}_j \langle \mathbf{u}_i, \mathbf{u}_j \rangle$$

$$= \sum_{i=1}^{n} \lambda_i |c_i|^2,$$

and

$$\sum_{i=1}^{n} |c_i|^2 = \|\mathbf{x}\|^2 = 1.$$

Therefore, $W(A) \subseteq \operatorname{conv} \sigma(A)$ and hence, as the opposite inclusion (40.1) is already known to be in force, the proof is complete. $\square$

40.3. The Gauss-Lucas theorem

Exercise 40.1. Verify the inclusion $\operatorname{conv} \sigma(A) \subseteq W(A)$ for normal matrices $A \in \mathbb{C}^{n \times n}$ by checking directly that every convex combination $\sum_{i=1}^{n} t_i \lambda_i$ of the eigenvalues $\lambda_1, \ldots, \lambda_n$ of A belongs to $W(A)$. [HINT:
$$\sum_{i=1}^{n} t_i \lambda_i = \sum_{i=1}^{n} t_i \langle A\mathbf{u}_i, \mathbf{u}_i \rangle = \left\langle A \sum_{i=1}^{n} \sqrt{t_i} \mathbf{u}_i, \sum_{j=1}^{n} \sqrt{t_j} \mathbf{u}_j \right\rangle.]$$

Exercise 40.2. Find the numerical range of the matrix $\begin{bmatrix} 0 & 0 & i \\ 1 & 0 & 0 \\ 0 & 1 & 0 \end{bmatrix}$.

40.3. The Gauss-Lucas theorem

Theorem 40.4. *Let $f(\lambda) = a_0 + a_1 \lambda + \cdots + a_{n-1} \lambda^{n-1} + \lambda^n$ be a polynomial of degree $n \geq 1$ with coefficients $a_i \in \mathbb{C}$ for $i = 0, \ldots, n-1$. Then the roots of the derivative $f'(\lambda)$ lie in the convex hull of the roots of $f(\lambda)$.*

Proof. The proof exploits two general facts for matrices $A \in \mathbb{C}^{n \times n}$ that are expressed in terms of the notation $\varphi(\lambda) = \det(\lambda I_n - A)$ and the block decomposition $A = \begin{bmatrix} A_{11} & A_{12} \\ A_{21} & A_{22} \end{bmatrix}$ with $A_{11} \in \mathbb{C}$ and $A_{22} \in \mathbb{C}^{(n-1) \times (n-1)}$ as:

(40.2) $$\frac{\varphi'(\lambda)}{\varphi(\lambda)} = \operatorname{trace}(\lambda I_n - A)^{-1} \quad \text{for } \lambda \notin \sigma(A)$$

and

(40.3) $$\frac{\det(\lambda I_{n-1} - A_{22})}{\varphi(\lambda)} = \mathbf{e}_1^T (\lambda I_n - A)^{-1} \mathbf{e}_1 \quad \text{for } \lambda \notin \sigma(A).$$

Let $\mu_1, \ldots, \mu_n$ denote the roots of $f(\lambda)$, allowing repetitions as needed, and set $A = UDU^H$, $D = \operatorname{diag}\{\mu_1, \ldots, \mu_n\}$ with $U \in \mathbb{C}^{n \times n}$ unitary, and $U^H \mathbf{e}_1 = \dfrac{\begin{bmatrix} 1 & 1 & \cdots & 1 \end{bmatrix}^T}{\sqrt{n}}$. Then

$$\mathbf{e}_1^T (\lambda I_n - A)^{-1} \mathbf{e}_1 = \mathbf{e}_1^T U (\lambda I_n - D)^{-1} U^H \mathbf{e}_1 = \frac{1}{n} \sum_{i=1}^{n} \frac{1}{\lambda - \mu_i}$$
$$= \frac{1}{n} \operatorname{trace}(\lambda I_n - D)^{-1} = \frac{1}{n} \operatorname{trace}(\lambda I_n - A)^{-1}$$

for $\lambda \notin \sigma(A)$. Thus, in view of (40.2) and (40.3),
$$\frac{\varphi'(\lambda)}{\varphi(\lambda)} = n \frac{\det(\lambda I_{n-1} - A_{22})}{\varphi(\lambda)}.$$

Consequently,
$$f'(\lambda) = \varphi'(\lambda) = n \det(\lambda I_{n-1} - A_{22}).$$

The last formula serves to identify the eigenvalues of A_{22} with the roots of $f'(\lambda)$. Moreover, in view of (40.1),

$$\begin{aligned}
\operatorname{conv} \sigma(A_{22}) &\subseteq W(A_{22}) = \{\mathbf{x}^H A_{22} \mathbf{x} : \mathbf{x} \in \mathbb{C}^{n-1} \text{ and } \|\mathbf{x}\| = 1\} \\
&= \left\{ \begin{bmatrix} 0 & \mathbf{x} \end{bmatrix}^H A \begin{bmatrix} 0 \\ \mathbf{x} \end{bmatrix} : \mathbf{x} \in \mathbb{C}^{n-1} \text{ and } \|\mathbf{x}\| = 1 \right\} \\
&\subseteq \{\mathbf{y}^H A \mathbf{y} : \mathbf{y} \in \mathbb{C}^n \text{ and } \|\mathbf{y}\| = 1\} = W(A) = \operatorname{conv} \sigma(A);
\end{aligned}$$

the last equality follows from Theorem 40.3, since A is normal. □

40.4. The Heinz inequality

Lemma 40.5. *If $A = A^H \in \mathbb{C}^{p \times p}$ and $B = B^H \in \mathbb{C}^{q \times q}$ are both invertible and $X \in \mathbb{C}^{p \times q}$, then*

$$(40.4) \qquad 2\|X\| \leq \|AXB^{-1} + A^{-1}XB\|.$$

Proof. Suppose first that $p = q$ and $X = X^H$, and let $\lambda \in \sigma(X)$. Then $\lambda \in \sigma(A^{-1}XA)$, and hence there exists a unit vector $\mathbf{x} \in \mathbb{C}^p$ such that

$$\lambda = \langle AXA^{-1}\mathbf{x}, \mathbf{x} \rangle$$

and

$$\overline{\lambda} = \langle \mathbf{x}, AXA^{-1}\mathbf{x} \rangle = \langle A^{-1}XA\mathbf{x}, \mathbf{x} \rangle.$$

Therefore, since $\lambda = \overline{\lambda}$,

$$|2\lambda| = |\langle (AXA^{-1} + A^{-1}XA)\mathbf{x}, \mathbf{x} \rangle| \leq \|AXA^{-1} + A^{-1}XA\|,$$

which leads easily to the inequality

$$(40.5) \qquad 2\|X\| \leq \|AXA^{-1} + A^{-1}XA\|.$$

To extend this inequality to matrices $X \in \mathbb{C}^{p \times p}$ that are not necessarily Hermitian, apply it to the matrices

$$\mathcal{X} = \begin{bmatrix} O & X \\ X^H & O \end{bmatrix} \quad \text{and} \quad \mathcal{A} = \begin{bmatrix} A & O \\ O & A \end{bmatrix}$$

and note that

$$\|\mathcal{X}\| = \|X\| \quad \text{and} \quad \|\mathcal{A}\mathcal{X}\mathcal{A}^{-1} + \mathcal{A}^{-1}\mathcal{X}\mathcal{A}\| = \|AXA^{-1} + A^{-1}XA\|.$$

Finally, (40.4) follows from (40.5) applied to the square matrices

$$\begin{bmatrix} O_{p \times p} & X \\ O_{q \times p} & O_{q \times q} \end{bmatrix} \quad \text{in place of } X \quad \text{and} \quad \begin{bmatrix} A & O_{p \times q} \\ O_{q \times p} & B \end{bmatrix} \quad \text{in place of } A. \quad \square$$

The **fractional power** A^t, $0 \leq t \leq 1$, of a matrix $A \succeq O$ with singular value decomposition $A = VSV^H$ is defined as $A^t = VS^tV^H$. We shall also make use of the fact that if f is continuous on the interval $[0,1]$ and $f((a+b)/2) \leq [f(a) + f(b)]/2$, then f is convex on $[0,1]$.

40.4. The Heinz inequality

Exercise 40.3. Show that if $A \in \mathbb{C}^{n \times n}$ and $A \succeq O$, then $f(t) = \|A^t\|$ is continuous on $[0, 1]$.

Theorem 40.6 (Heinz). *If $A \in \mathbb{C}^{p \times p}$ and $B \in \mathbb{C}^{q \times q}$ are both positive semidefinite and $X \in \mathbb{C}^{p \times q}$, then*

(40.6) $$2\|AXB\| \leq \|A^2 X + X B^2\|$$

and

(40.7) $$\|A^t X B^{1-t} + A^{1-t} X B^t\| \leq \|AX + XB\| \quad \text{for} \quad 0 \leq t \leq 1.$$

Proof. To verify (40.6), let $A_\varepsilon = A + \varepsilon I_p$ and $B_\varepsilon = B + \varepsilon I_q$ with $\varepsilon > 0$. Then, since A_ε and B_ε are invertible Hermitian matrices, we can invoke (40.4) to obtain the inequality

$$2\|A_\varepsilon X B_\varepsilon\| \leq \|A_\varepsilon^2 X B_\varepsilon B_\varepsilon^{-1} + A_\varepsilon^{-1} A_\varepsilon X B_\varepsilon^2\| = \|A_\varepsilon^2 X + X B_\varepsilon^2\|,$$

which tends to (40.6) as $\varepsilon \downarrow 0$.

Next, to verify (40.7), let $f(t) = \|A^t X B^{1-t} + A^{1-t} X B^t\|$, let $0 \leq a < b \leq 1$, and set $c = (a+b)/2$ and $d = (b-a)/2$. Then, as $c = a + d$ and $1 - c = 1 - b + d$, (40.6) implies that

$$\begin{aligned} f(c) &= \|A^c X B^{1-c} + A^{1-c} X B^c\| = \left\|A^d \left(A^a X B^{1-b} + A^{1-b} X B^a\right) B^d\right\| \\ &\leq \frac{1}{2}\left\|A^{2d}\left(A^a X B^{1-b} + A^{1-b} X B^a\right) + \left(A^a X B^{1-b} + A^{1-b} X B^a\right) B^{2d}\right\| \\ &= \frac{1}{2}\|A^b X B^{1-b} + A^{1-a} X B^a + A^a X B^{1-a} + A^{1-b} X B^b\| \\ &\leq \frac{f(a) + f(b)}{2}; \end{aligned}$$

i.e., $f(t)$ is a convex function on the interval $0 \leq t \leq 1$. Thus, as $f(0) = f(1) = \|AX + XB\|$,

$$f(t) = f((1-t)0 + t1) \leq (1-t)f(0) + tf(1) = (1-t)f(0) + tf(0) = f(0)$$

for every point t in the interval $0 \leq t \leq 1$, which is equivalent to (40.7). $\square$

Theorem 40.7. *Let $A \in \mathbb{C}^{p \times p}$, $B \in \mathbb{C}^{p \times p}$ and suppose that $A \succ O$ and $B \succ O$. Then*

(40.8) $$\|A^s B^s\| \leq \|AB\|^s \quad \text{for} \quad 0 \leq s \leq 1.$$

Proof. Let
$$Q = \{u \in [0,1] : \|A^u B^u\| \le \|AB\|^u\}$$
and let s and t be a pair of points in Q. Then, with the help of the auxiliary inequality
$$\begin{aligned}
\|A^{(s+t)/2} B^{(s+t)/2}\|^2 &= \|B^{(s+t)/2} A^{s+t} B^{(s+t)/2}\| \\
&= r_\sigma(B^{(s+t)/2} A^{s+t} B^{(s+t)/2}) \\
&= r_\sigma(B^s A^{s+t} B^t) \le \|B^s A^{s+t} B^t\| \le \|B^s A^s\| \, \|A^t B^t\| \\
&= \|A^s B^s\| \, \|A^t B^t\| \le \|AB\|^{s+t},
\end{aligned}$$
it is readily checked that $(s+t)/2 \in Q$ and hence that Q is convex. The proof is easily completed, since $0 \in Q$ and $1 \in Q$. $\square$

Exercise 40.4. Verify each of the assertions that lead to the auxiliary inequality in the proof of Theorem 40.7.

Exercise 40.5. Show that if A and B are as in Theorem 40.7, then $\varphi(s) = \|A^s B^s\|^{1/s}$ is an increasing function of s for $s > 0$.

40.5. Supplementary notes

This chapter is adapted from Chapter 22 of [30]. The presented proof of the convexity of numerical range is based on an argument that is sketched briefly in [46]. Halmos credits it to C. W. R. de Boor. The presented proof works also for bounded operators in Hilbert space; see also McIntosh [61] for another very attractive approach. The proof of the Heinz inequality is taken from the beautiful short paper [37] by Fujii, Fujii, Furuta, and Nakomoto that establishes the Heinz inequality (40.7) for bounded operators in Hilbert space and sketches the history. The elegant passage from (40.6) to (40.7) is credited to an unpublished paper of A. McIntosh. The proof of Theorem 40.7 is adapted from a paper by Furuta [38].

The proof of the Gauss-Lucas theorem in Section 40.3 is adapted from an exercise in the monograph [8] by Bakonyi and Woerdeman. A complete description of the numerical range $W(A)$ of a matrix $A \in \mathbb{C}^{n \times n}$ is presented in Helton and Spitkovsky [48].

Chapter 41

Riccati equations

In this chapter we shall investigate the existence and uniqueness of solutions $X \in \mathbb{C}^{n \times n}$ to the **Riccati equation**

(41.1) $\quad A^H X + XA + XRX + Q = O \quad$ when $\quad R = R^H, \quad Q = Q^H,$

and $A, R, Q \in \mathbb{C}^{n \times n}$. This class of equations has important applications, one of which (the LQR problem) will be discussed in Section 41.3.

Exercise 41.1. Show that if the Riccati equation (41.1) has exactly one solution $X \in \mathbb{C}^{n \times n}$, then $X = X^H$.

41.1. Riccati equations

The study of the Riccati equation (41.1) is intimately connected with the invariant subspaces of the matrix

(41.2) $\quad G = \begin{bmatrix} A & R \\ -Q & -A^H \end{bmatrix} \quad$ with $\quad R = R^H \quad$ and $\quad Q = Q^H,$

which is often referred to as the **Hamiltonian matrix** in the control theory literature. The first order of business is to verify that the eigenvalues of G are symmetrically distributed with respect to the imaginary axis $i\mathbb{R}$:

Lemma 41.1. *The roots of the polynomial $p(\lambda) = \det(\lambda I_{2n} - G)$ are symmetrically distributed with respect to $i\mathbb{R}$.*

Proof. This is a simple consequence of the identity

(41.3) $\qquad\qquad\qquad SGS^{-1} = -G^H,$

in terms of the orthogonal matrix

(41.4) $$S = \begin{bmatrix} O & -I_n \\ I_n & O \end{bmatrix}.$$ □

Exercise 41.2. Verify the identity $SGS^{-1} = -G^H$ and the assertion of Lemma 41.1.

Exercise 41.3. Show that if $E = \begin{bmatrix} A & B \\ C & A \end{bmatrix}$ is a $2p \times 2p$ matrix with $p \times p$ blocks $A = A^H$, $B = B^H$, and $C = C^H$, then $\lambda \in \sigma(E) \iff \overline{\lambda} \in \sigma(E)$. [HINT: It suffices to show that E is similar to E^H.]

If $\sigma(G) \cap i\mathbb{R} = \emptyset$, then Lemma 41.1 guarantees that G admits a Jordan decomposition of the form

(41.5) $$G = U \begin{bmatrix} J_1 & O \\ O & J_2 \end{bmatrix} U^{-1},$$

where $J_1, J_2 \in \mathbb{C}^{n \times n}$, $\sigma(J_1) \subset \mathbb{C}_L$, the open left half-plane, and $\sigma(J_2) \subset \mathbb{C}_R$, the open right half-plane.

It turns out that the upper left-hand $n \times n$ corner X_1 of the matrix U will play a central role in the subsequent analysis; i.e., upon writing

$$U \begin{bmatrix} I_n \\ O \end{bmatrix} = \begin{bmatrix} X_1 \\ X_2 \end{bmatrix} \quad \text{and} \quad \Lambda = J_1$$

so that

(41.6) $$G \begin{bmatrix} X_1 \\ X_2 \end{bmatrix} = \begin{bmatrix} X_1 \\ X_2 \end{bmatrix} \Lambda \quad \text{and} \quad \sigma(\Lambda) \subset \mathbb{C}_L,$$

the case in which X_1 is invertible will be particularly significant.

Lemma 41.2. If $X \in \mathbb{C}^{n \times n}$ is a solution of the Riccati equation (41.1) such that $\sigma(A + RX) \subset \mathbb{C}_L$, then $X = X^H$.

Proof. If X is a solution of (41.1), then

(41.7) $$A^H X^H + X^H A + X^H R X^H + Q = O.$$

Therefore,

$$A^H (X - X^H) + (X - X^H) A + XRX - X^H R X^H = O$$

and hence

$$(A^H + X^H R)(X - X^H) + (X - X^H)(A + RX) = O.$$

Consequently, upon setting $\Lambda = A + RX$, we see that the matrix $Z = X - X^H$ is a solution of the equation

(41.8) $$Z\Lambda + \Lambda^H Z = O.$$

However, since $\sigma(\Lambda) \subset \mathbb{C}_L \implies \sigma(\Lambda^H) \subset \mathbb{C}_L$, Theorem 37.6 ensures that $Z = O$ is the only solution of (41.8). Therefore, $X = X^H$. □

Exercise 41.4. Show that if $\sigma(G) \cap i\mathbb{R} = \emptyset$, then (41.6) is in force for some matrix $\Lambda \in \mathbb{C}^{n \times n}$ (that is not necessarily in Jordan form) and

(41.9) $$X_1^H X_2 = X_2^H X_1 .$$

[HINT: Use (41.3) to obtain (41.8), but with $Z = X_2^H X_1 - X_1^H X_2$.]

Theorem 41.3. *If $\sigma(G) \cap i\mathbb{R} = \emptyset$ and the matrix X_1 in (41.6) is invertible, then the matrix $X = X_2 X_1^{-1}$ enjoys the following properties:*

(1) X *is a solution of the Riccati equation* (41.1).
(2) $\sigma(A + RX) \subset \mathbb{C}_L$.
(3) $X = X^H$.

Proof. If X_1 is invertible and $X = X_2 X_1^{-1}$, then formula (41.6) implies that

$$G \begin{bmatrix} I_n \\ X \end{bmatrix} = \begin{bmatrix} I_n \\ X \end{bmatrix} X_1 \Lambda X_1^{-1}$$

and hence, upon filling in the block entries in G and writing this out in detail, that

$$A + RX = X_1 \Lambda X_1^{-1},$$
$$-Q - A^H X = X(X_1 \Lambda X_1^{-1}).$$

Therefore,

$$-Q - A^H X = X(A + RX),$$

which serves to verify (1).

Assertion (2) follows from the formula $A + RX = X_1 \Lambda X_1^{-1}$ and the fact that $\sigma(\Lambda) \subset \mathbb{C}_L$; (3) now follows from (1), (2), and Lemma 41.2. □

Exercise 41.5. Show that if X is a solution of the Riccati equation (41.1) such that $\sigma(A + RX) \subset \mathbb{C}_L$, then $\sigma(G) \cap i\mathbb{R} = \emptyset$, the matrix X_1 in (41.6), is invertible. [REMARK: This is a converse to Theorem 41.3.]

Theorem 41.4. *The Riccati equation* (41.1) *has at most one solution* $X \in \mathbb{C}^{n \times n}$ *such that* $\sigma(A + RX) \subset \mathbb{C}_L$.

Proof. Let X and Y be a pair of solutions of the Riccati equation (41.1) such that $\sigma(A + RX) \subset \mathbb{C}_L$ and $\sigma(A + RY) \subset \mathbb{C}_L$. Then, since

$$A^H X + XA + XRX + Q = O$$

and

$$A^H Y + YA + YRY + Q = O,$$

it is clear that
$$A^H(X-Y) + (X-Y)A + XRX - YRY = O .$$
However, as $Y = Y^H$ by Lemma 41.2 this last equation can also be reexpressed as
$$(A+RY)^H(X-Y) + (X-Y)(A+RX) = O ,$$
which exhibits $X - Y$ as the solution of an equation of the form
$$BZ + ZC = O$$
with $\sigma(B) \subset \mathbb{C}_L$ and $\sigma(C) \subset \mathbb{C}_L$. Theorem 37.6 ensures that this equation has at most one solution. Thus, as $Z = O_{n \times n}$ is a solution, it is in fact the only solution. Therefore $X = Y$, as claimed. □

The next theorem provides conditions under which the constraints imposed on the Hamiltonian matrix G in Theorem 41.3 are satisfied when $R = -BB^H$ and $Q = C^H C$.

Theorem 41.5. *Let $A \in \mathbb{C}^{n \times n}$, $B \in \mathbb{C}^{n \times k}$, $C \in \mathbb{C}^{r \times n}$ and suppose that*

(41.10) $\qquad \operatorname{rank} \begin{bmatrix} A - \lambda I_n \\ C \end{bmatrix} = n \quad \textit{for every point } \lambda \in i\mathbb{R}$

and

(41.11) $\qquad \operatorname{rank} \begin{bmatrix} A - \lambda I_n & B \end{bmatrix} = n \quad \textit{for every point } \lambda \in \overline{\mathbb{C}_R} .$

Then there exists exactly one solution $X \in \mathbb{C}^{n \times n}$ of the Riccati equation

(41.12) $\qquad A^H X + XA - XBB^H X + C^H C = O$

such that $\sigma(A - BB^H X) \subset \mathbb{C}_L$. Moreover, this solution X is positive semidefinite, and if A, B, and C are real matrices, then $X \in \mathbb{R}^{n \times n}$.

Proof. If $R = -BB^H$ and $Q = C^H C$, then
$$G = \begin{bmatrix} A & -BB^H \\ -C^H C & -A^H \end{bmatrix} .$$
The proof is divided into steps.

1. *If (41.10) and (41.11) are in force, then $\sigma(G) \cap i\mathbb{R} = \emptyset$.*

If
$$\begin{bmatrix} A & -BB^H \\ -C^H C & -A^H \end{bmatrix} \begin{bmatrix} \mathbf{x} \\ \mathbf{y} \end{bmatrix} = \lambda \begin{bmatrix} \mathbf{x} \\ \mathbf{y} \end{bmatrix}$$
for some choice of $\mathbf{x} \in \mathbb{C}^n$, $\mathbf{y} \in \mathbb{C}^n$, and $\lambda \in \mathbb{C}$, then
$$(A - \lambda I_n)\mathbf{x} = BB^H \mathbf{y}$$

41.1. Riccati equations

and
$$(A^H + \lambda I_n)\mathbf{y} = -C^H C \mathbf{x}.$$
Therefore,
$$\langle (A - \lambda I_n)\mathbf{x}, \mathbf{y}\rangle = \langle BB^H \mathbf{y}, \mathbf{y}\rangle = \|B^H \mathbf{y}\|_2^2$$
and
$$\langle (A + \overline{\lambda} I_n)\mathbf{x}, \mathbf{y}\rangle = \langle \mathbf{x}, (A^H + \lambda I_n)\mathbf{y}\rangle = -\langle \mathbf{x}, C^H C \mathbf{x}\rangle = -\|C\mathbf{x}\|_2^2.$$
Thus,
$$\begin{aligned}-(\lambda + \overline{\lambda})\langle \mathbf{x}, \mathbf{y}\rangle &= \langle (A - \lambda I_n)\mathbf{x}, \mathbf{y}\rangle - \langle (A + \overline{\lambda} I_n)\mathbf{x}, \mathbf{y}\rangle \\ &= \|B^H \mathbf{y}\|_2^2 + \|C\mathbf{x}\|_2^2\end{aligned}$$
and hence
$$\lambda + \overline{\lambda} = 0 \implies B^H \mathbf{y} = \mathbf{0} \quad \text{and} \quad C\mathbf{x} = \mathbf{0},$$
which in turn implies that
$$\begin{bmatrix} A - \lambda I_n \\ C \end{bmatrix}\mathbf{x} = \mathbf{0} \quad \text{and} \quad \mathbf{y}^H [A + \overline{\lambda} I_n \quad B] = \mathbf{0}$$
when $\lambda \in i\mathbb{R}$. However, in view of (41.10) and (41.11), this is viable only if $\mathbf{x} = \mathbf{0}$ and $\mathbf{y} = \mathbf{0}$. Consequently, $\sigma(G) \cap i\mathbb{R} = \emptyset$.

2. *If (41.10) and (41.11) are in force, then*
$$\begin{bmatrix} A & -BB^H \\ -C^H C & -A^H \end{bmatrix}\begin{bmatrix} X_1 \\ X_2 \end{bmatrix} = \begin{bmatrix} X_1 \\ X_2 \end{bmatrix}\Lambda \quad \text{and} \quad \mathrm{rank}\begin{bmatrix} X_1 \\ X_2 \end{bmatrix} = n,$$
where $X_1, X_2, \Lambda \in \mathbb{C}^{n \times n}$, $\sigma(\Lambda) \subset \mathbb{C}_L$, and X_1 is invertible.

The displayed formula and the inclusion $\sigma(\Lambda) \subset \mathbb{C}_L$ follow from step 1. The main task is to show that X_1 is invertible. Towards this end, suppose that $\mathbf{u} \in \mathcal{N}_{X_1}$. Then
$$-BB^H X_2 \mathbf{u} = X_1 \Lambda \mathbf{u},$$
and hence, as $X_2^H X_1 = X_1^H X_2$ by Exercise 41.4,
$$\begin{aligned}-\|B^H X_2 \mathbf{u}\|^2 &= \langle X_1 \Lambda \mathbf{u}, X_2 \mathbf{u}\rangle = \langle \Lambda \mathbf{u}, X_1^H X_2 \mathbf{u}\rangle \\ &= \langle \Lambda \mathbf{u}, X_2^H X_1 \mathbf{u}\rangle = \langle \Lambda \mathbf{u}, \mathbf{0}\rangle = 0.\end{aligned}$$
Thus, $B^H X_2 \mathbf{u} = \mathbf{0}$ and
$$X_1 \Lambda \mathbf{u} = -BB^H X_2 \mathbf{u} = \mathbf{0},$$
which means that $\mathcal{N}_{X_1}$ is invariant under Λ and hence that either $\mathcal{N}_{X_1} = \{\mathbf{0}\}$ or that $\Lambda \mathbf{v} = \lambda \mathbf{v}$ for some point $\lambda \in \mathbb{C}_L$ and some nonzero vector $\mathbf{v} \in \mathcal{N}_{X_1}$. In the latter case,
$$-BB^H X_2 \mathbf{v} = X_1 \Lambda \mathbf{v} = \lambda X_1 \mathbf{v} = \mathbf{0}$$

and
$$-A^H X_2 \mathbf{v} = X_2 \Lambda \mathbf{v} = \lambda X_2 \mathbf{v},$$
i.e.,
$$\mathbf{v}^H X_2^H [A + \overline{\lambda} I_n \quad B] = \mathbf{0}^H$$
for some point $\lambda \in \mathbb{C}_L$. Therefore, since $-\overline{\lambda} \in \mathbb{C}_R$, assumption (41.11) implies that $X_2 \mathbf{v} = \mathbf{0}$. Consequently,

$$\begin{bmatrix} X_1 \\ X_2 \end{bmatrix} \mathbf{v} = \mathbf{0} \implies \mathbf{v} = \mathbf{0} \implies \mathcal{N}_{X_1} = \{\mathbf{0}\} \implies X_1 \text{ is invertible}.$$

3. *Completing the proof*

Theorem 41.3 ensures that there exists at least one solution X of the Riccati equation (41.12) such that $\sigma(A - BB^H X) \subset \mathbb{C}_L$, whereas Theorem 41.4 ensures that there is at most one such solution. Therefore, there is exactly one such solution.

To verify that this solution X is positive semidefinite, it is convenient to express the Riccati equation
$$A^H X + XA - XBB^H X + C^H C = O$$
as
$$(A - BB^H X)^H X + X(A - BB^H X) = -C^H C - XBB^H X,$$
which is of the form
$$A_1^H X + XA_1 = Q,$$
where
$$\sigma(A_1) \subset \mathbb{C}_L \text{ and } -Q \succeq O.$$
The desired result then follows by invoking Lemma 37.7.

Finally, if the matrices A, B, and C are real, then the matrix $\overline{X}$ is also a Hermitian solution of the Riccati equation (41.12) such that $\sigma(A - BB^H \overline{X}) \subset \mathbb{C}_L$. Therefore, $X = \overline{X}$, i.e., $X \in \mathbb{R}^{n \times n}$. □

Exercise 41.6. Let $A \in \mathbb{C}^{n \times n}$, $B \in \mathbb{C}^{n \times k}$. Show that if $\sigma(A) \cap i\mathbb{R} = \emptyset$ and (41.11) holds, then there exists exactly one Hermitian solution X of the Riccati equation $A^H X + XA - XBB^H X = O$ such that $\sigma(A - BB^H X) \subset \mathbb{C}_L$.

For future applications, it will be convenient to have another variant of Theorem 41.5.

41.2. Two lemmas

Theorem 41.6. Let $A \in \mathbb{C}^{n \times n}$, $B \in \mathbb{C}^{n \times k}$, $Q \in \mathbb{C}^{n \times n}$, $R \in \mathbb{C}^{k \times k}$; and suppose that $Q \succeq O$, $R \succ O$,

(41.13) $\qquad \operatorname{rank} \begin{bmatrix} A - \lambda I_n \\ Q \end{bmatrix} = n \quad \text{for every point } \lambda \in i\mathbb{R},$

and (41.11) holds. Then there exists exactly one solution $X \in \mathbb{C}^{n \times n}$ of the Riccati equation

$$A^H X + XA - XBR^{-1}B^H X + Q = O$$

such that $\sigma(A - BR^{-1}B^H X) \subset \mathbb{C}_{\mathrm{L}}$. Moreover, this solution X is positive semidefinite, and if A, B, and Q are real matrices, then $X \in \mathbb{R}^{n \times n}$.

Proof. Since $Q \succeq O$, there exists a matrix $C \in \mathbb{C}^{r \times n}$ such that $C^H C = Q$ and $\operatorname{rank} C = \operatorname{rank} Q = r$. Thus, upon setting $B_1 = BR^{-1/2}$, we see that the matrix

$$\begin{bmatrix} A & -BR^{-1}B^H \\ -Q & -A^H \end{bmatrix} = \begin{bmatrix} A & -B_1 B_1^H \\ -C^H C & -A^H \end{bmatrix}$$

is of the form considered in Theorem 41.5. Moreover, since

$$\begin{bmatrix} A - \lambda I_n \\ C \end{bmatrix} \mathbf{u} = \mathbf{0} \iff \begin{bmatrix} A - \lambda I_n \\ Q \end{bmatrix} \mathbf{u} = \mathbf{0},$$

condition (41.13) implies that

$$\operatorname{rank} \begin{bmatrix} A - \lambda I_n \\ C \end{bmatrix} = n \quad \text{for every point } \lambda \in i\mathbb{R}.$$

Furthermore, as

$$\operatorname{rank} [A - \lambda I_n \quad B] = \operatorname{rank} [A - \lambda I_n \quad B(R^{1/2})^{-1}],$$

assumption (41.11) guarantees that

$$\operatorname{rank} [A - \lambda I_n \quad B_1] = n \text{ for every point } \lambda \in \overline{\mathbb{C}_{\mathrm{R}}}.$$

The asserted conclusion now follows from Theorems 41.5 and 13.7. $\qquad\square$

41.2. Two lemmas

The two lemmas in this section are prepared for use in the next section.

Lemma 41.7. Let $A, Q \in \mathbb{C}^{n \times n}$, $B, L \in \mathbb{C}^{n \times k}$, $R \in \mathbb{C}^{k \times k}$,

$$E = \begin{bmatrix} Q & L \\ L^H & R \end{bmatrix}$$

and suppose that $E \succeq O$, $R \succ O$ and that

(41.14) $\qquad \operatorname{rank} E = \operatorname{rank} Q + \operatorname{rank} R.$

Then the formulas

(41.15) $$\operatorname{rank} \begin{bmatrix} \widetilde{A} - \lambda I_n \\ \widetilde{Q} \end{bmatrix} = \operatorname{rank} \begin{bmatrix} A - \lambda I_n \\ Q \end{bmatrix}$$

and

(41.16) $$\operatorname{rank}[\widetilde{A} - \lambda I_n \quad B] = \operatorname{rank}[A - \lambda I_n \quad B]$$

are valid for the matrices

(41.17) $$\widetilde{A} = A - BR^{-1}L^H \quad \text{and} \quad \widetilde{Q} = Q - LR^{-1}L^H$$

and every point $\lambda \in \mathbb{C}$.

Proof. The Schur complement formula

$$\begin{bmatrix} Q & L \\ L^H & R \end{bmatrix} = \begin{bmatrix} I_n & LR^{-1} \\ O & I_k \end{bmatrix} \begin{bmatrix} \widetilde{Q} & O \\ O & R \end{bmatrix} \begin{bmatrix} I_n & O \\ R^{-1}L^H & I_n \end{bmatrix}$$

implies that
$$\operatorname{rank} E = \operatorname{rank} \widetilde{Q} + \operatorname{rank} R \quad \text{and} \quad \widetilde{Q} \succeq O.$$

Thus, in view of assumption (41.14),
$$\operatorname{rank} Q = \operatorname{rank} \widetilde{Q}$$

and, since $Q = \widetilde{Q} + LR^{-1}L^H$ is the sum of two positive semidefinite matrices,
$$\mathcal{N}_Q = \mathcal{N}_{\widetilde{Q}} \cap \mathcal{N}_{L^H} \subseteq \mathcal{N}_{\widetilde{Q}}.$$

However, since
$$\operatorname{rank} Q = \operatorname{rank} \widetilde{Q} \Longrightarrow \dim \mathcal{N}_Q = \dim \mathcal{N}_{\widetilde{Q}},$$

the last inclusion is in fact an equality:
$$\mathcal{N}_Q = \mathcal{N}_{\widetilde{Q}} \quad \text{and} \quad \mathcal{N}_{\widetilde{Q}} \subseteq \mathcal{N}_{L^H}$$

and hence,
$$\begin{bmatrix} \widetilde{A} - \lambda I_n \\ \widetilde{Q} \end{bmatrix} \mathbf{u} = \mathbf{0} \iff \begin{bmatrix} A - \lambda I_n \\ Q \end{bmatrix} \mathbf{u} = \mathbf{0}.$$

The conclusion (41.15) now follows easily from the principle of conservation of dimension.

The second conclusion (41.16) is immediate from the identity

$$[\widetilde{A} - \lambda I_n \quad B] = [A - \lambda I_n \quad B] \begin{bmatrix} I_n & O \\ -R^{-1}L^H & I_k \end{bmatrix},$$

since the last matrix on the right is invertible. $\square$

Lemma 41.8. *Assume that the matrices A, $\widetilde{A}$, Q, $\widetilde{Q}$, B, L, R, and E are as in Lemma 41.7 and that (41.14), (41.13), and (41.11) are in force. Then there exists exactly one solution $X \in \mathbb{C}^{n \times n}$ of the Riccati equation*

(41.18) $$\widetilde{A}^H X + X \widetilde{A} - XBR^{-1}B^H X + \widetilde{Q} = O$$

such that $\sigma(\widetilde{A} - BR^{-1}B^H X) \subset \mathbb{C}_L$. Moreover, this solution X is positive semidefinite, and if the matrices A, B, Q, L, and R are real, then $X \in \mathbb{R}^{n \times n}$.

Proof. Under the given assumptions, Lemma 41.7 guarantees that

$$\mathrm{rank}\begin{bmatrix} \widetilde{A} - \lambda I_n \\ \widetilde{Q} \end{bmatrix} = n \text{ for every point } \lambda \in i\mathbb{R}$$

and

$$\mathrm{rank}\,[\widetilde{A} - \lambda I_n \quad B] = n \text{ for every point } \lambda \in \overline{\mathbb{C}_R}.$$

Therefore, Theorem 41.6 is applicable with $\widetilde{A}$ in place of A and $\widetilde{Q}$ in place of Q. □

41.3. The LQR problem

Let $A \in \mathbb{R}^{n \times n}$ and $B \in \mathbb{R}^{n \times k}$ and let

$$\mathbf{x}(t) = e^{tA}\mathbf{x}(0) + \int_0^t e^{(t-s)A}B\mathbf{u}(s)ds, \quad 0 \leq t < \infty,$$

be the solution of the first-order vector system of equations

$$\mathbf{x}'(t) = A\mathbf{x}(t) + B\mathbf{u}(t), \; t \geq 0,$$

in which the vector $\mathbf{x}(0) \in \mathbb{R}^n$ and the vector-valued function $\mathbf{u}(t) \in \mathbb{R}^k$, $t \geq 0$, are specified. The **LQR (linear quadratic regulator) problem** in control engineering is to choose $\mathbf{u}$ to minimize the value of the integral

(41.19) $$Z(t) = \int_0^t [\mathbf{x}(s)^T \; \mathbf{u}(s)^T] \begin{bmatrix} Q & L \\ L^T & R \end{bmatrix} \begin{bmatrix} \mathbf{x}(s) \\ \mathbf{u}(s) \end{bmatrix} ds$$

when $Q = Q^T \in \mathbb{R}^{n \times n}$, $L \in \mathbb{R}^{n \times k}$, $R = R^T \in \mathbb{R}^{k \times k}$,

$$\begin{bmatrix} Q & L \\ L^T & R \end{bmatrix} \succeq O,$$

and R is assumed to be invertible.

The first step in the analysis of this problem is to express it in simpler form by invoking the Schur complement formula:

$$\begin{bmatrix} Q & L \\ L^T & R \end{bmatrix} = \begin{bmatrix} I_n & LR^{-1} \\ O & I_k \end{bmatrix} \begin{bmatrix} Q - LR^{-1}L^T & O \\ O & R \end{bmatrix} \begin{bmatrix} I_n & O \\ R^{-1}L^T & I_k \end{bmatrix}.$$

Then, upon setting
$$\widetilde{A} = A - BR^{-1}L^T, \quad \widetilde{Q} = Q - LR^{-1}L^T,$$
and
$$\mathbf{v}(s) = R^{-1}L^T\mathbf{x}(s) + \mathbf{u}(s),$$
the integral (41.19) can be reexpressed more conveniently as

$$(41.20) \qquad Z(t) = \int_0^t [\mathbf{x}(s)^T \ \mathbf{v}(s)^T] \begin{bmatrix} \widetilde{Q} & O \\ O & R \end{bmatrix} \begin{bmatrix} \mathbf{x}(s) \\ \mathbf{v}(s) \end{bmatrix} ds,$$

where the vectors $\mathbf{x}(s)$ and $\mathbf{v}(s)$ are linked by the equation

$$(41.21) \qquad \mathbf{x}'(s) = \widetilde{A}\mathbf{x}(s) + B\mathbf{v}(s),$$

i.e.,

$$(41.22) \qquad \mathbf{x}(t) = e^{t\widetilde{A}}\mathbf{x}(0) + \int_0^t e^{(t-s)\widetilde{A}} B\mathbf{v}(s)ds.$$

Theorem 41.9. *Let X be the unique solution of the Riccati equation (41.18) based on matrices A, B, Q, L, and R with real entries such that $\sigma(\widetilde{A} - BR^{-1}B^TX) \subset \mathbb{C}_L$. Then:*

(1) $X \in \mathbb{R}^{n \times n}$ and $Z(t)$ can be expressed in terms of the solution $\mathbf{x}(t)$ of (41.21) and the function $\varphi(s) = \mathbf{x}(s)^T X \mathbf{x}(s)$ as

$$(41.23) \qquad Z(t) = \varphi(0) - \varphi(t) + \int_0^t \|R^{-1/2}(B^T X \mathbf{x}(s) + R\mathbf{v}(s))\|_2^2 \, ds.$$

(2) $Z(t) \geq \varphi(0) - \varphi(t)$ with equality if $\mathbf{v}(s) = -R^{-1}B^T X \mathbf{x}(s)$ for $0 \leq s \leq t$.

(3) *If* $\mathbf{v}(s) = -R^{-1}B^T X \mathbf{x}(s)$ *for* $0 \leq s < \infty$, *then* $Z(\infty) = \varphi(0)$.

Proof. The proof is broken into parts.

1. *Verification of* (1).

In view of (41.21)
$$\begin{aligned}
\varphi'(s) &= \mathbf{x}'(s)^T X \mathbf{x}(s) + \mathbf{x}(s)^T X \mathbf{x}'(s) \\
&= (\widetilde{A}\mathbf{x}(s) + B\mathbf{v}(s))^T X \mathbf{x}(s) + \mathbf{x}(s)^T X(\widetilde{A}\mathbf{x}(s) + B\mathbf{v}(s)) \\
&= \mathbf{x}(s)^T (\widetilde{A}^T X + X\widetilde{A})\mathbf{x}(s) + \mathbf{v}(s)^T B^T X \mathbf{x}(s) + \mathbf{x}(s)^T X B \mathbf{v}(s) \\
&= \mathbf{x}(s)^T (XBR^{-1}B^T X - \widetilde{Q})\mathbf{x}(s) + \mathbf{v}(s)^T B^T X \mathbf{x}(s) + \mathbf{x}(s)^T X B \mathbf{v}(s) \\
&= (\mathbf{x}(s)^T XB + \mathbf{v}(s)^T R)R^{-1}(B^T X \mathbf{x}(s) + R\mathbf{v}(s)) \\
&\quad -\mathbf{x}(s)^T \widetilde{Q}\mathbf{x}(s) - \mathbf{v}(s)^T R\mathbf{v}(s).
\end{aligned}$$

Therefore,
$$\begin{aligned} Z(t) &= \int_0^t \{\mathbf{x}(s)^T \widetilde{Q}\mathbf{x}(s) + \mathbf{v}(s)^T R\mathbf{v}(s)\}ds \\ &= -\int_0^t \frac{d}{ds}\{\mathbf{x}(s)^T X\mathbf{x}(s)\}ds \\ &\quad + \int_0^t (B^T X\mathbf{x}(s) + R\mathbf{v}(s))^T R^{-1}(B^T X\mathbf{x}(s) + R\mathbf{v}(s))ds, \end{aligned}$$
which is equivalent to (41.23).

2. *Verification of* (2) *and* (3).

Assertion (2) is immediate from formula (41.23). Moreover, if $\mathbf{v}(s)$ is chosen as specified in assertion (3), then $\mathbf{x}(t)$ is a solution of the vector differential equation
$$\mathbf{x}'(t) = (\widetilde{A} - BR^{-1}B^T X)\mathbf{x}(t),$$
and hence, as $\sigma(\widetilde{A} - BR^{-1}B^T X) \subset \mathbb{C}_\mathrm{L}$, $\mathbf{x}(t) \to 0$ as $t \uparrow \infty$. □

41.4. Supplementary notes

This chapter is adapted from Chapter 18 in [**30**]. The discussion of Riccati equations and the LQR problem therein was partially adapted from the monograph Zhou, Doyle, and Glover [**77**], which is an excellent source of supplementary information on both of these topics. The monograph [**54**] by Lancaster and Rodman is recommended for more advanced studies.

Chapter 42

Supplementary topics

This chapter is devoted to four distinct topics: Gaussian quadrature, Bezoutians and resultants, general QR factorization, and the QR algorithm. The first, third, and fourth are presented in reasonable detail. The treatment of the second is limited to a brief introduction to the properties and significance of Bezoutians and resultants and is less complete. Companion matrices play a significant role in Gaussian quadrature and the theory of Bezoutians.

42.1. Gaussian quadrature

Let $w(x)$ denote a positive continuous function on a finite interval $a \leq x \leq b$ and let $\mathcal{U}$ denote the space of continuous complex-valued functions on this interval, equipped with the inner product

$$\langle f, g \rangle_{\mathcal{U}} = \int_a^b \overline{g(x)} w(x) f(x) dx \,.$$

Let $f_j(x) = x^{j-1}$ for $j = 1, 2, \ldots$ and let $\mathcal{P}_n = \mathrm{span}\{f_1, \ldots, f_n\}$ denote the n-dimensional subspace of polynomials of degree less than or equal to $n-1$ (with complex coefficients), and let

$$\mathfrak{P}_n = \Pi_n M|_{\mathcal{P}_n} \quad \text{for} \quad n = 0, 1 \ldots,$$

where Π_n denotes the orthogonal projection from $\mathcal{U}$ onto $\mathcal{P}_n$ and M denotes the linear transformation on $\mathcal{U}$ of multiplication by the independent variable, i.e., $(Mf)(x) = xf(x)$. Then,

$$\mathfrak{P}_n f_j = \begin{cases} f_{j+1} & \text{if } j = 1, \ldots, n-1, \\ c_1 f_1 + \cdots + c_n f_n & \text{if } j = n, \end{cases}$$

449

where the coefficients $c_1,\ldots,c_n$ are chosen so that

(42.1) $$\left\langle f_{n+1} - \sum_{j=1}^{n} c_j f_j, f_i \right\rangle_{\mathcal{U}} = 0 \quad \text{for } i = 1,\ldots,n.$$

Since the $n \times n$ matrix G with entries $g_{ij} = \langle f_j, f_i \rangle_{\mathcal{U}}$ for $i,j = 1,\ldots,n$ is invertible, the condition in (42.1) can be expressed as

(42.2) $$\mathbf{c} = \begin{bmatrix} c_1 \\ \vdots \\ c_n \end{bmatrix} = G^{-1} \begin{bmatrix} g_{1,n+1} \\ \vdots \\ g_{n,n+1} \end{bmatrix} \quad \text{with } g_{i,n+1} = \langle f_{n+1}, f_i \rangle_{\mathcal{U}}.$$

Now let $\mathcal{G} = \mathbb{C}^n$ equipped with the inner product $\langle \mathbf{x}, \mathbf{y} \rangle_{\mathcal{G}} = \langle G\mathbf{x}, \mathbf{y} \rangle_{\mathrm{st}}$. Then the linear operator T_n that maps $f_j \in \mathcal{P}_n$ to $\mathbf{e}_j \in \mathcal{G}$ for $j = 1,\ldots,n$ is unitary: T is invertible and

$$\langle T f_j, T f_i \rangle_{\mathcal{G}} = \langle \mathbf{e}_j, \mathbf{e}_i \rangle_{\mathcal{G}} = g_{ij} = \langle f_j, f_i \rangle_{\mathcal{U}},$$

and $\mathfrak{P}_n$ is unitarily equivalent to multiplication by the matrix

(42.3) $$A = \begin{bmatrix} \mathbf{e}_2 & \cdots & \mathbf{e}_n & \mathbf{c} \end{bmatrix} = \begin{bmatrix} 0 & 0 & \cdots & 0 & c_1 \\ 1 & 0 & \cdots & 0 & c_2 \\ \vdots & & & & \vdots \\ 0 & 0 & \cdots & 1 & c_n \end{bmatrix},$$

in $\mathcal{G}$, i.e.,

$$T\mathfrak{P}_n f_j = A\mathbf{e}_j = AT f_j \quad \text{for } j = 1,\ldots,n.$$

Lemma 42.1. *If A is the matrix that is defined by the formulas in (42.2) and (42.3), then:*

(1) *Multiplication by A is a selfadjoint linear transformation in $\mathcal{G}$, i.e.,*

(42.4) $$\langle A\mathbf{u}, \mathbf{v} \rangle_{\mathcal{G}} = \langle \mathbf{u}, A\mathbf{v} \rangle_{\mathcal{G}} \quad \text{for every choice of } \mathbf{u}, \mathbf{v} \in \mathbb{C}^n.$$

(2) *A has n distinct real eigenvalues $\lambda_1,\ldots,\lambda_n$ and a corresponding set $\{\mathbf{v}_1,\ldots,\mathbf{v}_n\}$ of eigenvectors that are orthonormal in $\mathcal{G}$.*

(3) *The coefficients in the expansion of $\mathbf{e}_j$ in terms of the orthonormal eigenvectors $\{\mathbf{v}_1,\ldots,\mathbf{v}_n\}$ of A are*

(42.5) $$\langle \mathbf{e}_j, \mathbf{v}_s \rangle_{\mathcal{G}} = \lambda_s^{j-1} \langle \mathbf{e}_1, \mathbf{v}_s \rangle_{\mathcal{G}} \quad \text{for } j = 1,\ldots,n$$

and

(42.6) $$\langle \mathbf{e}_j, \mathbf{e}_i \rangle_{\mathcal{G}} = \sum_{s=1}^{n} \lambda_s^{i+j-2} |\langle \mathbf{e}_1, \mathbf{v}_s \rangle_{\mathcal{G}}|^2 \quad \text{for } i,j = 1,\ldots,n.$$

42.1. Gaussian quadrature

Proof. Let $h_j = \langle f_j, f_2 \rangle_{\mathcal{U}}$. Then $g_{ij} = h_{i+j-2}$ for $i, j = 1, 2, \ldots$, and

$$GA = G \begin{bmatrix} \mathbf{e}_2 & \cdots & \mathbf{e}_n & \mathbf{c} \end{bmatrix} = \begin{bmatrix} g_{12} & \cdots & g_{1n} & g_{1,n+1} \\ \vdots & & & \vdots \\ g_{n2} & \cdots & g_{nn} & g_{n,n+1} \end{bmatrix}$$

$$= \begin{bmatrix} h_1 & h_2 & \cdots & h_n \\ h_2 & h_3 & \cdots & h_{n+1} \\ \vdots & & & \vdots \\ h_n & h_{n+1} & \cdots & h_{2n-1} \end{bmatrix} = (GA)^H = A^H G,$$

since $h_j \in \mathbb{R}$. Thus,

$$\langle A\mathbf{x}, \mathbf{y} \rangle_{\mathcal{G}} = \langle GA\mathbf{x}, \mathbf{y} \rangle_{\mathrm{st}} = \langle A^H G\mathbf{x}, \mathbf{y} \rangle_{\mathrm{st}} = \langle \mathbf{x}, A\mathbf{y} \rangle_{\mathcal{G}}$$

for every pair of vectors $\mathbf{x}, \mathbf{y} \in \mathbb{C}^n$, i.e., (1) holds.

Suppose next that A has k distinct eigenvalues $\lambda_1, \ldots, \lambda_k$. Then it follows readily from (42.4) that $\lambda_i \in \mathbb{R}$ for $i = 1, \ldots, k$ and that if $A\mathbf{v}_i = \lambda_i \mathbf{v}_i$, then $\langle \mathbf{v}_i, \mathbf{v}_j \rangle_{\mathcal{G}} = 0$ if $i \neq j$. Moreover, if $(A - \lambda_j I_n)^2 \mathbf{u} = \mathbf{0}$ for some vector $\mathbf{u} \in \mathbb{C}^n$, then

$$0 = \langle (A - \lambda_j I_n)^2 \mathbf{u}, \mathbf{u} \rangle_{\mathcal{G}} = \langle (A - \lambda_j I_n)\mathbf{u}, (A - \lambda_j I_n)\mathbf{u} \rangle_{\mathcal{G}}.$$

Therefore, $\mathcal{N}_{(A-\lambda_j I_n)^2} = \mathcal{N}_{(A-\lambda_j I_n)}$, and hence, the algebraic multiplicity α_j of each eigenvalue is equal to its geometric multiplicity γ_j, which is equal to one, since $\sigma(A) = \sigma(A^T)$ and A^T is a companion matrix. Thus, $k = \alpha_1 + \cdots + \alpha_k = n$. Consequently, (2) holds.

Finally, (42.5) follows from the observation that

$$\langle \mathbf{e}_j, \mathbf{v}_s \rangle_{\mathcal{G}} = \langle A\mathbf{e}_{j-1}, \mathbf{v}_s \rangle_{\mathcal{G}} = \langle \mathbf{e}_{j-1}, A\mathbf{v}_s \rangle_{\mathcal{G}} = \lambda_s \langle \mathbf{e}_{j-1}, \mathbf{v}_s \rangle_{\mathcal{G}}$$

for $j = 2, \ldots, n$; and (42.6) is obtained by substituting the expansions

$$\mathbf{e}_j = \sum_{s=1}^{n} \langle \mathbf{e}_j, \mathbf{v}_s \rangle_{\mathcal{G}} \mathbf{v}_s \quad \text{and} \quad \mathbf{e}_i = \sum_{t=1}^{n} \langle \mathbf{e}_i, \mathbf{v}_t \rangle_{\mathcal{G}} \mathbf{v}_t$$

into $\langle \mathbf{e}_j, \mathbf{e}_i \rangle_{\mathcal{G}}$ and then invoking (42.5). □

Theorem 42.2. *Let $W_j = |\langle \mathbf{e}_1, \mathbf{v}_j \rangle_{\mathcal{G}}|^2$ for $j = 1, \ldots, n$. Then the formula*

(42.7) $$\int_a^b f(x) w(x) dx = \sum_{i=1}^{n} W_i f(\lambda_i)$$

is valid for every polynomial $f(x)$ of degree less than or equal to $2n - 1$ with complex coefficients.

Proof. It suffices to verify this formula for $f(x) = f_k(x)$, $k = 1, \ldots, 2n$. Consider first the case $k = i + j$ with $i, j = 1, \ldots, n-1$. Then, by (42.6),

$$\int_a^b f_{i+j}(x) w(x) dx = \langle f_j, f_{i+1} \rangle_{\mathcal{U}} = \langle \mathbf{e}_j, \mathbf{e}_{i+1} \rangle_{\mathcal{G}}$$

$$= \sum_{s=1}^n \lambda_s^{i+j-1} W_s = \sum_{s=1}^n f_{i+j}(\lambda_s) W_s \,.$$

To complete the proof, it remains only to check that

(42.8) $$\int_a^b x^{2n-1} w(x) dx = \sum_{s=1}^n \lambda_s^{2n-1} W_s \,.$$

But, as $\langle \mathbf{e}_j, \mathbf{e}_i \rangle_{\mathcal{G}} = \sum_{s=1}^n \lambda_s^{i+j-2} W_s$ for $i, j = 1, \ldots, n$,

$$\int_a^b x^{2n-1} w(x) dx = \langle f_{n+1}, f_n \rangle_{\mathcal{U}} = \langle M f_n, f_n \rangle_{\mathcal{U}} = \langle \mathfrak{P}_n f_n, f_n \rangle_{\mathcal{U}}$$

$$= \langle A \mathbf{e}_n, \mathbf{e}_n \rangle_{\mathcal{G}} = \sum_{j=1}^n c_j \langle \mathbf{e}_j, \mathbf{e}_n \rangle_{\mathcal{G}}$$

$$= \sum_{j=1}^n c_j \left\{ \sum_{s=1}^n \lambda_s^{n+j-2} W_s \right\}$$

$$= \sum_{s=1}^n \left\{ \lambda_s^{n-1} W_s \sum_{j=1}^n c_j \lambda_s^{j-1} \right\} = \sum_{s=1}^n \lambda_s^{2n-1} W_s \,,$$

since

(42.9) $$\sum_{j=1}^n c_j \lambda_s^{j-1} = \lambda_s^n \quad \text{for } s = 1, \ldots, n \,,$$

because $\sigma(A) = \sigma(A^T)$ and A^T is a companion matrix. $\square$

Exercise 42.1. Verify formula (42.9). [HINT: $(\lambda - \lambda_1) \cdots (\lambda - \lambda_n) = \det(\lambda I_n - A) = \det(\lambda I_n - A^T)$ and A^T is a companion matrix.]

Finite sums like (42.7) that serve to approximate definite integrals, with equality for a reasonable class of functions, are termed **quadrature formulas**.

42.2. Bezoutians

Let
$$f(\lambda) = f_0 + f_1 \lambda + \cdots + f_n \lambda^n \,, \quad f_n \neq 0 \,,$$
be a polynomial of degree n with coefficients $f_0, \ldots, f_n \in \mathbb{C}$ and let
$$g(\lambda) = g_0 + g_1 \lambda + \cdots + g_n \lambda^n$$

42.2. Bezoutians

be a polynomial of degree less than or equal to n with coefficients $g_0, \ldots, g_n \in \mathbb{C}$, at least one of which is nonzero. Then the matrix $B \in \mathbb{C}^{n \times n}$ with entries b_{ij}, $i, j = 0, \ldots, n-1$, that is uniquely defined by the formulas

$$(42.10) \qquad \frac{f(\lambda)g(\mu) - g(\lambda)f(\mu)}{\lambda - \mu} = \sum_{i,j=0}^{n-1} \lambda^i b_{ij} \mu^j = \mathbf{v}(\lambda)^T B \mathbf{v}(\mu)$$

and

$$\mathbf{v}(\lambda)^T = \begin{bmatrix} 1 & \lambda & \cdots & \lambda^{n-1} \end{bmatrix}$$

is called the **Bezoutian** of the polynomials $f(\lambda)$ and $g(\lambda)$ and will be denoted by the symbol $B(f, g)$ (or just plain B if the polynomials are clear from the context).

It is clear from formula (42.10) that if $f(\alpha) = 0$ and $g(\alpha) = 0$, then

$$\mathbf{v}(\lambda)^T B \mathbf{v}(\alpha) = 0$$

for every point $\lambda \in \mathbb{C}$, and hence that

$$B\mathbf{v}(\alpha) = 0.$$

Moreover, if $f(\alpha) = f'(\alpha) = 0$ and $g(\alpha) = g'(\alpha) = 0$, then the identity

$$\frac{f(\lambda)g'(\mu) - g(\lambda)f'(\mu)}{\lambda - \mu} + \frac{f(\lambda)g(\mu) - g(\lambda)f(\mu)}{(\lambda - \mu)^2} = \mathbf{v}(\lambda)^T B \mathbf{v}'(\mu),$$

which is obtained by differentiating both sides of formula (42.10) with respect to μ, implies that

$$(42.11) \qquad \mathbf{v}(\lambda)^T B \mathbf{v}'(\alpha) = 0$$

for every point $\lambda \in \mathbb{C}$. Therefore, $\dim \mathcal{N}_B \geq 2$, since the vectors $\mathbf{v}(\alpha)$ and $\mathbf{v}'(\alpha)$ both belong to $\mathcal{N}_B$ and are linearly independent.

Much the same sort of reasoning leads rapidly to the conclusion that if

$$f(\alpha) = f^{(1)}(\alpha) = \cdots = f^{(k-1)}(\alpha) = 0 = g(\alpha) = g^{(1)}(\alpha) = \cdots g^{(k-1)}(\alpha),$$

then the vectors $\mathbf{v}(\alpha), \ldots, \mathbf{v}^{(k-1)}(\alpha)$ all belong to $\mathcal{N}_B$. Thus, as these vectors are linearly independent if $k \leq n$, $\dim \mathcal{N}_B \geq k$. Moreover, if $\alpha \neq \beta$ and

$$f(\beta) = f^{(1)}(\beta) = \cdots = f^{(j-1)}(\beta) = 0 = g(\beta) = g^{(1)}(\beta) = \cdots = g^{(j-1)}(\beta),$$

then the vectors $\mathbf{v}(\beta), \ldots, \mathbf{v}^{(j-1)}(\beta)$ all belong to $\mathcal{N}_B$. Therefore, since this set of vectors is linearly independent of the set $\mathbf{v}(\alpha), \ldots, \mathbf{v}^{(k-1)}(\alpha)$ when $\alpha \neq \beta$ (as was shown in Theorem 8.3), $\dim \mathcal{N}_B \geq k + j$. Proceeding this way, it is readily seen that $\dim \mathcal{N}_{B(f,g)} \geq \nu(f, g)$, where

$\nu(f, g) =$ the number of common roots of the polynomials $f(\lambda)$
and $g(\lambda)$, counting multiplicities.

It is also easy to show that $\dim \mathcal{N}_{B(f,g)} = \nu(f, g)$ when f has distinct roots, as is spelled out in Exercises 42.6 and 42.5 below. In fact equality holds even if f does not have distinct roots, but the proof is not so simple; it rests on a highly nontrivial identity and the following lemma:

Lemma 42.3. *If $g(\lambda)$ is a polynomial and $N = C_0^{(p)}$, then*

$$(42.12) \quad g(C_\lambda^{(p)}) = \sum_{j=0}^{p-1} \frac{g^{(j)}(\lambda)}{j!} N^j = \begin{bmatrix} g(\lambda) & \frac{g^{(1)}(\lambda)}{1!} & \cdots & \frac{g^{(p-1)}(\lambda)}{(p-1)!} \\ 0 & g(\lambda) & \cdots & \frac{g^{(p-2)}(\lambda)}{(p-2)!} \\ \vdots & & \ddots & \vdots \\ 0 & 0 & \cdots & g(\lambda) \end{bmatrix}$$

and

$$\operatorname{rank} g(C_\lambda^{(p)}) = \begin{cases} p & \text{if } g(\lambda) \neq 0, \\ p-k & \text{if } g(\lambda) = \cdots = g^{(k-1)}(\lambda) = 0 \text{ but } g^{(k)}(\lambda) \neq 0, \end{cases}$$

where, in the last line, k is an integer such that $1 \leq k \leq p$.

Proof. Let Γ denote a circle of radius $R > |\lambda|$ that is centered at the origin and is directed counterclockwise. Then, by Cauchy's formula,

$$\begin{aligned} g(\lambda I_p + N) &= \frac{1}{2\pi i} \int_\Gamma g(\zeta)(\zeta I_p - \lambda I_p - N)^{-1} d\zeta \\ &= \sum_{j=0}^{p-1} \frac{1}{2\pi i} \int_\Gamma \frac{g(\zeta)}{(\zeta - \lambda)^{j+1}} N^j d\zeta \\ &= \sum_{j=0}^{p-1} \frac{g^{(j)}(\lambda)}{j!} N^j . \end{aligned}$$

The formula for the rank of $g(C_\lambda^{(p)})$ is clear from formula (42.12), which exhibits $g(C_\lambda^{(p)})$ as an upper triangular matrix that is constant on diagonals, i.e., as an upper triangular Toeplitz matrix. $\square$

Example 42.1. If $p = 3$, then

$$g(C_\lambda^{(3)}) = g(\lambda I_3 + N) = \begin{bmatrix} g(\lambda) & g^{(1)}(\lambda) & g^{(2)}(\lambda)/2! \\ 0 & g(\lambda) & g^{(1)}(\lambda) \\ 0 & 0 & g(\lambda) \end{bmatrix}.$$

But this clearly exhibits the fact that

$$\operatorname{rank} g(\lambda I_3 + N) = \begin{cases} 3 & \text{if } g(\lambda) \neq 0, \\ 2 & \text{if } g(\lambda) = 0 \text{ and } g^{(1)}(\lambda) \neq 0, \\ 1 & \text{if } g(\lambda) = g^{(1)}(\lambda) = 0 \text{ and } g^{(2)}(\lambda) \neq 0. \end{cases}$$

42.2. Bezoutians

Exercise 42.2. Confirm formula (42.12) for the polynomial
$$g(\lambda) = \sum_{k=0}^{n} g_k \lambda^k \quad \text{by writing} \quad g(\lambda I_p + N) = \sum_{k=0}^{n} g_k (\lambda I_p + N)^k$$
and invoking the binomial formula. [REMARK: This is a good exercise in manipulating formulas, but it's a lot more work than the proof of Lemma 42.3 that was presented above.]

Theorem 42.4. *If $f(\lambda)$ is a polynomial of degree n (i.e., $f_n \neq 0$) and $g(\lambda)$ is a polynomial of degree $\leq n$ with at least one nonzero coefficient, then*

(42.13) $$\dim \mathcal{N}_{B(f,g)} = \nu(f,g).$$

Discussion. The proof rests on the **Barnett identity**

(42.14) $$B(f,g) = H_f\, g(K_f),$$

in which K_f is the companion matrix based on the polynomial f (see (8.19) and (8.20)) and H_f is an invertible Hankel matrix.

To understand why (42.14) yields (42.13), consider the case $f(\lambda) = (\lambda - \lambda_1)^3 (\lambda - \lambda_2)^2$ with $\lambda_1 \neq \lambda_2$. Then

$$K_f = V \begin{bmatrix} C_{\lambda_1}^{(3)} & O \\ O & C_{\lambda_2}^{(2)} \end{bmatrix} V^{-1} \quad \text{and} \quad g(K_f) = V \begin{bmatrix} g(C_{\lambda_1}^{(3)}) & O \\ O & g(C_{\lambda_2}^{(2)}) \end{bmatrix} V^{-1}.$$

Therefore,
$$\dim \mathcal{N}_{B(f,g)} = \dim \mathcal{N}_{g(K_f)} = \dim \mathcal{N}_{g(C_{\lambda_1}^{(3)})} + \dim \mathcal{N}_{g(C_{\lambda_2}^{(2)})}.$$

Thus, if $g(\lambda) = (\lambda - \lambda_1)^2 (\lambda - \lambda_2)(\lambda - \lambda_3)^2$ has three distinct roots, then $\dim \mathcal{N}_{g(C_{\lambda_1}^{(3)})} = 2$ and $\dim \mathcal{N}_{g(C_{\lambda_2}^{(2)})} = 1$. □

Exercise 42.3. Show that if $A \in \mathbb{C}^{n \times n}$, $B \in \mathbb{C}^{n \times n}$, and $AB = BA$, then
$$g(B) = \sum_{k=0}^{n} \frac{g^{(k)}(A)}{k!} (B - A)^k$$
for every polynomial $g(\lambda)$ of degree $\leq n$.

Exercise 42.4. Use Theorem 42.4 and formula (42.14) to calculate the number of common roots of the polynomials $f(x) = 2 - 3x + x^3$ and $g(x) = -2 + x + x^2$.

Exercise 42.5. Show that if $f(\lambda)$ is a polynomial of degree n with n distinct roots $\lambda_1, \ldots, \lambda_n$, $g(\lambda)$ is a polynomial of degree $\leq n$, and V is the Vandermonde matrix with columns $\mathbf{v}(\lambda_1), \ldots, \mathbf{v}(\lambda_n)$, then

(42.15) $$V^T B(f,g) V = \text{diag}\{f'(\lambda_1) g(\lambda_1), \ldots, f'(\lambda_n) g(\lambda_n)\}.$$

[HINT: Compute $\mathbf{v}(\lambda_j)^T B(f,g) \mathbf{v}(\lambda_k)$, for $j = k$ and $j \neq k$ via (42.10).]

Exercise 42.6. Show that (42.13) holds in the setting of Exercise 42.5. [HINT: Since V is invertible when f has n distinct roots, formula (42.15) implies that $\dim \mathcal{N}_{B(f,g)}$ is equal to the number of points α_j at which $g(\alpha_j) = 0$.]

42.3. Resultants

There is another formula for computing the number of common roots of a pair of polynomials $f(\lambda)$ and $g(\lambda)$ that is easier to write down, since it is expressed in terms of the $2n \times 2n$ matrix

$$R(f,g) = \begin{bmatrix} f_0 & f_1 & \cdots & f_{n-1} & f_n & 0 & \cdots & 0 \\ 0 & f_0 & \cdots & f_{n-2} & f_{n-1} & f_n & \cdots & 0 \\ \vdots & \vdots & & \vdots & \vdots & \vdots & & \vdots \\ 0 & 0 & \cdots & f_0 & f_1 & f_2 & \cdots & f_n \\ \hdashline g_0 & g_1 & \cdots & g_{n-1} & g_n & 0 & \cdots & 0 \\ 0 & g_0 & \cdots & g_{n-2} & g_{n-1} & g_n & \cdots & 0 \\ \vdots & \vdots & & \vdots & \vdots & \vdots & & \vdots \\ 0 & 0 & \cdots & g_0 & g_1 & g_2 & \cdots & g_n \end{bmatrix}$$

based on the coefficients of the polynomials $f(\lambda) = f_0 + f_1\lambda + \cdots + f_n\lambda^n$ and $g(\lambda) = g_0 + g_1\lambda + \cdots + g_n\lambda^n$. The matrix $R(f,g)$ is called the **resultant** of $f(\lambda)$ and $g(\lambda)$.

Theorem 42.5. *If $f(\lambda)$ is a polynomial of degree n (i.e., $f_n \neq 0$) and $g(\lambda)$ is a polynomial of degree $\leq n$ with at least one nonzero coefficient, then*

(42.16) $$\dim \mathcal{N}_{R(f,g)} = \nu(f,g).$$

Proof. See, e.g., Theorem 21.9 in [**30**]. □

The next exercise is a good way to check that the theorem is correct in at least one case.

Exercise 42.7. Show that if f is a polynomial of degree 3 with 3 distinct roots $\lambda_1, \lambda_2, \lambda_3$, g is a polynomial of degree 2 with 2 distinct roots μ_1, μ_2, and these five points are distinct, then $R(f,g)$ is invertible. [HINT: It's enough to show that if $\alpha \in \mathbb{C} \setminus \{\lambda_1, \lambda_2, \lambda_3, \mu_1, \mu_2\}$ and $\nu(f,g) = 0$, then the matrix

$$R(f,g) \begin{bmatrix} \mathbf{v}(\mu_1) & \mathbf{v}(\mu_2) & \mathbf{v}(\alpha) & \mathbf{v}(\lambda_1) & \mathbf{v}(\lambda_2) & \mathbf{v}(\lambda_3) \end{bmatrix}$$

with $\mathbf{v}(\lambda) = \begin{bmatrix} 1 & \lambda & \cdots & \lambda^5 \end{bmatrix}^T$ is invertible.]

42.4. General QR factorization

In this section we consider QR factorization for matrices $A \in \mathbb{C}^{p \times q}$ that are not assumed to be left invertible.

Theorem 42.6. *If $A \in \mathbb{C}^{p \times q}$ and $\operatorname{rank} A = r \geq 1$, then $A = QR$, where $Q \in \mathbb{C}^{p \times q}$ is isometric and $R \in \mathbb{C}^{q \times q}$ is upper triangular with r positive entries on the diagonal and $q - r$ zero entries on the diagonal.*

Discussion. The idea underlying the proof is best conveyed by example. The general case is established in exactly the same way, but with more elaborate bookkeeping.

Let $A = \begin{bmatrix} \mathbf{a}_1 & \mathbf{a}_2 & \mathbf{a}_3 & \mathbf{a}_4 \end{bmatrix}$ and suppose that $\mathbf{a}_1$ and $\mathbf{a}_3$ are the pivot columns of A. Then $\operatorname{rank} A = 2$, $\mathcal{R}_A = \operatorname{span}\{\mathbf{a}_1, \mathbf{a}_3\}$, and standard QR factorization ensures that

$$\begin{bmatrix} \mathbf{a}_1 & \mathbf{a}_3 \end{bmatrix} = \begin{bmatrix} \mathbf{q}_1 & \mathbf{q}_3 \end{bmatrix} \begin{bmatrix} a & b \\ 0 & c \end{bmatrix} \quad \text{with } \mathbf{q}_1, \mathbf{q}_2 \text{ orthonormal}, a > 0, \text{ and } c > 0.$$

Thus,

$$\begin{bmatrix} \mathbf{a}_1 & \mathbf{0} & \mathbf{a}_3 & \mathbf{0} \end{bmatrix} = \begin{bmatrix} \mathbf{q}_1 & \mathbf{0} & \mathbf{q}_3 & \mathbf{0} \end{bmatrix} \begin{bmatrix} a & 0 & b & 0 \\ 0 & 0 & 0 & 0 \\ 0 & 0 & c & 0 \\ 0 & 0 & 0 & 0 \end{bmatrix}$$

$$= \begin{bmatrix} \mathbf{q}_1 & \mathbf{q}_2 & \mathbf{q}_3 & \mathbf{q}_4 \end{bmatrix} \begin{bmatrix} a & 0 & b & 0 \\ 0 & 0 & 0 & 0 \\ 0 & 0 & c & 0 \\ 0 & 0 & 0 & 0 \end{bmatrix} = QB,$$

where $\mathbf{q}_2$ and $\mathbf{q}_4$ are orthonormal vectors such that $Q = \begin{bmatrix} \mathbf{q}_1 & \mathbf{q}_2 & \mathbf{q}_3 & \mathbf{q}_4 \end{bmatrix}$ is isometric and the 4×4 matrix B is upper triangular with nonnegative entries on the diagonal, two of which are positive.

Next, since $\begin{bmatrix} \mathbf{a}_2 & \mathbf{a}_4 \end{bmatrix} = \begin{bmatrix} \mathbf{a}_1 & \mathbf{a}_3 \end{bmatrix} \begin{bmatrix} d & e \\ 0 & f \end{bmatrix}$,

$$\begin{bmatrix} \mathbf{0} & \mathbf{a}_2 & \mathbf{0} & \mathbf{a}_4 \end{bmatrix} = \begin{bmatrix} \mathbf{a}_1 & \mathbf{0} & \mathbf{a}_3 & \mathbf{0} \end{bmatrix} \begin{bmatrix} 0 & d & 0 & e \\ 0 & 0 & 0 & 0 \\ 0 & 0 & 0 & f \\ 0 & 0 & 0 & 0 \end{bmatrix} = \begin{bmatrix} \mathbf{a}_1 & \mathbf{0} & \mathbf{a}_3 & \mathbf{0} \end{bmatrix} C,$$

where the 4×4 matrix C is strictly upper triangular. Consequently,

$$A = QB + QBC = QB(I_4 + C) = QR \quad \text{with } R = B(I_4 + C).$$

Thus, as B is upper triangular and $I_4 + C$ is upper triangular with ones on the diagonal, the diagonal entries of R coincide with the diagonal entries of B, two of which are positive and two are equal to zero. □

As an application we consider a subcase of the **CS decomposition**:

Theorem 42.7. *If $A, B \in \mathbb{C}^{k \times k}$ and $A^H A + B^H B = I_k$, then there exists a set of unitary matrices $W_1, W_2 \in \mathbb{C}^{2k \times 2k}$ such that*

$$W_1 \begin{bmatrix} I_k & A \\ O & B \end{bmatrix} W_2 = \begin{bmatrix} I_k & C \\ O & S \end{bmatrix},$$

where $C, S \in \mathbb{R}^{k \times k}$ are diagonal matrices such that $c_{11} \leq \cdots \leq c_{kk}$, $s_{11} \geq \cdots \geq s_{kk}$, and $C^2 + S^2 = I_k$.

Proof. By reordering the entries in the singular value decomposition of A, we obtain a pair of unitary matrices $V_1, U_1 \in \mathbb{C}^{k \times k}$ such that $A = V_1 C U_1^H$. Theorem 42.6 ensures that there exists a unitary matrix $Q \in \mathbb{C}^{k \times k}$ and an upper triangular matrix $R \in \mathbb{C}^{k \times k}$ with nonnegative entries on the diagonal such that $BU_1 = QR$. Thus, upon setting $W_1 = \text{diag}\{V_1^H, Q^H\}$ and $W_2 = \text{diag}\{V_1, U_1\}$, it is readily checked that

$$\begin{bmatrix} V_1^H & O \\ O & Q^H \end{bmatrix} \begin{bmatrix} I_k & A \\ O & B \end{bmatrix} \begin{bmatrix} V_1 & O \\ O & U_1 \end{bmatrix} = \begin{bmatrix} I_k & V_1^H A U_1 \\ O & Q^H B U_1 \end{bmatrix} = \begin{bmatrix} I_k & C \\ O & R \end{bmatrix}.$$

It remains to show that the upper triangular matrix R is actually a diagonal matrix. This follows from the fact that $C^H C + R^H R = I_k$. If $C = I_k$, then $R = O$. If $C \neq I_k$, then $0 \leq c_{11} < 1$. Consequently, $r_{11} > 0$, and hence, as the columns of $\begin{bmatrix} C & R \end{bmatrix}^T$ are orthonormal, $r_{1j} = 0$ for $j = 2, \ldots, k$. Similarly, if $r_{ii} > 0$, then $r_{ij} = 0$ for $j = i+1, \ldots, k$. □

42.5. The QR algorithm

Recall that if $A \in \mathbb{C}^{p \times q}$ and $\text{rank } A = q$, then there exists exactly one isometric matrix $Q \in \mathbb{C}^{p \times q}$ and exactly one upper triangular matrix $R \in \mathbb{C}^{k \times k}$ with positive entries on the diagonal such that $A = QR$. The **QR algorithm** defines a sequence of invertible matrices $A_j \in \mathbb{C}^{n \times n}$ by successive applications of QR factorization starting with $A = A_1$ by setting $A_{m+1} = R_m Q_m$ when $A_m = Q_m R_m$ and then applying QR factorization to A_{m+1} to obtain $A_{m+1} = Q_{m+1} R_{m+1}$ for $m = 1, 2, \ldots$. Thus,

$$A_1 = Q_1 R_1, \quad A_2 = R_1 Q_1 = Q_2 R_2, \quad A_3 = R_2 Q_2 = Q_3 A_3, \ldots,$$

where Q_m is unitary and R_m is upper triangular with positive diagonal entries for $m = 1, 2, \ldots$. The **notation**

(42.17) $$\mathbb{Q}_m = Q_1 Q_2 \cdots Q_m \quad \text{and} \quad \mathbb{R}_m = R_m R_{m-1} \cdots R_1$$

will be convenient; it will be used in this section only.

42.5. The QR algorithm

Exercise 42.8. Show that the QR factorization of A^m is equal to $\mathbb{Q}_m \mathbb{R}_m$, i.e.,

(42.18) $$A^m = \mathbb{Q}_m \mathbb{R}_m.$$

[HINT: $A^3 = Q_1 R_1 Q_1 R_1 Q_1 R_1 = Q_1(R_1 Q_1 R_1 Q_1) R_1 = Q_1 Q_2 R_2 Q_2 R_2 R_1$.]

Exercise 42.9. Show that

(42.19) $$A_{m+1} = \mathbb{Q}_m^H A \mathbb{Q}_m.$$

[HINT: $A_{m+1} = R_m Q_m \implies Q_m A_{m+1} Q_m^H = A_m$.]

Theorem 42.8. *If $A \in \mathbb{C}^{n \times n}$ is invertible with Jordan decomposition $A = XDX^{-1}$, where*

(1) $D = \mathrm{diag}\{\lambda_1, \ldots, \lambda_n\}$ *with* $|\lambda_1| > |\lambda_2| > \cdots |\lambda_n|$ *and*
(2) *the $k \times k$ upper left-hand corners of X^{-1} are invertible for $k = 1, \ldots, n$,*

then the absolute values of $(A_m)_{ij}$, the ij entry in A_m (the m'th matrix in the QR algorithm), tend to a limit when $i < j$ and

$$\lim_{m \uparrow \infty} (A_m)_{ij} = \begin{cases} 0 & \text{if } i > j, \\ \lambda_j & \text{if } i = j. \end{cases}$$

Proof. Under the given assumptions, Theorem 16.3 ensures that the matrix $Y = X^{-1}$ admits a factorization of the form $Y = L_Y U_Y$, where L_Y is lower triangular with ones on the diagonal and U_Y is upper triangular. Then, in view of formula (42.18),

$$\mathbb{Q}_m \mathbb{R}_m = A^m = X D^m L_Y U_Y = X C_m D^m U_Y \quad \text{with } C_m = D^m L_Y D^{-m}.$$

Since (in self-evident notation) $(C_m)_{ij} = \lambda_i^m (L_Y)_{ij} \lambda_j^{-m}$ and L_Y is lower triangular, $(C_m)_{ij} = 0$ if $j > i$. Furthermore, as $\lambda_i^m \lambda_j^{-m} \to 0$ when $m \uparrow \infty$ if $j < i$,

$$\lim_{m \uparrow \infty} (C_m)_{ij} = \lim_{m \uparrow \infty} \lambda_i^m (L_Y)_{ij} \lambda_j^{-m} = \begin{cases} 0 & \text{if } i \neq j, \\ 1 & \text{if } i = j. \end{cases}$$

Thus, C_m is an invertible matrix that tends to I_n as $m \uparrow \infty$.

Consequently, upon invoking the QR factorization $X = Q_X R_X$,

$$\mathbb{Q}_m \mathbb{R}_m = X C_m D^m U_Y = Q_X R_X C_m D^m U_Y$$
$$= Q_X (R_X C_m R_X^{-1}) R_X D^m U_Y = Q_X G_m R_X D^m U_Y$$
$$= Q_X \mathcal{Q}_m \mathcal{R}_m R_X D^m U_Y = Q_X \mathcal{Q}_m \Delta_m \Delta_m^H \mathcal{R}_m R_X D^m U_Y,$$

where $\mathcal{Q}_m$ is the unitary factor and $\mathcal{R}_m$ is the upper triangular factor with positive entries on the diagonal in the QR factorization of the invertible matrix $G_m = R_X C_m R_X^{-1}$, and $\Delta_m = \mathrm{diag}\{(\mu_m)_1, \ldots, (\mu_m)_n\}$ is a unitary matrix that is chosen so that the diagonal entries of the upper triangular matrix $\Delta_m^H \mathcal{R}_m R_X D^m U_Y$ are all positive. Therefore, by the uniqueness of QR factorization, $\mathbb{Q}_m = Q_X \mathcal{Q}_m \Delta_m$. The rest of the proof is divided into parts.

1. $\mathcal{Q}_m \to I_n$ *as* $m \uparrow \infty$

It is easily checked that the entries in the matrices $\mathcal{Q}_m$ and $\mathcal{R}_m$ are bounded and hence that:

(1) A subsequence of $\{\mathcal{Q}_1, \mathcal{Q}_2, \ldots\}$ converges to a unitary matrix $\mathcal{Q}_\infty$.
(2) A subsequence of the matrices $\{\mathcal{R}_1, \mathcal{R}_2, \ldots\}$ converges to an upper triangular matrix $\mathcal{R}_\infty$ with nonnegative entries on the diagonal.
(3) $\mathcal{Q}_\infty \mathcal{R}_\infty = I_n$.

Consequently, the unitary matrix $\mathcal{Q}_\infty = \mathcal{R}_\infty^H$ is a lower triangular matrix with nonnegative entries on the diagonal, which is only possible if $\mathcal{Q}_\infty = I_n$. Since the same conclusions hold for all convergent subsequences, it follows that the original sequences $\mathcal{Q}_m \to I_n$ and $\mathcal{R}_m \to I_n$ as $m \uparrow \infty$.

2. *Completion of the proof*

In view of (42.19) and the formulas $\mathbb{Q}_m = Q_X \mathcal{Q}_m \Delta_m$ and $X = Q_X R_X$,
$$A_{m+1} = \mathbb{Q}_m^H A \mathbb{Q}_m = \Delta_m^H \mathcal{Q}_m^H Q_X^H X D X^{-1} Q_X \mathcal{Q}_m \Delta_m$$
$$= \Delta_m^H \mathcal{Q}_m^H R_X D R_X^{-1} \mathcal{Q}_m \Delta_m \, .$$
The stated conclusions of the theorem are now easily read off the formula
$$(A_{m+1})_{ij} = \overline{(\mu_m)_i} (\mathcal{Q}_m^H R_X D R_X^{-1} \mathcal{Q}_m)_{ij} (\mu_m)_j \, ,$$
since $\mathcal{Q}_m^H R_X D R_X^{-1} \mathcal{Q}_m \to R_X D R_X^{-1}$ as $m \uparrow \infty$ and $R_X D R_X^{-1}$ is upper triangular with $(R_X D R_X^{-1})_{ii} = \lambda_i$ for $i = 1, \ldots, n$. $\square$

42.6. Supplementary notes

The treatment of Gaussian quadrature is partially adapted from the discussion in Section 8.14 of [30], which was adapted from the Ph.D. thesis of Ilan Degani [22]. The present version is based on first showing that the operator $\mathfrak{P}_n$ in the space of polynomials of degree $\leq n-1$ is unitarily equivalent to multiplication by a companion matrix in $\mathbb{C}^n$. The sections on Bezoutians and resultants are much abbreviated versions of the presentations in Chapter 21 of [30]. A neat way to establish the properties of Bezoutians and resultants that avoids the use of the Barnett identity is presented in Curgus and Dijksma [20]. The last two sections are new. Section 42.4 is adapted from [9]. The presented proof of Theorem 42.8 in the last section is taken from the paper [76] by J. N. Wilkinson. For additional insight and references, see the review article [73] by D. Watkins, and, for additional perspective [74].

Chapter 43

Toeplitz, Hankel, and de Branges

In this chapter we shall develop some elementary properties of finite-dimensional reproducing kernel Hilbert spaces and shall then use these properties to identify positive definite Hankel matrices as the Gram matrices of a set of polynomials in a very important class of reproducing kernel Hilbert spaces of the kind introduced and intensively studied by L. de Branges. Analogous conclusions will be presented first for positive definite Toeplitz matrices. This serves to identify the densities that define the inner products in these two spaces as solutions of a pair of truncated **moment problems**.

Recall that a matrix $A \in \mathbb{C}^{n \times n}$ with entries a_{ij}, $i,j = 1, \ldots, n$, is a:

- **Toeplitz matrix** if $a_{ij} = t_{i-j}$ for $i,j = 1, \ldots, n$ and some set of $2n-1$ numbers $t_{-n+1}, \ldots, t_{n-1}$,
- **Hankel matrix** if $a_{ij} = h_{i+j-2}$ for $i,j = 1, \ldots, n$ and some set of $2n-1$ numbers $h_0, \ldots, h_{2n-2}$.

Thus, for example, if $n = 3$,

$$B = \begin{bmatrix} t_0 & t_{-1} & t_{-2} \\ t_1 & t_0 & t_{-1} \\ t_2 & t_1 & t_0 \end{bmatrix}, \quad \text{and} \quad C = \begin{bmatrix} h_0 & h_1 & h_2 \\ h_1 & h_2 & h_3 \\ h_2 & h_3 & h_4 \end{bmatrix},$$

then B is a Toeplitz matrix and C is a Hankel matrix.

43.1. Reproducing kernel Hilbert spaces

A finite-dimensional inner product space $\mathcal{H}$ of complex-valued functions that are defined on a nonempty subset Ω of $\mathbb{C}$ is said to be a **reproducing kernel**

Hilbert space if there exists a function $K_\omega(\lambda)$ that is defined on $\Omega \times \Omega$ such that for every choice of $\omega \in \Omega$ the following two conditions are fulfilled:

(1) $K_\omega \in \mathcal{H}$ (as a function of λ).

(2) $\langle f, K_\omega \rangle_\mathcal{H} = f(\omega)$ for every point $\omega \in \Omega$ and every function $f \in \mathcal{H}$.

A function $K_\omega(\lambda)$ that meets these two conditions is called a **reproducing kernel** for $\mathcal{H}$.

Theorem 43.1. *If $K_\omega(\lambda)$ is a reproducing kernel for a finite-dimensional reproducing kernel Hilbert space $\mathcal{H}$ of complex-valued functions defined on a subset Ω of $\mathbb{C}$, then:*

(1) $K_\omega(\lambda)$ *is a positive kernel, i.e.,*

$$\sum_{i,j=1}^{m} \overline{c_i} K_{\omega_j}(\omega_i) c_j \geq 0 \tag{43.1}$$

for every positive integer m and every choice of points $\omega_1, \ldots, \omega_m$ in Ω and coefficients $c_1, \ldots, c_m \in \mathbb{C}$.

(2) $\overline{K_\alpha(\beta)} = K_\beta(\alpha)$ *for every pair of points $\alpha, \beta \in \Omega$.*

(3) $\mathcal{H}$ *has exactly one reproducing kernel (though it may be expressed in more than one way).*

Proof. Let $f = \sum_{j=1}^{m} c_j K_{\omega_j}$. Then the evaluation

$$\|f\|_\mathcal{H}^2 = \sum_{i,j=1}^{n} \langle c_j K_{\omega_j}, c_i K_{\omega_i} \rangle_\mathcal{H} = \sum_{i,j=1}^{m} \overline{c_i} K_{\omega_j}(\omega_i) c_j \tag{43.2}$$

clearly serves to establish (1), since $\|f\|_\mathcal{H}^2 \geq 0$.

The special case of (43.2) with $m = 2$, $\omega_1 = \alpha$, and $\omega_2 = \beta$ implies that

$$\begin{bmatrix} \overline{c_1} & \overline{c_2} \end{bmatrix} \begin{bmatrix} K_\alpha(\alpha) & K_\beta(\alpha) \\ K_\alpha(\beta) & K_\beta(\beta) \end{bmatrix} \begin{bmatrix} c_1 \\ c_2 \end{bmatrix} \geq 0$$

for every choice of $c_1, c_2 \in \mathbb{C}$ and hence that the 2×2 matrix in the preceding display is positive semidefinite. Therefore, it is Hermitian and (2) holds.

Suppose next that $L_\omega(\lambda)$ is also a reproducing kernel for $\mathcal{H}$. Then

$$L_\alpha(\beta) = \langle L_\alpha, K_\beta \rangle_\mathcal{H} = \overline{\langle K_\beta, L_\alpha \rangle_\mathcal{H}} = \overline{K_\beta(\alpha)} = K_\alpha(\beta)$$

for every pair of points $\alpha, \beta \in \Omega$. Therefore, (3) holds. $\square$

Example 43.1. Let $\mathcal{H}$ be an n-dimensional inner product space of complex-valued functions that are defined on a nonempty subset Ω of $\mathbb{C}$, and let $f_1, \ldots, f_n$ be a basis for $\mathcal{H}$ with Gram matrix $G \in \mathbb{C}^{n \times n}$, i.e.,

$$g_{ij} = \langle f_j, f_i \rangle_\mathcal{H} \quad \text{for } i, j = 1, \ldots, n \,.$$

43.2. de Branges spaces

Then

(43.3) $$K_\omega(\lambda) = \sum_{i,j=1}^n f_i(\lambda)(G^{-1})_{ij}\overline{f_j(\omega)}$$

is the reproducing kernel for $\mathcal{H}$.

It is clear that $K_\omega(\lambda)$ belongs to $\mathcal{H}$ (as a function of λ) for every choice of $\omega \in \Omega$. Moreover,

$$\langle f_k, K_\omega\rangle_\mathcal{H} = \left\langle f_k, \sum_{i,j=1}^n f_i(G^{-1})_{ij}\overline{f_j(\omega)}\right\rangle_\mathcal{H} = \sum_{i,j=1}^n (G^{-1})_{ji}f_j(\omega)\langle f_k, f_i\rangle_\mathcal{H}$$
$$= \sum_{i,j=1}^n (G^{-1})_{ji}f_j(\omega)g_{ik} = f_k(\omega),$$

since $\sum_{i=1}^n (G^{-1})_{ji}g_{ik}$ is equal to the jk entry of I_n. Therefore, by linearity,

$$\langle f, K_\omega\rangle_\mathcal{H} = f(\omega) \quad \text{for every } f \in \mathcal{H}$$

and hence $K_\omega(\lambda)$ is the reproducing kernel for $\mathcal{H}$. $\diamond$

Exercise 43.1. Show that in the setting of Example 43.1, the reproducing kernel $K_\omega(\lambda)$ defined by formula (43.3) can also be expressed as

$$K_\omega(\lambda) = F(\lambda)G^{-1}F(\omega)^H = -(\det G)^{-1}\det\begin{bmatrix} G & F(\omega)^H \\ F(\lambda) & 0 \end{bmatrix}$$

with $F(\lambda) = \begin{bmatrix} f_1(\lambda) & \cdots & f_n(\lambda)\end{bmatrix}$.

Exercise 43.2. Show that if the space $\mathcal{M}$ considered in Example 43.1 is a proper subspace of an inner product space $\mathcal{U}$ and $\langle g, h\rangle_\mathcal{M} = \langle g, h\rangle_\mathcal{U}$ for every choice of $g, h \in \mathcal{M}$, then the orthogonal projection $\Pi_\mathcal{M} f$ of $f \in \mathcal{U}$ at $\omega \in \Omega$ is equal to

$$(\Pi_\mathcal{M} f)(\omega) = \langle f, K_\omega\rangle_\mathcal{U}.$$

43.2. de Branges spaces

Classical **de Branges spaces** are reproducing kernel Hilbert spaces of **entire functions** (i.e., functions that are holomorphic on the full complex plane) with reproducing kernels of the form

(43.4) $$K_\omega(\lambda) = \frac{E_+(\lambda)\overline{E_+(\omega)} - E_-(\lambda)\overline{E_-(\omega)}}{-2\pi i(\lambda - \overline{\omega})} \quad \text{for } \lambda \neq \overline{\omega}$$

and inner product

(43.5) $$\langle f, g\rangle_{\Delta_\mathfrak{E}} = \int_{-\infty}^\infty \overline{g(\mu)}\Delta_\mathfrak{E}(\mu)f(\mu)d\mu \quad \text{with } \Delta_\mathfrak{E}(\mu) = |E_+(\mu)|^{-2},$$

where $E_\pm$ are entire functions that enjoy the following properties:

(1) $|E_+(\lambda)| > 0$ for all points λ in the closed upper half-plane.
(2) $|(E_+^{-1}E_-)(\lambda)| \leq 1$ for all points λ in the closed upper half-plane with equality for $\lambda \in \mathbb{R}$.

Analogous conditions can be formulated for the unit disc $\mathbb{D} = \{\lambda \in \mathbb{C} : |\lambda| < 1\}$. The reproducing kernel $K_\omega(\lambda)$ is then of the form

$$(43.6) \quad K_\omega(\lambda) = \frac{E_+(\lambda)\overline{E_+(\omega)} - \lambda\overline{\omega}\, E_-(\lambda)\overline{E_-(\omega)}}{1 - \lambda\overline{\omega}} \quad \text{for } \lambda\overline{\omega} \neq 1$$

where the functions $E_\pm(\lambda)$ are subject to the following constraints:

(3) $|E_+(\lambda)| > 0$ for all points λ in the closed unit disc.
(4) $|(E_+^{-1}E_-)(\lambda)| \leq 1$ for all points λ in the closed unit disc with equality for $|\lambda| = 1$.

The inner product in this setting is

$$(43.7) \quad \langle f, g \rangle_{\Delta_{\mathcal{E}}} = \frac{1}{2\pi} \int_0^{2\pi} \overline{g(e^{i\theta})} |E_+(e^{i\theta})|^{-2} f(e^{i\theta}) d\theta$$

To distinguish between these two cases, we shall refer to the space in the first (resp., second) setting as a de Branges space with respect to $\mathbb{C}_+$ (resp., $\mathbb{D}$).

43.3. The space of polynomials of degree $\leq n - 1$

For the rest of this chapter we shall focus on the special case that $\mathcal{H}$ is equal to the n-dimensional vector space of polynomials of degree $\leq n-1$, equipped with an inner product and $\Omega = \mathbb{C}$.

This is a nice space to work with because the vectors in $\mathcal{H}$ are entire functions and $\mathcal{H}$ is invariant under the action of the **generalized backward shift operator** R_α, which is defined by the formula

$$(43.8) \quad (R_\alpha f)(\lambda) = \begin{cases} \dfrac{f(\lambda) - f(\alpha)}{\lambda - \alpha} & \text{if } \lambda \neq \alpha, \\ f'(\alpha) & \text{if } \lambda = \alpha \end{cases}$$

for every $\alpha \in \mathbb{C}$: If $f(\lambda) = c_0 + c_1\lambda + \cdots + c_{n-1}\lambda^{n-1}$, then

$$\frac{f(\lambda) - f(\alpha)}{\lambda - \alpha} = \sum_{j=1}^{n-1} c_j \frac{\lambda^j - \alpha^j}{\lambda - \alpha} = \sum_{j=1}^{n-1} c_j \sum_{i=1}^{j-1} \alpha^{j-1-i}\lambda^i,$$

which is a polynomial in λ of degree $\leq n - 2$.

It is convenient to describe the space $\mathcal{H}$ in terms of the matrix-valued function $F(\lambda) = \begin{bmatrix} 1 & \lambda & \cdots & \lambda^{n-1} \end{bmatrix}$:

$$(43.9) \quad \mathcal{H} = \{F(\lambda)\mathbf{u} : \mathbf{u} \in \mathbb{C}^n\}.$$

43.3. The space of polynomials of degree $\leq n-1$

In particular

$$(R_0 F)(\lambda)\mathbf{u} = \frac{F(\lambda) - F(0)}{\lambda}\mathbf{u} = \begin{bmatrix} 0 & 1 & \cdots & \lambda^{n-2} \end{bmatrix}\mathbf{u} = F(\lambda)A\mathbf{u}$$

with $A = C_0^{(n)}$, the $n \times n$ Jordan cell with 0 on the diagonal. Consequently,

$$F(\lambda)(I_n - \lambda A) = F(0),$$

and hence, upon setting $F(0) = C$, we obtain the formulas

(43.10) $$F(\lambda) = C(I_n - \lambda A)^{-1}$$

and

(43.11) $$(R_\alpha F)(\lambda) = F(\lambda)A(I_n - \alpha A)^{-1}.$$

The formula

(43.12) $$\langle F\mathbf{u}, F\mathbf{v}\rangle_\mathcal{H} = \mathbf{v}^H G \mathbf{u}$$

defines an inner product on $\mathcal{H}$ for every $G \in \mathbb{C}^{n \times n}$ that is positive definite.

Lemma 43.2. *Let $\mathcal{H}$ denote the n-dimensional inner product space of polynomials of degree $\leq n-1$ equipped with the inner product (43.12). Then $\mathcal{H}$ is a reproducing kernel Hilbert space with reproducing kernel*

(43.13) $$K_\omega(\lambda) = F(\lambda)G^{-1}F(\omega)^H.$$

Proof. This is a particular case of Example 43.1. □

For each choice of $\alpha \in \mathbb{C}$, let

(43.14) $$\mathcal{H}_\alpha = \{f \in \mathcal{H} : f(\alpha) = 0\}.$$

Lemma 43.3. *The space $\mathcal{H}_\alpha$ is a reproducing kernel Hilbert space with inner product (43.12) and reproducing kernel*

(43.15) $$K_\omega^{(\alpha)}(\lambda) = K_\omega(\lambda) - K_\alpha(\lambda)K_\alpha(\alpha)^{-1}K_\omega(\alpha)$$

for each point $\alpha \in \mathbb{C}$.

Proof. Since $K_\alpha^{(\alpha)}(\alpha) = 0$, it is clear that $K_\alpha^{(\alpha)} \in \mathcal{H}_\alpha$. Moreover, if $f \in \mathcal{H}_\alpha$, then

$$\langle f, K_\omega^{(\alpha)}\rangle_\mathcal{H} = \langle f, K_\omega\rangle_\mathcal{H} - \langle f, K_\alpha K_\alpha(\alpha)^{-1}K_\omega(\alpha)\rangle_\mathcal{H}$$
$$= f(\omega) - K_\alpha(\omega)K_\alpha(\alpha)^{-1}f(\alpha) = f(\omega),$$

since $f(\alpha) = 0$ when $f \in \mathcal{H}_\alpha$. □

43.4. Two subspaces

In this section, we shall have special interest in two subspaces of the inner product space $\mathcal{H}$ considered in Lemma 43.2:

(43.16) $\quad \mathcal{H}_0 = \{f \in \mathcal{H} : f(0) = 0\} \quad \text{and} \quad \mathcal{H}_\bullet = \{R_0 f : f \in \mathcal{H}_0\}.$

To describe these spaces and their reproducing kernels, it is convenient to let

$$\Gamma = G^{-1} \quad \text{and} \quad \mathbf{e}_1, \ldots, \mathbf{e}_n \quad \text{denote the standard basis for } \mathbb{C}^n.$$

Then

(43.17) $\quad \mathcal{H}_0 = \{F\mathbf{v} : \mathbf{v} \in \mathbb{C}^n \text{ and } \mathbf{e}_1^H \mathbf{v} = 0\}$

and, in view of formula (43.15), the reproducing kernel for $\mathcal{H}_0$ is

(43.18) $\quad K_\omega^{(0)}(\lambda) = K_\omega(\lambda) - F(\lambda)\Gamma\mathbf{e}_1 \left(\mathbf{e}_1^H \Gamma \mathbf{e}_1\right)^{-1} \mathbf{e}_1^H \Gamma F(\omega)^H.$

Lemma 43.4. *The space* $\mathcal{H}_\bullet = \{R_0 f : f \in \mathcal{H}_0\}$ *can be described as*

(43.19) $\quad \mathcal{H}_\bullet = \{F\mathbf{v} : \mathbf{v} \in \mathbb{C}^n \text{ and } \mathbf{e}_n^H \mathbf{v} = 0\};$

it is a reproducing kernel Hilbert space with reproducing kernel

(43.20) $\quad K_\omega^{(\bullet)}(\lambda) = K_\omega(\lambda) - F(\lambda)\Gamma\mathbf{e}_n \left(\mathbf{e}_n^H \Gamma \mathbf{e}_n\right)^{-1} \mathbf{e}_n^H \Gamma F(\omega)^H.$

Proof. Observe first that $f = F\mathbf{v} \in \mathcal{H}_0$ if and only if $f(0) = F(0)\mathbf{v} = \mathbf{e}_1^H \mathbf{v} = 0$. Thus,

$$\mathcal{H}_0 = \left\{ F \begin{bmatrix} 0 \\ \mathbf{u} \end{bmatrix} : \mathbf{u} \in \mathbb{C}^{n-1} \right\}$$

and, as

$$R_0 F \begin{bmatrix} 0 \\ \mathbf{u} \end{bmatrix} = FA \begin{bmatrix} 0 \\ \mathbf{u} \end{bmatrix} = F \begin{bmatrix} \mathbf{u} \\ 0 \end{bmatrix},$$

we see that

$$\mathcal{H}_\bullet = \left\{ F \begin{bmatrix} \mathbf{u} \\ 0 \end{bmatrix} : \mathbf{u} \in \mathbb{C}^{n-1} \right\} = \{F\mathbf{v} : \mathbf{v} \in \mathbb{C}^n \text{ and } \mathbf{e}_n^H \mathbf{v} = 0\}.$$

Formula (43.20) implies that

$$K_\omega^{(\bullet)}(\lambda) = F(\lambda) B F(\omega)^H \quad \text{with} \quad B = \Gamma - \Gamma\mathbf{e}_n \left(\mathbf{e}_n^H \Gamma \mathbf{e}_n\right)^{-1} \mathbf{e}_n^H \Gamma.$$

Thus, as $\mathbf{e}_n^H B F(\omega)^H = 0$ for every point $\omega \in \mathbb{C}$, it follows that $K_\omega^{(\bullet)} \in \mathcal{H}_\bullet$. Moreover, if $f \in \mathcal{H}_\bullet$, then

$$\langle F\mathbf{v}, K_\omega^{(\bullet)} \rangle_\mathcal{H} = F(\omega) B^H G\mathbf{v} = F(\omega) BG\mathbf{v}$$
$$= F(\omega)(I_n - \Gamma\mathbf{e}_n \left(\mathbf{e}_n^H \Gamma \mathbf{e}_n\right)^{-1} \mathbf{e}_n^H)\mathbf{v} = F(\omega)\mathbf{v},$$

since $\mathbf{e}_n^H \mathbf{v} = 0$. $\quad\square$

43.5. G is a Toeplitz matrix

In this section we shall identify $\mathcal{H}$ as a de Branges space with respect to $\mathbb{D}$ when the matrix G is a Toeplitz matrix. The **notation**

$$(43.21) \qquad G_{[k,m]} = \begin{bmatrix} g_{kk} & \cdots & g_{km} \\ \vdots & & \vdots \\ g_{mk} & \cdots & g_{mm} \end{bmatrix} \quad \text{for } 1 \leq k \leq m \leq n$$

(which appeared earlier in (16.5)) and

$$(43.22) \qquad R_\bullet : f \in \mathcal{H}_\bullet \mapsto \lambda f(\lambda)$$

will be useful.

Lemma 43.5. *The operator R_0 maps $\mathcal{H}_0$ isometrically onto $\mathcal{H}_\bullet$ if and only if the positive definite matrix G that defines the inner product (43.12) in $\mathcal{H}$ is a Toeplitz matrix.*

Proof. It is readily checked that $R_\bullet R_0 f = f$ for every $f \in \mathcal{H}_0$. Moreover, R_0 maps $\mathcal{H}_0$ onto $\mathcal{H}_\bullet$, because if $f \in \mathcal{H}_\bullet$, then $R_\bullet f \in \mathcal{H}_0$ and $R_0 R_\bullet f = f$.

If $f, g \in \mathcal{H}_0$, then $f = F\mathbf{v}$, $g = F\mathbf{w}$ for vectors $\mathbf{v}, \mathbf{w} \in \mathbb{C}^n$ that are subject to the constraints $\mathbf{e}_1^H \mathbf{v} = 0$ and $\mathbf{e}_1^H \mathbf{w} = 0$. Thus, if $\mathbf{v} = \begin{bmatrix} 0 \\ \mathbf{x} \end{bmatrix}$ and $\mathbf{w} = \begin{bmatrix} 0 \\ \mathbf{y} \end{bmatrix}$ with $\mathbf{x}, \mathbf{y} \in \mathbb{C}^{n-1}$, then, in terms of the notation (43.21),

$$\langle F\mathbf{v}, F\mathbf{w} \rangle_\mathcal{H} = \mathbf{w}^H G \mathbf{v} = \mathbf{y}^H G_{[2,n]} \mathbf{x}$$

and

$$\langle R_0 F\mathbf{v}, R_0 F\mathbf{w} \rangle_\mathcal{H} = \langle FA\mathbf{v}, FA\mathbf{w} \rangle_\mathcal{H} = \mathbf{w}^H A^H G A \mathbf{v} = \mathbf{y}^H G_{[1,n-1]} \mathbf{x}.$$

Thus, R_0 maps $\mathcal{H}_0$ isometrically onto $\mathcal{H}_\bullet$ if and only if $G_{[2,n]} = G_{[1,n-1]}$, i.e., if and only if $g_{ij} = g_{i+1,j+1}$ for $i,j = 1, \ldots, n-1$, i.e., if and only if G is a Toeplitz matrix. $\square$

Exercise 43.3. Show that the operator S_α that is defined by the formula

$$(43.23) \qquad (S_\alpha f)(\lambda) = \frac{1 - \lambda \overline{\alpha}}{\lambda - \alpha} f(\lambda) \quad \text{for } |\alpha| \neq 1$$

maps $\mathcal{H}_\alpha$ isometrically onto $\mathcal{H}_{1/\overline{\alpha}}$ if and only if the positive definite matrix G that defines the inner product (43.12) in $\mathcal{H}$ is a Toeplitz matrix.

We remark that $R_0 = S_0 = \lim_{\alpha \downarrow 0} S_\alpha$ and $R_\bullet = \lim_{\alpha \downarrow 0} S_{1/\overline{\alpha}}$ (i.e., α tends to 0 through positive values).

Lemma 43.6. *If the positive definite matrix G that defines the inner product (43.12) in $\mathcal{H}$ is a Toeplitz matrix, then*

$$(43.24) \qquad \lambda\overline{\omega}\, K_\omega^{(\bullet)}(\lambda) = K_\omega^{(0)}(\lambda)$$

for every pair of points $\lambda, \omega \in \mathbb{C}$ and the reproducing kernel for $\mathcal{H}$ admits the representation

$$(43.25)\quad K_\omega(\lambda) = F(\lambda)\Gamma\overline{F(\omega)} = \frac{E_+(\lambda)\overline{E_+(\omega)} - \lambda\overline{\omega}\, E_-(\lambda)\overline{E_-(\omega)}}{1 - \lambda\overline{\omega}},\ \lambda\overline{\omega} \neq 1,$$

in terms of the matrix polynomials

$$(43.26)\qquad E_+(\lambda) = \frac{F(\lambda)\Gamma\mathbf{e}_1}{(\mathbf{e}_1^H\Gamma\mathbf{e}_1)^{1/2}} \quad \text{and} \quad E_-(\lambda) = \frac{F(\lambda)\Gamma\mathbf{e}_n}{(\mathbf{e}_n^H\Gamma\mathbf{e}_n)^{1/2}}.$$

Proof. Formula (43.24) follows easily from the evaluations

$$\langle R_0 K_\lambda^{(0)}, \overline{\omega} K_\omega^{(\bullet)}\rangle_\mathcal{H} = \omega(R_0 K_\lambda^{(0)})(\omega) = K_\lambda^{(0)}(\omega)$$

and

$$\langle \overline{\omega} K_\omega^{(\bullet)}, R_0 K_\lambda^{(0)}\rangle_\mathcal{H} = \langle \overline{\omega} R_\bullet K_\omega^{(\bullet)}, K_\lambda^{(0)}\rangle_\mathcal{H} = \lambda\overline{\omega} K_\omega^{(\bullet)}(\lambda)\,.$$

Formulas (43.25) and (43.26) are obtained by substituting the expressions (43.18) and (43.20) into (43.24) and then solving for $K_\omega(\lambda)$, the reproducing kernel for the full space $\mathcal{H}$. $\square$

Exercise 43.4. Show that if $G \in \mathbb{C}^{n\times n}$ is a positive definite Toeplitz matrix and $\Gamma = G^{-1}$, then

$$(43.27) \qquad \mathbf{e}_1^H\Gamma\mathbf{e}_1 = \mathbf{e}_n^H\Gamma\mathbf{e}_n\,.$$

Theorem 43.7. *If the positive definite matrix G that defines the inner product (43.12) in $\mathcal{H}$ is a Toeplitz matrix, then the space $\mathcal{H}$ is a de Branges space with respect to $\mathbb{D}$: Its reproducing kernel can be expressed in the form (43.25) with entries $E_\pm(\lambda)$ specified by (43.26) and inner product*

$$(43.28)\qquad \langle f, g\rangle_{\Delta_\mathfrak{E}} = \frac{1}{2\pi}\int_0^{2\pi} \overline{g(e^{i\theta})}|E_+(e^{i\theta})|^{-2} f(e^{i\theta})d\theta = \langle f, g\rangle_\mathcal{H}$$

for every pair of functions $f, g \in \mathcal{H}$, i.e.,

$$(43.29)\ \langle F\mathbf{u}, F\mathbf{v}\rangle_{\Delta_\mathfrak{E}} = \frac{1}{2\pi}\int_0^{2\pi} \mathbf{v}^H F(e^{i\theta})^H\, |E_+(e^{i\theta})|^{-2}\, F(e^{i\theta})\mathbf{u}\, d\theta = \mathbf{v}^H G\mathbf{u}$$

for every choice of $\mathbf{u}, \mathbf{v} \in \mathbb{C}^n$.

Proof. We have already obtained the basic identity (43.25) in tems of the polynomials $E_\pm$ in (43.26). It remains to show that E_+ and $E_+^{-1}E_-$ meet the constraints (3) and (4) in Section 43.2. The proof is divided into parts.

1. $|E_+(\omega)| > 0$ if $|\omega| \le 1$ and $|E_+(\omega)| = |E_-(\omega)|$ if $|\omega| = 1$.

Formula (43.25) clearly implies that
$$|E_+(\omega)|^2 = |\omega|^2 |E_-(\omega)|^2 + (1 - |\omega|^2) K_\omega(\omega) > 0$$
for every point $\omega \in \mathbb{D}$, since $K_\omega(\omega) > 0$, and that
$$|E_+(\omega)|^2 = |\omega|^2 |E_-(\omega)|^2 \quad \text{if } |\omega| = 1\,.$$
Consequently, $|E_+(\omega)| > 0$ when $|\omega| = 1$, because otherwise (by (43.25)),
$$F(\lambda) \Gamma F(\omega)^H = 0 \quad \text{for every point } \lambda \in \mathbb{D} \text{ when } |\omega| = 1\,.$$
But this in turn would imply that $\Gamma F(\omega)^H = \mathbf{0}$ when $|\omega| = 1$ and hence, as Γ is invertible, that $F(\omega)^H = \mathbf{0}$ when $|\omega| = 1$, which is false.

The inequality $|E_+(\omega)| > 0$ when $|\omega| = 1$ ensures that the inner product (43.28) is well-defined.

2. $\langle f, K_\omega \rangle_{\Delta_{\mathfrak{E}}} = f(\omega)$ for $f \in \mathcal{H}$.

Let $\rho_\omega(\lambda) = 1 - \lambda\overline{\omega}$ and then, using formula (43.25) for the reproducing kernel, check that for every $f \in \mathcal{H}$ and every $\omega \in \mathbb{D}$
$$\left\langle f, \frac{E_+ \overline{E_+(\omega)}}{\rho_\omega} \right\rangle_{\Delta_{\mathfrak{E}}} = \frac{1}{2\pi} \int_0^{2\pi} \frac{f(e^{i\theta})}{E_+(e^{i\theta})} \frac{1}{(1 - e^{-i\theta}\omega)} E_+(\omega) d\theta = f(\omega)$$
and (as $|E_+(e^{i\theta})| = |E_-(e^{i\theta})|$)
$$\left\langle f, \frac{(1 - \rho_\omega) E_- \overline{E_-(\omega)}^H}{\rho_\omega} \right\rangle_{\Delta_{\mathfrak{E}}} = \frac{1}{2\pi} \int_0^{2\pi} \frac{f(e^{i\theta})}{E_-(e^{i\theta})} \frac{e^{-i\theta}\omega}{(1 - e^{-i\theta}\omega)} E_-(\omega) d\theta = 0\,.$$
These two integrals can be evaluated either by using Cauchy's formula from complex function theory or by using the fact that $\int_0^{2\pi} e^{ik\theta} d\theta = 0$ for every nonzero integer k. This justifies the formula
$$f(\omega) = \frac{1}{2\pi} \int_0^{2\pi} f(e^{i\theta}) \Delta_{\mathfrak{E}}(e^{i\theta}) F(\omega) \Gamma F(e^{i\theta})^H d\theta$$
for $f \in \mathcal{H}$ and $\omega \in \mathbb{D}$. However, since both sides are entire functions of ω, the formula is valid for all points $\omega \in \mathbb{C}$.

3. *Verification of* (43.28).

Let Q denote the Gram matrix of the functions $f_j(\lambda) = \lambda^j$, $j = 0, \ldots, n-1$, with respect to the inner product (43.28). Then, in view of step 2,
$$F(\omega)\mathbf{u} = F(\omega)\Gamma G \mathbf{u} = \langle F\mathbf{u}, K_\omega \rangle_\mathcal{H} = \langle F\mathbf{u}, F\Gamma F(\omega)^H \rangle_\mathcal{H}$$
$$= \langle F\mathbf{u}, K_\omega \rangle_{\Delta_{\mathfrak{E}}} = \langle F\mathbf{u}, F\Gamma F(\omega)^H \rangle_{\Delta_{\mathfrak{E}}} = F(\omega)\Gamma Q \mathbf{u}$$
for every choice of $\omega \in \mathbb{C}$ and $\mathbf{u} \in \mathbb{C}^n$. Therefore, $\Gamma G = \Gamma Q$, and hence $G = Q$, since Γ is invertible. Therefore, (43.28) holds. $\square$

43.6. G is a Hankel matrix

In this section we shall identify $\mathcal{H}$ as a de Branges space with respect to $\mathbb{C}_+$ when the matrix G that defines the inner product (43.12) is a Hankel matrix. The analysis exploits the interplay between the operators

$$(43.30) \qquad (T_i f)(\lambda) = \frac{\lambda + i}{\lambda - i} f(\lambda) \quad \text{and} \quad (T_{-i} f)(\lambda) = \frac{\lambda - i}{\lambda + i} f(\lambda)$$

for $f \in \mathcal{H}_i$ and $f \in \mathcal{H}_{-i}$, respectively

Lemma 43.8. *The operator T_i maps $\mathcal{H}_i$ isometrically onto $\mathcal{H}_{-i}$ if and only if the positive definite matrix G that defines the inner product (43.12) in $\mathcal{H}$ is a Hankel matrix.*

Proof. Since

$$\frac{\lambda + i}{\lambda - i} f(\lambda) = f(\lambda) + 2i \frac{f(\lambda)}{\lambda - i} = f(\lambda) + 2i (R_i f)(\lambda)$$

for $f \in \mathcal{H}_i$, it is readily checked with the help of (43.11) that

$$T_i F \mathbf{u} = F \mathbf{v} \quad \text{with} \quad \mathbf{v} = (I_n + iA)(I_n - iA)^{-1} \mathbf{u}.$$

Consequently,

$$\langle F\mathbf{u}, F\mathbf{u} \rangle_{\mathcal{H}} - \langle T_i F\mathbf{u}, T_i F\mathbf{u} \rangle_{\mathcal{H}} = \mathbf{u}^H G \mathbf{u} - \mathbf{v}^H G \mathbf{v}$$
$$= \mathbf{u}^H (I_n + iA^H)^{-1} X (I_n - iA)^{-1} \mathbf{u},$$

with

$$X = (I_n + iA^H) G (I_n - iA) - (I_n - iA^H) G (I_n + iA)$$
$$= 2i(A^H G - GA).$$

Thus, isometry holds for all $f \in \mathcal{H}_i$ if and only if

$$\mathbf{u}^H (I_n + iA^H)^{-1} (A^H G - GA)(I_n - iA)^{-1} \mathbf{u} = 0$$

for all vectors $\mathbf{u} \in \mathbb{C}^n$ such that $\mathbf{e}_1^H (I_n - iA)^{-1} \mathbf{u} = 0$, i.e., if and only if

$$\begin{bmatrix} 0 & \mathbf{w}^H \end{bmatrix} (A^H G - GA) \begin{bmatrix} 0 \\ \mathbf{w} \end{bmatrix} \quad \text{for every } \mathbf{w} \in \mathbb{C}^{n-1}.$$

In view of Exercise 11.24, the last condition holds if and only if

$$\left[A^H G - GA \right]_{[2,n]} = O,$$

i.e., if and only if $g_{ij} = g_{i+1,j-1}$ for $i = 1, \ldots, n-1$ and $j = 2, \ldots, n$. Thus, the isometry holds if and only if G is a Hankel matrix. $\square$

Exercise 43.5. Show that the operator T_α that is defined by the formula
$$(T_\alpha f)(\lambda) = \frac{\lambda - \overline{\alpha}}{\lambda - \alpha} f(\lambda)$$
maps $\mathcal{H}_\alpha$ isometrically onto $\mathcal{H}_{\overline{\alpha}}$ if and only if the positive definite matrix G that defines the inner product (43.12) in $\mathcal{H}$ is a Hankel matrix. [REMARK: The assertion is correct for every $\alpha \in \mathbb{C}$, but only of interest for $\alpha \notin \mathbb{R}$.]

Exercise 43.6. Show that $T_{\overline{\alpha}}$ acting from $\mathcal{H}_{\overline{\alpha}}$ to $\mathcal{H}_\alpha$ is equal to the adjoint T_α^* of the operator T_α acting from $\mathcal{H}_\alpha$ to $\mathcal{H}_{\overline{\alpha}}$.

Lemma 43.9. *If the positive definite matrix G that defines the inner product (43.12) in $\mathcal{H}$ is a Hankel matrix, then*

(43.31) $$\frac{\lambda - i}{\lambda + i} K_\omega^{(-i)}(\lambda) = \frac{\overline{\omega} - i}{\overline{\omega} + i} K_\omega^{(i)}(\lambda).$$

Proof. Since $T_i K_\lambda^{(i)} \in \mathcal{H}_{-i}$ and $T_{-i} K_\lambda^{(-i)} \in \mathcal{H}_i$,
$$\langle T_i K_\lambda^{(i)}, K_\omega^{(-i)} \rangle_\mathcal{H} = (T_i K_\lambda^{(i)})(\omega) = \frac{\omega + i}{\omega - i} K_\lambda^{(i)}(\omega)$$
and
$$\langle T_{-i} K_\omega^{(-i)}, K_\lambda^{(i)} \rangle_\mathcal{H} = (T_{-i} K_\omega^{(-i)})(\lambda) = \frac{\lambda - i}{\lambda + i} K_\omega^{(-i)}(\lambda).$$
However, since
$$\langle T_{-i} K_\omega^{(-i)}, K_\lambda^{(i)} \rangle_\mathcal{H} = \langle K_\omega^{(-i)}, T_i K_\lambda^{(i)} \rangle_\mathcal{H} = \overline{\langle T_i K_\lambda^{(i)}, K_\omega^{(-i)} \rangle_\mathcal{H}},$$
it is readily seen that (43.31) holds. □

Theorem 43.10. *If the positive definite matrix G that defines the inner product (43.12) in $\mathcal{H}$ is a Hankel matrix, then*

(43.32) $$K_\omega(\lambda) = \frac{E_+(\lambda)\overline{E_+(\omega)} - E_-(\lambda)\overline{E_-(\omega)}}{-2\pi i(\lambda - \overline{\omega})} \quad \text{for } \lambda \neq \overline{\omega},$$

where

(43.33) $$\begin{aligned} E_+(\lambda) &= \pi(\lambda + i)K_i(\lambda)(\pi K_i(i))^{-1/2} \quad \text{and} \\ E_-(\lambda) &= \pi(\lambda - i)K_{-i}(\lambda)(\pi K_{-i}(-i))^{-1/2}. \end{aligned}$$

Proof. Substituting (43.15) with $\alpha = \pm i$ into (43.31) yields the identity
$$\frac{\lambda - i}{\lambda + i} \left(K_\omega(\lambda) - K_{-i}(\lambda) K_{-i}(-i)^{-1} K_\omega(-i) \right)$$
$$= \frac{\overline{\omega} - i}{\overline{\omega} + i} \left(K_\omega(\lambda) - K_i(\lambda) K_i(i)^{-1} K_\omega(i) \right).$$

Formulas (43.32) and (43.33) are then obtained by straightforward computation. □

Theorem 43.11. *If the positive definite matrix G that defines the inner product (43.12) in $\mathcal{H}$ is a Hankel matrix, then $\mathcal{H}$ is a de Branges space with respect to $\mathbb{C}_+$: Its reproducing kernel can be expressed in the form (43.32) with entries $E_\pm(\lambda)$ specified by (43.33) and inner product*

$$\langle f, g \rangle_{\Delta_\mathfrak{E}} = \int_{-\infty}^{\infty} \overline{g(\mu)} |E_+(\mu)|^{-2} f(\mu) d\mu \tag{43.34}$$

for every pair of functions $f, g \in \mathcal{H}$, i.e.,

$$\langle F\mathbf{u}, F\mathbf{v} \rangle_{\Delta_\mathfrak{E}} = \int_{\infty}^{\infty} \mathbf{v}^H F(\mu)^H |E_+(\mu)|^{-2} F(\mu) \mathbf{u} d\mu = \mathbf{v}^H G \mathbf{u} \tag{43.35}$$

for every choice of $\mathbf{u}, \mathbf{v} \in \mathbb{C}^n$.

Proof. We have already obtained the basic identity (43.32). The rest of the proof is divided into parts.

1. $|E_+(\omega)| > 0$ *if* $\Im \omega \geq 0$ *and* $|E_+(\omega)| = |E_-(\omega)|$ *if* $\omega \in \mathbb{R}$.

Formula (43.32) clearly implies that

$$|E_+(\omega)|^2 - |E_-(\omega)|^2 = -2\pi i (\omega - \overline{\omega}) K_\omega(\omega) > 0$$

for every point $\omega \in \mathbb{C}_+$, since $K_\omega(\omega) > 0$, and that

$$|E_+(\omega)| = |E_-(\omega)| \quad \text{if } \omega \in \mathbb{R}.$$

Consequently, $|E_+(\omega)| > 0$ when $\omega \in \mathbb{R}$, because otherwise,

$$F(\lambda) \Gamma F(\omega)^H = 0 \quad \text{for every point } \lambda \in \mathbb{C} \text{ when } \omega \in \mathbb{R}.$$

But this in turn implies that $\Gamma F(\omega)^H = \mathbf{0}$ when $\omega \in \mathbb{R}$ and hence, as Γ is invertible, that $F(\omega)^H = \mathbf{0}$ when $\omega \in \mathbb{R}$, which is false.

The last inequality ensures that the inner product (43.34) is well-defined.

2. $E_\pm$ *are polynomials of degree n and* $|E_-(\omega)| < 0$ *if* $\Im \omega < 0$.

It is clear from the formulas in (43.33) and the identity $|E_+(\omega)| = |E_-(\omega)|$ for $\omega \in \mathbb{R}$ that $E_\pm$ are both polynomials of degree m and $m \leq n$. On the other hand, in view of formulas (43.13) with $\Gamma = G^{-1}$ and (43.32),

$$\lim_{\nu \uparrow \infty} \frac{|E_+(i\nu)|^2 - |E_-(i\nu)|^2}{\nu^{2n-1}} = 4\pi \mathbf{e}_n^T \Gamma \mathbf{e}_n > 0.$$

But this is only viable if $m = n$ (and $\lim_{\nu \uparrow \infty} \nu^{-n}\{|E_+(i\nu)| - |E_-(i\nu)|\} = 0$). The proof that $|E_-(\omega)| < 0$ if $\Im \omega < 0$ is similar to the justification of step 1 and is left to the reader.

3. $\langle f, K_\omega \rangle_{\Delta_\mathfrak{E}} = f(\omega)$ *for* $f \in \mathcal{H}$.

Let $\rho_\omega(\lambda) = -2\pi i(\lambda - \overline{\omega})$ and then, using formula (43.32) for the reproducing kernel, check that for every $f \in \mathcal{H}$ and every $\omega \in \mathbb{C}_+$

$$\left\langle f, \frac{E_+ \overline{E_+(\omega)}}{\rho_\omega} \right\rangle_{\Delta_\mathfrak{E}} = \frac{1}{2\pi i} \int_{-\infty}^\infty \frac{f(\mu)}{E_+(\mu)} \frac{1}{(\mu - \omega)} E_+(\omega) d\mu = f(\omega)$$

and

$$\left\langle f, \frac{E_- \overline{E_-(\omega)}}{\rho_\omega} \right\rangle_{\Delta_\mathfrak{E}} = \frac{1}{2\pi i} \int_{-\infty}^\infty \frac{f(\mu)}{E_-(\mu)} \frac{1}{(\mu - \omega)} E_-(\omega) d\mu = 0\,.$$

These two integrals can be evaluated by using Cauchy's formula from complex function theory to evaluate the indicated integrals with limits $\pm R$ and then letting $R \uparrow \infty$. This justifies the formula

$$f(\omega) = \int_{-\infty}^\infty f(\mu) \Delta_\mathfrak{E}(\mu) F(\omega) \Gamma F(\mu)^H d\mu$$

for $f \in \mathcal{H}$ and $\omega \in \mathbb{C}_+$. However, since both sides are entire functions of ω, the formula is valid for all points $\omega \in \mathbb{C}$.

4. *Verification of* (43.35).

Let Q denote the Gram matrix of the functions $f_j(\lambda) = \lambda^j$, $j = 0, \ldots, n-1$, with respect to the inner product (43.34). Then, in view of step 2,

$$F(\omega)\mathbf{u} = F(\omega)\Gamma G \mathbf{u} = \langle F\mathbf{u}, F\Gamma F(\omega)^H \rangle_\mathcal{H} = \langle F\mathbf{u}, K_\omega \rangle_\mathcal{H}$$
$$= \langle F\mathbf{u}, K_\omega \rangle_{\Delta_\mathfrak{E}} = \langle F\mathbf{u}, F\Gamma F(\omega)^H \rangle_{\Delta_\mathfrak{E}} = F(\omega)\Gamma Q \mathbf{u}$$

for every choice of $\omega \in \mathbb{C}$ and $\mathbf{u} \in \mathbb{C}^n$. Therefore, $\Gamma G = \Gamma Q$, and hence $G = Q$, since Γ is invertible. Therefore, (43.35) holds. □

43.7. Supplementary notes

Formula (43.31) is a special case of the general formula

$$(43.36) \qquad \frac{\lambda - \alpha}{\lambda - \overline{\alpha}} K_\omega^{(\overline{\alpha})}(\lambda) = \frac{\overline{\omega} - \alpha}{\overline{\omega} - \overline{\alpha}} K_\omega^{(\alpha)}(\lambda)$$

which is valid for every point $\alpha \notin \mathbb{R}$ when the operator

$$T_\alpha : f(\lambda) \mapsto \frac{\lambda - \overline{\alpha}}{\lambda - \alpha} f(\lambda) = ((I + (\alpha - \overline{\alpha})R_\alpha)f)(\lambda)$$

maps $\mathcal{H}_\alpha$ isometrically onto $\mathcal{H}_{\overline{\alpha}}$. This identity is due to L. de Branges [21] and was used by him in conjunction with (43.15) to characterize reproducing kernel Hilbert spaces of entire functions that admit reproducing kernels $K_\omega(\lambda)$ of the form (43.32), i.e., in the terminology of this chapter, reproducing kernel Hilbert spaces of entire functions that are de Branges spaces with respect to $\mathbb{C}_+$. It extends to spaces of vector-valued functions; see, e.g., [**34**] and, for more information on such spaces, [**6**]. Analogous characterizations of vector-valued de Branges spaces with respect to $\mathbb{D}$ are furnished in [**31**]. To ease the reading, the analysis in this chapter was carried out for inner products based on positive definite matrices G. However, many of the basic identities remain valid when G is only assumed to be an invertible Hermitian matrix such that $\mathbf{e}_n^H G^{-1} \mathbf{e}_n \neq 0$. This in turn yields formulas that connect the number of zeros of the polynomials $E_\pm(\lambda)$ (or $\det E_\pm(\lambda)$ in the vector case) with the inertia of G; see, e.g., [**27**] and [**29**].

In the setting of Theorem 43.7 (wherein G is a Toeplitz matrix with $g_{jk} = t_{j-k}$ for $j,k = 1,\ldots,n$), formula (43.29) exhibits $(2\pi)^{-1}|E_+(e^{i\theta})|^{-2}$ as a solution of the **truncated trigonometric moment problem**:

$$\frac{1}{2\pi}\int_0^{2\pi} e^{ik\theta}|E_+(e^{i\theta})|^{-2}d\theta = t_k \quad \text{for } k = -(n-1),\ldots,(n-1).$$

Analogously, in the setting of Theorem 43.11 (wherein G is a Hankel matrix with $g_{jk} = h_{j+k-2}$ for $j,k = 1,\ldots,n$), formula (43.35) exhibits $|E_+(\mu)|^{-2}d\mu$ as a solution of the **truncated Hamburger moment problem**:

$$\int_{-\infty}^{\infty} \mu^k |E_+(\mu)|^{-2} d\mu = h_k \quad \text{for } k = 0,\ldots,2n-2.$$

Bibliography

[1] John G. Aiken, John A. Erdos, and Jerome A. Goldstein, *Unitary approximation of positive operators*, Illinois J. Math. **24** (1980), no. 1, 61–72. MR550652

[2] Alkiviadis G. Akritas, Evgenia K. Akritas, and Genadii I. Malaschonok, *Various proofs of Sylvester's (determinant) identity*, Symbolic computation, new trends and developments (Lille, 1993), Math. Comput. Simulation **42** (1996), no. 4-6, 585–593, DOI 10.1016/S0378-4754(96)00035-3. MR1430843

[3] T. Ando, *Majorizations and inequalities in matrix theory*, Linear Algebra Appl. **199** (1994), 17–67, DOI 10.1016/0024-3795(94)90341-7. MR1274407

[4] Dorin Andrica, *A new problem*, Eur. Math. Soc. Mag. **125** (2022), 53.

[5] Tom M. Apostol, *Mathematical analysis: a modern approach to advanced calculus*, Addison-Wesley Publishing Co., Inc., Reading, Mass., 1957. MR0087718

[6] Damir Z. Arov and Harry Dym, *Multivariate prediction, de Branges spaces, and related extension and inverse problems*, Operator Theory: Advances and Applications, vol. 266, Birkhäuser/Springer, Cham, 2018, DOI 10.1007/978-3-319-70262-9. MR3793176

[7] Sheldon Axler, *Down with determinants!*, Amer. Math. Monthly **102** (1995), no. 2, 139–154, DOI 10.2307/2975348. MR1315593

[8] Mihály Bakonyi and Hugo J. Woerdeman, *Matrix completions, moments, and sums of Hermitian squares*, Princeton University Press, Princeton, NJ, 2011, DOI 10.1515/9781400840595. MR2807419

[9] Rajendra Bhatia, *Matrix analysis*, Graduate Texts in Mathematics, vol. 169, Springer-Verlag, New York, 1997, DOI 10.1007/978-1-4612-0653-8. MR1477662

[10] Rajendra Bhatia, *Perturbation bounds for matrix eigenvalues*, Classics in Applied Mathematics, vol. 53, reprint of the 1987 original, Society for Industrial and Applied Mathematics (SIAM), Philadelphia, PA, 2007, DOI 10.1137/1.9780898719079. MR2325304

[11] Rajendra Bhatia, *Min matrices and mean matrices*, Math. Intelligencer **33** (2011), no. 2, 22–28, DOI 10.1007/s00283-010-9194-z. MR2813259

[12] Béla Bollobás, *Linear analysis: An introductory course*, 2nd ed., Cambridge University Press, Cambridge, 1999, DOI 10.1017/CBO9781139168472. MR1711398

[13] Jonathan M. Borwein and Adrian S. Lewis, *Convex analysis and nonlinear optimization: Theory and examples*, 2nd ed., CMS Books in Mathematics/Ouvrages de Mathématiques de la SMC, vol. 3, Springer, New York, 2006, DOI 10.1007/978-0-387-31256-9. MR2184742

[14] Richard A. Brualdi and Hans Schneider, *Determinantal identities: Gauss, Schur, Cauchy, Sylvester, Kronecker, Jacobi, Binet, Laplace, Muir, and Cayley*, Linear Algebra Appl. **52/53** (1983), 769–791, DOI 10.1016/0024-3795(83)80049-4. MR1500275

[15] John P. Burg, Maximum entropy spectral analysis, in *Modern Spectrum Analysis*, Donald G. Childers, ed., John Wiley & Sons, 1978. pp. 34–41.

[16] John P. Burg, *Maximum entropy spectral analysis*, Ph.D. Thesis, 1975.

[17] Chandler Davis, W. M. Kahan, and H. F. Weinberger, *Norm-preserving dilations and their applications to optimal error bounds*, SIAM J. Numer. Anal. **19** (1982), no. 3, 445–469, DOI 10.1137/0719029. MR656462

[18] Raymond Cheng and Yuesheng Xu, *Minimum norm interpolation in the $\ell_1(\mathbb{N})$ space*, Anal. Appl. (Singap.) **19** (2021), no. 1, 21–42, DOI 10.1142/S0219530520400059. MR4178411

[19] John B. Conway, *A course in functional analysis*, 2nd ed., Graduate Texts in Mathematics, vol. 96, Springer-Verlag, New York, 1990. MR1070713

[20] Branko Ćurgus and Aad Dijksma, *A proof of the main theorem on Bezoutians*, Elem. Math. **69** (2014), no. 1, 33–39, DOI 10.4171/EM/243. MR3182264

[21] Louis de Branges, *Hilbert spaces of entire functions*, Prentice-Hall, Inc., Englewood Cliffs, N.J., 1968. MR0229011

[22] Ilan Degani, *RCMS - right correction Magnus schemes for oscillatory ode's and cubature formulas and oscillatory extensions*, Ph.D. Thesis, 2005.

[23] Emeric Deutsch and Hans Schneider, *Bounded groups and norm-Hermitian matrices*, Linear Algebra Appl. **9** (1974), 9–27, DOI 10.1016/0024-3795(74)90022-6. MR382315

[24] Klaus Diepold, *Intersection of subspaces*, Patrick Dewilde Workshop on Algebra, Networks, Signal Processing and System Theory, Waasenaar (2008).

[25] R. G. Douglas, *On majorization, factorization, and range inclusion of operators on Hilbert space*, Proc. Amer. Math. Soc. **17** (1966), 413–415, DOI 10.2307/2035178. MR203464

[26] Peter Duren, *Invitation to classical analysis*, Pure and Applied Undergraduate Texts, vol. 17, American Mathematical Society, Providence, RI, 2012. MR2933135

[27] Harry Dym, *Hermitian block Toeplitz matrices, orthogonal polynomials, reproducing kernel Pontryagin spaces, interpolation and extension*, Orthogonal matrix-valued polynomials and applications (Tel Aviv, 1987), Oper. Theory Adv. Appl., vol. 34, Birkhäuser, Basel, 1988, pp. 79–135, DOI 10.1007/978-3-0348-5472-6_5. MR1021062

[28] Harry Dym, *J contractive matrix functions, reproducing kernel Hilbert spaces and interpolation*, CBMS Regional Conference Series in Mathematics, vol. 71, Published for the Conference Board of the Mathematical Sciences, Washington, DC; by the American Mathematical Society, Providence, RI, 1989, DOI 10.1090/cbms/071. MR1004239

[29] Harry Dym, *On Hermitian block Hankel matrices, matrix polynomials, the Hamburger moment problem, interpolation and maximum entropy*, Integral Equations Operator Theory **12** (1989), no. 6, 757–812, DOI 10.1007/BF01196878. MR1018213

[30] Harry Dym, *Linear algebra in action*, 2nd ed., Graduate Studies in Mathematics, vol. 78, American Mathematical Society, Providence, RI, 2013, DOI 10.1090/gsm/078. MR3154813

[31] Harry Dym, *Two classes of vector valued de Branges spaces*, J. Funct. Anal. **284** (2023), no. 3, Paper No. 109758, 31, DOI 10.1016/j.jfa.2022.109758. MR4513110

[32] Harry Dym and Israel Gohberg, *Extensions of band matrices with band inverses*, Linear Algebra Appl. **36** (1981), 1–24, DOI 10.1016/0024-3795(81)90215-9. MR604325

[33] Harry Dym and Israel Gohberg, *Extensions of kernels of Fredholm operators*, J. Analyse Math. **42** (1982/83), 51–97, DOI 10.1007/BF02786871. MR729402

[34] Harry Dym and Santanu Sarkar, *Multiplication operators with deficiency indices (p,p) and sampling formulas in reproducing kernel Hilbert spaces of entire vector valued functions*, J. Funct. Anal. **273** (2017), no. 12, 3671–3718, DOI 10.1016/j.jfa.2017.09.007. MR3711878

Bibliography

[35] Carl Eckart and Gale Young, *The approximation of one matrix by another of lower rank*, Psychometrika **1** (1936), 211–218.

[36] Avraham Feintuch, *Robust control theory in Hilbert space*, Applied Mathematical Sciences, vol. 130, Springer-Verlag, New York, 1998, DOI 10.1007/978-1-4612-0591-3. MR1482802

[37] Junichi Fujii, Masatoshi Fujii, Takayuki Furuta, and Ritsuo Nakamoto, *Norm inequalities equivalent to Heinz inequality*, Proc. Amer. Math. Soc. **118** (1993), no. 3, 827–830, DOI 10.2307/2160128. MR1132412

[38] Takayuki Furuta, *Norm inequalities equivalent to Löwner-Heinz theorem*, Rev. Math. Phys. **1** (1989), no. 1, 135–137, DOI 10.1142/S0129055X89000079. MR1041534

[39] I. M. Glazman and Ju. I. Ljubič, *Finite-dimensional linear analysis: a systematic presentation in problem form*, translated from the Russian and edited by G. P. Barker and G. Kuerti, The M.I.T. Press, Cambridge, Mass.-London, 1974. MR0354718

[40] Jochen Glück, *Evolution equations with eventually positive solutions*, Eur. Math. Soc. Mag. **123** (2022), 4–11, DOI 10.4171/mag-65. MR4429067

[41] W. Glunt, T. L. Hayden, Charles R. Johnson, and P. Tarazaga, *Positive definite completions and determinant maximization*, Linear Algebra Appl. **288** (1999), no. 1-3, 1–10, DOI 10.1016/S0024-3795(98)10211-2. MR1670594

[42] I. Gohberg, M. A. Kaashoek, and H. J. Woerdeman, *The band method for positive and contractive extension problems*, J. Operator Theory **22** (1989), no. 1, 109–155. MR1026078

[43] I. C. Gohberg and M. G. Kreĭn, *Introduction to the theory of linear nonselfadjoint operators*, translated from the Russian by A. Feinstein, Translations of Mathematical Monographs, Vol. 18, American Mathematical Society, Providence, R.I., 1969. MR0246142

[44] G. H. Golub, Alan Hoffman, and G. W. Stewart, *A generalization of the Eckart-Young-Mirsky matrix approximation theorem*, Linear Algebra Appl. **88/89** (1987), 317–327, DOI 10.1016/0024-3795(87)90114-5. MR882452

[45] Robert Grone, Charles R. Johnson, Eduardo M. de Sá, and Henry Wolkowicz, *Positive definite completions of partial Hermitian matrices*, Linear Algebra Appl. **58** (1984), 109–124, DOI 10.1016/0024-3795(84)90207-6. MR739282

[46] Paul R. Halmos, *A Hilbert space problem book*, D. Van Nostrand Co., Inc., Princeton, N.J.-Toronto, Ont.-London, 1967. MR0208368

[47] G. H. Hardy, J. E. Littlewood, and G. Pólya, *Inequalities*, 2nd ed., Cambridge, at the University Press, 1952. MR0046395

[48] J. William Helton and I. M. Spitkovsky, *The possible shapes of numerical ranges*, Oper. Matrices **6** (2012), no. 3, 607–611, DOI 10.7153/oam-06-41. MR2987030

[49] Michael Heymann, *The pole shifting theorem revisited*, IEEE Trans. Automat. Control **24** (1979), no. 3, 479–480, DOI 10.1109/TAC.1979.1102057. MR533402

[50] Charles R. Johnson, *Matrix completion problems: a survey*, Matrix theory and applications (Phoenix, AZ, 1989), Proc. Sympos. Appl. Math., vol. 40, Amer. Math. Soc., Providence, RI, 1990, pp. 171–198, DOI 10.1090/psapm/040/1059486. MR1059486

[51] Tosio Kato, *Perturbation theory for linear operators*, reprint of the 1980 edition, Classics in Mathematics, Springer-Verlag, Berlin, 1995. MR1335452

[52] Steven G. Krantz and Harold R. Parks, *The implicit function theorem: History, theory, and applications*, reprint of the 2003 edition, Modern Birkhäuser Classics, Birkhäuser/Springer, New York, 2013, DOI 10.1007/978-1-4614-5981-1. MR2977424

[53] M. G. Kreĭn and I. M. Spitkovskiĭ, *Some generalizations of Szegő's first limit theorem* (Russian, with English summary), Anal. Math. **9** (1983), no. 1, 23–41, DOI 10.1007/BF01903988. MR705805

[54] Peter Lancaster and Leiba Rodman, *Algebraic Riccati equations*, Oxford Science Publications, The Clarendon Press, Oxford University Press, New York, 1995. MR1367089

[55] Peter Lancaster and Miron Tismenetsky, *The theory of matrices*, 2nd ed., Computer Science and Applied Mathematics, Academic Press, Inc., Orlando, FL, 1985. MR792300

[56] Eliahu Levy and Orr Moshe Shalit, *Dilation theory in finite dimensions: the possible, the impossible and the unknown*, Rocky Mountain J. Math. **44** (2014), no. 1, 203–221, DOI 10.1216/RMJ-2014-44-1-203. MR3216017

[57] Chi-Kwong Li and Fuzhen Zhang, *Eigenvalue continuity and Geršgorin's theorem*, Electron. J. Linear Algebra **35** (2019), 619–625, DOI 10.13001/1081-3810.4123. MR4044371

[58] Ross A. Lippert, *Fixing two eigenvalues by a minimal perturbation*, Linear Algebra Appl. **406** (2005), 177–200, DOI 10.1016/j.laa.2005.04.004. MR2156435

[59] David G. Luenberger, *Optimization by vector space methods*, John Wiley & Sons, Inc., New York-London-Sydney, 1969. MR0238472

[60] John E. McCarthy and Orr Moshe Shalit, *Unitary N-dilations for tuples of commuting matrices*, Proc. Amer. Math. Soc. **141** (2013), no. 2, 563–571, DOI 10.1090/S0002-9939-2012-11714-9. MR2996961

[61] Alan McIntosh, *The Toeplitz-Hausdorff theorem and ellipticity conditions*, Amer. Math. Monthly **85** (1978), no. 6, 475–477, DOI 10.2307/2320069. MR506368

[62] L. Mirsky, *Symmetric gauge functions and unitarily invariant norms*, Quart. J. Math. Oxford Ser. (2) **11** (1960), 50–59, DOI 10.1093/qmath/11.1.50. MR114821

[63] F. Ninio, *A simple proof of the Perron-Frobenius theorem for positive symmetric matrices*, J. Phys. A **9** (1976), no. 8, 1281–1282. MR409523

[64] Vladimir V. Peller, *Hankel operators and their applications*, Springer Monographs in Mathematics, Springer-Verlag, New York, 2003, DOI 10.1007/978-0-387-21681-2. MR1949210

[65] Elijah Polak, *Optimization: Algorithms and consistent approximations*, Applied Mathematical Sciences, vol. 124, Springer-Verlag, New York, 1997, DOI 10.1007/978-1-4612-0663-7. MR1454128

[66] Walter Rudin, *Real and complex analysis*, McGraw-Hill Book Co., New York-Toronto-London, 1966. MR0210528

[67] Thomas L. Saaty and Joseph Bram, *Nonlinear mathematics*, reprint of the 1964 original, Dover Publications, Inc., New York, 1981. MR662681

[68] Jonathan R. Shewchuk, *An introduction to the conjugate gradient method without the agonizing pain*, https://www.cs.cmu.edu/ quake-papers/painless-conjugate-gradient.pdf (1994), 1–58.

[69] Alan Shuchat, *Generalized least squares and eigenvalues*, Amer. Math. Monthly **92** (1985), no. 9, 656–659, DOI 10.2307/2323714. MR810663

[70] Barry Simon, *Convexity: An analytic viewpoint*, Cambridge Tracts in Mathematics, vol. 187, Cambridge University Press, Cambridge, 2011, DOI 10.1017/CBO9780511910135. MR2814377

[71] Terence Tao, *Topics in random matrix theory*, Graduate Studies in Mathematics, vol. 132, American Mathematical Society, Providence, RI, 2012, DOI 10.1090/gsm/132. MR2906465

[72] Lloyd N. Trefethen and David Bau, III, *Numerical linear algebra*, SIAM, 1997.

[73] David S. Watkins, *The QR algorithm revisited*, SIAM Rev. **50** (2008), no. 1, 133–145, DOI 10.1137/060659454. MR2403061

[74] David S. Watkins, *Francis's algorithm*, Amer. Math. Monthly **118** (2011), no. 5, 387–403, DOI 10.4169/amer.math.monthly.118.05.387. MR2805025

[75] Roger Webster, *Convexity*, Oxford Science Publications, The Clarendon Press, Oxford University Press, New York, 1994. MR1443208

[76] J. H. Wilkinson, *Convergence of the LR, QR, and related algorithms*, Comput. J. **8** (1965), 77–84, DOI 10.1093/comjnl/8.3.273. MR183108

[77] Kemin Zhou, John C. Doyle, and Keith Glover, *Robust and optimal control*, Prentice Hall, 1996.

Notation index

$\Delta(\alpha;r)$, 393
$\Pi_\mathcal{V}$, 130
$\dotplus$, 43
$\oplus$, 126
α_j, 37
γ_j, 36
$\sigma(A)$, 37

$(A;\mathcal{X})_{\max}$, 305
$(A;\mathcal{X})_{\min}$, 305
$A \succ B$, 171
$A \succeq B$, 171
A^*, 118
A^H, 4
A^T, 4
$A^{1/2}$, 179
$A^\dagger$, 164
$A_{[j,k]}$, 173
$A_{\{I;J\}}$, 194
$A_{\{i,j,k;r,s,t\}}$, 197
$A_{\{i;j\}}$, 69
$\|A\|_s$, 101
$\|A\|_{s,t}$, 102
$\mathbb{A}$, 330
$|A|$, 65

$C_\mu^{(p)}$, 40
$\mathbb{C}$, 1
$\mathbb{C}^{p \times q}$, 1
$\mathbb{C}^p$, 1
$\mathbb{C}_\mathrm{L}$, 403
$\mathbb{C}_\mathrm{R}$, 403
$\mathcal{C}(Q)$, 235

$\mathcal{C}^k(Q)$, 88, 235

$(D_\mathbf{u}f)$, 263
$\det A$, 65
$\dim \mathcal{V}$, 3

$\mathcal{E}_+(A)$, 313, 409
$\mathcal{E}_-(A)$, 313, 409
$\mathcal{E}_0(A)$, 313, 409

(∇f), 237

$G_{[k,m]}$, 467

$H_f(\mathbf{x})$, 237
$\mathcal{H}_\alpha$, 465
$\mathcal{H}_\bullet$, 466

I_n, 3
$I_\mathcal{U}$, 7
$\int_\Gamma f(\lambda)d\lambda$, 382

$J_\mathbf{f}$, 239

$K_\omega^{(\alpha)}$, 465
$K_\omega^{(\bullet)}$, 466

$\mathcal{N}_T$, 7
$\mathcal{N}_A$, 7

$O_{p \times q}$, 3

P_Γ^A, 386
$P_\mathcal{V}^\mathcal{W}$, 128

R_α, 464

479

$R_\bullet$, 467
$\mathbb{R}$, 1
$\mathbb{R}^{p \times q}$, 1
$\mathbb{R}^{p}$, 1
$\mathbb{R}^{p \times q}_{>}$, 281
$\mathbb{R}^{p \times q}_{\geq}$, 281
$\mathcal{R}_T$, 7
$\mathcal{R}_A$, 7
$r_\sigma(A)$, 120

$\mathcal{S}_\mathbf{b}$, 359

T^*, 118
$T_k(x)$, 376
$\|T\|_{\mathcal{U},\mathcal{V}}$, 103

$\langle \mathbf{u}, \mathbf{v} \rangle_A$, 371
$\langle \mathbf{u}, \mathbf{v} \rangle_\mathcal{U}$, 113

$\mathcal{V}^\perp$, 126

$W(A)$, 429

$\mathcal{X}'$, 345
$\|\mathbf{x}\|_s$, 100
$\langle \mathbf{x}, \mathbf{y} \rangle_{\mathrm{st}}$, 114

$\mathcal{Z}_\Omega$, 413

Subject index

adjoint, 118
Aiken, John G., 344
Akritas, Alkiviadis G., 198
Akritas, Evgenia K., 198
algebraic multiplicity, 37
algorithm, 231
analytic, 381
Ando, Tsuyoshi, 323
Andrica, Dorin, 97
angle between subspaces, 154
Apostol, Tom M., 235, 236
approximate solutions, 169
approximating by unitary matrices, 337
area of parallelogram, 156
arithmetic mean, 90
Arov, Damir Z., 474

Banach space, 111
Barnett identity, 455
basis, 3
Bau, III, David, 379
Belevich, Vitold, 402
Bessel's inequality, 131
best approximation, 166, 167
Bezoutian, 453
Bhatia, Rajendra, 303, 314, 320, 323, 460
Binet-Cauchy formula, 189
binomial formula, 9
Birkhoff-von Neumann theorem, 297
block Gaussian elimination, 32
block multiplication, 6
block triangular matrices, 18

Bollobás, Béla, 97, 357
Borwein, Jonathan M., 122
bounded, 349
Bram, Joseph, 280
Brouwer fixed point theorem, 246
Brualdi, Richard A., 198
Burg, John P., 422

Carathéodory, Constantin, 96
Cassels, John W. S., 401
Cauchy interlacing theorem, 308
Cauchy sequence, 111
Cauchy-Riemann equations, 260
Cauchy-Schwarz inequality, 93, 115, 157
Cayley-Hamilton theorem, 55, 73
characteristic polynomial, 56, 72, 76, 188
Chebyshev polynomial, 376
circulant, 79
companion matrix, 75
complementary space, 45
completing the square, 200
complex vector space, 1
conjugate gradient recursion, 374
conjugate gradients, 369
conservation of dimension, 13
continuity of eigenvalues, 387
contour integral, 382
contractive fixed point theorem, 243
contractive matrix, 208
controllable pair, 399
convergence estimates, 375
convex function, 85, 241, 434

481

convex hull, 432
convex set, 85
Conway, John B., 357
cosine, 154
Courant-Fischer theorem, 305
Cramer's rule, 71
CS decomposition, 458
Curgus, Branko, 460
cyclic matrix, 138
cyclic vector, 83

Davis, Chandler, 428
de Boor, Carl W. R., 436
de Branges space, 463
de Branges, Louis, 473
de Sá, Eduardo M., 422
Degani, Ilan, 460
determinant, 65
determinant identities, 193
Deutsch, Emeric, 422
diagonalizable, 37
Diepold, Klaus, 160
difference equations, 214, 217
differentiability of eigenvalues, 312
differentiating determinants, 185
differentiation of matrix-valued
    functions, 222
Dijksma, Aad, 460
dimension, 3
direct sum decomposition, 43, 45, 46
directional derivative, 263
discrete dynamical system, 211
doubly stochastic matrix, 297
Douglas, Ron G., 428
Doyle, John C., 428, 447
dual extremal problems, 360
Duren, Peter L., 249

Eckhart, Carl, 170
eigenvalue, 35, 41, 72
eigenvector, 35, 41
entire function, 463
equivalent norms, 100
Erdos, John A., 344
exponential of a matrix, 222
extremal problems, 263, 266
extreme point, 96, 297

Fan, Ky, 300, 310
Farkas's lemma, 353
Feintuch, Avraham, 428
Fejér, Leopold, 401

Fibonacci sequence, 217
fitting a line, 204, 205
fixed point, 243
fixed point theorems, 246
Fourier matrix, 81
fractional powers of positive definite
    matrices, 401, 434
Frobenius norm, 114, 166, 338
Fuglede, Bert, 145
Fujii, J. I., 436
Fujii, M., 436
Furuta, T., 436

Gauss, Carl F., 249
Gauss-Lucas theorem, 433
Gauss-Seidel method, 30
Gaussian elimination, 21
Gaussian quadrature, 449
generalized backward shift, 464
generalized eigenvector, 37
generalized Vandermonde matrix, 78
generic, 390
geometric mean, 90
geometric multiplicity, 36
Geršgorin disks, 393
Glüch, Jochen, 290
Glazman, Israel N., 160
Glover, Keith, 428, 447
Glunt, William, 422
Gohberg, Israel, 170, 323, 422
Goldstein, Jerome A., 344
Golub, Gene H., 170
Google matrix, 292
gradient, 237
Gram matrix, 116, 117, 130
Gram-Schmidt method, 133
Grone, Robert, 422
Gronwall's inequality, 226

Hadamard's inequality, 191
Hadamard, Jacques, 145
Hahn-Banach extension theorem, 349
Halmos, Paul, 436
Hamiltonian matrix, 437
Hankel matrix, 82, 408, 461
Hardy, Godfrey H., 323, 401
Hautus, Malo L. J., 402
Hayden, Thomas L., 422
Heinz inequality, 435
Heinz, E., 436
Helton, J. William, 436
Hermitian matrix, 137, 140, 144

Subject index

Hermitian transpose, 4
Hessian, 237, 241
Heymann, Michael, 402
Hilbert matrix, 400
Hilbert space, 114
Hoffman, Alan, 170
Hölder's inequality, 91
holomorphic, 381
homogeneous system, 211
hyperplane, 348

identity matrix, 3
implicit function theorem, 255
inequalities for determinants, 191, 192
inertia, 410
inner product, 113
inner product space, 113
integration of matrix-valued functions, 222
invariant subspace, 39, 41
inverse, 6
inverse function theorem, 251
invertible, 6
irreducible matrix, 281, 285
isometric matrix, 137, 161
isospectral, 226

Jacobi matrix, 308
Jacobi's determinant identity, 195
Jacobi's formula, 187
Jacobi, Carl G., 198
Jacobian, 239
Jensen's inequality, 85
Johnson, Charles R., 422
Jordan cells, 40
Jordan chain, 55, 56
Jordan decomposition, 53

Kaashoek, Marinus A., 422
Kahan, William M., 428
Kantorovich, Leonid V., 280
Kato, Tosio, 391
keep in mind, 13–15, 28, 65, 122, 137, 140, 162, 168, 176, 182, 185
Krantz, Steven G., 261
Krein, Mark G., 160, 170, 323
Krein-Milman theorem, 96
Krylov subspace, 377
Ky Fan's inequality, 310
Ky Fan's maximum principle, 309

Lagrange multipliers, 266
Lancaster, Peter, 411, 447

Lax pair, 226
left invertible, 5, 14, 110
Leibniz's rule, 77
Leray-Schauder theorem, 246
Leslie matrices, 294
Levy, Eliahu, 209
Lewis, Adrian S., 122
Li, C-K., 391
Lidskii's inequality, 310
Lidskii, Viktor B., 310
linear combinations, 2
linear dependence, 2
linear functional, 345
linear independence, 2
linear mapping, 7
Lippert, Robert, 314
Littlewood, John E., 323, 401
Ljubic, Ju. L., 160
lower triangular, 8
LQR problem, 445
LU factorization, 174
Luenberger, David G., 379
Lyapunov equation, 407

Malaschonok, Genadii I., 198
mappings, 6
matrices with nonnegative entries, 281
matrices with positive entries, 281
matrix completion, 413
matrix multiplication, 4
maximum entropy completion, 417
McIntosh, Alan, 436
mean value theorem, 235, 238
Mihaly, Bakonyi, 436
minimal norm completion, 423, 424, 426
minimal polynomial, 55, 73
minimum matrix, 295
Minkowski's inequality, 93
minor, 69
Mirsky, Leonid, 170
moment problem, 422, 461, 474
Moore Penrose inverse, 164

Nakomoto, R., 436
negative definite, 171
negative semidefinite, 171
Newton recursion, 277
Newton step, 247, 277
Newton's method, 247, 275
Ninio, F., 290
nonhomogeneous system, 214
norm, 99

normal matrix, 137, 140, 142, 143
normal transformation, 145
normed linear space, 99
notation, 1, 114, 128, 171, 173, 235, 251, 284, 305, 313, 347, 371, 403, 418, 467
nullspace, 7
numerical range, 429

observable, 402
open mapping theorem, 253
operator norm, 103
orthogonal, 125
orthogonal complement, 126, 130
orthogonal decomposition, 126
orthogonal family, 125
orthogonal matrix, 4, 137, 140
orthogonal projection, 129, 130, 201
orthonormal expansion, 126
orthostochastic matrix, 300
othonormal family, 126

parallelogram law, 116
Parks, Harold R., 261
Parrott's lemma, 424
partial isometry, 161, 180
Peller, Vladimir V., 401
permutation matrix, 4, 63, 137
Perron-Frobenius theorem, 285
pivot column, 22
pivot variables, 22
pivots, 22
Polak, Elijah, 280
polar form, 180
polarization identity, 116
Polya, George, 323, 401
Popov, Vasilie M., 402
positive definite, 149, 171, 176
positive semidefinite, 171
principal submatrix, 188
products of singular values, 167
projection, 127
projection by iteration, 147
projection formulas, 151, 153
Putnam, Calvin R., 145

QR factorization, 134, 457
quadrature formulas, 452

range, 7
rank, 13, 14
real Jordan forms, 62
real vector space, 1

reproducing kernel, 462
reproducing kernel Hilbert space, 462
resultants, 456
Riccati equation, 437
Riesz projection, 386
Riesz, Frigyes, 401
right invertible, 6, 14, 110
Rodman, Leiba, 447
roots of polynomials, 258
Rudin, Walter, 122

Saaty, Thomas L., 280
Santanu, Sartu, 474
Schneider, Hans, 198, 422
Schur complements, 34, 200, 207
Schur's theorem, 141
Schur, Issai, 141, 145
selfadjoint transformation, 145
Shalit, Orr M., 209
Sherman Morrison formula, 74
Shewchuk, Jonathan R., 379
shifting eigenvalues, 400
Shuchat, Alan, 209
similar, 37
Simon, Barry, 97
simple curve, 382
simple permutation matrix, 63
sine, 155
singular value decomposition, 162, 164
singular values, 164, 165, 168
skew-Hermitian matrix, 137
smooth, 235
span, 2
spectral mapping, 74, 398
spectral radius, 120, 121, 396
spectrum, 37
Spitkovsky, Ilya M., 160, 436
square root, 178, 179
standard inner product, 114
Stein equation, 405
Stewart, G. W., 170
stochastic matrix, 291
strictly convex, 241, 370
strictly convex function, 85, 90, 201, 241
strictly convex normed linear space, 201
sublinear functional, 348
subspace, 1
sum of subspaces, 43
sums of singular values, 167
svd, 164
Sylvester equation, 405
Sylvester's determinant identity, 196

Subject index

Sylvester's law of inertia, 314
symmetric gauge function, 320
symmetric matrix, 140

Tao, Terence, 198
Tarazaga, Pablo, 422
Taylor's formula, 236
Tismenetsky, Miron, 411
Toeplitz matrix, 212, 405, 461
Toeplitz-Hausdorff theorem, 430
trace, 73, 90, 114, 158
transpose, 4
Trefethen, Lloyd N., 379
triangle inequality, 99
triangular, 8
triangular factorization, 173
triangular matrices, 18

UL factorization, 174, 200
unitarily invariant norm, 319
unitary matrix, 137, 140, 161
unitary transformation, 145
upper echelon matrix, 22
upper triangular, 8

Vandermonde matrix, 78
volume of parallelepiped, 157
von Neumann's inequality, 208
von Neumann's trace inequality, 301
von Neumann, John, 301, 320

warning, 7, 122, 137, 161, 172, 201, 353
warnings, 264, 265
Watkins, David, 460
Webster, Roger, 357
Weinberger, Hans F., 428
Weyl's inequalities, 310
Weyl, Herman, 310, 323
Wilkinson, John N., 460
Woerdeman, Hugo J., 422, 436
Wolkowicz, Henry, 422
Wronskian, 232

Young's inequality, 95
Young, Gale, 170

zero matrix, 3
Zhang, F., 391
Zhou, Kemin, 428, 447

SELECTED PUBLISHED TITLES IN THIS SERIES

232 **Harry Dym,** Linear Algebra in Action, Third Edition, 2023
231 **Luís Barreira and Yakov Pesin,** Introduction to Smooth Ergodic Theory, Second Edition, 2023
229 **Giovanni Leoni,** A First Course in Fractional Sobolev Spaces, 2023
228 **Henk Bruin,** Topological and Ergodic Theory of Symbolic Dynamics, 2022
227 **William M. Goldman,** Geometric Structures on Manifolds, 2022
226 **Milivoje Lukić,** A First Course in Spectral Theory, 2022
225 **Jacob Bedrossian and Vlad Vicol,** The Mathematical Analysis of the Incompressible Euler and Navier-Stokes Equations, 2022
224 **Ben Krause,** Discrete Analogues in Harmonic Analysis, 2022
223 **Volodymyr Nekrashevych,** Groups and Topological Dynamics, 2022
222 **Michael Artin,** Algebraic Geometry, 2022
221 **David Damanik and Jake Fillman,** One-Dimensional Ergodic Schrödinger Operators, 2022
220 **Isaac Goldbring,** Ultrafilters Throughout Mathematics, 2022
219 **Michael Joswig,** Essentials of Tropical Combinatorics, 2021
218 **Riccardo Benedetti,** Lectures on Differential Topology, 2021
217 **Marius Crainic, Rui Loja Fernandes, and Ioan Mărcuț,** Lectures on Poisson Geometry, 2021
216 **Brian Osserman,** A Concise Introduction to Algebraic Varieties, 2021
215 **Tai-Ping Liu,** Shock Waves, 2021
214 **Ioannis Karatzas and Constantinos Kardaras,** Portfolio Theory and Arbitrage, 2021
213 **Hung Vinh Tran,** Hamilton–Jacobi Equations, 2021
212 **Marcelo Viana and José M. Espinar,** Differential Equations, 2021
211 **Mateusz Michałek and Bernd Sturmfels,** Invitation to Nonlinear Algebra, 2021
210 **Bruce E. Sagan,** Combinatorics: The Art of Counting, 2020
209 **Jessica S. Purcell,** Hyperbolic Knot Theory, 2020
208 **Vicente Muñoz, Ángel González-Prieto, and Juan Ángel Rojo,** Geometry and Topology of Manifolds, 2020
207 **Dmitry N. Kozlov,** Organized Collapse: An Introduction to Discrete Morse Theory, 2020
206 **Ben Andrews, Bennett Chow, Christine Guenther, and Mat Langford,** Extrinsic Geometric Flows, 2020
205 **Mikhail Shubin,** Invitation to Partial Differential Equations, 2020
204 **Sarah J. Witherspoon,** Hochschild Cohomology for Algebras, 2019
203 **Dimitris Koukoulopoulos,** The Distribution of Prime Numbers, 2019
202 **Michael E. Taylor,** Introduction to Complex Analysis, 2019
201 **Dan A. Lee,** Geometric Relativity, 2019
200 **Semyon Dyatlov and Maciej Zworski,** Mathematical Theory of Scattering Resonances, 2019
199 **Weinan E, Tiejun Li, and Eric Vanden-Eijnden,** Applied Stochastic Analysis, 2019
198 **Robert L. Benedetto,** Dynamics in One Non-Archimedean Variable, 2019
197 **Walter Craig,** A Course on Partial Differential Equations, 2018
196 **Martin Stynes and David Stynes,** Convection-Diffusion Problems, 2018
195 **Matthias Beck and Raman Sanyal,** Combinatorial Reciprocity Theorems, 2018
194 **Seth Sullivant,** Algebraic Statistics, 2018
193 **Martin Lorenz,** A Tour of Representation Theory, 2018

For a complete list of titles in this series, visit the
AMS Bookstore at www.ams.org/bookstore/gsmseries/.